SPACE PROPULSION
ANALYSIS AND DESIGN

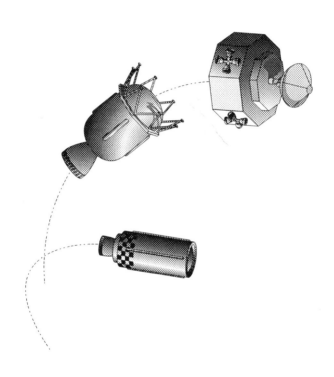

SPACE PROPULSION ANALYSIS AND DESIGN

REVISED

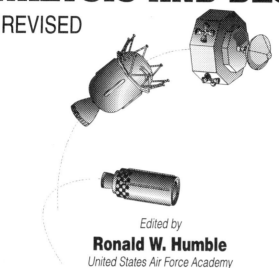

Edited by
Ronald W. Humble
United States Air Force Academy
University of Colorado

Gary N. Henry
United States Air Force Academy

Wiley J. Larson
United States Air Force Academy
University of Colorado
International Space University

This book is published as part of the
Space Technology Series, a cooperative activity
of the United States Department of Defense and the
National Aeronautics and Space Administration.

The McGraw-Hill Companies, Inc.
Primis Custom Publishing

New York St. Louis San Francisco Auckland Bogotá
Caracas Lisbon London Madrid Mexico Milan Montreal
New Delhi Paris San Juan Singapore Sydney Tokyo Toronto

SPACE PROPULSION ANALYSIS AND DESIGN

Copyright © 1995 by The McGraw-Hill Companies, Inc. All rights reserved. Printed in the United States of America. Except as permitted under the United States Copyright Act of 1976, no part of this publication may be reproduced or distributed in any form or by any means, or stored in a data base retrieval system, without prior written permission of the publisher.

   10 CCWCCW 11

First Edition—REVISED

ISBN-13: 978-0-07-031320-0

ISBN-10: 0-07-031320-2

Library of Congress Catalog Card Number: 95-79331

Editor: M. A. Hollander
Cover Design: Dale Gay
Text Design: Linda Pranke

Table of Contents

Chapter			Page
	List of Authors and Editors		viii
	Preface		xv
1	**INTRODUCTION TO SPACE PROPULSION**		**1**
	1.1 Rocket Fundamentals		6
		1.1.1 Thrust Equation	7
		1.1.2 Specific Impulse	10
		1.1.3 Effective Exhaust Velocity	11
		1.1.4 Ideal Rocket Equation	12
		1.1.5 Inert Mass Fraction	14
		1.1.6 Total Impulse	16
		1.1.7 Thrust Level	17
	1.2 The Design Process		19
		1.2.1 Define Mission Requirements	20
		1.2.2 Develop Criteria for Evaluating and Selecting Systems	22
		1.2.3 Develop Alternative Mission Concepts	23
		1.2.4 Define the Vehicle System and Select Potential Technologies	24
		1.2.5 Develop Preliminary Propulsion System Designs	29
		1.2.6 Assess Designs and Configurations	29
		1.2.7 Compare Designs and Baseline the Best Option(s)	30
		1.2.8 Iterate and Document Reasons for Choices	30
2	**MISSION ANALYSIS**		**31**
	2.1 Keplerian Orbits		32
		2.1.1 Satellite Equations of Motion	32
		2.1.2 Constants of Motion	35
		2.1.3 Classical Orbital Elements	36
		2.1.4 Time-of-Flight in an Elliptical Orbit	38
	2.2 Orbit Perturbations		40
		2.2.1 Third-Body Perturbations	41
		2.2.2 Perturbations Because of a Nonspherical Earth	42
		2.2.3 Perturbations from Atmospheric Drag	44
		2.2.4 Perturbations from Solar Radiation	46
	2.3 Orbit Maneuvering		47
		2.3.1 Coplanar Orbit Transfers	47
		2.3.2 Orbit-Plane Changes	52
		2.3.3 Orbit Rendezvous	54
	2.4 Launch Windows		55
	2.5 Orbit Maintenance		58
	2.6 Earth to Orbit		61
		2.6.1 The Effect of Vehicle Staging	62
		2.6.2 Velocity Budget to Low-Earth Orbit (LEO)	64

Chapter				Page
	2.6.3	Steering the Launch Vehicle		69
	2.6.4	Flight-Simulation Programs		74
3	**THERMODYNAMICS OF FLUID FLOW**			**77**
	3.1	Mass Transfer		78
	3.2	Thermodynamic Relations (Energy and Entropy)		81
		3.2.1	First Law for a System	82
		3.2.2	First Law for a Control Volume	83
		3.2.3	Perfect-gas Law and Specific Heats	87
		3.2.4	Second Law	93
		3.2.5	Isentropic Flow in One Dimension	95
	3.3	Thrust Equations		107
		3.3.1	Thrust Equation	109
		3.3.2	Coefficient of Thrust and Characteristic Exhaust Velocity	111
	3.4	Heat Addition		120
		3.4.1	Adding Heat to a Constant Volume with No Flow	120
		3.4.2	Steady Flow with Heat Addition	122
		3.4.3	Heat Addition to a Constant-Area Duct with Steady Flow	123
	3.5	Heat Transfer		128
		3.5.1	Conductive Heat Transfer	128
		3.5.2	Convective Heat Transfer	130
		3.5.3	Radiative Heat Transfer	134
		3.5.4	Numerical Modeling of Heat Transfer	137
	3.6	Design Example—Cold-Gas Thruster		138
		3.6.1	First Concerns	138
		3.6.2	First Design	140
		3.6.3	Predicting Performance	144
4	**THERMOCHEMISTRY**			**149**
	4.1	The Chemical Heat Source: Bond Energy		150
	4.2	Thermochemistry Basics		153
		4.2.1	Absolute and Relative Enthalpies	154
		4.2.2	Heats of Formation and Reaction	156
		4.2.3	Heat of Fusion and Vaporization	160
	4.3	Products of Combustion		161
		4.3.1	The Equilibrium Constant Method	162
		4.3.2	Minimization of Free Energy Method	166
	4.4	Flame Temperature: The Available-Heat Method		168
		4.4.1	Multiphase Flowfields	171
		4.4.2	Monopropellant Decomposition	171

Chapter			Page
	4.5	Chemical Kinetics: The Speed of the Chemical Reactions	172
	4.6	Combustion of Liquids vs. Solids	173
	4.7	Propellant Characteristics and Their Implications	174
	4.8	Key Thermochemical Parameters: The Bottom Line	176
5	**LIQUID ROCKET PROPULSION SYSTEMS**		**179**
	5.1	History	189
	5.2	Design Process	192
	5.3	Preliminary Design Decisions	194
		5.3.1 Estimate System Mass and Envelope	195
		5.3.2 Choose Propellants	199
		5.3.3 Determine the Engine Cycle	200
		5.3.4 Determine the Cooling Approach	202
		5.3.5 Determine Pressure Levels for the Engine & Feed System	203
		5.3.6 Estimate Propellant Mass and Size Tanks	214
	5.4	System Sizing, Design, and Trade-offs	217
		5.4.1 Thrust Chamber	217
		5.4.2 Configuring the Propellant-Feed System	244
		5.4.3 Turbomachinery	247
		5.4.4 Propellant Storage System	268
		5.4.5 Designing the Tank Pressurization System	275
		5.4.6 Thrust Vector and Attitude Control	280
		5.4.7 Structural Mounts	281
	5.5	Case Study	282
		5.5.1 Requirements and Design Considerations	282
		5.5.2 Preliminary Design Decisions	283
		5.5.3 System Sizing, Design, and Trade-offs	286
		5.5.4 Baseline Design	289
6	**SOLID ROCKET MOTORS**		**295**
	6.1	Background	297
	6.2	Design Process	302
	6.3	Preliminary Sizing	306
		6.3.1 Preliminary Estimate of Propellant Mass	307
		6.3.2 Techniques for Sizing Components	309
		6.3.3 Sizing Motor Cases	310
		6.3.4 Thrust Skirts and Polar Bosses	312
		6.3.5 Sizing Igniters	313
		6.3.6 Sizing Internal Insulation	314
		6.3.7 Sizing Nozzles	316

viii TABLE OF CONTENTS

Chapter				Page
		6.3.8	Sizing Systems for Thrust-Vector Control (TVC)	320
	6.4	Solid Rocket Propellants		323
		6.4.1	Fuels	324
		6.4.2	Oxidizers	325
		6.4.3	Binders	325
		6.4.4	Minor Ingredients	326
		6.4.5	Propellant Burning Rate	327
	6.5	Performance Prediction		331
		6.5.1	Lumped-Parameter Methods	334
		6.5.2	Ballistics with Variations in Spatial Pressure	342
		6.5.3	Calculating Specific Impulse, Mass Flow, and Thrust	348
	6.6	Case Study		352
		6.6.1	Requirements for Velocity and Propellant Mass	352
		6.6.2	Preliminary Design of the Motor Case	354
		6.6.3	Preliminary Design of the Thrust Skirt and Polar Boss	355
		6.6.4	Preliminary Design of the Insulation	356
		6.6.5	Designing the Nozzle and Igniter	356
		6.6.6	Preliminary Ballistics Analysis	357
		6.6.7	Preliminary Prediction of Performance	360
		6.6.8	Conclusions	360
7	**HYBRID ROCKET PROPULSION SYSTEMS**			**365**
	7.1	History		368
	7.2	Hybrid-Motor Ballistics		371
		7.2.1	Interior Ballistics Model	372
		7.2.2	The Burning-Rate Equation	374
		7.2.3	Alternate Expressions for Burning Rate	383
		7.2.4	Variations of Ballistic Parameters during a Burn	384
		7.2.5	Effect of Pressure on Radiation and Kinetics	392
		7.2.6	Correlating Experimental Data	397
	7.3	Design Process		403
	7.4	Preliminary Design Decisions		404
		7.4.1	Choose Propellants	404
		7.4.2	Determine Pressure Levels for the Engine and Feed System	408
		7.4.3	Size System	409
		7.4.4	Determine Requirements for the Initial Propellant Flow	410
		7.4.5	Configure Combustion Ports	411
	7.5	Performance Estimate		419
		7.5.1	Analytical Solution for Circular Ports with Constant Oxidizer Flow Rate	419
		7.5.2	Generic Simulation Algorithm	421

Chapter			Page
7.6	Preliminary Component Design		424
	7.6.1	Combustion-Chamber Design	424
	7.6.2	Injector Assembly	424
	7.6.3	Aft Mixing Chamber	425
	7.6.4	Nozzle Design	426
	7.6.5	Thrust-Vector Control	426
	7.6.6	Summary of Relationships between Mass Estimates for Components	426
7.7	Case Study		427
	7.7.1	Summarize Requirements	427
	7.7.2	Make Preliminary Design Decisions	428
	7.7.3	Estimate Performance	433
	7.7.4	Size and Configure Components	436
	7.7.5	Iterate	439
8	**NUCLEAR ROCKET PROPULSION SYSTEMS**		**443**
8.1	Introduction		444
	8.1.1	System Configuration and Operation	448
	8.1.2	Concepts	452
8.2	Design Process		455
8.3	Preliminary Design Decisions		459
	8.3.1	Evaluate Thermochemistry	459
	8.3.2	Sizing the System	461
	8.3.3	Determine the Required Reactor Power	461
	8.3.4	Determine System Pressure Levels	462
8.4	Size the Reactor		463
	8.4.1	Nuclear Physics	464
	8.4.2	Size the NERVA-Type Reactor	476
	8.4.3	Size the Particle-Bed Reactor Core	486
	8.4.4	Size the CERMET / Fast-Reactor Core	488
	8.4.5	Estimate the Reactor-Core Mass	488
	8.4.6	Thermal Hydraulics of the Reactor	491
	8.4.7	Reactor Materials	492
8.5	Size the Radiation Shield		493
8.6	Evaluate Vehicle Operation		498
8.7	Case Study		499
	8.7.1	Requirements and Design Considerations	499
	8.7.2	Make Preliminary Design Decisions	500
	8.7.3	Size the Reactor	503
	8.7.4	Size the Remaining Systems	503
	8.7.5	Baseline Design	506

Table of Contents

Chapter			Page
9	**ELECTRIC ROCKET PROPULSION SYSTEMS**		**509**
	9.1	History and Status	512
	9.2	Design Process	513
	9.3	Specify the Mission	514
	9.4	Select an Electric Thruster	523
		9.4.1 Overview of Electric Thrusters	523
		9.4.2 Background Physics for Electric Thrusters	535
		9.4.3 Details of Selected Thrusters	553
	9.5	Select Space Power	582
		9.5.1 Matrix of Power Technology for Space	582
		9.5.2 Power Conditioning	583
		9.5.3 Thermal Management	584
	9.6	Assess System Performance	585
	9.7	Evaluate the System	589
	9.8	Case Study	591
		9.8.1 Specify the Mission	591
		9.8.2 Select the Thruster	592
		9.8.3 Select the Power Source	593
		9.8.4 Power Conditioning Mass	593
		9.8.5 Thermal Management System Mass	593
		9.8.6 Propellant Tank and Power System Masses	594
		9.8.7 Assess System Performance	594
		9.8.8 Mission Constraints	594
		9.8.9 Evaluate the System	595
10	**MISSION DESIGN CASE STUDY**		**599**
	10.1	Define Mission Requirements	602
	10.2	Develop Criteria to Evaluate and Select a System	603
	10.3	Develop Alternative Mission Concepts	604
	10.4	Define the Vehicle System and Select Potential Technologies	612
		10.4.1 Size the On-board Propulsion System	613
		10.4.2 Size Potential Orbit-Transfer Stages	614
	10.5	Develop Preliminary Designs for the Propulsion System	622
	10.6	Assess Designs and Configurations	625
	10.7	Compare Designs and Choose the Best Option	627
11	**ADVANCED PROPULSION SYSTEMS**		**631**
	11.1	Air-Augmented Rockets	631
	11.2	Rocket Advancements	636

Chapter			Page
	11.2.1	Chemical Fuels with High-Energy Density	636
	11.2.2	Unconventional Propulsion Systems Using Nuclear Rockets	640
	11.2.3	Propulsion by Annihilating Antimatter	642
	11.2.4	Magnetic Thrust Chambers	647
11.3	Nonrocket Advancements		647
	11.3.1	Systems That Collect Solar Power	648
	11.3.2	Propulsion Systems Using External Beamed Power	653
	11.3.3	Catapult-Propulsion Systems	656
	11.3.4	Tether-Propulsion Systems	662
11.4	Interstellar Flight		668
	11.4.1	Relativistic Mechanics	668
	11.4.2	Astronomical Data and Nearest Stellar Systems	671
	11.4.3	Time, Acceleration, Velocity, and Energy Requirements	671
	11.4.4	Proposed Methods for Interstellar Flight	674
App. A	**Units and Conversion Factors**		685
App. B	**Thermochemical Data for Selected Propellants**		695
App. C	**Launch Vehicles and Staging**		715
Index			727

List of Authors and Editors

David Altman. Senior Vice President (ret.), Chemical Systems Division of United Technologies, San Jose, California. Senior Consultant, American Rocket Co., Ventura, California. Ph.D. (Physical Chemistry), University of California at Berkeley; A.B. (Arts and Science), Cornell University. Chapter 7—*Hybrid Rocket Propulsion Systems*

David Baker. President, DAB Engineering, Inc., Denver, Colorado. B.S. (Aerospace Engineering), University of Colorado. Chapter 10—*Mission Design Case Study*

William Bissell. Technical Staff Engineer (ret.), Advanced Programs, Rocketdyne Division of Rockwell International, Canoga Park, California. M.S. (Mechanical Engineering), Rensselaer Polytechnic Institute; B.S. (Mechanical Engineering), North Dakota State University at Fargo. Chapter 5—*Liquid Rocket Propulsion Systems - Turbomachinery*

Daryl G. Boden. Visiting Professor, U.S. Air Force Academy, Colorado Springs, Colorado. Ph.D. (Aeronautical and Astronautical Engineering), University of Illinois; M.S. (Astronautical Engineering), B.S. (Aerospace Engineering), University of Colorado. Chapter 2—*Mission Analysis*.

J. W. Erickson. Senior Staff Engineer, Atlas Flight Design, Lockheed-Martin, San Diego, California. B.S. (Aeronautical Engineering), University of Minnesota. Chapter 2—*Mission Analysis - Earth to Orbit*.

Robert Forward. Aerospace Consultant, Forward Unlimited, Clinton, Washington. Ph.D. and B.S. (Physics), University of Maryland. M.S. (Applied Physics), University of California at Los Angeles. Chapter 11—*Advanced Propulsion*.

Ronald Furstenau. Professor of Chemistry, U.S. Air Force Academy, Colorado Springs, Colorado. Ph.D. (Physical Chemistry), Montana State University in Bozeman; M.S. (Chemistry), University of Nebraska at Lincoln; B.S. (Chemistry), U.S. Air Force Academy. Chapter 4—*Thermochemistry*.

Stephen Heister. Associate Professor, School of Aeronautics and Astronautics, Purdue University, West Lafayette, Indiana. Ph.D. (Aerospace Engineering) University of California at Los Angeles. M.S. and B.S. (Aerospace Engineering), University of Michigan. Chapter 6—*Solid Rocket Propulsion Systems*.

Gary N. Henry. Chief of Flight Dynamics Branch, 418th Flight Test Squadron, Edwards Air Force Base, California. M.S. (Aeronautics and Astronautics), Stanford University; B.S. (Astronautics), U.S. Air Force Academy. *Editor;* Chapter 1—*Introduction to Space Propulsion*.

Ronald W. Humble. Visiting Professor, U.S. Air Force Academy, Colorado Springs, Colorado. Ph.D. and M.S. (Aerospace Engineering) University of Texas at Austin; B.S. (Aeronautics and Astronautics) University of Washington. *Editor;*

Chapter 1—*Introduction to Space Propulsion*; Chapter 4—*Thermochemistry*; Chapter 5—*Liquid Rocket Propulsion Systems*; Chapter 7—*Hybrid Rocket Propulsion Systems*; Chapter 8—*Nuclear Rocket Propulsion Systems*.

Wiley J. Larson. Visiting Professor, U.S. Air Force Academy, Colorado Springs, Colorado. D.E. (Spacecraft Design), Texas A&M University, M.S. (Electrical Engineering), University of Michigan; B.S. (Electrical Engineering), University of Michigan. *Editor*; Chapter 1—*Introduction to Space Propulsion*.

Timothy J. Lawrence. Instructor, U.S. Air Force Academy, Colorado Springs, Colorado. S.M. (Nuclear Engineering), Massachusetts Institute of Technology; B.S. (Mathematical Sciences), U.S. Air Force Academy. Chapter 8—*Nuclear Rocket Propulsion Systems*.

David Lewis. Senior Systems Engineer, Fluid and Thermophysics Department, Space and Electronics Group, TRW, Inc., Redondo Beach, California Ph.D. and M.S. (Mechanical Engineering), Massachusetts Institute of Technology. B.S. (Mechanical Engineering), University of California at Berkeley. Chapter 5—*Liquid Rocket Propulsion Systems*.

Perry D. Luckett. Director and Chief Consultant, Executive Writing Associates, Colorado Springs, Colorado. Ph.D. (American Studies), University of North Carolina at Chapel Hill; M.A. and B.A. (English), Florida State University. *Technical Editor*.

Linda K. Pranke. President, LK Editorial Services, Warrenton, Virginia. M.A. (International Affairs), George Washington University; B.A. (French), Carleton College. *Administrative Editor*.

Robert Sackheim. Manager, Propulsion and Combustion Center, TRW, Inc., Redondo Beach, California. M.S. (Chemical Engineering), B.S. (Chemical Engineering), Columbia University. Chapter 5—*Liquid Rocket Propulsion Systems*.

Peter J. Turchi. Professor, Department of Aerospace Engineering, Applied Mechanics and Aviation, The Ohio State University, Columbus, Ohio. Ph.D., M.A., B.S.E. (Aerospace and Mechanical Sciences), Princeton University. Chapter 9—*Electric Rocket Propulsion Systems*.

Gary Wirsig. Chief, Airborne Laser Technology, Lasers and Imaging Directorate, Phillips Laboratory, Albuquerque, New Mexico. M.S. (Astronautical Engineering), Air Force Institute of Technology (AFIT). B.S. (Aeronautical Engineering), AFIT. Chapter 3—*Thermodynamics of Fluid Flow*.

Jonathan K. Witter. Nuclear Engineer, Knolls Atomic Power Laboratory, Schenectady, New York. Ph.D. (Nuclear Engineering) Massachusetts Institute of Technology; M.S. and B.S. (Nuclear Engineering), Rensselaer Polytechnic Institute. Chapter 8—*Nuclear Rocket Propulsion Systems*.

Preface

The goal of this book is to answer the question: how do we design rockets? Of course, the complete answer to this question would probably fill a small library! Therefore, we limit our discussion to the preliminary or conceptual level of design. By staying at this level, we can rephrase our original question: given a mission objective and basic mission requirements, how can we create a preliminary propulsion-system design that allows us to do the mission? The answer usually includes a basic system configuration, a mass estimate, and an estimate of the system performance.

Over the years, people have developed a fair understanding of rocket design. Much of this knowledge is in the form of complex computer codes or in the heads of retiring engineers, who take their understanding of rocket design with them. Computer codes can tell you only whether your existing design will work. Experience defines the configurations to be analyzed and separates practical designs from the impractical.

The main problem for the editors and authors was to consolidate the accumulated knowledge into a comprehensible format, a process akin to extracting teeth through your ear. The result is a return to the basics of thermodynamics, thermochemistry, and historical precedent (experience is just historical data and the knowledge of what can be done with that data). We rely a lot on process charts. This approach allows the novice designer to move through the basic design algorithm and helps the experienced designer to remain rigorous and not forget the basics. Of course, these processes are intended only as a guide. We expect a designer to modify them, depending on design requirements. Otherwise, we could simply allocate design to a computer.

Another problem we had to wrestle with is how to decide what is reasonable and supportable at the preliminary design level. As already mentioned, one answer to this question is to rely on experience, but the rookie usually has little experience. To resolve this problem, we present historical numbers that include data such as typical chamber pressures or thrust-to-weight ratios. But this approach can give a designer the wrong impression. These numbers indicate only what has been done in the past and do not necessarily represent a fundamental physical limit. For example, a typical monopropellant thruster may have a thrust-to-weight ratio of 10. But for very specialized applications (exceptionally short burn durations), thrust-to-weight ratios as high as 500 have been demonstrated. By quoting the lower number, we run the risk of implying that thrust-to-weight ratios much above ten are impossible. Very high ratios are not impossible, but are just not typical. To quote one of our more colorful reviewers: "Please, we don't need a generation of stillborn rocket engineers. I'm begging you!"

One of the ground rules we set for this book is that it contain everything needed for the preliminary design of propulsion systems. Although designers are encouraged to look elsewhere for additional data, our approach allows us to do preliminary rocket design in a snow cave on Ellesmere Island if we have to. An example of this is the thermochemical data in Appendix B. The "industry standard" approach to thermochemical analysis is to run the "computer code," assuming equilibrium flow. But this method is not conducive to the "design chart" approach necessary for a book or to quick preliminary design of new rocket concepts. Thus, we use the frozen-flow approach (which can give equally valid results, especially in a snow cave!). Using frozen flow allows us to look at a wide variety of engine configurations quickly, without going back to the computer. It also keeps the book self-contained.

Any book requires serial presentation. Many of the equations and design concepts for various technologies are in earlier chapters. Thermodynamic and thermochemical analysis is key to much of the analysis in later chapters. In addition, many design considerations apply to different technologies. For example, we discuss propellant tank sizing in Chap. 5 (liquids), but we need to design propellant tanks for hybrids (Chap. 7), nuclear rockets (Chap. 8), and electric rockets (Chap. 9). In these later chapters, we simply refer to the earlier chapters so we can avoid repeating ourselves and shorten the book. We highly recommend you go through the book sequentially, but we do refer you to other sections. The first four chapters contain the basics. Chapters 5 through 9 have information on the major technology areas (liquids, solids, hybrids, nuclear, and electric). Chapter 10 gives a mission design example, and Chap. 11 discusses some future possibilities for space propulsion.

The target audience for this books includes practicing systems and propulsion engineers, engineering managers, graduate students, and advanced undergraduates in engineering. Draft versions of this book have been used to teach graduate-level classes at the University of Colorado and undergraduate classes at the U.S. Air Force Academy. Although you may cover the material in a single semester, a two-semester sequence is much more comfortable.

Despite the fact that most of the U.S. aerospace and rocket industry still uses English units, we decided to use SI (International System—its abbreviation comes from its French name, Le Système International) or metric units. We do this for several reasons. First, SI is the system of choice almost everywhere on the globe. Second, NASA, the DOD, and other agencies are trying to convert everyone in the U.S. to SI. Third, and most important to the neophyte and to the editors, the SI system has far fewer pitfalls. Appendix A gives conversion factors.

Many organizations sponsored and funded this book's development. Over the past three and a half years, the project was managed by the U.S. Air Force Academy's Department of Astronautics. Funding during this period was provided by the Air Force Space and Missile Center, Phillips Laboratory (Air Force), Air Force Space Command, Naval Research Laboratory, Office of Naval Research, Army Laboratory Command, Goddard Space Flight Center, Lewis Research Center, and the Advanced Research Projects Agency.

Preface

Our thanks go to the many people who made this book possible—in particular,

- Colonel Bob Giffen, Astronautics Department Head at the Air Force Academy, who furnished the leadership and the resources to complete the book.
- Perry Luckett, who made the text much more concise, grammatically correct, and readable.
- Linda Pranke, who designed the book and together with Anita Shute, formatted and edited the manuscript, interpreted and incorporated our obscure scribblings, and prepared drafts and camera-ready copy.
- Joan Aug and Connie Bryant, who cheerfully provided administrative support at the Academy.
- The people who spent many hours reviewing the book: Ellis Landsbaum (Aerospace Corporation); Jim Powell, George Maise, and Hans Ludewig (Brookhaven National Laboratory); Lee Lunsford (Lockheed, ret.); Al Martinez (Rocketdyne, ret.); John Whitehead, Preston Carter, and George Sutton (Lawrence Livermore National Laboratory); Peter Wilhelm and Tom Wilson (Naval Research Laboratory); Fazal Kauser (California State Polytechnic University); James Penick, Lee Meyer, Mark Mueller, Curt Selph, Richard LeClaire, Kevin Mahaffey, Dennis Tilley, Joel Beckman, Erin Durham, Dave Suzuki (Phillips Laboratory); Pete McQuade, Chuck Wood, Mike Lydon, Jerry Sellers, Scott Weston (U.S. Air Force Academy); and to anyone that we inadvertently left out, with our apologies.
- Our families for their support despite the craziness.

Through the many iterations of reviewing, editing, and revising, we have tried to eliminate all ambiguities and errors. If you find any, please contact us.

Ronald W. Humble　　HQ USAFA/DFAS
Gary N. Henry　　　　2354 Fairchild Dr., Suite 6J71
Wiley J. Larson　　　　USAF Academy, CO 80840-6224
　　　　　　　　　　　FAX: (719) 472-3723

CHAPTER 1
INTRODUCTION TO SPACE PROPULSION

Ronald W. Humble, *U.S. Air Force Academy*

Gary N. Henry, *U.S. Air Force Academy*

Wiley J. Larson, *U.S. Air Force Academy*

    1.1   Rocket Fundamentals
    1.2   The Design Process

    As early as A.D. 160, when the Greek satirist Lucian wrote of a ship caught in a fierce storm and transported to the Moon by a huge waterspout, dreams of flight through space captivated human imagination. Nearly eighteen hundred years were to pass before those dreams became reality. But as we look toward developments possible in the 21st century, we have come to acknowledge that the limitations of space propulsion reside more in the effective allocation of resources than in the laws of physics. We can get to the stars and beyond, but the choices we make determine how soon, and at what cost, we reach our destination.

    This book is about rocket design, a topic of potentially mammoth proportions. We intend to cover the preliminary or conceptual level of design: how do we start with a broad mission objective and a blank sheet of paper (or computer screen) and end up with a conceptual design for a propulsion system that meets the objective? By preliminary or conceptual design, we mean a basic system configuration (layout and dimensions), an estimate of the mass, and an estimate of the expected performance (thrust level and propellant usage). Doing this first step in the overall development is of utmost importance, if the final result is to be successful.

We provide a systematic approach to developing space propulsion systems. This approach relies heavily on defining representative design processes that can be used to walk through a design. We use

- Numerous process charts that help us tackle problems systematically
- Descriptions of the key technologies used for space propulsion, including both theory and practice
- Key ideas, concepts, and lessons learned from people in industry, government, and academia

The thrust of this book is propulsion, but we touch on other topics needed to develop a propulsion system. These topics include structures, thermodynamics, thermochemistry, control systems, and broader areas such as reliability and produceability.

In its simplest form, a propulsion system accelerates matter to provide a force of thrust that moves a vehicle or rotates it about its center of mass. Over the years, functions have been defined to more accurately describe what the propulsion system does. The main ones are

- *Launch*—accelerating a vehicle from Earth, or near Earth, through the atmosphere to a desired orbit
- *Orbit insertion*—moving a vehicle from an initial orbit to a mission orbit
- *Orbit maintenance and maneuvering*—keeping the space vehicle in the desired mission orbit or moving it to another desired orbit
- *Attitude control*—providing torque to help keep a spacecraft pointed in the desired direction

The first three functions—launch, orbit insertion, and orbit maintenance and maneuvering—provide the change in velocity (Δv) needed to *translate* the center of mass. The last function—attitude control—provides torque to *rotate* a vehicle about its center of mass.

The main system used for space propulsion is the *rocket*—a device that stores its own propellant mass and expels this mass to provide a force. Figure 1.1 shows the six basic elements of a rocket system discussed below:

- Propellant—A propellant can take the form of a gas, liquid, or solid. The propellant mainly provides the source for momentum transfer (see Sec. 1.1.1) to generate the thrust. In many systems propellants are also the source of energy.
- Propellant Storage—This system stores the propellant as a gas, liquid, or solid until it is used to provide thrust. It maintains the propellant in the proper conditions so it is usable when needed. Gases and liquids are normally maintained at particular pressures and temperatures for later use. For solid rockets, propellant storage is combined with energy conversion and accelerator hardware so that a feed system is not required.

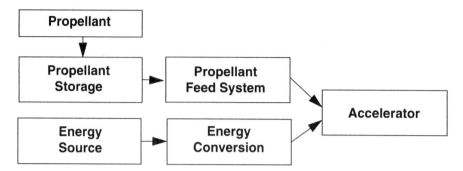

Fig. 1.1. **Generic Block Diagram for a Rocket System.** All of these elements are present in a given propulsion system, although they may be combined and named differently.

- Propellant Feed System—This system is used only in liquid and gaseous systems because we need some way to get the propellants from the storage tanks to the accelerator. In typical solid systems, the feed is part of the combustion process, so it overlaps the block for energy conversion.
- Energy Source—The most common energy sources are chemical reaction (of the propellants), gas pressure, electrical, and nuclear.
- Energy Conversion—The energy source determines how energy converts to thrust. Chemical systems rely on a chemical reaction to release energy in the form of heat and pressure. Pressure energy converts to thrust by allowing the gas to expand and accelerate, thus overlapping the "accelerator" block. Electrical energy creates thermal energy or electromagnetic force fields. Nuclear energy converts to thermal energy with subsequent heat transfer to a propellant.
- Accelerator—Accelerators take the propellant mass particles to the desired velocity and direction to provide thrust. They are basically thermodynamic or electromagnetic. Thermodynamic accelerators expand a gas to take advantage of thermal and pressure energy. Electromagnetic accelerators increase the speed of charged particles or plasmas in an electromagnetic field.

Thrust needed to change velocity or provide torque comes from a propulsion system made of one or a combination of the technologies shown in Table 1.1. These technologies provide the basic system capabilities (from Fig. 1.1) differently. Some of the more useful technologies include:

Cold Gas—A *cold-gas propulsion system* uses the pre-stored energy of a compressed gas to develop thrust. A typical system configuration is in Fig. 3.20. The gas, stored at high pressure, releases through a feed system and accelerates to a high velocity through a conventional converging/diverging nozzle. The advantages of this technology are that it is simple, safe to operate, and typically does not release contaminants into the space environment. The major disadvantage is that performance is quite low because stored pressure energy is usually far less than what is available from a chemical reaction. For this reason, cold-gas systems apply mainly to attitude control and minor orbital maneuvers.

Table 1.1. Key Propulsion Technologies and What They Do. The X's identify what these systems do or what they could conceivably do. This table does not imply that certain functions cannot be done with a given technology. We only indicate what is typical.

Propulsion Technology	Functions				Comments
	Launch	Orbit Insertion	Orbit Maintenance and Maneuver	Attitude Control	
Cold Gas N_2 H_2 He			X	X	Low-thrust performance and a high system mass make cold-gas inappropriate for launch and orbit insertion.

Chemical
Liquid
  Monopropellant
  Bipropellant
Solid
Hybrid
	
	X
	X
	X
		
		X
		X
		X
			
			X
			X
			
			X
				
				X
				X
					• Monopropellant systems are usually not used for launch and orbit insertion because of poor thrust performance
					• It is difficult to restart and modulate solid-rocket thrust for orbit maintenance and maneuver and attitude control
					• Precise thrust modulation of hybrids makes attitude control difficult

Nuclear Solid core Liquid core Gas core		X	X	X	• Solid-core reactors represent near-term possibilities • Environmental and political considerations make launch applications unlikely
Electric Electrothermal Electromagnetic Electrostatic		X	X		• Insufficient thrust makes launch impractical • Long thrust durations characterize missions • Attitude control is possible but has not been typically used.
Future Technologies	X	X	X	X	Chapter 11 discusses several possibilities

Liquid—A *liquid rocket propulsion system* (LRPS) stores its propellant (fuel and oxidizer for a *bipropellant* system) in tanks and feeds this propellant to a combustion chamber, in which energy (heat) is released through a chemical reaction. In some cases, a single liquid reacts alone by chemical decomposition. This is called a *monopropellant* system. We discuss and show typical configurations of the LRPS in Chap. 5. The main difference between candidate LRPS systems is the type of feed system. The pressure required to transport the propellant comes from either tank pressure or pump pressure. Once the heat is released (thermochemistry is discussed in Chap. 4), the hot combustion gases accelerate in a converging/diverging nozzle (discussed in Chap. 5). The advantages of a liquid system are that it can have the highest performance of any conventional chemical system and that it is highly controllable in terms of thrust modulation. Disadvantages can include complexity and development cost. Usually, monopropellant systems are less complex than bipropellant systems but have lower performance.

Solid—A *solid rocket propulsion system* (SRPS) simplifies the system illustrated in Fig. 1.1 by combining the feed system with the storage system. We discuss solid rockets and show a schematic of a typical SRPS in Chap. 6. Propellants are mixed before flight and stored in solid form in the SRPS's combustion chamber. This highly reactive and potentially explosive mixture ignites and burns until the propellant is exhausted. (It is difficult to stop the reaction once it starts.) Chemical combustion adds heat, and the combustion gases accelerate through a converging/diverging nozzle. Solid rockets are relatively simple to operate and relatively small (high propellant-packing density), but they are difficult to manufacture, handle, and throttle. Performance is relatively low compared with what is possible with a liquid rocket, and the exhaust products are often toxic.

Hybrid—The *hybrid rocket propulsion system* (HRPS) combines liquids and solids. We discuss HRPS technology and show a typical system configuration in Chap. 7. Usually, a hybrid system has a solid fuel stored in the combustion chamber. A liquid or gaseous oxidizer feeds into the combustion chamber with a feed system similar to a conventional liquid rocket. The solid fuel vaporizes with heat from the combustion process and mixes with oxidizer vapor to produce combustion. This process is very similar to what occurs in a wood stove or with a wax candle. The hot gases then exhaust through a conventional converging/diverging nozzle. Hybrid systems are simpler than bipropellant liquid systems, can have higher performance than solid rockets, usually are safer than other systems, and emit nontoxic exhaust products. Their disadvantages include lower packing density than solid rockets and poorer performance than liquid rockets.

Nuclear—A *nuclear rocket propulsion system* (NRPS) is similar to a liquid system except for the mechanism that adds heat. We discuss NRPS technology and show a typical configuration in Chap. 8. A single propellant, usually hydrogen, resides in a tank. A system similar to that in liquid rockets feeds the propellant to the heat-addition section. A nuclear fission reaction supplies heat to the propellant, which runs directly through a heat exchanger or over the heat-producing/fissioning material. The hot gases then expand in a conventional converging/diverging nozzle. The main advantage of this system is high performance. The disadvantages include system complexity and (at present) political opposition. But these systems can be made safe. In fact, using nuclear-propulsion systems for interplanetary travel can actually reduce radiation exposure as compared with chemical systems, because nuclear systems can reduce trip times, thus reducing exposure to cosmic radiation.

Electrical—An *electrical rocket propulsion system* (ERPS) uses electricity to add energy to a propellant. We discuss this technology and show a system schematic in Chap. 9. To add energy, the ERPS heats the propellant by using solid resistive elements or an arc discharge (electrothermal), or by ionizing the propellant and accelerating the ions or plasma in an electrostatic or electromagnetic field. In the first approach, the system is similar to an LRPS except for the heat source—electric heat instead of combustion (in some systems, combustion and electric heating are combined). In the second approach, the energy-addition block partially combines with the acceleration block. Some energy is added to ionize the propellant but most is added directly as kinetic energy. The advantage of an ERPS is specific impulse performance (see Sec. 1.1.2). An electrical system can have considerably higher performance than a chemical system. The principal disadvantages include low thrust levels and the large mass of the electrical power source.

Now, let us discuss a few more important points. The main difference between *rockets* and *jet engines* is that a rocket stores all of the mass it eventually expels. In contrast, a jet engine uses stored propellant with external atmospheric propellant (air) to provide thrust. Rocket motors and rocket engines are also distinct. In a typical chemical propulsion system, propellants chemically react to produce the expelled mass. The area where combustion occurs is called the *combustion chamber*.* If all of the propellant to be expelled is within the combustion chamber, the rocket is called a *motor*. If the propellant resides outside the combustion chamber, such as in a separate tank, the rocket is called an *engine*.

Although rocket systems are common to space propulsion, some propulsion concepts do not require mass expulsion. These include solar and magnetic sails, tethers, gravity assists, and aerobrakes. These systems involve the exchange of momentum, as discussed in Sec. 1.1.1, but they do not typically carry their own propellant. We discuss some of these systems in Chap. 11.

1.1 Rocket Fundamentals

Before discussing how to design a propulsion system, we need to introduce some rocket fundamentals. A general understanding of how rockets work, how vehicles affect the propulsion system (and vice-versa), and some performance parameters help us meet our main objective—designing propulsion systems.

To understand rocket propulsion and its effect on overall vehicle design, we need to introduce three fundamental ideas:

---

* The terms "combustion chamber" and "thrust chamber" are sometimes used interchangeably. Here, thrust chamber refers to the complete assembly of the combustion chamber, injector/manifold, and exhaust nozzle.

- How thrust is produced (Sec. 1.1.1)
- How efficiently thrust can be produced (Secs. 1.1.2 and 1.1.3)
- How thrust and thrust efficiency affect the vehicle mass in terms of propellant and inert mass required to hold the propellant and provide the thrust (Secs. 1.1.4 and 1.1.5)

1.1.1 Thrust Equation

In Chap. 3, we develop the equation for rocket thrust in a rigorous manner, but for now, we want to develop it more intuitively. The major source of thrust for most space propulsion systems comes from the exchange of momentum. For rockets, mass is expelled at a certain velocity and therefore has momentum:

$$P_{mom} = m v \qquad (1.1)$$

where P_{mom} = momentum (kg·m/s)
m = mass (kg)
v = velocity (m/s)

The total momentum of a system must remain constant, so if mass is expelled backward from a rocket with a certain momentum, the rocket must increase its forward momentum by an equal amount. Now, resorting to some calculus, if we eject a small mass (dm) from a rocket at an exit velocity (v_e), the change in the rocket's momentum (dP_{mom}) is

$$dP_{mom} = dm \, v_e \qquad (1.2)$$

so the change in momentum over a period of time (dt) is

$$\frac{dP_{mom}}{dt} = \frac{dm}{dt} v_e \qquad (1.3)$$

From Newton's Second Law, we know that the force on an object is equal to the rate of change of momentum (dP_{mom}/dt), so the *momentum thrust* is

$$F_m = \frac{dP_{mom}}{dt} = \frac{dm}{dt} v_e = \dot{m} v_e \qquad (1.4)$$

where F_m = momentum thrust magnitude (N)
$\dot{m}$ = mass flow rate of the propellant (kg/s)
v_e = exit or exhaust velocity* of the propellant (m/s)

---

* Exit velocity and exhaust velocity are used interchangeably. Do not confuse exhaust velocity with the effective exhaust velocity defined in Eq. (1.8).

The cartoon in Fig. 1.2 shows how this works. If you sit in a wagon and hurl rocks straight out the back, conservation of momentum propels the wagon in a direction opposite to the direction of each rock's velocity. The faster you throw the rock, or the bigger the rock, the more the wagon is propelled.

Fig. 1.2. **One-Person Rocket.** An astronaut throwing rocks out the back of a wagon is a simple example of a rocket. The astronaut uses his muscles to accelerate the rocks in one direction, leading to an equal but opposite force on the wagon that pushes it in the opposite direction. Note: if you try this experiment, friction on the wagon's wheels could keep the propulsive force from actually moving it. (Adapted from *Understanding Space: An Introduction to Astronautics* [Sellers, 1994].)

Some rocket systems have an additional source of thrust due to pressure.* Figure 1.3 shows all of the external pressures acting on a typical thrust chamber. When these pressure forces are added, all of them cancel except for the pressure effects at the nozzle exit:

$$\sum_{surface} F_p = (p_e - p_a) A_e \tag{1.5}$$

where F_p = pressure thrust (N)
 p_e = nozzle exit pressure (Pa)
 p_a = ambient pressure (Pa)
 A_e = nozzle exit cross-sectional area (m^2)

Combining the momentum thrust and the pressure thrust gives

---

* To be rigorous, all of the thrust generated by a rocket comes from integrating the pressure over all of the rocket's surfaces. However, the effect of most of this pressure is to provide a momentum exchange as described by Eq. (1.4). The momentum-exchange development misses only the pressure at the nozzle exit.

$$F = \lambda\left[\dot{m}v_e + (p_e - p_a)A_e\right] \tag{1.6}$$

where F = thrust magnitude (N)
$\dot{m}$ = mass flow rate of the propellant (kg/s)
v_e = exit or exhaust velocity (m/s)
λ = nozzle efficiency, typical range: 0.85-0.98

Notice that we have inserted an efficiency (λ—typical range 0.85 to 0.98) into this equation. With this parameter, we can account for various "real" effects, such as the nozzle flow not exiting the nozzle in a perfectly straight line.

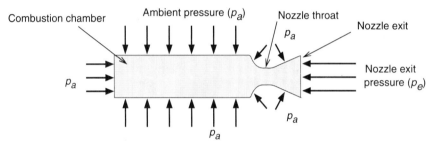

Fig. 1.3. **Pressure Thrust Schematic.** The figure shows the pressure acting on a thrust chamber (including the combustion chamber and the converging/diverging nozzle). Summing the pressure over the entire surface and taking into account the nozzle opening gives us the pressure thrust.

Equation (1.6) gives us the impression that we can maximize thrust by maximizing the exit pressure and velocity. This would be true if exit velocity and exit pressure were uncoupled. But the nozzle exit pressure and velocity are very closely and adversely coupled through the amount of nozzle expansion. If we increase nozzle expansion (by using a greater exit area for a given throat area), we find that exit velocity increases and exit pressure decreases by a corresponding amount. If we decrease the nozzle expansion, exit pressure increases but exit velocity decreases. We find, either through analytical proof or numerical experiments, that driving the exit pressure as close to the ambient pressure as possible (for a given combustion chamber pressure) maximizes the thrust for a given propellant flow rate.

It may seem counterintuitive that increasing the expansion increases the flow velocity. Our everyday experience with garden hoses indicates that pinching the hose (decreasing expansion) increases the flow velocity. However, in a typical rocket nozzle, the diverging part of the nozzle is supersonic (the flow is faster than pressure disturbances can travel through the flow). Forcing the flow to be supersonic in the expanding part of the nozzle (and sonic at the throat) allows us to substantially increase the exit velocity. From Eq. (1.6), we see that we want to increase the flow velocity. We talk about the physical reasons for this strange supersonic effect in Chap. 3.

For space applications, ambient pressure is zero. Our goal is to drive the exit pressure to zero, which requires us to increase our nozzle expansion (to get zero exit pressure requires infinite expansion). There is usually a practical limit to how much we can expand the flow, based on allowable size for the exhaust nozzle.

For launch systems or boosters, the ambient pressure can vary over the ascent (moving from sea-level pressure to a vacuum). We usually design our nozzle to give us the best average performance over the integrated ascent. The rule of thumb for preliminary design is to choose a design ambient pressure at the two-thirds point of the ascent. In other words, the exit pressure is below ambient at the start and is higher than ambient at the end of the ascent. A major design concern is avoiding flow separation inside the nozzle when the vehicle is at low altitude (see Chap. 3). The usual design goal is an exit pressure of between 15,000 Pa and 45,000 Pa for an engine operating from sea-level to a vacuum. The lower number is much more aggressive and more difficult to get than the higher number.

A very important consideration for nozzle design is combustion-chamber pressure. Given a required thrust level, increasing the chamber pressure reduces the size of our exhaust nozzle. This usually means we can reduce the mass of our engine. However, increasing the chamber pressure increases the pressure requirements of our propellant storage and feed systems. Usually, we have to make an important trade on chamber pressure to optimize our complete propulsion system.

1.1.2 Specific Impulse

A common performance parameter used for propulsion systems is *specific impulse* (I_{sp}). This parameter compares the thrust derived from a system as a function of the propellant mass flow rate:

$$I_{sp} = \frac{F}{\dot{m} g_0} \qquad (1.7)$$

where I_{sp} = specific impulse (s)
 F = thrust magnitude (N)
 $\dot{m}$ = propellant mass flow rate (kg/s)
 g_0 = 9.807 (m/s^2)

The g_0 term is thrown in to make "seconds" the unit for I_{sp} and eliminate other dimensions, so we can use the same I_{sp} number with English or SI units. It is common to leave the g_0 term out of this equation. Doing so gives us specific impulse units of velocity which are equivalent to the effective exhaust velocity discussed below. Typical numbers for the technologies in Table 1.1 are shown in Table 1.2.

Table 1.2. Performance for Key Technologies. The table shows performance and advantages of our more common rocket technologies. Higher I_{sp}s are desirable [Larson and Wertz, 1992].

Technology	I_{sp} (s)	Thrust (N)	Advantages	Disadvantages
Cold Gas N_2 H_2	 60 250	0.1–50	• simplicity • safe • low contamination	• low specific impulse

Chemical
Liquid
  Monopropellant
  Bipropellant
Solid
Hybrid
	
	140–235
	320–460
	260–300
	290–350
		0.1–12,000,000	• high thrust
			• heritage
				• moderate performance
				• combustion complications
				• safety concerns

Nuclear Solid core Liquid core Gas core	 800–1100 3000 6000	up to 12,000,000	• high specific impulse	• unproven • politically unattractive • expensive • low thrust/weight
Electric Electrothermal Electromagnetic Electrostatic	 500–1000 1000–7000 2000–10,000	0.0001–20	• very high specific impulse	• high system mass • low thrust levels • limited heritage

As we shall see in Chap. 3, increasing the temperature of the propellant or decreasing the molecular mass of the propellant improves specific impulse. Higher values of specific impulse are desirable because we can get more thrust for a given mass flow.

1.1.3 Effective Exhaust Velocity

The *effective exhaust velocity* (c) also frequently defines rocket performance:

$$c = v_e + \frac{(p_e - p_a) A_e}{\dot{m}} \tag{1.8}$$

If we compare this with Eq. (1.6), thrust (F) becomes

$$F = \dot{m}c \tag{1.9}$$

where we have left the efficiency term off for simplicity. Comparing this with Eq. (1.7) for specific impulse gives us

$$c = I_{sp} g_0 \tag{1.10}$$

This notation may replace the exhaust velocity (v_e) in the ideal rocket equations given in the next section.

1.1.4 Ideal Rocket Equation

The ideal rocket equation* allows us to size a rocket vehicle by estimating the amount of propellant required to change the vehicle's velocity (Δv) by a given amount. We start by assuming no external forces like gravity or drag are acting on the vehicle. Later we include these as Δv losses that increase our velocity change requirements. Without any external forces on the vehicle, the total momentum of the system (P_{system}), which includes the vehicle and the propellant, remains constant:

$$\frac{dP_{system}}{dt} = 0 \quad (1.11)$$

However, the momentum of the expelled propellant can be exchanged with the vehicle. To illustrate this, Fig. 1.4 shows a rocket traveling forward with propellant being ejected backward at some relative velocity.

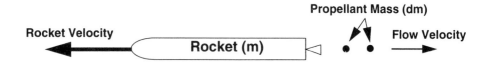

Fig. 1.4. **Schematic of Momentum Exchange for Developing the Rocket Equation.** Propellant masses are ejected to the right and, as a result, the vehicle travels to the left. As thrust continues, the vehicle mass decreases.

Because we are interested only in the change of momentum (not in the magnitude), we assume a reference frame which moves at the vehicle's velocity before thrust begins. Then, as the thrust accelerates the vehicle, the vehicle attains some velocity (v) relative to this reference frame. If an infinitesimal mass of propellant (dm) is expelled in a direction opposite to that of the vehicle's motion at a relative velocity of (v_e), the momentum imparted to the vehicle is equal to the propellant's momentum:

$$(m - dm)\,dv = -dm\,v_e \quad (1.12)$$

where m = mass of the vehicle before expulsion of mass (kg)

---

* The original derivation of this equation is usually attributed to Konstantin Tsiolkovsky (1857–1935). Tsiolkovsky was a Russian schoolteacher, and all of his works were published in Russian scientific journals. His two papers, "Investigating Space with Reaction Devices" and "Maximum Velocity of a Rocket" are key to our discussion [Tsiolkovsky, 1962].

1.1 Rocket Fundamentals

dm = infinitesimal propellant mass (kg)
dv = infinitesimal change in relative vehicle velocity (m/s)
v_e = exit or exhaust velocity of the propellant mass (m/s)

Multiplying through and recognizing that "$dm\,dv = 0$" gives us

$$dv = -v_e \frac{dm}{m} \quad (1.13)$$

Integrating this expression, we get

$$\int_0^{\Delta v} dv = -v_e \int_{m_i}^{m_f} \frac{dm}{m} \quad (1.14)$$

where Δv = change in velocity (m/s)
m_i = initial vehicle mass (kg)
m_f = final vehicle mass (kg)

$$\Delta v = -v_e \ln\left(\frac{m_f}{m_i}\right) \quad (1.15)$$

This equation is known as the *ideal rocket equation*. By rearranging terms, we can write alternate forms of the ideal rocket equation as

$$m_f = m_i e^{-\Delta v / v_e} \quad (1.16)$$

$$m_i = m_f e^{\Delta v / v_e} \quad (1.17)$$

$$m_{prop} = m_i - m_f \quad (1.18)$$

where m_{prop} = the mass of the propellant consumed (kg).

The propellant required to provide a change of velocity (Δv) is

$$m_{prop} = m_i \left(1 - e^{-\Delta v / v_e}\right) \quad (1.19)$$

$$m_{prop} = m_f \left(e^{\Delta v / v_e} - 1\right) \quad (1.20)$$

Notice that these equations do not include the pressure thrust effect. This effect can be added by simply exchanging effective exhaust velocity [c from Eq. (1.8)] with v_e. There is also a relationship between c and specific impulse, given by Eq. (1.10).

1.1.5 Inert Mass Fraction

In the previous section, we developed an expression for sizing a rocket vehicle given a velocity change requirement. Thus, if we know the initial mass we can find the final mass or vice versa. Usually, we know what the payload mass is, but without knowing the propellant mass or volume required, we do not know the tank's structural mass.

For different propellant types, historical or empirical data can relate a vehicle's inert mass to the propellant mass required. To do this, we use the concept of the *inert mass fraction* (typical numbers for liquid and solid rockets are given in Chaps. 5 and 6, respectively)[*]:

$$f_{inert} = \frac{m_{inert}}{m_{prop} + m_{inert}} \tag{1.21}$$

where f_{inert} = inert mass fraction (typical range 0.08–0.7)
m_{inert} = mass of the vehicle excluding propellant and payload[†] (kg)
m_{prop} = mass of the required propellant (kg)

It is important to remember two things. First, we use historical data to predict a typical inert mass fraction. Historical numbers only predict what has been done in the past. New technologies, new technology applications, or new requirements may change the number drastically. The second important thing to remember is that Eq. (1.21) assumes that the inert mass is only a function of propellant mass. However, such factors as the thrust level and pressures (i.e., propellant storage pressure, combustion chamber pressure, pump pressure) can significantly affect the mass of the structure and, in particular, the mass of the engine. For electric propulsion systems (Chap. 9), we find that the propulsion system mass is more sensitive to required power than to the propellant mass, but both have an effect. With all of this, the inert mass fraction approach is quite powerful and can give us very good preliminary design results.

An equivalent parameter that is also used is the *propellant mass fraction* (f_{prop}):

$$f_{prop} = \frac{m_{prop}}{m_{prop} + m_{inert}} \tag{1.22}$$

$$f_{prop} = 1 - f_{inert} \tag{1.23}$$

---

[*] Figs. 5.21 and 5.22 give typical mass fractions for liquid rockets. Table 6.2 and Figs. 6.9 and 6.10 give typical numbers (f_{prop}) for solid rockets.
[†] A vehicle's inert mass includes everything except the payload and propellant masses. The inert mass usually includes tank structure, support structure, engines, the propellant feed system, fairings, electronics, and any number of other nonreactive (inert) components.

Solving this expression for inert or structural mass:

$$m_{inert} = \frac{f_{inert}}{1-f_{inert}} m_{prop} \qquad (1.24)$$

The inert mass fraction is a performance parameter showing how well the structure is engineered. A small number indicates a high-performance structure whereas a higher number indicates lower performance and more inert mass. Initial mass and final mass are related as follows:

$$m_f = m_{pay} + m_{inert} \qquad (1.25)$$

$$m_i = m_{pay} + m_{inert} + m_{prop} = m_f + m_{prop} \qquad (1.26)$$

where m_i = initial vehicle mass (kg)
  m_f = vehicle mass at end of maneuver (kg)
  m_{pay} = mass of the vehicle payload* (kg)

Combining these equations with the ideal rocket equation, we get

$$m_{prop} = \frac{m_{pay}\left(e^{\left(\frac{\Delta v}{I_{sp}g_0}\right)} - 1\right)(1-f_{inert})}{1 - f_{inert} e^{\left(\frac{\Delta v}{I_{sp}g_0}\right)}} \qquad (1.27)$$

Notice that this equation gives a fundamental limit on vehicle performance. If the denominator is less than or equal to zero, it is impossible to build the vehicle as conceived. When the denominator goes to zero, the propellant required is infinite. If the denominator is negative, the propellant mass is negative. Neither of these possibilities is feasible:

$$1 - f_{inert} e^{\frac{\Delta v}{I_{sp}g_0}} \leq 0 \qquad (1.28)$$

Solving this equation, we get a "nonfeasible condition" for our mission:

$$I_{sp} \leq \frac{\Delta v}{\ln\left(\frac{1}{f_{inert}}\right) g_0} \qquad (1.29)$$

---

* The term "payload" means different things to different people. To a spacecraft designer, the payload could be a telescope or communications system attached to a spacecraft bus. To a launch vehicle designer, the payload is the entire spacecraft (bus and instrument).

This equation says that for a given mission (Δv) and a given technology (f_{inert}), the specific impulse must be above a certain value for the system to work. Notice that the payload, initial, or final masses do not show up in this equation, implying that the equation limits the ratio of masses (for example, m_{pay}/m_i) and that the vehicle's absolute size has no fundamental physical limit.

1.1.6 Total Impulse

Another useful way to describe the amount of propellant required, particularly in attitude-control applications, is *total impulse* (I). We determine this parameter by integrating the thrust magnitude (F) over time (t):

$$I = \int_0^{t_b} F\,dt \tag{1.30}$$

where I = total impulse (N·s)
  F = thrust magnitude at any time (N)
  t_b = thrust duration or burn time (s)

For a system with constant thrust magnitude,

$$I = F\,t_b \tag{1.31}$$

Total impulse and Δv are related to each other. Frequently, we need to convert one requirement to another ($\Delta v \rightarrow I$ or $I \rightarrow \Delta v$). We do this as follows:

$$I_{sp} = \frac{I}{m_{prop}\,g_0} \tag{1.32}$$

We can then solve this equation to get

$$m_{prop} = \frac{I}{I_{sp}g_0} \tag{1.33}$$

and substitute this result into Eq. (1.27):

$$I = \frac{I_{sp}g_0 m_{pay}\left(e^{\left(\frac{\Delta v}{I_{sp}g_0}\right)} - 1\right)(1 - f_{inert})}{1 - f_{inert} e^{\left(\frac{\Delta v}{I_{sp}g_0}\right)}} \tag{1.34}$$

We then solve this equation for Δv:

$$\Delta v = I_{sp}g_0 \ln\left[\frac{I + I_{sp}g_0 m_{pay}(1 - f_{inert})}{I f_{inert} + I_{sp}g_0 m_{pay}(1 - f_{inert})}\right] \tag{1.35}$$

where I = total impulse (N·s)
I_{sp} = specific impulse (s)
g_0 = 9.807 (m/s²)
m_{pay} = payload mass (kg)
f_{inert} = inert mass fraction
Δv = change in velocity (m/s)

The significance of these last two equations is that we have the ability to relate two key mission requirements, impulse and Δv, to each other.

1.1.7 Thrust Level

How much thrust is required? To remove dimensions from discussion of thrust level, we typically use the *thrust-to-weight ratio*:

$$F/W = \frac{F}{mg_0} \quad (1.36)$$

where F/W = thrust-to-weight ratio
F = thrust magnitude (N)
m = mass (kg)

If we can choose an appropriate F/W, and if we know the vehicle's mass, we can easily calculate the required thrust level.

We can usually define a lower level of allowable thrust-to-weight ratio (F/W). For example, launch vehicles must have an initial F/W that is greater than 1.0 or the vehicle cannot get off the ground. For orbital-transfer vehicles, the F/W for chemical systems is usually greater than 0.2. For orbit-maintenance systems and attitude-control systems, the F/W is usually much lower than 0.1.

A required maneuver usually has losses that can increase the Δv above the ideal change in velocity. The effective Δv to complete a particular maneuver is defined from the ideal rocket equation, Eq. (1.15). Increases in Δv come from gravity losses, drag losses, and steering losses. For example, to go from the surface of the Earth to low-Earth orbit, the actual change in velocity is about 7.5 km/s, whereas the Δv required to achieve orbit is more like 9 km/s. Another example is the extreme case of a launch vehicle with F/W less than 1.0. In this case, the gravity loss is the entire Δv. We need more propellant to overcome the time spent thrusting against gravity and against the drag force while moving through the atmosphere, as well as from vectoring the thrust to get our desired trajectory. These losses are discussed in Sec. 2.6.

Most launch-vehicle F/Ws lie in a very tight range from 1.2 to 1.5, with a few stragglers above 1.5. Numbers for existing and historical vehicles are shown in Fig. 1.5. Some systems have higher F/Ws to reduce the Δv loss. A higher F/W reduces the time that the vehicle spends thrusting against the gravity vector (gravity loss).

Of course, the higher F/W means that we go faster through the atmosphere, increasing the drag loss. But the savings in gravity loss usually outweighs the increased drag loss. Thrust-to-weight ratios for smaller vehicles, such as the Zenit (about 6.5), can be much higher, and missile systems can have F/W numbers with double or even triple digits.

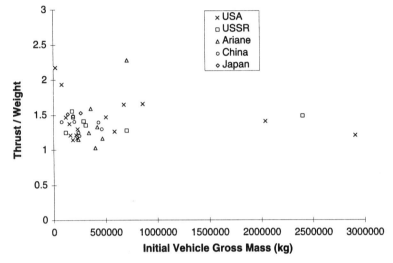

Fig. 1.5. **Initial Thrust-to-Weight Ratios for Launch Vehicles.** Numbers range from 1.2 up to 7 (Zenit, not shown), with the majority lying below 1.5 (data drawn from Isakowitz [1991]).

Thrust-to-weight ratios for space engines (upper stages) can range from 10^{-6} for electric systems to 30 for solid rockets at burn out. To avoid excessive gravity losses, F/W should be greater than 0.2. How much greater than 0.2 depends on how much acceleration the vehicle can withstand. As the F/W decreases below 0.2, the gravity losses can get excessive. In fact, Δv can as much as double for transfers using very low thrust.

Determining the thrust levels for orbit maintenance and attitude-control depends on how fast the maneuver needs to be completed and on the size of the spacecraft. Typical thrust magnitudes range from millinewtons to tens of newtons.

Thrust histories or profiles vary somewhat, depending on the technology:

- **Liquids**—thrust level is usually constant. Liquid rockets can be easily throttled to give almost any desired thrust history.
- **Solids**—thrust variations can be quite difficult to design and control. Solid-rocket combustion encourages increasing thrust levels over time, but it is possible to get level or slightly decreasing thrust profiles. However, as the mass of the vehicle decreases through propellant depletion,

the vehicle's acceleration level can get quite high. Solids are usually not used for orbit maintenance or attitude control because precise thrust management is very difficult.
- **Hybrids**—typical hybrid rocket designs give us decreasing thrust levels. However, the propellant mass usually decreases faster than the thrust level, so the vehicle acceleration increases. But the effect is not as drastic as with solids. Hybrids can be throttled, but it is difficult to get the small, highly controlled, impulses required for attitude control.
- **Nuclear**—thrust level is usually constant except at start-up and shutdown, when reactor-control transients dominate the operation. Nuclear rockets are throttleable and can be used for orbit maintenance and attitude-control—typically in addition to some other major function, such as electrical-power generation or orbital transfer. The thrust-to-weight ratio of the nuclear rocket engine alone is very poor when compared to chemical systems.
- **Electric**—electric rockets typically have very low thrust and very high specific impulse. Low thrust levels make using electric thrusters impractical for launch vehicles. Electric rockets provide significant advantages for orbit-insertion and orbit-maintenance applications because of their high specific impulse. The very low F/W tends to increase the required Δv (high gravity losses), but the high specific impulse more than makes up for this disadvantage.

1.2 The Design Process

Designing propulsion systems means different things to different people—the main distinction being the level of detail. In this book, we want to begin with the mission-level requirements and develop a preliminary (or conceptual) design of a propulsion system that fulfills these mission requirements. The goal is a baseline configuration for a propulsion system which includes an estimate of performance, mass, and envelope. This baseline configuration should have enough information to develop detailed requirements for the designers of individual components such as propellant tanks or pumps.

Preliminary design uses engineering principles but also relies heavily on creativity, art, intuition, and experience. We present a basic process for developing a preliminary design knowing that, once we understand what the design entails, we can modify the process as we see fit. This process is iterative. As we make decisions and do analysis, we hope to gain a greater understanding of the problem and the solution. As our understanding increases, we can refine the design and process to get a "better" solution.

Table 1.3 shows the overall process for preliminary design. We must use this or a similar approach to develop a high-quality propulsion system which meets mission needs. We discuss each of these steps in the following sections. Chapter 10 shows how to carry out the process for a particular example—an orbital transfer vehicle.

1.2.1 Define Mission Requirements

This first step is simply defining the mission. What do we want to do? What are the constraints and limitations? It is extremely important at this point to avoid the common error of trying to design the system. We are simply setting down, hopefully in relatively simple language, the mission's ultimate goals and limitations. If we inject design considerations at this point, we run the risk of biasing the rest of the process away from a "best" solution. The outputs of this step are as follows:

Mission Objective Statement—The first output is a clear statement of the mission objective—what do we want to do? A well defined mission objective should focus the development effort and avoid overly restrictive or excessive requirements and constraints. The importance of this step cannot be overemphasized!

List of Mission Requirements and Constraints—Coming up with a good list of mission requirements and constraints is the next step. Again, we do not want to overly restrict the solutions; on the other hand, we want to define a system that works to meet the mission objectives. We need to consider

- **Cost constraints**—In many cases, spending has (or should have) an upper limit, and this limit can drastically affect some of the major decisions in a project. For example, pumps on many liquid rockets can improve performance but they typically increase development and recurring hardware costs. A cost constraint could drive the design toward a system without a pump when a decision based strictly on performance would tend toward using a pump.
- **Schedule**—A key consideration. A "tight" schedule typically forces a project down a more conservative line. Development of new or more advanced technology usually implies a greater risk of having schedule problems. On the other hand, a looser schedule allows developers to look at more solutions with less "schedule risk." Many developers advocate a very tight schedule because "time is money." This is usually a good idea, but caution should be used.
- **Acceptable risk level**—An important constraint. Can we do the mission with off-the-shelf, proven technology? Do we need to increase performance above the usual level? What are the safety requirements? Typically, technical risk directly affects cost or schedule. Higher performance requirements usually mean higher technical risk, which in turn drives up the cost. The goal is to minimize technical risk, but this does NOT mean we should avoid new technology or new approaches. In many cases, new

Table 1.3. Process for Preliminary Design of a Propulsion System. The process allows us to start with a blank sheet of paper and methodically develop propulsion systems to best meet mission requirements.

Step	Outputs	Comments
1) Define mission requirements Section 1.2.1 Section 10.1 - example	• Mission objectives statement • List of mission requirements & constraints • Consideration of political, economic, and institutional environment	• Mission objectives include characteristics of the orbit, payload, and vehicle • Technical risk level & schedule determine level of technical aggressiveness • Safety concerns: safety factors, reliability, range safety
2) Develop criteria for evaluating and selecting systems Section 1.2.2 Section 10.2 - example	• Feasibility / acceptability criteria • Quantitative figures of merit • Qualitative figures of merit	• Predetermine the criteria by which the propulsion system assessment is done
3) Develop alternative mission concepts Section 1.2.3 Section 10.3 - example	• Assessment of past solutions • Launch site selections • Orbits / mission profiles • Low-thrust or high-thrust maneuvers, and where they are required • Δv or impulse budget	• Include several concepts that scope out the possibilities • Avoid vehicle design at this point
4) Define the vehicle system & select potential technologies Section 1.2.4 Section 10.4 - example	• Parameters of vehicle performance • Configuration of vehicle • Assessment of past solutions • Propulsion system requirements • List of potential technologies	• Vehicle configuration includes the number of engines, engine thrust levels, size, mass
5) Develop preliminary designs for the propulsion system Section 1.2.5 Chapters 5–9 - examples Section 10.5 - summary	• Assessment of "off the shelf" hardware applicability • System mass estimate • System configuration • System performance prediction	• Chapters 5 through 9 discuss this process • Chapters 3 and 4 support this analysis
6) Assess designs and configurations Section 1.2.6 Section 10.6 - example	• Feasibility assessment • Ranking of feasible options	• Assess the options to determine which one best meets mission objectives • Use step 2's ranking criteria
7) Compare designs and choose the best option(s) Section 1.2.7 Section 10.7 - example	• Baseline the best system • Documentation of reasons	• We need to decide whether we iterate or baseline a particular concept and set of options
8) Iterate and document reasons for choices Section 1.2.8 Section 10.8 - example	• Documentation of baseline if no iteration is required • Documentation of reason for iteration	• Iteration is required for all designs • The iteration can go back to any of the steps listed above, depending on the problem

technology can reduce technical risk. Increasing risk in a particular area can often reduce the overall risk in system development.
- **Environmental impacts**—An increasingly important constraint. Problems using toxic chemicals and radioactive materials are numerous and are becoming even more significant. The design team must decide ahead of time how much they want to spend on environmental concerns. The environmental impacts of conventional chemical fuels are much simpler and less costly than uranium. However, many missions become tenable only when they use less "friendly" materials.
- **Acceptable flight environment**—Flight-environment limitations for payloads could include the vibration environment, acceleration loads, contamination restrictions, or a dimensional envelope (length, breadth, and mass). These limitations can severely affect the propulsion-system design, so they must be well understood and specified.
- **The "ilities"**—This is the tough one! The "ilities" include reliability, produceability, transportability, storability, testability, and others. In many cases, these considerations are the key discriminators between a fair system and a good system. If we design a vehicle that is not transportable or requires a new transportation approach, costs tend to increase drastically. If we design a system that cannot be maintained, we are in big trouble. Careful definition of the "ilities" is a must.
- **Political, economic, and institutional environment**—Is the project a purely commercial venture, or is government involved? Is cost the bottom line, or is technology enhancement a key consideration? Is national security an issue? Could "turf" issues cloud and affect the development? Is the management and decision-making structure clear, or can we expect problems making important decisions in a correct and timely manner? Could "special interest groups" inhibit or enhance the project? All of these considerations, and more, are the intangibles that typically make or break an otherwise "good" project. Understanding these "considerations" early in the process lessens the inherent danger of these intangibles.

1.2.2 Develop Criteria for Evaluating and Selecting Systems

After we have developed several preliminary designs for our propulsion system, we need to evaluate them based on criteria that differentiate among the individual designs. However, it is important to determine what the criteria are ahead of time so we can avoid clouding the decision process later on. We should consider three basic types of criteria:
- **Feasibility and acceptability**—This first step is usually the easiest. Based on the "hard" requirements listed in the previous section, we need to evaluate the designs to ensure we have met these requirements. This is usually

a yes/no decision—does the system meet the requirement? Does the system fit within the defined envelope? Is the system mass less than the maximum allowable? We need to distinguish between "hard" and "soft" requirements. *Hard requirements* are similar to the ones just listed. These include requirements we can quantify. By contrast, *soft requirements* include such statements as "maximize performance" or "minimize costs." We try to avoid such "soft requirements" because they are usually difficult or impossible to evaluate, making them a nuisance at best and potentially destructive.

- **Quantitative figures of merit**—In an ideal world, we would like to create a "figure of merit" that takes into consideration all of the important aspects (cost, performance, schedule) of a project and quantifies them into a single unit of measure. An obvious example is cost. Suppose we could take considerations such as performance, development cost, schedule, life-cycle costs, transportability, and maintainability and combine them into some mathematical function that quantifies a total dollar cost. We could then easily evaluate different systems, rank them, and choose the least expensive. We use this approach in Chap. 10 for evaluating our particular example. Unfortunately, we can rarely do this because we usually must consider other "qualitative" measures.

- **Qualitative figures of merit**—In many "real life" situations quantifying all considerations is difficult or impossible. This truth is unfortunate because now we must rely on intuition or "gut feel" to make decisions. If more than one person is involved in the process, decisions are more difficult. Having said this, we must at least document the considerations (such as simplicity, high performance, and maintainability) and try to rank their relative importance.

Ultimately, design decisions are value judgments. However, documenting the value basis early on allows us, the designers, to move toward a "good" solution.

1.2.3 Develop Alternative Mission Concepts

At this point, some design bias enters the process. Based on experience and history, we must develop several mission scenarios that could meet the mission requirements. These usually include conservative and aggressive approaches. Hopefully, the remaining design process determines which of these concepts works best. Of course, as the design proceeds, we may learn other approaches to meeting the mission requirements, forcing us to come back to this point. We must develop several options which we can gradually eliminate until we reach the best one. But the number of options must not become so large that the remaining tasks in the process become unwieldy.

It may be appropriate to look at propulsion technology now, but we need to look at propulsion concepts from a high level. This usually means deciding whether low continuous thrust is part of the concept or whether high thrust is desirable. We should avoid deciding on the particular technology approach (for example, nuclear versus liquid). This may restrict the design space too much. The key outputs to this step are:

Orbits and mission profiles—Typically, the mission statement specifies the desired orbit or at least constrains the possibilities. In this step we must determine how we get from the launch site to our final mission orbit. Chapter 2 discusses orbits in detail.

Determine where propulsion is required—This is an obvious step. Once we have outlined the mission concept at a high level, we must identify all of the systems needed to achieve the mission objectives. A subset of these systems is the set of propulsion systems. Is a propulsion system required for the launch vehicle? Is a space engine required? Is an attitude-control system required? This step basically entails making a list of the propulsion systems requiring further analysis.

Low- or high-thrust maneuvers—At this point, we still do not want to specify the type of propulsion system. However, from the orbit decisions above, we know whether we need high- or low-thrust maneuvers. If we have chosen high thrust, we must either use a chemical or nuclear thermal system. If we have chosen low thrust, we normally use some form of electrical propulsion system or a low-performance chemical system.

Δv or impulse budget—Over the years we have used the concept of velocity change (Δv) or total impulse to define the vehicle's propulsion requirements. The main purpose of Chap. 2 is to help us come up with a good estimate of the mission Δv, so we can size the vehicle and the propulsion system. Another form of this requirement is total impulse. Basic sizing involves using the ideal rocket equation discussed in Sec. 1.1. Determining the Δv allows us to go on to the next step and establish parameters for the vehicle's overall performance.

Launch-site selection—Launch sites are located around the globe. Choice of the launch site is usually a political decision, but performance is an important consideration. If we are doing a mission to geostationary orbit, we want a launch site as close to the equator as we can get. For missions with higher inclinations, the performance savings of the low-latitude launch site are not as important.

List and assessment of past solutions—We need to look at past solutions, for two reasons. First, if an existing solution meets the requirements, we are finished! We simply implement that solution. Second, if previous solutions do not meet the requirements, they may indicate an approach that could work. Perhaps of more importance, the previous solution may reveal a direction NOT to take.

1.2.4 Define the Vehicle System and Select Potential Technologies

The ultimate goal of this process is to determine requirements for the propulsion system, so we should do the following steps in enough detail to accurately define these requirements. Once we have developed the basic concepts, we need to estimate the overall vehicle configuration that allows us to implement the particular concepts. We usually do this by trading various parameters. Notice that many of the requirements listed below are similar to the mission-level requirements. The main difference is that the mission requirements must flow down to the propulsion-system level, so the propulsion ones are much more specific.

Look at parameters for vehicle performance—The first step is to look at the various parameters defining a vehicle's performance, evaluate what is feasible, and determine the range of parameters that work. Typical parameters include number of stages, number of engines, inert mass fraction, specific impulse, initial mass, inert mass, transportable mass, or any other parameter of importance to the particular problem. Figure 1.6 shows an example of how we may do this for the particular problem of a single-stage-to-orbit (SSTO) launch vehicle. The requirements for the vehicle are as follows:

- Δv for the ascent is assumed to be 9 km/s
- Maximum inert mass is 2000 kg, for transportability
- Maximum launch pad limitation is 23,000 kg
- Specific-impulse limit is 430 seconds (a low-risk, oxygen/hydrogen liquid engine)
- Inert mass fraction limit is 8% (a very difficult task)

The construction of this figure is based on the rocket equation developed in Sec. 1.1 and is placed here simply to show how we can establish vehicle parameters. Chapter 10 shows how a similar figure is actually developed.

Configure the vehicle—From the parametric analysis we have a basic idea of what the vehicle looks like, but now we need to create a basic configuration for the vehicle. This configuration should include envelope estimates, mass estimates, tank configurations, engine configurations, and vehicle stacking and staging decisions. Including these values allows us to accurately define requirements for the individual propulsion systems. Again, Chap. 10 shows an example of this step.

List and assessment of past solutions—As in the previous section, we need to consider past solutions. At this point, however, we are looking at specific propulsion-system solutions. Does "off the shelf" hardware fit the bill? If the answer is yes, we do not need to design a new propulsion system. If the answer is no, we at least know what does NOT work.

Define the propulsion-system requirements—The process up to this point should be sufficient to define the propulsion-system requirements. The key ones are

- **Operating environment**—The operating environment includes such things as specifying the pressure and temperature of the environment. These affect thrust performance as well as many of the system's operational parameters, such as the propellant storage requirements and structural limitations. We must also determine the vehicle-specific

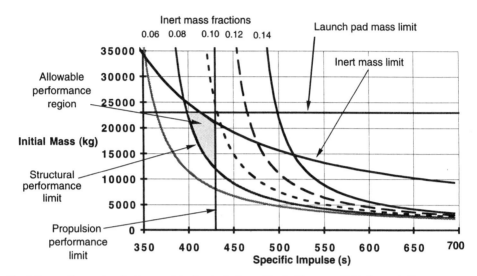

Fig. 1.6. **Example of a Parametric Analysis for a Single-Stage-to-Orbit Launch Vehicle.** This figure shows initial vehicle mass versus engine specific impulse for various possible inert mass fractions. The indicated limit curves are due to various mission constraints such as allowable launch pad mass (23,000 kg), allowable inert mass (2000 kg), and structural and propulsion technology limits (I_{sp} = 430 s and f_{inert} = 0.08). The shaded area indicates the performance region that allows us to meet the mission requirements (payload mass = 500 kg).

requirements such as vibration environment, acoustic environment, and external accelerations due to launch or transportation. All of these factors greatly influence the system design.

- **Performance requirements**—Performance requirements can include such things as initial vehicle mass, payload mass to a specified orbit, and Δv or impulse. Studying the relationships between these parameters shows what is possible, particularly when the next several requirements are being evaluated. Included in this effort are decisions on staging, number of engines per stage, and answers to other questions on vehicle design.
- **Cost constraints**—Second only to meeting the performance requirements is cost. Cost and performance are usually closely linked. We want to meet the performance requirement at the minimum cost. But as we discussed in Sec. 1.2.1, there are typically cost limitations. We must flow these down to the propulsion system level and determine the allowable costs for the propulsion system itself.

- **Technology risk level**—We have already made a decision for the mission technical risk (Sec. 1.2.1). We must now decide how this decision specifically affects the propulsion system.
- **Safety requirements**—Safety requirements usually involve redundancy and allowable design limits—typically based on mission safety considerations. For example, structural safety factors are usually different for manned versus unmanned vehicles.
- **Envelope constraints**—Many systems have physical size limitations. If the propulsion system must go on a particular launch vehicle, we must ensure the dimensions do not exceed the payload fairing's limitations. If we wish to ship the structure by truck, we must ensure that it fits on a truck. Depending on the particular development situation, system size has practical limits.
- **The "ilities"**—Again, we are looking at specific requirements for the propulsion system that flow down from the mission "ilities."
- **Thrust history**—The thrust history (see Fig. 1.7), or thrust profile, is one of the more important requirements. The thrust history defines the maximum thrust level (F_{max}), the thrust duration, propulsion system lifetime, throttling requirements, vehicle acceleration limits, cycling or the number and size of thrust pulses required, and directional control requirements.

List of potential technologies—Up to this point, we have tried hard to avoid making decisions about technology for each propulsion system specified above. Now, we need to list all of the necessary functions (e.g., attitude control, stationkeeping, orbital transfer) and select the propulsion-system technologies that meet the requirements for each. Table 1.1 begins to show which of the different technologies support which functions. In deciding on technology type, we usually consider

- Simplicity, reliability, and cost
- Δv or impulse required
- Maximum thrust magnitude
- Development time

For large-thrust, large-Δv maneuvers, we usually choose chemical rockets (solids and bipropellant liquids). The choice between one or the other is usually a trade between system complexity and performance. Liquid bipropellants have higher specific impulse than solids, though they can be more complex and costly. But if we need the performance or if we need good thrust controllability, liquids are our choice.

Typically, we use low-thrust maneuvers to take advantage of the high specific-impulse characteristics of electric rockets. This avoids high acceleration levels. If we select a low-thrust, large-Δv maneuver, we invariably select electric propulsion.

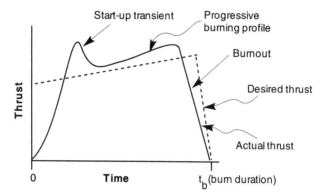

Fig. 1.7. **Thrust History for a Typical Solid Rocket.** The figure shows an example that compares achieved thrust (solid line) with desired thrust (dashed line).

Attitude control and many orbital maintenance systems use low thrust levels and have impulse requirements that are low compared to those for major orbital maneuvers. This characteristic allows us to trade simplicity for performance. The propellant tanks are usually not that large, so why not use a simpler system with slightly larger and heavier tanks? The most common solution for low-impulse situations is to use cold gas or monopropellant liquid systems. Electric thrusters have also been applied to orbit maintenance and attitude control, particularly with hydrazine arcjets.

Although hybrid rockets have not seen a lot of operational use, they offer much of the simplicity of monopropellant systems and better performance than either monopropellants or solids. They are also much more controllable, less expensive, and are typically safer to operate than are the other chemical options.

For high-Δv, high-thrust missions that are not overly constrained by allowable volume, nuclear propulsion has significant advantages over chemical systems, mainly because it can produce a high specific impulse (1000 seconds versus 450 seconds for liquids) at a high thrust level. Although using nuclear systems is unlikely for Earth-to-orbit applications, they are an enabling technology for manned interplanetary missions and could significantly enhance other orbital high-thrust, high-impulse missions.

In Chap. 11, we discuss various "advanced" technologies that should not be ignored. Although much of the development has not been done on these systems, in many cases they could be enabling or enhancing. For example, simple analysis shows that although single-stage-to-orbit (SSTO) missions are possible with chemical propulsion (specific impulse of 450 seconds or less), advanced technologies yielding specific impulses of 800 seconds or greater make SSTO vehicles more appealing. In addition, interstellar missions or manned missions to the outer planets really require the higher performance capabilities of these advanced systems.

1.2.5 Develop Preliminary Propulsion System Designs

Now that we have determined the propulsion system requirements, the next step is to come up with a preliminary design for each of the propulsion systems identified, based on these requirements. Chapters 5 through 9 each contain the process and information specific to a particular technology. However, each has a basic path:

- **Make preliminary decisions and trades**—Design is not a deterministic process. We must make many decisions before we can determine propulsion-system configuration and performance. Examples of these decisions include combustion-chamber pressure, type of propellant, and configuration of the propellant feed system. There are usually guidelines based on expected performance and historical precedent. Each chapter shows a way to determine these results for particular technologies.
- **Size the overall system**—Once we make some of the preliminary decisions, we need to come up with a basic idea of system configuration, mass, and performance. This estimate is usually based on empirical and historical data. The value of doing this step is twofold. First, this simplified analysis gives us an inkling of what the final system configuration is going to be. This insight allows us to visualize the system during the more detailed analysis and allows us to make a "sanity" check to ensure we are not making errors. For example, once we know the propellant type, the operational environment, and the required thrust history, we can easily determine the required propellant quantity based upon the ideal rocket equation (see Sec. 1.1). Second, we can then estimate the propulsion hardware's mass by comparing the desired system requirements with other systems. We use parameters such as thrust-to-weight (F/W) ratios to make these estimates.
- **Determine the configuration**—The next step is to come up with a configuration that meets the system requirements. This includes such items as propellant tank geometry, layout of the propellant feed system, and thruster geometry. We use this analysis later in the design process to define the baseline system.
- **Estimate performance**—Once we have a configuration, we need to determine whether it meets the system performance requirements such as thrust history and mass limits.

1.2.6 Assess Designs and Configurations

Assessing the particular systems we have designed requires that we

- **Determine feasibility**—In the previous steps, we developed various candidate propulsion system designs. Hopefully, we came up with several possibilities which satisfy the mission objective, requirements, and

constraints. If a design does not satisfy them, we can immediately throw it out or iterate on it to develop a system that meets the requirements.
- **Rank order feasible options**—If several possibilities satisfy the mission objective, we must rank them and determine which system is "best." In step 2 (Sec. 1.2.2) we decided on the evaluation criteria to use (quantitative and qualitative figures of merit)—now, we simply apply them.

1.2.7 Compare Designs and Baseline the Best Option(s)

Once we assess the relative merits of each candidate design for our propulsion system, the hard part is done. The next step is to choose or baseline the "best" system. But it is also appropriate at this point to ask: can we improve the designs? Also, which design has the highest likelihood of improvement? If we decide the selected "best" system is adequate, the preliminary design is complete. If we decide the knowledge we have gained in this iteration indicates further improvement in some area, we may need to do another iteration.

Documentation of the evaluation process and the reasons for rejecting specific designs is a key step for ensuring that all players in the project understand the decision process. This understanding becomes particularly important later on when anyone questions certain decisions. Chapter 10 gives an example of the required documentation.

1.2.8 Iterate and Document Reasons for Choices

Finally, we need to document the design effort of this particular iteration. What did we do? What are the system choices? Why did we make these choices? What are the configurations, masses, and performance? Can we improve things in another iteration and why do we think so?

References

Isakowitz, Steven J. 1991. *International Reference Guide to Space Launch Systems.* Washington, DC: American Institute of Aeronautics and Astronautics.

Larson, Wiley J. and James R. Wertz, eds. 1992. *Space Mission Analysis and Design.* 2nd edition. Norwell, MA: Kluwer Academic Publishers and Torrance, CA: Microcosm, Inc.

Sellers, Jerry J. 1994. *Understanding Space: An Introduction to Astronautics.* New York: McGraw-Hill.

Sutton, George P. 1992. *Rocket Propulsion Elements.* 6th edition. New York: John Wiley and Sons.

Tsiolkovsky, Konstantin E. 1962. *Selected Works* (in Russian). Moskva: Academy of Sciences of the USSR.

CHAPTER 2
MISSION ANALYSIS*

Daryl G. Boden, *U. S. Air Force Academy*

J. W. Erickson, *Lockheed Martin*

2.1 Keplerian Orbits
2.2 Orbit Perturbations
2.3 Orbit Maneuvering
2.4 Launch Windows
2.5 Orbit Maintenance
2.6 Earth to Orbit

Astrodynamics is the study of a satellite's trajectory or orbit: its path through space. The satellite *ephemeris* is a table listing its position and velocity over time. The first section below explains the terms used to describe satellite orbits, provides equations needed to calculate orbital elements from position and velocity, and shows how to predict a satellite's future position and velocity. This method is based on a simple, but accurate, model treating the Earth and the satellite as homogeneous, spherical masses. The next section discusses how forces other than the Newtonian gravitational force affect a satellite's orbit. The third section explains maneuvering strategies for changing the orbit. The next two sections discuss available launch times and ways to maintain satellite orbits. The final section describes a means to achieve low-Earth orbit.

Several textbooks discuss satellite orbits and celestial mechanics. Some of the most popular are Bate, Mueller, and White [1971], Battin [1987], Danby [1988], Escobal [1965], Kaplan [1976], Roy [1978], and Chobotov [1991].

---

* Sections 2.1 through 2.5 in this chapter have been adapted, with permission, from *Space Mission Analysis and Design*, 1992. 2nd ed. Wiley J. Larson and James R. Wertz, eds., Torrance, California, and Norwell, MA: Microcosm, Inc. and Kluwer Academic Publishers.

2.1 Keplerian Orbits

Explaining the motion of celestial bodies—especially the planets—has challenged observers for many centuries. The early Greeks believed celestial bodies moved in circles about the Earth. In 1543, Nicolaus Copernicus proposed a heliocentric (Sun-centered) system with the planets following circular orbits. Finally, with the help of Tycho Brahe's observational data, Johannes Kepler described elliptic planetary orbits about the Sun. Later, Isaac Newton mathematically solved this system using an inverse-square gravitational force.

Kepler spent several years reconciling the differences between Tycho Brahe's careful observations of the planets and their predicted motion based on previous theories. Having found that the data matched a geometric solution of elliptical orbits, he published his first two laws of planetary motion in 1609 and his third law in 1619. (These laws also apply to satellites orbiting the Earth.)

> **First Law**: The orbit of each planet is an ellipse, with the Sun at one focus.
>
> **Second Law**: The line joining the planet to the Sun sweeps out equal areas in equal times.
>
> **Third Law**: The square of the period of a planet is proportional to the cube of its mean distance from the Sun.

2.1.1 Satellite Equations of Motion

Figure 2.1 shows the key parameters of an elliptical orbit. The *eccentricity* (e) of the ellipse (not shown in the figure) is equal to c/a and measures the ellipse's deviation from a circle.

Isaac Newton explained mathematically why the planets (and satellites) follow elliptical orbits by combining his Second Law of Motion,

$$\vec{F} = m\ddot{\vec{r}} \tag{2.1}$$

with his Law of Universal Gravitation. Newton's law of gravitation states that any two bodies attract each other with a force proportional to the product of their masses and inversely proportional to the square of the distance between them. The equation for the magnitude of this force is

$$\begin{aligned} F &= \frac{-GMm}{r^2} \\ &\equiv \frac{-\mu m}{r^2} \end{aligned} \tag{2.2}$$

where F = magnitude of the force due to gravity (N)
       G = universal constant of gravitation (= 6.67×10^{-17} N·km/kg²)
       M = mass of the Earth (kg)

2.1 Keplerian Orbits

- m = mass of the satellite (kg)
- r = distance from the center of the Earth to the satellite (km)
- μ ≡ GM is the Earth's gravitational constant (= 398,600.5 km³·s⁻²).

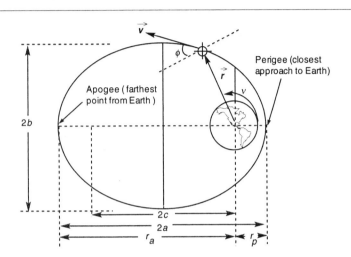

- $\vec{r}$: position vector of the satellite relative to the center of the Earth (km)
- $\vec{v}$: velocity vector of the satellite relative to the center of the Earth (km/s)
- ϕ : flight-path-angle, the angle between the velocity vector and a line perpendicular to the position vector (deg)
- a : semi-major axis of the ellipse (km)
- b : semi-minor axis of the ellipse (km)
- c : the distance from the center of the orbit to one of the foci (km)
- v : the polar angle of the ellipse, also called the true anomaly, measured in the direction of motion from the direction of perigee to the position vector (rad)
- r_a: radius of apogee, the distance from the center of the Earth to the farthest point on the ellipse (km)
- r_p: radius of perigee, the distance from the center of the Earth to the point of closest approach to the Earth (km)

Fig. 2.1. **Geometry of an Ellipse and Orbital Parameters.** Most of the orbits that we are interested in are ellipses.

Combining these two laws, we obtain an equation for the satellite's acceleration vector:

$$\ddot{\vec{r}} + \left(\frac{\mu}{r^3}\right)\vec{r} = \vec{0} \tag{2.3}$$

This equation, called the *two-body equation of motion*, is the relative equation of motion of a satellite's position vector as the satellite orbits the Earth. In deriving it, we assumed that gravity is the only force, the Earth is spherically symmetrical and homogeneous, the Earth's mass is much greater than the satellite's mass, and the Earth and the satellite are the only two bodies in the system.

A solution to the two-body equation of motion for a satellite orbiting the Earth is the *polar equation of a conic section*. It gives the magnitude of the position vector in terms of the vector's location in the orbit,

$$r = \frac{a\left(1-e^2\right)}{1+e\cos v} \tag{2.4}$$

where a = semi-major axis
       e = eccentricity
       v = polar angle or true anomaly

A *conic section* is a curve formed by passing a plane through a right circular cone. As Fig. 2.2 shows, the angular orientation of the plane relative to the cone determines whether the conic section is a *circle, ellipse, parabola,* or *hyperbola*. We can define all conic sections in terms of the eccentricity (e) in Eq. (2.4) above. The type of conic section is also related to the semi-major axis (a) and the energy (ε). Table 2.1 shows the relationships between energy, eccentricity, and semi-major axis and the type of conic section.

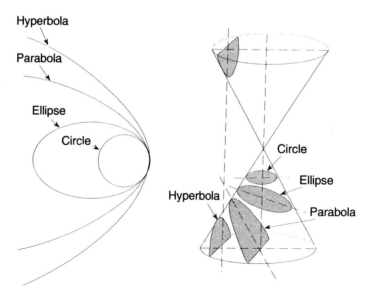

Fig. 2.2. **Conic Sections.** Satellite orbits can be any of four conic sections: a circle, an ellipse, a parabola, or a hyperbola.

2.1 Keplerian Orbits

Table 2.1. Conic Sections. The type of conic section has distinct regions of energy and eccentricity.

conic	energy, ε	semi-major axis, a	eccentricity, e
circle	< 0	= radius	0
ellipse	< 0	> 0	0 < e < 1
parabola	0	∞	1
hyperbola	> 0	< 0	> 1

2.1.2 Constants of Motion

Using the two-body equation of motion, we can derive several constants of motion for a satellite orbit. The first is

$$\varepsilon = \frac{v^2}{2} - \frac{\mu}{r} = \frac{-\mu}{2a} \qquad (2.5)$$

where ε is the total *specific mechanical energy*, or mechanical energy per unit mass, for the system and v is the magnitude of the velocity. It is the sum of the kinetic energy per unit mass and potential energy per unit mass. Equation (2.5) is known as the *energy equation*. The term for potential energy $(-\mu/r)$, defines the potential energy to be zero at infinity and negative at any radius less than infinity. Using this definition, we discover that the specific mechanical energy of elliptical orbits is always negative. As the energy increases (approaches zero), the ellipse gets larger and approaches a parabolic trajectory. From the energy equation (2.5), we find that the satellite moves fastest at perigee of the orbit and slowest at apogee.

We also know that for a circle the semi-major axis equals the radius, which is constant. Using the energy equation, we see that a satellite in a circular orbit has a velocity

$$\begin{aligned} v_c &= \sqrt{\frac{\mu}{r}} \\ &\cong 7.905\ 366 \left(\sqrt{\frac{R_E}{r}} \right) \\ &\cong 631.3481 \frac{1}{\sqrt{r}} \end{aligned} \qquad (2.6)$$

where v_c = circular velocity (km/s)
  R_E = radius of the Earth (km)
  r = orbit radius (km)

From Table 2.1, the energy of a parabolic trajectory is zero. A parabolic trajectory has the minimum energy needed to escape the Earth's gravitational attraction. Thus, we can calculate the velocity required to escape from the Earth at any distance (r) by setting energy equal to zero in Eq. (2.5) and solving for velocity.

$$V_{esc} = \sqrt{\frac{2\mu}{r}}$$
$$\cong 11.179\,88 \left(\sqrt{\frac{R_E}{r}}\right) \quad (2.7)$$
$$\cong 892.8611 \frac{1}{\sqrt{r}}$$

where V_{esc} is the escape velocity in km/s, and r is in km.

Another quantity associated with a satellite orbit is the *specific angular momentum* ($\vec{h}$), which is the satellite's total angular momentum divided by its mass (see Fig. 2.3). We can find it from the cross product of the position and velocity vectors:

$$\vec{h} = \vec{r} \times \vec{v} \quad (2.8)$$

We find from Kepler's second law that the angular momentum is constant in magnitude and direction for the two-body problem. Therefore, the plane of the orbit defined by the position and velocity vectors must remain fixed in inertial space.

2.1.3 Classical Orbital Elements

When solving the two-body equations of motion, we need six constants of integration (initial conditions) for the solution. Theoretically, the three components of position and velocity at any time could be found in terms of the position and velocity at any other time. We can also describe the orbit with five constants and one quantity that varies with time. These quantities, called *classical orbital elements*, are defined below and shown in Fig. 2.3. The coordinate frame in the figure is the geocentric inertial frame,* or GCI. Its origin is at the center of the Earth, with the x axis in the equatorial plane and pointing to the vernal equinox. Also, the z axis is parallel to the Earth's spin axis (the North Pole), and the y axis completes the right-hand set in the equatorial plane. The classical orbital elements are:

- a: *semi-major axis*: describes the size of the ellipse (km) (see Fig. 2.1)
- e: *eccentricity*: describes the shape of the ellipse
- i: *inclination:* the angle between the angular momentum vector and the Z axis (deg)
- Ω: *right ascension of ascending node:* the angle from the vernal equinox to

---

* A *sufficiently inertial coordinate frame* is a coordinate frame we can consider to be nonaccelerating for the particular application. The GCI frame is sufficiently inertial when considering Earth-orbiting satellites but is inadequate for interplanetary travel because of its rotational acceleration around the Sun.

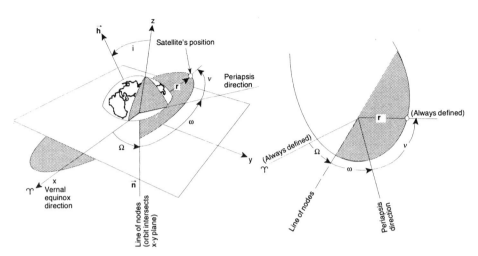

Fig. 2.3. **Defining the Keplerian Orbital Elements of a Satellite in an Elliptic Orbit.** Elements are defined relative to the GCI coordinate frame. We assume that the vernal equinox direction is fixed in space.

the ascending node. The *ascending node* is the point where the satellite passes through the equatorial plane moving from south to north. Right ascension is measured as a right-handed rotation about the pole, Z (deg)

ω: *argument of perigee*: the angle from the ascending node to the eccentricity vector measured in the direction of the satellite's motion. The *eccentricity vector* points from the center of the Earth to perigee with a magnitude equal to the eccentricity of the orbit (deg)

ν: *true anomaly*: the angle from the eccentricity vector to the satellite's position vector, measured in the direction of satellite motion. We can also use *time since perigee passage* (T) (deg)

Given these definitions, we can solve for the elements if we know the satellite's position and velocity vectors. Equations (2.5) and (2.8) allow us to solve for the energy and angular momentum vector. An equation for the *nodal vector* ($\vec{n}$) in the direction of the ascending node is

$$\vec{n} = \vec{Z} \times \vec{h} \tag{2.9}$$

We can calculate the eccentricity vector from the following equation:

$$\vec{e} = \left(\frac{1}{\mu}\right)\left\{\left(v^2 - \frac{\mu}{r}\right)\vec{r} - (\vec{r} \cdot \vec{v})\vec{v}\right\} \tag{2.10}$$

Table 2.2 lists equations to derive the classical orbital elements and related parameters for an elliptical orbit.

Table 2.2. Classical Orbital Elements. For the right ascension of ascending node, argument of perigee, and true anomaly, if the quantities in parentheses in the right-hand column (Decision) are positive, use the angle calculated. If the quantities are negative, use 360 deg minus the angle calculated.

Symbol	Name (units)	Equation	+/− Decision
a	semi-major axis (km)	$a = -\mu/(2\varepsilon) = (r_a + r_p)/2$	
e	eccentricity	$e = 1 - r_p/a = r_a/a - 1$	
i	inclination (deg)	$i = \cos^{-1}(h_z/h)$	
Ω	right ascension of ascending node (deg)	$\Omega = \cos^{-1}(n_x/n)$	$(n_y > 0)$
ω	argument of perigee (deg)	$\omega = \cos^{-1}[(\vec{n}\cdot\vec{e})/(n\cdot e)]$	$(e_z > 0)$
v	true anomaly (deg)	$v = \cos^{-1}[(\vec{e}\cdot\vec{r})/(e\cdot r)]$	$(\vec{r}\cdot\vec{v} > 0)$
r_p	radius of perigee (km)	$r_p = a(1-e)$	
r_a	radius of apogee (km)	$r_a = a(1+e)$	
P	period (min)	$P = 2\pi\sqrt{a^3/\mu}$ $\cong 84.489\sqrt{(a/R_E)^3}$ min $\cong \sqrt{0.000\ 165\ 87\ a^3}$ min, a in km	
ω_o	orbit frequency (rad/s)	$\omega_o = \sqrt{\mu/a^3}$ $\cong 631.348\ 16/\sqrt{a^3}$ rad/s, a in km	

2.1.4 Time-of-Flight in an Elliptical Orbit

By analyzing Brahe's observational data, Kepler was able to solve the problem of relating position in the orbit to the elapsed time $(t - t_0)$ or conversely, how long it takes to go from one point in an orbit to another. To do so, Kepler introduced the quantity M, called the *mean anomaly*. It is the fraction of an orbit period which has elapsed since perigee, expressed as an angle. The mean anomaly equals the true anomaly for a circular orbit. By definition,

$$M - M_0 \equiv n(t - t_0) \tag{2.11}$$

where M_0 is the mean anomaly at time t_0 and n is the *mean motion*, or average angular velocity, determined from the semi-major axis of the orbit:

$$n \equiv \sqrt{\frac{\mu}{a^3}}$$

$$\cong 36,173.585 \left(\frac{1}{\sqrt{a^3}}\right) \text{deg/s}$$

$$\cong 8,681,660.4 \left(\frac{1}{\sqrt{a^3}}\right) \text{rev/day} \quad (2.12)$$

$$\cong 3.125\,297\,7 \times 10^9 \left(\frac{1}{\sqrt{a^3}}\right) \text{deg/day}$$

where a is in km.

This solution gives the average position and velocity, but satellite orbits are elliptical, with a radius constantly varying in orbit. Because the satellite's velocity depends on this varying radius, it changes as well. To resolve this problem we can define an intermediate variable called *eccentric anomaly* (E) for elliptical orbits. Table 2.3 lists the equations necessary to relate time-of-flight to orbital position.

Table 2.3. Time-of-Flight in an Elliptical Orbit. All angular quantities are in radians.

Variable	Name	Equation
n	mean motion (rad/s)	$n = \sqrt{\mu/a^3}$ $\cong 631.34816/\sqrt{a^3}$ rad/s (a in km)
E	eccentric anomaly (rad)	$\cos E = (e + \cos v)/(1 + e \cos v)$
M	mean anomaly (rad)	$M = E - e \sin(E)$ (M in rad) $M = M_0 + n(t - t_0)$ (M in rad)
$t - t_0$	time-of-flight (s)	$t - t_0 = (M - M_0)/n$ ($t - t_0$ in s)
v	true anomaly (rad)	$v \approx M + 2e \sin M + 1.25 e^2 \sin(2M)$

As an example, we find the time it takes a satellite to go from perigee to an angle 90 degrees from perigee, for an orbit with a semi-major axis of 7000 km and an eccentricity of 0.1. For this example

$v_0 = E_0 = M_0 = 0.0$ rad $\quad t_0 = 0.0$ s
$v = 1.5708$ rad $\quad E = 1.4706$ rad
$M = 1.3711$ rad $\quad n = 0.001\,08$ rad/s
$t = 1271.88$ s

Finding the position in an orbit after a specified period is more complex. For this problem, we calculate the mean anomaly (*M*) using time-of-flight and the mean motion using Eq. (2.11). Next, we determine the true anomaly (*v*) using the series expansion shown in Table 2.3, a good approximation for small eccentricity (the error is of the order e^3). If we need greater accuracy, we must solve the equation in Table 2.3 relating mean anomaly to eccentric anomaly. Because this is a transcendental function, we must use an iterative solution to find the eccentric anomaly, after which we can calculate the true anomaly directly.

2.2 Orbit Perturbations

The Keplerian orbit discussed above provides an excellent reference, but other forces act on the satellite to perturb it away from the nominal orbit. We can classify these *perturbations*, or variations in the orbital elements, based on how they affect the Keplerian elements.

Figure 2.4 illustrates a typical variation in one of the orbital elements because of a perturbing force. *Secular variations* are linear, whereas *short-period variations* are periodic, with a period less than or equal to the orbital period. *Long-period variations* have a period greater than the orbital period. Because secular variations have long-term effects on orbit prediction (the orbital elements affected continue to increase or decrease), we discuss them in detail. If the satellite mission demands that we precisely determine the orbit, we must include the periodic variations as well. Battin [1987], Danby [1988], and Escobal [1965] describe ways to determine and predict orbits for non-Keplerian motion.

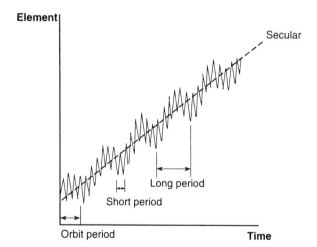

Fig. 2.4. **Secular and Periodic Variations of an Orbital Element.** Secular variations are linear. Short-period variations have a period less than the orbital period. Long-period variations have a period longer than the orbital period.

Unperturbed orbits have constant orbital elements. When perturbing forces are considered, the classical orbital elements vary with time. To predict the orbit we must determine this time variation using techniques of either special or general perturbations. *Special perturbations* employ direct numerical integration of the equations of motion, in which the accelerations are integrated directly to obtain velocity and position.

General perturbations are an approach to analytically solving some aspects of the motion of a satellite subjected to perturbing forces. For example, the polar equation of a conic applies to the two-body equations of motion. Unfortunately, most perturbing forces yield not to a direct analytic solution but to series expansions and approximations. Because the orbital elements are nearly constant, general perturbation techniques usually solve directly for the orbital elements rather than the inertial position and velocity. They are more difficult and approximate, but they allow us to better understand how the perturbations affect a large class of orbits. We can also obtain solutions much faster than with special perturbations.

The main forces that perturb a satellite's orbit arise from third bodies such as the Sun and the Moon, the mass distribution of the nonhomogeneous, nonspherical Earth, atmospheric drag, and solar radiation pressure.

2.2.1 Third-Body Perturbations

The gravitational forces of the Sun and the Moon cause periodic variations in all of the orbital elements, but only the right ascension of ascending node, argument of perigee, and mean anomaly experience secular variations. These secular variations arise when the satellite's orbit precesses about the ecliptic pole.* The secular variation in mean anomaly is much smaller than the mean motion and has little effect on the orbit; however, the secular variations in right ascension of the ascending node and argument of perigee are important, especially for high-altitude orbits.

For nearly circular orbits, e^2 is near zero and the resulting error is of order e^2. In this case, the equations for the secular rates of change of an Earth-centered satellite orbit resulting from the Sun and Moon are

right ascension of ascending node:

$$\dot{\Omega}_{Moon} = -0.00338 \frac{\cos i}{n} \qquad (2.13)$$

$$\dot{\Omega}_{Sun} = -0.00154 \frac{\cos i}{n} \qquad (2.14)$$

argument of perigee:

---

* The *ecliptic pole* is a vector perpendicular to the Earth's orbital plane around the Sun.

$$\dot{\omega}_{Moon} = 0.001\,69\,\frac{4-5\sin^2 i}{n} \qquad (2.15)$$

$$\dot{\omega}_{Sun} = 0.000\,77\,\frac{4-5\sin^2 i}{n} \qquad (2.16)$$

where i is the orbit inclination, n is the number of orbit revolutions per day, and $\dot{\Omega}$ and $\dot{\omega}$ are in deg/day. These equations are only approximate; they neglect the variation caused by the orbital plane's changing orientation with respect to the Moon's orbital plane and the ecliptic plane.

2.2.2 Perturbations Because of a Nonspherical Earth

When developing the two-body equations of motion, we assume the Earth is a spherically symmetrical, homogeneous mass. In fact, the Earth is neither homogeneous nor spherical. The most dominant features are a bulge at the equator, a slight pear shape, and flattening at the poles. For a potential function of the Earth, we can find a satellite's acceleration by taking the gradient of the potential function. One widely used form of the geopotential function is

$$\Phi = \frac{\mu}{r}\left[1 - \sum_{j=1}^{\infty} J_j\left(\frac{R_E}{r}\right)^j P_j \sin(jL)\right] \qquad (2.17)$$

where μ = GM, the Earth's gravitational constant (km^3/s^2)
R_E = Earth's equatorial radius (km)
P_j = Legendre polynomials
L = geocentric latitude (deg)
j = series counter

and J_j are the dimensionless geopotential coefficients, of which the first several are

$J_2 = 0.001\,082\,63$
$J_3 = -0.000\,002\,54$
$J_4 = -0.000\,001\,61$

This form of the geopotential function depends on latitude, and the geopotential coefficients (J_j) are called the *zonal coefficients*. Other, more general, expressions for the geopotential include *sectoral* and *tesseral terms* in the expansion. The sectoral terms divide the Earth into slices and depend only on longitude. The tesseral terms in the expansion depend on both longitude and latitude. They divide the Earth into a checkerboard pattern (*tesseral* is Latin for a tile pattern) of regions that alternately add to and subtract from the two-body potential.

The potential generated by the nonspherical Earth causes periodic variations in all orbital elements. The dominant effects, however, are secular variations in right ascension of ascending node and argument of perigee because of the Earth's oblateness, represented by the J_2 term in the geopotential expansion. The rates of change of Ω and ω due to J_2 are

$$\dot{\Omega}_{J_2} = -1.5 n J_2 \left(\frac{R_E}{a}\right)^2 \left(\frac{\cos i}{\left(1-e^2\right)^2}\right)$$
$$\cong -2.064\,74 \times 10^{14} \left(\frac{1}{\sqrt{a^7}}\right) \left(\frac{\cos i}{\left(1-e^2\right)^2}\right) \quad (2.18)$$

$$\dot{\omega}_{J_2} = 0.75 n J_2 \left(\frac{R_E}{a}\right)^2 \left(\frac{4 - 5\sin^2 i}{\left(1-e^2\right)^2}\right)$$
$$\cong 1.032\,37 \times 10^{14} \left(\frac{1}{\sqrt{a^7}}\right) \left(\frac{4 - 5\sin^2 i}{\left(1-e^2\right)^2}\right) \quad (2.19)$$

where n = mean motion (deg/day)
 R_E = Earth's equatorial radius (km)
 a = semi-major axis (km)
 e = eccentricity
 i = inclination (deg)

and $\dot{\Omega}$ and $\dot{\omega}$ are in deg/day. Table 2.4 compares the rates of change of right ascension of ascending node and argument of perigee resulting from the Earth's oblateness, the Sun, and the Moon. For satellites in geosynchronous orbit (GEO) and below, the J_2 perturbations dominate; for satellites above GEO, the Sun and Moon perturbations dominate.

Molniya orbits are highly eccentric ($e \cong 0.75$), with periods of about 12 hours (2 revolutions/day). We choose the orbital inclination so that the rate of change of perigee, Eq. (2.19), is zero. This condition occurs at inclinations of 63.4 deg and 116.6 deg. For these orbits the argument of perigee is typically in the southern hemisphere, so the satellite remains above the northern hemisphere near apogee for approximately 11 hours/orbit. We choose the perigee altitude to meet the satellite's mission constraints. Typical perigee altitudes vary from 200 to 1000 km. We can calculate the eccentricity and apogee altitude using the semi-major axis and perigee.

In a *Sun-synchronous orbit*, the satellite's orbital plane remains approximately fixed with respect to the Sun because the secular variation in the right ascension of ascending node [Eq. (2.18)] matches the Earth's rate of rotation around the Sun. A nodal precession rate of 0.9856 deg/day matches this rate. Because the rotation is positive, Sun-synchronous orbits must be retrograde. For a given semi-major axis

Table 2.4. Secular Variations in Right Ascension of the Ascending Node and Argument of Perigee. This table shows the perturbing effects of Earth oblateness (J_2) and "third-bodies" on the right ascension of the ascending node (Ω) and the argument of periapsis (ω) for typical orbits.

Orbit	Effect of J_2 [Eqs. (2.18), (2.19)] (deg/day)	Effect of Moon [Eqs. (2.13), (2.15)] (deg/day)	Effect of Sun [Eqs. (2.14), (2.16)] (deg/day)
Shuttle	a = 6700 km, e = 0, i = 28 deg		
$\Delta\Omega$	−7.35	−0.000 19	−0.000 08
$\Delta\omega$	12.05	0.002 42	0.001 10
Global Positioning System	a = 26,600 km, e = 0, i = 60.0 deg		
$\Delta\Omega$	−0.033	−0.000 85	−0.000 38
$\Delta\omega$	0.008	0.000 21	0.000 10
Molniya	a = 26,600 km, e = 0.75, i = 63.4 deg		
$\Delta\Omega$	−0.30	−0.000 76	−0.000 34
$\Delta\omega$	0.00	0.000 00	0.000 00
Geosynchronous	a = 42,160 km, e = 0, i = 0 deg		
$\Delta\Omega$	−0.013	−0.003 38	−0.001 54
$\Delta\omega$	0.025	0.006 76	0.003 07

(a) and eccentricity (e), we can use Eq. (2.18) to find the inclination needed to make the orbit Sun-synchronous.

2.2.3 Perturbations from Atmospheric Drag

The main nongravitational force acting on satellites in low-Earth orbit is atmospheric drag. Drag acts in a direction opposite to that of the velocity vector and removes energy from the orbit. Decreasing energy causes the orbit to shrink, leading to further increases in drag. Eventually, the orbit's altitude becomes so small that the satellite reenters the atmosphere.

The equation for acceleration due to drag on a satellite is

$$a_D = -\frac{1}{2}\rho\left(\frac{C_D A}{m}\right)v^2 \qquad (2.20)$$

where ρ = atmospheric density (kg/m^3)
C_D = coefficient of drag ≈ 2.2
A = satellite cross-sectional area (m^2)
m = satellite mass (kg)

v = satellite's velocity with respect to the atmosphere (m/s)

We can approximate the changes in semi-major axis and eccentricity per revolution and the satellite's lifetime in a circular orbit using the following equations:

$$\Delta a_{rev} = -2\pi \left(\frac{C_D A}{m}\right) a^2 \rho_p e^{-c[I_0 + 2eI_1]} \tag{2.21}$$

$$\Delta e_{rev} = -2\pi \left(\frac{C_D A}{m}\right) a \rho_p e^{-c\left[I_1 + \frac{e}{2}(I_0 + I_2)\right]} \tag{2.22}$$

where Δa_{rev} = change in semi-major axis per orbit revolution (km)
Δe_{rev} = change in eccentricity per orbit revolution (km)
a = semi-major axis (m)
ρ_p = atmospheric density at perigee (kg/m³)
c ≡ ae/H
H = density scale height (m)
I_i = Modified Bessel Functions* of order i and argument c

The term $m/(C_D A)$, or *ballistic coefficient*, is modelled as a constant for most satellites.

For circular orbits, we can use these equations to derive the much simpler expressions:

$$\Delta a_{rev} = -2\pi \frac{C_D A}{m} \rho a^2 \tag{2.23}$$

$$\Delta P_{rev} = -6\pi^2 \frac{C_D A}{m} \frac{\rho a^2}{v} \tag{2.24}$$

$$\Delta v_{rev} = \pi \frac{C_D A}{m} \rho a v \tag{2.25}$$

$$\Delta e_{rev} = 0 \tag{2.26}$$

where P is the orbit period and v is the satellite's velocity.

We can roughly estimate the satellite's lifetime (L) due to drag from

$$L \approx \frac{-H}{\Delta a_{rev}} \tag{2.27}$$

where, as above, H is the atmospheric density scale height (Table 2.5).

---

* Many standard mathematical tables contain values for I_i.

Table 2.5. Atmosphere-Related Parameters for Earth Satellites. The atmospheric density numbers can be used in Eqs. (2.20) through (2.27) to estimate drag perturbation effects.

Altitude (km)	Atmospheric Scale Height (km)	Atmospheric Density		Altitude (km)	Atmospheric Scale Height (km)	Atmospheric Density	
		Mean (kg/m^3)	Maximum (kg/m^3)			Mean (kg/m^3)	Maximum (kg/m^3)
0	8.4	1.2×10^0	1.2×10^0	1500	516	2.79×10^{-16}	1.16×10^{-15}
100	5.9	5.25×10^{-7}	5.75×10^{-7}	2000	829	9.09×10^{-17}	3.80×10^{-16}
150	25.5	1.73×10^{-9}	1.99×10^{-9}	2500	1220	4.23×10^{-17}	1.54×10^{-16}
200	37.5	2.41×10^{-10}	3.65×10^{-10}	3000	1590	2.54×10^{-17}	7.09×10^{-17}
250	44.8	5.97×10^{-11}	1.20×10^{-10}	3500	1900	1.77×10^{-17}	3.67×10^{-17}
300	50.3	1.87×10^{-11}	4.84×10^{-11}	4000	2180	1.34×10^{-17}	2.11×10^{-17}
350	54.8	6.66×10^{-12}	2.18×10^{-11}	4500	2430	1.06×10^{-17}	1.34×10^{-17}
400	58.2	2.62×10^{-12}	1.05×10^{-11}	5000	2690	8.62×10^{-18}	9.30×10^{-18}
450	61.3	1.09×10^{-12}	5.35×10^{-12}	6000	3200	6.09×10^{-18}	5.41×10^{-18}
500	64.5	4.76×10^{-13}	2.82×10^{-12}	7000	3750	4.56×10^{-18}	3.74×10^{-18}
550	68.7	2.14×10^{-13}	1.53×10^{-12}	8000	4340	3.56×10^{-18}	2.87×10^{-18}
600	74.8	9.89×10^{-14}	8.46×10^{-13}	9000	4970	2.87×10^{-18}	2.34×10^{-18}
650	84.4	4.73×10^{-14}	4.77×10^{-13}	10000	5630	2.37×10^{-18}	1.98×10^{-18}
700	99.3	2.36×10^{-14}	2.73×10^{-13}	15000	9600	1.21×10^{-18}	1.16×10^{-18}
750	121	1.24×10^{-14}	1.59×10^{-13}	20000	14600	7.92×10^{-19}	8.42×10^{-19}
800	151	6.95×10^{-15}	9.41×10^{-14}	20184	14600	7.92×10^{-19}	8.42×10^{-19}
850	188	4.22×10^{-15}	5.67×10^{-14}	25000	20700	5.95×10^{-19}	6.81×10^{-19}
900	226	2.78×10^{-15}	3.49×10^{-14}	30000	27800	4.83×10^{-19}	5.84×10^{-19}
950	263	1.98×10^{-15}	2.21×10^{-14}	35000	36000	4.13×10^{-19}	5.21×10^{-19}
1000	296	1.49×10^{-15}	1.43×10^{-14}	35786	37300	4.04×10^{-19}	5.12×10^{-19}
1250	408	5.70×10^{-16}	2.82×10^{-15}				

2.2.4 Perturbations from Solar Radiation

Solar radiation pressure causes periodic variations in all orbital elements. Its effect is strongest for satellites with low ballistic coefficients: light vehicles with large frontal areas. The magnitude of the acceleration in m/s^2 arising from solar radiation pressure is approximately

$$a_R \approx \frac{-4.5 \times (10)^{-6} A}{m} \quad (2.28)$$

where A is the cross-sectional area of the satellite exposed to the Sun in m^2 and m is the satellite's mass in kg. For satellites below 800 km altitude, acceleration from atmospheric drag is greater than that from solar radiation pressure; above 800 km, acceleration from solar radiation pressure is greater.

2.3 Orbit Maneuvering

At some point during the lifetime of most satellites, we must change one or more of the orbital elements. For example, we may need to transfer from an initial parking orbit to the final mission orbit, rendezvous with or intercept another satellite, or adjust for the perturbations discussed in the previous section. Most frequently, we must change the orbit's altitude, plane, or both. To change a satellite's orbit, we have to change the satellite's velocity vector in magnitude or direction. Most propulsion systems operate for only a short time compared to the orbital period, so we can treat the maneuver as an impulsive change in the velocity while the position remains fixed. For this reason, any maneuver changing the orbit of a satellite must occur at a point where the old orbit intersects the new orbit. If the two orbits do not intersect, we must use an intermediate orbit that intersects both. In this case, the total maneuver requires at least two propulsive burns.

In general, the change in the velocity vector to go from one orbit to another is given by

$$\Delta \vec{v} = \vec{v}_{Need} - \vec{v}_{Current} \tag{2.29}$$

We can find the current and needed velocity vectors from the orbital elements, keeping in mind that the position vector does not change much during impulsive burns.

2.3.1 Coplanar Orbit Transfers

The most common type of in-plane maneuver changes the orbit's size and energy, usually from a low-altitude parking orbit to a higher-altitude mission orbit such as a geosynchronous orbit. Because the initial and final orbit do not intersect (see Fig. 2.5), the maneuver requires a transfer orbit. Figure 2.5 shows a Hohmann[*] Transfer Orbit. In this case, the transfer orbit's ellipse is tangent to the initial and final circular orbits at the transfer orbit's perigee and apogee, respectively. The orbits are tangential, so the velocity vectors are collinear, and the Hohmann Transfer represents the most efficient transfer between two circular, coplanar orbits. For transfers from a smaller orbit to a larger one, the change in velocity is in the direction of motion; for transfers from a larger orbit to a smaller one, the change of velocity is opposite to the direction of motion.

The total change in velocity required for the transfer is the sum of the velocity changes at perigee and apogee of the transfer ellipse. Because the velocity vectors

---

[*] Walter Hohmann, a German engineer and architect, wrote *The Attainability of Celestial Bodies* [1925], which mathematically discusses the conditions for leaving and returning to Earth.

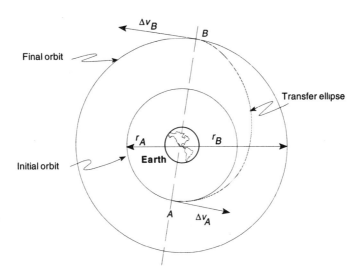

Fig. 2.5. Hohmann Transfer Ellipse Showing Orbit Transfer between Two Circular, Coplanar Orbits. The initial burn occurs at point A, putting us onto the transfer ellipse. The final burn, at point B, puts us into the final circular orbit. This sequence can be reversed to lower the altitude.

are collinear, the velocity changes are simply the differences in magnitudes of the velocities in each orbit. We can find these differences from the energy equation if we know the size of each orbit. If we know the initial and final orbits (r_A and r_B), we can calculate the semi-major axis of the transfer ellipse (a_{tx}) and the total velocity change (the sum of the velocity changes required at points A and B) using the following algorithm. Table 2.6 below illustrates transferring from an initial circular orbit of 6567 km to a final circular orbit of 42,160 km.

We can also write the total Δv required for a two-burn transfer between circular orbits at altitudes r_A and r_B as:

$$\Delta v_{Total} \equiv \Delta v_A + \Delta v_B \tag{2.30}$$

$$= \sqrt{\mu} \left[\left| \left(\frac{2}{r_A} - \frac{1}{a_{tx}}\right)^{\frac{1}{2}} - \left(\frac{1}{r_A}\right)^{\frac{1}{2}} \right| + \left| \left(\frac{2}{r_B} - \frac{1}{a_{tx}}\right)^{\frac{1}{2}} - \left(\frac{1}{r_B}\right)^{\frac{1}{2}} \right| \right] \tag{2.31}$$

2.3 Orbit Maneuvering

Table 2.6. Hohmann Transfer Equations. Given the initial and final circular orbit radii, we can calculate the velocity changes required to transfer between the circular orbits.

Step	Equations		Example
1	a_{tx}	$=(r_A + r_B)/2$	= 24,364 km
2	v_{iA}	$=\sqrt{\mu/r_A} = 631.3481/\sqrt{r_A}$	= 7.79 km/s
3	v_{fB}	$=\sqrt{\mu/r_B} = 631.3481/\sqrt{r_B}$	= 3.08 km/s
4	v_{txA}	$=\sqrt{\mu(2/r_A) - (1/a_{tx})}$ $= 631.3481\sqrt{(2/r_A) - (1/a_{tx})}$	= 10.25 km/s
5	v_{txB}	$=\sqrt{\mu(2/r_B) - (1/a_{tx})}$ $= 631.3481\sqrt{(2/r_B) - (1/a_{tx})}$	= 1.59 km/s
6	Δv_A	$= |v_{txA} - v_{iA}|$	= 2.46 km/s
7	Δv_B	$= |v_{fB} - v_{txB}|$	= 1.49 km/s
8	Δv_{Total}	$= \Delta v_A + \Delta v_B$	= 3.95 km/s
9	Time of transfer = $P/2$		= 5 hrs 15 min

where $\sqrt{\mu}$ = 631.3481 when Δv is in km/s and all of the semi-major axes are in km. As in step 1 above, $a_{tx} = (r_A + r_B)/2$.

The above expression applies to any coplanar Hohmann transfer. In the case of small transfers (that is, r_A close to r_B), we can conveniently approximate Δv in two forms:

$$\Delta v \approx v_{iA} - v_{fB} \tag{2.32}$$

$$\Delta v \approx 0.5(\Delta r/r) v_{A/B} \tag{2.33}$$

where

$$\Delta r \equiv r_B - r_A \tag{2.34}$$

and

$$r \approx r_A \approx r_B \qquad v_{A/B} \approx v_{iA} \approx v_{fB} \tag{2.35}$$

The two small burns are of nearly equal magnitude.

The result in Eq. (2.32) is more unusual than it may seem. Assume that a satellite is in a circular orbit with velocity v_{iA}. In two burns we *increase* the velocity by an amount Δv. The result is that the satellite is higher and traveling *slower* by the amount Δv. An example clarifies this result. Consider a satellite in a circular orbit at 400 km such that $r_A = 6778$ km and $v_{iA} = 7700$ m/s. We apply a total Δv of 20 m/s (= 0.26% of v_{iA}) in two burns of 10 m/s each. From Eq. (2.33) the total Δr is 0.52% of 6778 km or 35 km. Thus, the final orbit is circular at an altitude of 6813 km. Immediately following the first burn of 10 m/s the spacecraft is at the perigee of the transfer orbit with a velocity of 7710 m/s. When the spacecraft reaches apogee at 6813 km, it has slowed according to Kepler's second law by 0.52% to 7670 m/s. We then apply the second burn of 10 m/s to circularize the orbit at 7680 m/s, which is 20 m/s slower than its original velocity. By adding energy to the spacecraft, we raised the orbit and lowered kinetic energy but added enough potential energy to make up for the reduced speed and the added Δv.

Sometimes we may need to transfer a satellite between orbits in less time than that required for a Hohmann transfer. Fig. 2.6 shows a faster transfer called the *One-Tangent Burn*. In this instance the transfer orbit is tangential to the initial orbit. It intersects the final orbit at an angle equal to the transfer orbit's flight-path angle at the point of intersection. An infinite number of transfer orbits are tangential to the initial orbit and intersect the final orbit at some angle. Thus, we may choose the transfer orbit by specifying its size, the angular change of the transfer, or the time needed to complete the transfer. We can then define the transfer orbit and calculate the required velocities.

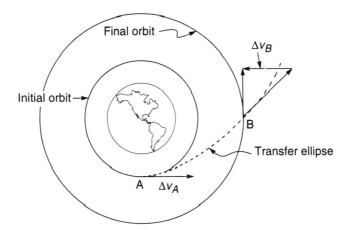

Fig. 2.6. **Transfer Orbit Using a One-Tangent Burn between Two Circular, Coplanar Orbits.** The first burn is made at point A, tangent to the initial orbit. The second burn, at point B, circularizes at the final orbit but the burn is not tangent to any orbit. The time for this transfer is shorter than with a Hohmann transfer but requires more Δv.

For example, we may specify the size of the transfer orbit, choosing any semi-major axis that is greater than the semi-major axis of the Hohmann transfer ellipse. Once we know the semi-major axis of the ellipse (a_{tx}), we can calculate the eccentricity, angular distance travelled in the transfer, the velocity change required for the transfer, and, using the equations in Table 2.7, the time required to complete the transfer.

Table 2.7. Computations for One-Tangent Burn Orbit Transfer. These equations can be used to calculate the velocity change requirements and transfer orbit parameters.

Quantity	Equation
eccentricity	$e = 1 - r_A / a_{tx}$
true anomaly at second burn (deg)	$v = \cos^{-1}[(a(1 - e^2) / (r_B - 1)) / e]$
flight-path angle at second burn (deg)	$\phi = \tan^{-1}[e \sin v / (1 + e \cos v)]$
initial velocity (km/s)	$v_{iA} = 631.3481 / \sqrt{r_A}$
velocity on transfer orbit at initial orbit (km/s)	$v_{txA} = 631.3481 \sqrt{(2/r_A) - (1/a_{tx})}$
initial velocity change (km/s)	$\Delta v_A = |v_{txA} - v_{iA}|$
final velocity (km/s)	$v_{fB} = 631.3481 / \sqrt{r_B}$
velocity on transfer orbit at final orbit (km/s)	$v_{txB} = 631.3481 \sqrt{(2/r_B) - (1/a_{tx})}$
final velocity change (km/s)	$\Delta v_B = \sqrt{v_{fB}^2 + v_{txB}^2 - 2 v_{fB} v_{txB} \cos v}$
total velocity change (km/s)	$\Delta v_T = \Delta v_A + \Delta v_B$
eccentric anomaly (rad)	$E = \tan^{-1}\left[\sqrt{1-e^2} \sin v / (e + \cos v)\right]$
time-of-flight (min)	$TOF = 631.34816 / \sqrt{a^3} \; (E - e \sin E)$, E in rads

Table 2.8 compares the total velocity change required and time-of-flight for a Hohmann transfer and a one-tangent burn transfer from a low-altitude parking orbit to geosynchronous orbit.

Table 2.8. Comparison of Coplanar Orbit Transfers from LEO to Geosynchronous Orbit. The one-tangent burn is faster than the Hohmann transfer. However, the transfer ellipse is larger and the velocity requirements are greater.

Variable	Hohmann Transfer	One-Tangent-Burn
r_A	6570 km	6570 km
r_B	42,200 km	42,200 km
a_{tx}	24,385 km	28,633 km
ΔV_T	3.935 km/s	4.699 km/s
TOF	5.256 hr	3.457 hr

Another way to change an orbit's size is to use a constant low-thrust burn, which results in a *spiral transfer*. We can approximate the velocity change for this type of orbit transfer by

$$\Delta v = |v_2 - v_1| \qquad (2.36)$$

where v_1 and v_2 are the circular velocities of the two orbits. Using a spiral transfer for the previous example, we find the total velocity change required to go from a low-Earth orbit to geosynchronous orbit is 4.71 km/s. Subtracting the results of step 3 from the results of step 2 in Table 2.6 gives us this value.

2.3.2 Orbit-Plane Changes

To change the orientation of the satellite's orbital plane (typically the inclination), we must change the direction of the velocity vector (Fig. 2.7). This maneuver requires a component of Δv to be perpendicular to the orbital plane and, therefore, perpendicular to the initial velocity vector. If the size of the orbit remains constant, the maneuver is called a *simple plane change* (Fig. 2.7a). We can find the required change in velocity by using the law of cosines. For the case in which v_f is equal to v, this expression reduces to

$$\Delta v = 2v_i \sin\left(\frac{\theta}{2}\right) \qquad (2.37)$$

where v_i is the velocity before and after the burn, and θ is the required angle change. Let us suppose that we are transferring from a low-altitude ($h = 185$ km) inclined ($i = 28$ deg) orbit to an equatorial orbit ($i = 0$) at the same altitude. The required change in velocity (Δv) is 3.77 km/s, with $r = 6563$ km and $v_i = 7.79$ km/s.

From Eq. (2.37) we see that if the angular change is 60 deg, the required change in velocity is equal to current velocity. Plane changes are very expensive in terms of fuel usage. To save fuel, we should change the plane when a satellite's velocity is at a minimum: at apogee for an elliptical orbit. In some cases, it may even be cheaper to boost the satellite into a higher orbit, change the orbit plane at apogee, and return the satellite to its original orbit.

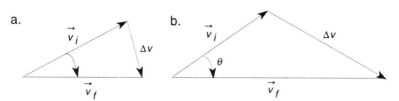

Fig. 2.7. **Vector Representation of Simple and Combined Changes in Velocity Direction.** For the simple plane change, the initial and final velocities are equal in magnitude.

Typically, orbital transfers require changes in the orbit's size and plane. An example would be transferring from an inclined parking orbit at low altitude to a zero-inclination orbit at geosynchronous altitude (see Chap. 10). We can do this transfer in two steps: a Hohmann transfer to change the orbit's size and a simple plane change to make the orbit equatorial. A more efficient method (less total change in velocity) would be to combine the plane change with a tangential burn at apogee of the transfer orbit (Fig. 2.7b). As we must change the velocity vector's magnitude and direction, we can find the required change in velocity using the law of cosines:

$$\Delta v = \sqrt{\left(v_i^2 + v_f^2 - 2v_iv_f \cos\theta\right)} \qquad (2.38)$$

where v_i is the initial velocity, v_f is the final velocity, and θ is the angle change required.

For example, we find the total change in velocity to transfer from a Shuttle parking orbit to a geosynchronous equatorial orbit as follows:

$r_i = 6563$ km $\qquad r_f = 42{,}159$ km
$i_i = 28$ deg $\qquad i_f = 0$ deg
$v_i = 7.79$ km/s $\qquad v_f = 3.08$ km/s
$\Delta v_A = 2.46$ km/s $\qquad \Delta v_B = 1.83$ km/s
$\Delta v_{Total} = 4.29$ km/s

Completing a Hohmann transfer followed by a simple plane change would require a velocity change of 5.44 km/s, so the Hohmann transfer with a combined plane change at apogee of the transfer orbit saves 1.15 km/s. As Eq. (2.38) shows, a small plane change ($\theta \geq 0$) can be combined with an energy change for almost no cost in Δv or propellant. Thus, standard practice is to do a geosynchronous transfer with a small plane change at perigee and most of the plane change at apogee.

Another option is to complete the maneuver using three burns. The first burn is a coplanar maneuver placing the satellite into a transfer orbit with an apogee much higher than the final orbit. When the satellite reaches apogee of the transfer orbit, a combined plane-change maneuver places the satellite in a second transfer orbit. This second orbit is coplanar with the final orbit and has a perigee altitude equal to the altitude of the final orbit. Finally, when the satellite reaches perigee of the second transfer orbit, another coplanar maneuver places the satellite into the final orbit. This three-burn maneuver may save fuel, but the fuel savings comes at the expense of the total time required to complete the maneuver.

2.3.3 Orbit Rendezvous

Orbital transfer becomes more complicated when the objective is to rendezvous with or intercept another object in space: both the interceptor and target must arrive at the rendezvous point at the same time. This precision demands a *phasing orbit*—any orbit which results in the interceptor's achieving the desired geometry relative to the target to start a Hohmann transfer. If the initial and final orbits are circular, coplanar, and of different sizes, the phasing orbit is simply the initial interceptor orbit (Fig. 2.8). The interceptor remains in the initial orbit until the relative motion between the interceptor and target results in the desired geometry. At that point, we would inject the interceptor into a Hohmann transfer orbit. The equation to solve for the wait time in the initial orbit is

$$\text{Wait Time} = \frac{(\phi_i - \phi_f + 2k\pi)}{(\omega_{int} - \omega_{tgt})} \qquad (2.39)$$

where ϕ_f = phase angle (angular separation of target and interceptor) needed for rendezvous (rad)
ϕ_i = initial phase angle (rad)
k = number of rendezvous opportunities, (for the first opportunity, $k = 0$)
ω_{int} = angular velocity of the interceptor (rad/s)
ω_{tgt} = angular velocity of the target (rad/s)

We calculate the lead angle (α_L) by multiplying ω_{tgt} by the time of flight for the Hohmann transfer; ϕ_f is π radians minus α_L.

The total time to rendezvous is equal to the wait time from Eq. (2.39) plus the time-of-flight of the Hohmann transfer orbit.

The denominator in Eq. (2.39) represents the relative motion between the interceptor and target. As the size of the interceptor orbit approaches the size of the target orbit, the relative motion approaches zero, and the wait time approaches infinity. If the two orbits are exactly the same, the interceptor must enter a new phasing orbit to rendezvous with the target (Fig. 2.9). The rendezvous occurs at the point where the interceptor enters the phasing orbit. The period of the phasing orbit is equal to the time it takes the target to get to the rendezvous point. Once we know the period, we can calculate the semi-major axis. The two orbits are tangential at their point of intersection, so the change in velocity is again the difference in magnitudes of the two velocities at the point of intersection of the two orbits.

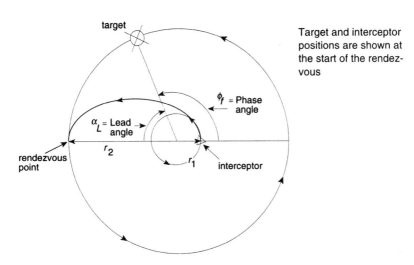

Fig. 2.8. Geometry Depicting Rendezvous between Two Circular, Coplanar Orbits. The phase angle is the angular separation between the target and interceptor at the start of the rendezvous. The lead angle is the distance the target travels from the start until rendezvous occurs.

Because we know the size of the two orbits, and therefore, the energies, we can use the energy equation (2.5) to solve for the current and needed velocity.

Satellites in circular orbits often must adjust their relative phasing in the orbit. We can adjust this phasing by making the satellite drift relative to its initial position. The *drift rate* in deg/orbit is given by

$$drift\ rate = 1080\left(\frac{\Delta v}{v}\right) \qquad (2.40)$$

where v is the nominal orbit velocity (km/s) and Δv is the velocity change (km/s) required to start or stop the drift.

2.4 Launch Windows

Similar to the rendezvous problem is the launch-window problem, or determining the appropriate time to launch from the Earth's surface into the desired orbital plane. Because the orbital plane is fixed in inertial space, the launch window is the time duration when the launch site on the Earth's surface rotates through the orbital plane. As Fig. 2.10 shows, the time of the launch depends on the launch site's latitude and longitude and the satellite orbit's inclination and right ascension of ascending node.

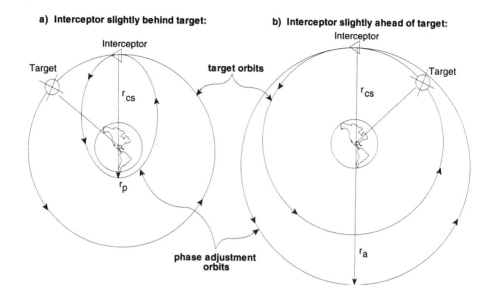

Fig. 2.9. Rendezvous from Same Orbit Showing the Target Leading and Trailing the Interceptor. By decreasing (a) or increasing (b) the interceptor's orbit period, we can rendezvous with the target.

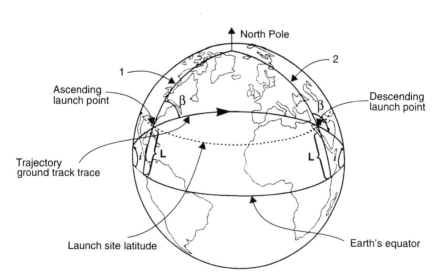

Fig. 2.10. Geometry for Launches on the Ascending Trajectory (1) and Descending Trajectory (2). The angles shown are the orbit inclination (i), launch-site latitude (L), and launch azimuth (β). δ is the equatorial angle between the launch site longitude and the trajectory node or intersection with the equator.

For a launch window to exist, the launch site must pass through the orbital plane, placing restrictions on the orbital inclinations (i) possible from a given launch latitude (L):

- No launch windows exist if $L > i$ for direct orbit or $L > 180$ deg $- i$ for retrograde orbits.
- One launch window exists if $L = i$ or $L = 180$ deg $- i$.
- Two launch windows exist if $L < i$ or $L < 180$ deg $- i$.

The *launch azimuth* (β) is the angle measured clockwise from north to the velocity vector. If a launch window exists, the launch azimuth required to achieve an inclination (i) from a given launch latitude (L) is

$$\beta = \beta_I \pm \gamma \approx \beta_I \tag{2.41a}$$

where

$$\sin \beta_I = \frac{\cos i}{\cos L} \tag{2.41b}$$

and

$$\tan \gamma = \frac{v_L \cos \beta_I}{v_0 - v_{eq} \cos i} \approx \left(\frac{v_L}{v_0}\right) \cos \beta_I \tag{2.41c}$$

where v_L = inertial velocity of the launch site given by Eq. (2.45) below (km/s)
v_{eq} = 464.5 m/s is the velocity of the Earth's rotation at the equator
$v_0 \approx$ 7800 m/s is the velocity of the satellite immediately after launch
β_I = inertial launch azimuth (deg)
γ = a small correction to account for the velocity added because of the Earth's rotation (deg)

For launches to low-Earth orbit, γ ranges from 0 for a due east launch to 3.0 deg for launch to a polar orbit. The approximation for γ in Eq. (2.41c) is good to better than 0.1 deg for low-Earth orbits. For launches near the ascending node, β is in the first or fourth quadrant, and the plus sign applies in Eq. (2.41a). For launches near the descending node, β is the second or third quadrant, and the minus sign applies in Eq. (2.41a).

If we let δ, shown in Fig. 2.10, be the angle in the equatorial plane from the nearest node to the longitude of the launch site, we can determine δ from

$$\sin \beta_I = \frac{\cos \beta}{\sin i} \tag{2.42}$$

where δ is positive for direct orbits and negative for retrograde orbits. Finally, the *local sidereal time (LST)* of launch is the angle at the time of launch from the vernal equinox to the longitude of the launch site:

$$LST = \Omega + \delta \quad \text{(launch at ascending node)}$$
$$= \Omega + 180 \text{ deg} - \delta \quad \text{(launch at descending node)} \quad (2.43)$$

where Ω is the right ascension of the ascending node of the resulting orbit.

Having calculated the launch azimuth required to achieve the desired orbit, we can now calculate the velocity needed to accelerate the payload from rest at the launch site to the required burnout velocity. To do so, we would use *topocentric-horizon* coordinates with the origin on the Earth's surface and velocity components v_S, (southward component), v_E (eastward component), and v_Z (northward component):

$$v_S = -v_{bo} \cos \phi \cos \beta_b$$
$$v_E = v_{bo} \cos \phi \sin \beta_b - v_L$$
$$v_Z = v_{bo} \sin \phi \quad (2.44)$$

where v_{bo} is the velocity at burnout (usually equal to the circular orbit velocity at the prescribed altitude), ϕ is the flight-path angle at burnout (see Fig. 2.1), β_b is the launch azimuth at burnout, and v_L is the velocity of the launch site on the Earth at a given latitude (L) as given by

$$v_L = 464.5 \cos L \text{ (in m/s)} \quad (2.45)$$

Equation (2.44) does not include losses in the velocity of the launch vehicle because of atmospheric drag and gravity—approximately 1500 m/s for a typical launch vehicle (see Table 2.10). Also, in Eq. (2.44) we assume that azimuth at launch and the azimuth at burnout are the same. Changes in latitude and longitude of the launch vehicle during powered flight introduce small errors into the calculation of the burnout conditions. We can calculate the velocity required at burnout from the energy equation if we know the semi-major axis and radius of burnout of the orbit.

2.5 Orbit Maintenance

Once in their mission orbits, many satellites need no additional orbit adjustments. On the other hand, mission requirements may demand that we maneuver the satellite to correct the orbital elements when perturbing forces have changed them. Two particular cases of note are satellites requiring repeating ground tracks and geosynchronous equatorial satellites.

Using two-body equations of motion, we can show that a satellite has a repeating ground track if it has exactly an integer number of revolutions per integer number of days. Its period must therefore be

$$P = \frac{m \text{ (sidereal days)}}{k \text{ (revolutions)}} \quad (2.46)$$

where m and k are integers, and 1 sidereal day = 1436.068 min. For example, a satellite orbiting the Earth exactly 16 times per day has a period of 89.75 min and a semi-major axis of 6640 km.

Next, we modify the satellite's period to account for the drift in the orbital plane caused by the Earth's oblateness (J_2). We can calculate the rate of change of the right ascension of ascending node, $\Delta\Omega$, because of J_2 from the two-body orbital elements. In this case the new period is

$$P_{New} = P_{Two-body} + \frac{\Delta\Omega}{\omega_{Earth}} \qquad (2.47)$$

Because the nodal drift is based on the two-body orbital elements, we must iterate to find the new orbital period and semi-major axis. Continuing with the previous example, we assume a perigee altitude of 120 km and an inclination of 45 degrees. In this case, the compensated period is 88.20 min, and the new semi-major axis is 6563 km.

Table 2.9 lists several examples of spacecraft placed in orbits with repeating ground tracks.

Table 2.9. Examples of Repeating Ground Tracks. These types of orbits occur when the orbital period divides into the length of a day (or an integer number of days), giving us an exact integer.

Satellite	Semi-major axis (km)	Orbit Revs per Repetition	Repeat Period (Days)
SEASAT	7168.3	43	3
LANDSAT 4/5	7077.8	233	16
GEOSAT	7173.6	244	17

The Earth's oblateness also causes the direction of perigee to rotate around the orbit. If the orbit is noncircular and the mission places limits on the altitude over specific targets, we must control the location of perigee. One possibility is to select the inclination of the orbit to be at the critical inclination (at 63.4 deg for a direct orbit and 116.4 deg for a retrograde orbit where the rotation of perigee is zero), so the location of perigee is fixed. If other constraints make this selection impossible, we must maintain the orbit through orbital maneuvers. We can change the location of perigee by changing the flight-path angle by an angle θ. Only the direction of the velocity vector is changing, so we can find the change in velocity from the equation for a simple plane change; see Eq. (2.37).

A final consideration for a low-altitude orbit with repeating ground tracks is the change in the semi-major axis and eccentricity due to atmospheric drag. Drag causes the orbit to become smaller. As the orbit becomes smaller, the period is also reduced, causing the ground track to appear to shift eastward. If some tolerance is specified, such as a maximum distance between the actual and desired ground track, the satellite must periodically maneuver to maintain the desired orbit.

We can use Eq. (2.21) to calculate the change in semi-major axis (Δa) per revolution of the orbit. Given the change in the orbit's size, we can also determine the change in the period (ΔP in seconds per orbit):

$$\Delta P = 3\pi \Delta a / (na) \tag{2.48}$$

If constraints exist for either the period or semi-major axis of the orbit, we can use Eqs. (2.21) and (2.48) to keep track of the period and semi-major axis until the orbit needs to be corrected. Applying a tangential velocity change at perigee adjusts the semi-major axis when required. Again, we can find the current and needed velocities from the energy equation (2.5) because we know the size, and therefore the energy, of the two orbits.

Geosynchronous equatorial orbits also require orbital-maintenance maneuvers. Satellites in these orbits drift when perturbations occur from the nonspherical Earth and from third-body interactions with the Sun and Moon. Matching the period of a geostationary orbit with the Earth's rotational velocity results in a resonance with the first sectoral harmonic term in the geopotential [see Eq. (2.17)]. This resonance term results in a transverse acceleration—an acceleration in the orbital plane—that causes the satellite to drift back and forth in longitude (*East-West drift*). The Sun and the Moon cause out-of-plane accelerations which make the satellite drift in latitude (*North-South drift*).

North-South stationkeeping is necessary when mission requirements limit either the latitude drift or inclination drift. If not corrected, the inclination of the orbit varies between 0 and 15 degrees with a period of approximately 55 years. The approximate equations to solve for the worst-case change in velocity are

$$\Delta v_{Moon} = 102.67 \cos\alpha \sin\alpha \quad \text{(m/s per year)} \tag{2.49}$$
$$\approx 36.93 \text{ m/s per year, for } i = 0$$

$$\Delta v_{Sun} = 40.17 \cos\gamma \sin\gamma \quad \text{(m/s per year)} \tag{2.50}$$
$$\approx 14.45 \text{ m/s per year, for } i = 0$$

where α is the angle between the orbital plane and the Moon's orbit, and γ is the angle between the orbital plane and ecliptic.

The transverse acceleration caused by resonance with the first sectoral harmonic term results in periodic motion about either of two stable longitudes at approximately 75 deg and 255 deg east longitude. If a satellite is placed at any other longitude, it will tend to orbit the closest of these two longitudes, resulting in east-west drift of up to 180 deg with periods of up to 900 days. Let us suppose a mission for a geostationary satellite specifies a required longitude (l_D). The change in velocity required to compensate for the drift and keep the satellite near the specified longitude is

$$\Delta v = 1.715 \sin(2||l_D - l_s|) \text{ (m/s per year)} \tag{2.51}$$

where l_D is the desired longitude and l_s is the closest stable longitude.

For example, if the desired longitude is 60° west, we find the velocity change required for one year as

$$l_D = -60° \qquad l_s = 255°$$

$$\Delta v = 1.715 \text{ m/s per year}$$

After the mission of the satellite is complete, several options exist, depending on the orbit. We may allow low-altitude orbits to decay and reenter the atmosphere or use a velocity change to speed up the process. We may also boost satellites at all altitudes into benign orbits to reduce the probability of collision with active payloads, especially at synchronous altitudes. Because coplanar velocity changes are more efficient than plane changes, we would normally apply tangential changes in velocity. Their magnitude would depend on the difference in energy of the two orbits. For example, the velocity change required to deorbit (drop perigee altitude to 0 km) a satellite in a circular orbit at radius (r) and velocity (v) is

$$\Delta v_{deorbit} \approx v \left(\frac{2R_E}{R_E + r} \right) \tag{2.52}$$

We need not reduce perigee altitude to 0 km. If we choose a less conservative deorbit altitude ($H_{deorbit}$) we can determine the deorbit Δv from Eq. (2.52) by replacing R_E with ($R_E + H_{deorbit}$). Note that only perigee is reduced in the deorbit burn. Reducing perigee to 100–150 km could result in several orbits over which apogee is reduced before the spacecraft reenters and might not allow adequate control of the deorbit conditions.

2.6 Earth to Orbit

In considering a new launch system that can deliver a given payload mass to a required orbit, a design team finds that getting the payload into low-Earth orbit (LEO)[*] forces them to focus on the launch vehicle's configuration. Various designs share some common features, dictated by the imposed environment, the operational constraints, and the velocity requirements to achieve LEO.

---

[*] We do not have to obtain an established low-Earth orbit. Planetary missions have used direct ascent: a continuous burn from sub-orbital velocity to the Earth-escape hyperbola, which becomes the elliptic transfer to the target planet. An example is the Mariner-Mars '69 mission.

We must recognize how much our early decisions affect the basic features of the propulsion and stage designs. For example, how many stages should we use to achieve LEO? How many, if any, stages should we recover?—complex and difficult questions that we merely mention here. Our point is that we can profitably work with the vehicle designers to balance the partially conflicting goals of performance, reliability, cost, and flexibility.

The lower stages (or stage) of the launch vehicle must place the upper stages into LEO. The size, shape, and mass of this "LEO payload" depend on its mission and the design adopted to achieve it. We must make sure these solutions integrate with the overall launch system. The vehicle configuration (stack-up) must withstand the dangers of moving through the atmosphere and other key events:

- Lift-off transient: mechanical vibration and acoustic noise
- Transonic flight (0.75 < Mach number < 1.2) buffet, acoustic noise, compartment venting transients, some pressure loads at aerodynamic limits
- Maximum dynamic pressure, (q_{max}) region (1.4 < M < 1.9, typical)
- Maximum aerodynamic heating region: region depends strongly on the vehicle's velocity and altitude history (20 km < alt < 70 km, typical)
- Engine start and shutdown transients: mechanical vibration and control issues associated with unsymmetrical starting or shutdown (multi-engine stages, primarily)
- Jettison transients: jettisoning solid rockets, lower stages, or payload fairings; shock and vibration, engine-exhaust blowback.

We need to address all applicable issues mentioned above to reliably place a payload into the desired Earth orbit. Other design issues, unique to a new vehicle, will probably arise. Section 2.6.2 discusses how to determine the propulsive Δv a launch system must deliver during the ascent stages.

2.6.1 The Effect of Vehicle Staging

Section 1.1.5 and Eq. (1.28) demonstrate that certain missions are not impossible, depending on what we can achieve in terms of specific impulse and the vehicle's inert-mass fraction. We also find that certain missions become impractical if performance is outside a certain envelope.* One way to overcome this problem is to use staging. Staging allows us to discard inert vehicle mass as it becomes unnecessary for our mission. This strategy allows us to reduce the effec-

---

* In Fig. 1.5, we see that, at lower specific impulse and a given inert-mass fraction, initial vehicle mass can grow very quickly. Small changes to a design can make huge changes in the complete system's mass, a very unstable and uncomfortable design situation. For example, with an inert-mass fraction of 0.08, we would like to keep specific impulse above 430 s to avoid the near-vertical section of the initial mass curve.

tive inert-mass fraction, enabling missions that are otherwise impossible or impractical.

Ideally, we would like to dispose of inert mass continuously. As propellant is consumed, an increasing amount of the propellant tank becomes unnecessary.* Because this method of disposal is usually impractical, we allow the maneuver to proceed for a time and then discard a large chunk of inert mass, which we call a *stage*.

So, how do we decide on the number of stages? We must do an important trade to answer this question. On one hand, increasing the number of stages decreases the mass of the vehicle (although the amount of payoff for succeeding stages decreases asymptotically). On the other hand, increasing the number of stages increases cost and complexity while decreasing reliability.

As an example, consider a simplified system in which the specific impulse and inert-mass fraction for each stage are the same. Further, assume we divide the Δv for the maneuver evenly between stages. Figure 2.11 shows how the initial vehicle mass decreases with increasing numbers of stages. The mass payoff becomes less significant as the number of stages increases. We have removed dimensions from the result by dividing the initial mass by the payload mass.

The bottom line is, as usual, not as concrete as we would like! Staging can greatly reduce our vehicle mass at the expense of cost, complexity, and reliability. An analysis similar to that shown in Fig. 2.11 shows us the mass savings, but we need to weigh these savings against the comfort or cost level of the additional stages. A direct relationship exists between complexity or cost and the number of stages required, so we want to choose the minimum number of stages that make our system practical.

What about parallel stages? In numerous existing and proposed launch vehicles, boosters (strap-ons) have been added to a core first stage to increase either payload or final Δv to orbit. Additional serial stages or increased capability of the upper stage usually benefits total mass more than a parallel stage. However, modifying the upper stage is usually more complicated than adding strap-ons. Although the particular reasons for selecting a parallel stage are numerous, we usually do so to enhance an existing or already baselined system for which we wish to minimize modification cost and complexity.

Another reason for adding a parallel stage is to increase the system's thrust-to-weight ratio for a short period at the start of the ascent. Examples of this technique are the Space Shuttle and the Atlas (the Shuttle uses solid strap-ons, and the Atlas uses parallel liquid boost-engines). The main or sustainer engines provide enough thrust to act as a sustainer after some of the propellant mass has been expended but have insufficient thrust to lift the entire vehicle mass off the launch pad. This approach has two benefits: we increase the initial thrust-to-weight ratio and we

---

* The ultimate situation occurs when propellant also serves as structure. In this extreme case, the effective inert-mass fraction is zero.

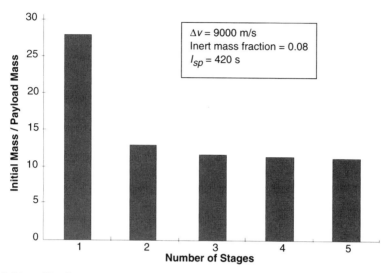

Fig. 2.11.   Varying the Vehicle's Initial Mass versus the Number of Stages. Mass decreases greatly (54%) as we shift from one to two stages. The initial mass continues to decrease as we increase the number of stages, but the payoff is not as significant.

can dispose of, or stage, the boosters shortly after lift-off, increasing our effective inert-mass fraction for the stage.

2.6.2   Velocity Budget to Low-Earth Orbit (LEO)

The following discussion focuses on the requirements placed on the propulsion systems used to achieve an Earth (or other central-body) orbit from a surface launch. We simply want to show how the Δv budget is allocated but are not implying that typical launch-trajectory analyses duplicate this approach. Trajectory analysis is discussed in detail in Sec. 2.6.4.

Consider a launch vehicle at an arbitrary point in its ascent trajectory but still in the sensible atmosphere* (Fig. 2.12). We must apply propulsive forces and control steering so the LEO payload, which may include the last used stage, achieves its desired orbit. The on-orbit arrival must occur within all inherent vehicle constraints and externally imposed limits, such as flight corridors that meet range-safety requirements. The integrals shown below are evaluated in a rotating coor-

---

* The *sensible* atmosphere has no strict meaning but conveys the sense of "still important to vehicle design." The altitudes where concerns disappear vary widely for different disciplines. Free molecular heating of spacecraft can still be important at 120 km or higher, whereas vacuum thrust is effectively achieved by 31 km, where the pressure has dropped below 1% of the sea-level value.

dinate system and give the change of magnitude of the velocities. In this section, we assume the Earth is nonrotating (rotation effects shown in Table 2.10).

$$V_{LEO} = \int_{ign}^{bo} \left[\frac{F \cos \alpha' - D}{m} - g_l \sin \gamma \right] dt \qquad (2.53)$$

Rewriting the equation slightly will isolate the primary role of the propulsion system(s):

$$V_{LEO} = \int_{ign}^{bo} \frac{F}{m} dt - \int_{ign}^{bo} \frac{F}{m}(1 - \cos \alpha') dt - \int_{ign}^{bo} \frac{D}{m} dt - \int_{ign}^{bo} g_l \sin \gamma \, dt \qquad (2.54)$$

$$(\Delta V_{prop}) \quad (\Delta V_{steering}) \quad (\Delta V_{drag}) \quad (\Delta V_{gravity})$$

or

$$\Delta V_{prop} = V_{LEO} + \Delta V_{grav} + \Delta V_{drag} + \Delta V_{steering} \qquad (2.55)$$

where F = time history of thrust, which depends on altitude in the sensible atmosphere and on possible throttling schedule (N)
m = time history of mass of the launch vehicle (kg)
D = time history of drag (N)
v = velocity of the center of mass (m/s)
g_l = local gravity acceleration at $r(t)$, (m/s^2)
r = radius from center of Earth to vehicle (m)
t = time (s)
α = pitch angle of attack (rad)
δ = thrust deflection angle (rad) (typically near zero)
α' = $\alpha + \delta$, angle between thrust vector and current velocity (rad)
γ = local flight-path angle, angle from local horizontal to velocity (rad)

Table 2.10 gives representative values of these velocities for several launch systems.

Ideal Propulsive Velocity—If multiple starts and shutdowns occur for a given stage, we must evaluate the equation over the appropriate intervals, simplifying wherever practical (Sec. 1.1). Note that expending some or all of the velocity capability does not necessarily change the magnitude of the vehicle's velocity. The expenditure may merely compensate for some dissipative acceleration such as drag or change the velocity direction. Typically, using it for several purposes simultaneously is most effective. The propulsive velocity does not depend only on a stage's engine or the propellant characteristics and mass. It also depends on the mass of all subsequent stages and the ascent profile up to full vacuum thrust.

Velocity at Orbit Injection (v_{LEO})—The net inertial velocity (m/s) to achieve a circular orbit about a central body is

$$V_{circ} = \sqrt{\frac{\mu}{r}} \qquad (2.56)$$

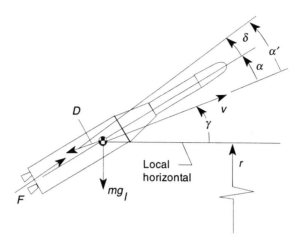

Fig. 2.12. Forces on a Vehicle in Planar Ascent Trajectory. The figure shows the various forces on a vehicle and the angles relating them to the vehicle's geometry.

Table 2.10. Velocity Budgets to Low-Earth Orbits for Selected Launch Vehicles. Ascent ΔVs vary between 8.8 and 9.3 km/s. Gravity, drag, and steering losses increase the effective velocity change above the velocity required in low-Earth orbit. The rotation of the Earth can be an advantage.

Vehicle	Orbit: $h_p \times h_a$ inclination (deg)	v_{LEO}	Δv_{grav}	$\Delta v_{steering}$	Δv_{drag}	Δv_{rot}*	$\Sigma \Delta v = \Delta v_{prop}$
Ariane A-44L	170 × 170† 7.0	7802	1576	38	135	−413	9138
Atlas I	149 × 607 27.4	7946	1395	167	110	−375	9243
Delta 7925	175 × 319 33.9	7842	1150	33	136	−347	8814
Space Shuttle	−196 × 278 28.5	7794‡	1222	358	107	−395	9086**
Saturn V	176 × 176 28.5	7798	1534	243	40	−348	9267
Titan IV/ Centaur	157 × 463 28.6	7896	1442	65	156	−352	9207

*Negative sign indicates beneficial effect of rotation.
†The third stage of Ariane 44L uses a continuous burn into a geosynchronous transfer orbit. We have arbitrarily terminated this burn at ≈ 170 km to give a better comparison with other vehicles.
‡Injection occurs at ≈ 111 km.
**An additional ΔV = 144 m/s is required to circularize at apogee.

where v_{circ} = circular orbit velocity (m/s)
  μ = central body's gravitational constant—the product of the central body mass and Newton's gravitational constant. $\mu = 3.98600 \times 10^{14}$ (m³/s²) for Earth
  r = distance from center of the body to the orbiting object. For Earth, $r = (R_E + h_{orbit}) \times 10^3$ (m)
  $R_E = 6378.14$ km, h_{orbit} = orbital altitude (km)

If the injection is into an elliptical orbit of semi-major axis (a in m), the injection velocity is

$$v_{inj} = \sqrt{\mu\left(\frac{2}{r} - \frac{1}{a}\right)} \qquad (2.57)$$

where v_{inj} = orbit injection velocity (m/s)
  r = the radius at which v_{inj} corresponds (m)
  a = semi-major axis of the elliptical orbit (m)

Assume the orbit altitude is 300 km. v_{circ} at this altitude is 7726 m/s. The propulsive stages must produce more ideal velocity than this, however, because of losses due to gravity, steering, and drag. A trajectory designer tries to minimize the sum of these losses within relatively tight constraints.

Gravity Loss—Gravity loss is typically the largest of the three Δv loss terms. During the intervals between firings, for positive flight-path angles, gravity continues to retard the vehicle, and the gravity-loss term increases. In some cases, however, significant time passes between firings. An example would be transferring from near perigee to apogee before circularizing the orbit. If so, we can more easily find the velocity states by using the standard methods of orbital mechanics than by evaluating the gravity-loss term. By qualitatively inspecting the integral, we see that shallower flight-path angles and shorter burn times (higher accelerations) produce smaller gravity losses. But these features may be in conflict with other aspects of vehicle operation. Limiting the acceleration levels for manned flight is a notable example. The gradual decay of the local gravity acceleration as a function of altitude only slightly affects the gravity-loss integral. The local g at 300 km is 91% of the reference value at the Earth's surface.

Drag Loss—The aerodynamic drag usually opposes the thrust, although as seen in Eq.(2.53), only when $\alpha' = 0.0$ are the two forces diametrically opposite. The drag force* is expressed in terms of a dimensionless coefficient (C_D), the flight dynamic pressure (q), and a reference area for the launch vehicle (or for each stage subjected to drag). The drag, expressed in newtons, is

---

* In missile aerodynamics it is customary to express the aerodynamic forces in the body axis system. For the planar case, axial and normal forces parallel and perpendicular to the longitudinal axis are resolved into the wind axes implied here. The wind axes drag term is used because all losses are evaluated along the vehicle velocity axis.

where
$$D = C_D\, q\, A_{ref} \qquad (2.58)$$

$$q = 1/2\, \rho\, v^2 \qquad (2.59)$$

and where ρ = atmospheric density (kg/m³)
A_{ref} = reference area (m²)

We do not provide the analysis techniques for estimating the drag of launch vehicles.* Fig. 2.13 gives a representative drag-coefficient variation with Mach number. Although the C_D function is unique to each launch vehicle, typical configurations have some "family resemblance." Figure 2.13† represents the configuration sketched in Fig. 2.12. Most of the drag loss accumulates in the first stage flight; the C_D function's importance dwindles as q diminishes in the trajectory.

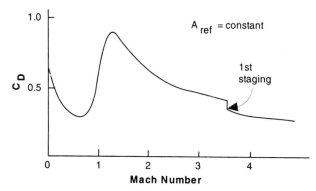

Fig. 2.13.   Typical Variation of Drag Coefficient with Mach Number. Drag coefficient increases near the sonic velocity and then decreases to Mach 4. The coefficient is essentially constant above Mach 4 (hypersonic speeds). This figure can be used as a first-order approximation for vehicles similar to that shown in Fig. 2.12.

Table 2.10 gives some representative values of drag loss. These values are relatively modest in terms of the overall velocity budget required. A larger vehicle has less drag loss than a smaller one with similar geometric shape, acceleration profile, and dynamic pressure history. This results because the drag is proportional to L^2, where L is some characteristic dimension of the vehicle, and the mass is proportional to L^3. The ratio of drag to mass is therefore proportional to $1/L$.

Steering Losses—Steering losses, like other loss terms, are usually unavoidable, but we can decrease them by using efficient steering strategies. With proper

---

* See Anderson [1982], Nielson [1961], and Hoerner [1965].
†The values of Fig. 2.13 differ from those recommended in Sec. 2.2.3 because of the different *aerodynamic regimes* involved. Section 2.2.3 deals with rarified fluid flow.

initial pitchover and control of the vehicle's acceleration history, a vehicle can fly directly into a circular or elliptical Earth orbit with essentially zero steering loss (α' = 0 deg, for nearly all flight times). This pure gravity turn (see Sec. 2.6.3) is far from an efficient steering strategy, however, because the gravity-loss term is much greater than the combined gravity and steering losses of more efficient methods.

So far we have assumed a planar ascent to simplify the discussion, but various missions use yaw steering (nonplanar ascents) to improve overall performance. If possible, yaw steering losses are incurred relatively early in flight to avoid or minimize greater losses later. The required parking-orbit inclination and possible range-safety constraints often dictate nonplanar ascents. The steering-loss term for this case has the same form as that for pitch in Eq. (2.54). To the $(1 - \cos \alpha')$ term, we add the additional yaw steering-loss term of

$$\Delta v_{y\ loss} = \int_{ign}^{bo} \frac{F}{m}(1 - \cos \beta') dt \qquad (2.60)$$

where β' = $\beta + \delta_y$ (rad)
β = yaw angle of attack (rad)
δ_y = thrust deflection in yaw (rad)

For similar acceleration histories, steering losses increase proportionately to the square of the characteristic α' and β' values (small α' and β'). This is evident from the series expansion for the cosine function: $\cos\beta' = 1 - \beta'^2/2 + ...$ (radian measure). Therefore, $(1 - \cos\beta') \approx \beta'^2/2$ (similarly for pitch). If constraints allow, we start yaw steering as early as possible, because we can change velocity direction most efficiently when the magnitude of the velocity is small.

Earth Rotation—The effect of Earth, or other central body, rotation typically affects the velocity budget required for orbital insertion. For Earth, the easterly inertial velocity at the launch site is given by Eq. (2.45). For launch sites at Cape Canaveral, with geocentric latitudes ≈ 28.4 deg N, the easterly velocity available at the launch site $v_{l/s} \approx 409$ m/s, of which a significant fraction can benefit a vehicle's velocity budget.

2.6.3 Steering the Launch Vehicle

Launch-vehicle steering involves controlling the pointing history of the propulsion vector throughout the trajectory to achieve the required orbit, within vehicle constraints. This aspect of steering is distinguished from the much tighter time loop of vehicle control, which for many vehicle designs requires part of the thrust vector to counteract aerodynamic overturning moments. Unless actively controlled, these moments cause the vehicle to tumble. This vitally important control function often contributes almost nothing to the vehicle's velocity budget. Three common steering methods are outlined briefly below.

Gravity Turn—A launch vehicle typically begins its ascent with a short vertical rise, during which it starts, and usually completes, a roll from its launch-pad azimuth to the desired flight azimuth. The vehicle then starts a pitch-over, during which the pitch angle of attack departs from zero because the angular change of the vehicle's longitudinal axis proceeds more rapidly than the change of the velocity vector from the vertical. After this initial pitchover phase, by appropriately controlling the vehicle's attitude we can make the longitudinal axis (the roll axis) and the velocity vectors coincident. Continuing this alignment leads to flight known as the gravity turn. Launch vehicles often use a close approximation to the gravity turn to keep the angle of attack near zero in the trajectory during the high aerodynamic loading region.*

Referring to Fig. 2.14,

$$-d\gamma = \frac{g_l \cos \gamma \, dt}{v + higher \ order \ terms} \quad \text{(rad)} \quad (2.61)$$

or

$$\frac{d\gamma}{dt} = -\frac{g_l \cos \gamma}{v} \quad \text{(rad/s)} \quad (2.62)$$

This equation states the pitch rate of change of the roll axis (the thrust axis for $\alpha' = 0.0$) we must provide if the vehicle is to be kept at zero angle of attack. The rate is measured in the local-horizontal, local-up coordinate system. The rate that these axes are turning in an Earth-centered, nonrotating system is

$$\frac{d\theta}{dt} = -\frac{v \cos \gamma}{r} \quad \text{(rad/s)} \quad (2.63)$$

where θ is the angle in the flight plane from the local horizontal at lift-off to the local horizontal at the time of interest (the positive direction is above the reference horizon). The vehicle's overall turning rate of the inertial pitch angle is $d\gamma/dt + d\theta/dt$.

To help understand the concept of a gravity turn, we superimpose onto the vehicle's trajectory at any instant in time the effects of vehicle accelerations due to engine thrust acting along the launch vehicle's longitudinal axis. At low vehicle velocities, the magnitude of the pitch change due to gravity is greatest, requiring the largest pitch rates. These rates rapidly decrease as the vehicle continues to accelerate to higher velocities.

---

* For simplicity we assume atmospheric winds are zero. In practice, winds are often the principal contributor to the vehicle's overall bending moments and therefore may dominate structural requirements for the stack-up. Steering methods to reduce the adverse effects of winds are beyond the scope of this discussion. These methods, which reduce wind-induced angles of attack, can often be considered perturbations to a basic gravity turn.

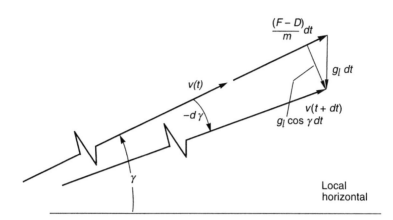

Fig. 2.14. Velocity Diagram for the Gravity Turn. The gravity force causes the trajectory to curve, shown here by a change in flight-path angle ($d\gamma$).

Controlled Angle of Attack—The gravity turn is not the best way to minimize overall ascent Δv losses. But it does decrease aerodynamic loads and localized heating in the region of ascent with high dynamic pressure. In the next section, we discuss linear-tangent steering, a steering scheme that best reduces Δv losses. As a bridge between these methods, the trajectory designer may select steering using controlled angle of attack. This approach reduces subsequent steering losses but meets aerodynamic loading or heating constraints in the transitional period. Figure 2.15 shows its main features.

The decisive element is the trajectory's decreasing dynamic pressure (q), which allows some nonzero angles of attack but still restricts these angles because of other vehicle constraints. Vehicle control, bending moments, and increased windward-side aerodynamic heating are roughly proportional to the product αq. Some nonzero α is usually acceptable, however, and a moderately increasing α history typically benefits performance, when we properly integrate it into the steering strategy.

The simplest form is a linearly ramped α schedule, as shown in the figure. A somewhat more sophisticated approach uses a constant αq constraint ($C_{\alpha q}$) after some appropriate time in the trajectory, such that $\alpha = C_{\alpha q}/q$. This form implies a step change in α after t_0. The step could be modified to a ramp, with a much shorter interval than t_r. This constant αq method begins to lose its validity when q decreases until α becomes larger than some moderate value: roughly $\alpha > 10°$. When constraints of atmospheric flight no longer dominate, we can minimize Δv losses by transitioning to linear-tangent steering.

Linear-Tangent Steering—This approach best uses the remaining propulsive Δv capability to reach a low-Earth or transfer orbit, once the launch vehicle enters

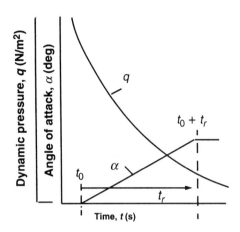

Fig. 2.15. Ramped Angle of Attack. We specify the vehicle's angle of attack (α) as a function of time. The vehicle is guided to this α. Although steering losses increase, overall losses decrease.

the part of its trajectory where atmospheric constraints no longer dominate. Linear-tangent steering becomes the best method as the drag becomes negligible and the propulsive specific impulse (I_{sp}) is effectively constant. We briefly overview this important method as it applies to planar flight (no yaw steering).

Figure 2.16 shows a typical time history of the thrust vector's steering angle, θ, together with a graphical definition of the steering angle in the inset. For configurations that are nearly axi-symmetrical, the thrust vector and the vehicle's longitudinal axis essentially coincide because the aerodynamic moments are near zero ($\delta = 0$). For unsymmetrical designs, a bias angle separates the axis and the thrust. We will assume $\delta = 0$ for this development. Linear-tangent steering derives its name from the following relationship:

$$\tan \theta = A + Bt \tag{2.64}$$

Our overall objective for linear-tangent steering is to minimize the Δv expended from the start of this steering to its completion (desired orbital end conditions are reached). We complete the steering by driving θ to the desired value at a given time. Note from Fig. 2.16 that θ is the sum of the flight-path angle (γ) and the angle of attack (α) at a given instant in time (t). From Eq. (2.64), the sign and magnitude (slope) of B will determine the amount θ is increasing or decreasing. The slope B is usually negative (steering beginning with higher pointing attitudes), and is then "washed out" to smaller (and possibly negative) values as the flight progresses to the point where thrust ends (t_f). The steering-attitude angle θ is approximately linear in time if $\tan \theta \approx \theta$ (radians), but we do not need the small-angle approximation. For example, during insertion into circular orbits, the final flight-path angle (γ_f) must be zero, but θ_f need not be zero.

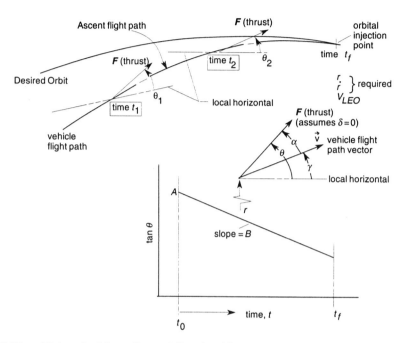

Fig. 2.16. History for Linear-Tangent Steering. We specify the tangent of the thrust angle as a linear function of time.

The guidance system must compute A, B, and t_f, given the required orbit definition. The system translates this data into a usable form for steering control and must command engine shutdown at the appropriate time. The vehicle requires a system to measure acceleration. It also needs a navigation module to process and integrate acceleration measurements into velocity $\vec{v}$ and to integrate the velocity and obtain position $\bar{R}$. The system must provide information that tracks the vehicle's attitude with respect to a reference coordinate system. These periodically updated vectors and attitudes are the result of the flight history to date, including all planned (and unplanned) steering, staging, and propulsion sequences and all externally imposed forces.*

Using this up-to-date information, the guidance system proceeds generally as follows. It needs a simple but adequate mathematical model of vehicle propulsion, jettison(s), and sequencing for the guided stage(s). It uses some initial-guess estimates for the constants in the steering law, A_1 and B_1, and for the final-stage

---

* The accelerometers do not measure gravity acceleration. We therefore need a high-quality mathematical model of the gravity vector as a function of position over the central body in the navigation routines.

shutdown time t_{f1}. From the initial conditions $\vec{r}_0$ and $\vec{v}_0$ at time t_0, the mathematical model simulates the vehicle's flight to its state at t_{f1}. But the computed orbital state typically does not match the desired orbit parameters. By numerically incrementing A and B independently, and twice resimulating the trajectory to its state at t_{f1}, we can obtain the partial derivatives $\frac{\partial r}{\partial A}$, $\frac{\partial \dot{r}}{\partial A}$, and $\frac{\partial r}{\partial B}$, $\frac{\partial \dot{r}}{\partial B}$.* The primary control of the velocity at the final state is through the shutdown time t_f, which we adjust as required. From this information the system updates the original A_1, B_1, and t_{f1} guesses to provide the values used at first in the steering control. As the flight proceeds and new information comes in, we update these values to drive the orbital injection errors toward zero. As the actual time nears the predicted shutdown time, we must prevent overly active steering which has little practical effect on injection accuracy.

2.6.4 Flight-Simulation Programs

The core of a computer program that simulates a launch vehicle's trajectory is a method that numerically integrates the equations of motion for the flight under study. In contrast to the classical methods of celestial mechanics, these equations typically do not have closed-form solutions because of the complex nature of the force and steering functions. Linked to the basic routine for numerical integration are other modules that model:

    a. the gravity vector of the central body
    b. the geometry of the central body (spherical or oblate)
    c. atmospheric properties, including winds, if required
    d. propulsion of the required stages
    e. time histories of the mass properties, including moments of inertia, if required. Simulation of these properties is typically linked to propulsion models that track the propellant levels in the respective tanks
    f. aerodynamic characteristics as a function of configuration, Mach number, and vehicle orientation with respect to the relative velocity
    g. vehicle steering

The complexity in modeling these functions varies widely among simulation programs. Indeed, various levels of complexity are typically available within a particular simulation program to match the varying needs of its users. The numeric-integration routines themselves are typically selectable, with options for integration methods, size of the integration step, and size variations to match different regions of the trajectory, where the acceleration histories vary in complexity. Such options can change the amount of computer time required to simulate a tra-

---

* The magnitudes of r and $\dot{r}$ are desired (not the vectors). Note that $\dot{r}/V = \sin \gamma$.

jectory. The trajectory simulation program may be coupled to constrained optimization schemes, for which the number of separate, simulated trajectories required to solve the problem is of the order of the square of the number of control variables used in the optimization. Under such use, efficient integration schemes can have a large payoff, allowing us to get solutions in minutes rather than hours.

The numeric integration is usually made in a nonrotating axis system, whose origin is fixed at the center of the central body. We label the velocities in this system *inertial*, and we must subtract the effects of the central body's rotation from the inertial velocity to obtain the velocity relative to the rotating central body. When transitioning a vehicle to an interplanetary trajectory, we must vectorially add the inertial velocity of the central body with respect to the Sun to the vehicle's inertial velocity with respect to the central body.

We may couple flight-simulation programs to sophisticated models of the vehicle's autopilot and guidance-steering routines. Very simple models of these functions often give good general results. Thus, we can use simpler ways to develop trajectories, especially in predesign. Of course, we still must analyze the autopilot and guidance systems before completing the design.

The data from flight-simulation programs can range from simple to vast. We normally can select how often we output such data as

> **time:** independent variable, different starting references optional, e.g., from lift-off, Greenwich mean time (GMT), or time from a key event, such as second-stage ignition
>
> **position:** altitude, range from pad, latitude and longitude of the sub-vehicle point, vector components in various coordinate systems
>
> **velocity**: Earth-relative and inertial, flight path angles, azimuth angles, vector components in various coordinate systems
>
> **acceleration**: vector components, load factors
>
> **attitude:** direction cosine matrices for vehicle orientation, particularly the matrix of the vehicle body axes on the reference integration axes
>
> **propulsion:** thrusts, flow rates, specific impulses, mixture ratios, propellant levels, pressures. Vehicle mass properties may be tracked through the propulsion routines.
>
> **aerodynamic parameters**: Mach number, dynamic pressure, angle of attack (pitch and yaw), aerodynamic coefficients, forces, and moments
>
> **tracking data:** (from selectable tracker locations) azimuth and elevation angles, slant range, and the time rates of change of these variables, orientation of the tracker-to-vehicle vector on the vehicle body axes
>
> **orbital data:** perigee, apogee, inclination, longitude of the ascending node, argument of perigee, true anomaly, period, perigee and apogee velocities, time to apsis, nodal regression, escape-hyperbola parameters.

References

Anderson, John D., Jr. 1982. *Modern Compressible Flow with Historical Perspective.* New York: McGraw-Hill Book Company.

Bate, Roger R., Donald D. Mueller, and Jerry E. White. 1971. *Fundamentals of Astrodynamics.* New York: Dover Publications.

Battin, Richard H. 1987. *An Introduction to the Mathematics and Methods of Astrodynamics.* New York: AIAA Education Series.

Cefola, P. J. 1987. The Long-Term Orbital Motion of the Desynchronized Westar II. AAS Paper 87-446 presented at the AAS/AIAA Astrodynamics Specialist Conference.

Chobotov, V. A. ed. 1991. *Orbital Mechanics.* Washington, DC: American Institute of Aeronautics and Astronautics.

Danby, J. M. A. 1988. *Fundamentals of Celestial Mechanics.* 2nd ed. Richmond, VA: Willmann-Bell.

Escobal, Pedro R. 1965. *Methods of Orbit Determination.* Malabar, FL: Robert E. Krieger Publishing Co.

Hoerner, Sigard F. 1965. *Fluid-Dynamic Drag.* Midland Park, NJ: published by the author.

Kaplan, Marshall H. 1976. *Modern Spacecraft Dynamics and Control.* New York: Wiley and Sons.

King-Hele, D. 1964. *Theory of Satellite Orbits in an Atmosphere.* London: Butterworths Mathematical Texts.

Larson, Wiley J. and James R. Wertz, eds. 1992. *Space Mission Analysis and Design,* 2nd ed., Torrance, California, and Norwell, MA: Microcosm, Inc. and Kluwer Academic Publishers.

Nielsen, Jack N. 1961. *Missile Aerodynamics.* New York: McGraw-Hill Book Company, Inc.

Pocha, J. J. 1987. *An Introduction to Mission Design for Geostationary Satellites.* Boston: D. Reidel Publishing Company.

Roy, A. E. 1978. *Orbital Motion.* Bristol and Philadelphia: Adam Hilger.

Tai, Frank and Peter D. Noerdlinger. 1989. A Low Cost Autonomous Navigation System, Paper No. AAS 89-001 presented to the 12th Annual AAS Guidance and Control Conference. Keystone, Colorado, Feb. 4-8.

Wiesel, William E. 1989. *Spaceflight Dynamics.* New York: McGraw-Hill Book Company.

CHAPTER 3
THERMODYNAMICS OF FLUID FLOW

Gary Wirsig, *U.S. Air Force Academy*

    3.1   Mass Transfer
    3.2   Thermodynamic Relations (Energy and Entropy)
    3.3   Thrust Equations
    3.4   Heat Addition
    3.5   Heat Transfer
    3.6   Design Example—Cold-Gas Thruster

Liquid, solid, and hybrid rockets all depend on fluid flow to provide propulsive thrust, so we must understand its mechanical and thermodynamical principles to analyze how well rocket motors perform. Thus, we ask questions such as:

- How much thrust does added energy or heat produce?
- How much energy goes to raising the fluid temperature as opposed to accelerating the flow?
- What are the pressure and temperature?
- How do fluid properties such as pressure, temperature, and density vary inside a rocket chamber and nozzle?

To answer these questions, we depend on four major equations which are the basis of this chapter: the laws of conservation of mass, momentum, and energy, plus the perfect-gas law. We use these equations to develop other helpful relationships.

This chapter first discusses conservation of mass and energy to develop the concept of isentropic flow and its applications. We then turn to conservation of momentum to develop the thrust equation and its related applications. Finally, we take an elementary look at heat transfer and see how it applies to rocket technology (see Fig. 3.1).

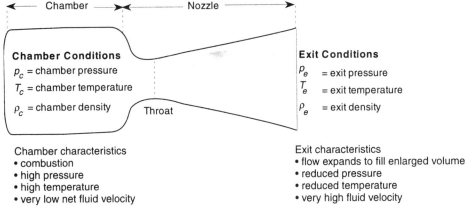

Fig. 3.1. **Generalized View of a Rocket Thrust Chamber.** Understanding the thermodynamic conditions in a rocket is key to understanding its performance.

3.1 Mass Transfer

Basically, rockets work by transferring or flowing mass in one direction through a tube or duct. Because a flowing mass has velocity, momentum transfers with it. Further, we must account for the forms of energy (internal, chemical, kinetic, and others) in the flow. We investigate mass transfer in this section and analyze momentum and energy transfer later.

The rate of mass flow perpendicular to a duct of cross-sectional area (A) is

$$\dot{m} = \rho v A \tag{3.1}$$

where $\dot{m}$ = mass flow rate (kg/s)
 ρ = fluid density (kg/m³)
 v = fluid velocity (m/s)
 A = cross-sectional area of duct (m²)

The density is mass per unit volume, and the velocity and area terms provide the amount of volume "filled" per unit time. Multiplied together, the terms express the amount of mass passing through a surface of cross-sectional area (A) per unit time.

With mass flow rate described, we discover the best way to analyze mass transfer by using either the control volume or the system approach.

A *control volume* is a region with constant shape and size that stays fixed in space. When analyzing a process using a control volume, we must carefully define rigid physical boundaries which are stationary relative to a fluid flow. Then, we determine the amount of mass crossing the boundaries, along with any associated energy and momentum. We must keep focused on a defined volume which is fixed in space as matter passes in and out. Such an *open system* is especially helpful for analyzing steady flow through a rocket nozzle.

The *system approach* (also known as *control mass approach*) allows us to focus on a fixed amount of matter rather than a fixed volume in space. The envelope containing the matter may change its size and shape, as well as its location, but the original matter in the system remains constant. A good example of such a *closed system* is a fixed amount of gas which changes volume as it passes from a high-pressure to a low-pressure tank.

We use the control-volume approach to derive the major equations in this chapter, beginning with the continuity equation, which follows. Because mass is neither created nor destroyed (known as *conservation of mass*) in conventional propulsion systems, total mass in a system must be conserved. In the most general case, this means that whatever mass enters a control volume must either be stored in that control volume or exit. We can rephrase this statement as: "the rate of change of mass in a control volume is equal to the net influx of mass." In differential form, this statement becomes

$$\frac{d}{dt}(m_{cv}) = \dot{m}_{in} - \dot{m}_{out} \tag{3.2}$$

where m_{cv} = amount of mass in the control volume (kg)

$\dot{m}_{in}$ = mass flow rate into the control volume (kg/s)

$\dot{m}_{out}$ = mass flow rate out of the control volume (kg/s)

Rearranging yields:

$$\frac{d}{dt}(m_{cv}) + \dot{m}_{out} - \dot{m}_{in} = 0 \tag{3.3}$$

The elemental mass flow rate ($d\dot{m}$) across the area (dA) can be expressed as $\rho v dA$. Once we have added all the elemental areas along the entire control surface in Fig. 3.2, the total mass flow is

$$\dot{m} = \int_{cs} \rho v \, dA \tag{3.4}$$

where $\dot{m}$ = total mass flow through control surface (kg/s)
subscript *cs* designates control surface

ρ = fluid density in the flow (kg/m³)
v = velocity of particles in flow (m/s)
dA = elemental area in control surface (m²)

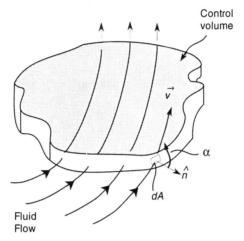

Fig. 3.2. Control Volume for Continuity. This illustration defines the geometry parameters used to develop the continuity relations. The control surface is the unshaded surface portion of the control volume.

To distinguish incoming mass flow from exiting mass flow in a concise form, we use the dot product between the velocity vector of a streamline ($\vec{v}$) and the outward normal ($\hat{n}$) of the control surface at the point of streamline intersection. If mass is flowing in, the dot product is negative, causing $\dot{m}_{in}$ always to be less than zero. Conversely, if mass is flowing out, the dot product is positive; thus, $\dot{m}_{out}$ is greater than zero. Additionally, the dot product corrects the magnitude of mass flow when the flow is not perpendicular to the part of the control surface area under consideration. Another expression for the dot product $(\vec{v} \cdot \hat{n})$ is $v \cos \alpha$, where α is the angle between the velocity vector and the unit outward normal vector for the elemental area under consideration. Therefore, for total mass flow rates in and out of a control volume,

$$\dot{m}_{out} - \dot{m}_{in} = \int_{cs} \rho(\vec{v} \cdot \hat{n}) \, dA \tag{3.5}$$

where $\vec{v}$ = velocity vector of flow through an elemental area on the control surface (m/s)
$\hat{n}$ = outward normal (unit vector) for the elemental area
$(\vec{v} \cdot \hat{n})$ = $v \cos \alpha$

Now we replace the mass flow terms of Eq. (3.3) with expression (3.5) and rewrite the time variant term as

$$\frac{d}{dt}(m_{cv}) = \frac{d}{dt}\int_{cv} \rho \, dV \tag{3.6}$$

where dV = infinitesimal element of the control volume (cv).
Combining all of the terms, we may now express continuity as

$$\frac{d}{dt}\int_{cv} \rho \, dV + \int_{cs} \rho(\vec{v}\cdot\hat{n}) dA = 0 \tag{3.7}$$

For a steady-state system, the amount of mass in the control volume remains constant, so the time variant term of Eq. (3.7) is zero. Analyzing the rest of the equation, we determine that the mass entering the control volume is equal to the mass leaving it. By expressing $(\vec{v}\cdot\hat{n})$ as v_n, we can write Eq. (3.7) as

$$(\rho v_n A)_{in} = (\rho v_n A)_{out} \tag{3.8}$$

where ρ = fluid density in the flow (kg/m³)
v_n = part of fluid velocity normal to the cross-sectional area through which the fluid is flowing (m/s)
A = cross-sectional area through which the fluid is flowing (m²)

Assumption: Steady flow

The final result for continuity in Eq. (3.8) may seem superfluous at first because it appears to bring us right back to $\dot{m} = \rho vA$, where we began in Eq. (3.1). But Eqs. (3.7) and (3.8), which include the possibility of nonaxial flow, are of particular value later when we develop an expression for nonaxial thrust.

3.2 Thermodynamic Relations (Energy and Entropy)

Why do we want to analyze the types and amounts of energy in a fluid flow? In the particular case of flow through a rocket nozzle, knowing how much energy the flow contains helps us determine the flow velocity and the rocket's thrust. We eventually develop the equations necessary for this analysis, but we first need to understand the types of energy to explore.

Energy falls into two general categories: (1) energy within a system of particles, called *system energy* (E) and (2) energy which transfers in and out of that system. This second type of energy can transfer through work (W) or a heat exchange (Q) between the system of particles and its environment. Notice that this transferred energy increases or decreases the system energy.

3.2.1 First Law for a System

The most prominent forms of system energy are internal energy (U), potential energy (PE), kinetic energy (KE), and chemical energy (ChE). We define *internal energy* as the energy due to the state (such as velocity) of particles within the system. For the types of gases discussed in this chapter (see perfect gas discussion in Sec. 3.2.3), the amount of this internal energy depends *only on the temperature* of the material within the system. Electric, magnetic, and nuclear energy are much less prominent forms but are used where appropriate.

Suppose we have a closed system consisting of an amount of gas enclosed in a cylinder, with a movable piston at one end as in Fig. 3.3. We assume that the gas is at a uniform temperature, thus fixing the internal energy. The internal energy can be changed by transferring outside energy into the system in either of two ways. First, we can compress the cylinder, doing work (W) on the system and increasing the internal energy. Second, we can transfer heat to the container. As the heat conducts from the container to the gas, the internal energy again increases. This careful accounting for all system energy during a physical process is called *conservation of energy*.

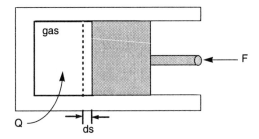

Fig. 3.3. Work and Heat Addition to Gas in a Cylinder. Heat added to a gas increases its internal energy. Work done by the gas decreases its internal energy.

The first law of thermodynamics allows us to quantify conservation of energy[*]:

$$Q = \Delta E + W \tag{3.9}$$

where Q = heat transferred into the system (J)
ΔE = change in system energy (J)
W = work done by the system (J)

Note that the form of Eq. (3.9) defines heat transferred to the system and work done by the system as positive quantities. The following illustrations clarify the

---

[*] This equation is empirically derived, meaning that we have created this equation by observation and not from some fundamental underlying principle.

definitions. If we do no work ($W = 0$) and increase Q (add heat to the system), ΔE must also increase. Similarly, if no heat transfers, positive work (done by the system) causes ΔE to decrease.

One more word of clarification helps explain the concepts of internal energy, heat, and work. The strict definition of work is

$$W = \int_{path} \vec{F} \cdot d\vec{s} \tag{3.10}$$

where W = work done on a system (J)
 $\vec{F}$ = force acting on the system (N)
 $d\vec{s}$ = infinitesimal portion of the path through which the force acts (m)

Because the force does not always line up perfectly with the path, losses usually accompany any work done. Known as path dependence, these losses are characteristic of both work and heat addition to a system. *Path dependence* simply means that different ways of doing work and transferring heat have different forms of losses, so we must carefully account for these losses in each case. Internal energy, on the other hand, has been experimentally proven to be a property of the system and therefore not does not depend on path [Holman, 1980].

3.2.2   First Law for a Control Volume

We would like to develop a general expression for energy which can vary over time and which we can adapt to various control volume shapes. We therefore take derivatives of the terms in Eq. (3.9) with respect to time to obtain

$$\frac{\delta Q}{dt} = \frac{dE}{dt} + \frac{\delta W}{dt} \tag{3.11}$$

(A)(B)(C)

The different notation for the derivatives in Eq. (3.11) shows that although energy is a property of the system, heat and work depend on the path. Term (A), expressed as $\dot{Q}$, is simply the rate at which heat enters or leaves the system.

In Fig. 3.4, volume I represents a defined control volume. Volume III contains mass which is moving into volume I, and volume II contains mass which is exiting volume I. Thus, at any time (t) our system of matter is in volumes I and III, but at a later time ($t + dt$) all the matter is in I and II. Using this convention, we can express term (B) in Eq. (3.11) as

$$\frac{dE}{dt} = \frac{\{E_{I_{(t+dt)}} + E_{II_{(t+dt)}}\} - \{E_{I_{(t)}} + E_{III_{(t)}}\}}{dt} \tag{3.12}$$

Rearranging Eq. (3.12) gives us

$$\frac{dE}{dt} = \frac{E_{I_{(t+dt)}} - E_{I_{(t)}}}{dt} + \frac{E_{II_{(t+dt)}}}{dt} - \frac{E_{III_{(t)}}}{dt} \quad (3.13)$$

$$\quad\quad\quad\quad\quad (1) \quad\quad\quad (2) \quad\quad (3)$$

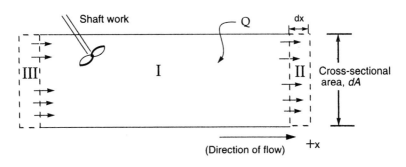

Fig. 3.4. **Control Volume for Energy Equation.** Volumes indicated by I, II, and III. This geometry is used to derive the first law for a control volume.

As dt approaches zero, and volumes II and III become inconsequential, expression (1) simply represents the time rate of change of energy in volume I:

$$\left(\frac{\partial E}{\partial t}\right)_{cv} \quad \text{or} \quad \frac{\partial}{\partial t}\int_{cv} e\rho dV \quad (3.14)$$

where   subscript cv indicates integration throughout the control volume
  e = specific system energy (independent of mass) (J/kg)
  ρ = system fluid density (kg/m^3)
  dV = infinitesimal volume element (m^3)

What kind of energy change within the control volume are we talking about in Eq. (3.14)? In the absence of energy entering or leaving the system, this energy change could be due to such things as internal chemical changes, nuclear exchanges between mass and energy, kinetic energy, or potential energy. But usually this term represents the difference between incoming and outgoing energy during unsteady flow conditions [Shapiro, 1953].

Expression (2) of Eq. (3.13) represents energy of the system leaving the control volume and may be rewritten as

$$\frac{\int e\, dm_{II_{(t+dt)}}}{dt} \quad (3.15)$$

(where dm is an infinitesimal amount of mass in volume II), or

$$\int e\, d\dot{m}_{II(t+dt)} \tag{3.16}$$

Because this is the energy flowing out of the control volume, we rewrite it as $\int e\, d\dot{m}_{out}$. But from continuity, $\dot{m} = \rho v A$ and therefore $d\dot{m} = \rho v\, dA$, assuming ρ and v to be constant across the differential area element (dA). So, finally, we may rewrite outgoing energy as

$$\left(\int e\rho v\, dA\right)_{out} \tag{3.17}$$

Expression (3) from Eq. (3.13), which represents energy entering the control volume, becomes

$$\left(-\int e\rho v\, dA\right)_{in} \tag{3.18}$$

Using the dot product as we did in Eq. (3.5)—to distinguish incoming energy from outgoing energy, expressions (2) and (3) combine as

$$\int e\rho(\vec{v}\cdot\hat{n})\, dA \tag{3.19}$$

Now we can use our new expressions for terms (1), (2), and (3) of Eq. (3.13), which represent expression (B) from our original equation (3.11). Thus, we have

$$\frac{dE}{dt} = \frac{\partial}{\partial t}\int_{cv} e\rho\, dV + \int_{cs} e\rho(\vec{v}\cdot\hat{n})\, dA \tag{3.20}$$

To summarize—the total change of energy with respect to time within a control volume is equal to the time rate of change of system energy in the control volume due to chemical or other state changes plus any energy entering the system through the control surface. (Remember, energy leaving is negative.)

Term (C) in Eq. (3.11), or rate of work done by the system, can split into two parts:

- Rates of shaft and shear work, designated P_s.
- Rate of *flow work* from pressure forces acting perpendicular to the control surface.

To understand the mechanism of flow work, look at Fig. 3.4 again and observe what happens to flow leaving the control volume. The amount of work done by pressure as the flow exits the control volume at an area element (dA) is $p(dA)(dx)$, where dx is the distance moved perpendicular to dA, and p is the pressure forcing the flow out of the control volume. But $(dA)(dx)$ is just the volume of mass in volume II at time $t + dt$, or $dV_{II(t+dt)}$, or again $v\, dm_{II(t+dt)}$. Note that v is specific volume, or volume per unit mass. Thus, $v dm$ equals dV. We can similarly analyze flow entering the system from volume III and represent the total flow work rate around a control surface as

$$\frac{\delta W}{\delta t} = \frac{\int p\upsilon \, dm_{II}}{dt} - \frac{\int p\upsilon \, dm_{III}}{dt} = \left(\int p\upsilon \, d\dot{m}\right)_{out} - \left(\int p\upsilon \, d\dot{m}\right)_{in} \quad (3.21)$$

If we treat the $d\dot{m}$ term as we did for Eq. (3.17) and apply our standard dot product, then

$$\frac{\delta W}{\delta t} = \int_{cs} p\,\upsilon \rho (\vec{v} \cdot \hat{n}) dA \quad (3.22)$$

Combining terms (A), (B), and (C) in Eq. (3.11), we can express conservation of energy as

$$\dot{Q} = P_s + \frac{\partial}{\partial t}\int_{cv} e\rho \, dV + \int_{cs} e\rho (\vec{v} \cdot \hat{n}) dA + \int_{cs} p\,\upsilon \rho (\vec{v} \cdot \hat{n}) dA \quad (3.23)$$

We can now further combine terms. Note that terms in Eq. (3.23) are in lower case, meaning they do not depend on mass. These are called *specific energy* terms. Kinetic energy per unit mass is expressed as $v^2/2$. Potential energy changes because of a gravitational field are not usually significant for flow inside a rocket. Finally, we define a term called *enthalpy*,

$$h = u + p\upsilon \quad (3.24)$$

We discuss enthalpy in more detail later; for now, simply note that it is helpful to combine internal energy (u) and $p\upsilon$ (a term useful in determining "flow work") to get our enthalpy term.

Using the definitions in the paragraph above and combining the last two terms of Eq. (3.23), we write the first law of thermodynamics for a control volume:

$$\dot{Q} = P_s + \frac{\partial}{\partial t}\int_{cv} e\rho \, dV + \int_{cs}\left(h + pe + \frac{v^2}{2}\right)\rho (\vec{v} \cdot \hat{n}) \, dA \quad (3.25)$$

(a) (b) (c)

Note that the terms in the so-called energy equation actually represent energy rates, or power.

Most of our analysis of rocket motors applies to steady-state, steady-flow conditions. In such cases Eq. (3.25) becomes greatly simplified. Terms depending on time, such as expression (b), vanish. We may usually ignore heat, on the left-hand side, because it is small compared to the terms for enthalpy and kinetic energy. As stated earlier, potential energy changes do not usually apply to flow in rockets. Shaft power and losses from shear forces, term (a), are insignificant for many rocket applications. Shaft power is important to air-breathing engines with compressor and turbine machinery and to rocket engines with pumps. From these simplifications of Eq. (3.25), and using continuity (mass flow rate is constant in steady flow), we can reduce the rest of Eq. (3.25) to

$$h_1 + \frac{1}{2}v_1^2 = h_2 + \frac{1}{2}v_2^2 \qquad (3.26)$$

h = specific enthalpy (J/kg)
v = flow velocity (m/s)

Assumptions:
1. Steady flow
2. Adiabatic process (no heat transfer)
3. No significant changes in potential energy
4. No shaft work or shear work done

Note that subscripts 1 and 2 indicate any two points in a flow passage, such as a rocket nozzle.

Equation (3.26) greatly simplifies flow analysis for a nozzle. Remember that $h = u + pv$ and that internal energy (u) is a function of temperature only. As we see in the next section, pv also depends solely on temperature. Therefore, for rocket engines, the energy equation reduces to a few essentials. It tells us that although flow velocities and enthalpies depending on temperature may change because cross-sectional nozzle areas are changing, the sum of the specific enthalpy and the specific kinetic energy remains constant for a given flow.

3.2.3 Perfect-gas Law and Specific Heats

Perfect Gas. A *perfect gas* (or an *ideal gas*) is defined by kinetic theory as a gas whose particles fulfill at least two conditions. First, the particles display motion with only three translational degrees of freedom (spin energy is negligible). Second, the gas molecules are spaced widely enough so that molecular force fields contribute little to the pressure exerted on the container's walls.

When gas pressures become so great that molecules "jam together," Van der Waals forces between molecules become significant in determining pressure because the gas no longer acts as a perfect gas. We assume the gases behave as perfect gases, but you can find in Holman [1980] an adaptation of the perfect-gas law which uses compressibility factors to account for Van der Waals effects at extremely high pressures.

We can simply define a perfect gas as one that obeys the perfect-gas law. The perfect-gas law (also called the ideal-gas law) relates the states of a gas as follows:

$$p = \rho R T \qquad (3.27)$$

where p = gas pressure (N/m^2)
ρ = gas density (kg/m^3)
R = specific gas constant (J/kg·K)

T = temperature (K)

The perfect-gas law may also take the form

$$p = \frac{\rho \mathcal{R}_u T}{\mathcal{M}} \qquad (3.28)$$

where $\mathcal{R}_u$ = universal gas constant (8314.41 J/kmol·K)
$\mathcal{M}$ = the molecular mass of the gas (kg/kmol)

note that

$$R = \frac{\mathcal{R}_u}{\mathcal{M}} \qquad (3.29)$$

$$pv = RT \qquad (3.30)$$

$$pV = \eta \, \mathcal{R}_u T \qquad (3.31)$$

where v = $1/\rho$ = specific volume (m³/kg)
V = volume (m³)
η = number of moles of the gas = $\dfrac{m}{\mathcal{M}}$ (kg·mol)

What about a perfect gas which is a mixture of gases, each with its unique molecular mass (for example, air or exhaust constituents in a rocket nozzle)? The following section (in smaller type) treats mixtures of several gases. We have included derivations and used the primary equations in an example problem.

Assuming the mixture temperature is uniform throughout, we can use two laws (Dalton's and Amagat's) to analyze the mixture. Dalton's Law relates partial pressures of the constituents of a gas mixture to the total mixture pressure. The partial pressure of a constituent gas is the pressure it would exert on a pressure vessel if it were the only gas occupying that vessel at the temperature of the mixture. Dalton's Law is stated as follows:

$$p = \sum_{i=1}^{n} p_i \qquad (3.32)$$

where p = mixture pressure (N/m²)
p_i = partial pressure of the i-th constituent gas (N/m²)
n = number of gases making up the mixture

Whereas Dalton assumed each gas occupied the original volume at mixture temperature to determine the partial pressure of the gas, Amagat reasoned that if each gas existed at the mixture pressure and temperature, one could determine the volume that gas would occupy, or the partial volume. Adding up all the partial volumes yields the total mixture volume. Amagat's Law is

$$V = \sum_{i=1}^{n} V_i \qquad (3.33)$$

We can add the masses of the gases making up the mixture to determine the total mass of the mixture:

$$m = \sum_{i=1}^{n} m_i \qquad (3.34)$$

It is helpful to define a mass fraction (C_i) and mole fraction (X_i). The *mass fraction* of an individual gas in a mixture is the gas mass divided by the mixture mass, or

$$C_i = \frac{m_i}{m} \qquad (3.35)$$

A *mole* is the quantity of a substance that has a mass numerically equal to the molecular mass. The *mole fraction* of an individual gas in a mixture is the number of moles of that gas divided by the number of moles of the gas mixture, or

$$X_i = \frac{\eta_i}{\eta} \qquad (3.36)$$

Because the density of a constituent gas may be expressed as

$$\rho_i = \frac{m_i}{V} \qquad (3.37)$$

and because the density of the mixture may be expressed as

$$\rho = \frac{\sum m_i}{V} \qquad (3.38)$$

we know

$$\rho = \sum_{i=1}^{n} \rho_i \qquad (3.39)$$

Using Amagat's Law, we can express the perfect-gas law for each constituent as

$$pV_i = \eta_i \mathcal{R}_u T \qquad (3.40)$$

Using Dalton's Law similarly, we have

$$p_i V = \eta_i \mathcal{R}_u T \qquad (3.41)$$

Because Eq. (3.31) applies for the entire mixture, we divide Eq. (3.40) by Eq. (3.31) and then divide Eq. (3.41) by Eq. (3.31) to get the following expressions for the mole fraction:

$$\frac{p_i}{p} = \frac{V_i}{V} = \frac{n_i}{n} = X_i \qquad (3.42)$$

We now determine the specific gas constant (R) for a gas mixture. We begin with the perfect-gas law for a constituent:

$$p_i = \rho_i R_i T \qquad (3.43)$$

This equation can be viewed as an empirical result or can be derived by substituting Eqs. (3.35), (3.27), and (3.29) into Dalton's Eq. (3.39). By writing ρ_i as $\frac{m_i}{V}$, we have

$$p_i = \frac{m_i}{V} R_i T = \frac{mT}{V} C_i R_i \qquad (3.44)$$

Substituting Eq. (3.32) into Eq. (3.44), we have

$$p = \frac{mT}{V} \sum_{i=1}^{n} C_i R_i = \rho T \sum_{i=1}^{n} C_i R_i \qquad (3.45)$$

Dividing by ρT and noting that $p/\rho T = R$, we arrive at

$$R = \sum_{i=1}^{n} C_i R_i \qquad (3.46)$$

Because $M = \mathcal{R}_u / R$, the molecular mass of the mixture is easy to determine once we have found the gas constant for a particular gas.

Example

A rigid tank contains 20 kg of nitrogen at 2×10^6 N/m² and 100°C. Enough oxygen is added to bring the pressure to 3×10^6 N/m² while the temperature remains at 100°C. What was the mass of oxygen added?

Solution

The number of moles of nitrogen remains the same,

$$n_{N_2} = \frac{m}{M} = \frac{20 \text{ kg}}{28 \text{ kg/kmol}} = 0.7143 \text{ kmol}$$

With added oxygen, the partial pressure of nitrogen is 2×10^6 N/m². Therefore,

$$p_{O_2} = p_{total} - p_{N_2}$$
$$= 3 \times 10^6 - 2 \times 10^6$$
$$= 10^6 \text{ N/m}^2$$

From Eq. (3.41):

$$\frac{n_{O_2}}{n_{N_2}} = \frac{p_{O_2}}{p_{N_2}}$$

$$n_{O_2} = \frac{(0.07143)(10^6)}{(2 \times 10^6)} = 0.3572 \text{ kmol}$$

Now we can find the mass of oxygen.

$$m_{O_2} = n_{O_2} M_{O_2} = (0.3572 \text{ kmol})(32 \text{ kg/kmol})$$
$$= 11.429 \text{ kg}$$

Specific Heat. At one time, in trying to measure heat, scientists hypothesized that bodies contained a substance called caloric. They thought that caloric was a property of the body and caloric could be "poured" from one body to another, similar to pouring water from one beaker to another. We have given caloric a new name, *heat*, and now understand that, although heat is not a system property, it can transfer from one body to another or from a system to its surroundings and vice versa. In these types of transfers we can measure heat not as a system property but as an energy transfer.

The concept of specific heat gives us a way to determine how much energy a system gains or loses as a function of temperature change. Thus *specific heat* is defined as

$$c = \frac{\delta q}{dT} \tag{3.47}$$

where c = specific heat (J/kg·K)
 δ – indicates path dependence of derivative
 q = heat which is transferred into a system (J/kg)
 T = temperature (K)

Because specific heat depends on the path, we use two typical processes (paths) to help define specific heat: constant volume and a constant pressure. First, consider a constant-volume process, such as heat added to a gas in a fixed volume. If the only form of system energy is the internal energy of the gas, the differential form of the energy equation for this process may be written as

$$\delta q = du \tag{3.48}$$

where u = specific internal energy of gas (J/kg)

Note that no work term appears in the equation because the volume does not change. We may substitute Eq. (3.48) directly into Eq. (3.47) to obtain the specific heat for constant volume:

$$c_v = \left(\frac{\partial u}{\partial T}\right)_v \tag{3.49}$$

where c_v = specific heat for a constant-volume process (J/kg·K).

A similar analysis gives the specific heat for a constant-pressure process:

$$c_p = \left(\frac{\partial h}{\partial T}\right)_p \tag{3.50}$$

where c_p = specific heat at constant pressure (J/kg·K)
h = specific enthalpy (J/kg)

The specific heat represents the change of internal energy (u) or enthalpy (h) as a function of temperature. As seen in Fig. 3.5, c_p is not constant, but rather is a function of temperature. However, for physical processes that take place within temperature ranges for which the changes in specific heat are small, the specific heat may be treated as constant. In this case, a gas is said to be *calorically perfect*. For these processes, we can use a variation from Eq. (3.49):

$$\Delta u = c_v \int_{T_1}^{T_2} dT \tag{3.51}$$

and from Eq. (3.50):

$$\Delta h = c_p \int_{T_1}^{T_2} dT \tag{3.52}$$

resulting in

$$\Delta u = c_v \Delta T \tag{3.53}$$

and

$$\Delta h = c_p \Delta T \tag{3.54}$$

For nonconstant specific heats, we need to integrate Eqs. (3.47) and (3.48) with specific heat as a function of temperature. Note that we do not usually talk about enthalpy in absolute terms because we are interested in the difference in enthalpy between end states of a process, not absolute enthalpy. Therefore, if we use gas tables to determine enthalpy at a given temperature, we do not see an absolute value. Rather, we find the difference in enthalpy between the given temperature and a reference temperature— typically 0° Celsius (273 K).

Differentiating the expression for enthalpy (Eq. 3.24) and assuming a perfect gas, we relate specific heat to the specific gas constant as follows:

$$dh = du + p\, dv + v\, dp \tag{3.55}$$

But in Eq. (3.30) $RT = pv$ and therefore

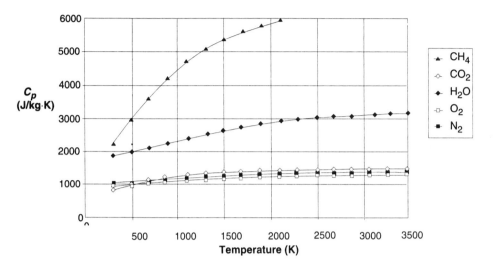

Fig. 3.5. C_p **as a Function of Temperature for Various Gases.** As the temperature of a gas increases, so does the specific heat.

$$R \, dT = p \, dv + v \, dp \tag{3.56}$$

Combining the previous two expressions, we have

$$dh = du + R \, dT \tag{3.57}$$

Dividing this expression by dT yields

$$c_p = c_v + R \tag{3.58}$$

The *ratio of specific heats* appears in many equations in this chapter. To use concise notation we define it as

$$\gamma = \frac{c_p}{c_v} \tag{3.59}$$

Table 3.1 contains values of molecular mass, specific heats, gas constant, and ratio of specific heats for various gases at low pressure and room temperature.

Assuming a perfect gas, the equipartition theorem of classical statistical mechanics predicts that γ lies in a range of 1 to 5/3. For a monatomic gas, the theory predicts a value of 5/3. As molecular complexity increases, the value of γ approaches one [Thompson, 1984].

3.2.4 Second Law

Entropy is the amount of disorder in a system. For any process, the following definition for entropy (S) holds true:

Table 3.1. Properties of Perfect Gases. These parameters are quoted at standard temperature and pressure (298.16 K and atmosphere).

Gas	Molecular Mass	c_p (kJ/kg·K)	c_v (kJ/kg·K)	R (kJ/kg·K)	γ
Air	28.97	1.005	0.718	0.287	1.4
Hydrogen, H_2	2.016	14.32	10.17	4.127	1.41
Helium, He	4.003	5.234	3.14	2.078	1.66
Nitrogen, N_2	28.02	1.038	0.741	0.296	1.40
Oxygen, O_2	32.00	0.917	0.653	0.260	1.40
Argon, Ar	39.94	0.515	0.310	0.208	1.67
Carbon Dioxide, CO_2	44.01	0.846	0.653	0.189	1.3

$$dS \geq \frac{\delta Q}{T} \tag{3.60}$$

where dS = the change in entropy of the system (J/K)
  δQ = the change in the heat energy transferred to the system (J)
  T = system temperature (K)

Now let us briefly define a reversible process. A process is *reversible* if the system undergoing the process can be restored to its initial state with no observable effects on the system or its surroundings. Irreversibilities usually result from friction, viscosity, or unrestrained expansion. If a process is reversible, the inequality of Eq. (3.60) vanishes:

$$dS = \left(\frac{\delta Q}{T}\right)_{rev} \tag{3.61}$$

In Eq. (3.61), we observe that, for a process which is reversible and *adiabatic* (no heat transfer, or $\delta Q = 0$), the change of entropy is zero. The process is called *isentropic*, meaning that the entropy of the system does not change. We assume all processes to be isentropic unless stated otherwise, so we can analyze without regard to friction and viscous effects. For large mass flows through rockets, this is a good assumption (see assumptions section below).

We consider entropy in terms of how it changes during a process rather than as a property of a system. Because the amount of change in entropy depends only on the system's end states, the entropy change is independent of path.

3.2.5 Isentropic Flow in One Dimension

We can now apply some of the concepts we have developed to determine specific characteristics of fluid flow in a duct, such as temperature, pressure, or density. Fluid flow in a duct applies to gaseous propellant in a nozzle, liquid or gaseous propellant in fuel plumbing, and coolant through a nozzle's cooling tubes.

Simplifying Assumptions. To reduce the many variables of rocket propulsion to a "real world" problem we can readily analyze, we assume:

1. *Isentropic flow,* i.e., reversible and adiabatic, from the chamber (after combustion of propellant) to the nozzle exit. *Adiabatic flow* means heat transfer does not dissipate energy from the flow. Although heat transfer through the chamber walls is significant, it is a relatively small percentage of the total energy generated. This assumption also neglects the effects of friction and fluid viscosity and does not apply to shock waves.
2. Flow is in one dimension. The larger the nozzle cone half angle, the less acceptable this assumption becomes. For most nozzles, this assumption causes less than 5% error. This issue is discussed later in the chapter.
3. Products of combustion constitute a perfect gas.
4. Flow is "frozen." Once established in the chamber, the products of combustion do not change in chemical composition while traversing the nozzle. This assumption is discussed extensively in Chap. 4.
5. Flow is steady. Therefore, the time-variant terms of the continuity, momentum, and energy equations are zero.

Isentropic Relations. If heat is added to a constant-pressure system, such as a gas-filled cylinder with a movable piston at one end, the volume slowly expands as the piston moves. This type of process is called *quasistatic*—consisting of a series of equilibrium states—and governed by the differential energy equation,

$$dq = du + p\,dv \qquad (3.62)$$

where dq = specific heat added (J/kg)
du = the increase of specific internal energy due to temperature rise (J/kg)
$p\,dv$ = flow work done on one kg of fluid while moving the piston (J/kg)

Assuming a reversible process, from Eq. (3.61) we have

$$ds = \frac{dq}{T} \qquad (3.63)$$

Incorporating Eq. (3.62),

$$ds = \frac{du}{T} + \frac{p}{T}dv \tag{3.64}$$

From the perfect-gas law and the definition of specific heat,

$$ds = c_v \frac{dT}{T} + R\frac{dv}{v} \tag{3.65}$$

If we now assume an isentropic process, then $ds = 0$, and

$$c_v \frac{dT}{T} = -R\frac{dv}{v} \tag{3.66}$$

Assume c_v and R are constant (calorically perfect), recall that $v = 1/\rho$, and integrate Eq. (3.66) arbitrarily, using any points in the process as endpoints. If we also use Eqs. (3.55) and (3.56) to simplify expressions, we arrive at

$$\frac{T_2}{T_1} = \left(\frac{\rho_2}{\rho_1}\right)^{(\gamma-1)} \tag{3.67}$$

$$\frac{p_2}{p_1} = \left(\frac{T_2}{T_1}\right)^{\frac{\gamma}{\gamma-1}} \tag{3.68}$$

$$\frac{p_2}{p_1} = \left(\frac{\rho_2}{\rho_1}\right)^{\gamma} \tag{3.69}$$

where p = pressure (N/m²)
  T = temperature (K)
  ρ = density (kg/m³)
Assumptions:
  1. Isentropic flow
  2. One-dimensional flow
  3. Steady flow
  4. Calorically perfect gas

If we use the previous assumptions of isentropic flow, these equations apply to any two points in a rocket from the chamber to the nozzle exit, as seen in Fig. 3.6. These equations have little use by themselves. But if we know nozzle geometry and can determine the Mach number of the flow through the nozzle, they become powerful tools for analyzing flow. Therefore, we must now explain the correlation between nozzle geometry and Mach number.

3.2 Thermodynamic Relations (Energy and Entropy)

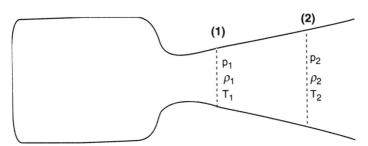

Fig. 3.6. **Arbitrary Points—Isentropic Flow Through a Nozzle.** The station numbers "1" or "2" refer to any point in a gas flow.

Mach Number. The *Mach number* (M) of a gaseous flow is the velocity of the flow (relative to a stationary observer) divided by the acoustic velocity of the flow. The *acoustic velocity* in a fluid is the velocity at which a small pressure disturbance (such as a compression or sound wave) can propagate through the fluid.* It is designated by a. We now begin a development which gives us a useful expression for the acoustic velocity.

Figure 3.7 depicts a sound wave traveling to the right through a tube filled with a perfect gas [Zucrow and Hoffman, 1976]. In part (a), we are stationary observers watching the wave travel through the tube at the velocity of sound. We note that the thermodynamic properties change by differential amounts as the wave passes. To conserve energy, the gas acquires velocity (dV) as the enthalpy changes by dh. In part (b), we are traveling to the right with the wave, noting that the gas approaches at velocity a and departs at velocity $a - dV$.

Applying continuity from one side of the wave to the other,

$$\rho A a = (\rho + d\rho) A (a - dV) \qquad (3.70)$$

where A = cross-sectional area of tube.

Doing the algebra and remembering that $d\rho \, dV = 0$,

$$\rho \, dV = a \, d\rho \qquad (3.71)$$

Now rearranging Eq. (3.64) and substituting the derivative of Eq. (3.24),

$$T \, ds = dh - \frac{dp}{\rho} \qquad (3.72)$$

Assuming an isentropic gas ($ds = 0$),

---

* Large disturbances create "shock waves" that can travel through a fluid at velocities higher than the speed of sound. These are not isentropic. Examples of large disturbances are supersonic combustion (detonation) waves and disturbances caused by vehicles traveling faster than the speed of sound (supersonic).

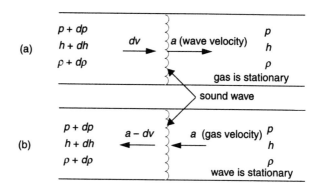

Fig. 3.7. Sound Waves Traveling Through a Gas in an Enclosed Tube. This figure defines the geometry and parameters used to derive the speed-of-sound relationships. Figure (a) shows a wave travelling past a stationary observer. Figure (b) shows the equivalent gas flow past a stationary wave.

$$dh = \frac{dp}{\rho} \tag{3.73}$$

To balance energy on either side of the sound wave, we use Eq. (3.25)

$$h + \frac{a^2}{2} = h + dh + \frac{(a-dv)^2}{2} \tag{3.74}$$

After combining terms and simplifying,

$$dh - a\,dv = 0 \tag{3.75}$$

Combining Eqs. (3.73) and (3.75),

$$\frac{dp}{\rho} = a\,dv \tag{3.76}$$

Now combining Eqs. (3.71) and (3.76) and solving for $dp/d\rho$,

$$\frac{dp}{d\rho} = a^2 \tag{3.77}$$

Having obtained the expression in Eq. (3.77), we must pause to develop another thermodynamic relation which helps us get a final expression for the acoustic velocity. We show that, for an isentropic process involving a perfect gas, pv^γ is a constant. We then differentiate the expression according to Eq. (3.77) to arrive at our equation for acoustic velocity.

We start with the differential energy equation (3.62) developed earlier for a thermodynamic process and assume an isentropic process (adiabatic and reversible $dq = 0$), so

$$du + p\,dv = 0 \tag{3.78}$$

According to our earlier definition of enthalpy ($h = u + pv$). By differentiating, we get

$$dh = du + p\, dv + v\, dp \tag{3.79}$$

But after substituting Eq. (3.78) into (3.79),

$$dh = v\, dp \tag{3.80}$$

Using our definitions of specific heat, we rewrite Eqs. (3.78) and (3.80) as

$$du = c_v\, dT = -p\, dv \tag{3.81}$$

and

$$dh = c_p\, dT = v\, dp \tag{3.82}$$

Dividing Eq. (3.82) by (3.81),

$$\frac{c_p}{c_v} = \gamma = -\frac{v\, dp}{p\, dv} \tag{3.83}$$

Rewriting, we have

$$\frac{dp}{p} + \gamma \frac{dv}{v} = 0 \tag{3.84}$$

Integrating Eq. (3.84),

$$\ln(p) + \gamma \ln(v) = C' \text{ (a constant)} \tag{3.85}$$

Making both sides of the equation exponents of e and combining terms, we finally arrive at

$$pv^\gamma = Cc = e^{C'} \tag{3.86}$$

or

$$p = C\rho^\gamma \tag{3.87}$$

We then look again at Eq. (3.77) and recall that the purpose for deriving Eq. (3.87) is to get a suitable expression for pressure. Finding $dp/d\rho$ for Eq. (3.87),

$$\frac{dp}{d\rho} = C\gamma\rho^{(\gamma-1)} = \frac{C\gamma\rho^\gamma}{\rho} = \frac{\gamma p}{\rho} \tag{3.88}$$

But $p/\rho = RT$, and we can apply Eq. (3.74)

$$\frac{dp}{d\rho} = a^2 = \gamma RT \tag{3.89}$$

This development gives us the following expression for acoustic velocity:

$$a = \sqrt{\gamma R T} \tag{3.90}$$

where γ = ratio of specific heats for the gas
  R = gas constant for the particular gas (J/kg·K)
  T = gas temperature (K)

Assumptions:
1. Continuity equation applies (steady flow)
2. One-dimensional flow
3. Perfect gas
4. Isentropic process (adiabatic and reversible)

Note that the acoustic velocity depends only on the square root of gas temperature. We know gas-particle velocities at the molecular level are functions of the square root of gas temperature. Thus, we see that disturbances which propagate through a gas because of molecular collisions also travel with a velocity which is a function of $T^{1/2}$.

As mentioned earlier, the *Mach number* (M) of a flow is simply the velocity of the flow (relative to a stationary observer) divided by the acoustic velocity of the flow. Because the *acoustic velocity* is the velocity at which a disturbance propagates through the flow, the Mach number shows whether the upstream flow can "sense" what is happening downstream and adjust itself accordingly. A Mach number greater than one shows the flow is moving at a rate faster than the acoustic velocity. Discussion in the section entitled "Throat Mach Number," which appears later in this chapter, deals with the important effect of upstream propagations on the flow. Because

$$M = \frac{V}{a} \tag{3.91}$$

the Mach number depends only on flow velocity and fluid temperature once we know the working fluid's makeup (γ and R).

Stagnation Conditions. For an isentropic, steady-state flow with no shaft work, we see that the energy equation simplifies to the following expression (3.26), which applies to any two arbitrary points in the flow:

$$h_1 + \frac{1}{2}v_1^2 = h_2 + \frac{1}{2}v_2^2$$

The enthalpy (h) at point 1 or 2 of Eq. (3.26) is called the *static enthalpy*. If the flow can be brought to rest adiabatically, reversibly, and without work, the velocity goes to zero (flow stagnation), and the enthalpy term goes to a maximum value. This value (h_0) is known as *stagnation enthalpy*. It occurs when the working fluid in

3.2 Thermodynamic Relations (Energy and Entropy)

a rocket chamber has very low axial velocity. For steady-state conditions this value is a constant. Now we can rewrite Eq. (3.26) as

$$h_0 = h_1 + \frac{1}{2}v_1^2 = h_2 + \frac{1}{2}v_2^2 \tag{3.92}$$

Stagnation conditions also apply to temperature (T_0), pressure (p_0), and density (ρ_0). Because $\Delta h = c_p \Delta T$, we substitute into Eq. (3.92) as it applies to the chamber* and some other arbitrary point:

$$c_p T_0 = c_p T + \frac{1}{2}v^2 \tag{3.93}$$

Dividing by $c_p T$, and using $v = M\sqrt{\gamma RT}$ and $\dfrac{R}{c_p} = \dfrac{\gamma-1}{\gamma}$, we discover

$$\frac{T_0}{T} = 1 + \frac{\gamma-1}{2}M^2 \tag{3.94}$$

Equation (3.91) now relates stagnation (or chamber) temperature to the temperature and Mach number at any other point in the nozzle. Using the isentropic relations developed earlier [Eqs. (3.64), (3.65), and (3.66)], we find

$$\frac{p_0}{p} = \left(1 + \frac{\gamma-1}{2}M^2\right)^{\frac{\gamma}{\gamma-1}} \tag{3.95}$$

and

$$\frac{\rho_0}{\rho} = \left(1 + \frac{\gamma-1}{2}M^2\right)^{\frac{1}{\gamma-1}} \tag{3.96}$$

where T_0 = stagnation temperature (K)
M = Mach number
T = static temperature (K) at a point with known Mach number
p_0 = stagnation pressure (N/m²)
p = static pressure (N/m²) at a point with known Mach number
ρ_0 = stagnation density (kg/m³)
ρ = static density (kg/m³) at a point with known Mach number
γ = isentropic parameter (ratio of specific heats)

---

* Temperature in a combustion chamber (flame temperature) is defined almost exclusively by the combustion chemistry (thermochemistry, see Chap. 4). The flow velocities in the combustion chamber are typically small (Mach number less than 0.3). Thus, we usually assume the combustion flame temperature is also the stagnation temperature.

Assumptions:
1. Isentropic flow
2. One-dimensional flow
3. Steady flow
4. Calorically perfect gas

If we know the temperature, pressure, and density within a combustion chamber—usually assumed to be the stagnation values—we may determine the values of those variables at any other point in the nozzle, provided we know the Mach number at that point. Now the challenge is to find the Mach number analytically for any point in the nozzle. To do so, we need a relationship between the size of the nozzle (known from a given nozzle geometry) and the Mach number for the flow at that point.

Area Ratio. The continuity equation for one-dimensional flow in a nozzle gives us the following relationship for any two points (1 and 2) in the nozzle:

$$(\rho A v)_1 = (\rho A v)_2 \tag{3.97}$$

Rearranging to solve for A_2/A_1 and using $v = M\sqrt{\gamma RT}$, we have

$$\frac{A_2}{A_1} = \frac{M_1}{M_2}\sqrt{\frac{T_1 \rho_1^2}{T_2 \rho_2^2}} \tag{3.98}$$

Using Eqs. (3.94) and (3.96) to substitute for the terms inside the radical, we arrive at an expression which relates the area ratio for two nozzle locations to the Mach numbers at those locations:

$$\frac{A_2}{A_1} = \frac{M_1}{M_2}\sqrt{\left\{\frac{1+\frac{\gamma-1}{2}M_2^2}{1+\frac{\gamma-1}{2}M_1^2}\right\}^{\left(\frac{\gamma+1}{\gamma-1}\right)}} \tag{3.99}$$

Normally the area ratio of greatest concern in a rocket nozzle is the ratio of the exit area to the throat area (A_e/A_t), commonly called the expansion ratio (ε). In the following section we prove that the Mach number at the nozzle throat is 1. If we use this fact in advance, Eq. (3.99) becomes

$$\varepsilon = \frac{A_e}{A_t} = \frac{1}{M_e}\sqrt{\left\{\frac{2}{\gamma+1}\left(1+\frac{\gamma-1}{2}M_e^2\right)\right\}^{\left(\frac{\gamma+1}{\gamma-1}\right)}} \tag{3.100}$$

where A_e = cross-sectional area of nozzle exit (m²)
       A_t = cross-sectional area of nozzle throat (m²)
       M_e = Mach number at exit
       γ = ratio of specific heats for the fluid

Assumptions:
1. Isentropic flow
2. One-dimensional flow
3. Steady flow
4. Perfect gas
5. Choked flow

Figure 3.8, based on Eq. (3.100), reveals that for any given area ratio (except 1) two distinct Mach numbers satisfy the equation. One is *subsonic* (Mach number less than 1.0), and the other is *supersonic* (Mach number greater than 1.0). In our nozzle analysis we are interested in the supersonic value so we can achieve the highest possible flow velocity and thrust. Substituting a throat Mach number of 1 into Eq. (3.95), we see that a supersonic nozzle containing a gas with $\gamma = 1.4$ must have a ratio of chamber pressure to throat pressure of at least 1.89. Equation (3.97) clearly shows that the exit Mach number depends on the expansion ratio. This observation highlights an important trade-off in nozzle design: we must pay the penalty of a larger, heavier nozzle to get the higher exit Mach number we want.

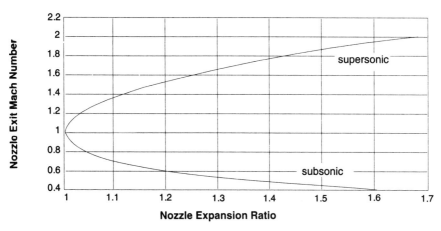

Fig. 3.8. **Mach Number as a Function of Area Ratio.** As flow expands after a nozzle throat (M = 1) it either accelerates ($M > 1$) or decelerates ($M < 1$) depending on the downstream ambient pressure ($\gamma = 1.4$).

Throat Mach Number. Now that we know how to determine the Mach number anywhere in the nozzle just by knowing the cross-sectional area there, we need to investigate why the Mach number at the throat is 1. For one-dimensional, isentropic flow, we may develop an expression relating velocity changes to changes in the nozzle's cross-sectional area for varying Mach numbers.

From continuity we know

$$\rho A v = \text{constant} \tag{3.8}$$

Differentiating Eq. (3.8) and dividing all terms by $\rho A v$, we have

$$\frac{d\rho}{\rho} + \frac{dA}{A} + \frac{dv}{v} = 0 \tag{3.101}$$

From the energy equation,

$$h + \frac{v^2}{2} = \text{constant} \tag{3.102}$$

so differentiating produces

$$dh + v\, dv = 0 \tag{3.103}$$

Substituting Eq. (3.73) into (3.103) and solving for dp,

$$dp = -\rho v\, dv \tag{3.104}$$

Rearranging Eq. (3.89),

$$d\rho = \frac{dp}{a^2} \tag{3.105}$$

Substituting Eq. (3.105) into Eq. (3.101),

$$\frac{dp}{a^2 \rho} + \frac{dA}{A} + \frac{dv}{v} = 0 \tag{3.106}$$

Substituting (3.104) into (3.106),

$$-\frac{v^2 dv}{a^2 v} + \frac{dA}{A} + \frac{dv}{v} = 0 \tag{3.107}$$

Recalling the definition of the Mach number and rearranging Eq. (3.107),

$$\frac{dA}{A} = \left(M^2 - 1\right)\frac{dv}{v} \tag{3.108}$$

where A = cross-sectional area of rocket nozzle (m²)
 M = Mach number of flow
 v = flow velocity (m/s)

3.2 Thermodynamic Relations (Energy and Entropy)

Assumptions:
1. Steady flow
2. Calorically perfect gas
3. Isentropic flow
4. One-dimensional flow

We may apply this equation to diverging and converging nozzle sections as shown in Fig. 3.9.

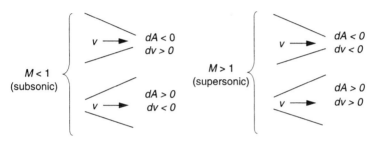

Fig. 3.9. Flow Through Converging and Diverging Nozzle Sections. Depending on whether the flow is initially subsonic or supersonic, the effect of a converging or diverging duct differs. If initially subsonic, decreasing the area increases flow velocity. If initially supersonic, increasing the area increases the flow velocity.

To satisfy the governing equation above for subsonic flow, where the Mach number is less than 1, the velocity must increase at nozzle constrictions ($dA < 0$ and $dv > 0$ as the area decreases) and decrease at expansions. For supersonic flow, the velocity must decrease (with accompanying shocks) at constrictions and increase at expansions. When $M = 1$, clearly $dA = 0$, so the Mach number equals 1 only where the nozzle area is neither increasing nor decreasing—the throat.

We can explain the transition from subsonic flow through sonic to supersonic by using the analogy of an aircraft approaching an air molecule. Because perturbations through the intervening air travel at the speed of sound, a subsonic plane does not reach the molecule until after the sound waves have contacted the molecule, "warning" it to get out of the way. Conversely, a supersonic jet strikes the molecule before it gets any warning. Now, as subsonic flow in a nozzle approaches the restriction at the throat, pressure waves traveling upstream cause the flow to adjust itself and increase velocity as cross-sectional area decreases, according to the equation above. If the upstream pressure increases, causing the velocity to increase further, the flow continues to adjust until it reaches Mach 1 at some point. This point must be the throat, because subsonic flow travels fastest at the smallest constriction. At this point, the flow upstream of the throat receives no information about the flow downstream and cannot adjust further. Thus, once a sufficient pressure ratio (p_o/p_e) is established, the flow approaching the throat is subsonic, flow at the throat is sonic, and the flow leaving the throat is supersonic (assuming the ambient pressure is not excessive).

Because information about the nozzle downstream of the throat cannot flow back into the chamber, we can analyze the chamber and throat separately from the nozzle. We see more about this analytical approach in our discussions of the thrust coefficient and characteristic exhaust velocity.

Nozzle Exit Velocity. Beginning with Eq. (3.26) and solving for velocity v_2, we have

$$v_2 = \sqrt{2(h_1 - h_2) + v_1^2} \tag{3.109}$$

Now we allow point 1 to be in the chamber and we use subscript 0 to indicate stagnation conditions. Note that v_1 in Eq. (3.109) is now the velocity in the chamber, which we assume to be zero (stagnation). Point 2 is at the nozzle exit (we use the subscript "e" to indicate nozzle conditions). Using the definition of enthalpy for constant $c_p (\Delta h = c_p \Delta T)$, we have

$$v_e = \sqrt{2c_p(T_0 - T_e)} \tag{3.110}$$

Factoring out the T_0 term and using $c_p = \gamma R/(\gamma - 1)$,

$$v_e = \sqrt{\frac{2\gamma R T_0}{\gamma - 1}\left(1 - \frac{T_e}{T_0}\right)} \tag{3.111}$$

Using Eq. (3.68) and the definition of the universal gas constant, we have the following expression for the nozzle exit velocity:

$$v_e = \sqrt{\frac{2\gamma \mathcal{R}_u T_0}{(\gamma - 1)\mathcal{M}}\left\{1 - \left(\frac{p_e}{p_0}\right)^{\frac{\gamma-1}{\gamma}}\right\}} \tag{3.112}$$

where v_e = nozzle exit velocity (m/s)
$\mathcal{R}_u$ = universal gas constant (8314.41 J/kmol·K)
T_0 = chamber temperature (K)
p_e = exit pressure (N/m^2)
p_0 = chamber pressure (N/m^2)
$\mathcal{M}$ = molecular mass of gas (kg/kmol)
γ = ratio of specific heat

Assumptions: 1. Same as for Eq. (3.25)
2. Constant c_p over the temperature range involved
3. Chamber represents stagnation conditions

Notice in Eq. (3.112) that two key parameters strongly affect exit velocities. Higher chamber temperatures and lower molecular masses result in higher exhaust velocities, which lead to higher specific impulses. Also, a small pressure ratio (p_e/p_0) yields a higher exhaust velocity [Sutton, 1986].

3.3 Thrust Equations

The thrust equation derives from the more general momentum equation, which starts with Newton's Second Law: the sum of the forces on an object equals the time rate of change of momentum of that object. We can express this law as

$$\sum F_x = \frac{d}{dt}(mv_x) \qquad (3.113)$$

The main concern of this section is how different forces affect the motion of a rocket. The main force is caused by propellant flowing through the nozzle, so we concentrate on it here before developing the thrust equation in the next section. In Eq. (3.113) we limit the flow through our system to one direction (in x), but we may readily generalize this result to include all directions.

Using Fig. 3.10 as in Shapiro [1953] and the definition of a derivative as we did for the energy equation, we see that as dt becomes very small, the right-hand side of Eq. (3.113) may be rewritten as follows:

$$\frac{d}{dt}(mv_x) = \frac{\{(mv_x)_{I_{(t+dt)}} + (mv_x)_{II_{(t+dt)}}\} - \{(mv_x)_{I_{(t)}} + (mv_x)_{III_{(t)}}\}}{dt} \qquad (3.114)$$

or

$$\frac{d}{dt}(mv_x) = \underbrace{\frac{(mv_x)_{I_{(t+dt)}} - (mv_x)_{I_{(t)}}}{dt}}_{(A)} + \underbrace{\frac{(mv_x)_{II_{(t+dt)}}}{dt}}_{(B)} - \underbrace{\frac{(mv_x)_{III_{(t)}}}{dt}}_{(C)} \qquad (3.115)$$

Note that the entire right-hand side of Eq. (3.115) represents the time rate of change of momentum of the fluid flowing through the control volume. As dt approaches zero and volumes II and III become insignificant, expression (A) simply represents the time rate of change of momentum in volume I, or

$$\frac{\partial}{\partial t}(mv_x)_{cv}$$

But mass (m) is just ρV, so we can rewrite expression (A) as

Expression (B) becomes

$$\frac{\partial}{\partial t}\int_{cv}\rho v_x\, dV$$

or

$$\frac{\int_{cv} v_x\, dm_{II(t+dt)}}{dt}$$

$$\int_{cv} v_x\, d\dot{m}_{II(t+dt)}$$

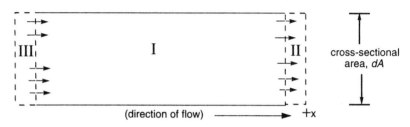

Fig. 3.10. Control Volume for Momentum Equation. This figure defines the geometry used to derive the "thrust" equation.

Because this is the mass flowing out of the control volume, we rewrite it as $\int_{cv} v_x\, d\dot{m}_{out}$. But $\dot{m} = \rho v_n A$, and therefore $d\dot{m} = \rho v_n\, dA$, assuming ρ and v_n are constant across the differential area (dA). Finally, we rewrite expression (B) again to get

$$\left(\int_{cs} v_x \rho v_n\, dA\right)_{out}$$

Similarly, rewriting expression (C) gives us $\left(\int v_x \rho\, v_n\, dA\right)_{in}$. Using the dot product (as we did for the continuity equation) to distinguish incoming momentum from outgoing momentum, we combine expressions (B) and (C):

$$\int_{cs} v_x \rho (\vec{v}\cdot\hat{n})\, dA$$

Now the momentum equation in the x direction *due to fluid flow* becomes

$$F_{x_{thrust}} = \frac{\partial}{\partial t}\int_{cv}\rho v_x\, dV + \int_{cs} v_x \rho(\vec{v}\cdot\hat{n})\, dA \qquad (3.116)$$

or generally

3.3 Thrust Equations

$$\vec{F}_{thrust} = \frac{\partial}{\partial t}\int_{cv}\rho\vec{v}\,dV + \int_{cs}\vec{v}\rho(\vec{v}\cdot\hat{n})\,dA \qquad (3.117)$$

where $\vec{F}_{thrust}$ = thrust force on the rocket (N)
 ρ = density of fluid in the flow (kg/m³)
 $\vec{v}$ = flow velocity (m/s)
 dV = differential volume element (m³)
 $\hat{n}$ = outward normal unit vector
 dA = differential area element (m²)

We read this expression as "the sum of the forces acting on a control volume equals the time rate of change of momentum in the control volume plus the net momentum flux across the control surface of the system." Again, note that Eq. (3.117) considers only the effect of fluid moving through the system, so its left-hand side does not include electrical, magnetic, or other forces.

3.3.1 Thrust Equation

We can now adapt the momentum equation to develop a more usable expression for rocket thrust [Hill and Peterson, 1970]. Figure 3.11 illustrates the major forces acting on a rocket mounted on a test stand. Momentum thrust from ejected propellant acts on the rocket in the positive x direction. Pressure at the exit (p_e) also acts on the rocket in the positive x direction, whereas ambient pressure (p_a) acts everywhere else around the rocket. The resultant produces a pressure thrust equal to ($p_e - p_a)A_e$. All other pressure forces, normal to the control surface, are symmetrical and cancel each other. The force F must perfectly balance momentum thrust and pressure thrust to keep the rocket stationary. Because it is exactly equal and opposite to the thrust force,

$$F = \text{momentum thrust} + \text{pressure thrust}$$

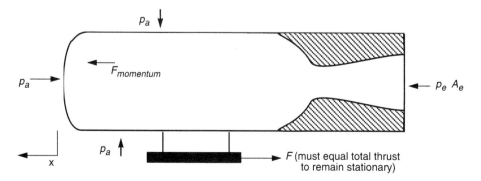

Fig. 3.11. **Forces Acting on a Rocket Motor.** Thrust is a combination of momentum exchange and exit pressure. Ambient pressure acts to reduce the thrust level.

Substituting for momentum thrust from the momentum equation derived earlier, we have the following expression for thrust:

$$F = \frac{\partial}{\partial t}\int_{cv} v\rho \, dV + \int_{cs} v\rho(\vec{v}\cdot\hat{n})dA + (p_e - p_a)A_e \qquad (3.118)$$

where we dropped most of the vector notation because the forces combine in the x direction. The flow is steady in the combustion chamber, so the time-dependent term disappears. As seen in the derivation of continuity, the term $\vec{v}\cdot\hat{n}$ simply equals v_n. Adopting this convention for the present analysis assumes one-dimensional (1-D) flow. In other words, we assume all streamlines in the flow are parallel throughout the nozzle. Thus,

$$F = v_n \rho_e v_e A_e + (p_e - p_a)A_e \qquad (3.119)$$

or

$$F = \dot{m}v_e + (p_e - p_a)A_e \qquad (3.120)$$

where F = thrust force (N)
 $\dot{m}$ = fluid mass flow rate (kg/s)
 v_e = exhaust velocity (m/s)
 p_e = nozzle exit pressure (N/m^2)
 p_a = ambient pressure (N/m^2)
 A_e = nozzle exit area (m^2)

Assumptions:
 1. Steady flow
 2. One-dimensional flow through nozzle

Effective Exhaust Velocity. Equation (3.120) shows the pressure-thrust term increases or decreases a rocket's thrust velocity, depending on the values of exit and ambient pressure. We may combine this effect with gaseous exhaust velocity (v_e) to obtain an *effective exhaust velocity (c)*. By dividing both sides of Eq. (3.120) by $\dot{m}$ and setting $F/\dot{m}$ equal to c, we discover

$$c = v_e + \frac{A_e}{\dot{m}}(p_e - p_a) \qquad (3.121)$$

where c = effective exhaust velocity (m/s)
 v_e = actual exhaust velocity (m/s)
 A_e = nozzle exit area (m^2)
 $\dot{m}$ = mass flow rate through nozzle (kg/s)
 p_e = nozzle exit pressure (N/m^2)
 p_a = ambient pressure (N/m^2)

Assumptions: Same as for Eq. (3.120)

3.3.2 Coefficient of Thrust and Characteristic Exhaust Velocity

The *coefficient of thrust* (c_F), is defined as

$$c_F = \frac{F}{A_t p_0} \quad (3.122)$$

where F = thrust (N)
A_t = cross-sectional area at the throat (m²)
p_0 = chamber pressure (N/m²)

The thrust equation developed above [Eq. (3.120)] may be rearranged using Eq. (3.112) (nozzle exit velocity) and $\dot{m} = \rho_t A_t v_t$. Because we know flow velocity is equal to the acoustic velocity at the throat for a supersonic nozzle, we have

$$v_t = \sqrt{\gamma R T_t} \quad (3.123)$$

If we rearrange Eq. (3.96) for sonic conditions at the throat,

$$p_t = p_0 \left(\frac{2}{\gamma+1} \right)^{\frac{1}{\gamma-1}} \quad (3.124)$$

Combining our thrust equation [Eq. (3.120)] with Eqs. (3.112), (3.123), and (3.124), we obtain the following expression for thrust:

$$F = A_t p_0 \gamma R \left\{ \frac{2 T_t T_0}{\gamma-1} \left(\frac{2}{\gamma+1} \right)^{\frac{1}{\gamma-1}} \left[1 - \left(\frac{p_e}{p_0} \right)^{\frac{\gamma-1}{\gamma}} \right] \right\}^{\frac{1}{2}} + (p_e - p_a) A_e \quad (3.125)$$

Now, using the isentropic flow relations

$$T_t T_0 = \left(\frac{p_0}{\rho_t R} \right)^2 \left(\frac{2}{\gamma+1} \right)^{\left(\frac{\gamma+1}{\gamma-1}\right)} \quad (3.126)$$

and

$$\frac{1}{\rho_t^2} \left(\frac{2}{\gamma+1} \right)^{\frac{2}{\gamma-1}} = \frac{1}{\rho_0^2} \quad (3.127)$$

we arrive at the following expression for thrust:

$$F = A_t p_0 \gamma \left\{ \frac{2}{\gamma-1}\left(\frac{2}{\gamma+1}\right)^{\frac{\gamma+1}{\gamma-1}} \left[1-\left(\frac{p_e}{p_0}\right)^{\frac{\gamma-1}{\gamma}}\right]\right\}^{\frac{1}{2}} + (p_e - p_a)A_e \tag{3.128}$$

Using the earlier definition for thrust coefficient [Eq. (3.122)], we develop the following expression:

$$c_F = \left\{ \frac{2\gamma^2}{\gamma-1}\left(\frac{2}{\gamma+1}\right)^{\frac{\gamma+1}{\gamma-1}} \left[1-\left(\frac{p_e}{p_0}\right)^{\frac{\gamma-1}{\gamma}}\right]\right\}^{\frac{1}{2}} + \frac{(p_e - p_a)A_e}{p_0 A_t} \tag{3.129}$$

Note that expression (3.129) depends entirely on nozzle characteristics (except for γ). Suppose a laboratory is testing a particular nozzle setup with known expansion ratio on an engine with known chamber conditions using a propellant of known ratio of specific heats (γ). In this case, they can closely estimate the thrust to help choose a load cell for the test. Experimenting with different nozzle sizes changes only area and pressure ratios, so the thrust is easily and accurately predicted from the thrust coefficient.

Characteristic Exhaust Velocity. Just as the coefficient of thrust allows analysis of nozzles independent of the rest of the rocket, the *characteristic exhaust velocity* (c^*) allows us to focus on propellant and chamber performance independent of the nozzle. Development of c^* begins with the expression used earlier for mass flow rate at the throat ($\dot{m} = (\rho A v)_t$).

Now we would like to get all of the variables except for throat area expressed in terms of chamber conditions. Using throat velocity ($v_t = \sqrt{\gamma R T_t}$) Eqs. (3.94), (3.96), and the perfect-gas law, we get mass flow rate

$$\dot{m} = \frac{A_t p_0}{\sqrt{\gamma R T_0}} \left\{ \gamma \left(\frac{2}{\gamma+1}\right)^{\frac{\gamma+1}{2(\gamma-1)}} \right\} = \frac{A_t p_0}{\sqrt{\gamma R T_0}} \Gamma' \tag{3.130}$$

We designate the expression in the large brackets as Γ' and note that $a_0 = \sqrt{\gamma R T_0}$ is the value of acoustic velocity in the chamber, so

$$\dot{m} = \frac{\Gamma'}{a_0} A_t p_0 \tag{3.131}$$

We now define the characteristic exhaust velocity (c^*):

$$c^* = \frac{a_0}{\Gamma'} \tag{3.132}$$

where a_0 = acoustic velocity in the chamber (m/s)
Γ' = function of γ as defined above (Eq. (3.130),

From Eq. (3.131) we may also develop the following expression:

$$c^* = \frac{A_t\, p_0}{\dot{m}} \tag{3.133}$$

where A_t = throat cross-sectional area (m²)
p_0 = chamber pressure (N/m²)
$\dot{m}$ = mass flow rate through the rocket (kg/s)

Assumptions:
1. Isentropic flow
2. Perfect gas
3. One-dimensional, steady flow

Equation (3.132) shows that c^* is a function only of temperature and gas properties in the combustion chamber (γ, R, and T_0). These are all determined *from the thermochemical reaction in the combustion chamber*. In other words, we can use c^* to analyze the chamber and throat independent of the nozzle. Usually, after choosing a propellant for testing, we use Eq. (3.132) to calculate a theoretical value for c^*. After testing, we substitute empirical values into Eq. (3.133) to determine an experimental value of c^*. Based on experience, the actual value of c^* should be 96% to 98% of the theoretical value. Incomplete combustion causes this reduction.

Conditions for Maximum Thrust. We can also apply Eq. (3.128) to determine the pressure conditions that provide the most thrust for a nozzle. A qualitative look at the equation suggests the momentum-thrust term is largest when exit pressure is zero and the pressure-thrust term is largest when ambient pressure is zero. But we need a nozzle of infinite expansion ratio to get an exit pressure of zero. In addition, rocket propulsion of launch vehicles largely concerns performance in the atmosphere where ambient pressure is not zero.

The conditions for maximum thrust may be determined analytically by taking the derivative of Eq. (3.128) with respect to the ratio p_e/p_0. Setting the result equal to zero to find the maximum thrust, we find the relation which satisfies this condition is $p_e = p_a$. Thus, the maximum thrust for a given altitude is obtained by a nozzle which expands the flow to the precise ambient pressure for that altitude. This maximum-thrust condition, in which the nozzle-exit pressure is equal to the ambient pressure, is called *ideal expansion*.

We must be careful how we state the results of this analysis. Our first inclination may be to say that a given rocket nozzle achieves maximum thrust when $p_e = p_a$. However, this is false, as an investigation of Eq. (3.128) reveals. We must say that for a given altitude and a given γ, the rocket which achieves maximum thrust is the one whose nozzle expands the flow to the precise ambient pressure for that altitude. This conclusion can be verified by choosing γ, p_c, and p_a (fixing the altitude) and solving Eq. (3.129) for various values of p_e. We then discover the maximum thrust coefficient results when $p_e = p_a$. On the other hand, a rocket motor with constant p_0, p_e, and γ always produces higher thrust as it increases in altitude, even above the altitude for which $p_e = p_a$. This statement can be verified by inspection of Eq. (3.128). However, that motor's thrust at an altitude where $p_e > p_a$ always is less than a similar motor for which $p_e = p_a$ at that altitude.

As a result, the best nozzle for a launch vehicle must have an adjustable exit area, match exit pressure with ambient pressure throughout the vehicle's ascent, and thus provide maximum thrust at all altitudes. Because mechanical difficulties usually disallow such a nozzle, we design a rocket so each of its stages produces the best thrust at some nominal point in its operating altitude. This point is normally about two-thirds of the maximum atmospheric height for the stage.

Figure 3.12 depicts thrust curves as a function of altitude. The solid curve on the right shows altitude vs. thrust for ideal expansion (exit pressure equals ambient pressure) at all altitudes. The other two curves represent performance for engines designed for ideal expansion at only one altitude each. We want the curve closest to the solid curve—usually from a nozzle that achieves ideal expansion at two-thirds of its maximum operating altitude.

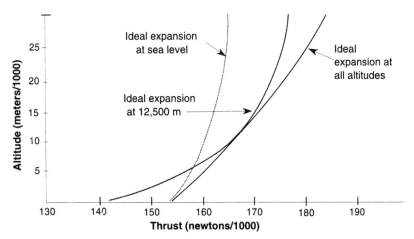

Fig. 3.12. **Thrust vs. Altitude for Nozzles with Various Expansion Ratios.** The thrust level increases with altitude. However, nozzle geometry affects how it increases.

Overexpansion and Underexpansion. For steady supersonic flow (in which the flow is not separated from the nozzle at any point), the exit pressure is constant for a given engine/nozzle system. Thus, ideal expansion ($p_e = p_a$) occurs at only one altitude. At any other altitude, the flow is either overexpanded or underexpanded.

If a rocket is operating below its optimum altitude, ambient pressure is greater than exit pressure. As exhaust products flow toward the nozzle exit, the nozzle increases in area and the flow expands to fill it, causing the pressure in the flow to decrease. Because the flow has expanded to a pressure below the ambient pressure, the flow is said to be *overexpanded*. Conversely, the flow in a rocket operating above its best altitude is *underexpanded*. A televised Space Shuttle launch reveals both conditions. When the main engines ignite on the launch pad, the overexpanded flow is restricted to a narrow cylinder by the higher ambient pressure. As the Shuttle reaches high altitudes, the now underexpanded flow fans out in the greatly reduced ambient pressure.

Although both conditions contribute to lower thrust, the overexpansion is not always as harmful as it would first appear to be. In moderate to extreme overexpansion the flow separates from the wall near the exit because of the high ambient pressure. As a result, the effective exit area shrinks and, as seen in Eq. (3.128), the negative effect of high ambient pressure lessens.

Summerfield [1954] (also see Oates [1984]) developed a way to estimate the altitude at which flow separation occurs for a given nozzle. As Fig. 3.13 shows, when a rocket operates above the design altitude (h_d where the nozzle is ideally expanded), the flow crosses expansion fans which expand it to the lower ambient pressure. If we imagine the rocket descending below the design altitude, the flow crosses oblique shock waves forming at the nozzle lip. These waves boost flow pressure to ambient pressure. As altitude decreases, the shocks become stronger until the rocket reaches the separation altitude (h_{sep}). At this point the shocks are as strong as they can get, if they are to remain attached to the nozzle lip. If ambient pressure increases, the shock waves move inside the nozzle, causing flow to separate from the wall downstream of the shock wave. In separated flow, as seen in Fig. 3.14, the pressure just before the shock (p_s) is the pressure for that part of the nozzle if it were operating normally. This condition occurs because effects of the shock cannot travel upstream in supersonic flow.

Oates observes from Summerfield's experimentation with separated flow that the pressure ratio p_a/p_s (which he terms "K") is somewhere between 2.5 and 3.0 for his configuration. A good "rule of thumb" value for K, regardless of the position of the oblique shock in the nozzle, is 2.75.

We gain from this value the ability to calculate thrust when the flow separates. Because the flow separates from the wall at some point inside the nozzle, the exit area effectively shrinks. Using $K = 2.75$, and knowing the ambient pressure (p_a) for a given altitude, we can find the flow pressure (p_s) just before the shock. We may then use Eqs. (3.95) and (3.99) to determine the effective exit area and subsequently use Eq. (3.120) to calculate the thrust.

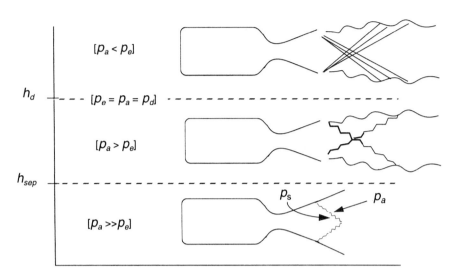

Fig. 3.13. Flow Regimes in a Nozzle as a Function of Altitude. At high altitudes, the nozzle is underexpanded. At lower altitudes, the nozzle is overexpanded, possibly allowing the flow to separate from the nozzle.

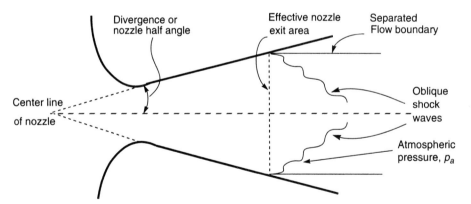

Fig. 3.14. Representation of Separated Flow in a Nozzle. At the point of flow separation, oblique shock waves radiate into the flow. These shocks act to compress the flow to a pressure close to ambient.

Oates has developed an expression to predict the altitude at which the flow separates. Following his method, we can model atmospheric pressure to within 3500 Pa using

$$\frac{p}{p_{SL}} = e^{\frac{-h}{h_{scl}}} \qquad (3.134)$$

where p = ambient pressure at a given altitude above sea level (Pa)
p_{SL} = pressure at sea level (N/m²)
h = altitude of concern above sea level (m)
h_{scl} = scale height for the model = 7010.4 m

Rearranging Eq. (3.134),

$$h = -h_{scl} \ln\left(\frac{p}{p_{SL}}\right) \qquad (3.135)$$

Applying Eq. (3.135) to conditions at the design altitude (h_d) and the separation altitude (h_{sep}), we have design altitude

$$h_d = -h_{scl} \ln\left(\frac{p_d}{p_{SL}}\right) \qquad (3.136)$$

where p_d is the pressure at the design altitude. At the separation altitude,

$$h_{sep} = -h_{scl} \ln\left(\frac{p_{sep}}{p_{SL}}\right) \qquad (3.137)$$

After we combine Eqs. (3.136) and (3.137),

$$h_{sep} = h_d - h_{scl} \ln\left(\frac{p_{sep}}{p_d}\right) \qquad (3.138)$$

At the separation altitude, $p_d = p_s$ because the shock is at the nozzle exit plane. Separation pressure $p_{sep} = p_a$ for that altitude. Therefore,

$$\frac{p_{sep}}{p_d} = \frac{p_a}{p_s} = K \qquad (3.139)$$

which is Summerfield's pressure ratio for separated flow. Substituting this value into Eq. (3.138), we have

$$h_{sep} = h_d - h_{scl} \ln K \qquad (3.140)$$

Equation (3.140) is a good approximation for the altitude above sea level at which the flow separates inside a nozzle.

Real Nozzles. Up to now we have analyzed ideal nozzles assuming one-dimensional flow, which is not really possible. The flow at the exit of a conical nozzle is nonaxial, with a variance that depends on the half angle (θ_{cn}) of the nozzle [see Fig. 3.15 (a)]. Thus, we need to relate ideal one-dimensional thrust to nonaxial thrust.

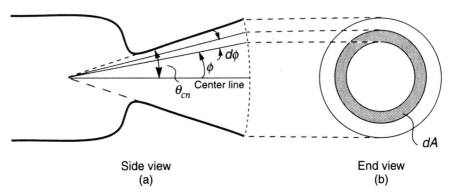

Fig. 3.15. **Views of a Nonaxial Nozzle.** This geometry is used to develop the efficiency relationship to account for nonaxial flow from the nozzle.

As seen earlier in the development of the thrust equation, the thrust force equals momentum thrust plus pressure thrust. For a rocket operating at steady state, we combine Eqs. (3.116) and (3.120) to write the momentum equation for the rocket pictured in Fig. 3.15:

$$F_x = \int_{cs} v_x \rho(\vec{v} \cdot \hat{n})dA + (p_e - p_a)A_e \tag{3.141}$$

We can no longer assume a planar exit area for this nozzle. Rather the area is a section of a sphere of radius R. To find the exit area we must add up all the areas of the concentric rings of infinitesimal width on the surface as seen in Fig. 3.15 (b). The width of the ring pictured is

$$dA = 2\pi R^2 (\sin \phi) \, d\phi \tag{3.142}$$

Because $\vec{v} \cdot \hat{n} = v_e$ and $v_x = v_e \cos \phi$, we may rewrite the thrust equation as

$$F = \int_0^{\theta_{cn}} 2\pi R^2 \rho v_e^2 \sin \phi \cos \phi \, d\phi + (p_e - p_a)A_e \tag{3.143}$$

Integrating from 0 to θ_{cn}, we have

$$F = 2\pi R^2 \rho v_e^2 \left(\frac{1 - \cos^2 \theta_{cn}}{2}\right) + (p_e - p_a)A_e \tag{3.144}$$

By designating the curved exit area as A_e', we can express the mass flow rate as $\dot{m} = \rho v_e A_e'$. Because

$$A_e' = \int_0^{\theta_{cn}} 2\pi R^2 \sin \phi \, d\phi = 2\pi R^2 (1 - \cos \theta_{cn}) \tag{3.145}$$

3.3 Thrust Equations

the mass flow rate becomes

$$\dot{m} = 2\pi R^2 \rho v_e (1 - \cos\theta_{cn}) \quad (3.146)$$

Substituting into the thrust equation above,

$$F = \dot{m} v_e \left(\frac{1+\cos\theta_{cn}}{2}\right) + (p_e - p_a) A_e \quad (3.147)$$

The traditional planar exit area $A_e = \pi R^2 \sin^2\theta = \pi R^2 (1 - \cos^2\theta)$, allows us to rewrite the exit area as

$$A_e = A_e' \left(\frac{1+\cos\theta_{cn}}{2}\right) \quad (3.148)$$

Substituting once again into the thrust equation,

$$F = \left(\frac{1+\cos\theta_{cn}}{2}\right)\left[\dot{m} v_e + (p_e - p_a) A_e'\right] \quad (3.149)$$

In the expression above, the term $(p_e - p_a)$ is usually small compared to the momentum-thrust term. In addition, the difference between A_e and A_e' is very small for all conventional nozzles. Because the product of the two terms is even smaller, we substitute A_e for A_e' with little change to the overall thrust equation. We may finally write the thrust equation as

$$F = \left(\frac{1+\cos\theta_{cn}}{2}\right)\left[\dot{m} v_e + (p_e - p_a) A_e\right] \quad (3.150)$$

and we can express the ratio of actual thrust in a conical nozzle to thrust in an ideal nozzle as

$$\lambda = \frac{\text{actual thrust}}{\text{ideal thrust}} = \frac{1}{2}(1 + \cos\theta_{cn}) \quad (3.151)$$

where λ = ratio of actual over ideal thrust or thrust efficiency
θ_{cn} = nozzle half-angle

Equation (3.150) shows that, as θ_{cn} decreases, the actual thrust increases. Unfortunately, as θ_{cn} decreases, we need a longer nozzle to expand the flow adequately. Therefore, using a conical nozzle involves the trade-off mentioned before: a compact nozzle with lower thrust or a longer, heavier nozzle with higher thrust. Table 3.2 shows values for λ for nozzles of various half-angles.

Table 3.2. **Thrust Efficiency as a Function of the Nozzle's Half-angle.** As the nozzle half angle increases the thrust efficiency decreases. Numbers are derived from Eq. (3.151).

Nozzle Half-angle (θ_{cn})	Thrust Efficiency (λ)
0	1.000
4	0.999
8	0.995
12	0.989
16	0.981
20	0.970
24	0.957

The most widely used nozzle is bell-shaped. Just past the throat, a rapid increase in volume allows the flow to expand rapidly, which produces weak expansion shocks. As the bell shape brings the flow back to the axial direction, weak compression shocks form. Shaping the bell so that the compression shocks coincide with and "cancel" the expansion shocks is the major challenge for engineers who design nozzles. The bell-shaped nozzle can achieve near-axial flow at the exit with adequate expansion in a relatively compact size [Hill and Peterson, 1970]. This nozzle is discussed in Sec. 5.4.

3.4 Heat Addition

Heat addition is the main way energy transfers to the rocket's gas flow, providing thrust. To add heat, we can use

- Exothermic chemical reactions (combustion)
- Atomic fission or fusion
- Electric elements (resistojets)
- Electric arcs (arcjets)
- Concentrated solar energy
- Other heat sources.

We need to know how adding heat affects the behavior of gases and liquids. When we presented the first law, we said heat addition and work effects depend on the process path. We consider here how they affect rocket propulsion.

3.4.1 Adding Heat to a Constant Volume with No Flow

A typical example of this problem is adding heat to a closed pressure vessel, such as a propellant storage tank. We can analyze it by considering the gas in the tank either as a "control mass" or as a "control volume," with the mass flow rates set to zero. Results in both cases are the same. We use the first law for a system:

3.4 Heat Addition

$$\frac{\delta Q}{dt} = \frac{dE}{dt} + \frac{\delta W}{dt} \tag{3.152}$$

where $\frac{\delta Q}{dt}$ = rate of heat addition (path dependent) (W)

$\frac{dE}{dt}$ = rate of change of system energy (W)

$\frac{\delta W}{dt}$ = rate of work done (path dependent) (W)

We assume no irreversible work is done. The change in reversible work,

$$\delta W_{rev} = p \, dV \tag{3.153}$$

where p = system pressure (Pa)
dV = change in system volume (m^3)

but because we have prescribed a constant volume,

$$dV = 0 \Rightarrow \delta W_{rev} = 0 \tag{3.154}$$

Substituting these values leaves the differential equation

$$dQ = dE \tag{3.155}$$

Having defined the process (no work done), we can change the derivative sign in front of the Q term and then integrate it to give

$$\Delta Q = E_2 - E_1 \tag{3.156}$$

where the ΔQ term refers to the amount of heat added to the system. If mass has no net motion, corresponding to a gas or fluid in a storage tank, the kinetic or potential energy does not change. This equation becomes

$$\Delta Q = U_2 - U_1 \tag{3.157}$$

which reduces to

$$\Delta Q = m \int_{T_1}^{T_2} c_v(T) \, dT \tag{3.158}$$

where T_1 = initial temperature (K)
T_2 = final temperature (K)
$c_v(T)$ = constant volume specific heat (J/ kg·K)
m = mass of the system (kg)

For a calorically perfect gas,

$$\Delta Q = mc_v(T_2 - T_1) \tag{3.159}$$

3.4.2 Steady Flow with Heat Addition

A rocket combustion chamber and nozzle operating in a steady state is equivalent to steady flow with heat addition. This implies that the mass entering the chamber is equal to the mass exiting (no mass accumulation), and that the energy distribution of the mass within the control volume is also constant. Assuming no irreversible work is done and changes in flow potential energy are negligible, the first law for a control volume is

$$\dot{Q} = \dot{m}(h_{0e} - h_{0i}) \tag{3.160}$$

where h_{0e} = total specific enthalpy at control volume exit (J/kg·K)
h_{0i} = total specific enthalpy at control volume inlet (J/kg·K)

This equation still has a a minor problem. Total enthalpy (h_0) includes both static enthalpy (h) and kinetic energy:

$$h_0 = h + \frac{1}{2}v^2 \tag{3.161}$$

where h = static, specific enthalpy (J/kg·K)
v = the flow velocity (m/s)

We need more information to determine how the added heat divides between kinetic energy and static enthalpy.

For rocket-propulsion systems, we usually assume the propellants are injected into the control volume (combustion chamber) at a relatively low velocity (Mach number). We also assume heat is added very quickly at stagnation, so the exit condition is also at relatively low velocity. These assumptions are valid only if the area of the combustion chamber is large enough compared with the nozzle throat, ensuring low Mach numbers (less than 0.4) throughout the combustion chamber. From these assumptions the total enthalpy in the above equations reduces to static enthalpy, giving us

$$\dot{Q} = \dot{m}(h_e - h_i) \tag{3.162}$$

where $\dot{Q}$ = heat flow rate (W)
$\dot{m}$ = mass flow rate (kg/s)
h_e = static enthalpy at exit (J/kg·K)
h_i = static enthalpy at intake (J/kg·K)

Of course the question now is, "What does 'large enough' mean?" To determine the answer, first assume Eq. (3.162) is correct. When a design is complete, check the result using the following method (Sec. 3.4.3), assuming heat addition in steady flow. Despite the assumptions, Eq. (3.162) gives remarkably good results, particularly for liquid-rocket systems. In the next section, we discuss one way to determine the distribution of enthalpy and kinetic energy.

3.4.3 Heat Addition to a Constant-Area Duct with Steady Flow

Most propulsion systems have combustion chambers or heat exchangers with a constant cross-sectional area. Let us look at the flow characteristics through a constant-area duct when adding heat. The main assumptions are still reversible flow with no potential energy change. Figure 3.16 shows a schematic of a constant-area duct with all of the heat added between stations 1 and 2 over the total length L.

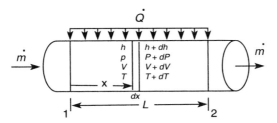

Fig. 3.16.   Model Diagram for Heat Added to a Constant-Area Steady Flow. This figure helps us define the parameters used to derive the effect of adding heat to a flowing fluid.

To analyze this problem, we must look at the three key equations: energy, continuity, and momentum. The energy equation is given by Eq. (3.160), for which we assumed nothing concerning velocities or Mach numbers. The continuity equation for one-dimensional flow is from Eq. (3.1):

$$\dot{m} = \rho V A \tag{3.163}$$

where $\dot{m}$ = gas mass flow rate through the duct (kg/s)
  ρ = gas density (kg/m³)
  V = gas flow velocity (m/s)
  A = duct cross-sectional area (m²)

In this case, assuming steady flow ensures the mass flow rate ($\dot{m}$) is constant, and assuming a constant-area duct makes the cross-sectional area of the duct (A) constant. We can write this equation as

$$G = \rho V = \frac{\dot{m}}{A} \to \text{constant} \tag{3.164}$$

where G is the mass flux rate and has a constant value. Another form is

$$\rho_1 V_1 = \rho_2 V_2 \tag{3.165}$$

where the subscripts refer to the two duct cross sections shown in Fig. 3.16. The one-dimensional momentum equation, from Eq. (3.104), is

$$dp + \rho V \, dV = 0 \tag{3.166}$$

where p = static flow pressure (Pa)
ρ = static flow density (kg/m³)
v = flow velocity (m/s)

Substituting Eq. (3.164) into this expression gives us

$$dp = -G\,dv, \qquad (3.167)$$

an equation easily integrated between stations 1 and 2 because G is constant:

$$\int_{p_1}^{p_2} dp = -G \int_{v_1}^{v_2} dv \qquad (3.168)$$

$$p_2 - p_1 = G(v_1 - v_2) \qquad (3.169)$$

The problem now is to find the fluid state at one station given the state at the other. Assume we know the fluid state at station 1, as well as the mass and heat flow rates. We then want to determine p_2, ρ_2, T_2, and M_2. Taking Eq. (3.164) and substituting the perfect-gas law, the momentum Eq. (3.169), and a form of the energy Eq. (3.158) and (3.159), we have three equations and three unknowns.

Summary of equations:

$$G = \frac{p_2 v_2}{R_2 T_2} \qquad (3.170)$$

$$p_2 - p_1 = G\,(v_1 - v_2) \qquad (3.165)$$

$$q = \frac{\dot{Q}}{\dot{m}} = \left[h_2(T_2) + \frac{1}{2}v_2^2\right] + \left[-h_1(T_1) + \frac{1}{2}v_1^2\right] \qquad (3.171)$$

where q is the heat flow per unit mass, and h_1 and h_2 are the relative static enthalpies at stations 1 and 2 respectively. The T in parentheses indicates that h is a function only of T. Typically, we do not know the absolute enthalpy values because we do not know c_p at low temperatures. Instead, we evaluate relative enthalpies from a standard temperature, usually 298.16 K. This method allows us to determine the difference in the absolute enthalpies, which is the same as the difference in enthalpies relative to the arbitrary 298.16 K point:

$$h_2 = \int_{298.16}^{T_2} c_p(T)\,dT + \Delta h_f \qquad (3.172)$$

$$h_1 = \int_{298.16}^{T_1} c_p(T)\,dT + \Delta h_f \qquad (3.173)$$

where the Δh_f (heats of formation at 298.16 K) terms account for any heat added from a chemical reaction or change of state (liquid to gas). See Chapter 4, "Thermochemistry" for more details on how we account for chemical reactions.

The three equations (3.171) through (3.173) contain three unknowns (p_2, T_2, and v_2), making this set of algebraic equations solvable, at least in principle. However, these three equations are highly nonlinear and require a numerical solution. Many commercial "codes" are available to solve them. The codes typically rely on an initial guess for p_2, T_2, and v_2 and iterate using some numerical-solution technique, such as Newtonian iteration or bisection. When the changes in the iterated values get below some allowable error limit, the numerical solution is complete.

Oates [1984] shows that, if the gas constants are constant through the duct, we can determine the Mach number (M) and total temperature ratio (T_0) from

$$\frac{T_{02}}{T_{01}} = 1 + \frac{q}{c_p T_{01}} = \frac{f(M_2^2)}{f(M_1^2)} \tag{3.174}$$

where

$$f(M_1^2) = M_1^2 \frac{1 + \frac{\gamma-1}{2} M_1^2}{\left(1 + \gamma M_1^2\right)^2} \tag{3.175}$$

But we also need to solve these equations numerically for M_2.

Analysis of heat addition for a constant-area duct shows that a gas entering one end of the duct at low Mach number exits at low Mach number, depending on the amount of heat added and other variables. This validates the assumption of a low Mach number, discussed in Sec. 3.4.2, where kinetic energy has little effect. However, if the inlet Mach number is greater than 0.4, constant-area heat addition gives different results.

Example

A large tank contains H_2 gas at a temperature of 500 K. A pressure regulator releases the H_2 at 300 kPa into a 1-cm diameter heat-exchanger pipe. If the heat transfer into the pipe is known to be 100 kwatts, and the pipe exhausts into a 200-kPa atmosphere, determine the state of the flow at the beginning (station 1) and end (station 2) of the pipe.

From the problem statement:

T_{01} = 500 K [total temperature of H_2 at station (1)]
P_1 = 300,000 Pa [static pressure of H_2 at station (1)]
$A_{pipe} = \pi \left(\frac{0.01}{2}\right)^2 = 0.008$ m² (area of pipe)
$\dot{Q}$ = 100,000 watts

For the H_2 gas we assume from Van Wylen and Sonntag (1971) that

$\mathcal{M}$ = 2.016 kg/kmol (molecular mass)

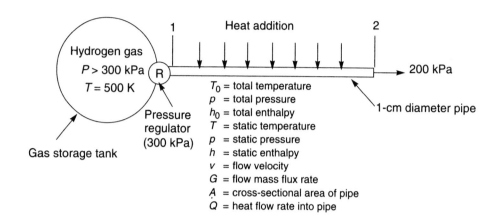

$$c_p = \frac{1000\left(56.505 - 702.79\left(\frac{T}{100}\right)^{-0.75} + 1165\left(\frac{T}{100}\right)^{-1} - 560.7\left(\frac{T}{100}\right)^{-1.5}\right)}{M}$$

$$\gamma = \frac{c_p}{c_p - R}$$

$$h = \int_{298.16}^{T} c_p(t)\, dT$$

At station (1), the total enthalpy is

$$h_{01} = 2{,}915{,}634.485 \text{ J/kg}$$

$$= \int_{298.16}^{500} c_p(T)\, dT$$

In this problem, we actually have five unknowns:

$\dot{m}$ = mass flow rate (kg/s)
T_1 = static temperature at station (1) (K)
T_2 = static temperature at station (2) (K)
v_1 = flow velocity at station (1) (m/s)
v_2 = flow velocity at station (2) (m/s)

and five nonlinear equations:

$$h_{01} = \int_{298.16}^{T_1} c_p(T)\, dT + \frac{1}{2}v_1^2 \quad \text{(definition of total enthalpy)}$$

$$G = \frac{p_2 v_2}{R T_2} \quad \text{(constant mass flux)}$$

$$G = \frac{p_1 v_1}{RT_1} \text{ (constant mass flux)}$$

$$p_2 - p_1 = G(v_1 - v_2) \text{ (momentum equation)}$$

$$\dot{Q} = (GA)\left[\left(h_2 + \frac{1}{2}v_2^2\right) - h_{01}\right] \text{ (energy equation)}$$

We can solve these equations numerically to yield

G = 57.625 kg/m²·s (mass flux rate)
T_1 = 494.696 K
T_2 = 1790.196 K
v_1 = 391.89 m/s
v_2 = 2127.25 m/s

which allows us to determine the ratio of specific heats (γ):

$c_p(T_1)$ = 14,478.11 J/kg·K
$c_p(T_2)$ = 16,584.07 J/kg·K
γ_1 = 1.398 32
γ_2 = 1.331

The flow Mach numbers are

$$M_1 = \frac{v_1}{\sqrt{\gamma_1 RT_1}} = 0.232\,02$$

$$M_2 = \frac{v_2}{\sqrt{\gamma_2 RT_2}} = 0.6786$$

The temperature at station 2 is

$$T_{02} = T_2\left(1 + \frac{\gamma_2 - 1}{2}M_2^2\right) = 1926.63 \text{ K}$$

The total pressures at stations 1 and 2 are

$$p_{01} = p_1\left(1 + \frac{\gamma_1 - 1}{2}M_1^2\right) = 311,444.13 \text{ Pa}$$

$$p_{02} = p_2\left(1 + \frac{\gamma_2 - 1}{2}M_2\right) = 268,716.79 \text{ Pa}$$

Notice that as total temperature increases from added heat, the total pressure actually drops along the pipe.

$$\Delta p = p_{02} - p_{01} = -42727.39 \text{ Pa}$$

3.5 Heat Transfer

Until now, we have assumed for propulsion that no heat transfers from the system. This assumption has given good results because heat transfer consumes little energy compared to thrust. But heat transfer is significant in designing a propulsion system, most obviously in the injectors, combustion chamber, and nozzle. Besides having to contend with the extremely high operating temperatures of these components, materials must also undergo potentially harmful thermal expansion and contraction. Table 3.3 lists characteristics of typical refractory materials used for the rocket applications mentioned above, including their response to high temperatures. Table 3.4 lists characteristics of typical materials used in rockets. Let us look at the principles of heat transfer, including the ways to transfer heat in propulsion systems and some numerical solutions for practical problems.

Table 3.3. Physical Properties of Typical Refractory Materials for Rocket Applications. This table shows values of several key parameters for several materials. We need these values for the discussion of heat transfer in this section.

Material	Typical Composition	Melting Point (K)	Linear Coefficient of Thermal Expansion (10^{-6} K^{-1})	Thermal Conductivity (W/m^2-K/m)	Specific Heat (kcal/kg·K)
Alumina	99.2% Al_2O_3 0.6% SiO_2	2322	4.0 at 283 to 1280 K	2.45 at 0 to 1273 K	0.28–0.30 at 290 to 2000 K
Carbon	Essentially C	Sublimates 3812	1.6 to 2.9 at 288 to 378 K	1.7 to 20 at 1375 K	0.29 at 15 to 1375 K
Graphite	Essentially C	2300	1.0 to 2.6 at 288 to 378 K	35 to 206 at 1373 K	0.29 at 15 to 1283 K
Silicon carbide	SiC with complex nitride bond	Decomposes at 2483	3.7 at 288 to 1773 K	16.3	0.23

3.5.1 Conductive Heat Transfer

Conduction is the transfer of heat between molecules from one part of a system (usually a solid) to another because of a temperature gradient. Fourier's Law defines it as

$$\dot{Q} = -kA \frac{dT}{dx} \tag{3.176}$$

where $\dot{Q}$ = rate of heat transfer (W)
k = thermal conductivity of the material, (W/(m·°C))
A = cross-sectional area perpendicular to x, the direction of heat flow (m^2)
$\frac{dT}{dx}$ = change of temperature along x (K/m)

Table 3.4. Properties of Materials at High Temperatures. Temperature changes affect the properties of materials as shown here.

Parameter	Commercially Pure Copper	Aluminum Alloy 24S-T	Low Carbon Steel SAE 1020	Alloy Steel SAE X4130	Stainless Steel AISI Type 302	Nickel Alloy, Inconel
Thermal Conductivity (W/m·°C)	0°C 381.9 200°C 371.9 400°C 362.9 600°C 344.9	17.8°C 225.9	0°C 52.0 200°C 49.0 400°C 43.0 800°C 25.9 1000°C 27.0 1100°C 28.0 1200°C 30.0	0°C 43.0 200°C 42.0 400°C 39.0 600°C 34.0 800°C 26.0 1000°C 28.0 1200°C 30.0	0°C 15.9 200°C 17.2 400°C 19.6 600°C 22.9 800°C 26.0 1000°C 28.0 1200°C 29.7	30°C 18.0 200°C 19.0 800°C 28.0 1200°C 36.0
Coefficient of Thermal Expansion, Average Value ((cm/cm·°C) × 10⁻⁶)	300°C 20.8 400°C 22.3 500°C 24.1 600°C 26.2 800°C 31.2	20–100°C 23.2 20–200°C 23.9 20–300°C 24.7	20–100°C 11.7 20–200°C 12.1 20–300°C 12.8 20–500°C 13.9 20–600°C 14.4 20–700°C 14.8	100°C 12.7 300°C 13.5 500°C 14.2 700°C 14.8 900°C 13.0 1000°C 13.9 1100°C 14.5	100°C 14.8 300°C 17.1 500°C 18.0 700°C 18.8 900°C 19.2 1000°C 19.4	10 10.1 93 12.3 204 14.4 316 15.6 427 16.7 538 18.2 760 18.4
Specific Heat (kJ/kg·°C)	23.9°C 0.384 200°C 0.402 400°C 0.419 800°C 0.456 1000°C 0.477	17.8–100°C 0.888	20°C 0.482 300°C 0.557 500°C 0.628 700°C 0.636 900°C 0.645 1100°C 0.653 1300°C 0.695	100°C 0.477 300°C 0.544 500°C 0.657 700°C 0.825 800°C 0.883 1000°C 0.601 1250°C 0.645	100°C 0.511 300°C 0.549 500°C 0.595 700°C 0.624 900°C 0.653 1100°C 0.662 1300°C 0.678	25.6–100°C 0.456

Figure 3.17 depicts a temperature gradient in a wall of thickness Δl. Writing Eq. (3.176) in finite terms and using q, or heat transfer per unit area, we have the following expression for conduction heat transfer:

$$\dot{q} = \frac{\dot{Q}}{A} = -k\left(\frac{T_2 - T_1}{\Delta l}\right) \quad (3.177)$$

where $\dot{Q}$ = rate of heat transfer (W)
k = thermal conductivity of the material, (W/(m·°C))
A = cross-sectional area perpendicular to x, the direction of heat flow
$\frac{dT}{dx}$ = change of temperature along x
Δl = distance through material (m, see Fig. 3.17)

Note that, because $T_2 < T_1$, $\dot{q}$ is positive in the x direction.

Two common methods use conduction to protect rocket components from absorbing excessive heat. The first uses a large heat sink with high conductivity to draw heat from the chamber and nozzle areas. The second uses insulating material with low conductivity to cover the inner surfaces of chamber and nozzle walls, so heat flow to the walls decreases.

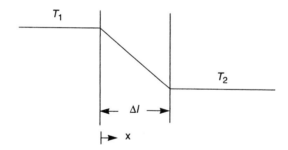

Fig. 3.17. **Heat Transfer by Conduction.** The temperature profile through a solid wall is essentially linear. This figure is used to define parameters in Eqs. (3.176) and (3.177).

3.5.2 Convective Heat Transfer

Convection is an energy exchange between a moving fluid and a solid surface. Propulsion systems therefore transfer heat through convection because they use high-speed flow through a nozzle. There is no clear-cut distinction between high-speed flow and low-speed flow in terms of a Mach number or Reynolds number.

3.5 Heat Transfer

We consider flow high-speed if elevated temperature and velocity cause irreversibilities through viscous dissipation in the boundary layer.

We can gain a first-order understanding of high-speed flow by assuming that the nozzle wall is adiabatic (does not conduct heat) as in Hill and Peterson [1970]. The left portion of Fig. 3.18 represents the free stream, characterized by high-speed flow and relatively low static temperature (T_∞). As we approach the wall, the velocity decreases, accompanied by an increase in T_∞. We would expect that the adiabatic wall temperature (T_{aw}) (where the velocity is zero) equals the free-stream stagnation temperature. But it does not because, in a rocket nozzle, the temperature difference between the free stream and the wall is so great that we get radiative heat transfer back into the free stream from the slower-moving hot fluid near the wall. Because of this heat loss, T_{aw} is lower than T_0. So how do we account for this energy loss? To satisfy conservation of energy for steady flow, a region near the wall has to have a higher stagnation temperature than the free stream to balance the region at the wall which has a stagnation temperature lower than the free stream, as seen in Fig. 3.18. Because of these complications, we cannot simply measure T_{aw} and then use the stagnation and isentropic relations to solve for the free-stream conditions. To account for the anomalies we define a *recovery factor*

$$r = \frac{T_{aw} - T_\infty}{T_0 - T_\infty} \qquad (3.178)$$

where T_{aw} = adiabatic wall temperature (K)
 T_∞ = free-stream static temperature (K)
 T_0 = stagnation (chamber) temperature (K)

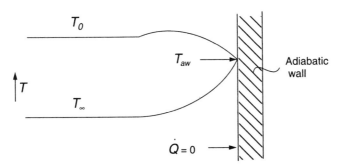

Fig. 3.18. **Temperature Profile Near an Adiabatic Wall for High-speed Flow.** The static temperature increases and stagnation temperature increases, then decreases as we get closer to a solid wall. These temperatures converge to a single number at the stagnation point (velocity = 0) on the wall. T_{aw} is less than T_0 because heat radiates back into the flow from the hot, slow-moving flow near the wall. This radiated heat increases stagnation temperature locally as shown.

For typical rocket propellants and compressible boundary layers in fluid flows up to Mach 4, experimental results show we can use $r = 0.91$. Applying this recovery factor, we can determine the adiabatic wall temperature and use it in the following expression for heat transfer by convection:

$$\dot{q} = h_g(T_{aw} - T_{wh}) \tag{3.179}$$

where h_g = heat-transfer coefficient for a hot gas (W/m²·K)
T_{wh} = actual wall temperature on the hot side of the wall (K)

Assumption: one-dimensional flow

At the throat, the adiabatic wall temperature (T_{aw}) is essentially equal to the stagnation temperature of the hot gas in the chamber (chamber temperature, T_{0g}) because the recovery factor (r) is one. Elsewhere, we must use the recovery factor [Eq. (3.178)] to determine the adiabatic wall temperature. Also note that for Eq. (3.179), if the wall temperature is equal to the theoretical adiabatic wall temperature, the heat transfer drops to zero.

The most common application of convection for large, modern, liquid rockets is regenerative cooling. The propellant, routed through small tubes which encircle the nozzle, absorbs heat from the nozzle wall by convection to provide the cooling. Because propellant moving through the tubes is not at high speed, we cannot use Eq. (3.179) to determine the amount of heat carried away by regenerative cooling. The convection equation for cooling is

$$\dot{q} = h_l(T_{wc} - T_l) \tag{3.180}$$

where h_l = heat transfer coefficient for the cooling liquid (W/m²·K)
T_{wc} = wall temperature on the cooled side of the wall (K)
T_l = temperature of the cooling liquid (K)

Assumption: one-dimensional flow

We can also use convection to cool a nozzle by injecting a film of fuel or a cool gas along the nozzle wall, creating a protective boundary layer.

Figure 3.19 shows a wall with hot gas on one side at temperature T_g and coolant on the other at temperature T_c. The dotted lines represent the change in temperature through the boundary layer on either side of the wall. From Eq. (3.175), $\dot{q}_1$ is the heat transfer from the gas to the wall,

$$\dot{q}_1 = h_g(T_{0g} - T_{wh}) \tag{3.181}$$

From Eq. (3.177), heat transfer through the wall is

$$\dot{q}_2 = \frac{-k}{\Delta l}(T_{wc} - T_{wh}) \tag{3.182}$$

Finally, from Eq. (3.180), heat transfer from the wall to the coolant is

$$\dot{q}_3 = h_l(T_{wc} - T_{cl}) \tag{3.183}$$

For steady state, all three heat-transfer rates are equal. If we know the stagnation gas temperature and coolant temperature, as well as the heat-transfer coefficients and conductivity, we may combine the three previous equations to get

$$\dot{q} = \frac{T_{0g} - T_{cl}}{\dfrac{1}{h_g} + \dfrac{\Delta l}{k} + \dfrac{1}{h_l}} \tag{3.184}$$

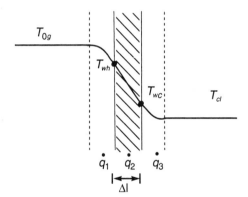

T_{0g} = stagnation temperature of the gases (K)

T_{wh} = static temperature at the hot side of the wall (K)

T_{wc} = static temperature at the cold side of the wall (K)

T_{cl} = static temperature of the cooling liquid (K)

Fig. 3.19. Nozzle Wall with Regenerative Cooling. Convection causes the temperature to decrease from the hot gas to the wall, followed by a linear decrease through the wall and, finally, a convective decrease through the coolant-flow boundary layer.

Heat-Transfer Coefficients. The heat-transfer coefficients (h_g and h_l) are functions of the fluid properties (conductivity, viscosity, specific heat, and density), as well as the flow characteristics (diameter of channel and flow velocity). Empirical results for high-speed flow of a hot gas in an enclosed channel yield the following expression:

$$h_g = 0.026k\left(\frac{\rho v}{\mu}\right)^{\frac{4}{5}}\left(\frac{1}{D}\right)^{\frac{1}{5}}\left(\frac{c_p\mu}{k}\right)^{\frac{2}{5}} \tag{3.185}$$

where h_g = heat transfer coefficient of a hot gas (W/m²·K)

ρ = density of gas (kg/m³)
v = flow velocity (m/s)
μ = dynamic viscosity of gas (kg/m·s)
D = diameter of flow channel (m)
c_p = specific heat (J/kg·K)
k = thermal conductivity (W/m·K)

Assumption: fully developed flow in a pipe of circular cross section

Note that the smallest diameter of the flow channel (D) corresponds to the largest heat-transfer coefficient (h_g), so heat transfer in a nozzle is greatest at the throat. Also note from Eq. (3.185) that h_g is a function only of geometry and gas properties.

Similarly, empirical results for a slower-moving coolant yield

$$h_l = 0.023 \left(\frac{c_p \dot{m}}{A} \right) \left(\frac{\mu}{D \rho v} \right)^{\frac{1}{5}} \left(\frac{k}{\mu c_p} \right)^{\frac{2}{3}} \quad (3.186)$$

where h_l = heat-transfer coefficient of a hot gas (W/m²·K)
c_p = specific heat (J/kg·K)
$\dot{m}$ = mass flow rate of coolant (kg/s)
A = cross-sectional area of tube (m²)
μ = dynamic viscosity of coolant (kg/m·s)
D = diameter of tube (m)
ρ = density of coolant (kg/m³)
k = thermal conductivity of coolant (W/m·K)
v = flow velocity (m/s)

Assumption: fully developed flow in a tube of circular cross section

To increase h_l and thus better cool a nozzle, we would increase the mass flow rate and decrease the channel diameter. Therefore, the best cooling system consists of many small-diameter tubes.

Because solid rockets cannot use regenerative cooling for their nozzles, another method is used which is not purely conduction, convection, or radiation. Usually the chamber, throat, and much of the nozzle are coated with an ablative material which absorbs a great amount of heat as it vaporizes and joins the fluid flow.

3.5.3 Radiative Heat Transfer

We can determine how much heat transfers by radiation from a hot gas to the nozzle and chamber walls through the expression

$$\dot{q} = \varepsilon \sigma T^4 \quad (3.187)$$

where ε = emissivity of the gas ($\varepsilon = 1$ for a perfect radiator)

σ = Stefan-Boltzmann constant = 5.669×10^{-8} W/m$^2 \cdot$K^4
T = gas temperature (K)

Obviously, the nozzle does not absorb all of the heat radiated by the gas. How much the nozzle absorbs depends on its reflectivity and absorptivity. Because solid rockets cannot be cooled regeneratively, they usually have nozzles constructed of a refractive material, such as graphite, with a high resistance to heat. As a result, the nozzles reject as much heat as possible. See Table 3.3 for characteristics of refractory materials typically used in rockets. In high-temperature rocket motors the heat radiated from the gas to the chamber and nozzle walls may amount to 5 to 35% of the total heat transfer. For a more complete treatment of radiation heat transfer, consult Penner and Weinbaum [1958], Logan [1958], and Pearce [1978].

Example

An experimental rocket's thrust chamber has an outside (cool) wall temperature of 150°C at the nozzle throat, with 3300°C chamber temperature. The heat-transfer rate at the throat is measured to be 1.4×10^7 W/m^2. The liquid-cooled throat wall is stainless steel 0.25 cm thick with k = 26 W/m·°C.

(a) Assuming the coolant's surface area is equal to the hot gas's surface area, what is the inner (hot) wall's temperature at the throat?

From Eq. (3.173) we have

$$1.4 \times 10^7 \frac{W}{m^2} = -26 \frac{W}{m \cdot °C} \left(\frac{150°C - T_{wh}}{0.0025 \text{ m}} \right)$$

where T_{wh} is the wall's temperature on the hot side.

Solving for T_{wh},

$$T_{wh} = \left(\frac{1.4 \times 10^7}{26} \right) (0.0025) + 150$$

$$T_{wh} = 1496°'C$$

(b) We expect this temperature to cause failure of the stainless steel in actual application. Therefore, the throat must be lined with a ceramic of conductivity k = 8.65 W/m·°C to protect the metal. Assuming all exhaust-gas properties from part (a) remain unchanged, find the thickness of ceramic required to reduce the peak metal temperature to 1100°C while the cool side remains at 150°C.

Because we know T_2 and T_3, we may solve for the heat-transfer rate through the wall:

$$\dot{q} = \frac{k_{ss}}{\Delta l_{ss}} (T_2 - T_3) = \frac{26}{0.0025} (1100 - 150)$$

$$= 9.88 \times 10^6 \text{W/m}^2$$

Referring to the diagram, $T_3 = 150°C$ and from the problem statement, $T_2 = 1100°C$.

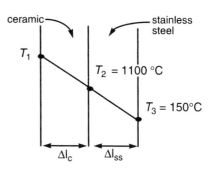

We assume all of this heat transfers from the gas to the wall through convection. Because gas properties are unchanged from part (a), we may solve for h_g from (a) and apply it to part (b):

$$\dot{q} = h_g(T_{0g} - T_{wh})$$

Here we have used the fact that $T_{aw} = T_{0g}$ at the throat:

$$h_g = \frac{1.4 \times 10^7}{(3300 - 1496)} = 7760.53 \left(\frac{W}{m^2 \cdot °C}\right)$$

Now, with h_g determined, we can find T_1, the temperature on the hot side of the ceramic:

$$\dot{q} = h_g(T_{0g} - T_1)$$

$$T_1 = T_{0g} - \frac{\dot{q}}{h_g} = 3300 - \frac{9.88 \times 10^6}{7760.53}$$

$$T_1 = 2026.89°C$$

And we can use T_1 to find Δl_c, the thickness of the ceramic:

$$\dot{q} = \frac{k_c}{\Delta l_c}(T_1 - T_2)$$

$$\Delta l_c = \frac{8.65}{9.88 \times 10^6}(2026.89 - 1100)$$

$$\Delta l_c = 0.0008 \text{ m} = 0.08 \text{ cm}$$

3.5.4 Numerical Modeling of Heat Transfer

Unfortunately, as with most aspects of engineering, at a certain point a problem becomes too unwieldy to solve it analytically. Some of the reasons are

- Complex hardware geometry
- Complex heat sources
- Time-dependent solutions
- Accuracy requirements which preclude simplifying assumptions
- Difficult and complex differential equations

If we run into this trouble, we must build hardware or software models of the system and run tests. Building hardware models is best but is usually very time-consuming and expensive. Software usually costs less time and money. To model heat-transfer problems, we can use finite differences or finite elements. Each approach has its own advantages and disadvantages.

Finite Differences

1. The general heat-transfer equation is a partial differential equation. To solve it, we must reduce it to its finite form, meaning that the differentials are no longer infinitesimally small.
2. Finite differences tend to "lump" parameters and then treat the geometry as a "thermal circuit." In this case, material becomes a thermal "capacitance" which can store heat, and different materials are connected by thermal "resistors" (inverse of conductance). The result is a "thermal circuit" we can solve like an electric circuit.
3. Finite-difference codes model radiation effects much better than finite elements do.
4. Finite differences handle problems with transient heat flow (time-dependent solutions) better than finite elements.
5. Usually, finite-difference codes are "generic" enough to solve somewhat unusual problems such as fluid flow, heat-pump operation, heat sources with unusual input or output properties, incorporation of other analysis results such as CFD output, and many other "real life" problems.
6. Finite differences do not give extremely precise steady-state temperature profiles over the solid geometry as do finite elements.

Finite Elements

1. Finite elements break the solid geometry up into small parts containing assumed temperature-distribution shapes. Compatibility between the elements is ensured by forcing heat flows and

temperatures between the elements to be consistent with geometry and with the governing differential equations.
2. Finite elements provide good information on how the temperatures change throughout the solid material.
3. Finite elements do not handle time-dependent problems and radiation problems very well.

Fortunately for the designer or analyst, a number of heat-transfer codes are on the market. Unfortunately, wading through them in search of the right code for your particular problem can be onerous. Usually, the agency requesting work prefers a code based on previous experience.

3.6 Design Example—Cold-Gas Thruster

3.6.1 First Concerns

This example shows how to use thermodynamic and flow equations to design a simple "cold-gas" thruster system. These systems are typically used for satellite attitude control, which requires

- A high degree of reliability
- Low system complexity, no combustion involved
- Low Δv
- Extremely safe operation
- No contamination of the satellite's external surfaces from exhaust gases

Fig. 3.20 shows a typical system architecture.

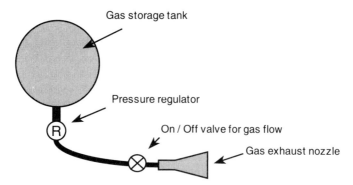

Fig. 3.20. **System Architecture for a Cold-Gas Thruster.** This drawing shows the basic systems required to operate a cold-gas thruster.

3.6 Design Example—Cold-Gas Thruster

The gas is stored in a high-pressure tank. Typically the gas feeds from the tank through a regulator which reduces the gas pressure. Although this feature is not necessary, it has two advantages:

1. The hardware outside the regulator operates at lower pressure and therefore has lower strength requirements.
2. The thruster operates at a constant pressure until the tank pressure drops below the regulator pressure; this feature ensures more consistent levels of thrust.

A valve turns the flow on and off, and the nozzle accelerates and expands the flow. Requirements for this propulsion system are

1. 40-N thrust level in a vacuum
2. 1000 seconds of thrust duration (40,000 N·s impulse)
3. use a low reactivity gas
4. intermittent system operation as needed
5. gas storage at 298.15 K

For our first design, we assume

1. no losses in feed lines, valve, or regulator
2. no pressure or mass flow losses in the nozzle
3. isentropic flow from the regulator to the nozzle exit
4. intermittent operation allowing the gas in the storage tank and the feed lines to be isothermal at a temperature of 298.15 K

The last assumption means that heat flows from the satellite into the storage tank as the tank pressure drops. If we assume an isentropic system, the tank temperature drops as the pressure decreases. However, the relatively slow decrease in tank pressure, from intermittent operation, allows the tank to maintain an isothermal state. We consider seven candidate gases from Table 3.1. The acoustic, or sonic, velocity in the gases is

$$a_0 = \sqrt{\gamma R T_0} \qquad (3.188)$$

and the characteristic velocity is

$$c^* = \frac{a_0}{\gamma \left(\frac{2}{\gamma+1}\right)^{\frac{\gamma+1}{2\gamma-2}}} \qquad (3.189)$$

The thrust equation is

$$F = A_t p_c \gamma \left[\left(\frac{2}{\gamma-1}\right)\left(\frac{2}{\gamma+1}\right)^{\frac{\gamma+1}{\gamma-1}}\left\{1-\left(\frac{p_e}{p_c}\right)^{\frac{\gamma-1}{\gamma}}\right\}\right]^{\frac{1}{2}} + (p_e - p_a)A_e \quad (3.190)$$

The ambient vacuum condition gives $p_a = 0$; the specified thrust level (F) is 40 N. Assume we have an infinite expansion ratio so the exit pressure (p_e) is zero. This assumption gives us the following equations:

$$\dot{m} = \frac{A_t p_c}{c^*} \quad (3.191)$$

Remember: These equations ONLY apply for infinite expansion, where exit pressure is zero

$$F = \dot{m} c^* \gamma \left[\left(\frac{2}{\gamma-1}\right)\left(\frac{2}{\gamma+1}\right)^{\frac{\gamma+1}{\gamma-1}}\right]^{\frac{1}{2}} \quad (3.192)$$

$$\dot{m} = \frac{F}{c^* \gamma}\left[\left(\frac{2}{\gamma-1}\right)\left(\frac{2}{\gamma+1}\right)^{\frac{\gamma+1}{\gamma-1}}\right]^{\frac{-1}{2}} \quad (3.193)$$

$$I_{sp} = \frac{F}{g_0 \dot{m}} = \frac{c^*}{g_0} \gamma \left[\left(\frac{2}{\gamma-1}\right)\left(\frac{2}{\gamma+1}\right)^{\frac{\gamma+1}{\gamma-1}}\right]^{\frac{1}{2}} \quad (3.194)$$

$$m_p = \dot{m}\,\Delta t, \quad \Delta t = 1000 \text{ s, total propellant mass} \quad (3.195)$$

$$V_p = \frac{m_p R T}{p_c}, \text{ total tank volume} \quad (3.196)$$

From Eqs. (3.188) through (3.196), we can construct Table 3.5, below.

Figure 3.21, generated from Eq. (3.196), shows the storage tank's volume as a function of its pressure.

3.6.2 First Design

Now we can choose a particular gas. We want one with a high specific impulse (I_{sp}) but low reactivity, so we choose helium (He). Now look again at the nozzle conditions—ambient pressure is still zero and exhaust pressure is greater than zero. If we assume pressure thrust is negligible, we can find parameters for pres-

3.6 Design Example—Cold-Gas Thruster

Table 3.5. Parameters for a Cold-Gas Propulsion System. Assuming an infinite expansion ratio and a storage temperature of 298.15 K, we can determine the key performance parameters for several candidate gases.

Gas	a_0 (m/s)	c^* (m/s)	$\dot{m}$ (kg/s)	I_{sp} (s)	m_p (kg)
Air	346.153	427.251	0.05168	78.933	51.678
Argon (Ar)	320.773	343.640	0.07153	57.031	71.525
CO_2	269.348	356.845	0.05635	72.384	56.354
He	1013.875	1085.608	0.02265	180.121	22.647
H_2	1314.394	1617.456	0.01369	297.866	13.695
N_2	351.976	434.439	0.05082	80.261	50.823
O_2	328.731	406.977	0.05408	75.434	54.076

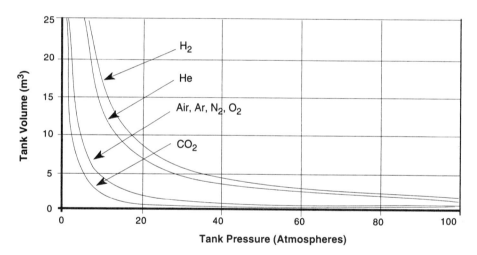

Fig. 3.21. Storage Tank's Volume vs. Its Pressure for Various Gases. If we assume infinite expansion in the exhaust nozzle, this figure shows how tank pressure affects the tank volume needed to meet our requirements.

sure ratios and expansion ratios. Let us plot specific impulse as a function of expansion ratio to determine a reasonable expansion ratio:

$$I_{sp} = \frac{c^*}{g_0} \gamma \left[\left(\frac{2}{\gamma-1} \right) \left(\frac{2}{\gamma+1} \right)^{\frac{\gamma+1}{\gamma-1}} \left\{ 1 - \left(\frac{p_e}{p_c} \right)^{\frac{\gamma-1}{\gamma}} \right\} \right]^{\frac{1}{2}} \quad (3.197)$$

where

$$\varepsilon = \frac{A_e}{A_t} = \frac{1}{M_e}\left[\left(\frac{2}{\gamma+1}\right)\left(1+\frac{\gamma-1}{2}M_e^2\right)\right]^{\frac{\gamma+1}{2\gamma-2}} \quad (3.198)$$

$$\frac{p_e}{p_c} = \left[1+\frac{\gamma-1}{2}M_e^2\right]^{\frac{\gamma}{1-\gamma}} \quad (3.199)$$

To create Fig. 3.22, we vary the expansion ratio (ε) over the desired range. For each ratio value, we must calculate an exhaust Mach number (M_e) from Eq. (3.198). By substituting this number into the pressure ratio, Eq. (3.199), we can use the result to solve for I_{sp} with Eq. (3.197).

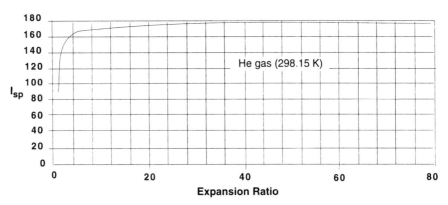

Fig. 3.22. **Specific Impulse vs. Nozzle Expansion Ratio for Helium (He) Gas.** Specific-impulse performance increases rapidly to an expansion ratio of 10. Above this value, the increase in marginal performance drops off.

From Fig. 3.22, we can see that the I_{sp} asymptotically approaches 180 seconds. However, at an expansion ratio of 25, the I_{sp} is close to 175 seconds. At this point we choose an expansion ratio of 25, so

M_e = 7.0157
p_e/p_c = 0.000 77
I_{sp} = 174.8 s
$\dot{m}$ = 0.023 3 kg/s

To determine the thruster's dimensions, we can consult Fig. 3.23, which shows the throat diameter (D_t) and nozzle-exit diameter (D_e) as a function of regulator pressure (p_r). The pressure regulator supplies gas to the nozzle at a specified static

3.6 Design Example—Cold-Gas Thruster

pressure. This pressure (p_r) equals the chamber pressure. We have created the figure by using

$$A_t = \frac{\dot{m}c^*}{p_r}$$

$$A_e = 25\, A_t$$

$$D_t = 2\sqrt{\frac{A_t}{\pi}}$$

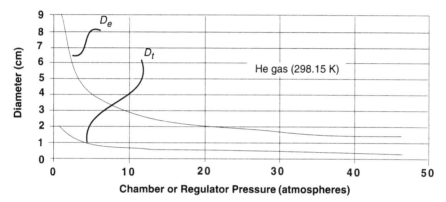

Fig. 3.23. Nozzle-Orifice Diameters vs. Regulator Pressure. This graph shows the required throat (D_t) and nozzle-exit (D_e) diameters needed to get the required thrust level as a function of our feed-system pressure. As the pressure increases, the size of our thruster decreases.

Now, choosing the regulator pressure determines the throat and exit dimensions. A choice of 12.7 atmospheres (1.29 MPa) gives reasonable thruster dimensions and still allows reasonably low pressures for tank storage:

p_r = 1,290,000 Pa
D_t = 0.5 cm
D_e = 2.5 cm

In addition, we choose a high-efficiency bell nozzle:

λ = 0.98

and a pressure of 40 MPa for the propellant tank.

3.6.3 Predicting Performance

So far in the design, our assumptions have made decisions a bit more straightforward. Now is the time to estimate the "true" performance without these assumptions. First, we look at the isothermal performance and size the propellant tank. Then, we look at performance based upon an isentropic assumption. From Eqs. (3.189), (3.198), and Table 3.5, we know the characteristic velocity and the exit Mach number:

$$c^* = 1085.608 \text{ m/s}$$
$$M_e = 7.0157$$

With these values we can determine the nozzle-exit pressure from Eq. (3.199):

$$p_c = p_r = 1.29 \text{ MPa}$$
$$p_e = 997.8 \text{ Pa}$$

We now have enough information to determine the amount of thrust from a form of Eq. (3.190):

$$F = \lambda \left\{ A_t p_c \gamma \left[\left(\frac{2}{\gamma-1}\right)\left(\frac{2}{\gamma+1}\right)^{\frac{\gamma+1}{\gamma-1}} \left\{1 - \left(\frac{p_e}{p_c}\right)^{\frac{\gamma-1}{\gamma}}\right\}\right]^{\frac{1}{2}} + (p_e - p_a)A_e \right\}$$

$$F = 39.6755 \text{ N}$$

The mass flow rate, given by Eq. (3.191), is

$$\dot{m} = 0.023\,334\,37 \text{ kg/s}$$

so the total propellant used over the 1000 seconds is

$$m_p = 23.3313 \text{ kg}$$

However, this amount does not include the propellant mass (m_r) left over when the tank pressure gets down to the regulator pressure. To determine the total propellant required, we must simultaneously solve two algebraic equations for tank volume (V) and residual propellant mass:

$$p_r V = m_r RT \quad (3.200)$$

$$p_{ti} V = (m_r + m_p) RT \quad (3.201)$$

where p_{ti} is the initial tank pressure of 40 MPa. The solution gives

$$V = 0.3735 \text{ m}^3$$

3.6 Design Example—Cold-Gas Thruster

$$m_r = 0.7775 \text{ kg}$$
$$m_t = m_r + m_p = 24.1088 \text{ kg}$$

The specific impulse is

$$I_{sp} = \frac{F}{g_0 \dot{m}} = \frac{39.68}{9.81\,(0.023334)} = 173.3 \text{ s}$$

This is lower than the value in Table 3.5 because we no longer have ideal expansion.

Although this system is isothermal, let us take a quick look at the performance if the thrust were continuous and isentropic. From the first law for a control volume [Eq. (3.23)], we get an equation for the rate of change in internal energy:

$$\dot{U} = -\dot{m} h_0 = -\dot{m} c_p T_t \qquad (3.202)$$

where T_t is the instantaneous temperature in the tank. We would expect the tank's temperature to drop as its internal energy decreases during the "blowdown" process. Internal energy

$$U = m_t c_p T_t \qquad (3.203)$$

where the "t" subscript on the mass term refers to the instantaneous propellant mass in the tank at any time during the "blowdown." The propellant's mass flow rate determines the mass in the tank as a function of time and is itself changing throughout the blowdown:

$$\dot{m} = \frac{A_t p_r}{c^*} \qquad (3.204)$$

where c^* is a function of the instantaneous tank pressure. Integrating Eqs. (3.201) and (3.203) numerically gives the time history of the thruster performance. This performance history is presented graphically in Figs. 3.24 through 3.27. They show the tank's mass, pressure, and temperature, as well as the thruster's specific impulse over time. Because we assumed a calorically perfect gas, thrust stays constant over the time, but mass flow rate changes to match the change in tank temperature. Notice that the integration is for only 500 seconds, so we can assume constant gas parameters.

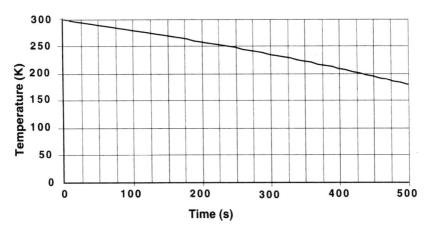

Fig. 3.24. **Isentropic Tank Temperature vs. Time.** This graph shows the decrease in the temperature of the gas stored in the tank if the tank is drained quickly.

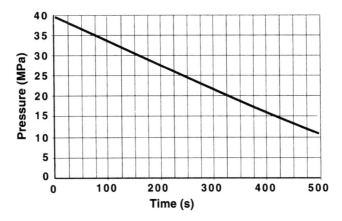

Fig. 3.25. **Isentropic Tank Pressure vs. Time.** This graph shows the drop in tank pressure if the tank is drained quickly.

3.6 Design Example—Cold-Gas Thruster

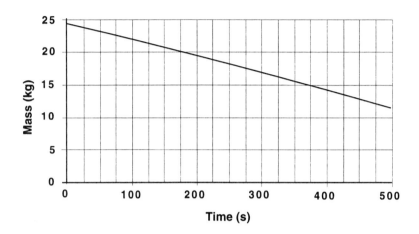

Fig. 3.26. **Isentropic Propellant Mass in Tank vs. Time.** This graph shows the decrease in propellant mass if the tank is drained quickly.

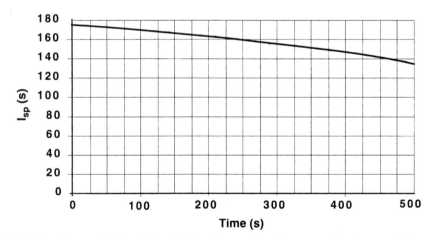

Fig. 3.27. **Isentropic Thrust Specific Impulse vs. Time.** Because of the decrease in the temperature of the propellant gas, the specific impulse of the thruster decreases. This drop occurs despite the constant pressure feed.

References

Altmann, D., J. M. Carter, S. S. Penner, and M. Summerfield. 1960. *Liquid Propellant Rockets*. Princeton, NJ: Princeton University Press.

Holman, J. P. 1980. *Thermodynamics*. New York: McGraw-Hill Book Company.

Hill, Philip G., and Carl R. Peterson. 1970. *Mechanics and Thermodynamics of Propulsion*. New York: Addison Wesley Publishing Company.

Langton, N. H. 1970. *Rocket Propulsion*. New York: American Elsevier Publishing Company.

Logan, J. G. 1958. Recent Advances in Determination of Radiative Properties of Gases at High Temperatures. *Jet Propulsion*. 28(12):795–798.

Oates, G. C. 1984. *Aerothermodynamics of Gas Turbine and Rocket Propulsion*. New York: American Institute of Aeronautics and Astronautics.

Pearce, B. E. 1978. Radiative Heat Transfer within a Solid Propellant Rocket Motor. *Journal of Spacecraft and Rockets*. 15(2):125-127.

Penner, S. S., and S. Weinbaum. 1958. Some Considerations of the Effect of Radiation on the Performance of Liquid Fuel Rockets. *Journal of the Optical Society of America*. 38(7).

Shapiro, A. H. 1953. *The Dynamics and Thermodynamics of Compressible Fluid Flow*. Vol. I. New York: John Wiley and Sons.

Summerfield, M., C. R. Foster, and W. C. Swan. Flow Separation in Overexpanded Supersonic Exhaust Nozzles. *Jet Propulsion*. 24:319–321.

Sutton, George P. 1986. *Rocket Propulsion Elements*. 5th ed. New York: John Wiley and Sons.

Thompson, Philip A. 1984. *Compressible-Fluid Dynamics*. New York: Maple Press Company.

Van Wylen, G. J., and R. E. Sonntag. 1971. *Introduction to Thermodynamics*. New York: John Wiley and Sons.

Zucrow, Maurice J., and Joe D. Hoffman. 1976. *Gas Dynamics*. Vol. I. New York: John Wiley and Sons.

CHAPTER 4
THERMOCHEMISTRY

Ronald Furstenau, *U.S. Air Force Academy*
Ronald Humble, *U.S. Air Force Academy*

4.1 The Chemical Heat Source: Bond Energy
4.2 Thermochemistry Basics
4.3 Products of Combustion
4.4 Flame Temperature: The Available-Heat Method
4.5 Chemical Kinetics: The Speed of the Chemical Reactions
4.6 Combustion of Liquids vs. Solids
4.7 Propellant Characteristics and Their Implications
4.8 Key Thermochemical Parameters: The Bottom Line

Thermochemistry is a crucial, fundamental discipline to understand when designing a rocket. It is the starting point for rocket design. Our choice of rocket propellants determines the temperature of the combustion chamber and the molar masses of the exhaust products, which in turn determine such critical parameters as exit velocity and specific impulse. The combustion temperature also dictates the types of materials used for rocket construction as well as the type of cooling which may be necessary. Even before the reaction chemistry occurs in the combustion chamber, the propellants' physical and chemical properties are important in determining the type and size of the tank and feed system. Nor is chemistry out of the picture once the exhaust leaves the exit plane of the nozzle. The chemistry occurring in the exhaust plume is crucial to a rocket's detectability and potential plume contamination of spacecraft surfaces and the atmosphere.

For rocket propulsion, we must solve two basic thermochemical problems:

- Given the combustion-chamber conditions (pressure and temperature), determine the chemical-reaction products
- Given the chemical-reaction products, determine the combustion-chamber temperature.

These two goals seem to be somewhat contradictory. We require results from one to get the other. To solve this problem, we must resort to iteration. For example, if we assume a reaction temperature, we can estimate the reaction products, which in turn allow us to evaluate the reaction temperature. The big question then remains: does the process converge? The answer: we have enough experimental and analytical experience to estimate well enough so iteration almost always converges to the "correct" answer.

In presenting the various concepts involved in thermochemistry, we run into a similar problem. We need to discuss several topics that are somewhat disconnected from each other but which require information from all of the topics. We start by discussing the fundamental source of heat in a chemical reaction: the making and breaking of chemical bonds. Then, after discussing some fundamental concepts, we introduce two approaches to determining the products of combustion. This discussion leads us to the "available heat method" for determining the combustion flame temperature, plus several other short but important topics.

From this type of analysis, we can predict combustion-chamber temperature (T_c), isentropic parameter (γ), and the molecular mass ($\mathcal{M}$) of the combustion gases as a function of the proportions of injected propellant (oxidizer-to-fuel ratio—O/F). In this chapter we show how to do this analysis, but we usually need fairly sophisticated computer codes to actually carry it out (see reference list). Appendix B contains the thermochemical data for many of the more popular propellant combinations.

4.1 The Chemical Heat Source: Bond Energy

From a chemical viewpoint, the energy associated with a chemical reaction has its roots in the strength of the chemical bonds of the reactants and products. A common rocket reaction involves the reaction between liquid hydrogen and liquid oxygen:

$$\underset{\text{reactants}}{\underset{\text{2 moles}\quad\text{1 mole}}{2H_2\,(l) + O_2\,(l)}} \to \underset{\text{products}}{\underset{\text{2 moles}}{2H_2O\,(g)}} \tag{4.1}$$

We need to point out several key things about this simple reaction. First of all, the numbers in front of the chemical species are known as *stoichiometric coefficients*. These numbers represent the number of moles of each chemical specie involved in the reaction. From basic chemistry, a mole is Avogadro's number (6.022×10^{23}) of molecules or atoms. The letters in parentheses following the chemical symbols are *state symbols* and represent the physical state of the chemical species. State symbols are crucial in determining the heat given off or taken in by a reaction. The chemical species on the left side of the arrow are the *reactants*. The chemical species on the right side of the arrow are the *products*. With these brief definitions, we can read the above equation as follows: "Two moles of liquid hydrogen react with one mole of liquid oxygen to form two moles of water vapor (or gaseous water)." But as the opening sentence to this paragraph stated, energy changes accompany chemical reactions, and these changes concern us most when designing a rocket.

From a chemical standpoint, it makes sense that the energy accompanying a chemical reaction should be related to the strengths of the bonds in the reactants as compared to the strength of the bonds of the products. The sharing of electrons is holding the atoms of the molecule together by forming a chemical bond. A measure of the strength of these chemical bonds is known as the *bond energy*. When designing a rocket, we want to use propellant molecules (reactants) that have relatively weak chemical bonds. A weak chemical bond means that it takes little energy to tear an atom away from another atom with which it is sharing electrons. When the products of the combustion reaction are formed, we want to form chemical products that have relatively strong chemical bonds. The essential point is this: breaking weak chemical bonds takes relatively little energy; when strong chemical bonds form, they release relatively large amounts of energy. We want to use propellants with weak chemical bonds and form products with strong chemical bonds.

Using bond energies, we can estimate the amount of energy given off by a chemical reaction by finding the difference between the energy of all of the chemical bonds in the reactants (left side of a chemical reaction) and the energy of all of the chemical bonds in the products (right side). This energy difference is known as the *heat of reaction*.

By convention, bond energies are tabulated for chemical species in the gaseous phase. Phase changes are discussed in Sec. 4.2. Looking at the reaction between hydrogen and oxygen again (this time, all species are gases),

$$2H_2(g) + O_2(g) \rightarrow 2H_2O(g) \tag{4.2}$$

we notice that the relationship between atoms is different in the products than it is in the reactants. On the reactants' side, a hydrogen is bonded to a hydrogen and an oxygen is bonded to an oxygen. On the products' side, two hydrogens are bonded to one oxygen in water. In a general sense, if atom A is bonded to atom B, the *bond energy* is defined as the average enthalpy change for breaking an A-B bond in the gaseous phase. The word "average" is important. It should make sense that if we had a water molecule H—O—H, the energy required to break the first O—H bond would be different from the energy required to break the second O—H bond, once the first H atom was gone. Once a hydrogen atom is removed from the water molecule, the electron environment of the remaining molecule fragment is completely different. The average of these two energies is considered to be the bond energy of an O—H bond.

To calculate the approximate enthalpy of a reaction using bond energies, we add up the bond energies for the reactants (energy required to break the bonds in the reactants) and subtract the sum of the bond energies for the products (energy released when bonds are formed in the products). Table 4.1 is a compilation of average bond energies.

Table 4.1. **Bond Energies.** The upper matrix shows the average energy of a single bond. For example, it takes 386,000 joules of energy to break all of the N—H bonds in 1 mole (6.02 × 10^{23} bonds) of material. The lower level lists the energy of multiple bonds. A double bond is denoted by "=" and a triple bond is denoted by "≡".

Energies of Single Bond (kJ/mol)						
	H	C	N	O	F	Cl
H	432					
C	411	346				
N	386	305	167			
O	459	358	201	142		
F	565	485	283	190	151	
Cl	432	327	313	218	249	243
Energies of Multiple Bond (kJ/mol)						
C=C	620	C=N	615	C=O (in CO_2)	799	
C≡C	812	C≡N	887	C≡O	1072	
N=N	418	N=O	607			
N≡N	942	O=O	494			

Looking at Table 4.1, how do we know whether we have single, double, or triple bonds in our chemical species? The nature of the chemical bond can be easily determined by finding what is known as the *Lewis dot structure* of the molecule. This is a standard lesson in general chemistry, with Lewis dot structures determined by following a relatively simple set of rules. Any general chemistry textbook, including Brown and LeMay [1981], and Ebbing [1984], contains a section describing how to determine the bond structure in chemical compounds using Lewis dot diagrams.

Going back to the hydrogen/oxygen reaction [Eq. (4.2)], on the reactants' side, we are breaking 2 moles of H—H bonds and 1 mole of O=O bonds:

$$\text{Energy of bonds broken} = (2 \text{ mol})(432 \text{ kJ/mol}) + (1 \text{ mol})(494 \text{ kJ/mol})$$
$$= 1358 \text{ kJ}$$

On the products' side, we are forming 2 moles of two H—O bonds:

Energy of bonds formed = (2 mol) (2) (459 kJ/mol)
= 1836 kJ

The heat of reaction (ΔH_{rxn}) is then

ΔH_{rxn} = broken − formed = 1358 − 1836 = −478 kJ, or
= −239 kJ/mol of H_2O

Bond energies can provide a quick method for calculating heats of reaction. We are simply comparing the strength of the chemical bonds in the compounds we start with to the strength of the bonds in the compounds we end up with. Another method, using heats of formation data, is the preferred method by chemists and is more accurate than using bond energies when the appropriate data exists. Heat of formation is one of the topics in the next section.

4.2 Thermochemistry Basics

Chemical reactions are accompanied by energy changes. The energy changes in chemical reactions provide the link between chemistry and engineering of designing a rocket. Before going into the chemical thermodynamics details essential to designing a rocket, it is useful to first present some basic thermochemistry terms. It is probably best to start with the first law of thermodynamics. This is essentially the conservation of energy principle and can be stated many ways. One way to state this is that the energy change (ΔE) of a system for any process which takes the system from state 1 to state 2 can be expressed as

$$\Delta E = E_2 - E_1 \tag{4.3}$$

E is known as the *internal energy* of the system. We can also write E in terms of Q (the heat gained by the system) and W (the work done by the system).

$$\Delta E = Q - W \tag{4.4}$$

The work term (W) is considered to be the mechanical work associated with the force of expansion.* Because pressure is force per unit area, we can express the W term as an integral of the external pressure that drives the piston (p_{ext}) and the change in volume of the system (dV):

$$W = -\int_{V_1}^{V_2} p_{ext} \, dV \tag{4.5}$$

Equation (4.5) is considered to be "pV work."

---

* Such as work done by moving a piston by gas in a cylinder.

When dealing with chemical systems, two special conditions occur frequently. The first condition occurs when the *volume* of the system is kept constant and the second condition happens when the *pressure* of the system is kept constant.

In a *constant-volume* process, there is no pV work. There could be other work, however, such as electrical work. For a constant-volume process, $W = 0$ and Eq. (4.4) becomes

$$E = Q \tag{4.6}$$

where Q is the heat absorbed by the system.

In a *constant-pressure process*, if the system expands by ΔV against a constant external pressure, Eq. (4.4) becomes

$$\Delta E = Q - p\Delta V \tag{4.7}$$

If we lump together the changes in the system on one side of the equation, Eq. (4.7) becomes

$$\Delta E + p\Delta V = Q \tag{4.8}$$

At this point, we introduce the new energy term called *enthalpy*, which is defined as

$$H = E + pV \tag{4.9}$$

For any process, the change in enthalpy (ΔH) is

$$\Delta H = \Delta E + \Delta(pV) \tag{4.10}$$

For a constant pressure process, Eq. (4.10) becomes

$$\Delta H = \Delta E + p\Delta V \tag{4.11}$$

It then follows from Eq. (4.8) that for a *constant-pressure process with only pV work*,

$$\Delta H = Q \tag{4.12}$$

From an energy standpoint, then, it is more convenient to work with internal energy (E) for constant-volume processes [Eq. (4.6)]. However, for constant-pressure processes, it is more convenient to work with enthalpy [Eq. (4.12)]. Enthalpy is a function introduced for convenience and is the prime energy term we are concerned with in our discussion of the thermodynamics of rockets.

4.2.1 Absolute and Relative Enthalpies

The reaction between liquid oxygen and liquid hydrogen,

$$2H_2(l) + O_2(l) \rightarrow 2H_2O(g) + \text{heat} \tag{4.13}$$

4.2 Thermochemistry Basics

is accompanied by a heat flow of 460.38 kJ at 25°C at a pressure of 1 atm.* More specifically, when the reaction occurs, 460.38 kJ of energy is released. This heat flow is often called $Q_{reaction}$ and is equal to the enthalpy (H) difference between the products and the reactants. Recall that the change in enthalpy is the energy difference associated with a system at constant pressure, which is the normal condition for rocket combustion.

$$Q_{reaction} = \Delta H = H_{products} - H_{reactants} \qquad (4.14)$$

Enthalpy, or chemical heat content, is not something we can measure absolutely for a substance. The absolute enthalpy of a substance can be expressed in terms of an integral of the heat capacity at constant pressure (C_p). C_p is equal to $\left(\frac{\partial H}{\partial T}\right)_p$ and is determined empirically.

$$\Delta H_{abs} = \int_0^T C_p dT \qquad (4.15)$$

However, we do not know C_p at low temperatures. Therefore, it is more convenient to evaluate the integral based on a reference temperature (T_{ref}, which is usually 25°C).

$$\Delta H_r = \int_{T_{ref}}^T C_p dT \qquad (4.16)$$

Thus, an enthalpy change can be measured for a substance relative to something else or, in our specific case, for a chemical reaction. In our hydrogen-oxygen example, the heat of reaction (ΔH_{rxn}) is –460.38 kJ. Figure 4.1 illustrates what is going on in an energy sense for the hydrogen-oxygen reaction.

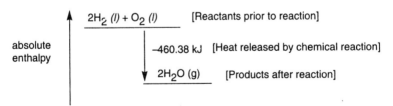

Fig. 4.1. **Enthalpy Diagram for an Isothermal and Isobaric Hydrogen-Oxygen Reaction.** Heat must be removed from the system to maintain a constant temperature. The heat removed exactly equals the enthalpy change for a constant-pressure (isobaric) reaction.

---

* Notice that this is an isothermal (constant temperature) reaction. A negative sign on the heat term means that heat is produced and must be removed to maintain a constant reaction temperature.

Chemical reactions which have a negative heat flow, or a negative change in enthalpy (ΔH) are considered to be *exothermic*, which means they give off heat. Reactions which have a positive heat flow, or positive change in enthalpy, are considered to be *endothermic*, which means they must take in heat to react. When designing a rocket, we are searching for propellant combinations which result in very exothermic reactions. The more exothermic the reaction, the higher the temperature we can expect to get for the combustion products, which leads to a higher exhaust velocity. It is worth mentioning that, as odd as it may sound, endothermic reactions can occur in rocket engines. Going back to our hydrogen-oxygen system, the following reaction, known as a *dissociation reaction*, can occur at high temperatures:

$$H_2(g) \rightarrow H(g) + H(g) \tag{4.17}$$

This reaction has a ΔH_{rxn} of +436.08 kJ. Because this reaction is endothermic, it removes valuable heat from the combustion gases. This problem is particularly pronounced in high-speed, air-breathing engines. Decelerating the air increases the temperature. Adding more heat increases the temperature for W, but there is a limit. Beyond a certain point, we just create ions and do not effectively increase the usable thermal energy.

4.2.2 Heats of Formation and Reaction

A useful concept concept is *standard heats (or enthalpies) of formation*. Data on standard heats of formation provides a convenient reference for evaluating the energy change of a reaction at constant pressure. The standard enthalpy of formation (ΔH_f°) of a substance is the enthalpy change for the formation of one mole of the substance from its elements at a standard pressure (1 atm) and a specified temperature (usually 25°C). The most stable form of an element under these conditions is said to be the reference state and is defined to have a $\Delta H_f^\circ = 0$. The superscript ° refers to standard conditions (25°C and one atmosphere). If we had conducted our hydrogen-oxygen reaction with gaseous hydrogen and oxygen,

$$2H_2(g) + O_2(g) \rightarrow 2H_2O(g) \tag{4.18}$$

The reaction is now using elemental hydrogen and oxygen in their most stable form at 25°C. Figure 4.2 shows the enthalpy diagram for this reaction.

Notice that the ΔH_{rxn}° for this reaction is −483.86 kJ. We can say from this reaction that the standard enthalpy of formation for $H_2O(g)$ is −483.86 kJ/2 mol or −241.93 kJ/mol, because $H_2O(g)$ is being formed from elemental hydrogen and oxygen in their most stable form under standard conditions. Notice also from Fig. 4.2 that the concept of standard heats of reaction has established a scale on our enthalpy axis. With the arbitrary definition of the heats for formation of elements being zero, the heat of formation of everything else can be measured relative to this reference point.

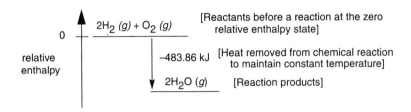

Fig. 4.2. **Enthalpy Diagram for a Gaseous H_2/O_2 Reaction at Constant Temperature and Pressure.** Notice that more heat is removed in this reaction than is shown in Fig. 4.1. This difference is attributed to the phase difference (gas vs. liquid) of the reactants.

Do we have to do an experiment to actually measure the heat of reaction of all reactions in our rocket system? Fortunately, the answer is "no." According to *Hess's Law*, if a reaction can be carried out in a series of steps, ΔH for the reaction is equal to the sum of the enthalpy changes for each step. Enthalpy is a state function, which means that it does not matter how we get from point A to point B, chemically. The energy change is always the same. Hess's Law allows us to use standard heats of formation to calculate ΔH_{rxn}° for any reaction. It also allows us to use the following mathematical relationship to determine the heat of reaction:

$$\Delta H_{rxn}^{\circ} = \sum n_p \Delta H_{f,\,products}^{\circ} - \sum n_r \Delta H_{f,\,reactants}^{\circ} \qquad (4.19)$$

where n represents the coefficient for each species in the balanced chemical equation. Determining the actual n values is the most difficult part of determining the combustion chamber temperature of a rocket. ΔH_f° values can be determined experimentally from thermal measurements or theoretically from spectroscopic measurements. Wilkins [1963] does a superb job in discussing methods of theoretically calculating thermodynamic functions such as heats of formation, heat capacity, enthalpy, and entropy. ΔH_f° values have been compiled for hundreds of chemical compounds and can be found in nearly any general chemistry book, thermodynamics book, and in the CRC *Handbook of Chemistry and Physics*. Table 4.2 lists the ΔH_f° for some of the chemical compounds commonly encountered when working with rocket propellants and their combustion products.

To illustrate how the data in Table 4.2 is used with Eq. (4.19), we can calculate the ΔH_{rxn}° for our original hydrogen/oxygen reaction:

$$2H_2\,(l) + O_2\,(l) \rightarrow 2H_2O\,(g) \qquad (4.20)$$

Table 4.2. Standard Heats of Formation. The right-hand columns show the energy required to form one mole of the substances in the left-hand columns from their stable element forms at a constant temperature (298 K) and pressure (1 atm). Notice that several elements (i.e., Al, H_2) are already in their stable form, so they require no heat of formation.

Compound	kJ/mol	Compound	kJ/mol
Al(s)	0.00	N_2(g)	0.00
Al_2O_3	−1670.53	N(g)	+472.71
C(s, graphite)	0.00	NH_3(l)	−65.63
CO(g)	−110.59	NH_3(g)	−46.21
CO_2(g)	−393.68	N_2H_4(l)	+50.48
C(g)	+718.70	N_2O_4(l)	−20.13
CH_4(g)	−74.90	NH_4ClO_4(s)	−290.58
CH_3OH(l)	−238.76	HF(g)	−268.73
C_2H_5OH(l)	−277.77	HCl(g)	−92.34
H_2(g)	0.00	F_2(l, 85 K)	−12.68
H_2(l, 20 K)	−7.03	O_2(g)	0.00
H(g)	+218.04	O_2(l, 90 K)	−9.42
H_2O(g)	−241.93		
H_2O(l)	−285.96		
H_2O_2(l)	−187.69		
HNO_3(l)	−173.08		

By writing the reaction in this manner, we are assuming that all of the reactants are converted only to the products shown. This makes calculating ΔH_{rxn}° quite straightforward.

$$\Delta H_{rxn}^\circ = [\,(2 \text{ mol})\,(\Delta H_f^\circ, H_2O\,(g))\,] \quad (4.21)$$
$$-[\,(2 \text{ mol})\,(\Delta H_f^\circ, H_2\,(l)) + (1 \text{ mol})\,(\Delta H_f^\circ, O_2\,(l))\,]$$

Note that we must choose the correct physical state (solid, liquid, or gas) of each chemical species.

$$\Delta H_{rxn}^\circ = [\,(2 \text{ mol})\,(-241.93 \text{ kJ/mol})\,] \quad (4.22)$$
$$-[\,(2 \text{ mol})\,(-7.03 \text{ kJ/mol}) + (1 \text{ mol})\,(-9.42 \text{ kJ/mol})\,]$$
$$\Delta H_{rxn}^\circ = -460.38 \text{ kJ}$$

Calculating the ΔH_{rxn}° determines rocket performance because it tells us how much energy we can expect the chemical reaction to give. Remember, we now have two methods for determining this crucial quantity: heats of formation and bond energies.

There may be some confusion about this whole concept of standard states—this superscript zero (°). After all, a rocket reaction certainly does not take place at 25° and 1 atm. Figure 4.3 illustrates the usefulness of the standard-state concept. Because enthalpy is a state function, it does not matter what path we take to get from the initial to the final conditions.

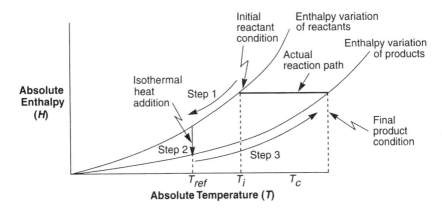

Fig. 4.3. **A Chemical Reaction Path for a Constant-Pressure Reaction.** Total enthalpy does not change during the reaction because heat added exactly equals the propellant enthalpy change. To evaluate the chamber temperature it is not necessary to follow the actual reaction path. Any appropriate path, such as the three-step path shown, suffices.

For example, if our propellant reactants start out at some temperature higher than the standard temperature, we can use the heat capacity of these reactants to calculate how much energy would be released to cool them down to standard conditions. Then, we react these propellants and obtain the products at standard conditions, liberating heat in the amount of ΔH_{rxn}°. Once we have obtained the products, we can use the heat gained from lowering the reactants to standard temperature and the heat gained from forming the products to heat the products up to some chamber temperature T_c, which depends on the heat capacity of the products. From a physical standpoint, of course, this is not what happens inside a rocket; the actual process is more direct. From a theoretical standpoint, however, this is a perfect, convenient bookkeeping tool which accounts for all of the enthalpy changes. We can do all of this because enthalpy is a state function. We use this state function property to calculate a combustion chamber temperature in Sec. 4.3.

Figure 4.3 illustrates how we are using the concept of a standard state to account for all energy changes in a chemical reaction. The ΔH_{rxn}° accounts for the energy difference between the reactants and products under a standard condition. In this illustration, the reactants are initially at a higher temperature (T_i) and higher enthalpy than the standard state. Thus in Step 1, they must lose energy to reach that standard state (T_{ref}, 25°C or 298 K). Likewise, energy is lost in the chemical reaction, which is illustrated as Step 2. Note that the energy loss in Step 2 is occurring under isothermal conditions. The energy removed in Steps 1 and 2 is now available to heat up the resulting products to some combustion chamber temperature, T_c. In other words, for i chemical species involved,

$$\text{Heat lost in Step 1 + Step 2 = Heat gained in Step 3} \quad (4.23)$$

$$\sum_i n_{i,\,reactants} \int_{T_i}^{298} C_{p,\,i}\, dT + \Delta H_{rxn}^\circ = -\sum_i n_{i,\,products} \int_{298}^{T_c} C_{p,\,i}\, dT \quad (4.24)$$

Essentially, we have used the fact that enthalpy is a state function to divide the energetics of a chemical reaction into a series of bookkeeping steps. Although these steps are not the actual reaction path, they take us from the initial reactant condition to the final product condition. From an energetics standpoint, this is all that matters!

4.2.3 Heat of Fusion and Vaporization

The ΔH_{rxn}° is different for the hydrogen-oxygen reaction depending upon whether or not we use liquid or gaseous reactants. When using $H_2(l)$ and $O_2(l)$, the ΔH_{rxn}° is -460.38 kJ. When using gases, the ΔH_{rxn}° is -483.86 kJ. This difference occurs because it takes energy to vaporize the liquid reactants, making the overall enthalpy of reaction less.

A reaction we are all familiar with, the melting of ice, requires 6.00 kJ/mol at 0°C. This is known as the *heat of fusion*.

$$H_2O(s) \rightarrow H_2O(l) \quad \Delta H = 6.00 \text{ kJ at } 0°C \quad (4.25)$$

The energy required for the phase change from a liquid to a gas is much greater. Water, for example, requires 40.7 kJ/mol at 100°C, its boiling point.

$$H_2O(l) \rightarrow H_2O(g) \quad \Delta H = 40.7 \text{ kJ at } 100°C \quad (4.26)$$

To no surprise, this *heat of vaporization* is greater at lower temperatures. At 25°C, the heat of vaporization of water is 44.9 kJ/mol. The overall enthalpy of a reaction depends upon the states of the reactants and products, as well as the chosen standard state. The bottom line: pay attention to the physical states of the reactants and products when determining the enthalpy of a reaction.

The importance of enthalpy to design lies in the fact that determining the enthalpy of a reaction dictates how much energy is available to heat up the reaction products. To determine the enthalpy, we first must determine the chemical species we are actually getting from the reaction. The concept of chemical equilibrium will allow us to do so.

4.3 Products of Combustion

The concept of chemical equilibrium plays an important role in rocket chemistry. Chemical equilibrium allows us to predict the chemical species which are present as products when using reactants at a given set of conditions.

When a chemical reaction is at *equilibrium*, there is no observable change in the amounts of reactants and products as time goes by. The rate of the forward reaction is the same as the rate of the reverse reaction (which implies the reaction is reversible). For example, if we look at the dissociation of hydrogen gas:

$$H_2(g) \rightleftarrows H(g) + H(g) \tag{4.27}$$

the double arrow implies that the reaction is reversible. When the above reaction is at equilibrium, it does NOT mean that the concentration of the products and the reactants are the same; it simply means that the concentration of the products and reactants are not observably changing. At room temperature, the amount of atomic hydrogen produced from molecular hydrogen is very small. At the much higher temperatures of a rocket engine, however, the amount of atomic hydrogen is much more significant. The energy produced by a rocket can produce enough energy to break a significant fraction of the H—H bonds. This lowers the amount of energy available for heating up the product molecules. The types of chemical species produced are essential in determining the amount of heat produced by the reaction.

But how do we know the quantity of products and reactants existing at chemical equilibrium? The answer lies in a quantity called the *equilibrium constant*. From empirical observations, we know that at any particular temperature, there is a constant relationship between the concentrations (or partial pressures of gases) of the products and reactants. For the general reaction

$$aA + bB \rightleftarrows cC + dD \tag{4.28}$$

the equilibrium constant can be expressed in terms of concentrations (K_c), where the brackets "[ ]" represent a molar concentration.

$$K_c = \frac{[C]^c [D]^d}{[A]^a [B]^b} \tag{4.29}$$

If all of the species are gaseous (which is the case for meaningful rocket combustion applications), it is more convenient to represent the equilibrium constant (K_p) in terms of the partial pressures (p) of the bases present.

$$K_p = \frac{p_C^c \, p_D^d}{p_A^a \, p_B^b} \tag{4.30}$$

For the dissociation of hydrogen in Eq. (4.27), this equilibrium constant is

$$K_p = \frac{p_H^1 \, p_H^1}{p_{H_2}^1} = \frac{p_H^2}{p_{H_2}} \tag{4.31}$$

In this expression, the equilibrium constant in terms of pressure (K_p) is expressed in terms of the product of the partial pressure of each of the product constituents raised to the power of their stoichiometric coefficients, divided by the product of the partial pressure of each of the reactant constituents raised to the power of their stoichiometric coefficients. p_H represents the partial pressure of monatomic hydrogen (H) and p_{H_2} represents the partial pressure of diatomic hydrogen (H_2). We can also see that K_p may be different depending on how the reaction is written; that is, changing the stoichiometric coefficients changes the numerical value for K_p. **No matter what the initial pressures of any of the constituents are, the relationship between the final partial pressures at equilibrium is always equal to K_p at any given temperature.** It is important to remember that K_p depends on temperature—the important clue for using the equilibrium constant in rocket design! To calculate performance parameters, we need to know the temperature of the combustion chamber. To calculate the temperature, we need to know how much heat is released. To calculate the heat released, we need to know the chemical species present. The equilibrium constant tells us what chemical species must be present at any particular temperature. A method which employs the equilibrium constant to calculate the chemical species present in a rocket and, thus, the combustion chamber's temperature, is known as the *equilibrium-constant method*.

4.3.1 The Equilibrium Constant Method

Taking the reaction

$$2H_2\,(g) + O_2\,(g) \rightarrow 2H_2O\,(g) \tag{4.32}$$

or, with "1" being the largest stoichiometric coefficient,

$$H_2\,(g) + 1/2\,O_2\,(g) \rightarrow H_2O\,(g) \tag{4.33}$$

we assume that the reactants go completely to products in the stoichiometric ratios indicated by the coefficients. Unfortunately, this complete conversion rarely occurs in rockets or in chemical reactions in general. At any particular pressure and temperature, the reaction tries to achieve an *equilibrium state*. To accomplish this, some secondary reactions will also occur during the combustion process. For

Eq. (4.33), the following significant reactions occur with the main reaction, achieving chemical equilibrium:

(a) $H_2 + 1/2\, O_2 \rightleftarrows H_2O$ (4.34)

(b) $1/2\, O_2 + 1/2\, H_2 \rightleftarrows OH$

(c) $1/2\, H_2 \rightleftarrows H$

(d) $1/2\, O_2 \rightleftarrows O$

How do we know that the above reactions are going to occur? How do we predict what species are actually going to be present before we find their relative amounts? There are two answers to these questions. First of all, we do not have to know the species which are present; we can guess. Any chemically feasible combination of atoms could exist. If we are able to calculate or find equilibrium-constant data for these "made-up" reactions, the method which follows tells us whether or not we actually get any of these chemical species. Secondly, we can use intuition to eliminate certain chemical species and thus eliminate some of the equilibrium expressions. Fortunately, sets of equilibrium data already exist for most chemical species in common combinations of rocket propellants. In other words, we are unlikely to have to figure out the possible chemical species. A computer program can do it for us!

As the number of elements and compounds increase for a chemical reaction (our example has only two different elements), the number of possible side reactions increases dramatically, particularly if carbon or nitrogen is present. But all of these side reactions are not important because chemical equilibrium allows us to determine which species actually form. Do not forget that the reason for using the concept of equilibrium is to find the number of moles of each species present at any particular temperature, so we may find the heat released by the reaction.

For the chemical reactions (4.34 a) through (4.34 d) above, we can write the following *equilibrium-constant expressions*. Equilibrium constants (K_p, equilibrium constants in terms of partial pressure) are known functions of temperature. Thermodynamic tables exist for equilibrium constants versus temperature for numerous equilibrium systems. The remainder of this discussion on the equilibrium-constant method follows the nomenclature used by Penner [1957].

The equilibrium constants for reactions (4.34) (a)–(d) are as follows:

(a) $K_{p,a} = \dfrac{p_{H_2O}}{p_{H_2}\, p_{O_2}^{1/2}}$ (4.35)

(b) $K_{p,b} = \dfrac{p_{OH}}{p_{O_2}^{1/2} p_{H_2}^{1/2}}$

(c) $K_{p,c} = \dfrac{p_H}{p_{H_2}^{1/2}}$

(d) $K_{p,d} = \dfrac{p_O}{p_{O_2}^{1/2}}$

Table 4.3, adapted from NBS Circular 500, *Selected Values of Chemical Thermodynamic Properties*, 1 February 1952, shows the values for these equilibrium constants at various temperatures.

Table 4.3. Equilibrium Constants for Equations (4.35) (a)–(d). The equilibrium constants for the four reactions vary with temperature as shown here. Each column corresponds to an individual equilibrium reaction.

T (K)	$K_{p,a}$	$K_{p,b}$	$K_{p,c}$	$K_{p,d}$
298	1.1143×10^{40}	2.8340×10^{-7}	2.4831×10^{-36}	4.8978×10^{-41}
300	6.1235×10^{39}	3.1311×10^{-7}	4.2560×10^{-36}	9.0157×10^{-41}
400	1.7418×10^{29}	2.1419×10^{-5}	1.3459×10^{-26}	5.5335×10^{-30}
500	7.6913×10^{22}	2.6984×10^{-4}	6.9984×10^{-21}	7.7140×10^{-23}
1000	1.1482×10^{10}	4.1295×10^{-2}	2.2693×10^{-9}	1.9055×10^{-10}
1500	5.2541×10^{5}	2.1324×10^{-1}	1.7575×10^{-5}	1.6315×10^{-5}
2000	3.3931×10^{3}	4.7664×10^{-1}	1.6233×10^{-3}	7.3350×10^{-4}
2500	1.6127×10^{2}	7.6648×10^{-1}	2.5090×10^{-2}	1.5574×10^{-2}
3000	2.0999×10^{1}	1.0478	1.5762×10^{-1}	1.2010×10^{-1}
3500	4.9295	1.3046	5.8993×10^{-1}	5.1807×10^{-1}
4000	1.6623	1.5315	1.5933	1.5528
4500	7.0307×10^{-1}	1.7322	3.4602	3.6521
5000	3.5465×10^{-1}	1.9067	6.4506	7.2395

What do the numbers in Table 4.3 actually mean? Notice that at 298 K (room temperature), Eq. (4.35) (a) has a value of 1.1143×10^{40}, whereas the other three reactions have K_p values far less than one. This means that the only reaction that appreciably occurs at room temperature for the hydrogen/oxygen set of reactions is reaction (4.35) (a). However, as the temperature goes up, the K_p values for the other reactions (4.35) (b–d) get larger and larger, meaning that the products of those reactions are becoming more appreciable. In other words, the equilibrium chemical composition of the hydrogen/oxygen system changes significantly with temperature.

What good are all of these pressures? We need moles of gas, not pressures! Remember, the ideal-gas law states that

$$pV = nRT, \quad \text{or} \quad n = pV/RT \tag{4.36}$$

so we can easily convert any pressure values into moles. (The ideal-gas law is a pretty good assumption for preliminary analysis of rocket engines!) At this point, even if we knew the K_p values at particular temperatures for Eqs. (4.35) (a–d), we **still** could not solve for the n values. We have six chemical species, but only four mathematical equations. We can now introduce equations governing the conservation of mass. For our two-element reaction system, the total number of moles of hydrogen (N_H) and the total number of moles of oxygen (N_O) must be the same at equilibrium as the number of moles of these elements that we injected into the rocket in the first place. Remembering the chemical species in our system, the following conservation of mass equations must hold true:

$$2\,n_{H_2O} + n_{OH} + 2\,n_{H_2} + n_H = N_H \tag{4.37}$$

$$n_{H_2O} + n_{OH} + 2\,n_{O_2} + n_O = N_O \tag{4.38}$$

Using the ideal-gas law, we can express Eqs. (4.37) and (4.38) in terms of partial pressures, which appear in the equilibrium-constant expressions:

$$2\,p_{H_2O} + p_{OH} + 2\,p_{H_2} + p_H = N_H(RT/V) \tag{4.39}$$

$$p_{H_2O} + p_{OH} + 2\,p_{O_2} + p_O = N_O(RT/V) \tag{4.40}$$

Remember, we know what N_H and N_O are because we know how much fuel and oxidizer we are putting into the rocket in the first place. Notice that we now have six equations [(4.35) (a–d) and (4.39, 4.40)] and six unknowns (partial pressures). We can express the partial pressures of all of the involved species in terms of the equilibrium constants and the reactants (H_2 and O_2):

$$p_{H_2O} = K_{p,a} p_{H_2} p_{O_2}^{1/2} \tag{4.41}$$

$$p_{OH} = K_{p,b} p_{H_2}^{1/2} p_{O_2}^{1/2} \tag{4.42}$$

$$p_{H} = K_{p,c} p_{H_2}^{1/2} \tag{4.43}$$

$$p_{O} = K_{p,d} p_{O_2}^{1/2} \tag{4.44}$$

Our goal at this point is to numerically solve six equations in six unknowns. This allows us to solve for the partial pressure of each gaseous specie and thus the number of moles of each specie.

4.3.2 Minimization of Free Energy Method

Another approach, known as the *minimization of free energy* method, can also be used to determine the amounts of chemical species at any particular temperature. Free energy is derived from the concept of entropy. *Entropy* is a measure of randomness or disorder. The second law of thermodynamics states that, for any spontaneous process (occurring on its own), the entropy of the universe increases. The change in the entropy of the universe (ΔS_{univ}) is equal to the change in entropy of the thermodynamic system (ΔS_{sys}) plus that of the surroundings (ΔS_{surr}):

$$\Delta S_{univ} = \Delta S_{sys} + \Delta S_{surr} \tag{4.45}$$

If a reaction causes the entropy of the universe to increase, it occurs spontaneously. If the change in the entropy of the universe is negative, the reaction does not occur spontaneously, but the reverse reaction does. If the change in the entropy of the universe is zero, the system is at equilibrium.

However, working with the entropy of the surroundings can make Eq. (4.45) a difficult equation to solve. We must focus on the chemical system, which is much easier to quantify. At constant pressure, the entropy of the surroundings is proportional to the heat transferred into or out of the system, or

$$\Delta S_{surr} = \frac{-\Delta H_{sys}}{T} \tag{4.46}$$

Equation (4.46)* allows us to express the entropy of the universe in terms of the system only, or

$$\Delta S_{univ} = \frac{-\Delta H_{sys}}{T} + \Delta S_{sys} \tag{4.47}$$

---

* This equation is the second law of thermodynamics (see Sec. 3.2.4) for a reversible system. For a constant-pressure process, the heat change is equal to the change in enthalpy (see Sec. 4.2.1).

multiplying both sides of (4.47) by $-T$,

$$-T\Delta S_{univ} = \Delta H_{sys} - T\Delta S_{sys} \qquad (4.48)$$

A new thermodynamic function, free energy (G, older references use F), or Gibbs free energy, is now introduced as a function of entropy (S) and enthalpy (H):

$$G = H - TS, \qquad (4.49)$$

or

$$\Delta G = \Delta H - T\Delta S \qquad (4.50)$$

where $\Delta G = -T\Delta S_{univ}$.

From a chemistry standpoint, free energy tells us whether or not the reaction occurs spontaneously—whether or not a reaction will even go under the set conditions. A negative free-energy change indicates that the reaction occurs spontaneously as written. A positive free-energy change indicates that the reaction is not spontaneous. For our rocket problem, however, the most important value for the free-energy change occurs when ΔG for a chemical reaction is equal to zero, the equilibrium condition.

At equilibrium, ΔG for a chemical system, that is, $G_{products} - G_{reactants}$, is equal to zero. This is the key to the "minimization of free energy method:"

$$\Delta G° = \sum_i n\Delta G°_{f,\,products} - \sum_i n\Delta G°_{f,\,reactants} \qquad (4.51)$$

ΔG values for a chemical species are tabulated as functions of temperature or can be calculated from enthalpy and entropy data or functions. These values are all empirically determined. In using the minimization of free energy approach, we assume a pressure, temperature, and initial product and reactant composition for the rocket combustion chamber. We then iterate to see what specific proportion (mole values) of products and reactants minimizes the free energy or approaches a ΔG of zero, which places the system at equilibrium. Because this method is also based on the concept of equilibrium, the final answer is the same as with the equilibrium-constant approach. The big advantage of the free-energy approach is that a computer program can converge very rapidly on a solution by iterative methods. In addition, the minimization of free energy approach can handle a mixture of condensed (solid or liquid) and gaseous species. Today, minimization of free energy is the preferred approach.

Neither the equilibrium-constant method nor the minimization of free energy method seem very friendly for a "back of the envelope" calculation of combustion-chamber temperature...and they are not! Many computer programs exist to do these calculations. Nearly all of them are based on the work of Gordon and McBride, some of whose work is listed in the reference section [1967, 1976]. We highly recommend these references if you want to know how the concepts discussed in this chapter are actually developed into the format of a computer program. Gordon and McBride's program, often referred to as the *Lewis code*, is still considered the standard rocket-thermodynamics code throughout the world, though some groups have made many changes in hopes of improving it (see, for example, Straub [1989]). The U.S. Air Force's I_{sp} code, developed by Curtis Selph, is another example of a performance program based on these methods.

At this point, we have shown only how to calculate the equilibrium concentrations of chemical species at a **particular temperature**. We have essentially "guessed" a temperature. How do we know the "correct" temperature for the combustion chamber of our rocket? In Sec. 4.4, we describe the procedure known as the *available-heat method* to calculate the adiabatic flame temperature of the combustion chamber. The available-heat method uses information about the chemical species from the equilibrium-constant method or the minimization of free energy method.

4.4 Flame Temperature: The Available-Heat Method

The available-heat method is the standard approach for calculating combustion-chamber temperatures. Wilkins [1963] gives perhaps the most detail regarding this method. The available-heat method follows these steps:

1. Assume a combustion-chamber temperature $T_c = T_c'$ at a particular chamber pressure.

2. Determine the equilibrium composition of the combustion gases at the assumed temperature (T_c') and at the assumed chamber pressure. This is by far the most complicated part of the available-heat method, because it involves using the equilibrium-constant approach or the minimization of free energy approach.

3. Using the calculated composition from step (2) at the assumed temperature (T_c'), determine the available heat ($Q_{available}$ or, simply Q) using

$$\Delta H_{rxn}^\circ + \sum_i n_{i,\, reactants} \int_{298}^{T_i} C_{p,\, i} dT = Q_{available} \qquad (4.52)$$

4. Calculate the heat absorbed by the combustion products when they are heated from the standard temperature (298 K) to T_c'. This heat absorbed is called Q' or $Q_{required}$ and can be found by using "enthalpy tables" which compile heat capacities of gases with respect to temperature, allowing us to solve the following equation:

$$Q' = Q_{required} = \sum_i n_{i,\, products} \int_{298}^{T_c} C_p\, dT \qquad (4.53)$$

More typically, heat capacity curve fits for gases are used. These curve-fit equations are usually in the form of

$$C_p = a + b(T/1000) + c(T/1000)^2 + d(T/1000)^3 \qquad (4.54)$$

where the constants a, b, c, and d are given in thermodynamic tables or included in computer programs. One must be careful when using such curve fits, because the constants are good only in a specified temperature range.

5. Compare Q' from step (4) with Q from step (3). If $Q < Q'$, the assumed value of T_c' from step (1) is too large. If $Q > Q'$, the assumed T_c' is too small. If $Q = Q'$ (within an allowable tolerance), we have found the correct temperature $(T_c = T_c')$!

6. Using the criteria in step (5), iterate until we find the correct T_c'. Again, except for step (2), this procedure is very straightforward.

It is perhaps easiest to see how the available-heat method works if we use an enthalpy table rather than the heat-capacity equations. Let us use a reaction we have used all along: the combustion of hydrogen and oxygen to form water vapor. For this example, we burn a fuel-rich reaction, which is more representative of conditions in a rocket:

$$5H_2(l) + O_2(l) \rightarrow 2H_2O(g) + 3H_2(g) \qquad (4.55)$$

To make the example simpler to follow, we assume that we get complete combustion to products. Now, let us step through the available heat method:

1. We can truly assume any combustion-chamber temperature. To begin the calculation, we choose a temperature (T_c') of 2500 K, which lies somewhere in the middle of our enthalpy table, Table 4.4.

2. Again, this is the most difficult step in the available-heat method. Because we are assuming the reaction goes completely to H_2O and H_2 with no side reactions, dissociation reactions, or reactants remaining, this calculation is easy. We simply use the stoichiometric coefficients to represent the chemical composition of the combustion gases.

3. Using Eq. (4.52), we can calculate $Q_{available}$, or Q, using the heats of formation data in Table 4.2. We assume that the reactants are starting out in their standard states, so the only component of $Q_{available}$ is ΔH_{rxn}°:

    $Q = -\Delta H_{rxn}^\circ = -\{[2 \text{ mol } (-241.93 \text{ kJ/mol}) - [3 \text{ mol } (0.00 \text{ kJ/mol})]$
    $- [5 \text{ mol } (-7.03 \text{ kJ/mol}) + 1 \text{ mol } (-9.42 \text{ kJ/mol})]\}$

    $Q = \underline{439.29 \text{ kJ}}$

4. We now use Table 4.4 to calculate how much heat is absorbed by the products (2 mol of water vapor, H_2O, and 3 mol of hydrogen, H_2) at T_c'. This value is Q'.

Table 4.4. Enthalpy Table. Enthalpy required to raise one mole of compound from 298 K to a temperature (T). T is in kelvins, all other values are kJ/mol.

T	H_2	H_2O	CO	CO_2	O_2	N_2
298	0.0	0.0	0.0	0.0	0.0	0.0
400	2.909	3.395	2.926	3.943	3.018	2.918
500	5.831	6.869	5.877	8.246	6.057	5.856
600	8.761	10.448	8.895	12.859	9.222	8.841
700	11.704	14.149	11.980	17.715	11.851	11.888
800	14.659	17.966	15.132	22.776	15.815	14.994
900	17.631	21.910	18.360	28.013	19.230	18.180
1000	20.641	25.987	21.646	33.388	22.692	21.424
1100	23.680	30.185	25.003	38.892	26.280	24.727
1200	26.761	34.522	28.402	44.493	29.867	28.084
1300	29.880	38.980	31.839	50.186	33.446	31.479
1400	33.048	43.530	35.317	55.954	37.034	34.920
1500	36.628	48.219	38.825	61.785	40.621	38.398
1600	39.512	52.995	42.362	67.679	44.338	41.898
1700	42.806	57.858	45.929	73.619	48.059	45.435
1800	46.142	62.815	49.516	79.605	51.777	48.989
1900	49.512	67.838	53.116	85.625	55.494	52.568
2000	52.928	72.937	56.737	91.682	59.215	56.164
2100	56.369	78.106	60.375	97.776	63.045	59.772
2200	59.839	83.339	64.021	103.896	66.867	63.388
2300	63.343	88.622	67.683	110.050	70.697	67.030
2400	66.871	93.959	71.350	116.203	74.519	70.672
2500	70.438	99.346	75.026	122.407	78.349	74.335
2600	74.021	104.780	78.726	128.632	82.293	78.010
2700	77.617	110.251	82.426	134.864	86.236	81.694
2800	81.250	115.768	86.131	141.118	90.179	85.390
2900	84.900	121.310	89.848	147.393	94.126	89.095
3000	88.576	126.899	93.570	153.676	98.065	92.804
3100	92.264	132.520	97.291	159.997	102.071	96.521
3200	95.972	138.163	101.033	166.326	106.090	100.242
3300	99.702	143.839	104.767	172.647	110.125	103.968
3400	103.444	149.536	108.514	179.010	114.177	107.702
3500	107.208	155.267	112.260	185.360	118.242	111.440
3600	110.979	161.019	116.015	191.752	122.311	115.182
3700	114.763	166.791	119.774	198.132	126.409	118.933
3800	118.573	172.593	123.537	204.536	130.511	122.679
3900	122.759	178.403	127.305	210.949	134.638	126.438
4000	126.220	184.226	131.076	217.379	138.778	130.201

T_c'	H_2	H_2O	$3 \times H_2$	$2 \times H_2O$	$Q_{required}$
2500 K	70.538	99.346	211.614	198.692	410.006 kJ

5. We can see that at 2500 K, $Q > Q'$, which means that the assumed temperature (T_c') is too small.
6. We now go back to step (1) with a new T_c'. This time, try 2700 K. Again, because we are assuming that the reaction goes to products as written, our next "working step" is (4):

T_c'	H_2	H_2O	$3 \times H_2$	$2 \times H_2O$	$Q_{required}$
2700 K	77.617	110.251	232.851	220.502	453.353 kJ

We can see that at 2700 K, T_c' is too large, because $Q < Q'$. We have now bracketed the correct temperature (T_c) between 2500 and 2700 K. Assuming the heat-capacity function is linear in this 220 K temperature range, we can do a linear interpolation to calculate T_c. T_c is interpolated as 2635 K. Again, we are assuming that no side reactions or dissociation reactions are occurring. Referring to Table 4.3 at this temperature, we see that small amounts of dissociation occur, lowering the actual temperature of the combustion chamber. However, this is a good initial estimate that is relatively simple to do.

4.4.1 Multiphase Flowfields

Rocket flowfields may certainly consist of more than a gas phase. Small droplets and solid particles may also be present. During the start-up and shutdown transients of liquid rocket engines, it is possible to eject unburned propellant droplets into the flowfield. In solid rocket motors, such as those employing aluminum powder as a fuel, metal-oxide particles can also form as combustion by-products. If the particles or droplets are very small, they have nearly the same velocity as the gas. Only when the droplets or particles attain a certain size (>0.002 mm) do they have a significant effect on performance because of drag and heat transfer [Sutton, 1992]. Computer programs developed for the U.S. Air Force [Hoffman, et al., 1982] employ a method-of-characteristics analysis which includes coupling terms for drag and heat transfer and applies to liquid droplets and solid particles.

4.4.2 Monopropellant Decomposition

The most common method of using a chemically-reacting monopropellant for rockets is to pass the liquid, usually H_2O_2 or N_2H_4, over a catalyst bed, which causes the liquid to thermally decompose. Catalyst beds, like the catalytic converter on an automobile, usually consist of metal-oxide pellets or fine grains, such as aluminum oxide, with a surface-active metal, such as iridium, imbedded into the metal oxide. A catalyst simply speeds up a chemical reaction. At equilibrium, the same K_p relationships hold whether or not we use a catalyst. However, that is the key—at equilibrium. In a hydrazine monopropellant engine, the chemical system never reaches equilibrium. The chemical products depend on a variety of factors, including the particle size in the catalyst bed, the inlet pressure of the propellant, and the temperature of the catalyst bed, all of which affect the rate of the reaction (see Sec. 4.5 on Chemical Kinetics). Semi-empirical methods were developed by Kesten at Hamilton Standard in the 1960s for predicting exhaust products from hydrazine monopropellants.

4.5 Chemical Kinetics: The Speed of the Chemical Reactions

The chemical species of combustion can be handled in three ways:

- The composition of the chemical species does not change from the combustion chamber to the nozzle exit plane. This is known as the *frozen flow* assumption. There are no chemical reactions occurring in such a flowfield.
- As the temperature and pressure change along the nozzle, a new, instantaneous equilibrium condition is achieved for the chemical species. This is known as the *shifting equilibrium* assumption. All chemical reactions are occurring infinitely fast. The chemical composition is different from the combustion chamber to the nozzle exit plane, and at every point in between.
- The flowfield is a chemically reacting gas mixture. Equilibrium may not occur in the flowfield gases. This is known as a *kinetics* assumption. This is essentially somewhere between frozen flow and shifting equilibrium and is the most effective method of predicting performance for rockets.

Chemical kinetics refers to the speed at which chemical reactions occur. To no surprise, the best performance analysis considers chemical kinetics, but this is much easier said than done. Data on rates of chemical reactions is very hard to get and the calculations using kinetics information are only as good as the rate data. According to Timnat [1987], performance losses due to reacting flowfields can range from 0.1% to 10% of the ideal performance calculated based on frozen flow. Kinetic losses can also occur in an engine operating at low pressure [Timnat, 1987]. The bottom line on kinetics: an equilibrium calculation is the most "optimistic" in terms of performance. Any type of kinetics calculation results in a lower performance. As for equilibrium calculations (in fact, even more so!), we want a computer to do the work for us when it comes to chemical kinetics! Almost any general chemistry book (for example, Brown and LeMay, [1981], or Ebbing, [1984]) discusses the basics of chemical kinetics, with more details in books on physical chemistry [Berry, Rice, and Ross, 1980]. Penner [1957] discusses reacting flows in nozzles in much more detail, as do Hoffman, et al. [1982].

Two main mathematical expressions are important to understand in chemical kinetics. The first is known as a *rate law*. For the general reaction ($aA + bB \rightarrow$ products), the rate of the reaction—the change in the concentration of reactants (or products) with respect to time—is proportional to the concentrations of the reacting species raised to some power which may or may not be equal to the stoichiometric coefficients. In other words,

$$\text{reaction rate} = k\,[A]^x[B]^y \quad (4.56)$$

The x, y, and k values are found experimentally. The rate law tells us how fast a particular reactant depletes with respect to time, in terms of the concentration of the reactants. The proportionality constant k is known as the *rate constant* and is expressed in our second kinetics mathematical expression as

$$k = A \exp(-E/\mathcal{R}\,T) \quad (4.57)$$

where A is the *frequency factor* representing how often the molecules or atoms are colliding in the reaction. E is the *activation energy*, which is the minimum amount of energy required to initiate a chemical reaction. $\mathcal{R}$ is the gas constant [see Eq. (3.29)] and T is the absolute temperature. Semi-empirical models exist which account for molecular speed, angle of collision, and many other variables. However, good kinetics information for complex systems depends on good experimental data. Kinetics calculations for rocket propulsion systems are limited by the amount of kinetics data available for the chemical system and are not done in most analyses.

4.6 Combustion of Liquids vs. Solids

Both Sutton [1992] and Timnat [1987] give excellent, detailed discussions on the differences between combustion processes in liquid- and solid-propellant rockets. The chemical principles we have discussed (equilibrium, kinetics, and thermodynamics) apply equally well for liquid and solid systems. The differences between liquid and solid combustion are really more engineering differences than chemical ones. In this section, we highlight only some of these differences.

With a liquid-propellant system, we are looking at spray combustion, which involves understanding how droplets burn. Fortunately, this process is well understood and was originally developed in the 1950s. Timnat [1987] discusses several computer models which take into account droplet concentration, density, momentum, and thermal energy while including similar gas-phase conservation equations. The U.S. Air Force developed a computer program in the late 1970s and early 1980s called the *Transient Performance Program* (TPP), which does an excellent job of predicting performance of liquid-bipropellant rocket engines.

In a solid rocket motor, the particle size, metal content, and orientation of the burning surface to the acceleration vector can all affect the burning rate of a solid propellant [Timnat, 1987]. It is also important to consider the mechanical properties of the propellant, which depend on

- The nature of the polymeric binder
- The concentration, size, and shape of the solid fuel and oxidizer particles
- The strength of the bond between the binder and the fuel/oxidizer particles [Siegel and Schieler, 1964]

Probably the best performance-prediction program for solid rocket motors is the *Solid Performance Program* (SPP), completed for the U.S. Air Force in the early 1980s. SPP includes models which account for two-phase flow, chemical kinetics, boundary layer losses, combustion efficiency, nozzle-throat erosion, and losses from impinging particles [Siegel and Schieler, 1964].

4.7 Propellant Characteristics and Their Implications

By far the easiest information to obtain about chemical propellants is their physical and chemical properties. One of the best early books on this topic is by Kit and Evered [1960]. From an engineering standpoint, we can get a lot of information from the physical and chemical properties of propellants, which should precede and may preclude a detailed calculation of their performance. This section discusses some of the major propellant characteristics to consider when choosing a propellant. In addition to Kit and Evered [1970], Siegel and Schieler [1964] and Sutton [1992] are excellent sources of information on this topic. Appendix B shows some properties for the more popular propellants.

Propellants as Hazardous Materials—To no surprise, many propellants are hazardous to handle. Besides the obvious hazards of being capable of causing explosions or fires under poor handling or storage conditions, propellants have other physical hazards which should concern us. Many propellants, such as those in the hydrazine family, are toxic to breathe and touch and must be handled with care. Regulations govern the "cradle to grave" handling of wastes generated when using such materials. We must also be aware of the materials compatibility of propellants when choosing tankage, piping, seals, and motor-casing materials for the rocket.

Vapor Pressure—When a propellant has a high vapor pressure at a particular temperature, an appreciable amount of the liquid (or solid) converts into a gas at that temperature. At the boiling point of a liquid, the vapor pressure becomes equal to the atmospheric pressure. Water, for example, has a vapor pressure of less than 7000 Pa at room temperature. Nitrogen tetroxide (N_2O_4), on the other hand, boils just at room temperature. At the boiling point, a liquid's vapor pressure equals the atmospheric pressure. Cryogenic propellants (very low boiling points) such as H_2 and O_2 must be stored in thick-walled, insulated tanks to keep the vapor pressure at a reasonable value. If a toxic propellant has a high vapor pressure, toxic vapors can more easily escape during handling procedures. From an engineering standpoint, propellants with a high vapor pressure cause cavitation in the pumps, leading to pressure oscillations and combustion instability. On the "good" side, these propellants can self-pressurize a tank. For solid rocket motors, a plasticizer with a high vapor pressure may cause shrinking and cracking of the propellant grain. In usual terms, propellants with low vapor pressures are "good;" propellants with high vapor pressures are "bad."

Heat-Transfer Properties—The heat-transfer properties of propellants are important to consider when an engine must be regeneratively cooled. Compounds with a high specific heat make good choices for regenerative cooling. Liquid hydrogen has a specific heat of 7320 J/kg·K at 20 K, compared to water's specific heat of 4217 J/kg·K at 273 K. It is very unusual for compounds to have specific heats greater than 4000 J/kg·K. The propellant, of course, must also be chemically stable over the temperature range encountered during regenerative cooling.

Viscosity—Viscosity is an important consideration when regeneratively cooling an engine. It is also a crucial variable when designing any pumping system. For comparison, water has a viscosity of 1.0 centipoise near its freezing point, whereas liquid hydrogen has a viscosity of 0.024 near its boiling point, and RP-1 has a viscosity of 0.75 near room temperature [Sutton, 1992]. The more viscous the liquid, the more difficult it is to pump. Viscosity is normally not a big consideration in propellant choice; however, it is an important design parameter to know!

Density—The density plays a critical role in the design of propellant tanks. Quite simply, the denser the propellant, the more mass of that propellant we can put in a given volume of tankage. Density is also referred to as *specific gravity*, which is the measurement of density relative to water (1.00). RP-1 has a specific gravity of 0.8 near room temperature. Liquid hydrogen has a very poor rating in this category, with a specific gravity of 0.07 near its boiling point [Sutton, 1992].

Chemical Reactivity—In solid rocket motors, the oxidizer and fuel components must be chemically compatible during storage. In liquid engines, the reactivity factor becomes important during the mixing of fuel and oxidizer droplets in the combustion chamber. *Hypergolicity*, spontaneous combustion upon mixing, is also a factor to consider. Hypergolic propellants eliminate the need for a separate ignition system. Monomethylhydrazine (CH_3NHNH_2) and nitrogen tetroxide (N_2H_4), used on the Space Shuttle Orbital Maneuvering System, is a hypergolic fuel/oxidizer combination. Hydrogen and oxygen, used on the Space Shuttle main engines, is not a hypergolic system.

Heat of Formation—A very telling piece of information about a propellant at first glance is its heat of formation, a familiar thermodynamic variable from Sec. 4.2. Because heats of reaction are calculated from "products – reactants," we want the heats of formation of the reactants (propellants) to be **as positive as possible**. The more positive the heats of formation for the reactants, the more negative the "products – reactants" are and the more energy the reaction gives off. In the world of chemistry, propellants with positive heats of formation are unusual. All of the hydrazine family fit the bill, along with hydrogen peroxide. Some of the more exotic fuels and oxidizers, such as pentaborane and oxygen difluoride, also fare well with positive heats of formation.

4.8 Key Thermochemical Parameters: The Bottom Line

The bottom line for thermochemical analysis is obtaining the variation of three parameters as a function of the input propellant proportions. The common parameter used to define the propellant proportions is the *oxidizer-to-fuel ratio* (O/F). This is defined as the ratio of the oxidizer mass flow rate ($\dot{m}_{ox}$) over the fuel mass flow rate ($\dot{m}_{fuel}$):

$$O/F = \frac{\dot{m}_{ox}}{\dot{m}_{fuel}} \quad (4.58)$$

and only applies to bipropellant systems.

The three parameters commonly used to describe the thermochemical properties of the combustion exhaust gases are

- *Isentropic parameter*—or the ratio of specific heats (C_p/C_v)
- *Molecular mass*—this is the mass of one mole of the exhaust gases
- *Flame temperature*—or combustion chamber temperature. This is the total temperature (Mach number = 0) due to the chemical reaction.

From these three parameters, we can determine all of the other thermodynamic parameters such as characteristic velocity, thrust coefficient, and specific impulse.

As mentioned in Sec. 4.5, the values of these parameters vary depending on our assumptions for analysis: frozen flow, shifting equilibrium, or chemical kinetics. Remember, the frozen-flow assumption gives us a conservative estimate of specific impulse; shifting equilibrium predicts a specific impulse that is better than we can achieve, and chemical kinetics is somewhere in between. However, we actually have more decisions to make:

- Where do we freeze the flow—in the combustion chamber ($M = 0$), in the throat ($M = 1$), or elsewhere?
- If the equilibrium is shifting, then the isentropic parameter and molecular mass are changing through the thrust chamber. How do we determine a single value?

Figure 4.4 shows how this works for the isentropic parameter (propellants are solid polybutadiene and liquid oxygen). The top curve is γ frozen in the chamber. The next lowest curve shows results for freezing the flow at the throat. The bottom

4.8 Key Thermochemical Parameters: The Bottom Line

curve shows results for shifting equilibrium. To determine the isentropic parameter, we use the isentropic flow relation [Eq. (3.68)]:

$$\frac{T_e}{T_c} = \left(\frac{p_e}{p_c}\right)^{\frac{\gamma-1}{\gamma}} \qquad (4.59)$$

where T_e = static temperature at nozzle exit (K)
T_c = total temperature ($M \ll 1$) in the combustion chamber (K)
p_e = nozzle exit static pressure (Pa)
p_c = combustion chamber total pressure (Pa)
γ = equilibrium flow isentropic parameter
   $\neq c_p / c_v$
c_p = constant pressure specific heat (J/kg·K)
c_v = constant volume specific heat (J/kg·K)

from Chap. 3. If we had taken a chemical-kinetics approach, the result would lie somewhere in the shaded area.

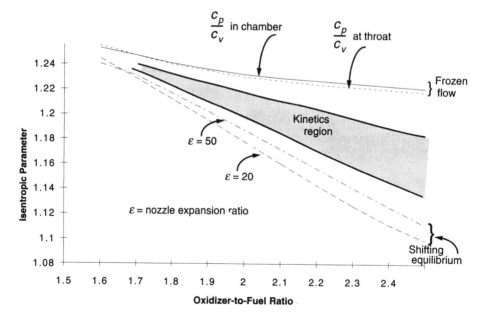

Fig. 4.4. Isentropic Parameter vs. Oxidizer-to-Fuel Ratio Based on Various Analysis Assumptions. The fuel is a solid polybutadiene and the oxidizer is liquid oxygen. Shifting equilibrium predicts γs that are much lower than the frozen-flow assumption. Shifting equilibrium γs vary depending on nozzle geometry whereas frozen flow fixes the value of γ at the combustion chamber.

For conceptual design, we recommend using the frozen-flow assumption, freezing the flow at the throat [all rockets have a throat ($M = 1$) but we have not yet determined the expansion ratio]. When we have finally made the basic decisions on thrust chamber geometry, we usually assume shifting equilibrium (the industry's standard approach). However, as Fig. 4.4 indicates, a more accurate solution is to average results from analyses using frozen flow and shifting equilibrium.

Appendix B gives thermochemical results for several typical propellant combinations.

References

Berry, R. S., S. A. Rice, and J. Ross. 1980. *Physical Chemistry*. New York: John Wiley & Sons.

Brown, Theodore L., and H. Eugene LeMay, Jr. 1981. *Chemistry: The Central Science*. 2nd ed. Englewood Cliffs, NJ: Prentice-Hall, Inc.

Ebbing, Darrell D. 1984. *General Chemistry*. Boston: Houghton Mifflin Company.

Goodger, E. M. 1975. *Hydrocarbon Fuels*. New York: John Wiley & Sons.

Gordon, S. and B. J. McBride. 1976. *Computer Program for Calculation of Complex Chemical Equilibrium Compositions, Rocket Performance, Incident and Reflected Shocks, and Chapman-Jouguet Detonations*. Interim Revision. NASA SP-273. Washington, DC: National Aeronautics and Space Administration.

Hoffman, R. J., M. A. Hetrick, Jr., G. R. Nickerson, and F. J. Jarossy. 1982. *Plume Contamination Effects Prediction, CONTAM III Computer Program: Volume II, Contaminant Transports and Chemical Kinetics*. AFRPL TR-82-033. U.S. Air Force Rocket Propulsion Laboratory.

Kit, Boris, and Douglas S. Evered. 1960. *Rocket Propellant Handbook*. New York: The Macmillan Company.

McBride, B. J. and S. Gordon. 1967. *Fortran IV Program for Calculation of Thermodynamic Data*. NASA TN D-4097. Washington, DC: National Aeronautics and Space Administration.

Penner. S. S. 1957. *Chemistry Problems in Jet Propulsion*. New York: Pergamon Press.

Siegel, Bernard, and Leroy Schieler. 1964. *Energetics of Propellant Chemistry*. New York: John Wiley & Sons, Inc.

Straub, Dieter. 1989. *Thermofluiddynamics of Optimized Rocket Propulsions: Extended Lewis Code Fundamentals*. Basel, Switzerland: Birkhauser Verlag.

Sutton, George P. 1992. *Rocket Propulsion Elements*. 6th ed. New York: John Wiley & Sons, Inc.

Timnat, Y. M. 1987. *Advanced Chemical Rocket Propulsion*. London: Academic Press.

Wilkens, Roger L. 1963. *Theoretical Evaluation of Chemical Propellants*. Englewood Cliffs, NJ: Prentice-Hall, Inc.

CHAPTER 5
LIQUID ROCKET PROPULSION SYSTEMS

Ronald W. Humble, *U.S. Air Force Academy*

David Lewis, *TRW, Inc.*

William Bissell, *Rockwell International, Rocketdyne Division (ret.)*

Robert Sackheim, *TRW, Inc.*

- 5.1 History
- 5.2 Design Process
- 5.3 Preliminary Design Decisions
- 5.4 System Sizing, Design, and Trade-offs
- 5.5 Case Study

Liquid Rocket Propulsion Systems (LRPSs) are the most popular form of rocket propulsion when relatively high specific impulse and high thrust levels are required. This popularity has encouraged research and development in many of the aspects of LRPSs, bringing our capabilities to a high level of readiness. As a result, we have many options for getting the "best" performance from a system. But sorting through these options can be a bit daunting for the uninitiated, or even for an experienced designer. To get us started, we introduce the basic layout of an LRPS and discuss its main features.

Figure 5.1 shows a simplified layout of a typical bipropellant LRPS, including boxes describing the operation of the system and some key considerations for components or subsystems. Figure 5.2 presents similar information for a monopropellant system. Figure 5.3 shows the schematic for a cold-gas thruster system. Although this system is not strictly an LRPS, much of the technology is very similar. Section 3.6 evaluates the performance of a cold-gas thruster and sizes it.

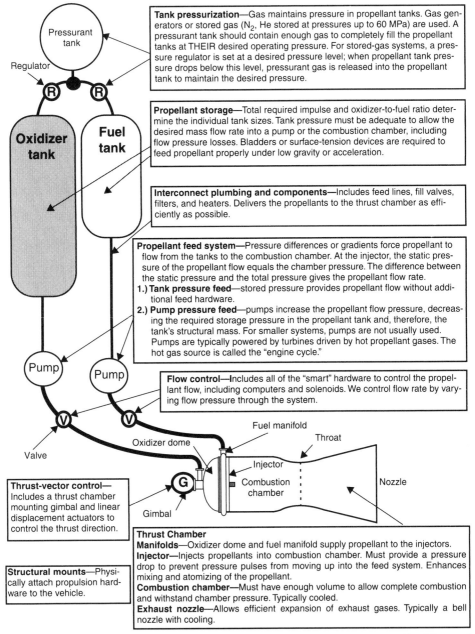

Fig. 5.1. **Operation of a Bipropellant Liquid Rocket Propulsion System.** Liquid oxidizer and liquid fuel feed into the thrust chamber where they mix and react chemically. The hot combustion gases then accelerate and are exhausted through a converging-diverging nozzle.

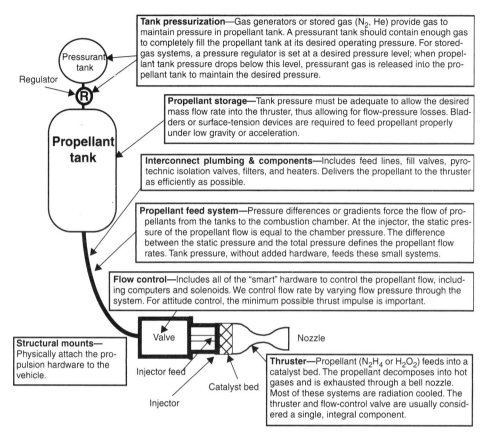

Fig. 5.2. Operation of a Monopropellant Liquid Rocket Propulsion System. A single liquid propellant feeds into a catalyst bed in the thruster. The propellant catalytically and thermally decomposes to create a hot gas that is exhausted through the nozzle.

Liquid Rocket Propulsion

LRPSs offer two advantages over the other types of chemical propulsion. They

- Generate the highest demonstrated performance (specific impulse) of any rocket system using chemical combustion
- Can precisely control impulse delivery and energy management, including:
  - throttling
  - pulse-mode operation (thousands of highly repeatable and accurate pulses) for precise control of a spacecraft's attitude
  - multiple starts and stops with extremely long off times between burns

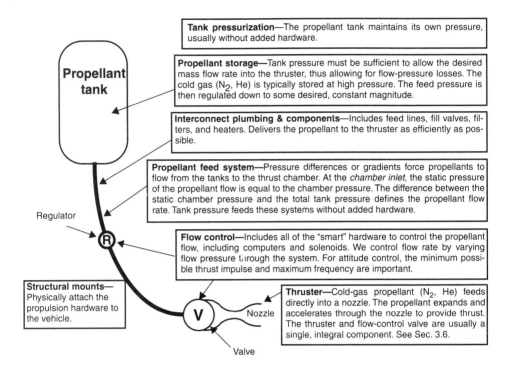

Fig. 5.3. Operation of a Cold-Gas Thruster System. A high-pressure gas passes through a nozzle and accelerates to create thrust. We do not discuss cold-gas systems specifically in this chapter, but many of the concepts still apply. Chapter 3 contains an example design for a cold-gas thruster.

Table 5.1 lists many of the advantages and disadvantages of liquid systems. We must track these capabilities against those of solid rockets and other propulsion systems to select the best candidate for each application.

Table 5.2 summarizes the flight applications for liquid propulsion systems. It shows that some form of liquid propulsion applies to any phase of space flight. Applications requiring large impulses (ascent, orbital transfer, and upper stages) typically use higher-performance (specific impulse) propellants with a pump-pressure-feed system to reduce the propellant tank's mass and reduce engine size. Lower-impulse applications, such as perigee or apogee kick motors and orbit-adjust systems, may or may not use pumps. The choice typically depends upon whether we can lower system mass, lower the parts count, or lower the cost. For applications needing even smaller impulse (stationkeeping and attitude control), system simplicity takes precedence over specific-impulse performance: we use tank pressure to feed low-specific-impulse propellants, such as monopropellant hydrazine, into the thruster. Electrically heated systems use electric heating (see

Chap. 9, electrothermal rockets) to increase the propellant's exhaust velocity and thereby the specific impulse. These systems have relatively low thrust and typically apply to orbit-maintenance applications.

Table 5.1. Considerations for Liquid Rocket Propulsion Systems. There are many advantages and disadvantages with liquid rockets. We must decide when advantages outweigh disadvantages.

Advantages	Disadvantages

• Highest achievable specific impulse for chemical rocket propulsion
• Controllability
  - On/off control
  - Easily restarted
  - Throttleable
  - Pulse mode (up to a million pulses demonstrated)
• Easy to package, enabling us to control and change a vehicle's center of gravity
• Amenable to realistic ground testing and pre-launch checkout
• Lower total system cost in many cases
• Higher total system reliability in many cases
• Can have little effect on the environment
• Enable component redundancy to virtually any level
• Most have relatively clean exhaust
	• Lower propellant density than solid rockets
	• Higher parts count than other chemical systems
	• More complicated in some cases
	• Concerns about leakage and associated system safety or personnel hazards
	• Concerns about propellant volatility and boil off
	• Concerns about combustion instability
	• Difficult to control propellant flow
	• Less reliable for some applications
	• Can be expensive for some applications
	• Need mixture ratio control
	• Need special design for a zero-g start

Table 5.2. Applications for Liquid Rockets. Different system types have traditionally applied to different missions. This table shows the breakdown. Tank-pressure-fed systems for launch vehicles usually apply to upper stages. A monopropellant orbital insertion system was used when Viking arrived at Mars.

System Type	Applications			
	Launch	Orbit Insertion	Orbit Maintenance and Maneuver	Attitude Control
Pump-pressure-fed bipropellant system	X	X		
Tank-pressure-fed bipropellant system	X	X	X	X
Monopropellant		X	X	X
Electrically heated monopropellant (Chap. 9)			X	
Cold gas (Sec. 3.6)				X

System Configuration

The operation schematics of Figs. 5.1 through 5.3 show an LRPS's eight major subsystems: propellant feed, propellant storage, tank pressurization, propellant-flow control, interconnect plumbing and components, thrust chamber, thrust-vector control (TVC), and structural mounts. We can group the first five subsystems into a *propellant handling system*. The following paragraphs briefly discuss each subsystem.

Propellant Feed

Historically, liquid systems have been categorized by the way propellant feeds from the propellant tanks to the thrust chamber. The two basic propellant-feed categories are

- Tank-pressure-feed (Fig. 5.4)
- Pump-pressure-feed (Fig. 5.5)

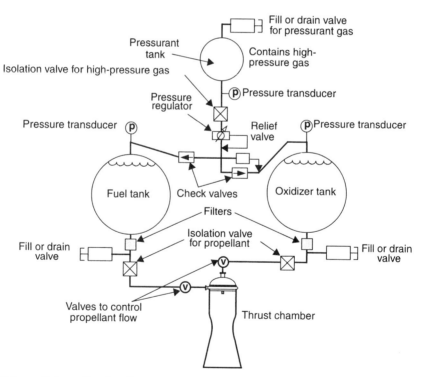

Fig. 5.4. Schematic of a Tank-pressure-fed Liquid Rocket. Pressure-regulated gas (pressurant) is used to keep the propellant tanks at their desired pressure. Tank pressure then forces the propellant into the thrust chamber as desired.

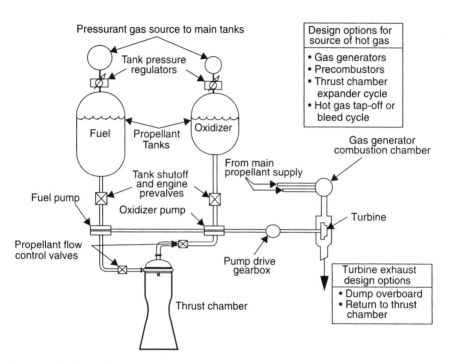

Fig. 5.5. Schematic of a Liquid Rocket with Turbopump-pressure Feed. The major differences between this and a tank-pressure-fed system are the pumps and pump drives. This figure illustrates a gas generator being used to drive the turbine.

Although both categories are "pressure-fed systems," the *tank-pressure-fed* category implies that the energy required to deliver propellant into the thrust chamber is added prior to flight by pressurizing the propellant tanks, so that propellant feeds to the thrust chamber under the influence of prestored pressure.* In contrast, the *pump-pressure-fed* system typically uses the chemical energy of the propellant to power a pump, which in turn pressurizes the propellant before feeding it into the thrust chamber. In practice, we call tank-pressure-fed systems simply *pressure-fed* and refer to pump-pressure-fed systems as *pump-fed*.

Distinguishing between pressure-fed and pump-fed systems trades complexity against system mass. Pressure-fed systems typically require heavier propellant

---

* In general, this is not strictly true. Several propulsion systems rely on other mechanisms for generating tank pressure at later times in a particular mission. For example, externally stored gas, gas generators (from either solid or liquid reactants), or tank heaters can raise tank pressure before engine ignition, to name just a few possibilities.

tanks than pump-fed because of higher storage pressures. On the other hand, pump systems are more complex than pressure-fed systems and can be heavier for lower-impulse systems. We must trade the relative merits of each to design a particular system. Pump-pressure-fed systems are usually used when large thrust impulses and large propellant tanks are required, as in launch-vehicle applications. If a relatively small impulse is required, the penalty for using heavier tanks for a tank-pressure-fed system can be justified.

The pump-fed system consists of two major components: the pump and the pump driver. Typically, the pump uses a centrifugal-flow configuration having an axial-flow inducer, with a very low rise in pressure, placed in front of the centrifugal pump, which has a much higher rise in pressure. In addition, recent work done with piston pumps for medium-impulse systems, such as orbital transfer vehicles, shows promise [Whitehead, 1994].

The most commonly used type of pump driver is a turbine powered by hot gas, which can be supplied from several sources:

- Unreacted gases or liquids heated up while cooling the thrust chamber (expander cycle)
- Hot combustion gases from a separate combustion chamber (gas generator)
- Hot combustion gases generated from a combustor ahead of the main combustion chamber (staged combustion or precombustion)
- Hot combustion gases taken from the main thrust chamber (bleed flow)

Each of these approaches has its advantages and disadvantages, all of which are discussed in Secs. 5.3 and 5.4.

Propellant Storage

The *propellant storage* system consists of the propellant-tank structure and the propellant-expulsion assembly. The tank structure consists mainly of the pressure vessel, the structural attachments, propellant slosh baffles, and thermal insulation. Typical tank materials are aluminum, stainless steel, and metal-lined, fiber-reinforced composite materials.

One of the more difficult problems to solve in using stored propellant is expelling it. Under positive gravity or thrust, acceleration can force propellants to a desired expulsion point. However, under low gravity—or unusual accelerations—we need a positive expulsion method, such as a collapsible bladder, piston, or surface-tension device.

Tank Pressurization

The *tank pressurization* system ensures that the propellant tanks maintain their desired pressure. As propellant is consumed, the pressure of the propellant tank drops unless a pressurant gas makes up the volume. A pressurant tank contains "pressurant" gas, typically helium (He) or nitrogen (N_2), at a very high pressure (up to 60 MPa). This gas is regulated down to the desired tank pressure and fed into the propellant tank. A pressurizing method is particularly important for pressure-fed systems but is also used for pump-fed systems. To maintain propellant-tank pressure, we may also use

- Electric heaters to heat the propellant and increase vapor pressure
- An external chemical reaction to generate pressurant gas
- High-pressure gases fed from the combustion chamber into the tanks
- A heat exchanger to heat some of the propellant

Stored gas pressurant is routine for spacecraft but is rarely used for boosters.

Propellant-Flow Control

This system includes on/off valves, directional flow or check valves, and transducers, as well as the computers necessary to command this hardware. This system is distinct from the interconnect plumbing in that it includes only the "smart" components in the flow system. The flow-control system also controls the propellant-feed system by affecting such parameters as pump speed.

Interconnect Plumbing and Components

The purpose of this system is evident from its name. It consists of all the feed lines, fill or drain valves, and other similar components required for propellant handling but excludes the flow-control components.

Thrust Chamber

The *thrust chamber* generates thrust by providing a volume for combustion and converting the thermal energy to kinetic energy. The thrust chamber is made up of

- *Propellant injectors and feed manifolds*—to feed the propellants into the combustion chamber. The injectors atomize the propellants as much as possible and spray them into the combustion chamber in a pattern that helps the propellant to mix and burn.
- *Igniter*—often, but not always, required to start combustion. After combustion begins, we do not need it again unless restarting is necessary. *Hypergolic propellants* ignite on contact with each other, so they do not need an igniter. Hypergolics are well suited for spacecraft applications requiring multiple thrust pulses. Typically, monopropellant thrusters react on contact with a solid catalyst. Although a heater may preheat the catalyst, igniters are not usually required.
- *Combustion chamber*—the area where most of the chemical reaction occurs. It must be long enough to allow complete combustion and wide enough to keep the mass flow velocity low. The chamber usually requires some kind of cooling to keep the high temperatures from damaging its walls.

- *Exhaust nozzle*—accelerates the exhaust gases up to high velocity and low pressure. Typical configurations include the conical, deLaval nozzle discussed in Sec. 3.3.3 or the bell nozzle discussed in Sec. 5.4.1. The nozzle normally requires cooling, particularly in the throat area, to avoid overheating the structure.
- *Structural cooling*—necessary to keep the structure below its maximum allowable temperature. Usually, the exothermic chemical reaction creates temperatures that damage most materials. Common cooling approaches include ablatives, radiative cooling, channel cooling, and film cooling.

Thrust-Vector Control (TVC)

TVC maintains the vehicle's correct attitude for the thrust duration by rotating (gimballing) the thrust chamber or by redirecting the exhaust-gas flow so the thrust generates a vehicle torque. Common methods for directing the exhaust gas are moveable vanes or injection of another fluid into the flow. TVC is typically used to control pitch and yaw attitude on boosters and upper stages (high-thrust systems). Spacecraft use dedicated attitude-control thrusters in lieu of TVC.

Structural Mounts

As the name suggests, this equipment includes all of the additional structure required to mount the LRPS systems to each other and to the vehicle.

Propellants

There are several levels of classification for propellants. At the first level we distinguish between monopropellants and bipropellants. *Monopropellants* feed into a chamber where they exothermically react or decompose by themselves, usually in the presence of a chemical catalyst. Monopropellant systems are undoubtedly the simplest form of rocket, particularly from an operational standpoint. The most popular monopropellant today is hydrazine (N_2H_4, a fuel). However, this propellant has a relatively low specific impulse (230 seconds), so we do not use it when we need large impulses. In the past, hydrogen peroxide (H_2O_2, an oxidizer) was also used as a monopropellant. However, it has even lower specific impulse (150 seconds) than hydrazine.

The liquid propellants that give the highest performance are bipropellants.[*] *Bipropellants* are fuel and oxidizer combinations. We select bipropellants to get the best possible specific impulse. As shown in Chap. 3, the smaller the exhaust-gas molecules, the higher the specific impulse. Higher specific impulse means we need less total propellant mass and, usually, a lower overall mass for the vehicle. This

---

[*] There are some advantages in using even more than two propellants. For example, tripropellant systems using LOx, H_2, and RP-1 have been investigated. By varying the proportions of the fuels (H_2 and RP-1), we can reduce a vehicle's overall mass.

may not always be the case. For example, hydrogen and oxygen provide a good specific impulse. However, the hydrogen molecule is so light that a given mass of hydrogen requires an extremely large and relatively massive storage tank.

Another important consideration in choosing a propellant combination is its storability. For launch vehicles, storability is not usually a problem, so hydrogen/oxygen cryogenics are often used. However, if the propellants must be stored for a long time, as in a long-duration mission in space, cryogenics may be a poor choice compared to *space storables*. These storables include propellants that can be stored at much higher temperatures, such as the hydrazines.

Toxicity of the propellants is another consideration. For example, the hydrogen/fluorine reaction provides higher specific impulse than hydrogen/oxygen. However, fluorine is HIGHLY toxic and HIGHLY corrosive. We can deal with this problem, but unfortunately, even the reaction products (HF) are toxic. Most propellants are toxic, including oxygen in many forms. The question to consider is the degree of toxicity we are willing to accept.

5.1 History

Modern liquid rocketry began in the United States in 1909 with the well-publicized work of Dr. Robert H. Goddard. Early workers in other countries, such as the deaf Russian schoolmaster, Konstantin Tsiolkovsky, and the German scientist and high-school teacher Hermann Oberth, also contributed much to the field.

Liquid rockets came into serious consideration as practical, high-speed propulsion devices around the beginning of the 20th century. As these systems evolved from theory to experimental demonstrations, the early rocket pioneers began earnestly to advocate their use for military weapons. During World War II the German V-2 rocket was the first—and most infamous—application of liquid propulsion and the first successful ballistic missile. Since then liquid-rocket technology has evolved, and the world has used it to explore and exploit space.

After World War II, liquids were first developed in the United States for the original family of intercontinental- and intermediate-range ballistic missiles (ICBMs and IRBMs). These earliest long-range, liquid-propellant missiles became known as the Redstone, Thor, Jupiter, Juno, Atlas, and Titan. These missiles gradually evolved into the United States' family of expendable space launch vehicles. The United States has developed two liquid launch vehicles—the Saturn and the Space Shuttle—expressly for manned launch systems. The final version of Saturn (Saturn-V) used five of the largest liquid rocket engines ever flown (Rocketdyne's F-1, using LOx and kerosene) for its first stage. It also used a high-thrust liquid-oxygen/liquid-hydrogen engine (the first LO_2/LH_2 engine to be "man-rated"), the Rocketdyne J-2, for its upper stages. The Space Shuttle orbiter's main engines also use LO_2/LH_2 with the basic designation of SSME (Space Shuttle Main Engine). In addition to being the best-performing (highest specific impulse) chemical rocket engine in the world today, the SSME is both man-rated and reusable.

Similar periods of development of liquid rockets, first for military purposes and then for space launchers, occurred in the Soviet Union (USSR) after World War II. The USSR had a large inventory of liquid rockets to use as military and space launchers. Other countries of the world, including Great Britain, France, Italy, Canada, Sweden, Japan, China, and India, also went on to develop liquid rocket propulsion during the years after World War II. Table 5.3 is a chronology of significant U.S. achievements in liquid propulsion.

Table 5.3. **Highlights in the Development of Liquid Propulsion in the United States.** The table lists the salient, historical milestones in the advancement of liquid rockets. Developments in other countries are similar.

Program	Flight Period	Propulsion Development
V-2	1942–1950	• Largest engine for its time • Earliest production liquid rocket (more than 800 were built) • Thrust = 249,000 N • Developed in Germany, but brought to U.S. following World War II
Jet Assisted Takeoff (JATO)	1942	• First use of a hypergolic propellant combination (red fuming nitric acid with aniline)
Bumper - WAC	1948–1950	• First system with a large second stage
X-15	1959–1968	• Ammonia/liquid oxygen XRL-99 propulsion
Earth to Orbit		
Redstone	1953–1958	• First large liquid rocket engine • Upgrade of the V-2 • Thrust = 311,000 N
Atlas	1957–present	• First U.S. ICBM • 1-1/2-stage vehicle • Pressure-stabilized airframe • Large tubular engines, 1,734,800-N thrust • First-stage walls (vehicle structure) served as tank walls
Titan II	1962–present	• Large, tubular, first- and second-stage engines • Storable N_2O_4/Aerozine-50 propellant • First-stage thrust = 2,000,000 N • Second-stage thrust = 400,000 N
Saturn I	1963–1965	• Eight H-1s, thrust = 800,000 N per engine, in first stage • Six LOx/LH_2 engines (RL-10) clustered in second stage
Saturn IB	1966–1975	• First high-thrust LH_2 engine • One J-2 on second stage, thrust = 1,023,000 N • Variable-thrust (throttled) cryogenic engine
Saturn IC	1967–1968	• Very large cluster of liquid rocket engines in first stage • 6,050,000-N thrust for five engines on first stage

Table 5.3 Highlights in the Development of Liquid Propulsion in the United States. (Continued)

Program	Flight Period	Propulsion Development
Saturn V	1967–1973	• Five biggest engines to date in first stage (F-1) • Thrust = 6,700,000 N per engine • Five J-2 engines clustered in second stage
Space Shuttle Main Engines	1981–present	• Reusable space engine • High chamber pressure (P_c = 22 MPa) • High turbomachinery power density (164 kW/kg) • Staged combustion cycle
Upper Stage and Spacecraft		
ABLE 4, 5	1959–1960	• First use of monopropellant hydrazine in space
Gemini	1964–1966	• Ablative RCS engines • First RCS to use a blend of bipropellant N_2O_4/hydrazine
Centaur	1963–present	• Large-scale use of hydrogen • Integral LOx/LH$_2$ tankage (common bulkhead tanks) • High specific impulse (444 s) expander cycle engine
Apollo Program	1966–1975	• RCS in service module: radiation-cooled N_2O_4/Aerozine 50 • Ablative-radiation engine with up to 50 starts (Apollo command and service modules, CSM) • Highly throttleable (10:1) engine for lunar descent • Supercritical helium pressurization in lunar module (Lunar excursion module, LEM)
Viking	1975–1976	• Throttleable engine using monopropellant hydrazine
Agena	1959 to present	• Zero-g retention devices in feed tank • Multistart IRFNA/UDMH bipropellant combination (IRFNA regeneratively cooled) • Small 2-N thrust hydrazine monopropellant engines for ACS
Shuttle RCS	1981 to present	• Tank feed system using a single, low-g/high-g propellant • Largest pulse-modulated N_2O_4/MMH RCS (3870-N thrust)
Space Shuttle Orbital Maneuvering Engine	1981 to present	• High-performance, pressure-fed engine for on-orbit maneuvering of the Space Shuttle. This engine is rated to support manned missions. Also reusable and Earth-storable.
Intelsat V	1980 to present	• First flight of electrothermally augmented thrusters using monopropellant hydrazine (1994)
Commercial COMSATS	1980 to present	• Storable, bipropellant, liquid apogee engines and RCS-class, storable, bipropellant thrusters. Produce high performance with low thrust (8-40 N).
Advanced high power, high performance COMSATS	1991 and beyond	• Hydrazine arcjets and other electric propulsion devices • Dual Mode (N_2O_4/N_2H_4) flight propulsion system (1991)

5.2 Design Process

We can establish a baseline design for an LRPS by thinking of it as an input/process/output methodology, presented graphically in Fig. 5.6. The elements are:

- **Inputs**—requirements provided to the propulsion system from the mission-level analysis (see Sec. 1.2). Propulsion performance requirements derive from the mission requirements and typically consist of the Δv required to meet the terminal velocity for each phase of the mission (boost phase, orbit transfer, stationkeeping, interplanetary); attitude control for the vehicle; and orientation accuracy and lifetime. In addition there are requirements or constraints on cost, limiting technology risk, geometric envelope, safety, operational environment, the "—ilities" (such as reliability, maintainability, manufacturability), and schedule.

- **Outputs**—include a baseline design configuration, performance estimate, operating sequence and timeline, an equipment list, plus estimates of mass, cost, reliability, and lifetime.

- **Preliminary design process**—has several parts, shown in Fig. 5.6. *Preliminary design decisions* (Sec. 5.3) incorporate the most fundamental decisions, including propellant selection and engine type. We base these selections on trades among potential candidates and evaluate them against the requirements for impulse, mass, and geometry requirements "input" to this part of the process. During this phase, we decide on the propellants, number and type of engine(s), engine operating conditions (primarily chamber pressure and oxidizer-to-fuel ratio), and whether to "make the engine or buy it." Using the selected propellants and engine parameters, we can then size, design, and do trades on the system, as well as do preliminary designs of the subsystems (*system sizing, design and trade-offs*: Sec. 5.4). Major subsystems are propellant feed, propellant storage, tank pressurization, and flow control. They also include interconnect plumbing and components, thrust chamber, thrust-vector control, and structural mounts. We iterate the sizing and design within and between subsystems to ensure that the LRPS is properly integrated and can meet its requirements. This is an "end-to-end" evaluation and incorporates previous flight data.

A "good" preliminary design sets the stage for a successful program, and a good design process can prevent potential pitfalls that may scuttle the mission. Early in the program we need to identify and correct these potential pitfalls, which include requirements that cannot be met at reasonable "cost" (in money, schedule, or risk). If we discover them during preliminary design, before hardware is built, we can negotiate them with little effect on the program. If the "stressing" requirements cannot be relaxed, we can at least understand the cost implications early on.

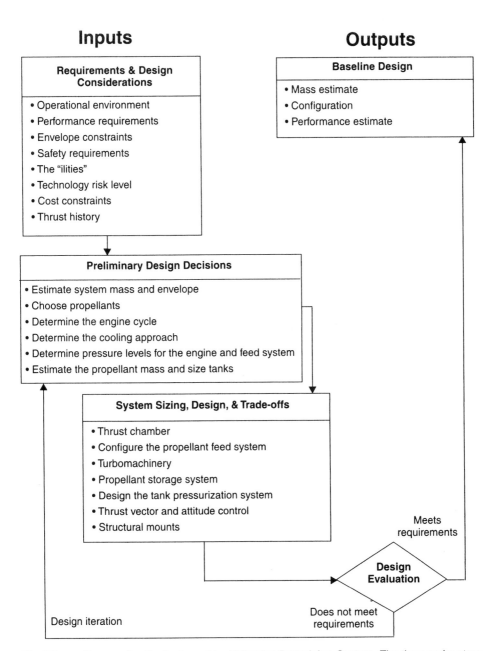

Fig. 5.6. Process for Designing a Liquid Rocket Propulsion System. The three major steps are shown. We leave out the many possible iteration paths to avoid cluttering the figure.

5.3 Preliminary Design Decisions

Before we can start designing the individual components, we must make some decisions that affect the overall configuration and design approach. Table 5.4 outlines them. Typically, many of these decisions are not "deterministic," so a certain amount of estimating or "guessing" is required. Sometimes the initial estimates are good, sometimes not. This leads us to an iterative approach.

Table 5.4. **Preliminary Design Decisions.** We need to make these decisions prior to designing the individual components.

Step	Outputs	Comments
1) Estimate system mass and envelope (Sec. 5.3.1)	• Engine mass estimate • Engine length • Engine diameter	• Use historical data • Roughly estimate system mass and size
2) Choose propellants (Sec. 5.3.2)	• Decision • Target O/F	• Performance • Storability • Toxicity • Cost
3) Determine the engine cycle (Sec. 5.3.3)	• Decision	• Determine how to drive the pump (if used) • Gas-generator cycle • Expander cycle • Staged combustion cycle • Hot-gas bleed from combustion chamber
4) Determine the cooling approach (Sec. 5.3.4)	• Decision	• Regenerative cooling • Film cooling • Radiation cooling • Ablative cooling
5) Determine engine and feed-system pressures (Sec. 5.3.5)	• Combustion chamber pressure • Nozzle expansion • Dynamic pressure • Pressure drop in feed system • Pressure drop in the injector • Propellant tank pressure • Engine balance • Pressurant system	• Trade chamber pressure & expansion ratio to get desired performance • Create a profile of system pressure
6) Estimate propellant mass and size tanks (Sec. 5.3.6)	• Propellant masses • Tank volumes	• Based on analysis of the rocket equation (Sec. 1.1)
7) Iterate	• New decision parameters	• Some later decisions affect earlier ones

5.3.1 Estimate System Mass and Envelope

The mass and dimensions of existing and historical liquid rocket systems (excluding tanks) correlate well with thrust magnitude. From mission-level analysis, we know how much thrust we need, so we can easily estimate system mass, length, and diameter. Figures 5.7, 5.8, and 5.9 show engine thrust-to-weight ratio versus vacuum thrust for launch-vehicle engines, bipropellant space engines, and monopropellant space engines, respectively. We have not distinguished among pressures, propellants, feed systems, or anything else, so these numbers give approximate curve fits. However, this rough estimate gives us a more intuitive feel for what we need in the following design sections.

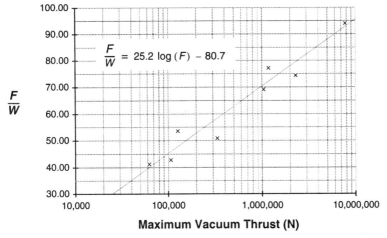

Fig. 5.7. **Thrust-to-Weight Ratio vs. Thrust for the Launch Vehicle's First-Stage Engines in an LRPS.** This figure indicates that as thrust level increases, so does the engine thrust-to-weight. We show a least-squares fit of the data.

Figures 5.7, 5.11, and 5.12 show engine length (L_E in cm) and diameter (D_E in cm, usually the nozzle exit) versus thrust magnitude (F in N). In each of Figs. 5.7 through 5.12, curve fit equations approximate the particular data plotted. Given a required thrust level, we can approximate engine mass, length, and diameter.

It is EXTREMELY important to recognize that these figures show representative historical numbers. They DO NOT indicate fundamental physical limitations. It is possible to design engine systems that deviate drastically from these numbers. But designers must understand why the design is different.

Launch-vehicle engines:

$$m_E = \frac{F}{g_0 \, (25.2 \, \log F - 80.7)} \tag{5.1}$$

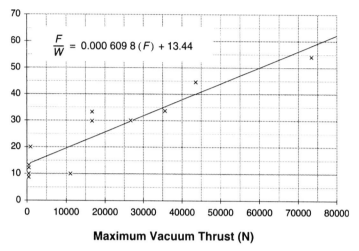

Fig. 5.8. **Thrust-to-Weight Ratio vs. Thrust for Bipropellant Engines Used in Space.** This figure shows that as thrust level increases, so does the engine thrust-to-weight. We show a least-squares fit of the data.

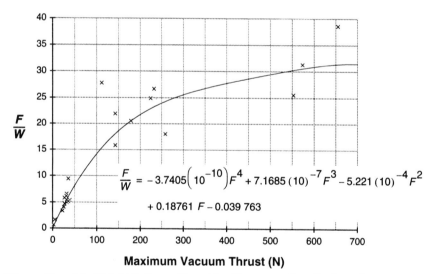

Fig. 5.9. **Thrust-to-Weight Ratio vs. Thrust for Monopropellant Engines.** These numbers include the thruster and valve. We show a least-squares, fourth-order polynomial fit of the data. The drop-off at the lower thrust levels indicates the difficulty of building miniature hardware. The trend to smaller spacecraft will change this aspect.

5.3 Preliminary Design Decisions

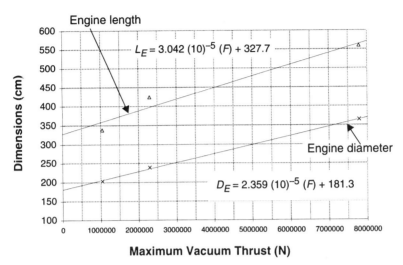

Fig. 5.10. **Dimensions vs. Thrust for a Launch Vehicle's First-Stage Engines in an LRPS.** This figure shows that as thrust level increases, so does the engine thrust-to-weight. Linear-regression analysis gives relationships between (L_E, diamonds) and diameters (D_E, x's)

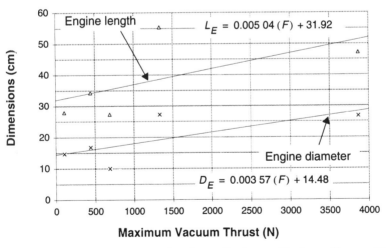

Fig. 5.11. **Dimensions of a Bipropellant Space Engine vs. Maximum Vacuum Thrust Level.** The figure shows a plot of engine length and diameter as a function of thrust level. Linear regression analysis gives relationships between length (L_E, diamonds) and diameter (D_E, x's).

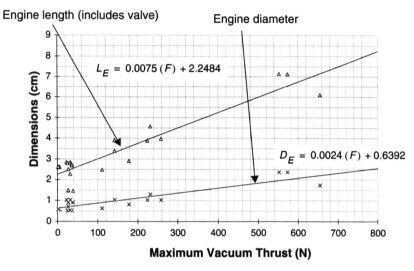

Fig. 5.12. Monopropellant Engine Dimensions versus Maximum Vacuum Thrust Level. The figure shows a plot of engine length and diameter as a function of thrust level. Linear regression analysis gives relationships between length (L_E, diamonds) and diameter (D_E, x's).

$$L_E = 0.000\,030\,42F + 327.7 \tag{5.2}$$

$$D_E = 0.000\,023\,59\,F + 181.3 \tag{5.3}$$

Space bipropellant engines:

$$m_E = \frac{F}{g_0\,(0.000\,609\,8\,F + 13.44)} \tag{5.4}$$

$$L_E = 0.0054\,F + 31.92 \tag{5.5}$$

$$D_E = 0.003\,57\,F + 14.48 \tag{5.6}$$

Monopropellant Engines:

$$m_E = \frac{F}{g_0 \begin{bmatrix} -3.7405\,(10)^{-10}F^4 + 7.1685\,(10)^{-7}F^3 + (-5.2221\,(10)^{-4}F^2) \\ + 0.18761F - 0.039763 \end{bmatrix}} \tag{5.7}$$

$$L_E = 0.0075\,F + 2.2484 \tag{5.8}$$

$$D_E = 0.0024 F + 0.6392 \tag{5.9}$$

where m_E = engine mass (kg)
L_E = length of the engine (cm)
D_E = diameter of engine (cm)
F = vacuum thrust of engine (N)
g_0 = 9.81 (m/s^2)
W = weight of engine in 1–g gravity (N)

5.3.2 Choose Propellants

We need to decide what propellants we are going to use, the O/F ratio, and the total quantity of propellant required. In many cases, the vehicle design specifies the propellants to be used, the desired O/F ratio, and the propellant quantity. However, the following discussion assumes a given specific impulse range and impulse (or Δv).

Appendix B gives the frozen-flow thermochemistry for several propellant combinations. It includes flame temperature (T_c), exhaust gas isentropic parameter (γ) frozen at the throat, exhaust gas molecular mass, and the vacuum specific impulse as a function of O/F. Table B.1 gives some of the more important parameters for these same propellants.

Determine the Oxidizer-to-Fuel Ratio. With a few exceptions, we want to choose an oxidizer-to-fuel ratio (O/F) that minimizes the total vehicle mass, which usually means the O/F that corresponds to the maximum specific impulse. In Appendix B, plots of specific impulse versus O/F allow us to do so. This best O/F is virtually independent of the operating condition, so these curves work for many possible conditions. However, the actual specific impulse changes depending on the operating condition.

In two cases, we do not choose the O/F corresponding to maximum specific impulse:

- The density of one propellant differs greatly from the other—the best example being the liquid hydrogen and liquid oxygen propellants. The optimum specific impulse is near an O/F of 3.5, whereas O/Fs of 5 to 6 are usually chosen. This is because lower O/Fs make the hydrogen tank prohibitively large, driving up the total system mass. To determine the best O/F for propellant combinations such as this, we must look at the overall vehicle design.[*]
- Hydrazine systems have an optimal O/F very near the condition for which the individual propellant tanks have equal volume. For mono-methyl hydrazine/nitrogen tetroxide systems, we choose an O/F of 1.65. For the

---

[*] A "stoichiometric" O/F of 8 gives us the maximum combustion temperature. However, maximum specific impulse occurs at a different O/F because I_{sp} is also a function of the combustion product's molecular mass and isentropic parameter (γ).

other hydrazines this number is slightly different and depends upon the particular propellant densities. This feature allows us to manufacture fuel tanks and oxidizer tanks of the same size and reduces the cost of hardware development.

5.3.3 Determine the Engine Cycle

If we have chosen to use a pump, we must have some way to power it—typically with a turbine driven by hot gases. The question is: what is the source of the hot gases? Fig. 5.13 illustrates three approaches:

- Expander cycle—pumped, high-pressure propellant feeds through a heat exchanger used to cool the thrust chamber structure. The exchanger heats and vaporizes the propellant, which then feeds into the turbine and, eventually, into the combustion chamber.
- Staged-combustion—small amounts of fuel feed into the oxidizer flow or small amounts of oxidizer feed into the fuel flow. This mixture then preburns to produce warm gas that drives the turbine or turbines. The partially burned stream or streams of gas then enter the combustion chamber to complete the combustion process.
- Gas generator—some of the fuel and oxidizer (2%–5% of total flow) feed into a separate combustor. The hot gases then move into the turbine and, typically, exhaust overboard.

Choosing one cycle over another usually involves a trade between cost and complexity on the one hand and simplicity and decreased performance on the other. The simplest cycle is the gas generator. However, by dumping the burned propellants overboard, we normally get a 2% to 5% drop in overall specific impulse at a given chamber pressure. Typically, the quantity of propellant needed for the pumping work is a few percent of the total propellant flow. The exact number depends on factors such as pressure, propellant density, gas density, and turbomachinery efficiencies. If we can take this hit, dumping is the best approach. If we cannot, then we should use either an expander or staged-combustion cycle. Although they both work in a similar manner, the expander operates at a lower turbine inlet temperature. This characteristic translates into lower turbine power and, consequently, lower chamber pressure. As a result, staged combustion yields higher performance for most applications. However, staged combustion requires more hardware, so it is more complex and costly.

The cycle strongly affects the engine and feed system pressure levels (Sec. 5.3.5). As shown in Fig. 5.13, the turbine is in parallel with the thrust chamber in a gas-generator cycle and is in series with the thrust chamber in both the staged-combustion and expander cycles. Therefore, the turbine pressure ratio does not raise the required pump pressure in a gas-generator cycle, but strongly affects the other two cycles. In addition, the preburner injector for the staged-combustion cycle is in series with the thrust chamber (with an associated 20% pressure drop). As a result,

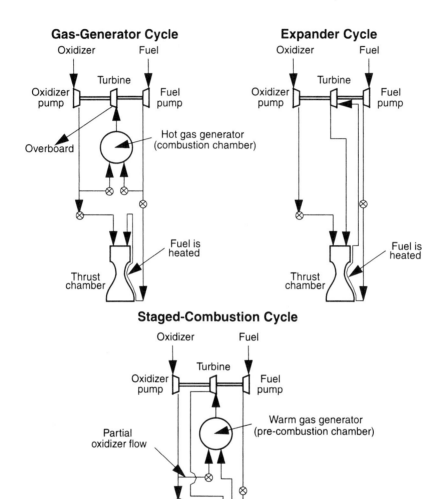

Fig. 5.13. Cycles of Candidate Rocket Engines. The gas-generator, expander, and staged-combustion cycles are the three cycles that most often drive propellant pumps. (Courtesy of Rocketdyne [Bissell, 1985])

the gas-generator and the staged-combustion cycles represent the two extremes in system and pump pressure levels. These extremes are shown in Fig. 5.14, which compares the sums of drops in fuel-system pressure for the two cycles at the same thrust-chamber pressure. Sec. 5.3.5 discusses the effect of system pressure levels.

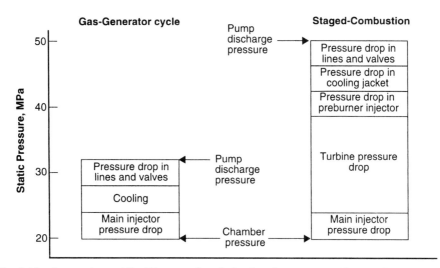

Fig. 5.14. Comparison of Fuel Pressure Levels for Gas-Generator and Staged-Combustion Cycles. Staged combustion requires much higher pump pressures than a gas generator but has better specific impulse performance. These numbers are for a typical booster engine [Bissell, 1985].

5.3.4 Determine the Cooling Approach

The thrust chamber typically operates at temperatures above the limits of structural materials. Therefore, to avoid a structural failure, we must cool the structure with one of four basic cooling approaches:

- *Radiation cooling*—the thrust chamber heats up during combustion until the structure gets red or white hot and heat radiates into space, keeping the chamber at a reasonable temperature. We use this approach in a vacuum. Sometimes, we combine this approach with film cooling at the throat. Radiation cooling is then appropriate at the nozzle exit where the temperature is somewhat lower.

- *Ablative cooling*—the thrust chamber's lining absorbs heat as it ablates. The expendable lining material also insulates the structural wall. This approach works well where simplicity is more important than low weight. However, the lifetime of the engine is restricted by the ablation rate of the expendable material.

- *Regenerative cooling*—cold propellant runs through a heat exchanger which is an integral part of the thrust-chamber wall. The propellant absorbs heat being transferred to the structure, allowing the structure to maintain a lower temperature. Predicting the performance of this arrangement is somewhat difficult and operating it is more complex than an ablative system. But if we have chosen an expander cycle, we must use this approach.
- *Film or boundary-layer cooling*—coolant, fuel, or oxidizer is injected close to the thrust-chamber wall. The mixture ratio here is much different from that in the main part of the chamber, allowing a lower flame temperature. Again, this system is more complex than an ablative one but is lighter. There is also a small loss in specific-impulse performance.

The choice of cooling approach balances complexity against performance. If we are operating a space motor with short-duration burns, we choose radiation cooling. If we require longer burns but do not want the complexity of other approaches and can handle a higher-weight system, we use ablation. For higher performance requirements, we could use regenerative cooling. If the regenerative approach is inadequate to handle the heat load, we can use film cooling. Radiation cooling is often used on the cooler part of a nozzle when the other approaches are used. In some situations, all cooling approaches are combined.

5.3.5 Determine Pressure Levels for the Engine & Feed System

One of the most critical considerations for liquid engine systems is pressure distribution. Pressure variations from the storage tank, through the feed system, and into the thrust chamber determine the propellant flow rate. Figures 5.15 and 5.16 show schematics of typical pressure distributions through tank-pressure-fed and pump-pressure-fed systems, respectively. To determine the pressures, we start the discussion at the thrust chamber to make sure that we get the required engine performance. We then determine the pressure changes back through the feed system to the propellant tank, ignoring, for a moment, any pumping system. The final step is to balance the engine pressures, add the turbopump, and make sure the power requirements for the pump match the turbine output. The actual process for coming up with the pressures is usually somewhat iterative rather than rigidly following this path.

Combustion Chamber Pressure and Nozzle Expansion Ratio

Chamber pressure affects the specific impulse of engines (operating in a sensible atmosphere) and the size of the thrust chamber. For a given required thrust, we can determine the required mass flow rate. We can then see the effect of chamber pressure on engine size from the characteristic velocity equation (see Sec. 3.3.2),

$$\dot{m} = \frac{p_c A_t}{c^*} \tag{5.10}$$

where $\dot{m}$ = mass flow rate (kg/s)
 p_c = chamber pressure (Pa)
 A_t = throat cross-sectional area (m²)
 c^* = propellant characteristic velocity (m/s)

For a given propellant and mass flow rate, this equation shows the direct relationship between chamber pressure and throat area (engine size). If we can increase chamber pressure, we can reduce the throat dimensions and, therefore, all other engine dimensions. This means that engine mass depends on the chamber pressure choice.

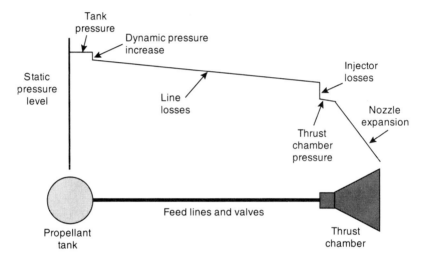

Fig. 5.15. Schematic Representation of the Pressure Levels in a Tank-pressure-fed System.
Pressure drops continuously from the tank to the exhaust nozzle.

Chamber pressure also improves the specific impulse possible for engines operating in a high-pressure atmosphere, the main consideration being the separation of the exhaust gas flow in the nozzle. Figure 5.17 shows chamber pressure versus nozzle expansion ratio (area ratio) for several engines. The data marked "x" is for space engines. We can see space engines do not have a fixed design exit pressure (and corresponding expansion ratio) because there is no concern with flow separation (ambient pressure = 0). However, there is a definite correlation for first-stage engines on launch vehicles (and other rockets operating in an atmosphere). The two curves show the relationship needed between chamber pressure and nozzle expansion to get a fixed nozzle exit pressure. The curve on the left is for a pressure of 45,000 Pa, and the curve on the right is for 15,000 Pa [Eqs. (5.14) and (5.15) are used to generate these curves with $\gamma = 1.238$]. There are several points on the

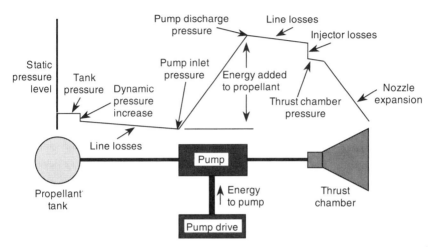

Fig. 5.16. Schematic of the Pressure Levels in a Pump-pressure-fed System. Pressure drops continuously to the pump inlet. The pump drastically drives up the pressure, which then drops continuously again. This schematic is for a gas-generator cycle. For expander and staged-combustion cycles, the turbine is after the pump and contributes a substantial pressure drop. (See Fig. 5.14.)

45,000 Pa curve indicating that aiming for this exit pressure is a conservative design goal for first-stage engines. The single point on the 15,000 Pa curve is the Space Shuttle Main Engine (SSME); it is the limit of what we can achieve in flow expansion without getting flow separation. Section 3.3.2 discusses flow separation. For first-stage engines, then, our design approach is to trade chamber pressure and expansion ratio to get an exit pressure between 15,000 Pa (aggressive) and 45,000 Pa (conservative). For space engines, we do not have this restriction.

Figure 5.18 shows a plot of specific impulse versus expansion ratio for various chamber pressures. The upper curve represents space-engine performance (zero ambient pressure) and is independent of chamber pressure. The other curves represent engines operating at sea level (1 atmosphere = 101,325 Pa). For engines operating in a sensible atmosphere, an increase in chamber pressure allows us to choose a larger expansion ratio without getting flow separation in the nozzle. This leads to higher possible specific impulses. The ends of these curves represent a 45,000-Pa exit pressure, and the maximum specific impulse corresponds to ideal expansion ($p_e = p_a$). As the engines ascend from sea level, the specific impulse increases (straight up at a constant expansion ratio) to the upper limit possible, represented by the vacuum line. Booster engines start operation at sea level ($p_a = 1$ atmosphere) and ascend into a virtual vacuum ($p_a = 0$). This means the effective specific impulse over the ascent is some average value.

The curves for Figs. 5.17 and 5.18 are developed from

$$I_{sp} = \lambda \left\{ \frac{c^* \gamma}{g_0} \sqrt{\left(\frac{2}{\gamma-1}\right)\left(\frac{2}{\gamma+1}\right)^{\frac{\gamma+1}{\gamma-1}} \left[1 - \left(\frac{p_e}{p_c}\right)^{\frac{\gamma-1}{\gamma}}\right]} + \frac{c^* \varepsilon}{g_0 p_c}(p_e - p_a) \right\} \quad (5.11)$$

$$c^* = \frac{\eta_{c^*} \sqrt{\gamma R T_c}}{\gamma \left(\frac{2}{\gamma+1}\right)^{\frac{\gamma+1}{2\gamma-2}}} \quad (5.12)$$

$$R = \frac{8314}{\mathcal{M}} \quad (5.13)$$

where λ = nozzle efficiency (0.9–0.98)
g_0 = acceleration due to gravity at sea level, 9.81 m/s^2
p_e = nozzle exit pressure (Pa)
p_c = combustion chamber pressure (Pa)
p_a = ambient pressure (Pa)
γ = isentropic parameter (ratio of specific heats)
c^* = characteristic exhaust velocity (m/s)
ε = nozzle expansion ratio
R = exhaust gas constant (J/kg·K)
T_c = flame temperature (K)
$\mathcal{M}$ = molecular mass (kg/kmol)
η_{c^*} = c^* combustion efficiency (frozen flow range 1.0–1.15; equilibrium flow range 0.9–0.98)

These equations require knowledge of the nozzle flow pressures, particularly p_e. Assuming isentropic flow [Eq. (3.95)]:

$$p_e = p_c \left[1 + \frac{\gamma-1}{2} M_e^2\right]^{\frac{\gamma}{1-\gamma}} \quad (5.14)$$

where we must determine the exit Mach number (M_e) from the implicit expansion-ratio equation [Eq. (3.100)]:

$$\varepsilon = \frac{1}{M_e}\left[\left(\frac{2}{\gamma+1}\right)\left(1 + \frac{\gamma-1}{2}M_e^2\right)\right]^{\frac{\gamma+1}{2\gamma-2}} \quad (5.15)$$

Table 5.5 outlines how to use these equations so we can generate plots similar to Fig. 5.18.

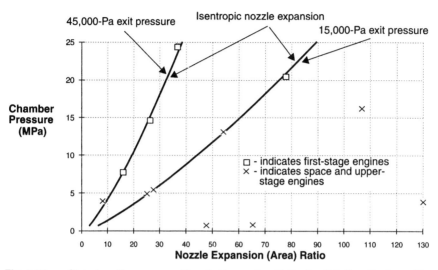

Fig. 5.17. Chamber Pressure vs. Nozzle Expansion Ratio for Existing Engines. There is no correlation for space engines, but first-stage engines must avoid nozzle flow separation. The left curve shows expansion to 45,000 Pa nozzle exit pressure and the right curve shows expansion to 15,000 Pa (frozen flow $\gamma = 1.238$).

Dynamic Pressure

As the propellant leaves the tank, its velocity goes from zero to the required flow velocity. To a first approximation, the total pressure remains constant, at least for the first short distance. Thus, the static pressure must drop to allow the increase in dynamic pressure. For incompressible flow, which includes all liquids to a good approximation, the pressure drop from the increase in dynamic pressure is given by Bernoulli's equation:

$$\Delta p_{dynamic} = \frac{1}{2}\rho v^2 \tag{5.16}$$

A typical number for flow velocity is 10 m/s. To determine the size of a flow channel that gives this velocity, use the continuity equation:

$$\dot{m} = \rho v A \tag{5.17}$$

where ρ = liquid propellant density (kg/m³)
 v = flow velocity (m/s)
 $\Delta p_{dynamic}$ = pressure drop due to dynamic pressure (Pa)
 $\dot{m}$ = mass flow rate through the particular channel (kg/s)
 A = flow channel cross-sectional area (m²)

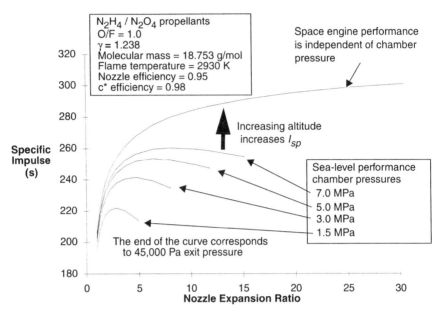

Fig. 5.18. **Specific-Impulse Performance of a Hydrazine Bipropellant System versus Expansion Ratio for Various Chamber Pressures.** The top curve is for engines operating in a vacuum and is independent of chamber pressure. The other curves are for performance at sea-level conditions. Expansion stops at the exit pressure of 45,000 Pa (see Table 5.5).

Table 5.5. **Process for Determining the Relationship between Specific Impulse and Nozzle Expansion Ratio.** By following this process, we can create figures similar to Fig. 5.18.

Action	Inputs	Outputs	Source
1. Evaluate thermochemistry	O/F, η_{c*}, propellants	γ, $\mathcal{M}$, T_c, R, c^*	Appendix B for γ, $\mathcal{M}$, T_c; Eqs. (5.12, 5.13)
2. Choose fixed parameters for each desired curve		p_c p_a λ	p_c - decision p_a - required environment λ - nozzle efficiency
3. Vary p_e over an applicable range		p_e	Horizontal axis coordinate
4. Evaluate nozzle exit Mach number	p_e, p_c, γ	M_e	Eq. (5.14)
5. Evaluate nozzle expansion ratio	γ, M_e	ε	Eq. (5.15)
6. Evaluate I_{sp}	λ, c^*, γ, p_e, p_c, ε	I_{sp}	λ - decision Eq. (5.11)

Pressure Drop in the Feed System

The goal in designing a feed line is to make the pressure drops (Δp_{feed}) as low as possible. Typically, we limit these losses to between 35,000 Pa and 50,000 Pa. For early decisions, we assume a number in this range. If we wish to be aggressive, or if the lines are relatively short, we choose the lower end. If we wish to be conservative, we choose the upper end.

$$\Delta p_{feed} = 35{,}000 \text{ to } 50{,}000 \text{ Pa} \tag{5.18}$$

For a regenerative cooling system, pressure drops in the cooling jacket (Δp_{cool}) can vary between 10% and 20% of chamber pressure (p_c),

$$\Delta p_{cool} = 0.15 p_c \tag{5.19}$$

Pressure Drop in the Injector

As a rule of thumb, this pressure drop (Δp_{inj}) should be about 20% or more of the chamber pressure (p_c) for unthrottled engines:

$$\Delta p_{inj} = 0.2\, p_c \quad \text{(unthrottled)} \tag{5.20}$$

This drop isolates chamber-pressure oscillations from the feed system, reducing pressure coupling between the combustion chamber and the feed system which could lead to instabilities or oscillations in the flow that are driven by variations in combustion. The actual pressure drop is also a function of throttling and a complex function of geometry. The rule of thumb for throttled systems is a 30% drop:

$$\Delta p_{inj} = 0.3\, p_c \quad \text{(throttled)} \tag{5.21}$$

However, for some pintle-type injectors, we can go as low as 5%.

The major source of pressure drop through the injectors is from an increase in dynamic pressure. Small injector orifices accelerate the flow so the high-speed propellant streams can atomize and vaporize more easily. However, some flow losses are associated with forcing the propellant through the small holes.

Propellant-Tank Pressure

The first big question to be answered is whether or not we should use a pump. Usually, big systems use pumps, and small systems do not. As the system size grows, decreasing the tank mass justifies the additional complexity and cost of pumps. The larger the tank, the bigger the mass savings.* Figure 5.19 plots tank pressure versus tank volume for various systems. The larger propellant-tank

---

* The main difference between pump-fed systems and pressure-fed systems is the inert mass fraction. Pumps usually allow us to lower the mass fraction. The rocket equation tells us that the inert mass fraction affects mass ratios such as *payload mass / initial mass*. So, strictly speaking, the mass ratio savings is independent of absolute mass number. However, a 10% savings on a million kilograms is greater than a 10% savings on 10 kilograms.

volume drives us to use pumps. The lack of pumps drives us to a higher tank pressure. If the system we are designing falls right in the overlap region (1–10 m³) of the figure, we typically go to a pressure-fed system. This choice makes sense because the system's complexity and cost decrease. In addition, tank materials and manufacturing technologies have improved since many of the tanks in this figure were built.

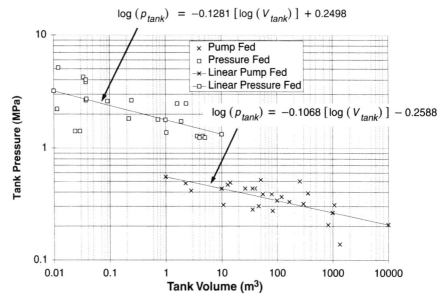

Fig. 5.19. Tank Pressure vs. Tank Volume for Different Feed Systems. Small tanks are tank-pressure-fed, with an overall trend of decreasing pressure as the tank gets bigger (see linear regression curve). For larger systems, we use pumps. (From NASA SP-8112 [1975])

Once we have answered the pump question, we can determine the tank pressure. Again, refer to Fig. 5.19. If we have decided on a pumped system, tank pressures vary from 0.2 MPa to about 0.5 MPa and average about 0.3 MPa:

$$p_{tank} = \left[10^{-0.1068 \, [\log \, (V_{tank}) \, - \, 0.2588]} \right] \times 10^6 \qquad (5.22)$$

Choosing the average is a good first estimate. If we have decided on a pressure-fed system, we can get a tank pressure based on the linear curve fit shown:

$$p_{tank} = \left[10^{-0.1281 \, [\log \, (V_{tank}) \, + \, 0.2498]} \right] \times 10^6 \qquad (5.23)$$

where p_{tank} = tank pressure (Pa)
       V_{tank} = tank volume (m³)

There are large dispersions around these curve fits.

For either the pump-fed or pressure-fed systems, we can bump the average or nominal pressure up or down a bit depending on whether we wish to be aggressive or conservative. If we are using a pressure-feed system, the required tank pressure is not quite so simply determined as Eq. (5.23) suggests. The tank pressure must equal the sum of the chamber pressure and all the pressure drops through the feed system:

$$p_{tank} = p_c + \Delta p_{dynamic} + \Delta p_{feed} + \Delta p_{cool} + \Delta p_{inj} \qquad (5.24)$$

If the pressure calculated in this equation turns out to be much higher than what is reasonable based on Fig. 5.19, we need to consider a pumped system. Remember, Fig. 5.19 shows only historical trends. Deviating from these numbers is feasible.

Engine Balance

If we have chosen to use a pump, the next step is to balance the engine-feed system in terms of pressures and the power required by the pump versus the power provided by the pump drive (turbine). This process is iterative. At this point we are grossly estimating the pressure changes and power requirements. When we have completed the component designs described in Sec. 5.4.3, we must make sure our original assumptions are accurate and adjust them as required. This iteration goes on for the entire life of a real engine development, as we better understand the design and its performance.

If we define the turbine pressure ratio as

$$p_{trat} = \frac{p_{ti}}{p_{td}} \qquad (5.25)$$

where p_{trat} = turbine pressure ratio
       p_{ti} = turbine inlet pressure (Pa)
       p_{td} = turbine discharge pressure (Pa)

we know from experience that the pressure ratio is approximately 1.5 for staged-combustion or expander cycles and can be as high as 20 for a gas generator. For a staged-combustion cycle, pressure also drops through the injector of the pre-combustor. In a preliminary estimate, we can use these pressure ratios to define the turbine and pump pressures (Fig. 5.14 indicates how we can do so).

To illustrate how to balance the flows, pressures, and powers, we need to discuss the power balance for a pump-fed engine because it involves all three major variables: the flow rates, pressures, and powers. Assuming a turbine drives each pump, the power of that turbine must equal the power used by that pump:

$$P_{req} = \frac{g_0 \dot{m} H_p}{\eta_p} = \eta_T \dot{m}_T c_p T_i \left[1 - \left(\frac{1}{P_{trat}}\right)^{\frac{\gamma-1}{\gamma}}\right] \quad (5.26)$$

where P_{req} = power required to drive the pump (W)
g_0 = 9.81 m/s^2
$\dot{m}$ = propellant mass flow rate through the pump (kg/s)
H_p = pump head rise (m) (see Eq. 5.59)
η_p = pump efficiency (see discussion below)
η_T = turbine efficiency (see discussion below)
$\dot{m}_T$ = mass flow rate through the turbine (kg/s)
c_p = specific heat of the turbine drive gases (J/kg·K)
T_i = turbine inlet temperature (K)
γ = isentropic parameter (ratio of specific heat) of the turbine drive gases

For a gas-generator cycle, the turbine is in parallel with the thrust chamber, and the drive gases are either dumped overboard or injected part way down the skirt of the thrust-chamber nozzle. Typically, about 2% to 5% of the propellant mass flow goes through the gas generator. We can usually generate about one-half the specific impulse with the resulting gases. This gives us an overall reduction of 1% to 2.5% in I_{sp}. In any event, the specific impulse of the turbine drive gas is less than that of the thrust chamber gas. Therefore, the design goal for this cycle is to minimize turbine flow rate. To do this, we want the highest pump efficiency, turbine efficiency, turbine inlet temperature, and turbine pressure ratio we can get. Turbine and pump efficiencies vary widely, depending on pump size and other requirements. Lower efficiencies have advantages in simplification and lower cost. Even though reduced efficiency drives up the turbomachinery mass, the effect on the overall system mass is usually negligible. For gas generators, reasonably high initial assumptions for these parameters are

- Pump efficiency—0.80 efficiency for pumps other than hydrogen, 0.75 efficiency for hydrogen pumps (see Fig. 5.35)
- Turbine efficiency—0.70 (see Fig. 5.39)
- Turbine inlet temperature—1100 K, which is close to typical material strength limits (see Fig. 5.37); we must choose the correct thermochemistry (O/F) to get this number
- Turbine pressure ratio—20; higher values can produce pressure-distribution problems elsewhere in the engine. In a liquid oxygen/liquid hydrogen engine with two turbines in series, the overall pressure ratio is the product of the two ratios, and the hydrogen pump's power is much greater than the oxygen pump. This arrangement can translate to turbine pressure ratios more like 8.0 and 2.5 for the hydrogen and oxygen turbines, respectively.

For staged-combustion and expander cycles, the fact that the turbine is in series with the thrust chamber removes the negative impact of turbine flow rate on engine specific impulse. However, as shown in Fig. 5.14, a rising pressure ratio greatly raises the required pump discharge pressure. Therefore, with these cycles, we want to have the lowest turbine pressure ratio, so that we get either a lower pump-pressure rise (thereby minimizing the size and mass of the turbopumps) or a higher pressure in the main combustion chamber downstream. Remember, high pressure in this chamber gives us a smaller thrust chamber [Eq. (5.33)] and lower thrust-chamber mass. For booster engines, high chamber pressure also leads to the largest possible nozzle area ratio and, consequently, engine specific impulse (Fig. 5.18).

Equation (5.26) calls for maximum pump and turbine efficiencies, as well as turbine inlet temperature. The pump efficiencies and the inlet temperatures for a staged-combustion turbine are usually about the same as for the gas-generator cycle. However, because these cycles have higher turbine velocity ratios, turbine efficiencies are more like 80%. Heat transfer from the thrust chamber's cooling jacket controls turbine inlet temperatures for the expander cycle. These temperatures are usually no more than 650 K and can be as low as 225 K.

The remaining independent variable in Eq. (5.26) is the turbine mass flow rate, which should also be maximized. Assuming we have fuel-rich turbine drive gases, we get the best flow rate by using all of the fuel to drive the turbines. In staged-combustion cycle engines, enough oxidizer is added in the preburner to raise the turbine inlet temperature to the desired limit. These procedures typically produce turbine pressure ratios of 1.5 to 2.0 at the power limit (when all independent variables are at maximum), with occasional values approaching 2.5.

Pressurant System

We need a pressurant system to keep the propellant tanks at the desired operating pressure. Typically, we choose from three approaches:

- External stored gas—a gas, such as helium or nitrogen, is stored in a separate tank at high pressure. As the propellant level drops, the gas feeds into the propellant tank through a regulator to maintain the desired pressure.
- Evaporation—pressurant is stored as a liquid. The pressurant, which could be the propellant itself, is heated so the vapor pressure is equal to the desired propellant tank's operating pressure.
- Chemical reaction—chemicals stored separately either react and feed into a tank to increase pressure or feed into the tank where they react with the propellant to maintain pressure.

Figure 5.20 shows a plot of pressurization systems as a function of the propellant tank's operating (ullage) pressure and tank volume. Pressurant systems using inert gas can be pressurized, initially, to as high as 60 MPa. Chemical reaction and evaporation systems operate close to the propellant-tank pressure.

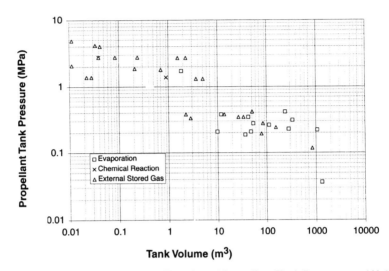

Fig. 5.20. Pressurization Types as a Function of Propellant Tank Pressure and Volume. This plot shows three types of pressurization. Larger systems (boosters) tend to rely on evaporation, whereas smaller systems (spacecraft) rely on inert gas. Chemical reaction is not widely used (From NASA SP-8112 [1975].)

The other option is to exclude a pressurization system. This is typically called a *blowdown* system. Unfortunately, as the propellant level drops, so does the tank pressure and the system thrust level. This is a very simple system, however, and is popular in small attitude-control systems for which we value simplicity above performance.

5.3.6 Estimate Propellant Mass and Size Tanks

From Sec. 1.1.5, we can estimate the total propellant mass required from the ideal rocket equation (Sec. 1.1):

$$m_{prop} = \frac{m_{pay}\left[e^{\frac{\Delta v}{I_{sp} g_0}} - 1\right](1 - f_{inert})}{1 - f_{inert} e^{\frac{\Delta v}{I_{sp} g_0}}} \quad (5.27)$$

where m_{prop} = total mass of propellant (kg)
m_{pay} = payload mass (kg)
Δv = effective velocity change required (m/s)

I_{sp} = specific impulse (s)
g_0 = standard acceleration of Earth gravity (9.81 m/s)
f_{inert} = inert mass fraction (see Figs. 5.21 and 5.22)

Previous analysis should have revealed all of these parameters except for mass fraction numbers. Figure 5.21 shows inert mass fractions for typical liquid-rocket systems. Remember, the inert mass fraction is defined as

$$f_{inert} = \frac{inert\ mass\ of\ stage}{total\ mass\ of\ stage} \tag{5.28}$$

and includes all of the stage mass, less the propellant.

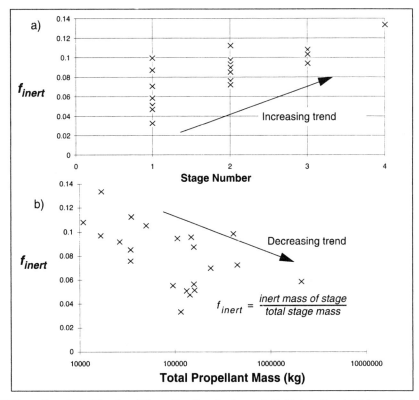

Fig. 5.21. **Trends in Structural Mass Fraction for Launch Vehicles.** Graph (a) is a plot of mass fractions versus stage number. Graph (b) is a plot of the same data as in (a) but is a function of propellant mass. These plots show that as propellant mass increases, the structural mass fraction decreases. As the stage number goes up, the mass fraction also tends to increase [Isakowitz, 1991].

We usually strive to achieve low inert mass (subject to the usual constraints such as cost), so our vehicle is lighter and, hopefully, less costly. Figure 5.21a shows selected historical mass fractions as a function of stage number, and Fig. 5.21b shows the same data as a function of total propellant mass. There is a fairly large dispersion in the mass fraction numbers. The choice of mass fraction depends on how conservative we wish to make the structure, the stage number, and the expected propellant mass and density, and on how innovative we are.

Figure 5.22 shows similar data for upper stages, transfer stages, and reaction-control systems. The data varies widely and reveals no clear trend. Reaction-control systems (attitude-control systems) are relatively small and have inert mass fractions near 0.7 (only 30% of the total system mass is propellant). Other space propulsion systems (upper stages, kick motors, and maneuvering systems) average a mass fraction of 0.17. To decide on an appropriate mass fraction, we need to look at our system and decide whether we want a conservative mass estimate (higher fraction) or an aggressive mass estimate (lower fraction).

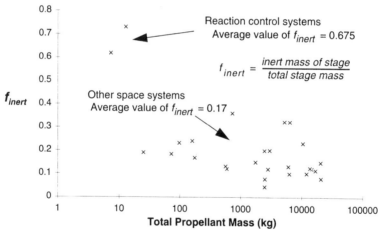

Fig. 5.22. Trends in Structural Mass Fraction for Space Engines. There is no clear trend here, so we choose numbers based on how aggressive we wish to be (Isakowitz [1991] and Larson and Wertz [1992]).

The other parameter in Eq. (5.27) is specific impulse (I_{sp}). Appendix B gives typical specific-impulse numbers for a given condition. If this is the desired condition, we have completed the preliminary design. If not, we need to do more analysis, as discussed in Sec. 5.3.5.

The following equations give us the fuel and oxidizer mass and volume:

$$m_{fuel} = \frac{m_{prop}}{1 + O/F} \qquad (5.29)$$

$$m_{ox} = m_{prop} \frac{O/F}{O/F + 1} \tag{5.30}$$

$$V_{fuel} = \frac{m_{fuel}}{\rho_{fuel}} \tag{5.31}$$

$$V_{ox} = \frac{m_{ox}}{\rho_{ox}} \tag{5.32}$$

where m_{fuel} = fuel mass (kg)
m_{ox} = oxidizer mass (kg)
m_{prop} = total propellant mass (kg)
ρ_{fuel} = fuel density (kg/m³)
ρ_{ox} = oxidizer density (kg/m³) (see Table B.1)
V_{fuel} = fuel volume (m³) (see Table B.1)
V_{ox} = oxidizer volume (m³)

5.4 System Sizing, Design, and Trade-offs

We have now made enough decisions to design individual components. This section discusses the basic technology and allows us to estimate the component configurations and masses.

5.4.1 Thrust Chamber

Figure 5.23 shows a sketch of a typical thrust chamber. It consists of three major sections:

- *Combustion chamber*—provides an area for proper mixing of propellants and enough length for the propellants to complete chemical combustion. This section is usually cooled to ensure sufficient structural strength in the high-pressure and high-temperature environment. We commonly include the converging part of the nozzle assembly in the combustion chamber.
- *Exhaust nozzle*—usually connected directly to the combustion chamber. In the nozzle, the enthalpy of the hot gases converts to kinetic energy and produces the engine's thrust. The hot combustion gases accelerate from Mach 1 at the throat to the desired exit pressure at supersonic velocities. The nozzle is usually cooled, particularly in the throat area where the most intense heat transfer occurs. At some station downstream of the throat, beyond Mach ~3–4, the cooling requirements decrease, and the nozzle may not need to be actively cooled.
- *Injector*—distributes the prescribed propellant mass flows to the combustion chamber. We must make sure the propellants properly atomize, vaporize, and mix. A large drop in the flow's static pressure (see Sec. 5.3.5)

must occur through the injector orifice to ensure that pressure oscillations in the combustion chamber do not travel into the propellant-feed system.

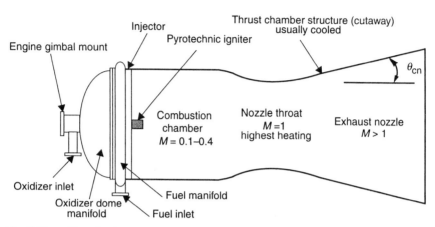

Fig. 5.23. Sketch of a Typical Thrust Chamber. The illustration shows the basic configuration of a thrust chamber, the various components, and the flow Mach number (*M*) conditions.

Although we design the thrust chamber, nozzle, and injector separately, we have to evaluate the operation and performance of the thrust-chamber assembly as a complete system.

At this point, we must decide a fundamental question: Given the mission requirements, do we need to build a new engine or can we use an existing one? The development of any new engine is expensive, so strong cost arguments support using an existing engine. If we select an existing engine, we know the engine mass and configuration. If we develop a new engine, we must estimate the system mass and configuration.

Dimensions of the Combustion Chamber

We have already chosen the chamber pressure, nozzle expansion ratio, propellant type, propellant mass flow rates, and the oxidizer-to-fuel ratio (*O/F* and associated chemistry). The throat area derives from conservation of mass and from knowing the flow is choked at the throat. Once we know the throat area, we can easily calculate the diameter [Eqs. (3.133) and (5.12)]:

$$A_t = \frac{\dot{m}\sqrt{\gamma R T_c}}{\eta_{c^*} p_c \gamma \left(\frac{2}{\gamma+1}\right)^{\frac{\gamma+1}{2\gamma-2}}} = \frac{\dot{m} c^*}{p_c} \qquad (5.33)$$

$$D_t = 2\sqrt{\frac{A_t}{\pi}} \qquad (5.34)$$

where A_t = throat cross-sectional area (m²)
$\dot{m}$ = mass flow rate (kg/s)
T_c = combustion flame temperature (K)
p_c = chamber pressure (Pa)
η_{c^*} = c* efficiency (see discussion below)
c^* = characteristic velocity (m/s)
γ = isentropic parameter of combustion gas
R = combustion gas constant (J/kg·K)
D_t = throat diameter (m)

The characteristic velocity efficiency (η_{c^*}) accounts for the assumptions we made when evaluating the thermochemistry. Remember from Chap. 4, the frozen-flow assumption is conservative, so η_{c^*} ranges from 1.0 to 1.15. The equilibrium-flow assumption is optimistic, so η_{c^*} ranges from 0.93 to 0.97.

Next, we need to establish the combustion chamber's dimensions by combining historical data and gas dynamics. The combustion chamber must be large enough to allow the propellants to completely mix, atomize, and burn. Detailed analysis of these coupled processes is complex and is still based on empirical data. The main drivers are propellant combination, the state of the propellants (gas, liquid, or gel), chamber geometry, injector geometry, and chamber pressure. It helps to define a combustion chamber's *characteristic length* (L^*) as

$$L^* = \frac{V_c}{A_t} \qquad (5.35)$$

where L^* = characteristic length (m)
V_c = total volume of the combustion chamber (m³)
A_t = nozzle throat area (m²)

Small values of L^* imply small engines, so to minimize size and mass, we want to make L^* as small as possible while maintaining adequate combustion efficiency. However, vaporizing, mixing, and burning of propellants require finite time and therefore volume. Historical data on vaporizing in the combustion chamber and engine mass efficiency suggest that L^* values on the order of 1.8 to 2.5 meters achieve good mass efficiency. But these engines are larger, have higher heat and pressure losses, and weigh more than short L^* engines. On the other hand, an engine with a low L^* of 0.3 m may vaporize, mix, and burn propellants poorly. Historical data suggests that L^* in the range of 0.8 to 1.0 m is common and depends on propellant choice. Table 5.6 shows ranges for L^*, depending on the propellant choice. An aggressive choice would be a lower value, whereas the conservative choice would be a higher value. It is important to be wary of these L^* numbers.

Recent propulsion system advances indicate that we can decrease the minimum values by a significant amount. As we have said before, these numbers only represent history and do not represent fundamental physical limits.*

Table 5.6. Numbers for Characteristic Lengths of Typical Propellant Combinations [Huzel and Huang, 1992].

Propellants	Characteristic Length (L^*)	
	Low (m)	High (m)
Liquid fluorine / hydrazine	0.61	0.71
Liquid fluorine / gaseous H_2	0.56	0.66
Liquid fluorine / liquid H_2	0.64	0.76
Nitric acid / hydrazine	0.76	0.89
N_2O_4 / hydrazine	0.60	0.89
Liquid O_2 / ammonia	0.76	1.02
Liquid O_2 / gaseous H_2	0.56	0.71
Liquid O_2 / liquid H_2	0.76	1.02
Liquid O_2 / RP-1	1.02	1.27
H_2O_2 / RP-1 (including catalyst)	1.52	1.78

Having determined throat area and selected L^*, we can uniquely define the required combustion chamber volume [Eq. (5.35)]. If we simplify things by assuming a simple circular cylinder,

$$V_c = A_c L_c = \frac{\pi}{4} D_c^2 L_c \qquad (5.36)$$

where A_c = chamber's average cross-sectional area (m²)
L_c = chamber's length (m)
D_c = chamber's diameter (m)

The question remains: how do we determine D_c and L_c?

---

* Recent research indicates that we can do better than Table 5.6 if we increase chamber pressure. For example, the Marquardt 150-N Leap-Divert engine has an L^* of 0.13 m and an injector-to-throat length (L') of 3.8 cm [Acampora and Wickman, 1992]. Similarly, the 2200-N, RS-59 engine made by Rocketdyne has an L^* = 0.18 m. Both of these systems use N_2H_4 and N_2O_4. By comparison, the L^* quoted in Table 5.6 is 0.6 m.

5.4 System Sizing, Design, and Trade-offs

A gas dynamic consideration in designing combustion chambers is that the Mach number be "small," so that the combustion-chamber pressure nearly equals the stagnation pressure. A small Mach number implies low velocity and therefore long residence time for mixing and chemical reaction. A common design choice is Mach numbers in the range of 0.1–0.6, with $M = 0.1$ being conservative. This fact implies that the combustion chamber area is some multiple of the throat area, as determined by the thermochemistry of the propellants. From Sec. 3.2.5 we can determine the chamber contraction ratio (A_c / A_t):

$$\frac{A_c}{A_t} = \frac{1}{M}\left[\left(\frac{2}{\gamma+1}\right)\left(1+\frac{\gamma-1}{2}M^2\right)\right]^{\frac{\gamma+1}{2(\gamma-1)}} \quad (5.37)$$

$$L = L^*\frac{A_t}{A_c} \quad (5.38)$$

where L = combustion chamber length (m)
L^* = chamber characteristic length (m)
M = Mach number in the combustion chamber (typical range 0.2– 0.4)
γ = isentropic parameter

This equation relates the combustion chamber's physical length directly to L^* if we assume a constant-cross-section cylinder for our combustion-chamber volume and use Eq. (5.35).

What is a good number to use for Mach number? Historically, smaller engines have a lower Mach number, and larger engines use a higher Mach number. Although thermodynamics explain this variance in Mach numbers, we basically want to avoid "pencils and pancakes"—chambers that are excessively long and narrow or excessively short and wide. Looking again at historical engines, we find that chamber length-to-diameter ratios (L_c / D_c) vary from 0.5 to 2.5. Again, this range is so large that it is difficult to choose a number. The good news is, however, that the historical data is "all over the map." This means that requirements other than combustion performance drive the decision, so it does not matter too much. Having said this, we note Huzel and Huang [1992] suggest a curve similar to the one in Fig. 5.24. Given a contraction ratio, we multiply by throat area to get chamber area, which in turn gives us chamber length and diameter.

The above method for determining the chamber dimensions is based on engine test data and simple gas-dynamic considerations. We can make this evaluation much more complicated by analyzing the individual processes which occur in the chamber, propellant impingement, stream atomization, vaporization, mixing, and combustion. We can then separate this set of simultaneous events into a series of uncoupled processes. Finally, we model these processes individually to arrive at a description of the engine design's end-to-end performance, including a thrust estimate. We can determine how each parameter of hardware design affects engine

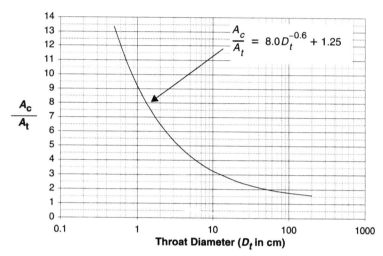

Fig. 5.24. Scaling Relations between the Combustion-Chamber Length and Contraction Ratio. If we know the nozzle throat's diameter, we can determine the chamber-contraction ratio. The contraction ratio gives us the chamber's diameter and, subsequently, its length. (adapted from Huzel and Huang [1992])

performance. Injector orifice diameters and gap widths, oxidizer and fuel impingement angles, stream velocities, and other parameters can all be determined. Computer codes can solve limited combinations of coupled processes, such as atomizing, vaporizing, and mixing. But we do not have one comprehensive, end-to-end, analytical or numerical tool that does not rely on some amount of empirical data. The section on injectors discusses this subject in more detail.

Dimensions of the Exhaust Nozzle

We determine the diameter of the nozzle exit geometry from the throat area and the desired expansion ratio:*

$$D_e = \sqrt{\frac{4\varepsilon A_t}{\pi}} \tag{5.39}$$

where D_e = nozzle exit diameter (m)
A_t = nozzle throat area (m²)
ε = nozzle expansion ratio

Selection of the nozzle contour offers several choices. Common nozzle geometries include conical, bell, deLaval and Rao (see Huzel and Huang [1992]). Conical

---

* The terms "expansion ratio" and "area ratio" are used interchangeably.

nozzles (see Sec. 3.3.3) are simple in concept and relatively easy to manufacture but they are not the most efficient in terms of thrust losses for a given length. Oblique shock formation causes a small loss in thrust when the flow expands from the throat area to the final expansion ratio. Also, because a radial velocity-component is at the nozzle exit plane, more thrust is lost. The deLaval nozzle* is designed to expand the flow to minimize thrust losses from oblique shocks. It has an axial velocity profile at the nozzle exit plane, so there is no radial thrust loss. But to maximize thrust, this nozzle is relatively long, resulting in a larger mass and geometric envelope. The Rao nozzle is a clever design which expands the flow and has an axial velocity profile with minimum length and thrust losses [Rao, 1958; Koelle, 1961; Frey and Nickerson, 1970]. Despite these fine qualities, it is not often used because fabricating its contour is difficult compared to other designs. The bell nozzle is a compromise. It has good performance, is simpler to design, and is both shorter and lighter. It turns and expands the flow with only small losses and produces a nearly axial flow at the nozzle exit plane. This combination of features makes it very popular.

For any of the nozzle concepts mentioned thus far, a fixed pressure profile at the nozzle exit plane limits expansion (see Sec. 3.3.2). But one family of nozzle concepts seems to achieve a variable-pressure profile at the nozzle exit plane and, therefore, efficient expansion over more altitudes. These nozzle designs include step, aero-spike, plug, expansion-deflection, radial flow, and horizontal flow [Koelle, 1961; Sutton, 1992]. To some degree, all of these nozzles attempt to match the nozzle exit plane pressure to the ambient pressure but suffer from increased thermal loading or drawbacks in carrying out the design, so we do not routinely use them.

The nozzle length depends on the nozzle design selected. For conical nozzles, the length depends on the divergence half angle (θ_{cn}). The greater the divergence angle, the shorter the nozzle for a given expansion ratio. However, an increased divergence angle means a greater thrust loss because the radial-velocity component increases. This effect is captured in the nozzle efficiency, which for an inviscid conical nozzle (see Sec. 3.3.2) is

$$\lambda = \frac{1}{2}(1 + \cos\theta_{cn}) \tag{5.40}$$

where λ = nozzle efficiency
 θ_{cn} = nozzle cone half angle (typical range 12°–18°)†

---

* Sometimes the bell nozzle is called an isentropic-ideal bell and the deLaval nozzle is referred to as a truncated-isentropic-ideal-bell nozzle. We use the shortened nomenclature here.
† The terms "cone half angle," "nozzle half angle," nozzle "divergence angle," and nozzle "expansion angle" are interchangeable. See Fig. 5.23 or Sec. 3.3.3.

For a conical nozzle with a circular cross section, the length of the diverging section is

$$L_n = \frac{D_e - D_t}{2 \tan \theta_{cn}} \qquad (5.41)$$

where L_n = conical nozzle length (m)
D_t = nozzle throat diameter (m)

We have assumed that the transition from the throat area to the cone is small and the effect on the nozzle length is negligible. Typically, cone half angles can range from 12° to 18°. The usual compromise is a half angle of 15°.

Bell nozzles offer better performance than 15° half-angle, conical nozzles of the same nozzle length. The exact analysis of bell nozzles requires complex computations using computational fluid dynamics based on the method of characteristics [Frey and Nickerson, 1970]. However, we can use parabolic-geometry approximations to the bell-nozzle dimensions. This technique works well enough so we seldom use the method of characteristics techniques for preliminary design.

A routinely used term for designing bell nozzles is the "percent bell": 80% bell, 70% bell, and so on [Huzel and Huang, 1992; Koelle, 1961]. The phrase refers to the length of the bell nozzle compared to a 15° half-angle, conical nozzle. Figure 5.25b shows the bell nozzle has an increased efficiency compared to conical nozzles for expansion ratios in the 10–40 range. For example, the 80% bell nozzle has an efficiency greater than 98.5%.

The parabolic design approximation for the bell nozzle is summarized in Fig. 5.25. The upstream throat contour is circular, with a radius 1.5 times the throat radius (r_t) terminating at the geometric throat (point T). The downstream throat radius is also circular with radius 0.382 r_t; it joins smoothly at the geometric throat with the upstream radius and continues to point N where the angle is θ_n. For a given nozzle length (L_n) defined by the percent bell, and expansion ratio (ε), the nozzle lip exit angle (θ_e) is specified as a function of expansion ratio and percent bell. This procedure locates point E on Fig. (5.25) and allows us to complete the design by drawing a smooth, parabolic curve between points N and E.

To use these charts:

1. Choose the fraction (L_f) from Fig. 5.25b that gives the desired nozzle efficiency. Usually, an 80% bell ($L_f = 0.8$) is used for an efficiency of 0.987.

2. Knowing L_f and the desired expansion ratio, use Fig. 5.25c to get the initial and final parabola angles.

3. Determine the nozzle configuration using basic geometry.

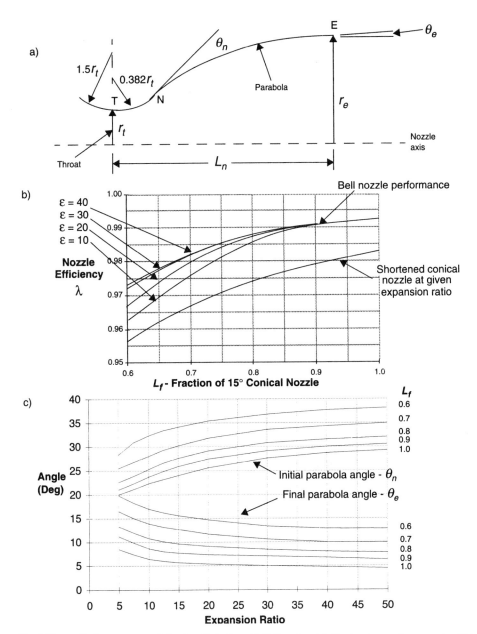

Fig. 5.25. Performance and Design Data for a Bell Nozzle. a) is the basic nozzle geometry; b) is the nozzle efficiency as a function of the percent of an equivalent 15°, conical nozzle; c) plots the initial and final parabolic angles versus desired nozzle expansion ratio (adapted from Rao, [1958]).

Estimating the Mass of the Thrust Chamber

Figure 5.26 summarizes mass fractions for select ablatively cooled engines.* The orbit-transfer engine, shown on the far left, is the Lunar Module Descent Engine (LMDE), a pressure-fed, 43,900-N thrust, ablatively cooled design that uses storable hypergolic propellants, N_2O_4 and Aerozine–50 [Giffoni, 1973/1974; TRW, 1967; TRW, 1970]. The family of boost engines, shown in the nine columns to the right of the LMDE, are pump-fed, ablatively cooled designs for use with LOx/H_2 or $LOx/RP-1$ propellants. The figure lists engine mass fractions of the injector, nozzle, chamber, ablative liner, and nozzle. Table 5.7 lists these parameters for each engine, including the chamber pressure, expansion ratio, throat diameter, and the total mass of the thrust chamber. We can use this table to generate subsystem mass estimates if other information is not available.

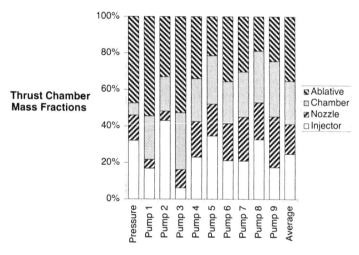

Fig. 5.26. Thrust Chamber Mass Fractions for Various Engine Components. Ten ablatively cooled engines and an average of these are shown; one is pressure-fed and the others are pump-fed. The mass percentage of the total thrust-chamber mass is broken down into the injector (including feed components), the nozzle structure, the chamber structure, and the ablative material. This data is shown in Table 5.7.

---

* Estimating the mass of engines with other cooling approaches is VERY difficult at the conceptual level. The numbers we use here are based on ablatively cooled engines because they are conservative. The main reason we would go to a different cooling approach is to reduce mass. Because other cooling methods increase complexity, we should not use them unless mass is severely limited. In this case, ablatively cooled engines define the upper limit on engine mass.

5.4 System Sizing, Design, and Trade-offs 227

Table 5.7. Mass Fraction (and Other Data) for Various Ablatively Cooled Engines. Knowing how the component masses vary as a fraction of the total engine mass, we can estimate the component masses if we know any particular component mass. The column on the right side shows a numerical average of the ten sets of data. We use this average for design (see Fig. 5.26). LMDE is the Lunar Module Descent Engine. The boost engine numbers are based on design studies done at TRW, Inc.

	Pressure-Fed	Boost Engines—Pump-Fed									
	LMDE	1	2	3	4	5	6	7	8	9	Avg
Thrust (kN)	43.9	178.3	178.3	245.2	1560	1560	1783	3343	3343	6687	--
Chamber pressure (kPa)	717	2068	6893	2068	2068	6893	2068	2068	6893	2068	--
Expansion ratio	47.4	160	180	60.7	7.2	16	7.3	7.2	16.6	8.1	--
Throat dia. (cm)	21.1	23.6	13.0	28.2	84.3	43.2	89.4	123.4	62.5	176.0	--
Mass (kg)	163	393	291	475	1082	1309	1263	2809	3193	7600	--
Injector fraction / misc. feed	0.184/ 0.137	0.169	0.43	0.062	0.231	0.347	0.212	0.210	0.327	0.175	0.249
Nozzle fraction	0.138	0.098	0.051	0.101	0.194	0.173	0.202	0.240	0.201	0.277	0.163
Chamber fraction	0.067	0.238	0.189	0.310	0.235	0.265	0.230	0.246	0.282	0.303	0.237
Ablative fraction Nozzle/chamber	0.241/ 0.233	0.545	0.330	0.526	0.340	0.215	0.356	0.303	0.189	0.245	0.352

Table 5.8 summarizes some analytical formulas that can be used to calculate the mass of simple combustion-chamber and nozzle designs. In these simple models, the combustion chamber is a cylinder with a conical subsonic contraction to the throat. We assume that the nozzle is conical with expansion half angle θ_{cn} and present results for nozzles with both constant and tapered wall thicknesses.

The thickness of the thrust-chamber wall is a function of the local static pressure and the structural strength of the chamber material. Assuming a cylindrical combustion chamber, the hoop stress in the wall (f_h) is given by Sarafin and Larson [1995]:

$$f_h = \frac{pr_c}{t_w} \qquad (5.42)$$

Table 5.8. Analytical Method for Estimating Thrust-Chamber Mass. We can approximate the material volume of the combustion chamber and exhaust nozzle, then multiply it by the material density. Typical values of material density are 8000 to 8600 kg/m³. The point of this chart is to allow us to estimate the thrust chamber's structural mass. We assume a cylindrical combustion chamber and conical contraction area, both with constant thickness. We also assume a conical nozzle, but it can have straight or tapered walls. Now we can use the mass-fraction data to determine the masses of other components.

Mass Component	Analytical Result	Sketch
Combustion chamber (m_{cc}) • Throat radius, r_t • Chamber radius, r_{cc} • Constant wall thickness, t_w • Constant contraction half angle, θ_{cc} • Cylinder length before contraction, L_{cc} • Wall density, ρ	$m_{cc} = \rho V_w$ $= \pi \rho t_w \left(2 r_{cc} L_{cc} + \left(\dfrac{r_{cc}^2 - r_t^2}{\tan \theta_{cc}} \right) \right)$	
Nozzle - constant thickness (m_n) • Nozzle half angle, θ_{cn} • Nozzle exit radius, r_e	$m_n = \pi \rho t_w L_n (r_e + r_t)$	
Nozzle - tapered wall (m_n) • Throat wall thickness, t_w • Nozzle exit wall thickness, t_e	$m_n = 2 \pi \rho L_n \left[\dfrac{1}{3} f_1 f_2 L_n^2 + \dfrac{1}{2}(f_1 r_t + f_2 t_t) L_n + r_t t_t \right]$ $f_1 = \dfrac{t_e - t_t}{L_n}$ $f_2 = \dfrac{r_e - r_t}{L_n}$	

where f_h = hoop stress in the circular cylinder wall (Pa)
  p = pressure in the cylinder (Pa)
  r_c = radius of the circular cylinder (m)
  t_w = thickness of the cylinder wall (m)

We want to determine the wall thickness. To find it, we define the *ultimate tensile strength* (F_{tu}) of the wall material and let this be the "allowable stress." The applied pressure (p, usually the chamber pressure) typically has a safety factor (f_s) of "2" applied. We call this factored pressure *burst pressure* (p_b):

$$p_b = f_s p \tag{5.43}$$

The resulting wall thickness is

$$t_w = \frac{p_b r_c}{F_{tu}} \tag{5.44}$$

The reality of structural design requires the actual wall to be 1.5 to 2.5 times thicker than this number to account for stress concentrations and other stresses not modeled in a simple hoop-stress analysis. Columbium is a typical material for thrust chambers. Nickel-based materials of this sort have densities in the range of 8000 to 8600 kg/m^3 and a practical tensile strength of around 310 MPa.

Static pressures through the thrust chamber quickly drop. At the throat, the static pressure is about 50% of the total (chamber) pressure. The pressure further drops to near ambient at the exit. In fact, overexpanded nozzles (operating at low altitude) actually have a compressive buckling problem near the nozzle exit. To account for changes in nozzle thickness, Table 5.8 includes relations for linearly tapered nozzles. In real systems, we typically taper the thickness faster than linear, but the linear assumption is conservative.

Once we determine the masses of the chamber and nozzle, we can use the mass fraction data (Fig. 5.26 and Table 5.7) to determine the ablative material mass and the injector mass.

Designing the Propellant Injector

The propellant injectors are the heart of the thrust chamber. A better injector design gives better engine performance. The injectors are the hardware that enable us to turn the analytic thermochemical predictions of specific impulse into reality. Designing liquid-propellant injectors is difficult and time consuming—difficult, because so many physical processes are occurring simultaneously; time consuming, because we must optimize many parameters of the injector design (gap widths, hole diameters, injection angles and velocities) through painstaking test programs.

Injectors fall into two broad categories: impinging and nonimpinging. The word "impinging" suggests that two or more streams strike one another to atomize the streams. By far the most popular approach is the impinging type. Consider two impinging streams. Those with two fuels or two oxidizers are "like;" those with one fuel and one oxidizer are "unlike." Fig. 5.27 introduces several injector types. A typical injector assembly incorporates dozens and even hundreds of individual injectors.[*] The sketches suggest we must specify many design parameters in an

---

[*] Pintle injector systems typically employ just one or, at most, a few elements.

injector design, including injector type (doublet, triplet, coaxial, pintle); which propellant is at the center of a repeating element (fuel centered, ox–fuel–ox, ox–ox or fuel–fuel); number of injectors; impingement angles of the propellant stream; flow rates; hole sizes; and evaluation of combustion stability [Huzel and Huang, 1992; NASA, 1976; Sutton, 1992; TRW, 1967, 1970; Elverum and Morey, 1959].

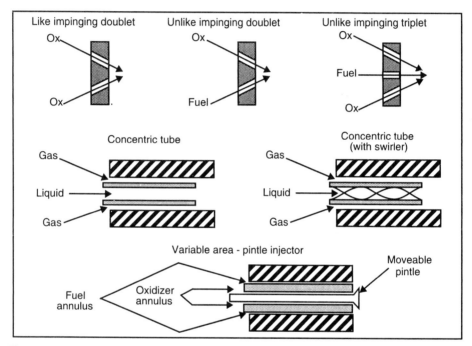

Fig. 5.27. **Typical Injector Assemblies.** These are the typical configurations used. They include doublets, triplets, concentric tubes, and the pintle injector.

Analysis of injector designs is complicated because many physical phenomena influence propellant injection, mixing, and combustion, all of which occur simultaneously. When the oxidizer and fuel enter the engine, the flow streams strike one another, shatter, atomize, vaporize, diffusively mix, exchange heat, and undergo violent combustion reactions. All of these processes occur virtually at once, in a relatively small space and short time. By analyzing each of the processes as a series of uncoupled actions, we make analysis workable. Uncoupled action means that each one of the physical phenomena, such as atomization, occurs separately and distinctly from all the other processes, such as vaporization, mixing, and reaction. Thus, we can analyze atomizing and vaporizing separately, making the total task of designing an injector much easier to complete. We assume uncoupled physical processes for two reasons: first, we want to make analyzing the injector design pos-

5.4 System Sizing, Design, and Trade-offs

sible because solving the entire coupled-processes problem cost effectively is too difficult. Second, we simplify interpreting test data from nonreacting flow experiments—data used in injector design.

The difficulty is that such simplification may not represent what is really happening in the rocket engine. Complex computer codes have been developed to try modeling the simultaneous occurrence of all the processes. This approach is limited by an incomplete understanding of the details of propellant-reaction chemistry and the multiphase fluid dynamics. In addition, doing such extensive calculations and repeating them many times while designing an injector can be expensive. Commonly, we compromise by analyzing experimental data and assuming uncoupled processes. We use the individual analyses to evaluate atomization, drop-size distributions, mixing, and set design parameters on the injector. Then, we project numerical performance to verify the injector design. Of course, we must still run rigorous tests to select the best design parameters and verify overall engine performance.

For preliminary design, we want to know just the basics: what type of injector elements do we use? How many injector elements do we need? What are the approximate element dimensions? We can estimate answers to these questions with relative ease, but verifying that they are correct is difficult and beyond the scope of our discussion here. We start by answering the simple questions.

Choose the Injector Type—We begin by selecting injector classes: impinging or nonimpinging, like or unlike. We then select the injector type: doublet, triplet, coaxial, or coaxial pintle. The selection criteria include

- Injector and engine performance under operating conditions similar to the conditions of our mission
- Unique mission requirements that might mandate or favor one type of injector over another
- The drop-size performance of the injector using the same or similar propellants
- The number of repeating elements required
- Hardware complexity
- The completeness and quality of the database on a particular injector type with the selected propellant for our application

Because final validation of an injector occurs in testing, we are motivated to use and build on databases of injector performance using the selected propellants. Various injector types have been used successfully for different missions. Doublets, triplets, coaxial pintle, and other injectors have been demonstrated in Earth-to-LEO pump-fed applications, and coaxial doublets, quadlets, and coaxial pintle injectors have worked for LEO-to-GEO pump- and pressure-fed engines.

Often, an engine-development team, contractor, or government may prefer a specific type of injector. If we are part of that team, we need to find out what has worked for similar mission applications and engine types. We should exploit the strength of the development team by selecting an injector type that has been successful for the team. We also need to consider the availability of test data for the injector type and propellants. Injector development is largely based on test data, and test data is usually expensive to obtain, so large databases mean reduced development cost.

Check the Injector Pressure Drop—In Sec. 5.3.5, we assume an injector pressure drop without knowing the injector type. The rule of thumb is that the pressure drop across the injectors is some percentage of the chamber pressure, typically ranging from 20% to 25%. However, this percentage can vary with injector type. For injectors that use unlike-impinging elements, 10% to 15% is typical, whereas like-impinging elements are in the 15% to 25% range. With concentric-tube elements, lower bounds of 5% have been recorded. Some amount of pressure drop is desirable to isolate the propellant feed system from combustion-driven fluctuations[*] in chamber pressure. At this point we may wish to change our estimate based on our choice of injector type. Because altering Δp also changes the system pressure levels, we need to iterate the design. Remember, only extensive analysis and testing can determine the "correct" answer.

Choose the Number and Size of Injectors—Once we have specified the desired overall injector pressure drop, we can determine the number of injector elements from the total propellant mass flow rate and the database on the mass flow rate per injector element. Most of the pressure drop (Δp) through the injector comes from the increase in the flow dynamic pressure[†] (assuming that the flow is initially stagnated [$v = 0$]—a reasonable assumption for large-diameter manifolds):

$$\Delta p = K \frac{1}{2} \rho v^2 \tag{5.45}$$

We can then use the continuity equation to determine the flow area for an individual injector:

$$\frac{\dot{m}}{N} = \rho v A_{inj} \tag{5.46}$$

Combining these results by equating velocity gives the relationship between injector pressure drop and the number of injectors:

$$A_{inj} = \frac{\dot{m}}{N} \sqrt{\frac{K}{2\rho \Delta p}} \tag{5.47}$$

---

[*] The usual definition for a "smoothly" running engine is a pressure fluctuation of 5% around the mean level of chamber pressure.
[†] We try to minimize the nonreversible losses so the maximum amount of pressure can be used to increase flow velocity. A higher velocity of the injected propellant makes the propellant atomize and mix better.

where N = number of injectors
  v = flow velocity through the injector (m/s)
  $\dot{m}$ = total mass flow rate (kg/s)
  A_{inj} = injector inlet cross-sectional area (m²)
  K = head loss coefficient (1.2 for radiused inlet, 1.7 for nonradiused)
  ρ = density of the propellant (kg/m³)
  Δp = desired pressure drop across the injector (Pa)

Notice we have added a factor K. This *head loss coefficient* takes nonreversible pressure losses into account. Typical numbers for K are 1.2 if the injector inlet has radiused edges and 1.7 if the edges are sharp; we get larger losses with sharp edges (see Huzel and Huang, [1992]). These "typical numbers" are for standard injector configurations with typical lengths and cross sections. For specific designs, it is a good idea to perform tests (usually with water) early in a program to accurately determine values for K.

Representative data for injector types is in Table 5.9. This data captures an initial set of trade-offs for injector design. Equation (5.47) implies that as the number of tubes or flow channels (N) increases, the mass flow in each channel decreases and the flow area (or diameter) decreases. For an injector to work well—that is, produce small droplets which atomize, vaporize, and mix rapidly—the tube diameter or channel width must be small. The velocity must be high so when the streams hit, the fluid inertia force is much greater than the surface tension forces in the fluid. Small drops result. However, to minimize fabrication complexity and cost, we want the number of injector elements and channels or tubes to be small. An injector type that requires a large number of elements to guarantee uniform mass flow through each injector element means that the manifold hardware for the propellant supply is complex and therefore costly. Reduced numbers of injector elements or flow channels imply less complex hardware that is easier to design and less expensive to make. Some injectors require larger numbers of injector elements. Doublets and triplets often require hundreds of elements across the head end of the combustion chamber, whereas the coaxial pintle injector requires only a few elements.

A method to select N (the number of injector elements), and therefore the number of flow tubes or channels, is as follows:

- Select N
- For the type of injector, calculate mass flow rate, flow area, diameter or gap width, and velocity
- Verify that these values are reasonable for the injector type selected.

Table 5.9 contains this information for several injector types. For more detail, consult Huzel and Huang [1992]; Oefelein and Yang [1992]; NASA [1976]; TRW [1967, 1970]; Sutton [1992], and Elverum, et al. [1967].

Table 5.9. Typical Injector Parameters. We can use this historical data to estimate values for new injectors.

Parameter	Engine Designation		
	F-1	RL-10	LMDE
Thrust (N)	6,770,000 (sea level) 7,776,400 (vacuum)	66,700 (vacuum)	43,800 (vacuum)
O/F	2.27 (engine) 2.40 (chamber)	5.0	1.60
I_{sp} (s)	265.4 (sea level) 304.8 (vacuum)	444 (vacuum)	304 (vacuum)
Chamber pressure (Pa)	6,768,786	2,757,143	716,857
Oxidizer	Liquid oxygen	Liquid oxygen	N_2O_4
Fuel	RP-1	Liquid H_2	Aerozine - 50
Injector type	Like doublet	Coaxial	Coaxial pintle (for throttling)
Oxidizer mass flow rate (kg/s)	1812 (engine) 1788 (chamber)	12.80	9.06
Fuel mass flow rate (kg/s)	798 (engine) 743.6 (chamber)	2.56	5.66
Number of injector elements (N)	714 (Ox) 702 (Fuel)	216	1
Total flow area (cm²), A	423.7 (Ox) 561.7 (Fuel)	5.16 (Ox) 15.48 (Fuel)	3.10 (Ox) 3.23 (Fuel)
Orifice diameter or gap width (cm)	0.615 (Ox) 0.714 (Fuel)	0.201 (Ox) 0.043 (Fuel-annulus)	Variable area for throttling
Injector impact half angle (deg)	20 (Ox) 15 (Fuel)	parallel	90
Injector flow velocities (m/s)	40.50 (Ox) 17.00 (Fuel)	25.91 (Ox) 43.65 (Fuel)	20.06 (Ox-max) 17.50 (Fuel-max)

Cooling the Thrust Chamber

Liquid bipropellant engines for boost, sustainment, and upper-stage applications require thrust-chamber cooling to control the wall temperature of the combustion chamber. Table 5.10 summarizes the commonly used cooling techniques and states when we usually use each method.

Two other techniques are frequently used for engine testing but not often for flight applications: heavyweight heat-sink cooling and water-jacket cooling. Both methods usually add too much mass to the engine to be considered seriously for

Table 5.10. Methods for Cooling the Thrust Chamber. We typically use four methods, listed here with their usual application.

Cooling Technique	Mission Phase	Comment
Regenerative	Boost, sustain, orbit transfer	• Sometimes combined with film cooling
Ablative	Boost, sustain, orbit transfer	• Often combined with film cooling
Radiation	Orbit transfer, stationkeeping	• Often combined with film cooling
Film/boundary layer	Orbit transfer, stationkeeping	• Often combined with radiation cooling

flight applications. The four different flight-system approaches are shown in Fig. 5.28 [NASA, 1972; Bartz, 1968; Huzel and Huang, 1992; Sutton, 1992].

Regenerative cooling is often used in large liquid engines. In *regenerative cooling*, one propellant, typically the fuel,[*] is forced through cooling passages that surround the combustion chamber and nozzle walls. The heat from the combustion gases conducts through the walls—mainly in the radial direction—and is convected away by the propellant flowing through the passages. In this way the propellant recaptures the "waste heat" by increasing the energy content. This cooling technique mainly applies in mid-to-high thrust levels because the heat flux to the chamber walls increases as the mass flow through the engine (thrust) increases. As a result, regenerative cooling is used predominantly with pumped-engine designs operating in steady-state types of duty cycles. Of course, manufacturing the propellant-flow passages increases the hardware complexity and therefore the cost of producing the engine.

A variation of regenerative cooling is *inter–regenerative* cooling, in which we create a cool propellant film at the chamber wall. To do this, we place annular passages inside the combustion chamber. The propellant from the injectors (typically the fuel) forces through the annular channel, creates a film at the walls, and reduces the wall temperature. But this propellant does not mix and react well with the rest of the propellants in the chamber, so this approach results in a slight loss of specific impulse.

The essential features of *ablative cooling* are shown in Fig. 5.28. As the engine operates, hot combustion gases vaporize the ablator material that lines the combustion chamber wall. These materials tend to be silica, quartz, or carbon cloth and resin composites attached to the thrust chamber. The thickness of the ablative creates enough mass in the layer to maintain cooling throughout the burn duration. In principle this technique works on engines of all thrust levels, but the operational designs using this technique have tended to be for engines with low- to mid-sized thrust levels. The Apollo space engines are typical examples of this type of cooling

---

[*] The oxidizer has been used successfully in various programs, such as the hydrogen peroxide-cooled Black Knight engines. However, there are additional concerns, such as "oxidizing" the coolant channels.

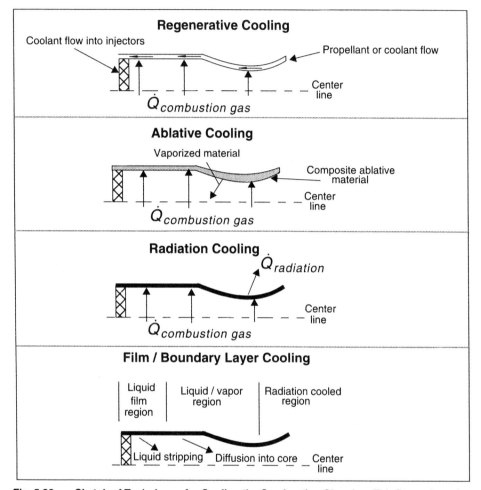

Fig. 5.28. Sketch of Techniques for Cooling the Combustion Chamber. This figure shows how each of the cooling techniques work [$\dot{Q}$ = heat flow (W)].

technology. Because the liner material on the combustion-chamber wall ablates away as the engine fires, these engines have limited life—typically fewer than 2,000 seconds of firing time. This technique for cooling the thrust chamber has a flexible duty cycle. That is, we can stop and restart the engine as long as ablative material is left on the chamber walls.

In *radiation cooling* of pressure-fed engines, the heat from the combustion gases transfers to the combustion chamber walls by convection and radiation. The heat conducts through the chamber walls mainly in the radial direction and radiates away to the surroundings. The main limitation on steady-state temperature is that when the walls reach this temperature, they must be thick enough so the material has the strength to contain the engine pressure.

The temperatures of the chamber walls are not uniform axially, because the flow speed of the combustion gases varies. The Mach number in the combustion chamber is low (0.1 to 0.4), so the convective heat transfer to the walls is relatively low even though the combustion gases are hot. The heat load to the wall is largest in the throat region, where the flow speeds are high, and the densities and temperatures are also near their stagnation values. Wall temperatures in the supersonic part of the nozzle are lower. In this region, the flow speed is high and the static temperatures are low because the energy has converted to kinetic energy.

Axial nonuniformities in temperature generally require a detailed thermal analysis of the combustion chamber, throat region, and nozzle to calculate the distribution of steady-state wall temperature. In a radiation-cooled engine, we want the material in the combustion chamber wall to have a high melting point. Thus, we often use refractory materials, such as columbium, with special coatings for oxidation protection, such as rhenium, on the inside of the chamber and maximum surface emissivity in the outside. This cooling technique has a flexible duty cycle, but we must analyze the time between restarts to assess the effects of initial wall temperature if the system has not cooled down completely. Within these operational constraints, the radiation-cooled engine can last a long time.

Film or *boundary-layer cooling,* sketched in Fig. 5.28, is rarely used by itself. Most often, we combine it with other techniques for cooling the combustion chamber. The sketch shows the injectors intentionally throwing a liquid film, typically the fuel, along the chamber wall. The liquid propellant forms a film along the chamber walls; as the cooler film flows along the walls of the chamber, it increases the thermal resistance and reduces the wall temperature. Near the throat the film is mostly vapor with little liquid remaining. The heat flux to the walls increases because of convection. The chamber walls must have a relatively high thermal conductivity so the heat flux in the throat region conducts quickly away in the axial direction, both upstream into the chamber and downstream into the supersonic nozzle. As shown in the sketch, we commonly use radiation cooling in the downstream part of the nozzle because the gas temperatures are dropping, the walls are relatively cool, and the area available for radiation cooling is increasing. If the nozzle walls have high thermal conductivity, heat from the throat moves along the nozzle walls and radiates away to the surroundings. Film cooling most often applies to low-thrust, pressure-fed engines with low chamber pressures. This type of cooling supports engine restarts, so it works for long missions.

We can analyze any of these cooling methods at varying levels of detail by using the energy conservation (first law of thermodynamics) equation:

$$\dot{Q}_{in} = \dot{Q}_{out} + \Delta h \qquad (5.48)$$

where $\dot{Q}_{in}$ = heat flow into the control volume (W)

$\dot{Q}_{out}$ = heat flow out of the control volume (W)

Δh = any change in enthalpy in the control volume (W)

The simplicity of these conservation equations is deceiving. During the detailed design of any real engine, we must analyze fully coupled mass, momentum, and energy reacting flow, as well as detailed thermal characteristics of the injector and thrust chamber. Below, we briefly discuss the key assumptions needed at a particular form of the energy equation for each technique to arrive.

Regenerative Cooling—For the regenerative system, the control volume (Fig. 5.29) is taken as a small length of the regenerative flow tubing. Convection and radiation from the thrust chamber add heat to the coolant flow. Radiation from the cooling channel wall and an increase in the coolant flow enthalpy remove this heat:

$$(\dot{Q}_{rad} + \dot{Q}_{con})_{in} = (\dot{Q}_{rad})_{out} + \dot{m}_{coolant} C (T_{out} - T_{in}) \tag{5.49}$$

where $\dot{Q}_{rad}$ = heat addition due to radiation (W)
$\dot{Q}_{con}$ = heat addition due to convection (W)
$\dot{m}_{coolant}$ = coolant mass flow rate (kg/s)
C = heat capacity of the fluid (J/kg·K)
T_{in} = temperature of the fluid entering the control volume (K)
T_{out} = temperature of the fluid leaving the control volume (K)

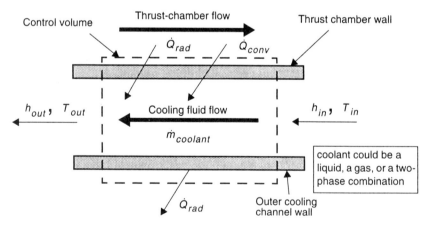

Fig. 5.29. Schematic of Analysis for Regenerative Cooling. Coolant moves through passages next to a hot wall and absorbs heat from the wall.

The total change in temperature of the cooling fluid is the sum of the individual temperature changes. Analytically, we set the length of each segment of fluid so the convective and radiative heat inputs are essentially constant over the length of the cooling tube. Typically, each tube segment is 1% to 10% of the total tube length. We assume the flow is steady and the cooling fluid has negligible changes in kinetic energy. We also assume the temperature drop across the tube wall is negligible. The radiative heat input from the combustion gases can be optically thick or thin depending on whether the combustion gases contain particles as products. The emissivity of the gases and particles can vary with wavelength. The combustion gas temperature varies with axial position. We assume it is constant and equal to the local value for each small tube segment. We can analyze in more detail by making the length of the individual tube segments smaller and smaller, passing to the limit of a differential tube length. At this level of complexity, we do a detailed thermal analysis at the same time, allowing for radial and axial heat transfer in the tube walls over the length of the cooling tubes.

Ablative Cooling—To analyze ablation-cooled engines, we use a different control volume (Fig. 5.30). The control volume is a small length of the combustion-chamber wall, taken so the heat input to the wall through radiation and convection balances the enthalpy of the ablated mass flow:

$$(\dot{Q}_{rad} + \dot{Q}_{con})_{in} = \dot{m}_{ablative} h_{ablative} \qquad (5.50)$$

where $\dot{m}_{ablative}$ = the mass flow of the ablative material (kg/s)
$h_{ablative}$ = enthalpy of material ablation (J/kg)

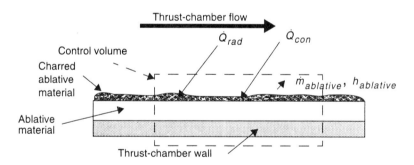

Fig. 5.30. Sketch for Analysis of Ablative Cooling. Heat from the combustion chamber ablates material on the wall, preventing excessive heat from reaching the wall.

Usually, this analysis neglects changes in the kinetic energy of the main flow of combustion gases as well as the ablated mass flow. The convective and radiative heat inputs can vary with axial position in the thrust chamber—the heat load into the control volume being largest in the throat region. Evaluation of the radiative heat input must address the possibility of upstream/downstream coupling. Sections of the chamber that "see" the throat must include the throat's contribution in the radiation heat. The throat wall is hot, and its shape and configuration may alter the total temperature of the other elements from radiation. A conductive heat flow out of the control volume is included in the energy-conservation equation, allowing for conductive heating of the unreacted ablative material. The resulting energy equation is either steady or unsteady, depending on whether we consider the heat conduction into the unreacted ablative material. We can best characterize the enthalpy of the ablated material using equilibrium thermochemistry for this relatively simple model. In more detailed analysis, we would include the actual kinetics of ablation. Also, if we include conduction heating of the unablated material, we should analyze temperatures in the thrust-chamber walls.

Radiation Cooling—The control volume (Fig. 5.31) for analyzing radiation-cooled engines consists of a small section of combustion chamber. The analysis is steady state, and changes in kinetic energy of the core combustion gases are neglected. The input to the control volume has two heat inputs: convection and radiation. We usually neglect the thermal resistance of the chamber walls,

$$(\dot{Q}_{rad} + \dot{Q}_{con})_{in} = \varepsilon A \sigma F \left(T_w^4 - T_{environment}^4 \right) \tag{5.51}$$

where ε = emissivity (range 0–1)
A = cross-sectional area of the wall (m²)
σ = Stefan-Boltzman constant $\left(5.67051 \times 10^{-8} \dfrac{W}{m^2 K^4} \right)$
F = shape factor
T_w = temperature of the thrust chamber wall (K)
$T_{environment}$ = temperature of the surroundings near the wall (K)

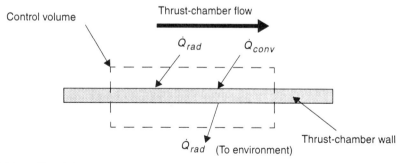

Fig. 5.31. **Sketch for Analysis of Radiative Cooling.** Heat is absorbed by the wall and then radiates again into space.

In determining the local wall temperature, we must include the possibility of upstream/downstream coupling for radiative heat output and input. Some sections of the chamber walls "see" the hot throat region. Likewise, heat from the walls radiates not only to cold space but also to the cooler areas of the nozzle. Depending on the geometry of the engine, these effects may be important.

Film/Boundary Layer Cooling—To analyze conservation of energy in film/boundary layer-cooled engines, we employ yet another control volume. The analysis is steady-state, neglects changes in kinetic energy, and includes radiative and convective heat inputs. The heat flows out of the film layer and into the chamber walls. We can evaluate the mass flow and enthalpy entering the control volume in several ways. One way is to assume the film is liquid only, and the enthalpy and mass flow correspond to that of the fuel or oxidizer, whichever is used for cooling the walls. Typically the fuel is selected to reduce corrosion wherever the chamber walls are hot. The conservation equation assumes that the propellant leaving the control volume fully vaporizes and then heats to some known temperature, such as the mean temperature of the core combustion gases at the end of the control volume:

$$(\dot{Q}_{rad} + \dot{Q}_{con})_{in} = \dot{Q}_{wall} + \dot{m}_{film}\left(\int_{T_{vap}}^{T_{out}} c_p dT + h_{fg} + c[T_{vap} - T_{in}]\right) \quad (5.52)$$

where c_p = heat capacity of vaporized gases at constant pressure (J/kg·K)
C = heat capacity of the prevaporized fluid (J/kg·K)
h_{fg} = heat of vaporization of the fluid (J/kg)
T_{vap} = vaporization temperature of the fluid (K)
$\dot{Q}_{wall}$ = heat flow into the wall (W)
$\dot{m}_{film}$ = mass-flow rate of the wall coolant (kg/s)
T_{in} = temperature of the coolant entering the control volume (K)
T_{out} = temperature of the vaporized coolant leaving the control volume (K)

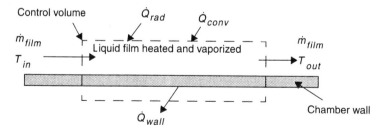

Fig. 5.32. Sketch for Analysis of Film Cooling. A thin fluid film near the wall is cooler than the engine core. This allows the wall to be cooler.

Other important issues in designing combustion-chamber hardware relate to materials of construction. In bipropellant engines the combustion-chamber walls must withstand corrosive oxidizers and high temperatures. This mandates compatible materials and extensive thermal analysis to make sure the combustion-chamber hardware survives during the firing. Current designs often use exotic metals, such as columbium, to exploit their high-temperature properties. Other designs incorporate coatings (such as rhenium) which can survive the corrosive action of bipropellant oxidizers.

Liquid Monopropellant Thrust Chamber

A *monopropellant* is a single substance (usually a liquid) that reacts or decomposes by itself to a gas, releasing energy. Monopropellants are reacted by metallic or thermal catalysts to a lower energy state while releasing stored chemical energy in the form of heat. They can be used in some propulsive applications requiring low energy expenditures and for gas generators. Two often-used monopropellants are hydrogen peroxide (H_2O_2), an oxidizer, and hydrazine (N_2H_4), a fuel.

Hydrogen peroxide usually comes with some percentage of water. Typical concentrations of H_2O_2 range from 85% to 98% by mass. In the catalytic decomposition of H_2O_2, we get hot oxygen gas and more water:

$$2\,H_2O_2\,(l) + n\,H_2O\,(l) \rightarrow (n+2)\,H_2O\,(g) + O_2\,(g) - 108{,}480\text{ joules}^* \quad (5.53)$$

Remember, by convention, this chemical equation is isothermal, so for an exothermic reaction (gains heat) we must remove heat to maintain the constant temperature; thus, there is a negative sign in front of the energy. Temperatures range from about 1500 K down to the injection temperature and are a function of the amount of water mixed with the peroxide. For complete decomposition, the catalyst bed is typically 5 cm long and the diameter is sufficient to give a "rule of thumb" mass flux of around 250 kg/m²·s.

Hydrazine (N_2H_4) is a bit more complicated. In the presence of a catalyst, it decomposes into ammonia (NH_3) and nitrogen gases:

$$3\,N_2H_4 \rightarrow 4\,NH_3 + N_2 - 336{,}280\text{ joules} \quad (5.54)$$

However, further decomposition of the ammonia is an endothermic reaction:

$$4\,NH_3 \rightarrow 2\,N_2 + 6\,H_2 + 184{,}400\text{ joules} \quad (5.55)$$

Thus, when we design the catalyst beds, some empirical art affects our choice of configuration. A certain configuration gives us complete hydrazine decomposition

---

* This equation is a bit misleading. The energy (108.48 kJ) does not include the vaporization of the n moles of H_2O. If we know the concentration (n) then, from Table 4.2, we can decrease the energy number by $n \times 44{,}030$ joules.

with the corresponding highest possible temperature. A slightly larger bed starts to allow the ammonia to decompose, lowering the maximum temperature. The highest temperature is usually desired for thrusters, and lower temperatures are used in gas-generator applications. The temperature can range from 1650 K down to 860 K. Schmidt [1984], Rocket Research Corporation [undated], Schmitz, et al. [1966], Sackheim [1973], and Huang and Sackheim [unpublished] give guidelines on how to design the "cat beds" to get the desired thermochemistry. Typical mass fluxes for hydrazine catalyst beds can be as high as 200 kg/m^2·s.

Monopropellant hydrazine is by far the most popular type of propulsion for controlling spacecraft attitude and small velocity changes because it can produce a higher specific impulse than H_2O_2 (230 s versus 150 s for H_2O_2). Its good handling characteristics, relative stability under normal storage conditions, and clean decomposition products have made it the standard. The general sequence of operation in a hydrazine thrust chamber is:

- When the attitude-control system signals for thruster operation, an electric-solenoid valve opens, allowing hydrazine to flow. This action may be pulsed (as short as 5 ms) or long duration (steady state).
- The pressure in the propellant tank forces liquid hydrazine into the injector. It enters as a spray into the thrust chamber and contacts the catalyst bed.
- The catalyst bed consists of alumina-ceramic pellets coated with iridium. The most widely used catalyst, manufactured by Shell Oil Company, is called Shell 405. Incoming liquid hydrazine heats to its vaporizing point by contact with the catalyst bed and with the hot gases, leaving the catalyst particles. The temperature of the hydrazine rises to a point where the rate of its decomposition becomes so high that the chemical reactions are self-sustaining.
- By controlling the flow variables and the geometry of the catalyst chamber, we can tailor the proportion of chemical products, the exhaust temperature, the molecular weight, and thus the enthalpy for a given application. In a thruster application for which specific impulse is paramount, we try to provide 30% to 40% ammonia dissociation, which is about the lowest percentage that can be maintained reliably. For a gas generator, in which we usually want lower-temperature gases, we provide higher levels of ammonia dissociation (up to 80% dissociation is achievable).
- Finally, in the thruster, the hydrazine decomposition products leave the catalyst bed and exit from the chamber through an exhaust nozzle to produce thrust.

At the beginning of this chapter, Figure 5.2 shows a basic schematic of a single monopropellant thruster. Figure 5.33, in the next section, is a schematic of the monopropellant-hydrazine system used for the NASA Tracking and Data Relay Satellite (TDRS) spacecraft. It is one of the larger hydrazine systems built and flown to date, containing about 680 kg of hydrazine. It operates in a blowdown mode, in which the propellant and pressurant gas reside in the same tank. As propellant expels from the tank, the pressure level drops. The system has 24 thrusters operating at a nominal thrust of 5 N for reaction control. These hydrazine thrusters are completely redundant for all functions. The system has two large positive-expulsion tanks with elastomeric diaphragms.

Latching isolation valves are used in series for fault protection and redundancy, to tolerate any single fault. They are called latching valves because they stay in whatever position they were last commanded to take (open or closed), therefore requiring power only for the command period of approximately 200 ms. The system also has manual fill-and-drain valves. Each tank has one propellant valve and one gas valve to load the hydrazine and gaseous nitrogen pressurizing gas on their respective sides of the positive-expulsion tank. The TDRS propulsion system also has miscellaneous components such as filters (to keep from contaminating the system), pressure transducers, temperature sensors, and thermostatically controlled heaters for the catalyst bed and lines. These heaters increase thruster lifetime (temperature gradients during repeated cold-starts cause the particles to break down) and keep the propellant from freezing.

5.4.2 Configuring the Propellant-Feed System

The propellant-feed system consists of all components between the propellant tanks and the thrust chamber, including pumps and pump drives. However, we discuss pumps and pump drives in Sec. 5.4.3 and limit our discussion here to the propellant-feed lines and the many valves, actuators, and fittings required for safe and reliable operation. Figure 5.33 sketches the feed configuration for a tank-pressure-fed system, showing most typical components.

As mentioned in Sec. 5.3.5, we want to keep line losses between 35,000 and 50,000 Pa. This, of course, does not include losses from a cooling jacket or injector. To meet this goal, propellant flow rates are usually about 10 m/s. We then use the continuity equation to determine the propellant channel area (A_c) and diameter (D_c), given the propellant mass flow rate ($\dot{m}$):

$$A_c = \frac{\dot{m}}{\rho v} = \frac{Q}{v} \qquad (5.56)$$

where $\dot{m}$ = propellant mass flow rate (kg/s)
ρ = propellant density (kg/m³)
v = velocity of the propellant flow (m/s)
Q = volume flow rate of propellant (m³/s)
A_c = cross-sectional area of the propellant channel (m²)
D_c = diameter of the propellant channel (m)

$$D_c = 2\sqrt{\frac{A_c}{\pi}} \qquad (5.57)$$

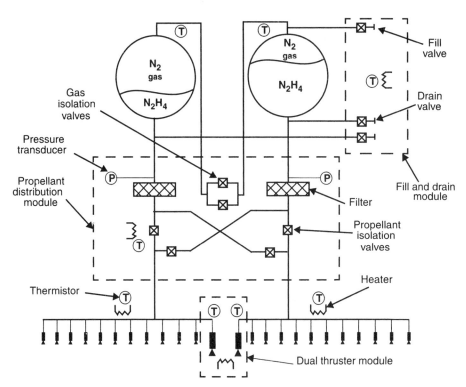

Fig. 5.33. TDRS Monopropellant Hydrazine Propulsion System Schematic. This shows the basic feed system configuration for the Tracking and Data Relay Satellite (courtesy of TRW).

To configure a generic propulsion system, we must integrate various standard components. A short list of the possibilities follows:

- Fluid ducting—the fluid ducting transfers the propellant from the storage tank through other components to the thruster. We want the shortest, least massive feed lines possible in a robust system. The ducts must be long enough and flexible enough to deal with vibration, thermal expansion and contraction, and component movement for operations such as gimballing to control thrust vectors. In addition, ducting must support the pressure and chemical-corrosion environments.
- Tank fill and drain valves—we must always fill a propellant tank prior to use and, sometimes, drain it. Valves for these operations are usually separate but can be combined for some systems. Typically, the valves attach directly to the tank but this depends on the system layout.

- On/off valves—in all systems, we must have some way to control whether the propellant is flowing or not. In nonthrottled systems, this involves a simple valve (ball, needle, gate) that can start or stop the flow. In throttled systems, the on/off valve can combine with the control valve, or we can use a separate valve, depending on redundancy requirements. For monopropellant thruster systems, the on/off valve is usually an integral part of the thruster assembly.
- Control valves—in many systems, we wish to control the propellant flow rates. We do this by varying the pressure distribution through the feed system. If we can reduce the feed pressure, we can reduce the flow rate, and vice versa—typically by placing a variable flow restriction device in the flow. By restricting the flow, we can create a pressure drop (through nonreversible losses) which, in turn, decreases the flow rate.
- Isolation valves—isolation valves are in a propulsion system's ducting to connect, disconnect, or bypass redundant subsystems. For example, in Fig. 5.33, a tank or thruster set can be isolated, in case of a failure, to save a mission.
- Check valves—similar to isolation valves. They keep fluids from going in the wrong direction, such as when pressure oscillations in the combustion chamber try to force hot propellants back up the feed system.
- Pressure regulator—pressure regulators maintain tanks at a desired constant pressure. These can be either mechanically controlled or computer controlled. As an example, consider the problem of maintaining a propellant tank at a constant pressure by using a stored gas pressurant. As the propellant level drops, so does the tank pressure. This reaction releases a spring mechanism (in a mechanical regulator) which allows gas pressurant into the propellant tank. As the tank pressure regains the desired level, the spring mechanism closes, restricting flow of the pressurant gas. We can use pressure regulators to control gas or liquid pressurant flows.
- Filters—no matter how clean we keep the environment, particulate contaminants are always possible. Placing a filter between the propellant tank and the components is a good idea.
- Transducers—pressure and temperature transducers routinely reside throughout the propulsion system to monitor for proper operation. We usually install temperature sensors with heaters to ensure that the propellant and hardware are at their correct operating temperature before operation. Pressure sensors in various locations diagnose problems.
- Heaters—many components cannot operate in the cold space environment. Hydrazine fuel can freeze, thruster catalyst beds are ineffective, and many materials are temperature sensitive. We install heaters with temperature sensors to maintain a desired temperature range.

5.4.3 Turbomachinery

Figure 5.34 is a cross-sectional (along the rotor axis) assembly drawing of a typical turbopump for a rocket engine. It shows the LOx turbopump used in the J-2 LOx / LH$_2$ engine used for the second stage of the Saturn V, which launched the Apollo vehicle that went to the Moon. In this turbopump, the LOx flow enters the impeller at the inducer inlet, where it is treated very gently by the inducer until enough static pressure has been added to suppress cavitation in the centrifugal pump. Then the flow passes into the centrifugal impeller, where most of the total pressure rise occurs. The impeller adds pressure by simultaneously adding kinetic energy and diffusing part of the resulting velocity head, yielding a flow that has more velocity head and more static pressure at the impeller discharge than at the inducer inlet. The rest of the velocity head is diffused in the volute (which is stationary and collects the flow as it leaves the impeller) and in the conical diffuser (not shown) located in the downstream discharge duct. The net result is a flow that has a static pressure rise of 7.6 MPa for the J-2 LOx pump.

This pump is driven by a turbine located on the other end of the shaft. The turbine is driven by hot combustion gases from a gas generator. These gases enter the turbine inlet manifold through the turbine inlet and pass from there into the turbine inlet nozzles. The gases are expanded through the nozzles to lower static pressure and are directed in a tangential direction. The result is a high gas velocity entering the first row of turbine rotor blades in a primarily tangential direction. The turbine rotor blades then turn the flow through a large angle, causing a change in gas momentum which produces a tangential force acting on the outside of the first turbine rotor disc. The gas then enters the stationary turning vanes (stators) located between the rows of turbine rotor blades, where it is realigned before entering the second row of turbine rotor blades. These blades turn the flow through a somewhat smaller angle, producing a tangential force on the outside of the second turbine rotor disc. At the design point, this second turbine rotor removes the rest of the tangential kinetic energy in the flow, resulting in a relatively low gas velocity in the axial direction at the turbine discharge.

The tangential forces on the two turbine discs are transmitted through those discs to the turbopump shaft, producing the shaft torque that drives the pump. The shaft is supported by two bearings that are lubricated by the propellant. One bearing is just inboard of the pump impeller, and one is inboard of the turbine rotors. The pump is a single-stage, centrifugal configuration. The turbine is a two-row, velocity-compounded configuration. Although other rocket turbopumps may differ in numbers of stages and types of axial turbines, they all share the features of pumps that add velocity head and then diffuse it, as well as turbines that expand gases to a high velocity and then deflect that high velocity flow to obtain a driving force.

We can do a preliminary design of turbomachinery in at least two ways. One method is to get right into the nitty-gritty of evaluating the flow physics through pumps and turbines, but this approach is way beyond the scope of this book or any

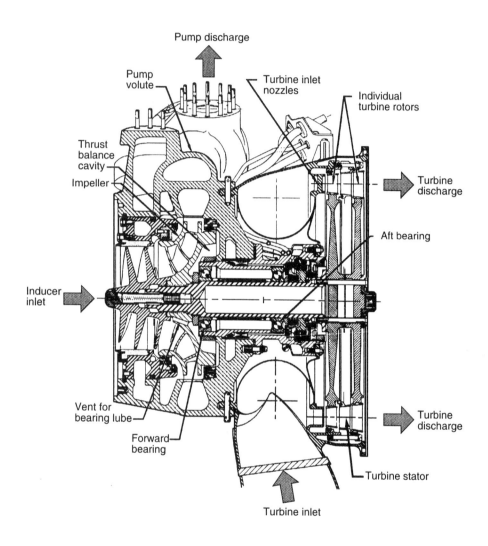

Fig. 5.34.   Cutaway View of a LOx Turbopump. This figure shows many of the typical components found in a turbopump and explains its operation. This is the turbopump for Rocketdyne's J-2 engine that flew on the Saturn-V second stage [NASA SP-8107, 1974].

reasonable preliminary design. A second approach is to use overall design parameters that are similar to those used in the past to get the best combinations of good performance, light weight, reliability, life, simplicity, and overall cost acceptability. This approach yields design concepts that are reasonable and understandable without bogging down in excessive detail.

Similarity Relationships

The following discussion of similarity relationships is intended to give a physical understanding for the various parameters we use. Experience or historical data defines the magnitudes of similarity parameters such as flow coefficients and specific speeds. Knowing the values of these parameters for similar machinery allows us to design and size a similar pump or turbine to a reasonably high accuracy.

Specific speed—Stage-specific speed [Eq. (5.61)] is a parameter that characterizes pumps and is calculated using very basic overall values of rotational speed, volume flow rate, and head rise for that stage. For centrifugal pumps, it indicates the ratio of an impeller's inlet and exit diameters. A low value of specific speed indicates a small inlet diameter relative to the exit diameter, which means the flow passages are small relative to the exit diameter. This situation results in high friction losses which translate into relatively low efficiencies, as shown (for example) at a stage-specific speed of 0.8 in Fig. 5.35. It then follows that a high value of specific speed indicates a large inlet diameter relative to the exit diameter, and flow passages that are large relative to the exit diameter. At very high specific speeds, we have to modify a centrifugal-pump stage so much to keep the exit diameter larger than the inlet diameter that it loses its efficiency, as also shown (for example) at a stage-specific speed of 8.0 in Fig. 5.35. At such a high specific speed, it is better to use an axial pump stage, in which the inlet diameter is equal to the exit diameter, and the throughflow direction is mainly axial rather than radial. In between these extremes, for specific speeds between 3 and 5, Fig. 5.35 shows that centrifugal pumps reach very respectable efficiencies (above 80%). Note that this comment, and Fig. 5.35, apply to centrifugal impellers greater than 10 cm in diameter. Smaller pumps have lower efficiencies.

Suction-specific speed—This parameter [Eq. (5.63) solved for u_{ss}] characterizes the pump's suction performance. It indicates the minimum net positive suction head (*NPSH*, inlet pressurization above vapor pressure expressed in meters) at which a pump, operating at rotational speed (N) and volume flow rate (Q), can operate without having cavitation affect its performance. A given pump-inlet geometry gives close to the same suction-specific speed no matter what its absolute values of size, rotational speed, or volume flow rate are. Pump-inlet geometry includes values for blade angles, diameter ratios, blade height-to-diameter ratios, and length-to-diameter ratio distributions. This fact is the basis of using similarity relationships to scale similar machinery.

A large value of suction-specific speed indicates an ability to operate at low inlet pressures. These large values are obtained by using

- Large diameters for the pump-inlet tip and small diameters for the inlet hub so as to minimize inlet axial velocity head (which is a big part of *NPSH*)
- Thin, gradually curved blades with sharp leading edges to get the minimum static pressure gradients on the blade surfaces.

Inducers (as shown in Fig. 5.34) have these features and, therefore, are used for most pumping applications in rocket engines.

Head coefficient—Stage head coefficient [Eq. (5.62) solved for ψ] expresses pump-stage head rise as a fraction of the impeller tip's speed squared over the gravitational constant. This value is based on the fact that, from Euler's equation for open (nonpositive displacement) turbomachinery elements, a pump stage's ideal head rise is equal to a geometrical factor times the impeller tip's speed squared over the gravitational constant. Pumps having similar stage-specific speeds and similar geometries (i.e., similar blade angles, L/D ratios) have similar head coefficients.

Inducer-inlet flow coefficient—This coefficient [Eq. (5.67) solved for ϕ] is the ratio of axial velocity to blade tangential velocity at the inducer inlet tip. It is the tangent of the flow angle approaching the inducer blade tip and is one of the main design parameters used to maximize suction performance. A low value of 0.07 usually yields close to the highest values of suction-specific speed.

Design Approach

Table 5.11 illustrates the conceptual design of turbomachinery. The discussion helps us understand the problems in designing rocket-engine turbopumps and gives us a way to make reasonable estimates. More detailed analyses are required, for example, to determine angles of fluid flow and velocity diagrams for the pump impeller, deal with the compressibility of hydrogen, and estimate height effects and corresponding stresses for the turbine blade. We discuss some of these considerations briefly at the end of the section to reflect some of the details we have overlooked in the interest of brevity.

Establish Design Goals

The first step is to establish the relative importance of the various design goals. High efficiency, high performance, and minimum mass are desirable because they help to increase the vehicle's payload. But designs that increase efficiency and decrease mass can also increase the pump-inlet pressure required to suppress cavitation which, in turn, can increase the propellant-tank pressure and mass. Also, these design processes can compromise design simplicity, reliability, life, and cost. We must carefully examine requirements for the mission and vehicle to determine the proper balance among design objectives. High efficiency, low mass, and simplicity usually result from minimizing the design margins relative to the component limits. Reliability and life usually improve when we increase those margins. Of course, all of these considerations affect the development and recurring costs.

Summarize Pump Requirements

The engine and vehicle requirements dictate the types, flow rates, and pressure levels of the propellants that the pumps deliver to the engine's thrust chamber. We

5.4 System Sizing, Design, and Trade-offs

Table 5.11. Process for Turbomachinery Design. We start by taking the decisions already made to this point and turn them into pump requirements. We then size the pump to meet these requirements and size the turbine required to drive the pump. We complete everything by configuring the turbopump system and estimating its mass.

Step	Inputs	Outputs
1.) *Establish design goals for turbomachinery*	• Propulsion system design goals	Define relative emphasis: • Performance • Mass • Simplicity • Reliability • Lifetime
2.) *Summarize pump requirements* (see Table 5.12 for a summary of details)	From Sec. 5.3: • Propellant choice • Propellant mass flow rate • Pump pressure rise • Pump inlet pressure	• Propellant density • Propellant vapor pressure • Volume flow rate • Pump head rise • Net positive suction head
3.) *Design and size pump* (see Table 5.13 for a summary of details)	• Propellant density • Propellant vapor pressure • Volume flow rate • Pump head rise • Net positive suction head	• Number of pump stages • Pump rotational speed • Impeller tip speed • Impeller tip diameter • Inducer tip diameter • Pump efficiency • Pump power required
4.) *Summarize turbine requirements*	• Drive gas type • Drive gas flow rate • Inlet temperature and pressure • Pressure ratio	• Summary of gas properties
5.) *Design and size turbine* (see Table 5.14 for a summary of details)	• Gas properties • Power required • Rotational speed (if direct driven)	• Spouting velocity • Pitchline velocity • Turbine type • Number of stages • Turbine efficiency • Turbine tip diameter • Rotational speed (if geared)
6.) *Select turbopump arrangement*	• Pump parameters • Turbine parameters	• Machinery configuration • System drawing
7.) *Estimate turbopump mass*	• Turbopump configuration	• Mass estimate

have already established these in Sec. 5.3. The pumps must deliver propellant without requiring an inlet pressure (to suppress cavitation) higher than that allowed by the propellant tanks and feed system. Table 5.12 shows how we do this. A given pump operating under a given condition (rotational speed and fluid angles of attack on the internal flow surfaces) produces a constant-volume flow rate and a constant head rise. Thus, the engine requirements of mass flow rates and pressure levels translate into pump requirements for volume flow rates and head levels (pressure levels expressed in meters of propellant being pumped). The volume flow rate is expressed by Eq. (5.58); the required head rise is expressed by Eq. (5.59); and the net positive suction head supplied by the system to the pump inlets is expressed by Eq. (5.60). These pump requirements are inputs to pump design.

Table 5.12. Summarizing Pump Requirements. The inputs, such as mass flow rate and required pressure rise, have already been established. Now we simply convert them to expressions common to the turbomachinery community.

Step	Calculation	Comments
1.) Summarize inputs (from Section 5.3)		• Propellant choice • Propellant mass flow rate, $\dot{m}$ (kg/s) • Pump pressure rise, Δp_p (Pa) • Pump inlet pressure, p_i (Pa) • Engine cycle choice
2.) Determine propellant density		• Taken from propellant properties Appendix B, ρ (kg/m³)
3.) Determine propellant vapor pressure		• Taken from propellant properties Appendix B, p_v (Pa)
4.) Calculate volume flow rate [Q, (m³/s)]	$Q = \dfrac{\dot{m}}{\rho}$	• Volume flow rate is customary in pump design instead of mass flow rate
5.) Calculate pump head rise [H_p, (m)]	$H_p = \dfrac{\Delta p_p}{g_0 \rho}$	• Head rise is the height a column of propellant would attain if under pressure • Head rise is a customary term used in place of pressure rise
6.) Calculate Net Positive Suction Head [$NPSH$, (m)]	$NPSH = \dfrac{p_i - p_v}{g_0 \rho}$	• NPSH is a measure of the margin allowed to ensure the propellant does not vaporize and cause pump cavitation

Traditionally, we use the concept of pressure head, in length units, as opposed to pressure rise in force per area. Head rise is the height that a column of water can be raised given a certain pressure. The *Net Positive Suction Head* (NPSH) is the limit on a pump system that keeps the propellant from vaporizing (*cavitation*)* during the pumping action. Vaporizing affects performance and, if severe enough, can cause erosion on surfaces of the pump's impeller blades.

$$Q = \frac{\dot{m}}{\rho} \qquad (5.58)$$

---

* Cavitation occurs when vapor bubbles form and then suddenly collapse, causing tiny shock waves with enough energy to erode the metal surfaces.

$$H_p = \frac{\Delta p_p}{g_0 \rho} \quad (5.59)$$

$$NPSH = \frac{p_i - p_v}{g_0 \rho} \quad (5.60)$$

where Q = volume flow rate (m³/s)
$\dot{m}$ = mass flow rate (kg/s)
ρ = propellant density (kg/m³)
H_p = pump's head pressure rise (m)
Δp_p = required rise in pump pressure (Pa)
g_0 = 9.81 m/s²
p_i = pump-inlet pressure (Pa)
p_v = propellant vapor pressure (Pa)

Design and Size Pump

Table 5.13 shows how we size a pump to meet our requirements. A similarity parameter used to characterize pumps is stage-specific speed (N_s). Equation (5.61) expresses this parameter in terms of the requirements established above. Figure 5.35 shows how this parameter influences the pump's hydraulic efficiency (η_p).

$$N_s = \frac{N_r \sqrt{Q}}{\left(\frac{H_p}{n}\right)^{0.75}} \quad (5.61)$$

where N_s = stage-specific speed $\left(\sqrt{m^3/s}/m^{0.75}\right)$
N_r = pump's rotational speed (rad/s)
n = number of pump stages

A centrifugal pump's stage operates at a higher specific speed than most other pumps. This means that, for a given volume flow rate and head rise, centrifugal pumps operate at higher rotational speeds, which results in smaller pump diameters and masses. Centrifugal pumps can have high efficiencies relative to other pump types, depending on design choices. Finally, they have excellent off-design operating characteristics, compared to axial pumps, which greatly simplifies engine start-up and shutdown.* Together, these attributes make centrifugal pumps highly favored for rocket engines. Axial pumps, which have slightly higher efficiencies and similar weights, have been used occasionally. But their limited off-

---

* Positive displacement pumps, such as piston or diaphragm pumps, usually have much better start-up and shutdown characteristics than either centrifugal or axial pumps [Whitehead, 1994].

Table 5.13. Designing and Sizing the Pump. Given the requirements, six basic steps are needed to size a pump.

Step	Calculation	Comments

1.) Determine the number of stages (n)	$n \geq \dfrac{\Delta p_p}{\Delta p_{ps}}$	• n is the next higher integer of this ratio
		• Δp_p is the required pump pressure rise (Pa)
		• Δp_{ps} is the allowable pressure rise over a single stage. Use:
		  16 MPa for liquid H_2
		  47 MPa for all others

2.) Determine pump rotational speed
N_r (rad/s)
N (RPM)
	$N_r = \dfrac{u_{ss} NPSH^{0.75}}{\sqrt{Q}}$
	
	$N_r = \dfrac{N_s \left(\dfrac{H_p}{n}\right)^{0.75}}{\sqrt{Q}}$
	
	$N = \dfrac{30 N_r}{\pi}$
		• Use the lesser of the two numbers for N_r if there is no boost pump
		• Use second N_r expression if there is a boost pump
		• u_{ss} is the suction specific speed. Use:
		  130 for liquid hydrogen
		  90 for other cryogenic liquids
		  70 for others
		• N_s is the stage specific speed. Use:
		  2.0 for liquid hydrogen
		  3.0 for all others

3.) Determine pump impeller tip speeds
u_t (m/s)
	$u_t = \sqrt{\dfrac{g_0 H_p}{n \psi}}$	• Where ψ is the pump head coefficient, use:
		  0.60 for liquid hydrogen
		  0.55 for all others

4.) Determine pump impeller diameters D_{2t} exit tip diameter (m) D_{1t} inlet tip diameter (m)	$D_{2t} = \dfrac{2 u_t}{N_r}$ $D_{1t} = \sqrt[3]{\dfrac{(4/\pi) Q}{\phi N_r \left(1 - L^2\right)}}$	• ϕ is the inducer-inlet flow coefficient, use: 0.10 • L is the inducer inlet hub-to-tip diameter ratio; use: 0.3 • Impeller inlet tip diameter is equivalent to the diameter of the inducer exit tip
5.) Determine pump efficiency (η_p)	Use Fig. 5.35	• Check the value of N_s using Eq. (5.61)
6.) Determine power required to drive the pump	$P_{req} = \dfrac{g_0 \dot{m} H_p}{\eta_p}$	• This is the power required from the turbine to drive the pump

design operating ranges complicate the start-up and shutdown sequences for chemical rocket engines, which typically require very short start durations.

The number of pump stages is generally the next largest integer value greater than the ratio of the pump-pressure rise to the maximum allowable stage-pressure

5.4 System Sizing, Design, and Trade-offs

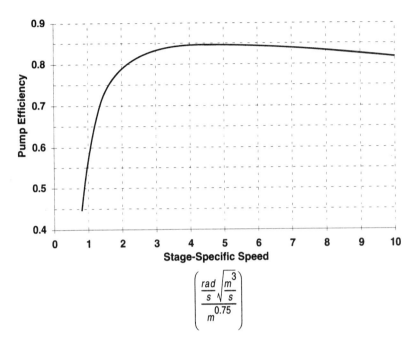

Fig. 5.35. Pump Efficiency Versus Stage-Specific Speed. Once we have determined N_s from Eq. (5.61), we can pick off the pump efficiency (data courtesy of Rocketdyne).

rise. This choice minimizes the stages subject to the limit of stage-pressure rise. Assuming the use of centrifugal pumps, the approximate limits on stage-pressure rise are 47 MPa for propellants other than liquid hydrogen and 16 MPa for liquid hydrogen. These limits are based on experience. Although the numbers can change a bit depending on pump size and flow rates, these numbers are adequate for preliminary design. At 47 MPa the impeller blades get so thick they begin to restrict the flow passages, thereby beginning to cause large losses. However, for liquid hydrogen, the 47-MPa limit is way beyond the impeller's structural limit, so we keep the stage-pressure rise under 16 MPa, which corresponds to the limit of 610 m/s on tip speed for a shrouded centrifugal impeller made from forged titanium.

All impellers are shrouded for best efficiency and fewest rotordynamic problems. We also assume each pump has an inducer at the inlet to its first impeller. This inducer is a low-pressure pumping device that treats the flow very gently until it adds enough pressure to suppress cavitation. Inducers allow lower pump-inlet pressures, which is a major factor in vehicles with large propellant tanks.

The next steps are to estimate the impeller-tip speeds, the rotational speeds, and, finally, the impeller sizes. We use Eq. (5.62) to estimate the impeller-tip speed using the head rise from Eq. (5.59), the number of pump stages just discussed above, and the typical stage head coefficient (ψ) of 0.60 for liquid hydrogen and 0.55 for other propellants.

$$u_t = \sqrt{\frac{g_0 H_p}{n \psi}} \qquad (5.62)$$

where ψ = the pump-stage head coefficient
(use 0.6 for liquid hydrogen and 0.55 for others)

As discussed earlier, we want the highest rotational speed, within the component limits, to get the lowest mass and highest performance. The pump imposes several rotational speed limits, which we can easily calculate early in the design procedure. However, several rotational speed limits can also be imposed by the turbine or the bearings and require the complete system to be defined before we can determine their impact. We can resolve this issue by assuming the pump sets the speed and then continue the design through to the bearings and the turbine. If the pump sets the speed at a lower level than either the bearings or the turbine, we can continue the design to completion. But if the bearings or the turbine set the speed at a lower level than the pump, some iteration is necessary.

To simplify our preliminary design, we treat only the two major pump limits of cavitation and stage-specific speed. This approach is not really all that cavalier because, from experience, our assumed limits on stage-specific speed usually keep the rotational speeds within the limits imposed by the bearing and turbine.

If there is no boost pump, and if there is a definite NPSH [Eq. (5.60)] limit, we can use Eq. (5.63) to estimate the pump's rotational speed. Suction-specific speeds (u_{ss}) of 130, 90, and 70 are reasonable maximums for liquid hydrogen, cryogenic propellants other than liquid hydrogen, and noncryogenic propellants, respectively.

$$N_r = \frac{u_{ss} NPSH^{0.75}}{\sqrt{Q}} \qquad (5.63)$$

$$N = \frac{30 N_r}{\pi} \qquad (5.64)$$

where N_r = pump's rotational speed (rad/s)
N = pump's rotational speed (RPM)

The limit on stage-specific speed is based on a value that yields close to the maximum efficiency for a centrifugal stage. As shown in Fig. 5.35, a specific speed of 3 does this. Higher values could yield slightly higher efficiencies, but the large inducer diameters of these configurations with low inlet pressure could complicate the resulting designs. For liquid hydrogen, experience shows that a stage-specific speed of 2 is more common because limits on turbine stress reduce possible speeds for liquid-hydrogen turbopumps. Therefore, we use Eq. (5.65) [which is Eq.

(5.57) solved for rotational speed], combined with specific speed limits of 2 and 3, to estimate the stage specific-speed limits for liquid-hydrogen pumps and for pumps for other propellants, respectively.

$$N_r = \frac{N_s \left(\frac{H_p}{n}\right)^{0.75}}{\sqrt{Q}} \tag{5.65}$$

The estimated rotational speed would then be the lesser of these two values [Eqs. (5.63), (5.65)]. If there is some flexibility in inlet pressure, or if there is a boost pump, we can set the speed by Eq. (5.65) and determine the resulting inlet-pressure requirement by solving Eqs. (5.63) and (5.60) for pressure at the pump inlet.

Once we know the rotational speed, we can estimate the diameters of the impeller outlet tip and the inducer inlet tip using

$$D_{2t} = \frac{u_t}{N_r} \tag{5.66}$$

$$D_{1t} = \sqrt[3]{\frac{4Q}{\pi \phi N_r (1 - L^2)}} \tag{5.67}$$

where D_{2t} = diameter of the impeller outlet (m)
D_{1t} = diameter of the impeller inlet (m)
L = ratio of hub diameter to tip diameter
ϕ = inducer-inlet flow coefficient

Equation (5.67) is based on the definition of the inducer-inlet flow coefficient. This coefficient is the ratio of axial-flow velocity over the speed of the blade tips at the inducer's inlet. We use a flow coefficient of 0.10 for these estimates. The other variable in the equation is the ratio (L) of the inducer inlet's hub and tip diameters, which we can assume to be 0.3. Minimizing the factors ϕ and L minimizes the pump inlet's axial velocity head we use, which, in turn, minimizes the NPSH required by the pump.

Summarize Turbine Requirements

The turbine requirements of gas type, gas flow rate, inlet temperature, inlet pressure, pressure ratio, and gas properties are the same as those used in the engine balance (Sec. 5.3.3). As a result, they require little or no translation in the "Design and Size Turbine" section below.

Design and Size Turbine

The usual pump-fed rocket uses a turbine to drive the pump; therefore, we assume the use of a turbine here. The turbine is driven by hot gases from a gas generator in a gas-generator cycle, by warm gases leaving the cooling jacket in an expander cycle, or by hot gases from a preburner in a staged-combustion cycle (Fig. 5.13). We assume a single turbine directly driving a single pump. Table 5.14 shows how to size the turbine. First, we estimate the isentropic spouting velocity, which is the velocity that the turbine gas flow would have if it were expanded isentropically from the turbine inlet conditions to the turbine exit static pressure:

$$C_o = \sqrt{2c_p T_i \left[1 - \{\frac{1}{P_{trat}}\}^{\frac{\gamma-1}{\gamma}} \right]} \quad (5.68)$$

where C_o = isentropic spouting velocity (m/s)
 c_p = constant pressure specific heat of the turbine drive gases (J/kg·K)
 T_i = turbine inlet temperature (K)
 γ = ratio of specific heats
 P_{trat}= turbine pressure ratio (20 for gas-generator,* 1.5–2.0 for staged-combustion and expander obtained from the engine balance from Sec. 5.3.3)

Second, we determine the allowable pitchline velocity; third, we select the turbine and its best efficiency. Figure 5.39 shows that turbine efficiency depends mostly on turbine velocity ratio (u_m/C_0), turbine type, and the number of stages† (or rotors for velocity-compounded turbines). The turbine types are defined in Figs. 5.36 and 5.38. High values of mean pitchline velocity (u_m) result in the best velocity ratio which, in turn, maximizes turbine efficiency. From Fig. 5.37, pitchline velocities of 450 to 550 m/s are close to the structural limits of high-temperature turbine rotors. If we place no more than 2 rotors on the turbine to avoid too much complexity and mass, we usually select either a 2-row, velocity-compounded turbine, or a 2-stage, pressure-compounded turbine for a gas-generator cycle. The velocity-compounded turbine has taller inlet blades, so we normally use it if the pressure-compounded turbine has inlet blade heights that are too small (less than 0.5 cm). Otherwise, we would select the pressure-compounded turbine because it can perform better. Also, note that other turbine losses reduce efficiencies to about 95% of those shown in Fig. 5.39. Small turbines (less than 10 cm in diameter) can be as much as 20% less efficient (75% of the efficiencies in Fig. 5.39).

---

* For turbines in series, this is the overall ratio. For LH₂/LOx turbines, the individual ratios are around 2.5 and 8.0, respectively. For RP-1/LOx, they are more like 4.0 and 5.0. This is because of the relative power required to pump each propellant.
† A *turbine stage* is characterized by a pressure drop through either a conventional nozzle or a stationary blade row. Although some turbines may have several rotating blade rows (rotors) or stationary blade rows (stators), if they have only a single pressure drop, they are single-stage turbines.

Table 5.14. Designing and Sizing the Turbine. This is a six-step process that depends on our required pump power and the nature of the available turbine drive gas.

Step	Calculation	Comments
1.) Determine the turbine's drive gas parameters c_p = constant pressure specific heat (J/kg·K) T_i = turbine inlet temperature (K) γ = ratio of specific heats	From Chaps. 3 and 4, thermodynamic and thermochemical analysis	• Turbine materials keep inlet temperature below 1100 K • see Fig. 5.37
2.) Determine the turbine's isentropic spouting velocity, C_o (m/s)	Eq. (5.68): $$C_o = \sqrt{2c_p T_i \left[1 - \left\{\frac{1}{P_{trat}}\right\}^{\frac{\gamma-1}{\gamma}}\right]}$$	• P_{trat} is the turbine pressure ratio from engine balance (Sec. 5.3.5) • for GGs maximum allowable value is 20 • for SC and expanders $1.5 < P_{trat} < 2.5$ • spouting velocity depends on chemistry and cycle
3.) Determine maximum turbine pitch velocity, u_m = pitch velocity (m/s)	From Fig. 5.37	• Limited by turbine materials and temperature
4.) Select turbine and best efficiency	From Fig. 5.39	• Limit number of rotors to two • Maximize efficiency • Velocity-compounded or pressure-compounded typical for GG cycle • Reaction typical for expander and staged-combustion cycles
5.) Compare turbopump efficiency with engine balance value	$\eta = \eta_t \times \eta_p$	• If different from engine balance value, redo engine balance and start over • For GG cycles, minimize the turbine mass flow rate • For expander and staged-combustion cycles, lower turbine P_{trat} to get lowest pump pressure rise or highest engine chamber pressure
6.) Determine the turbine's mean pitch diameter, D_m (m)	$D_m = \dfrac{2u_m}{N_r}$	• Assume the turbine's rotational speed is the same as the pump's, i.e., no gearbox

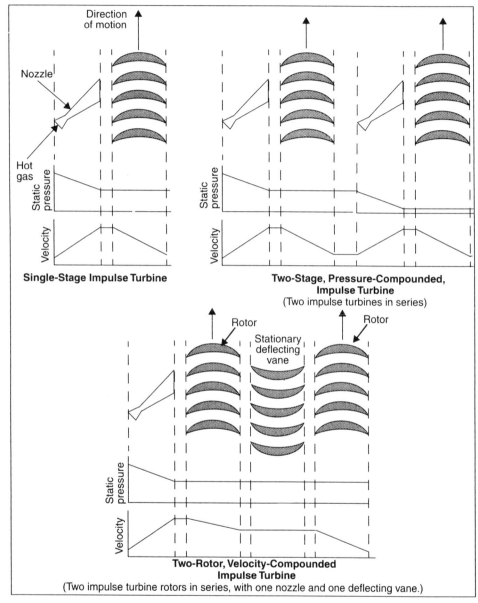

Fig. 5.36. Impulse Turbine Types. We use these turbines mainly to produce high-pressure ratios for a gas-generator cycle. We use a two-stage, pressure-compounded turbine unless the blade height gets too small. In that case, we go to a two-rotor, velocity-compounded configuration. Velocity-compounded turbines rely on a large pressure drop and corresponding velocity increase. We then need several rotors and stators to straighten the flow back out [Faires, 1947].

5.4 System Sizing, Design, and Trade-offs

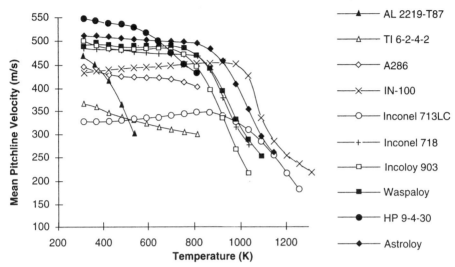

Fig. 5.37. Allowable Rotor Speeds versus Temperature for Various Turbine Materials. This figure shows the turbine blade's allowable tangential velocity at its midpoint (mean pitchline velocity) versus operating temperature for a 10-hour lifetime (data courtesy of Rocketdyne [Bissell, 1985]).

For expander and staged-combustion cycle turbines, the velocity ratios are much higher than those for the gas-generator cycle because the expander and the staged-combustion cycles have lower pressure ratios. As a result, two-stage reaction turbines are the primary candidates for these cycles. In some cases, we have to use pitchline velocities below the material limits to keep the design velocity ratio from going above the point of peak efficiency. Sometimes, this means reducing the diameter or using one stage.

After we have determined the turbine efficiency, we have to check whether the engine balance needs modifying by comparing the pump and turbine efficiency product with that used in the last engine balance. If it is reasonably close, we can continue the design. If it is not, we must update the engine balance using corrected values of turbopump efficiency. This update is necessary because changes in turbopump efficiency change such factors as engine performance and turbine pressure ratio. In turn, these factors change the flow rates and pressure distributions throughout components. Once we have agreement between the engine balance and mean pitchline velocity, we can estimate the turbine mean pitch diameter (D_m) if we assume the rotational speed is the same as the pump's (a direct-drive system with no gearing):

$$D_m = \frac{2u_m}{N_r} \quad (5.69)$$

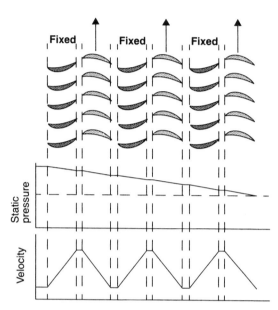

Fig. 5.38. Schematic of a Reaction Turbine. This is a three-stage turbine with a 50% reaction. (There is a pressure drop in all three fixed-blade rows.) The stator and a moving-blade row make up a stage. Reaction (50% in illustration) refers to the fraction of overall turbine pressure drop that occurs in moving-blade rows (rotors). We use this turbine mainly in expander and staged-combustion cycles, because we want to minimize the pressure drop between the pump and the combustion chamber [Faires, 1947].

Other parameters of turbine geometry, such as blade heights and blade-height distributions, require analyzing the turbine gas path, which is beyond the scope of this discussion. This type of analysis allows us to accurately predict how the turbine blade's centrifugal stress limits rotational speed, which is particularly important for liquid-hydrogen turbopumps. Also, the initial turbine in a gas-generator cycle and turbines for low-thrust engines (less than 90,000 N thrust), may have blade heights that are either below minimum limits (less than 0.5 cm) or above limits on end-wall loss (hub diameter/blade tip diameter > 0.93). We need to analyze the turbine gas path to evaluate these cases. The results of such an analysis could lead us to use partial-admission turbines, switch from pressure-compounded turbines to velocity-compounded turbines, or use geared turbines (discussed in the following section). Because we neglect these factors, the turbine efficiencies generated here may be optimistic, and the turbopump configurations may change as we do more analysis. This is not an unusual situation, because preliminary analyses cannot cover everything.

5.4 System Sizing, Design, and Trade-offs

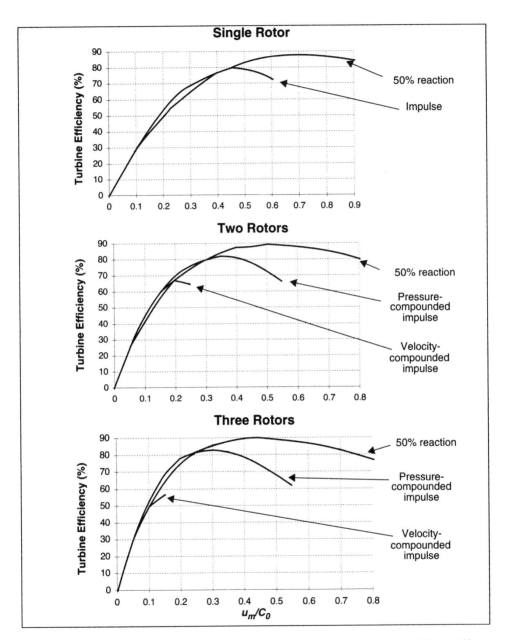

Fig. 5.39. Turbine Efficiencies. This chart shows the efficiencies for 50% reaction turbines and impulse turbines for 1, 2, and 3 rotors. We have plotted them versus the nondimensional ratio of mean pitchline velocity over spouting velocity. This ratio is very near constant for a given system. These values do not include viscous, friction, or leakage losses. Large turbines produce about 95% of these numbers; small turbines produce 75% [NASA SP-8107, 1974].

Select Turbopump Arrangement

Figure 5.40 shows the major options available for arranging the turbopumps on a rocket engine. Each option has two pumps—one for the oxidizer and one for the fuel. We need options because each application has different requirements for rotational speed, and we have to balance these requirements against such factors as complexity and reliability. In the case of the LOx/LH$_2$ propellant combination, the low density of LH$_2$ relative to LOx causes the LH$_2$ head rise to be much higher than that of the LOx [Eq. (5.59)]. As a result, the rotational speed needed to attain a reasonable design-stage-specific speed is much higher for LH$_2$ [Eq. (5.65)] than for LOx. To accommodate this disparity in pump design speeds, the J-2 rocket engine uses the dual-shaft arrangement with series turbines (arrangement "c"); the RL-10 rocket engine uses the single geared pump (arrangement "f"); and the SSME rocket engine uses dual shafts with parallel turbines (arrangement "g").

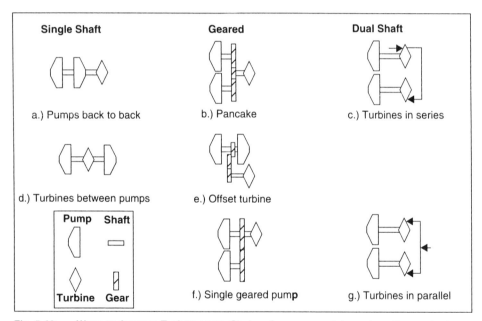

Fig. 5.40. Ways to Arrange Turbopumps. Choice of arrangement depends on the required pump and turbine speeds [NASA SP-8107, 1974].

The other major element that can cause a speed disparity is the effect of the turbine drive cycle on the optimum design speed for the turbine. For staged-combustion and expander cycles, the best design speed for the turbine is usually close enough to that of the pump so we can use a direct drive. However, for gas-generator cycles, the blade heights of direct-drive turbines are usually quite small and, as a result, we often want to raise turbine speeds to a level higher than those of the pumps. An example is the Atlas's turbopumps, for which the pump-flow densities and, consequently, the pump speeds are close enough to allow a direct connect between the pumps. The turbine is allowed to operate at a higher speed by gearing it to the main pump shaft (configuration "e"). Configuration "b" allows each major component to operate at its own best speed, an arrangement used for the engines in the Titan's first stage.

The trend has been to opt for simplicity and reliability by eliminating the gearbox. Examples are the F-1 engine with a gas-generator cycle and two dense propellants (configuration "a") and Rocketdyne's version of the XLR-132 research rocket engine (configuration "d"). The difference between configurations "a" and "d" is that configuration "d" allows each pump to have good suction performance by placing each one on the end of the shaft. But this approach requires placing the hot turbine in the middle, thereby requiring additional structure to guarantee satisfactory alignment of the bearing. Configuration "a" avoids these structural problems by placing the turbine on the end. Unfortunately, this approach compromises the suction performance of the pump whose inlet faces the turbine.

Except for the XLR-132 engine, which has a design thrust level of less than 17,000 N, most recent designs use dual shafts. Series turbines are the rule for gas-generator engine cycles (configuration "c"), and parallel turbines are common for staged combustion and expander engine cycles (configuration "g"). Series turbines yield higher performance because they have higher turbine-velocity ratios, and parallel turbines are more controllable by devices with high reliability and life requirements. Obviously, we often need to iterate between the pump and turbine designs and turbopump arrangement to meet each application's requirements.

Other arrangement options involve bearing placements, most of which are illustrated in Fig. 5.41. These configurations almost always use bearings that are cooled and lubricated by the propellants being pumped. The inboard arrangement (configuration "a") is usually favored because it is the simplest. However, this arrangement can compromise the design speed because the drive shaft must pass through both bearings, tending to overspeed the bearings. Also, it can cause critical speed problems for LH_2 turbopumps, which usually have several pump stages that can overhang too much. The J-2, LH_2 turbopump avoids overhang by placing the pump bearing outboard (configuration "e"). The SSME-LH_2 turbopump avoids both issues by placing both bearings outboard (configuration "c"). The outboard bearing on the turbine end does create a complication in that propellant must be supplied to the bearing for cooling. We can avoid many of these complications by using hydrostatic bearings, an advanced-technology element coming into favor which avoids the contact-fatigue problems of rolling-element bearings by eliminating the rolling elements. We discuss advanced components below under "Other Considerations."

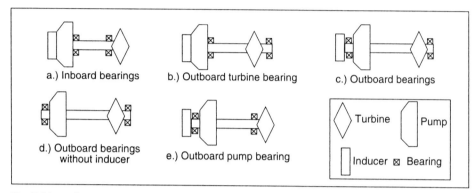

Fig. 5.41. Arrangements for Bearing Supports. The figure shows some of the typical turbine, pump, inducer, and bearing configurations [NASA SP-8107, 1974].

Estimate Turbopump Mass

Conventional scaling rules yield a mass relationship that is proportional to diameter cubed. We also know that flow (and engine thrust) are proportional to diameter squared. Combining these two relationships yields a turbopump mass that is proportional to engine thrust to the 1.5 power. However, experience has shown that this is too strong a relationship with thrust because bearings and seals do not scale the same as other subcomponents and materials reach minimum-gauge thicknesses. Actual masses of turbopump hardware for rocket engines correlate well with torque in the main pump shaft:

$$m_{tp} = A \tau^B \qquad (5.70)$$

$$\tau = \frac{P_{req}}{N_r} \qquad (5.71)$$

where m_{tp} = mass of the turbopump (kg)
  A = empirical coefficient (range 1.3–2.6)
  B = empirical exponent (range 0.6–0.667)
  τ = pump shaft torque (N·m)
  P_{req} = required pump power (W)
  N_r = pump rotational speed (rad/s)

Depending on the design emphasis, the coefficient A can vary between 1.3 and 2.6, and the exponent B can vary between 0.6 and 0.667. For conceptual design, we can use $A = 1.5$ and $B = 0.6$. Examining the resulting relationship reveals a mass that increases with power and decreases with increasing rotational speed, as we would expect.

Other Considerations

We have discussed an approach that yields reasonable design values for conceptual analyses of turbopumps in rocket engines. But special design situations could produce different configurations and eventually require more complex analyses.

As thrust decreases in a pump-fed rocket engine, blade heights become smaller and smaller in centrifugal pumps and axial turbines. Eventually, partial-emission centrifugal pumps and partial-admission turbines become necessary to achieve blade heights large enough to yield reasonable efficiencies. As engine thrusts drop further to values of 4500 N, design-specific speeds drop, and other pump and drive candidates (such as piston pumps) may become more efficient. But we must select carefully. For example, positive-displacement (piston) pumps require lubrication of their rubbing surfaces, so we have to consider compatibility with the propellant—particularly the oxidizer.

Engine cycles can be altered to enhance individual cycle performance. For example, we can use mixed preburners (fuel-rich for the fuel turbopump, oxidizer-rich for the oxidizer turbopump) to obtain more turbine drive gases for staged combustion cycles. This approach raises the attainable chamber pressure. Another alternative (used on the SSME) is to add a small stage to the LOx pump so only the small amount of LOx needed in the preburners is raised to the maximum pressure. This reduces the power requirements for the oxidizer pump, which reduces the pressure ratio of the drive turbine.

The compressibility of liquid hydrogen causes its density to rise with pressure, somewhat reducing the pumping power available. Also, pump inefficiencies tend to drop the pump flow density, thereby dropping the isentropic efficiency to a value lower than we can get from a purely incompressible fluid.

The shaft sizes required to avoid vibration problems (structural resonance occurs at critical speeds) are usually larger than the sizes required to transmit the torque based on material strength. This can cause bearing speed problems that we can solve by rearranging the bearing placement, as discussed before, but not without balancing the solution against issues of reliability and complexity.

To predict turbine blade stress, which often limits the rotational speed of liquid-hydrogen turbopumps, we must estimate the turbine blade heights. This concern also applies when minimum blade heights, or maximum blade hub-to-tip ratios, may be exceeded. These applications include turbines for gas-generator cycles and for small engines. To estimate these blade heights, we have to analyze the gas-flow path through the turbine.

The need to throttle an engine affects our choice of centrifugal pump. For stable pump operation, we need a negative slope on the pump head versus flow characteristic (the pump-performance map) to allow a wide throttling range. This slope should be negative down to low flow rates. However, this condition requires hydrodynamic modifications that reduce the pump design head coefficient, thereby raising requirements for the pump tip speed. These changes also reduce the stage pressure-rise capabilities to values below our standard values of 16 MPa for LH$_2$ and 47 MPa for other propellants. In some cases, this can raise the required number of pump stages, particularly for hydrogen pumps.

Technology is continually advancing, often influencing the designs or their performances. Hydrostatic bearings, which are becoming more popular, offer the attraction of increasing the bearing speed limits that exist with rolling contact bearings. Thus, we can use higher speeds for configurations with overhung pumps and turbines and simpler overhung configurations for some applications that conventionally use outboard bearings. Other technology advances that could influence turbopump designs are in the materials for the pump impellers and turbine rotors and blades. Better materials for pump impellers can reduce the number of pump stages required to deliver a given pressure level, which is very important to liquid hydrogen. Better turbine materials can allow higher turbine-inlet temperatures and higher turbine efficiencies (because of higher attainable velocity ratios), both of which enhance turbopump and engine performance.

5.4.4 Propellant Storage System

Tank Volume

In Sec. 5.3.2, we estimate the volume of propellant required, but the tank volumes need to be somewhat higher than this number. The total volume of a tank has four components:

1. Usable propellant volume (V_{pu})—The volume of tank required to hold the usable propellant mass (either m_{fuel} or m_{ox}, determined in Sec. 5.3.2). This also includes any design reserves needed to account for contingencies.
2. Ullage volume (V_{ull})—The volume of the tank left unfilled to allow for expansion of the propellant or contraction of the tank structure. This volume ranges from 1% to 3% of the total tank volume.
3. Boil-off volume (V_{bo})—This volume, for cryogenic propellants, allows for the propellant that boils off due to heat transfer into the tank between filling and draining.
4. Trapped volume (V_{trap})—The volume of unusable propellant left in the feed lines, valves, tank, and other components at the end of the engine operation. This is typically the volume of the feed system.

Therefore, the total tank volume (V_{tot}) is

$$V_{tot} = V_{pu} + V_{ull} + V_{bo} + V_{trap} \tag{5.72}$$

Tank Shape

The two most popular tank shapes are spherical and cylindrical. Spherical tanks offer the most volume for a given surface area, so they are usually lighter than cylindrical tanks. However, envelope constraints, particularly for launch vehicles, tend to drive the configuration to a cylindrical shape. Cylindrical tanks can also supply structural rigidity to longer vehicles without an additional shell structure. Figures 5.42, 5.43, and 5.44 show spherical tanks, tandem cylindrical tanks, and tandem cylindrical tanks with a common bulkhead, respectively. These are the most common tank configurations.

For propellant or pressurant storage tanks, pressure has the greatest effect on tank structural requirements, although acceleration, vibration, and handling loads are also important. For preliminary design, we consider only the pressure loads. The design burst pressure of a tank is

$$p_b = f_s MEOP \qquad (5.73)$$

where p_b = the structure's design burst pressure (Pa)
  f_s = a factor of safety to account for errors and contingencies (2.0 is typical for pressure vessels)
  $MEOP$ = Maximum Expected Operating Pressure of the tank (Pa)

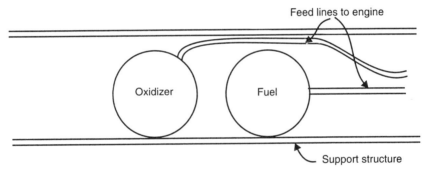

Fig. 5.42. Spherical Tanks. These tanks have the most volume for a given surface area and tend to be lightest. Envelope constraints restrict their use.

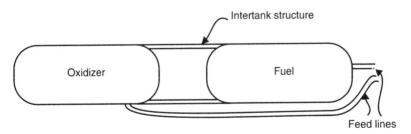

Fig. 5.43. Tandem Cylindrical Tanks. This is the simplest configuration when vehicle and tank diameter are limited.

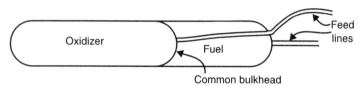

Fig. 5.44. Tandem Cylindrical Tanks with a Common Bulkhead. This configuration shortens the vehicle and reduces mass by eliminating one bulkhead. Manufacturing is more complex than for separate tanks, and leakage of one propellant to the other poses a safety issue.

Table 5.15 shows, for various materials, the strengths and properties we use in the following equations.

Table 5.15. Tank Materials. Quoted strengths for composite materials are for actual tanks using these materials. Note that these values are representative only; for more complete details, consult *MIL-Handbook 5*. The largest value of $F_{tu} / \rho\, g_0$ is the most structurally efficient material.

Material	ρ kg/m³ (lb/in³)	F_{tu} GPa (ksi)	$F_{tu} / (\rho\, g_0)$ km
2219 - Aluminum	2800 (0.101)	0.413, 0.214 welded (60.0, 31.0 welded)	15.04
Titanium	4460 (0.161)	1.23 (178.0)	28.81
4130 Steel	7830 (0.283)	0.862 (125.0)	11.23
Graphite*	1550 (0.056)	0.895 (130.0)	58.88

*Composite fiber materials using an epoxy resin matrix for structural stability. The numbers shown here are for an approximately isotropic layup. Individual fibers are much stronger.

Another primary concern is the choice of tank material. Typically, we wish to choose a material that gives us the most strength for a given mass. But the material must also be chemically compatible with the propellant, weldable, and formable—to consider just a few of its needed properties (as of the publication date, the use of composite tank materials has not been demonstrated for cryogenic propellants because of concerns for brittleness). Appendix B lists the compatibility of materials for different propellants. The materials in this list meet many of the requirements.

Spherical tanks—The equations we use for sizing spherical tanks are

$$V_s = \frac{4}{3}\pi r_s^3 \qquad (5.74)$$

$$A_s = 4\pi r_s^2 \qquad (5.75)$$

$$t_s = \frac{p_b r_s}{2 F_{all}} \tag{5.76}$$

$$m_s = A_s t_s \rho_{mat} \tag{5.77}$$

where r_s = radius of the sphere (m)
A_s = surface area of the sphere (m²)
V_s = volume of the sphere (m³)
t_s = wall thickness of the sphere (m)
p_b = design burst pressure (Pa) [from Eq. (5.73)]
F_{all} = allowable material strength (Pa) (F_{tu} from Table 5.15)
m_s = mass of a spherical tank (kg)
ρ_{mat} = density of the tank structure material (kg/m³) (ρ from Table 5.15)

Cylindrical Tanks—These are not quite as simple as the spherical ones, as there are more design decisions involved, including cylindrical radius, and total tank length. For the cylindrical section, the important equations are

$$V_c = \pi r_c^2 l_c \tag{5.78}$$

$$A_c = 2\pi r_c l_c \tag{5.79}$$

$$t_c = \frac{p_b r_c}{F_{all}} \tag{5.80}$$

$$m_c = A_c t_c \rho_{mat} \tag{5.81}$$

where r_c = radius of the cylindrical section (m)
l_c = length of the cylindrical section (m)
A_c = surface area of the cylindrical section (m²)
V_c = volume of the cylindrical section (m³)
p_b = design burst pressure (Pa) (from Eq. 5.73)
F_{all} = allowable material strength (Pa) (F_{tu} from Table 5.15)
ρ_{mat} = density of the tank structure material (kg/m³) (ρ from Table 5.15)
m_c = mass of the cylindrical tank section (kg)
t_c = thickness of cylinder wall (m)

The elliptical ends of the tank are illustrated in Fig. 5.45. The smaller the value of the crown height (h_e), the more compact the tank is. This characteristic is important if there are stack height limitations on launch vehicles or envelope limitations on spacecraft. However, an elliptical end is typically heavier than a spherical end ($h_e = r_c$) because of structural bending loads at the knuckle. To simplify our preliminary analysis of mass and configuration, we look only at spherical ends. Huzel and Huang [1992] thoroughly discuss the properties of elliptical ends. For the spherical ends of cylindrical tanks, the important equations are the same as for spherical tanks, except that the volume and surface area for *each* end is half of what it is for an entire sphere. Of course, if both ends of the cylindrical tank are spheres, the combination of both ends is a complete sphere.

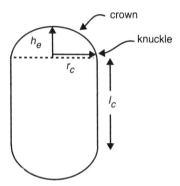

Fig. 5.45. Geometry of a Cylindrical Tank. The figure shows the geometrical parameters used in Eqs. (5.78) through (5.81).

The problem with this type of "hoop stress" analysis is that we have intentionally left out many of the complications that can substantially increase the tank mass. These complications include structural lugs, feed-system fittings, stress concentrations at the tank knuckles, and weld efficiencies, to name but a few. In addition, we have considered only the effect of loads from tank pressure. There are also acceleration (quasi-static) loads and dynamic (vibration and acoustic) loads. Typically, **a given tank weighs 2.0 to 2.5 times more** than what we calculate using the simplified approach outlined above.

Estimating Tank Mass with the pV/W Method

A purely empirical approach involves a tank mass factor [ϕ_{tank} (in meters)] based on tank burst pressure [p_b (in pascals)], tank volume [V_{tot} (m³)], and tank mass [m_{tank} (kg)]:

$$\phi_{tank} = \frac{p_b V_{tot}}{g_0 m_{tank}} \tag{5.82}$$

For completely metallic tanks, the tank factor is 2500 meters. For tanks using fiber-reinforced composite materials, we may be able to drive the mass factor as high as 10,000 meters. Given the burst pressure and required tank volume, we can solve Eq. (5.82) for the tank mass.

Insulation

For tanks storing cryogenics or propellants that may freeze on orbit, we must insulate the tanks against heat transfer. Insulations usually consist of a metallic foil covering a foam insulator. Another common approach uses nonmetallic honeycombs.

Propellant-Expulsion Devices

Tanks must not only store propellants reliably but also manage propellant usage throughout the mission profile. Part of this function entails supplying the engines with gas-free propellant and draining the maximum amount of loaded propellant out of the tanks to minimize the residuals. Also, the propellants in the tanks must not transmit forces and moments to the spacecraft structure that may overwhelm the attitude-control system (propellant slosh). Propellant management is complicated by the changing acceleration environment of the vehicle throughout the mission. The rocket on the launch pad starts in a 1-g field. During the boost and sustainer phase, accelerations can be several g's in any direction. Orbital insertion maneuvers typically have lower accelerations, often less than 1 g. Finally, on-orbit acceleration can be anywhere from ~0.1g to 10^{-6} g.

The crux of the propellant-supply problem is shown in Fig. 5.46. The sketch shows a simple tank with liquid propellant and a pressurant gas bubble. If, at some time in the mission profile, the acceleration is low, the liquid propellant and gas can reorient to a configuration as shown in the sketch, and the gas bubble can end up over the propellant outlet. If the engines fire at this time, the feed system ingests gas and the engines "burp," an undesirable situation. Thus, we need propellant-expulsion devices.

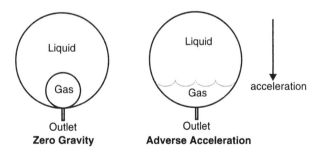

Fig. 5.46. **Sketch of Propellant Supply Problem Motivating Need to Manage Propellant.** In the event of adverse or zero acceleration, gas can accumulate at the outlet and keep engines from starting.

Propellant-expulsion devices may be positive or passive. Positive expulsion devices use physical barriers between the propellant and pressurant gas, such as bladders, pistons, diaphragms, and bellows. Fig. 5.47 illustrates positive-expulsion devices, and Table 5.16 lists their advantages and disadvantages. Using a spinning spacecraft or stage is also considered positive expulsion because a local acceleration field is intentionally developed to orient the liquid propellant over the outlet. Passive-expulsion devices use surface tension on the propellant to keep the fluid in contact with the propellant drain. Examples of passive-expulsion devices are vanes, porous sheets, and screens.

Expulsion devices must also deliver gas-free propellant even in the presence of lateral disturbance accelerations. During the boost and sustain mission phases, the vehicle acceleration mainly moves along the rocket's axis. Lateral accelerations can be as high as ~10% of the axial acceleration. In missions for high-speed military interceptors the lateral accelerations can be even higher. During the on-orbit phase of the mission, the accelerations can orient in any direction because the attitude-control system must react to and correct random disturbances. Positive-expulsion devices can "hold" propellant up to a few 10s of g's of lateral accelerations. While positive-expulsion devices lower the risk of delivering gas-free propellant, the "cost" of these devices is high in terms of system complexity and weight. Piston devices require more hardware elements and increase mass. Bladders must bond to the tank walls, and the bladder material must retain its properties throughout the mission life in the presence of corrosive propellants. Propulsion-system mass, complexity, and longevity, with their attendant dollar costs, are part of the propulsion engineer's trade-off criteria.

Passive-expulsion devices for managing propellant exploit surface tension to control the location of liquid-free surfaces. The defining surface-tension experiment is shown in Fig. 5.48, which demonstrates how surface-tension forces can support a pressure drop in the presence of a radius of curvature provided by boundary geometry.

In designing passive-expulsion devices, we apply this concept in reverse. By imposing a specified radius of curvature at the fluid boundary with vanes or screens, we can create a pressure difference between the liquid and gas. This allows separation of liquid propellant and pressurant gas and segregates the gas bubble from certain parts of the tank, such as the outlet. If we wish, we can use hardware that "holds" against one value of adverse or lateral accelerations and "flows" at another value. Sketches and quantitative details showing these ideas are in Fig. 5.49. The lateral accelerations that passive-expulsion devices hold against are much lower than those for positive-expulsion devices. We can use vanes up to about 0.001 g, porous sheets up to 0.01 g, and screens up to few g's of lateral accelerations. Once again, packing a propellant tank with screens gives gas-free propellant delivery at the cost of extra system mass. A common approach is to use screens in a limited volume near the tank drain as a "start basket," plus other expulsion devices, such as vanes, to deliver propellant to the start basket.

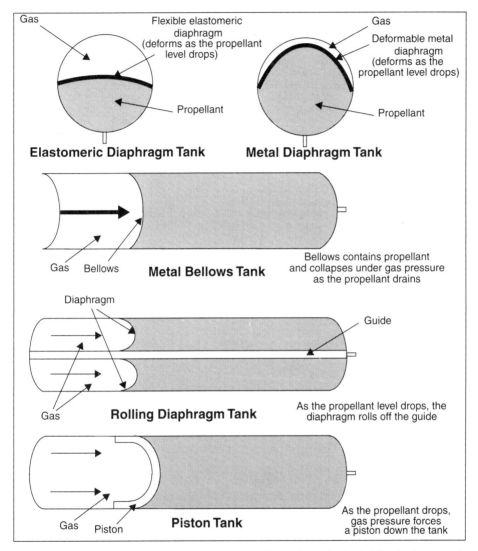

Fig. 5.47. Positive-Expulsion Devices to Manage Propellant. A physical barrier between the propellant and pressurant gas ensures that the gas is not inadvertently ingested.

5.4.5 Designing the Tank Pressurization System

The tanks store and feed propellants to the engines. Pressure in the tanks forces the flow out through the valves, filters, fittings, and supply piping and into the engine or pump. The tank pressure must be high enough to overcome the pressure

Table 5.16. Advantages and Disadvantages of Positive-Expulsion Devices for Propellant Management.

	Metal Diaphragm Tank	Rolling Diaphragm Tank	Piston Tank	Rubber Diaphragm Tank	Metal Bellows Tank
Advantages	• High volume efficiency • Good center of gravity control • No ullage volume • No sliding seals • Proven design	• Low mass • Low cost • Low Δp during expulsion	• Extensive database • Low Δp during expulsion • Design adapts easily to growth	• Extensive database • Low Δp during expulsion • Not cycle limited • Proven design • High expulsion efficiency	• No sliding seals • Good center-of-gravity control • Proven design • Good compatibility • Hermetically sealed
Disadvantages	• High mass • High cost • High-expulsion Δp • Optimizes only for special envelope	• Inspection of internal welds is difficult	• High cost • Low volumetric efficiency • Critical tolerance on shell • Sliding seals possible blow-by	• Compatibility limits on propellants	• High mass • High cost • Limited cycle capability • Low volumetric efficiency
Typical applications	• Spacecraft control & maneuvering • Launch vehicles • Upper stages • Missiles	• Missile interceptors • Maneuvering missiles	• High acceleration missiles	• Spacecraft control & maneuvering • Launch vehicles • Upper stages	• Missiles • Spacecraft • Launch vehicles

drops in the supply-system piping and force the liquid propellants to enter the engine at the design chamber pressure. As previously discussed, propulsion engineers can design the pressurization system in two ways: pump-fed and pressure-fed.

Pump-fed systems are either regulated Net Positive Suction Head (NPSH) or autogenous. Common pressure-fed pressurization systems are either regulated or blowdown. For pump-fed systems, regulated NPSH implies that an external pressurant gas maintains the pressure in the propellant tank at a constant value. The autogenous system uses combustion products to do the same thing. A tapped-off flow from the combustion chamber or a preburner can supply the combustion products. A related concept uses products of a gas generator for the same purpose. Even dumping small amounts of fuel into a hypergolic oxidizer has been suggested as another way to generate combustion products for pressurization but we do not often use this technique. To select a design, we must choose between the straightforward simplicity and increased mass of the regulated NPSH design vs. the potentially lighter, more complex flow routing and thermal management required for autogenous pressurization.

5.4 System Sizing, Design, and Trade-offs

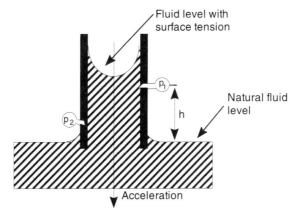

Fig. 5.48. Defining Experiment Involving Surface Tension for Passive-Expulsion Devices. Even though acceleration exists, surface tension allows an effective pressure head (h where $p_1 < p_2$) (adapted from NASA SP-106, [1966]).

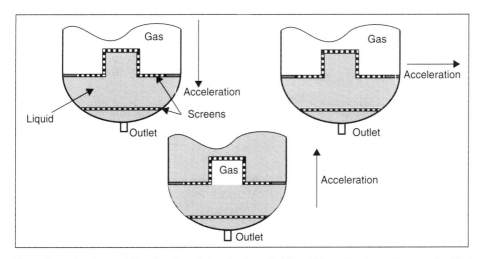

Fig. 5.49. Designs of Passive-Expulsion Devices Hold and Flow Configurations under Various Acceleration Conditions. In all cases, liquid "holds" near the outlet.

Similar choices must be made for a pressure-fed system. First, we decide between a pressure-regulated or blowdown system. The mission profile usually determines this selection. A pressure-regulated system supplies constant (or nearly constant) tank pressure, hence constant pressure to the engine throughout the engine's firing time. In a blowdown system, the tank pressure and the thrust continuously drop throughout the firing timeline. The blowdown system is simpler in design, has fewer parts, and usually weighs less than the regulated system. Fig. 5.50 shows schematic diagrams of blowdown and pressure-regulated systems.

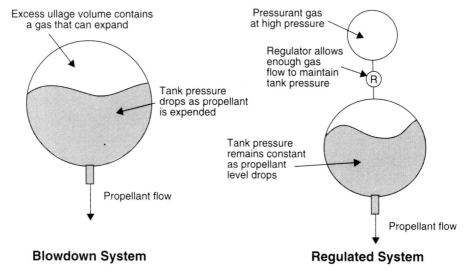

Fig. 5.50.   Sketch of Pressure-Regulated and Blowdown Systems. Blowdown systems are simpler than pressure-regulated systems but typically mean lower performance.

Stored-Gas Systems

Stored gases can be used to pressurize the propellant tank in either the blowdown mode or the pressure-regulated mode (see Fig. 5.50). The most popular pressurant gas is helium because it is inert and lightweight. However, it is a bit expensive. Other gases include nitrogen, hydrogen (for liquid H_2 propellants) and oxygen (for liquid oxidizer systems). There is some concern about the gas dissolving in particular liquid propellants. See NASA SP-8112 [1975] for guidelines on pressurant gas choices.

Usually, we know the initial (21,000,000 Pa is typical, but we can go up to 60,000,000 Pa) and final (usually tank) pressures of our pressurant system and the initial temperature (usually the vehicle temperature). For attitude-control applications, propellant is released slowly, and heat transfer through the structure allows the temperature to remain constant (isothermal). For longer thrust durations, there is insufficient time for heat to transfer into the propellant. For these systems, we assume an isentropic change in temperature [from Eq. (3.68)]:

5.4 System Sizing, Design, and Trade-offs

$$T_f = T_i \left(\frac{p_f}{p_i}\right)^{\frac{\gamma-1}{\gamma}} \tag{3.68}$$

where T_f = temperature at end of burn (K)
T_i = initial gas temperature (K)
p_f = final gas pressure (Pa)
p_i = initial gas pressure (Pa)
γ = isentropic parameter of the pressurant gas

We determine the required pressurant mass with the perfect gas law (see Sec. 3.2.3):

$$m_{press} = \frac{V_{press} p_f}{R T_f} \tag{5.83}$$

where V_{press} = the final volume of pressurant required (m³)
m_{press} = the mass of the pressurant required (kg)
R = pressurant gas constant (J/kg·K)
  = 8314/M
M = molecular mass of the pressurant (kg/kmol)

The final volume of pressurant required is usually equal to the volume of the propellant tank(s) plus the volume of the pressurant storage tank. Of course, not knowing the pressurant tank volume ahead of time requires us to iterate. One possible iteration scheme is as follows:

1. Assume the pressurant tank volume is zero
2. Let V_{press} = propellant tank volume plus pressurant tank volume
3. Calculate m_{press} using Eq. (5.83)
4. Calculate the required pressurant tank volume (V_{ptank}) to hold m_{press} at the initial storage pressure and temperature using the perfect gas law
5. Go to step 2 unless this algorithm has converged

To estimate the mass of the pressurant tank, we use the pV/W approach described in Sec. 5.4.4. Typical pressurant tanks use a titanium structure and have a tank-mass factor of 6350 m (see NASA SP-8112 [1975]). This factor already includes a burst-safety factor, so we can use the initial pressurant pressure:

$$m_{ptank} = \frac{p_i V_{ptank}}{g_0 (6350)} \tag{5.84}$$

Autogenous Systems

For autogenous pressurization, we need to provide enough combustion products to fill the propellant tank volume at the required pressure and temperature. Thermochemical analysis tells us the nature of the pressurant gases. Once we know these numbers, we can do analysis similar to that for the stored gas.

5.4.6 Thrust Vector and Attitude Control

A means of directing the engine thrust vector, or "steering" the vehicle, may be required. This system must account for vehicle directional charges commanded by the guidance and control subsystem. It must also allow for shifts in the vehicle's center of mass and account for errors that accrue when imprecise manufacturing tolerances stack up. If the disturbance torques resulting from a misaligned thrust vector are small, the vehicle's pulsing thrusters in pitch, roll, or yaw can overcome these perturbations. In other cases, the entire vehicle may be intentionally spun during firings of the main engine to cancel out any thrust misalignments. Other thrust-vector control (TVC) modes involve gimballing of the entire rocket engine through the use of actuators, swivel joints or gimbal bearings, and flex hoses. Alternatively, several rocket clusters may provide the main thrust, with rocket engines on the opposing axis turned off to briefly compensate for misalignment or disturbance torques induced by thrust imbalances. Similarly, we may use a rocket engine's throttling ability to modulate its imbalance-induced disturbances relative to any opposing rocket in the cluster. In general, TVC of the flight vehicle over its trajectory uses one of the modes listed in Table 5.17.

We use gimballed engines most often because of reliability and relatively high performance. All of the techniques—except for off-pulsing, throttling, RCS thruster control, and secondary injection—require actuators which may operate by hydraulic, pneumatic, or electromechanical means (or various combinations). For the remaining systems, we use flow regulation for TVC.

Thrust-vector control through secondary injection of matter into the thrust chamber nozzle (SITVC) has been successful in solid and hybrid motors. It has found only limited, mainly experimental, application in liquid-propulsion systems. It appears especially promising for upper-stage engines, in which the lateral forces required are smaller than with boosters. The main methods of secondary injection are

- Gas injection, using
  - Inert stored gas
  - Thrust chamber tapoff
  - Gas generator
- Liquid injection, using
  - Inert fluid
  - Propellants

In a gimballed thrust chamber, the side force is at the injector end. With an SITVC system, the applied side force is downstream of the nozzle throat and approximately at the point of injection, resulting in an increased moment arm which decreases the required side force. As a rule, a system requires four injection elements equally spaced on the main chamber's circumference, with no more than two adjacent ones operating at a given time. A logic and a servosystem similar to that of a gimbal actuator controls the required valves.

Table 5.17. **Strategies for Controlling the Thrust Vector.** We list here the usual methods of thrust vector control along with their control authority.

Method	Typical Range of Control Authority
Gimbals and thrust chamber or complete engine assembly	±7 degrees
Off-pulsing	20 to 40% of thrust
Throttling	±10 to 20% of thrust
Exhaust jet deflectors (jet tabs and jet vanes)	±10 degrees
RCS thruster control	Full range of attitude-control rates as determined by thrust level, moment arm (torque), and duty cycle
Gimballed thrust chamber nozzle (rare with liquid propellants)	±6 degrees
Secondary material injection into the thrust chamber	±5 degrees

5.4.7 Structural Mounts

Determining the configuration and mass of the structure required to hold all of the individual components in place and attach them to the rest of the vehicle is difficult at the preliminary design level. Getting accurate results usually requires a structural analysis, possibly using sophisticated computer modeling. For preliminary analysis, we usually assume that the mass of all the structural mounts and associated hardware is about 10% of the inert mass (stage mass minus propellant and pressurant).

Configuring the structural elements usually involves either a truss structure or a stressed skin/stringer construction. Again, choosing an approach is complex at the preliminary design level. See Larson and Wertz [1992] or Sarafin and Larson [1995] for guidance.

5.5 Case Study

The purpose of this section is to run through an example of the process for designing liquid rockets. The example is based upon the mission case study from Chap. 10— in particular, the first-stage liquid system for the "Option 4" mission.

5.5.1 Requirements and Design Considerations

Recall that for this mission, we are using a liquid propulsion system to take us from an initial 400-km circular orbit to a 5000-km altitude circular orbit. From there, an electric ion engine-based propulsion system is used to spiral up to the final orbit. The basic requirements at the propulsion-system level for this "stage" are as follows:

- Δv = 1721 m/s
- Payload mass = 4914 kg (includes 2600-kg payload plus the ion engine mass)
- Minimum thrust-to-weight ratio = 0.3
- Initial vehicle mass $\leq$ 12,000 kg
- Envelope limitations = 3-m diameter by 3-m length

Chapter 10 also indicates the propellants probably should be kerosene (RP-1) and liquid oxygen.

Because this is a "commercial" development, we do not have to get the highest possible performance from our system as long as it works. Therefore, we specify a "low risk" approach, which means we must try to keep the system as simple as possible and avoid having to develop complex systems such as pumps and fluid cooling systems.

Before we start configuring and designing our vehicle, we need to add some margin to the design. For example, from Chap. 10 we have a Δv requirement that gives us an initial vehicle mass of 9432 kg. However, what happens if we are not placed in the correct initial orbit? Or what happens if the engine performance is a few percent below the design point? Experience from many historical systems developments indicates that a 30% increase in vehicle inert mass during development is not unusual! All of these "intangible" considerations could severely affect our program down the road, so adding a pad at this point in the design makes sense.

The major area for padding is in the stage mass. The case study analysis (Chap. 10) indicates that we have a margin to eat into. However, if we do not design our tanks and other systems to handle additional propellant now, we can never take advantage of this pad if required. First of all, let us assume a 10% margin on the Δv requirement:

- Design Δv = 1.1 × 1721   = 1893 m/s

5.5.2 Preliminary Design Decisions

Estimate System Mass and Envelope

If we assume an initial mass of 12,000 kg (the upper limit) and an initial thrust-to-weight ratio (F/W) of 0.3, we can determine the required thrust level (F):

- $F = 0.3(9.81)(12{,}000) = 35{,}305$ N

Now, to get a rough size for the thrust chamber, we use Eqs. (5.4) through (5.6), which give us the mass, length, and diameter of "space bipropellant engines." Remember, this is a coarse estimate at best; there are large dispersions in these numbers.

- Engine mass   = 103 kg
- Engine length   = 2.23 m
- Engine diameter   = 1.41 m
- Engine thrust/weight = 35

Choose Propellants

The case study in Chap. 10 trades various propellant combinations and settles on kerosene and liquid oxygen (RP-1/LOx). This seems to be a reasonable choice, so we stick with it.

Determine Engine Cycle and Cooling Approach

In the interests of simplicity and low cost, we choose an ablative engine fed with tank pressure.

Determine Pressure Levels for the Engine and Feed System

Thrust Chamber—From Appendix B, we can determine the combustion parameters for RP-1/LOx. At the optimal O/F:

- O/F   = 2.3
- Flame temperature   = 3510 K
- Isentropic parameter = 1.225
- Molecular mass   = 22.1 kg/kmol

Assuming a combustion efficiency of 1.0 (remember, we are using frozen-flow thermochemistry, so this efficiency is conservative) and a nozzle efficiency of 0.98, we can use the process in Table 5.5 to determine the specific impulse as a function of nozzle expansion ratio. The result of this analysis is shown in Fig. 5.51. From this figure, we see that an expansion ratio of 100 gives us a specific impulse of 337 seconds (as we already assumed). Doubling the expansion ratio takes the specific impulse up to 342 seconds. This amount seems excessive, so we use a "diminishing returns" argument to choose an expansion ratio of 100.

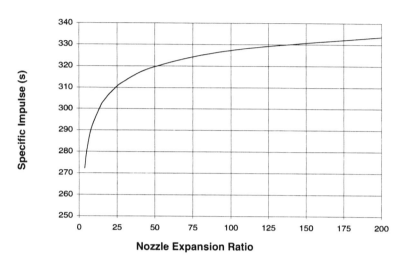

Fig. 5.51. Performance of the RP-1/LOx Space Engine. The graph shows the specific-impulse performance of our engine as a function of expansion ratio. Increasing the expansion ratio above 100 does not increase I_{sp} significantly (see Table 5.5).

For space engines, we are NOT required to drive up chamber pressure to get higher performance. If we go to a lower pressure, we can get equivalent performance by increasing the throat area. This means the thrust chamber gets larger. However, the lower pressure requirement drives down the pressure requirements and the mass of the propellant and storage feed system. Therefore, we choose a combustion-chamber pressure at an historically valid lower limit of 700,000 Pa (about 100 psi). In summary:

- Nozzle expansion ratio   = 100
- Combustion-chamber pressure = 700,000 Pa

Now, using the equations from Sec. 5.3.5, we can determine the pressure profile through the feed system. From the propellant tank volume and Fig. 5.19, we see that the question of pump feed versus pressure feed is unclear because we are right in the overlap area. However, we can improve the numbers somewhat by using composite tanks (at least for the RP-1, which is not cryogenic) which are not considered in the historical data. Using this argument and having a desire for system simplicity, we choose a pressure-fed system for a "first cut."

- Injector pressure drop $= 0.2 \times 700{,}000 = 140{,}000$ Pa
- Feed system pressure drop $= 50{,}000$ Pa
- Fuel dynamic pressure $= 0.5 \times 810 \times 10^2 = 40{,}500$ Pa
- Oxidizer dynamic pressure $= 0.5 \times 1142 \times 10^2 = 57{,}100$ Pa

- Fuel tank pressure = 700,000 + 140,000 + 50,000 + 40,500
  = 930,500 Pa
- Oxidizer tank pressure = 700,000 + 140,000 + 50,000 + 57,100
  = 947,100 Pa

From Fig. 5.20, we see that using a pressurant system that employs a stored inert gas is a good choice. We assume a fairly standard initial pressure in the pressurant tank of 21 MPa (see Sec. 5.4.5). The resulting propellant tank pressure corresponds quite well (perhaps a bit on the high side) with the numbers in Fig. 5.20.

Estimate Propellant Mass and Size Tanks

We assume a specific impulse of 337 seconds and an O/F of 2.3. If we take these values and apply them to the rocket equation, we find that an inert mass fraction of 0.261 gives us an initial mass for the complete vehicle of 12,000 kg. We solve the rocket equation assuming an initial mass of 12,000 kg to get this number. A summary of the ideal rocket analysis is as follows:

- Payload mass = 4914 kg (includes payload and ion engine stage)
- Structural mass fraction = 0.261 (conservative)
- Initial mass of the vehicle = 12,000 kg
- Specific impulse = 337 s
- Propellant mass = 5232.7 kg
- Stage inert mass = 1853.3 kg
- Thrust level = 35,305 N (for an initial F/W of 0.3)
- Thrust duration = 489.8 s
- Total impulse = 17,293,891 N·s
- Propellant mass flow rate = 10.683 kg/s (from specific impulse and thrust required)
- Fuel flow rate = 3.237 kg/s
- Oxidizer flow rate = 7.445 kg/s

From Fig. 5.22, we know that a good average value for inert mass fraction is 0.17. However, the dispersion in this number for space systems is quite large (0.08–0.37). If we are trying to use a simple pressure-fed system, our actual mass fraction could be higher than the 0.17 number. Choosing a mass fraction to hit the maximum allowable mass gives us some margin.

From Appendix B, we know that the density of the propellants is:

- RP-1 = 810 kg/m^3
- Liquid oxygen = 1142 kg/m^3

Assuming we have 10% ullage (Sec. 5.4.4), the required propellant tank volume is (mass = mass flow rate × burn duration):

- Mass of RP-1 = 3.237 × 489.8 = 1585.7 kg
- Mass of LOx = 7.445 × 489.8 = 3647.0 kg
- Volume of RP-1 tank   = 2.153 m^3
- Volume of LOx tank   = 3.513 m^3

If we assume individual spherical storage tanks, we can use Eq. (5.74) to determine the required tank radii:

- RP-1 tank radius   = 0.801 m
- LOx tank radius   = 0.943 m

5.5.3 System Sizing, Design, and Trade-offs

Thrust Chamber

We start the thrust-chamber design with the nozzle. From the thrust requirement and the predicted specific impulse, we have determined the required mass flow rate as 10.683 kg/s. From this flow rate, we can determine the required throat area (A_t) and corresponding exit area (A_e) by using the equation for characteristic velocity [Eq. (3.130)]. We can determine the characteristic velocity from the thermochemical data and Eq. (5.12):

- $c^* = 1759$ m/s
- $A_t = \dfrac{\dot{m} c^*}{p_c} = \dfrac{10.683\,(1759)}{700{,}000} = 0.026\,84$ m^2
- $A_e = \varepsilon A_t = 100\,(0.026\,84) = 2.6842$ m^2

Now the area diameters are

- Throat diameter   = 0.1849 m
- Exit diameter   = 1.8487 m

We use the Mach-number approach to determine the combustion chamber's dimensions. Assuming a chamber Mach number of 0.2 and a design L^* of 0.9 (this is a bit aggressive, but we have the additional nozzle contraction area as margin), we get the following dimensions by using Eqs.(5.37) and (5.38):

- Chamber length   = 0.298 m
- Chamber area   = 0.0810 m^2
- Chamber diameter   = 0.3211 m
- Contraction ratio   = 3.02 (a bit higher than indicated in Fig. 5.24)

We can now estimate the thrust chamber's mass. For the hoop-stress approach, we assume columbium as our material with a yield strength of 310 MPa. Applying a multiplication factor (ϕ) of 3 (includes a pressure factor of 2 and a mass factor of 1.5), we determine the chamber wall's thickness [Eq. (5.44)]:

- thickness $= \dfrac{P_c D_{chamber} \phi}{2 F_{tu}} = \dfrac{700{,}000 (0.3211) 3}{2(310{,}000{,}000)} = 0.00109$ m

This gives us a wall thickness of 1.09 millimeters in the chamber.

To determine the combustion-chamber mass, we must first estimate the surface area. To do this, we use the simplified, thin-wall-geometry approach given in Table 5.8 (we assume a 45° contraction angle for the combustion chamber):

- $m_{cc} = \pi \rho t_w \left[2 r_{cc} L_{cc} + \dfrac{\pi \left(r_{cc}^2 - r_t^2 \right)}{\tan \theta_{cc}} \right]$

- $m_{cc} = \pi (8500)(0.00109) \left[0.3211(0.298) + \dfrac{\left((0.3211/2)^2 - (0.1848/2)^2 \right)}{\tan 45°} \right]$

  $= 4.355$ kg

To determine the nozzle length, we use Eq. (5.41) for the length of a conical nozzle. Given the throat radius and the nozzle expansion ratio, for a 15° half angle nozzle:

- 15° nozzle length = 3.1047 m

From Fig. 5.25b, we can determine the fractional length for a bell nozzle, given the required nozzle efficiency (0.98):

- Fraction of 15° conical nozzle   = 0.675
- Bell nozzle length   = 0.675 × 3.1047 = 2.0957 m

To estimate the mass of this nozzle, we go back to the equations in Table 5.8. Assuming a constant-thickness nozzle is usually too conservative, so we assume a tapered nozzle. At the throat, the static pressure is about one half of the chamber pressure, so we assume a throat-wall thickness of one half the combustion-chamber thickness. For simplicity, we assume the nozzle tapers to a wall thickness of zero. Although this is not physically accurate, assuming a constant taper is very conservative. In addition, we assume the nozzle length is the bell nozzle length:

- Nozzle mass   = 24.0 kg

From Table 5.7, we know that the combustion chamber and nozzle masses make up about 40% (0.163 + 0.237) of the total thrust-chamber mass for ablatively cooled engines. Using this mass-fraction approach (with average values), we can deduce the other masses:

- Total engine mass   = 70.70 kg
- Injector mass = 0.249 × 70.70   = 17.60 kg
- Ablative material = 0.352 × 70.70 = 24.89 kg

Determine Engine Cycle and Cooling Approach

In the interest of simplicity, we avoid using pumps and thus remove the need for a decision on the engine cycle. Also in the interest of simplicity, we ablatively cool the thrust chamber.

Propellant Storage System

We use the tank-mass-factor approach to estimate the propellant tank masses. From previous analyses, we know the tank volumes and required storage pressures. For both tanks, we assume a conservative tank-mass factor of 2500 meters and a burst-pressure factor of 2. From Eq. (5.82):

$$m_{tank} = \frac{p_b V_{tank}}{\phi_{tank} g_0} \tag{5.85}$$

This gives tank masses as follows:

- RP-1 tank mass = 163.45 kg
- LOx tank mass   = 271.40 kg

Designing the Tank Pressurization System

We have assumed a constant pressure in the propellant tank as part of our design. One way to achieve this constant pressure is with a regulated, inert-gas pressurant. Because the operation of this propulsion system involves a single, relatively short-duration burn, we assume isentropic expansion of the pressurant gas into a constant-pressure propellant tank. We also choose to use helium because of its low molecular mass. The gas properties of helium (from Table 3.1) are:

- $\gamma = 1.66$
- Molecular mass = 4.003 kg/kmol

We assume an initial temperature of 273 K and an initial pressure of 21 MPa (as mentioned in Sec. 5.4.5). The pressurant expands into the propellant tanks, through regulators, at an average pressure of 938,800 Pa (this is just the average of the fuel and oxidizer tank pressures). The pressurant must occupy the entire volume of the propellant tanks and its own storage tank at this pressure at the end of the burn. However, we allow ourselves an additional 5% margin on the pressurant. Unfortunately, we do not know the pressurant tank volume (this is what we are trying to determine) ahead of time, so we must rely on an iteration to estimate the total volume. In addition, the temperature of the pressurant drops during the burn because of isentropic expansion. The final temperature is

$$T_f = T_i \left(\frac{p_f}{p_i}\right)^{\frac{\gamma-1}{\gamma}} = 273\left(\frac{938,000}{21,000,000}\right)^{\frac{0.66}{1.66}} = 79.4 \text{ K} \tag{5.86}$$

5.5 Case Study

As an initial guess for tank volume, we simply take the sum of the RP-1 and LOx tank volumes and add a 5% design margin. We can determine the required pressurant mass from the ideal gas law:

$$m_{press} = \frac{pVM}{R_u T_f} = \frac{(1.05)\,938{,}800\,(2.153 + 3.513)\,4.003}{8314\,(79.4)} = 33.889 \text{ kg} \quad (5.87)$$

We can then determine the pressurant tank volume, again using the ideal gas law:

$$V_{press} = \frac{m_{press} R_u T_i}{pM} = \frac{33.889\,(8314)\,(273)}{21{,}000{,}000\,(4.003)} = 5.666 \text{ m}^3 \quad (5.88)$$

We then add this volume to the previously assumed volume and iterate on the mass and pressurant tank volume until we get a converged solution. The result is:

- mass of pressurant = 40.416 kg (with a 5% margin)
- pressurant tank volume = 1.091 m³

To determine the pressurant tank mass, we use the structural-mass-factor approach as for the propellant tanks. We assume an operating pressure of 21,000,000 Pa and a tank factor of 6350 m:

$$m_{tank} = \frac{21{,}000{,}000\,(1.091)}{9.807\,(6350)} = 368.0 \text{ kg} \quad (5.89)$$

This is a very significant mass as compared to the propellant tank masses.

Thrust Vector Control

To estimate the masses of the feed system and the system for thrust-vector control, we simply subtract the calculated thrust chamber mass from our estimate of the system mass (based on historical data):

$$103 - 71 = 32 \text{ kg} \quad (5.90)$$

Structural Mounts

To estimate the support-structure mass, we assume a 10% increase in the overall inert (structural) mass of the system (see Sec. 5.4.7):

$$0.1 \times (71 + 435 + 368) = 87 \text{ kg} \quad (5.91)$$

5.5.4 Baseline Design

Figure 5.52 shows the assembled system configuration. The extremely tight envelope requires us to get a bit inventive in laying out our tankage. Spherical and cylindrical tanks simply do not fit into the envelope. Toroidal tanks have been used in several vehicles developed in the Soviet Union. We must resort to this configuration. Figure 5.53 shows the basic dimensions. The key dimensions for our tanks are (assuming a maximum dimension [OD] of 3 m):

- LOx: r = 0.403 m
- LOx: ID = 1.389 m
- RP-1: r = 0.302 m
- RP-1: ID = 1.793 m
- He: r = 0.2068 m
- He: ID = 2.173 m

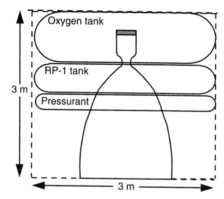

Fig. 5.52. System Configuration. This figure shows the layout of all of the individual components. All of the systems fit within the 3 m × 3 m cylinder.

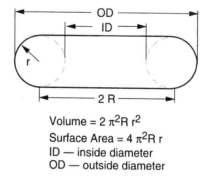

Volume = $2\pi^2 R r^2$
Surface Area = $4\pi^2 R r$
ID — inside diameter
OD — outside diameter

Fig. 5.53. Key Dimensions for Toroidal Tanks. This geometry sizes our propellant tanks.

The following tables summarize the design results. Table 5.18 summarizes the overall performance we can expect from the system. The final thrust-to-weight ratio (3.5), may be a bit excessive, depending on required thrust levels. If so, throttling may be required.

Table 5.18. **Summary of System Performance.** These performance parameters allow us to meet the propulsion system requirements.

Parameter Name	Value	Comments
Thrust magnitude	35,305 N	• initial thrust-to-weight ratio is 0.3 • final thrust-to-weight ratio is 3.5 (throttling may be required)
Thrust duration	490 s	• assuming a constant thrust magnitude
Specific impulse	337 s	• chamber pressure = 700,000 Pa • nozzle expansion ratio = 100
Impulse	17,293,891 N·s	• based on a Δv of 1893 m/s
Fuel flow rate	3.237 kg/s	• O/F = 2.3 • total mass flow rate = 10.683 kg/s
Oxidizer flow rate	7.445 kg/s	• $c^* = 1759$ m/s

Table 5.19 summarizes the system masses. The total inert mass is 1034 kg, which is below the amount allowed by the rocket-equation analysis by 820 kg. This result gives us an 80% margin, which should be adequate for even the most conservative of detailed design.

Table 5.19. **Summary of System Mass.** All of the individual components are listed along with their mass. The final mass is 13% lower than the initial assumed mass, giving us a healthy margin.

Component	Mass (kg)	Comments
RP-1 (fuel)	1586	• includes 10% Δv margin
LOx	3647	• includes 10% Δv margin
Helium pressurant	40	• includes a 5% mass margin
Thrust chamber	71	
RP-1 tank	163	
LOx tank	271	
Pressurant tank	368	
Feed system	32	• this is the difference between the system-level estimate of mass and the component-level estimate of thrust-chamber mass • 103 – 71 = 32 kg
Support structure	87	• this is 10% of the combined thrust-chamber and propellant-tank mass
Total	6366	• this gives us an actual structural-mass fraction of 16.5%

References

Acampora, K.J., and H. Wickman. 1992. *Component Development for Micro-Propulsion Systems*. AIAA-92-3255. Washington, DC: American Institute of Aeronautics and Astronautics.

Altman, D., J. M. Carter, S.S. Penner, and M. Summerfield. 1960. *Liquid Propellant Rockets*. Princeton, NJ: Princeton University Press.

Bartz, D. R. 1968. "Survey of Relationships between Theory and Experiment for Convective Heat Transfer in Rocket Combustion Gases." in *Advances in Rocket Propulsion*. S. S. Penner, ed. Manchester, England: AGARD, Technivision Services.

Bissell, W.R. 1985. *Rocket Engine Turbopump Tutorial*. Rocketdyne-RI/RD 85-245, USAF Contract F33657-82-C-0346 for Analysis of Foreign Chemical Rocket Propulsion Systems. Rockwell International, Rocketdyne Division.

Elverum, G. W., Jr. and T. F. Morey. 1959. *Criteria for Optimum Mixture Ratio Distribution Using Several Types of Impinging Stream Injector Elements*. JPL Memorandum No. 30-5. Pasadena, CA: Jet Propulsion Laboratory/California Institute of Technology.

Elverum, G. W., Jr., P. Staudhammer, J. Miller and A. Hoffman. 1967. The Descent Engine for the Lunar Module. AIAA Paper 67-521. Paper presented at the AIAA 3rd Propulsion Joint Specialist Conference. Washington, DC:American Institute of Aeronautics and Astronautics.

Faires, Virgil Moring. 1947. *Applied Thermodynamics*. New York: McMillan Co.

Frey, H. M. and G. R. Nickerson. 1970. (revised 1971). *TDK—Two Dimensional Kinetic Reference Program*. Developed for NASA Contract NAS9-10391. Irvine, CA: Dynamic Science.

Giffoni, S. F. 1973/74 (approx.). TRW Storable Bipropellant Engines for Space Applications. Unpublished manuscript.

Gordon, S. and B. McBride. 1973. *Computer Program for Calculation of Complex Chemical Equilibrium Compositions, Rocket Performance, Incident Shocks and Chapman-Jouget Detonations*. NASA SP-173. Washington, DC: National Aeronautics and Space Administration.

Hill, Philip G., and Carl R. Peterson. 1970. *Mechanics and Thermodynamics of Propulsion*. New York: Addison Wesley Publishing Company.

Huang, D. T. and R. L. Sackheim. *Monopropellant Hydrazine Spacecraft Propulsion Systems— 30 Years of Safe, Reliable, Flexible and Predictable Performance*. Unpublished technical manuscript.

Huzel, Dieter K., and David H. Huang. 1992. *Modern Engineering for Design of Liquid-Propellant Rocket Engines*. Revised Edition. AIAA Progress in Astronautics and Aeronautics, Vol. 147. Washington, DC: American Institute of Aeronautics and Astronautics.

Ingebo, R. D. 1958. *Drop-Size Distributions for Impinging-Jet Breakup in Airstreams Simulating the Velocity Conditions in Rocket Combustors*. NACA Technical Note TN 4222.

Washington, DC: National Advisory Committee on Aeronautics.

Isakowitz, Steven J. 1991. *International Reference Guide to Space Launch Systems*. Washington, DC: American Institute of Aeronautics and Astronautics.

Koelle, H. H. ed. 1961. *Handbook of Astronautical Engineering*. New York: McGraw-Hill, Inc.

Krishnan, A. and A. J. Przekwas. 1990. *Orbital Maneuvering Vehicle (OMV) Thrust Chamber Performance—Interim Report for Phase 1*. CFDRC Report 4075/1. Huntsville, AL: CFD Research Corporation.

Larson, W. J., and Wertz, J. R. eds. 1992. *Space Mission Analysis and Design* second edition. Norwell, MA and Torrance, CA: Kluwer Academic Publishers and Microcosm, Inc.

NASA SP-106. 1966. *The Dynamic Behavior of Liquids in Moving Containers*. Washington, DC: National Aeronautics and Space Administration.

NASA SP-8087. 1972. *Liquid Rocket Engine Fluid-Cooled Combustion Chambers*. Washington, DC: National Aeronautics and Space Administration.

NASA SP-8088. 1974. *Liquid Rocket Metal Tanks and Tank Components*. Washington, DC: National Aeronautics and Space Administration.

NASA SP-8089. 1976. *Liquid Rocket Engine Injectors*. Washington, DC: National Aeronautics and Space Administration.

NASA SP-8107. 1974. *Turbopump Systems for Liquid Rocket Engines*. Washington, DC: National Aeronautics and Space Administration.

NASA SP-8109. 1973. *Liquid Rocket Engine Centrifugal Flow Turbopumps*. Washington, DC: National Aeronautics and Space Administration.

NASA SP-8112. 1975. *Pressurization Systems for Liquid Rockets*. Washington, DC: National Aeronautics and Space Administration.

NASA SP-8113. 1974. *Liquid Rocket Engine Combustion Stabilization Devices*. Washington, DC: National Aeronautics and Space Administration.

NASA SP-8120. 1976. *Liquid Rocket Engine Nozzles*. Washington, DC: National Aeronautics and Space Administration.

Oates, G.C. 1984. *Aerothermodynamics of Gas Turbine and Rocket Propulsion*. New York: American Institute of Aeronautics and Astronautics. 1984.

Oefelein, J. C. and V. Yang. 1992. *A Comprehensive Review of Liquid Propellant Combustion Instabilities in F-1 Engines*. University Park, PA: The Pennsylvania State University, Department of Mechanical Engineering, Propulsion Engineering Research Center.

Priem, R. J. and M. F. Heidmann. 1959. Vaporization of Propellants in Rocket Engines. *ARS Journal*. November: 836

Priem, R. J. and M. F. Heidmann. 1960. *Propellant Vaporization as a Design Criterion for Rocket-Engine Combustion Chambers*. NASA Technical Report TR R-67. Cleveland, OH: NASA Lewis Research Center.

Rao, G. V. R. 1958. Exhaust Nozzle Contour for Optimum Thrust. *Jet Propulsion.* 28:377.

Rocket Research Corporation. *Monopropellant Hydrazine Design Data.* Seattle, WA: Rocket Research Corporation.

Sackheim, R. L. 1973. Survey of Space Applications of Monopropellant Hydrazine Propulsion Systems. Paper presented at Tenth International Symposium on Space Technology and Science. Tokyo, Japan.

Sarafin, Thomas P. and Wiley J. Larson, eds. 1995. *Spacecraft Structures and Mechanisms— from Concept to Launch.* Norwell, MA and Torrance, CA: Kluwer Academic Publishers and Microcosm, Inc.

Schmidt, E. W. 1984. *Hydrazine and Its Derivatives, Preparation, Properties, Applications.* New York: John Wiley and Sons.

Schmitz, B. W., D. A. Williams, W. W. Smith, and D. Maybee. 1966. Design and Scaling Criteria for Monopropellant Hydrazine Rocket Engines and Gas Generators Employing Shell 405 Catalyst. AIAA paper 66-594 presented at the AIAA Second Propulsion Joint Specialist Conference. Colorado Springs, CO: American Institute of Aeronautics and Astronautics.

Sutton, George P. 1992. *Rocket Propulsion Elements.* 6th ed. New York: John Wiley and Sons.

TRW, Inc. 1967. *Characteristics of the TRW Lunar Module Descent Engine,* Vol. I. TRW Report No. 01827–6119–T000. Prepared for Grumman Aircraft Engineering Corporation Under Contract PO 2-18843-C. Redondo Beach, CA: TRW, Inc.

TRW, Inc. 1970. *Apollo Lunar Module Descent Engine Performance Characterization Program Report.* Prepared for Grumman Aircraft Engineering Corporation Under Contract 2-18843-C. Redondo Beach, CA: TRW, Inc.

Whitehead, John, Lee Pittenger, and Nick Colella. 1994. *Design and Flight Testing of a Reciprocating Pump-Fed Rocket.* AIAA-94-3031. Washington, DC: American Institute of Aeronautics and Astronautics.

CHAPTER 6
SOLID ROCKET MOTORS

Stephen Heister, *Purdue University*

6.1 Background
6.2 Design Process
6.3 Preliminary Sizing
6.4 Solid Rocket Propellants
6.5 Performance Prediction
6.6 Case Study

One of the more common rocket-propulsion systems in use today is the Solid Rocket Motor (SRM). Solid rocket motors have appeared in

- Strap-on boosters for space-launch vehicles
- Upper-stage propulsion systems for orbital-transfer vehicles
- Spin and despin systems for spacecraft
- Strategic and tactical missile-propulsion systems
- Gas generators for starting liquid engines and pressurizing tanks

In fact, even the fundamentals employed in airbags for new automobiles essentially derive from SRM technologies.

Figure 6.1 shows a typical SRM with its major components highlighted. The pressure vessel enclosing the propellant grain is referred to as the *motor case*. The nozzle assembly usually bolts to the motor case through a fitting called a *polar boss*. Most motor cases contain polar bosses at both ends, although Fig. 6.1 shows only the aft boss. The motor case usually has a cylindrical extension called the *thrust skirt* which enables the motor to be attached to an upper stage or spacecraft interface. Motors may contain either zero, one, or two skirts (forward and aft) depending on the application.

The *igniter* starts the propellant burning within the chamber, while the propellant and *internal insulation* shield the motor case from hot propellant gases. The *nozzle* converts the thermal energy of propellant gases to thrust. Nozzles can be

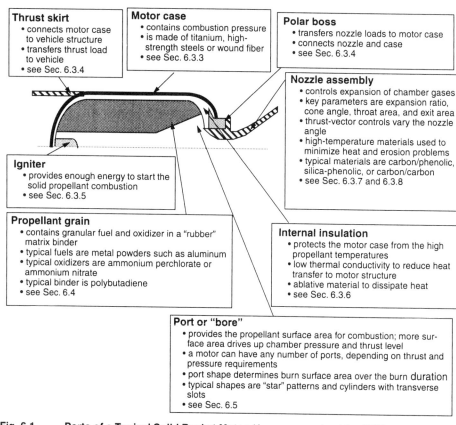

Fig. 6.1. Parts of a Typical Solid Rocket Motor. Key components of the SRM are shown along with key issues associated with each component.

quite complex and contain many parts, depending on whether or not they can gimbal to control the thrust vector. The mass of propellant is usually called a *propellant grain* after powder grains historically used for ammunition. The motor's thrust is proportional to the burning surface area (Sec. 6.5), so we can vary the geometry of the grain to obtain different thrust histories. Finally, the channel formed by the perforation in the propellant grain is called the *bore* or *port* of the motor.

An SRM is simple (very few moving parts), responds quickly (no need to fill tanks with liquid), and delivers very high thrust levels without significant design compromises. Unfortunately, an SRM cannot usually change thrust in response to current conditions or be test-fired without expending itself.

In this chapter, we discuss the analysis and design of SRMs used in space propulsion systems. As noted in the above list, the main uses for SRMs in space systems are as boosters for launch vehicles and upper-stage systems. For this reason, our discussion focuses on these areas, although the analytical approaches presented in this chapter apply to virtually all SRMs. We describe state-of-the-art systems and briefly discuss each of the major components listed above. After offering insight into the propellant formulations and theory of combustion, we close the chapter with a case study involving the design of a solid-rocket propulsion system for an orbital-transfer vehicle.

6.1 Background

Solid rocket motors often work as boosters for space-launch vehicles. In boosters, the SRM can provide very high thrust levels (more than 5,000,000 N) by simply having a nozzle large enough to expel the large mass flows required to attain these forces. In contrast, liquid rocket engines have limited thrust because of restrictions in the design of turbopumps needed to force propellants into the combustion chamber. Limits in materials keep these pumps relatively small, so SRMs produce higher thrust. Of course, a designer can use several liquid engines to achieve very high thrusts, but not without penalties such as increased inert mass.

Solid rocket motors are also used in upper stages, or orbital-transfer vehicles (OTVs), to raise a payload (normally a satellite) from low-Earth orbit to a higher orbit. Because the launch vehicle must transport the entire upper-stage propulsion system, high performance (specific impulse) and compact size are important. SRMs usually produce a lower specific impulse than liquid systems, but the high density of solid propellants permits a compact package. In fact, solid propellants have a higher density impulse (product of specific impulse and propellant density) than liquid systems, making them attractive for upper-stage propulsion. Although a solid propulsion OTV is usually heavier than the liquid OTV, the solid motor's lower volume has advantages in small spaces such as the bay of the Space Shuttle.

Finally, the high thrust of SRMs makes them a good source for first-stage propulsion in space-based missile systems. In this application, high thrust levels over short durations are required to accelerate missiles to high velocities. Solid systems are also simple and storable—added advantages for this application. Solid-propulsion systems are being investigated for several space-based missile concepts.

The SRMs' high thrust has led many designers to choose them as first-stage propulsion for launch vehicles. In fact, six U.S. launch systems (Scout, Delta, Titan, Atlas, Pegasus, and Space Shuttle) use solid propulsion as part of the first stage. Foreign launch vehicles such as the Ariane (French), the H-2 (Japanese), and the Long March CZ-2E (Chinese) also use solid propulsion systems on the first stage.

The Space Shuttle's boosters are designated the Redesigned Solid Rocket Motors (RSRM) because their design changed after the Challenger failure. A new booster, the Advanced Solid Rocket Motor (ASRM) allows the Space Shuttle a larger payload. Figure 6.2 highlights some of the improved technology in the ASRM design. Most notably, the ASRM has reduced the number of segments from four to three and has selected a new propellant system. The United States Air

Force's heavy-lift Titan has a long history of using solid rocket boosters. A schematic of the Titan IV booster, the newest version, is shown in Fig. 6.3. The Air Force has developed a new solid booster, the Solid Rocket Motor Upgrade (SRMU), to increase the Titan IV's payload.

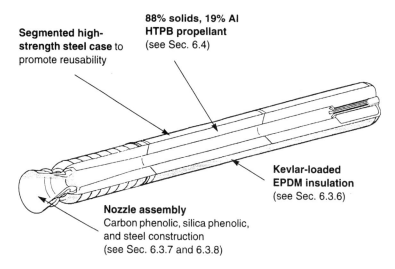

Fig. 6.2. Schematic of ASRM (Advanced Solid Rocket Motor). Courtesy of Aerojet Corporation.

The Delta vehicle has used the Castor family of solid motors since the early 1950s. The current vehicle (Delta II) employs the Graphite Epoxy Motor (GEM), which is an upgrade to the Castor IVA motors used on previous versions of the Delta vehicle. A schematic diagram of the GEM is shown in Fig. 6.4.

The smaller launch vehicles in the United States extensively use solid rocket motors. The Pegasus vehicle, which is launched from an aircraft, also uses only solid propulsion, with stages 1–3 using the Orion 50SAL, 50, and 38 motors, respectively. Figure 6.5 shows a schematic of the winged Pegasus vehicle.

The U.S. has three upper-stage systems that use solid propulsion. The Inertial Upper Stage and Payload Assist Module (PAM) rely completely on SRMs, whereas the Transfer Orbit Stage uses a solid stage and a liquid stage to achieve the desired impulse. The Inertial Upper Stage's first and second states are called ORBUS 21 and ORBUS 6E, respectively. Fig. 6.6 shows a schematic of this vehicle. Note that the second-stage ORBUS 6E motor incorporates an *extendible exit cone* which deploys before firing to give the nozzle a large expansion ratio in a minimal length.

The PAM-D uses a Star 48B motor and the newer PAM-DII uses the larger Star 63D motor. A schematic of the Star 48B is in Fig. 6.7. Finally, we note that the Star

6.1 Background

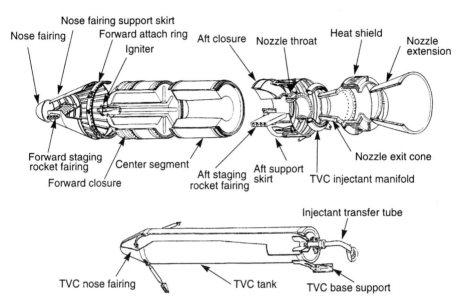

Fig. 6.3. Schematic of Titan IV Motor. (TVC = Thrust Vector Control) Courtesy of United Technologies Chemical Systems.

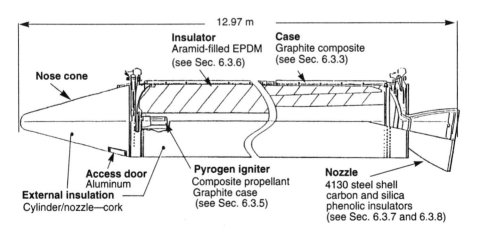

Fig. 6.4. Schematic of GEM (Graphite Epoxy Motor). Courtesy of Hercules Aerospace Company.

37XFP motor is also used frequently in space applications, mainly as an apogee kick motor for the Global Positioning System constellation of satellites.

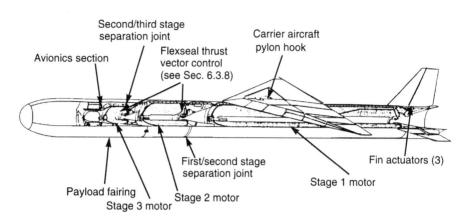

Fig. 6.5. Schematic of Pegasus Launch Vehicle. Courtesy of Hercules Aerospace Company.

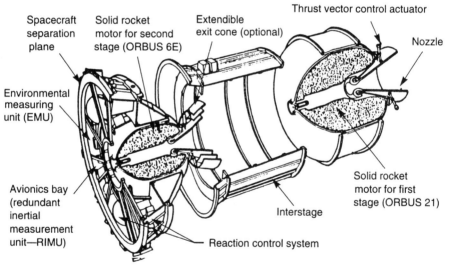

Fig. 6.6. The Inertial Upper Stage Vehicle using Solid Propulsion. Courtesy of United Technologies Chemical Systems.

Tables 6.1–6.3 summarize dimensions, mass, and performance for the family of boosters and upper-stage motors discussed above. As noted in Table 6.1, many of the larger boosters are segmented to ease transportation and handling. Newer designs have used fewer segments to reduce the number of possible leak paths (as well as inert mass) in the motor. Studies show that motors as long as 18 meters can

6.1 Background

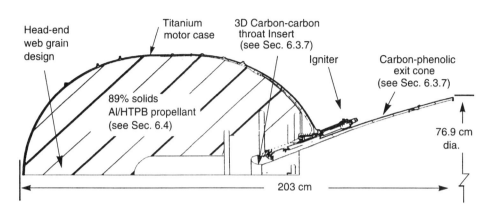

Fig. 6.7. The Star 48B Motor. Courtesy of Thiokol Corporation.

ride on double-length rail cars. The ORBUS 6E motor for the Inertial Upper Stage vehicle uses an extendible exit cone to achieve better packaging efficiency as shown in Fig. 6.6.

The mass summary in Table 6.2 shows that SRMs can have a wide range of propellant masses. Miscellaneous masses in this table include items such as a destruct system (used in case of malfunctions), electrical systems, motor attachments, and other minor parts. Nozzle masses include provisions for thrust-vector control where applicable except for the Titan IV nozzle, which uses liquid injectant thrust-vector control. Each Titan IV SRM carries several thousand kilograms of nitrogen tetroxide, which can be injected through 24 ports in the exit cone to generate side forces required to steer the vehicle. Finally, we note that these masses are intended for use as rough estimates; mass accounting systems differ somewhat from vehicle to vehicle.

The performance summary in Table 6.3 indicates that the upper-stage motors achieve roughly a 10% advantage in specific impulse by using higher expansion ratios and more energetic propellant formulations. The boosters are limited to lower expansion ratios because the flow may separate within a highly overexpanded nozzle (see Chapter 3). We can convert the table's values for specific impulse in a vacuum to an arbitrary ambient pressure using the correction:

$$I_{sp} = I_{sp_v} - \varepsilon \frac{p_a}{p_c} \frac{c^*}{g_0} \tag{6.1}$$

where I_{sp} = specific impulse at ambient conditions (s)
I_{sp_v} = specific impulse at vacuum conditions (s)
ε = nozzle expansion ratio
p_a = ambient atmospheric pressure (Pa)

p_c = chamber pressure (Pa)
c^* = propellant characteristic velocity (m/s)
g_0 = 9.81 m/s^2

Table 6.1. Summary of Dimensions for Current Space-Propulsion SRMs. All dimensions are in meters, and booster lengths do not include nosecones.

Vehicle	Motor Designation	Overall Length (m)	Motor Case Diameter (m)	No. of Segments	Nozzle Diameter (m)	
					Throat	Exit
Shuttle	RSRM	38.40	3.71	4	1.37	3.67
Shuttle	ASRM	38.40	3.81	3	1.38	3.78
Titan	Titan IV	34.40	3.05	9	1.01	3.20
Titan	SRMU	30.76	3.20	3	0.87	3.27
Delta	Castor IVA	9.19	1.02	1	0.28	0.82
Delta II	GEM	11.09	1.03	1	0.25	0.82
IUS	ORBUS 21	3.15	2.31	1	0.16	1.31
IUS	ORBUS 6E	1.98*	1.61	1	0.11	1.44
PAM	Star 48B	2.03	1.24	1	0.10	0.75
AKM	Star 37XFP	1.50	0.93	1	0.81	0.60
PAMD-II	Star 63D	1.78	1.60	1	0.11	0.55
Pegasus	Orion 50SAL	8.88	1.28	1	0.22	1.42
Pegasus	Orion 50	2.66	1.28	1	0.11	0.86
Pegasus	Orion 38	1.34	0.97	1	0.07	0.53

*With extendible exit cone in stowed position.

6.2 Design Process

As with any system, we can design an SRM at various levels. In *preliminary* or *conceptual* design, we use simplifying assumptions to see how the design meets top-level systems parameters and basic operating characteristics. At this point, we may simply estimate mass and crudely represent dimensions for various parts. We use these preliminary results for detailed engineering and manufacturing drawings required to make the hardware.

Figure 6.8 outlines how we do preliminary design for an SRM. In preliminary design, we wish to estimate the overall size and mass of all major parts. Because many of the component designs depend on the amount of propellant, we often need a rough estimate (or best guess) to get us started. Section 6.3.1 describes how we use historical data to estimate how much propellant we need.

Table 6.2. Mass Summary for Current Space-Propulsion SRMs. All masses are in kilograms. See text for explanation of miscellaneous masses.

Motor Designation	Propellant	Insulation	Case	Nozzle	Igniter	Misc.	Total	f_{prop}*
RSRM	501,809	11,177	44,793	10,860	227	670	568,536	0.883
ASRM	548,670	8587	45,114	8469†	199	2251	613,290	0.895
Titan IV	268,168	20,478	27,401	4315	128	8660	329,150	0.815
SRMU	313,130	6443	15,684	6739	91	4892	346,979	0.902
Castor IVA	10,101	234	749	225	10	276	11,595	0.871
GEM	11,767	312	372	242	7.9	291	12,992	0.906
ORBUS 21	9707	145	354	143	16	7	10,374	0.936
ORBUS 6E	2721	64.1	90.9	105.2	9.5	5.3	2996	0.908
Star 48B	2010	27.1	58.3	43.8	0.0‡	2.2	2141	0.939
Star 37XFP	884	12.7	26.3	31.7	0.0‡	1.3	956	0.915
Star 63D	3250	71.4	106.3	60.8	1.0	11.6	3501	0.928
Orion 50SAL	12,160	265.2	547.9	235.4	9.1	21.0	13,239	0.918
Orion 50	3024	75.6	133.4	118.7	5.3	9.9	3367	0.898
Orion 38	770.7	21.9	39.4	52.8	1.3	10.6	896.7	0.859

*$f_{prop} = \frac{\text{mass of propellant}}{\text{total mass}}$ (see Sec. 1.1.5)

† Excludes mass of actuation system which is included in miscellaneous mass.
‡ Igniter mass included in nozzle.

The propellant mass estimate, combined with many preliminary design decisions, underpin the design process. Preliminary design decisions include not only important performance parameters such as chamber pressure and burn duration but also the desired technology level (which determines the construction material). In addition, operational requirements such as operating temperature range and storage life can strongly affect the type of system selected. The design process can actually illuminate the best materials or operating conditions by comparing designs based on different assumptions.

Now, we can start designing components. Additional information required at this stage includes design variables, materials selections, and other system requirements. During the preliminary design, we want to determine the best chamber pressure, burning duration, nozzle expansion ratio, and geometry of the propellant grain. We also have to choose which materials to use in various components in order to get the best performance, mass, or cost. Additional systems issues might involve a maximum acceleration limit for our payload or a dynamic pressure limit for vehicles operating in the atmosphere.

Table 6.3. Performance Summary at Standard Temperature for Current Space-Propulsion SRMs. Standard performance is usually quoted at or near 16° C (60° F).

Motor Designation	p_c MPa	$p_{c\,max}$ MPa	ε	t_b (s)	F_v kN	I_{sp_v} (s)
RSRM	4.58	6.30	7.72	111.1	11,648	268.0
ASRM	4.12	6.60	7.48	138	10,573	269.2
Titan IV	4.34	6.00	10.0	120	6670	272.0
SRMU	6.14	8.14	15.7	140	7560	283.0
Castor IVA	4.76	5.10	8.58	52.5	494.7	267.0
GEM	5.63	7.36	10.75	55.1	502.8	274.0
ORBUS 21	4.52	5.87	63.9	141.6	199.8	295.5
ORBUS 6E	4.21	5.78	181	101.0	80.73	303.8
Star 48B	3.96	4.26	39.6	85.2	68.86	294.2
Star 37XFP	3.63	3.97	54.8	66.0	38.4	292.6
Star 63D	4.18	6.60	27.5	118	85.34	285.0
Orion 50SAL	5.84	7.38	35.2	72.4	488.9	294.4
Orion 50	5.83	6.76	50.3	73.3	118.4	292.3
Orion 38	3.78	4.58	46.1	68.4	32.12	289.0

p_c = Average chamber pressure (Pa)
$p_{c_{max}}$ = Maximum chamber pressure (Pa)
ε = Initial expansion ratio
t_b = Burning duration (s)
F_v = Average vacuum thrust (N)
I_{sp_v} = Average vacuum specific impulse (s)

Once we have made these choices, we can determine inert system masses and sizes. Now we can undertake more detailed ballistics and performance analyses to see if our vehicle meets system requirements for our assumed propellant mass. Our first estimate is usually wrong, so we have to iterate until we find an acceptable system. We can also iterate to determine the best values for our design variables and materials. Eventually, we have a fairly good idea what our SRM looks like, so we can generate detailed drawings of all components and detailed analyses to ensure the integrity of all components (including propellant). These detailed analyses often reveal deficiencies in the design and lead to changes.

When everyone is satisfied with the design, we can build test units for components, subsystems, and the integrated motor. Major component tests include hydrostatic proof tests of the motorcase, burning rate and structural tests of propellant samples, and subscale motor firings to assess how insulation and nozzle

6.2 Design Process

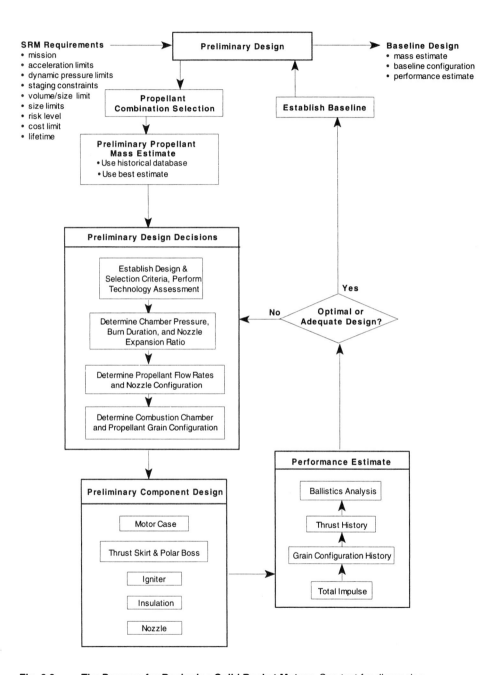

Fig. 6.8. **The Process for Designing Solid Rocket Motors.** See text for discussion.

materials respond to heat and cold. Initial full-scale firings are called development tests; they validate the design and manufacture of the motor. After the development phase, a series of full-scale qualification tests are conducted over the operating temperature range to certify the motor for flight. By analyzing data from full-scale tests, we determine whether the system meets all requirements or needs more design changes. Once the SRM passes all tests, it is ready to enter production.

6.3 Preliminary Sizing

In this section, we consider only the preliminary design process used in determining top-level characteristics of the system. Preliminary design falls into two categories. Our simplest representation of the motor assumes the system includes propellant and *inert* masses only. The inert masses represent any parts other than the propellant used in accelerating the vehicle in the intended direction. Good examples of inert components are the motor case, insulation, nozzle, and igniter. A second category of design includes more detail. We may model the SRM with major components and try to size those components based on past history and current requirements. We briefly discuss both of these common approaches.

By modelling the SRM as having only propellant and inert masses, we can rapidly estimate the propellant mass required to meet certain performance objectives (usually expressed as a velocity gain, or Δv, to be imparted to the payload). Moreover, we can use a wealth of historical data to improve the design. The easiest way to represent the historical data is in terms of the *propellant mass fraction* (f_{prop}) defined as the ratio of propellant mass (m_{prop}) to overall mass:

$$f_{prop} = \frac{m_{prop}}{m_i} = \frac{m_{prop}}{m_{prop} + m_{inert}} \tag{6.2}$$

where we have written the overall mass (m_i) as the sum of the propellant and the inert masses (m_{inert}) in the motor (see Sec. 1.1.5).

We want f_{prop} values as close to unity as possible, so we can reduce the amount of inert structure we need to accelerate along with our payload. It is easy to calculate f_{prop} values for existing systems in Tables 6.1–6.3, and we can plot the data as a function of propellant mass as shown in Fig. 6.9. Trends are difficult to pick out in Fig. 6.9 because we have a mixed database of many different types of motors. Boosters usually have a lower mass fraction than space motors because we have to include structures that attach the motor to the main vehicle. Segmented motors suffer because they need additional structure at joints. The Orion motors have slightly lower mass fractions than space motors because they are used on a winged vehicle, which adds loadings. The extendible exit cone on the ORBUS 6E decreases mass fraction but also increases specific impulse. As one can plainly see from Fig. 6.9, using propellant mass fractions is appropriate for preliminary estimates only; they do not include enough detail to be reliable for more extensive studies.

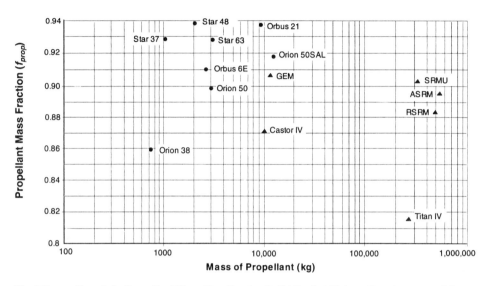

Fig. 6.9. Trends in Propellant Mass Fraction for Solid Rocket Motors. Booster motors (shown as triangles) usually have a lower propellant mass fraction than space motors (shown as solid circles).

Trends become more visible if we look at a database of only space motors. As Fig. 6.10 shows, propellant mass fraction increases with propellant mass. The behavior results from the fact that larger systems are more structurally efficient. In addition, some items (such as actuators and batteries) may not scale down in mass as the system gets smaller, thereby penalizing smaller motors in terms of inert mass.

On the large side, f_{prop} asymptotically approaches a value near unity. The current systems making up our database in Fig. 6.10 have a maximum value of about 0.95. It is very difficult to raise f_{prop} appreciably for these motors because the inert masses are such a small fraction of the motor mass. For example, to raise f_{prop} from 0.95 to 0.96 would require a 20% reduction in the inert mass of the motor.

6.3.1 Preliminary Estimate of Propellant Mass

If we know something about the motor's mass fraction, we can rapidly calculate the amount of propellant (and hence inert mass) needed for a given mission by using the *rocket equation* discussed in Sec. 1.1.4:

$$\Delta v = g_0 I_{sp} \ln\left(\frac{m_i}{m_f}\right) \qquad (6.3)$$

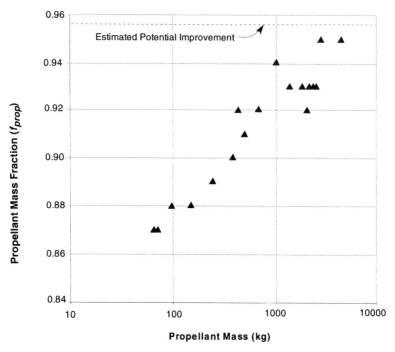

Fig. 6.10. Trends in Propellant Mass Fraction for Space Motors. This figure shows that mass fraction usually increases with propellant mass.

where m_i and m_f are the initial and final (burnout) masses of our rocket, and g_0 is the gravity acceleration at the Earth's surface (9.81 m/s^2). We assume the Δv in this expression includes the desired velocity increase plus any losses due to gravity, drag, and steering. In addition, we assume that we know something about the average I_{sp} of the system we have selected based on historical data. We can "solve" Eq. (6.3) for the *mass ratio*:

$$R = \frac{m_i}{m_f} = e^{\frac{\Delta v}{g_0 I_{sp}}} \tag{6.4}$$

To use Eq. (6.4) in solving for our propellant masses, note that we can write the initial and burnout masses:

$$m_i = m_{prop} + m_{inert} + m_{pay} \tag{6.5}$$

$$m_f = m_i - m_{prop} = m_{pay} + m_{inert} \tag{6.6}$$

where m_{pay} is the effective payload for the system under consideration, and m_{inert} is inert mass. Note that for a multistage vehicle, the effective payload of the lower stage includes the mass of the entire upper stage because this mass also must be accelerated by the lower-stage propulsion system. In addition, intermediate stages, guidance systems, payload fairings, and nosecones must be part of our payload mass because we normally do not include them in the motor mass fraction.

Introducing our definition of f_{prop}, we can combine Eqs. (6.2–6.6) to give

$$m_{prop} = m_{pay}\left(\frac{f_{prop}(R-1)}{1 - R(1 - f_{prop})}\right) \tag{6.7}$$

which gives us the propellant mass, assuming we know the payload mass and have an estimate of f_{prop}. We can then calculate the inert mass in the motor using the definition of propellant mass fraction:

$$m_{inert} = m_{prop}\left(\frac{1 - f_{prop}}{f_{prop}}\right) \tag{6.8}$$

6.3.2 Techniques for Sizing Components

The mass-fraction approach described above suits only studies of top-level systems because this simple technique tells us little about their components. The one useful piece of information we do get from this approach is an estimate of the propellant mass required for the mission. We can use this information, combined with some assumed values for basic design parameters, to actually "size" the major components in the motor and estimate their masses. By adding up these masses, we can re-estimate the required propellant mass because our calculated inert masses will not match our initial estimate using the propellant mass fraction. In addition, sizing components in more detail permits us to optimize fundamental design parameters.

Typical design parameters assumed for this procedure include the motor chamber's average pressure, the burning time, and the nozzle expansion ratio. We want to optimize the chamber pressure because it influences nearly all of the inert masses in the system. Although increasing the nozzle expansion ratio increases the motor's vacuum I_{sp}, it also increases the nozzle and interstage mass, so the best value balances these properties. In the case of a launch system, criteria for nozzle separation usually limit the expansion ratio for a selected chamber pressure.

Burning time is also an important parameter because it influences the nozzle's size. If we write the average mass-flow rate as the total propellant mass divided by the burn duration, we can calculate the nozzle throat area from [Eq. (3.133)]:

$$A_t = \frac{c^* m_{prop}}{t_b p_c} \quad (6.9)$$

where A_t = nozzle throat area (m²)
 c^* = characteristic velocity (m/s)
 m_{prop} = mass of propellant (kg)
 t_b = burn duration (s)
 p_c = chamber pressure (Pa)

We see from this relation that we can decrease the nozzle size (and hence mass) by extending the motor's burning time. Counterbalancing this trend is the tendency to increase insulation requirements (in both the case and nozzle) as the burning time increases. In addition, because thrust must be greater than weight for a launch system, burning time is limited. Finally, for launch systems, the drag interactions and gravity losses are both sensitive to burning time. Presuming we know propellant mass, chamber pressure, expansion ratio, and burning time, we can start sizing each of the motor's major components.

6.3.3 Sizing Motor Cases

Technology. The motor case acts as a pressure vessel to contain the high-pressure combustion processes occurring within the bore of the motor. Cases may be constructed from metal materials, from composite fiber materials, or from a combination of a metal "liner" with a composite overwrap. Table 6.4 summarizes some of the critical properties of various candidate materials; the variables ρ, F_{tu}, and E represent density, ultimate tensile strength, and Young's modulus for the material.

Table 6.4. SRM Motor-Case Materials. Quoted strengths for composite materials are for actual motors using these materials. Note that these values are representative only; for more complete details consult *Mil Handbook 5*.

Material	ρ kg/m³ (lb/in³)	F_{tu} GPa (ksi)	E GPa (msi)	$F_{tu} / (\rho g)$ km
2219 - Aluminum	2800 (0.101)	0.413, 0.214 welded (60.0, 31.0 welded)	68.9 (10.0)	15.04
Titanium	4460 (0.161)	1.23 (178.0)	103 (15.0)	28.81
D6aC Steel	7830 (0.283)	1.52 (220.0)	200 (29.0)	19.7
4130 Steel	7830 (0.283)	0.862 (125.0)	200 (29.0)	11.23
Graphite*	1550 (0.056)	0.965–1.72 (140–250)	69–140 (10–20)	63.5–113
Kevlar*	1380 (0.050)	0.827–1.10 (120–160)	41–55 (6.0–8.0)	61.0–81.3
Fiberglass*	1990 (0.072)	1.10 (160)	32.4 (4.7)	53.4

*Composite fiber materials using an epoxy resin matrix for structural stability.

Many lower-cost SRMs use metal designs that are simple to build, especially if the motor is quite small or has to be reusable (such as the Space Shuttle RSRM). Many small motors in air-launched missiles employ steel cases using either the 4130 alloy or the higher strength D6aC (or Maraging) alloy. Small and mid-sized space motors have often employed titanium cases. Aluminum is an attractive choice as a case material for only the smallest motors because of its low strength.

As highlighted in Table 6.4, the composite materials are very attractive case materials because of their high strength-to-weight ratios. Motor cases designed with these materials can be 20–30% lighter than conventional metal cases. Some of the mass savings is lost because these materials are not as stiff (they have lower modulus values) as titanium or steel, so large deflections are possible. For example, the composite motor case tested for the Space Shuttle SRB actually grew more than three inches under pressure. Often, stiffness concerns of this type override strength considerations, thus increasing the mass of the composite case above its minimum value solely to retain internal pressure. Designs of this type are said to be stiffness-critical rather than strength-critical.

Mass Estimate. The motor case must be able to withstand the maximum chamber pressure under any possible operating condition. This condition occurs at the upper limit of the SRM's operating temperature range and includes worst-case grain geometry, propellant burn rate, and nozzle geometry allowed by the motor specifications. The pressure calculated in this way is called the *maximum expected operating pressure* (*MEOP*), for the motor. This pressure must account for normal variations about the average pressure as well as the effect of manufacturing tolerances and hot-day conditions, which tend to enhance the burning rate slightly. The case is designed to burst at a pressure

$$p_b = f_s(MEOP) \tag{6.10}$$

where f_s is the *factor-of-safety* applied to account for case material and manufacturing variations, post-production damage, and overall safety margin for the design. Safety factors are usually 1.25 for unmanned systems, 1.4 for manned systems, and as high as 2.0 for small pressurization bottles or gas generators. Composite structures sometimes employ a higher factor-of-safety because they can be damaged easier than metal cases and because strength can degrade from temperature and humidity extremes as well as aging.

Using these ideas, we can calculate the case thickness for a given case material, burst pressure, and case size. For a cylindrical case, we apply the theory for thin-walled pressure vessels to get

$$t_{cs} = \frac{p_b r_{cs}}{F_{tu}} \tag{6.11}$$

where t_{cs} = motorcase thickness (m)
p_b = burst pressure (Pa)

r_{cs} = case radius (see Fig. 6.11, in m)
F_{tu} = ultimate tensile strength of case material (Pa)

We can use t_{cs} to estimate the case mass if we know the case length and dome geometry. Fig. 6.11 illustrates this geometry.

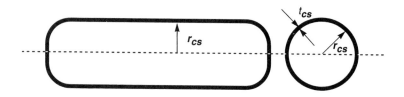

Fig. 6.11. Geometry for a Cylindrical Motor Case. We use these definitions in Eq. (6.11).

Often, the parameter $p_b V/W$ is used in the preliminary design of motor cases. In this expression, p_b refers to burst pressure, V is the enclosed case volume, and W is the actual weight of the case. This parameter reflects case performance because it measures the mass of material required to maintain the total pressure load ($p_b V$). Typical values lie between 5 and 15 km for metal cases and from 7 to nearly 20 km for composite cases. By knowing this parameter, we can quickly estimate case mass because we assume the propellant mass (and hence volume) and burst pressure at the beginning of a design.

The case volume is of course a function of the propellant volume needed for the mission. Often, we can get preliminary estimates by introducing a parameter called the *volumetric loading efficiency* (η_v). This parameter measures the percentage of enclosed case volume which the propellant uses:

$$\eta_v = \frac{V_{prop}}{V_{cs}} \qquad (6.12)$$

where V_{prop} and V_{cs} represent propellant and enclosed case volumes, respectively. Typical η_v values lie in the range of $0.8 < \eta_v < 0.98$. Of course, we like to have η_v be as close to unity as possible, but grain design and thrust history usually dictate these actual values. Space motors commonly have η_v values above 0.90, whereas boosters usually have values between 0.85 and 0.90.

6.3.4 Thrust Skirts and Polar Bosses

The critical design factors influencing the thickness and mass of thrust skirts are the compressive stresses and bending loads resulting from forces transmitted through this member. Because composite materials have lower strength in compression, skirts (and the region where the skirt is joined to the case) are usually thicker than the case material itself. Still, we often assume the skirt is the same thickness as the case when doing preliminary designs.

Polar bosses are designed to survive blowout loads from the internal chamber pressure and any torques the nozzle develops during vectoring operations. They also provide a surface for winding the fibers around the domes of cases made with composite materials. Note that at the aft boss, thrust loads through the nozzle relieve the outward bending of the boss caused by the chamber pressure. Bosses are nearly always made of metal (usually aluminum) and contribute little mass (usually less than 10%) compared to the case itself on many designs.

6.3.5 Sizing Igniters

Technology. Practically all igniters employed in SRMs are of the pyrogen variety. In *pyrogen igniters* the igniter itself operates as a small SRM to provide hot gases and high pressures for igniting the main propellant grain. A typical pyrogen igniter is in Fig. 6.12. An electrical signal starts a small charge which ignites the boron-potassium nitrate pellets at the head end of the igniter. Gases from the $BKNO_3$ charge start combustion in the main igniter grain.

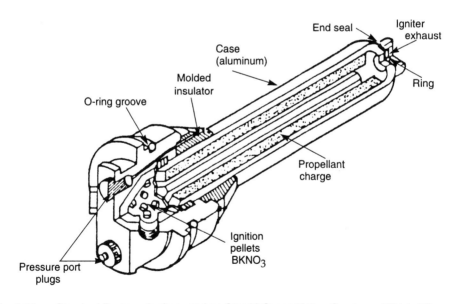

Fig. 6.12. Standard Pyrogen Igniter used on Star 20 Space Motor. Courtesy of Thiokol Corporation.

Smaller SRMs simply use the pellets themselves to provide adequate energy to start combustion. An electrical signal ignites the pellets to begin combustion.

Designing a pyrogen igniter is similar to designing a small rocket motor. The igniter case (or housing) must be able to withstand the predicted internal pressure while the igniter is burning. Igniters also must be well insulated externally because they normally have to withstand the chamber environment during the entire SRM firing. Recently, consumable igniters have been developed to eliminate insulation and slightly improve motor mass fraction. When the motor fires, the ablative materials in these igniters burn up.

Mass Estimate. The design (and mass) of the igniter is often neglected in preliminary analyses because the igniter mass is typically a few percent of the total inert mass. The propellant charge in pyrogen designs is counted in the total useful propellant load because it does exit the nozzle and contribute to thrust. One way to estimate igniter mass is based on the port volume within the main chamber at ignition. We can estimate this quantity using the volumetric loading efficiency:

$$V_{port} = V_{prop}\left(\frac{1}{\eta_v} - 1\right) \tag{6.13}$$

where V_{port} is the port volume and V_{prop} is the volume of propellant in the main chamber. Heister [1990] has correlated igniter masses for several motors with this parameter to give

$$m_{ig} = 0.0138 V_{port}^{0.571} \tag{6.14}$$

where m_{ig} is in kg and V_{port} is in cubic centimeters.

6.3.6 Sizing Internal Insulation

Technology. Internal insulation protects the motorcase when it is exposed to the high temperatures within the chamber. The insulator must match the propellant selected and provide strain capability to ensure the propellant remains bonded when the case deforms under pressure. In addition, we want the insulation to have low thermal conductivity so we can keep it as thin as possible. Low density is also desirable because it reduces inert mass.

These requirements usually lead to rubber-based materials with fibrous fillers to improve structural integrity. Ethylene Propylene Dimethyl Monomer or Natural Butadiene Rubber are popular rubber materials, usually mixed (loaded) with fibrous materials such as asbestos, silica, or Kevlar pulp.

The heat from chamber gases causes the insulation to decompose and ablate. Rubber material exposed to the flame decomposes to form pyrolysis gases, leaving a *char layer* composed mostly of carbon near the surface. The filler materials increase the strength of the char, which is normally quite brittle and easily removed. As the char layer builds up, it insulates the virgin rubber lying behind it because of its very low thermal conductivity. Thus, superior insulators have filler material that retains the char near the surface under stress and vibration during firing. Char loss varies widely, so ablated depth also reflects large variations for a

6.3 Preliminary Sizing

given controlled test condition. We must include this condition in the design, so we use the worst-case ablation depth to determine the design thickness.

Design of the insulation thickness anywhere within the motor is largely empirical. The required thickness is a function of the *exposure time* (t_{exp}) which is the amount of time a particular place on the insulator is exposed to the chamber environment. (Remember, the propellant itself shields the insulator in many locations for parts of the burn.) In addition, the thickness is proportional to the *ablation* or *erosion rate*, and a factor-of-safety,

$$t_{insul} = t_{exp} \dot{e} f_s \qquad (6.15)$$

where t_{insul} = insulation thickness (m)
t_{exp} = insulation exposure time (s)
$\dot{e}$ = insulation erosion rate (m/s)
f_s = safety factor

We can define the safety factor for insulator designs in many ways. Some designs use a factor of twice the average char depth at a given location as measured from test data, or 1.25 times the maximum depth at a given axial station. In some instances, a safety factor is implied in a requirement that the motor case itself remain below some maximum allowable temperature (either at burnout or at some specified interval after burnout). Often, initial tests of a new design employ large amounts of insulation to protect against temperature swings. Ablation data from these tests, in addition to experience gained from other motors, allow us to generate a flight-weight insulation design to be validated on later tests.

Mass Estimate. Predicting insulation mass for preliminary designs is also a challenge, because in this case we do not know the propellant grain design. Most preliminary design codes use correlations (regressions involving known masses from previous motors) to estimate insulation mass. One such correlation, based on data from space motors using silica-loaded EPDM insulation, is shown below:

$$m_{insul} = 1.788 \times 10^{-9} m_{prop}^{-1.33} t_b^{0.965} (L/D)^{0.144} L_{sub}^{0.058} A_w^{2.69} \qquad (6.16)$$

where m_{insul} = mass of insulation (kg)
m_{prop} = total mass of propellant (kg)
t_b = burn duration (s)
L = motor case length (see Fig. 6.13, in m)
D = diameter (see Fig. 6.13, in m)
L_{sub} = (see Fig. 6.13, in m) = $100X/L$
X = nozzle submerged length (see Fig. 6.13, in m)
A_w = surface area of wall in motor case exposed to hot gases (cm^2)

This equation predicts insulation masses within ±15% on the SRMs from the database (Tables 6.1 to 6.3). This accuracy is typical for insulation correlations because various motors have insulation thicknesses optimized to varying degrees.

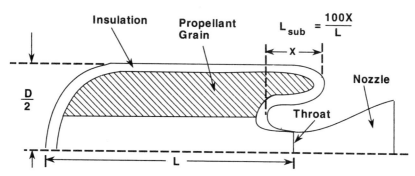

Fig. 6.13. **Nomenclature for Correlating Insulation Mass.** Nozzles are very often submerged in the case to shorten the engine. This figure defines several parameters used in Eq. (6.16).

6.3.7 Sizing Nozzles

Technology. Much of the design, analysis, and fabrication time for an SRM lies in developing the nozzle. The nozzle converts high thermal energy of chamber gases to kinetic energy and thrust. For this reason, the highest velocities, heat fluxes, and pressure gradients are in this part of the motor. **Studies of historical SRM failures show that roughly 50% of all failures occur from problems in the nozzle area.** The high failure rate in this area is attributed to the severe environment and the fact that nozzles can contain many parts and may have to vector to control the vehicle. Analysis of nozzle behavior is also difficult because both thermal and structural loads are important to the overall state of stress in this region.

SRM nozzles are constructed of composite materials, metals, and rubbers (or polymers). Current throat materials are primarily polycrystalline graphite or 3-D carbon-carbon. Carbon- and graphite-phenolic are used in larger throats where the higher ablation rate has only a small effect on the expansion ratio and hence the I_{sp}. Also, 3-D carbon-carbon is costly for large throats while the polycrystalline graphite is difficult to produce in very large sizes without defects that may cause failure. Exit cones are principally carbon-phenolic with silica-phenolic used in lower erosion areas. There are some 2-D carbon-carbon exit cones where the cost of obtaining a lower mass is justified. Entrance insulation is again carbon- or graphite-phenolic with silica phenolic in the lower heat flux zones. Structural materials are primarily metals with some fiberglass-epoxy.

The composite phenolics are woven cloth impregnated with phenolic and laid up on a mandrel, either as large flat pieces or as tape wrapped around the mandrel. The lay-up is cured under temperature and pressure and machined to the final configuration. 2-D carbon-carbon is produced similarly but is carbonized at high temperature. 3-D carbon-carbon is produced by weaving a three-dimensional billet, impregnating it with pitch, and then adding graphite to it at high temperature and pressure. These cycles of impregnation and adding graphite continue until a part reaches the desired density, typically about 1900 kg/m³.

Table 6.5 summarizes insulator and structural materials used in today's nozzles. Because properties of these materials often depend on temperature and orientation, the quantities in the table are only approximate. In the throat region, we want materials with low erosion, such as graphite or carbon/carbon, to lessen the loss in expansion ratio during the firing. Materials of this type are made of fibers wrapped axially and tangentially (2-D) or axially, tangentially, and radially (3-D). Here we note that actual erosion rates were scaled to the condition given at the bottom of Table 6.5, using the notions developed in the insulation section of this chapter. If the heat flux is proportional to the quantity $\rho v = \dot{m}/A$, it is easy to show that the highest heat fluxes are in the throat region (see Sec. 3.5.2).

Table 6.5. Materials for the SRM Nozzle Insulator and Supporting Structures. Note that properties of composite materials can vary with orientation in the material. Values listed here represent actual nozzle materials.

Material	ρ kg/m³	c_p cal/gm·K	κ W/m·K	F_{tu} MPa	$\dot{e}^*$ mm/s
Pyrolytic Graphite	2200	0.50	0.059	103	0.05
Polycrystalline Graphite	1700	0.60	26.0	48	0.10
2-D Carbon/Carbon	1400	0.54	13.8	110	-
3-D Carbon/Carbon	1900	0.50	31.5	186	0.10
Carbon/Phenolic	1400	0.36	1.00	72.4	0.18
Graphite/Phenolic	1400	0.39	1.59	52.4	0.28
Silica/Phenolic	1700	0.30	0.55	52.4	1.3
Glass/Phenolic	1900	0.22	0.028	414	1.5
Paper/Phenolic	1200	0.37	0.40	152	1.9

ρ = Material density
c_p = Material specific heat
κ = Material thermal conductivity
F_{tu} = Ultimate tensile strength
$\dot{e}$ = erosion rate

*Reference data:
p_c = chamber pressure = 6.9 MPa
D_t = throat diameter = 0.3 m
T_c = combustion temperature = 3030 K
Note that values are scaled using correlations for convective heat transfer (see Heister [1990]).

Typically, we require an insulating material behind the throat insert (as shown in Fig. 6.14) because of the high thermal conductivity, density, and cost associated with graphite and carbon/carbon materials. Low density (mass) and thermal conductivity are desirable characteristics for these "backup insulators." Carbon, graphite, glass, and silica phenolics have these characteristics. Glass phenolic has excellent insulating properties, but we seldom use it because it has poorer erosion resistance than the other insulators. Finally, a paper phenolic material has been proposed as a low-cost alternative for some applications.

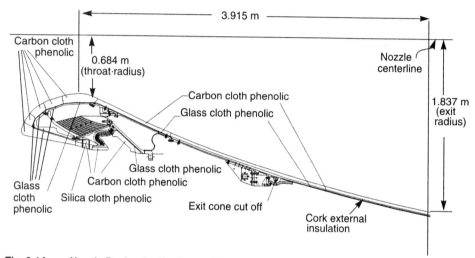

Fig. 6.14.   Nozzle Design for the Space Shuttle's Redesigned Solid Rocket Motor. The nozzle uses extensive metal backup to make it more reusable. (Courtesy of Thiokol)

In the exit cone, the thermal environment is less severe, so some of the insulators (carbon and silica phenolics) work as primary materials on the gas side. Two-dimensional carbon/carbon materials have also been used in this region, alone and with insulators. The thickness of exit cones is typically set by thrust (pressure) loads placed on the structure, so this region can be as low as 0.8–1.3 mm.

In addition to the so-called "plastic parts" used as nozzle insulators, throat, and exit cone materials, most nozzles include some metal parts. These parts are used as structural members to support the composite parts and to accept vectoring loads for distribution to the motor case. Aluminum, steel, and titanium are the most common materials for backup structures.

Figures 6.14–6.16 present nozzle designs for the Redesigned Solid Rocket Motor, the Castor IVA motor, and the Star 48B space motor. The RSRM nozzle uses carbon phenolic insulators (with glass phenolic backup) along internal aerodynamic surfaces. The entire nozzle, except for the aft exit cone, uses metals to back up its housing and make it more reusable. An explosive linear-shaped charge severs the aft exit cone before splashdown and recovery in the ocean, lowering splashdown loads on other nozzle components and allowing us to reuse metal parts and the flexbearing. All plastic parts are replaced after each mission for this nozzle.

The Castor IVA expendable motor (Fig. 6.15) is used as a portion of the first-stage propulsion for the Delta launch vehicle. Because the nozzle has no thrust-vector-control requirements, it is very simple in design, using graphite as a throat insert and a combination of graphite and carbon phenolic in regions upstream and downstream from the throat, respectively. This nozzle also uses steel for backup structure.

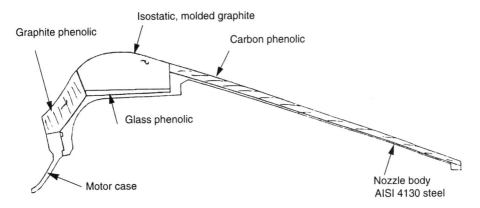

Fig. 6.15. **Nozzle Design for the Castor IVA SRM.** This is a very simple nonvectoring nozzle with a graphite throat insert. (Courtesy of Thiokol)

The Star 48B nozzle, shown in Fig. 6.16, uses a 3-D carbon/carbon throat insert with carbon phenolic backup insulators and primary insulators in regions of lower heat flux. The exit cone is overwrapped with a glass epoxy to provide meridional or "hoop" strength. Fibers wrapped tangentially provide enough strength to withstand the internal pressures developed within the exit cone during the firing. This nozzle uses little metal to reduce inert mass in this upper-stage motor.

Mass Estimate. It is difficult to predict the nozzle's mass in preliminary designs, because the nozzle contains many parts, a thrust-vector control system heavily influences the design, and historical designs vary in estimating material thicknesses. Nozzles built for applications within the atmosphere may be subject to buckling loads if they are highly overexpanded. For these reasons, it is difficult to develop accurate correlations of historical nozzle masses.

With these caveats in mind, we present a correlation that can provide an early estimate of nozzle masses:

$$m_{noz} = 0.256 \times 10^{-4} \left[\frac{(m_{prop} c^*)^{1.2} \varepsilon^{0.3}}{p_c^{0.8} t_b^{0.6} (\tan \theta_{cn})^{0.4}} \right]^{0.917} \qquad (6.17)$$

where m_{prop} is the propellant mass in kg, c^* is in m/s, p_c is in MPa, θ_{cn} is the nozzle included half-angle, and t_b is in seconds. This equation is developed using space motors and strap-on boosters without thrust-vector control and is accurate to only about 20% for the database employed. As mentioned above, we want to consider the actual detailed design of the nozzle (even in preliminary designs) in order to estimate mass more accurately. A design of this type (e.g., Heister [1990]) considers the size and mass of the major nozzle parts including insulators, throat insert, and appropriate backup structure.

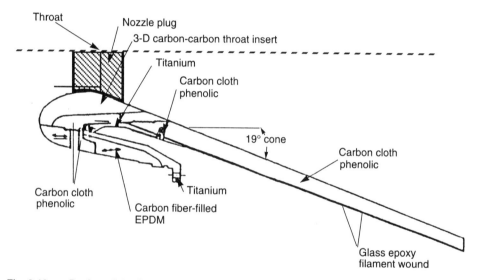

Fig. 6.16. **Design of the Star 48B Nozzle for Space Motors.** Note that the exit cone does not use metallic backup in this space motor. A nozzle plug keeps debris from entering the motor before use. (Courtesy of Thiokol)

6.3.8 Sizing Systems for Thrust-Vector Control (TVC)

Many SRMs contain thrust-vector-control (TVC) systems to help control the vehicle while the solid motor is operating. Including TVC tends to complicate the nozzle design because we now must include a means by which to vary the thrust vector. Studies using historical data show that including a TVC system increases nozzle mass by roughly 50%. The following list summarizes some of the more popular approaches to TVC.

Flexible bearing (Flexseal). This is probably the most popular TVC scheme in use today. A schematic of this system for the Space Shuttle's Solid Rocket Booster is in Fig. 6.17. The Flexseal is made up of alternating layers of *shims* (which stabilize the bearing) and *pads*—made of an elastomeric material with a low shear modulus and high shear elongation. A hydraulic actuator pushes the pads on the side of the nozzle to permit up to 8° of nozzle rotation.

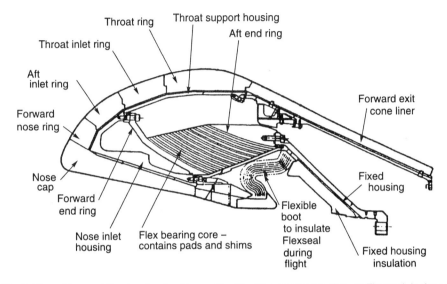

Fig. 6.17. Flexseal Design for the Space Shuttle. The flexible bearing, or Flexseal, is the most common scheme for controlling the thrust vector in current motors. (Courtesy of Thiokol)

Techroll seal. This system, developed by Chemical Systems Division of United Technologies, is used on the ORBUS-21 and ORBUS-6E motors (see Fig. 6.6). Fig. 6.18 shows a schematic of the system.

The Techroll seal provides a moveable nozzle needing very low torque or actuator forces (the Flexseal requires relatively high forces). The forward part of the nozzle actually rides on a rubber-lined, oil-filled reservoir. When the actuator force torques the nozzle to one side, the oil rushes around the annulus of the reservoir, allowing the nozzle to move. While the system has lower actuator requirements than a Flexseal, it is more complex.

Liquid Injection Thrust-Vector Control (LITVC). In this system, a liquid is injected at various places around the circumference of the motor nozzle's exit cone to provide the required control moments. Side forces develop in this system because of distributed pressure on the wall resulting from the shock structure formed by the injectant. A schematic of an LITVC system used on the Titan SRM is in Fig. 6.19. By using energetic liquids (such as nitrogen tetroxide), the injectant

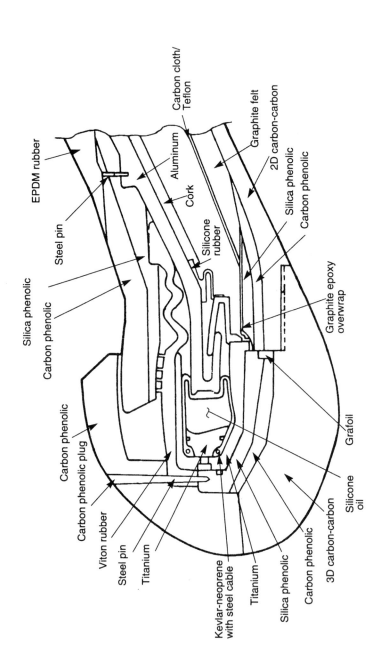

Fig. 6.18. Nozzle with Techroll Seal on the Inertial Upper Stage ORBUS 21. The Techroll seal permits nozzle actuation with very little torque. Courtesy of United Technologies Chemical Systems Division.

can actually add a bit of axial thrust during operation. LITVC systems require no actuators but are complicated by their liquid manifolding and many valves. Historically, these systems were used much earlier than Techroll or Flexseals because of the many problems in developing large moveable nozzles.

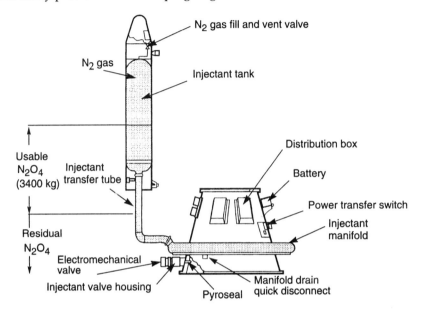

Fig. 6.19. Liquid Injection Thrust-Vector Control on a Titan SRM. Using Liquid Injector Thrust Vector Control (LITVC) permits thrust-vector control with a fixed nozzle. (Courtesy of United Technologies, Chemical Systems Division)

Jet Vanes and Pintles. These devices generate TVC forces by deflecting the flow within the nozzle without moving the nozzle itself. Schematics of the concepts are in Fig. 6.20. Both concepts require moving parts within the hot nozzle flow, so they normally apply to motors with short burning times. Both concepts introduce additional frictional losses even when not operating. No materials exist which can hold up through longer burns, so these systems are not as common as the others.

6.4 Solid Rocket Propellants

Solid propellants fall into two general categories: double base and composite. In *double-base* propellants, molecules of fuel and oxidizer mix. These propellants usually contain a nitrocellulose type of gun powder dissolved in nitroglycerine with other minor additives. Both of these primary ingredients are explosives containing fuel and oxidizer (oxygen) within their molecular structures. Double-base

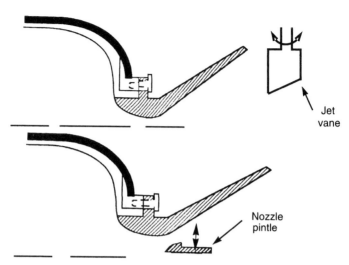

Fig. 6.20. Using a Jet Vane and Nozzle Pintle for Thrust-Vector Control. These concepts are rarely used in space-propulsion systems because long burning times develop high heat loads, which degrade the vane or pintle.

propellants have been used most often in military applications (mainly in ballistic missiles) but are becoming less common because the military has increased safety requirements for munitions.

The other general type of solid propellant is called a *composite propellant* to highlight the fact that it is a heterogeneous mixture of fuel, oxidizer, and binder. Fuels for composite propellants are usually metallic powders. Oxidizers are typically crystalline materials, whereas the binder is a rubber-based material which holds the powder/crystal mixture together in a cohesive grain. Note that the binder itself also acts as a fuel and is oxidized during combustion. In fact, low-smoke formulations used in tactical propulsion systems may incorporate very little (if any) metallic powder.

6.4.1 Fuels

Table 6.6 summarizes typical fuels used in composite propellants in ascending order of performance. The most common fuel today is aluminum, using anywhere from 2–21% Al. Magnesium is being used in some modern "clean" formulas with reduced exhaust emissions. The beryllium fuels are the most energetic available, but the beryllium oxide in the exhaust products is toxic. Repeated ingestion of beryllium oxide can lead to berylliosis, a condition associated with reduced lung capacity similar to the effects of long-term exposure to asbestos. For this reason, beryllium fuels are considered seriously only for space-based applications such as space interceptor weapons.

Table 6.6. Fuels used in Composite Propellants. Aluminum is the most common fuel in current motors. As we descend through the fuel list, the performance increases (i.e., boron is better than titanium).

Fuel	Chemical Symbol	Molecular Mass (kg/kmol)	Density (kg/m^3)	Remarks
Zirconium	Zr	91.22	6400	Low performance but high ρ
Titanium	Ti	47.90	4500	Low performance but high ρ
Magnesium	Mg	24.32	1750	Clean propellant applications
Aluminum	Al	26.98	2700	Low cost, common fuel
Aluminum Hydride	AlH$_3$	30.0	1420	Difficult to make
Beryllium	Be	9.01	2300	Toxic exhaust products
Boron	B	10.81	2400	Inefficient combustion
Beryllium Hydride	BeH$_2$	11.03	≈ 650	Toxic exhaust products

6.4.2 Oxidizers

Table 6.7 describes typical oxidizers used in composite propellants. High oxygen content is a desirable characteristic for an oxidizer and is one measure of its overall performance.

Ammonium perchlorate, or AP, is by far the most common oxidizer in use today because of its moderately low cost, moderate performance, and good processability. The one concern associated with AP is that the chlorine in this compound combines with hydrogen to yield hydrochloric acid in the exhaust products. The HCl cloud from a large solid motor can combine with water vapor to form acid rain in the area near the launch pad and can cause ozone depletion in the upper atmosphere. The industry is trying to develop new propellant formulas with less than 1% HCl in the exhaust [see McDonald, et al., (1991)].

Of the remaining oxidizers in Table 6.7, ammonium nitrate (AN) has the most applications. While AN is not as energetic as AP, it is less expensive and provides a "clean" exhaust (at least in terms of HCl). Problems with AN include its inherently low burn rate as well as susceptibility to a phase change near 30°C. The rest of the oxidizers in Table 6.7 are not used in today's motors.

6.4.3 Binders

The binder holds the entire formulation in a structurally sound grain which can withstand temperature variations as well as pressure and acceleration loads during flight. We want binders to have low density and energy of combustion. In addition, the best binder materials can provide needed structural integrity using a minimal binder volume. The *solids loading* in the propellant addresses this desired characteristic by expressing the total mass of fuel and oxidizer as a percentage of

Table 6.7. Oxidizers used in Composite Propellants. Ammonium perchlorate is now the most common oxidizer.

Compound	Chemical Formula	Oxygen Content, Mass %	Density (kg/m^3)	Remarks
Ammonium Perchlorate	NH_4ClO_4	59.5	1950	Moderate performance, cost
Ammonium Nitrate	NH_4NO_3	60.0	1730	Moderate performance, low cost
Sodium Nitrate	$NaNO_3$	56.4	2170	Moderate performance
Potassium Perchlorate	$KClO_4$	46.2	2520	Low regression rate, moderate performance
Potassium Nitrate	KNO_3	47.5	2110	Low cost and performance
Nitronium Perchlorate	NO_2ClO_4	66.0	2200	Very reactive, unstable

the total propellant mass. Composite propellants have solids loadings in the 84–90% range, which implies that only 10–16% of the propellant mass is made up of binder and minor ingredients.

Binders are usually long-chain polymers that can keep the propellant's powders and crystals in place by forming a continuous matrix through polymerizing and cross-linking. Cross-linking takes place by mixing a curative into the propellant just before *casting* (pouring) the mixture into the motor. The cast grain is cured at high temperature to promote cross-linking. Binder materials used in composite propellants for SRMs are summarized in Table 6.8. The first three compounds described in this table are older technology, whereas PBAN, CTPB, and particularly HTPB are used in many of today's applications. The new binder GAP has not yet been used in space applications, although designers have proposed it for tactical rockets.

6.4.4 Minor Ingredients

We mentioned above that a small amount of curative is required to promote polymerizing. A *cure catalyst* is often included with the curative to increase the rate of chemical reaction caused by adding this ingredient. Some propellants also use a *plasticizer* to improve physical properties at low temperatures. Many double-base formulas require a *darkening agent* (such as carbon black) to make the translucent propellant darker and avoid excessive thermal radiation through the propellant itself.

Because the propellant must be "tailored" to achieve the desired burn rate, it often needs small amounts of burn-rate catalysts or modifiers. These ingredients, such as iron oxide or copper chromite, tend to increase the propellant's burning rate. These additives can be quite important because variations between different lots of fuel, oxidizer, and binder materials can lead to significant changes in the burning rate. By doing subscale tests on new batches of propellant, we can deter-

Table 6.8. **Binders Used in Composite Propellants.** Hydroxy-terminated Polybutadiene (HTPB) is most common today.

Binder Designation	SRM Applications
Polysulfide (PS)	Older rockets
Polyether Polyurethane (PEPU)	Polaris Stg. 1
Polybutadiene Acrylic Acid (PBAA)	Older rockets
Polybutadiene Acrylonitrile (PBAN)	Titan and Shuttle SRMs
Nitrocellulose (Plasticized) (PNC)	Minuteman and Polaris
Carboxy-terminated Polybutadiene (CTPB)	Minuteman Stg. 2, 3
Hydroxy-terminated Polybutadiene (HTPB)	IUS, Peacekeeper, Star 48
Nitrate Ester Polyether (NEPE)	Peacekeeper, SICBM
Glycidal Azide Polymer (GAP)	New Binder

mine the amount of catalyst required to achieve the desired burning rate in the motor. Often, we require *antioxidants* to prevent binder oxidation reactions in the propellant over long storage periods. Finally, some propellants use a *bonding agent* to improve the bond between the oxidizer and binder. In some space-motor applications, HMX (Her Majesty's' Explosive) has been added to increase the energy content (I_{sp}) of the propellant. While the addition of HMX in small amounts is permissible, large amounts (greater than 12% by mass) make the propellant much easier to detonate.

6.4.5 Propellant Burning Rate

The propellant in an SRM not only must burn at the proper rate to maintain the desired chamber pressure but also must have enough structural continuity to withstand pressure and acceleration loads during flight. Because we cannot predict most physical characteristics (including burning rate) through analysis, we often rely heavily on test data.

The *propellant burning rate* (r_b) is the rate at which the exposed propellant surface is consumed; it is measured as the distance normal to the surface consumed in a given time. By theory and experiment, we know this quantity is a function of the motor-chamber pressure. In many cases, we write

$$r_b = a p_c^n \qquad (6.18)$$

where a is the *burn rate coefficient* and n is the *burn rate exponent*. Typical burn rates for solid propellants lie in the range of 0.1–5 cm/s, with most lying in the 0.6–1.3 cm/s range. In practice, the constants a and n are determined experimentally. Typical n values lie in the range $0.2 < n < 0.6$, whereas typical a values depend on the units of pressure and can be inferred from Eq. (6.18). For example, in metric units,

a is expressed in cm/s/MPa^n. The expression in Eq. (6.18) is most commonly known as St. Robert's Law.

To determine the burning rate with subscale testing, we can burn small strands of propellant in a constant pressure "bomb" or fire small ballistic test motors. Strands are more commonly used during research and early development studies. Most manufacturers prefer ballistic test motors, which are small SRMs with simple grain designs similar to that shown in Fig. 6.21 (without the restrictors). By firing the motors with different throat sizes, we can obtain burning rate data at different chamber pressures and plot the data in the form of $\ln(r)$ versus $\ln(p_c)$. The slope of this line represents n, whereas the y-axis intercept represents $\ln(a)$. Fig. 6.22 shows these values for a typical composite propellant.

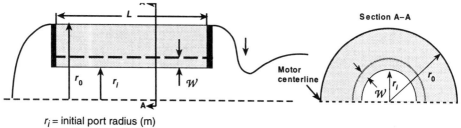

r_i = initial port radius (m)
r_0 = final port radius (m)
w = web thickness; thickness of propellant grain consumed as a function of time of burn (m)

Fig. 6.21. Sample Grain Geometry for a Simple Solid Rocket Motor. Restrictors (or inhibitors) are occasionally applied to some grain surfaces to inhibit burning.

Note that Fig. 6.22 is empirical in this representation; it does not address the complex thermochemical and combustion processes that occur when a propellant actually burns. St. Robert's Law cannot characterize the behavior of some propellants. Fig. 6.23 shows samples of observed burning rate behavior. In *plateau burning*, the burning rate can be totally insensitive to pressure ($n = 0$) over some pressure range, whereas in *mesa burning*, the burning rate can actually decrease with increasing pressure ($n < 0$). These behaviors are usually restricted to double-base propellants. The burning rate variation shown in curve (d) in Fig. 6.23 occurs in composite propellants, but this behavior requires the use of variable n for accurate modelling.

Fortunately, St. Robert's Law can quite accurately represent many composite propellants. Table 6.9 presents propellant characteristics for some of the current systems in Tables 6.1–6.3. The actual mass percentage of major ingredients is included below the table with minor ingredients lumped into the binder values. The highest aluminum loading is 21% as indicated in the TP-H-1202 formulation. This propellant uses 12% HMX to help this large amount of fuel react.

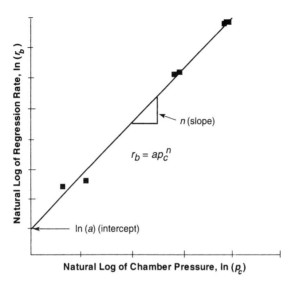

Fig. 6.22. Determining Burning Rate Coefficient (*a*) and Exponent (*n*) from Subscale Test Data. If our propellant obeys St. Robert's Law, plotting data on a log chart gives us a linear relation as shown. The regression rate exponent is the slope of the line and the coefficient is the intercept.

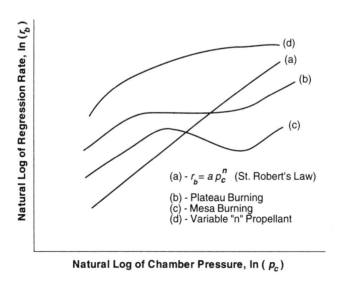

Fig. 6.23. Burning Rate Behavior of Solid Propellants. Most composite propellants behave as in curves (a) or (d) above.

Table 6.9. Characteristics of Composite Propellants used in Space Systems. We show typical thermochemical and regression rate parameters. The parameters *a* and *n* correspond to p_c units of MPa and regression rates in cm/s. Typical isentropic parameters (γ) for the exhaust gases are around 1.2.

Item	ρ_p kg/m³	c^* m/s	T_c K	Pressure Range MPa	$r_b = a p_c^n$, cm/s		π_k %/K
					a	n	
A	1840	1541	3636	2.75–7.0	0.415	0.31	0.18
B	1800	1527	3396	2.75–5.0	0.399	0.30	0.18
C	1760	1568	3392	2.75–7.0	0.561	0.35	0.20
D	1680	1511	3288	2.7–10.0	0.41	0.39	--

A – TP-H-1202, 21% Al, 57% AP, 12% HMX, 10% HTPB; Star 63D
B – TP-H-3340, 18% Al, 71% AP, 11% HTPB; Star 48, Star 37
C – TP-H-1148, 16% Al, 70% AP, 14% PBAN; RSRM
D – 22% Mg, 66% AP, 12% HTPB; Mg "clean" propellant

Temperature Effects. The variable π_k in Table 6.9 addresses the temperature sensitivity of the motor to changes in burning rate. At higher propellant temperatures, the increased internal energy within the propellant leads to small increases in burning rate as compared to normal temperature conditions. Most space systems operate over small temperature ranges which normally make this effect quite small (but not negligible). The parameters σ_p and π_k measure temperature sensitivity:

$$\sigma_p = \left.\frac{\partial \ln (r_b)}{\partial T}\right|_{p_c = \text{const}} \tag{6.19}$$

$$\pi_k = \left.\frac{\partial \ln (p_c)}{\partial T}\right|_{A_b/A_t = \text{const}} \tag{6.20}$$

where T is the temperature of the propellant grain before combustion.

For a propellant obeying St. Robert's law, we can show that

$$\pi_k = \frac{\sigma_p}{1-n} \tag{6.21}$$

Typical values for these parameters lie in the ranges $0.06 < \sigma_p < 0.18\%$ /K, and $0.16 < \pi_k < 0.30\%$ /K, so a temperature change of 25 K changes the burning rate only a few percent. Normally, the temperature change is described with a reference temperature of 15°C (59°F) as an assumed "standard" condition. Including temperature effects, the burning rate of the propellant becomes

$$r_b = e^{(\sigma_p \Delta T)} a\, p_c^n \qquad (6.22)$$

where we assume the constant a has been determined using data at a standard temperature, and ΔT is the difference in temperature from this standard condition.

Erosive Effects. If the motor is long, gases at the end of the grain can have high speed and momentum. The increased heat flux associated with this condition can enhance the burning rate through what is called *erosive burning*. In this case, the burning rate not only depends on pressure but also may vary according to the local Mach number or mass flux in the bore. Probably the best-known correlation for this effect is that of Lenoir and Robillard:

$$r_b = a p_c^n + \frac{\alpha G^{0.8}}{L^{0.2}} e^{\frac{-\beta \rho_p r_b}{G}} \qquad (6.23)$$

where α and β are new constants which must be determined experimentally, and L is the length of the grain. In addition, the parameter G is the bore mass flux (kg/m² ·s), and the 0.8 exponent comes from the effects of convective heat transfer. We consider the Lenoir-Robillard model as an erosion based on mass flux.

The other option, addressed by Green [1954] (and others), presumes that the erosive effects occur mainly because of compressibility. In this case, Mach number (M) influences the burning rate:

$$r_b = a p_c^n (1 + kM) \qquad (6.24)$$

where k is the new empirical constant which addresses the erosive effects.

Erosive effects are usually restricted to the early stages of the firing, when the port area is smallest. Designers try to avoid significant erosive burning because experimental evidence shows it can vary greatly. Often, designers limit bore Mach numbers (by increasing available port area) to avoid erosive effects early in the firing.

6.5 Performance Prediction

Internal ballistics analysis predicts the time history of the chamber pressure in the motor. The field of internal ballistics is concerned with the flowfield generated within the combustion chamber and the motor performance associated with this flowfield. The main outputs of a ballistics prediction are the chamber pressure and grain burnback histories for the SRM. To generate this data, we need information about the propellant burning rate (or burn rate), the exposed surface area of the propellant grain, the nozzle throat and exit areas, and the propellant formula.

If we know the burning rate, we can determine the amount of mass entering the chamber at a given instant. Recalling that for steady one-dimensional flow, mass-flow rate = density × velocity × flow area, or $\dot{m} = \rho v A$, we may write

$$\dot{m}_{in} = \rho_p r_b A_b \tag{6.25}$$

where now $\dot{m}_{in}$ is the mass flow into a control volume (kg/s), ρ_p is the *propellant density* (kg/m^3), the velocity corresponds to the burning rate (r_b [m/s]), and A_b is called the *propellant burn surface area* (m^2). From this expression, we can see that the amount of mass entering the chamber is directly proportional to the burn surface area exposed to the flame.

Consider the simple tubular grain shown in Fig. 6.21. The burn surface is simply the surface area of a cylinder with changing radius. The ends of the cylinder are assumed to be restricted from burning by bonding a rubber *restrictor* (or *inhibitor*) on these faces, so we can assume that no burning occurs on these faces. For the geometry shown, we can see that the initial surface area is simply $2\pi r_i l$ while the surface area just before the motor extinguishes is $2\pi r_o l$. Because $r_o > r_i$, we see that the burn surface in this case increases with time. Motors of this type are called *progressive burning* motors. If the burn surface is reasonably constant with time, the motor is said to be *neutral burning*, whereas a motor with a burn surface which decreases with time is said to be *regressive burning*.

Figure 6.24 gives a few examples of grain designs. End burners, segmented tubes, rods and tubes, star grains, and slotted tubes give us approximately neutral behavior. As we show in Fig. 6.27, the internal burning tube is a progressive burner.

Because both chamber pressure and thrust follow the burn surface trends (A_b), by designing the proper grain geometry, we can obtain the desired thrust history for the motor. We usually want either neutral or regressive thrust histories for real SRMs because a progressive history gives the highest thrust when the vehicle is lightest. This usually leads to structural problems in either the payload or the stages being boosted because of higher acceleration levels. A neutral burning behavior provides the design for minimum inert mass because structural components must withstand maximum pressure loads and the maximum and average pressures are very nearly the same for this situation.

Accurate ballistics prediction depends on the intended application. Although current computational fluid dynamics (CFD) codes can calculate the three-dimensional flowfield inside a SRM, this calculation is prohibitive because we need to repeat it tens or hundreds of times to determine the motor's overall thrust and chamber pressure history. Because a single calculation can take as much as 40–60 hours on the fastest computers available, this approach is clearly too expensive to be practical for preliminary design. In addition, accurate predictions can be

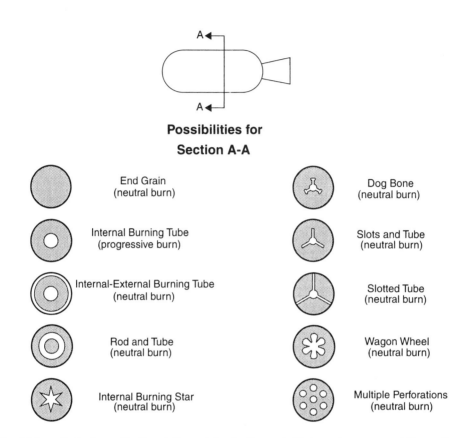

Fig. 6.24. Grain Cross-Sectional Geometries to Generate Various Pressure and Thrust Responses. By altering the grain design, we can achieve progressive or neutral burning. Shaded areas indicate propellant, blank areas indicate the port.

obtained without the use of a 3-D pressure field. For these reasons, we simplify the flowfield to make ballistics calculations efficient and easy to use.

Current ballistics calculations fall into two general categories. If the velocities inside the chamber are low compared to the local sound speed (as in many space motor designs), pressure variations within the chamber are necessarily small. In this case, we can represent the entire chamber with a single pressure. This approach, referred to as a *lumped-parameter method*, is very easy to use and requires little computation. In the case of a long motor (such as the boosters discussed in the previous section), pressure variations within the bore are important, so we must make the model spatially dependent. In this type of model, the bore is broken into a series of *ballistic elements* to which one-dimensional mass and momentum balances are applied. Two and three-dimensional effects are typically treated in an empirical fashion in this type of model, so we can compute efficiently.

6.5.1 Lumped-Parameter Methods

Steady-State Lumped-Parameter Methods

The control volume we employ for a lumped-parameter method is shown in Fig. 6.25 where all exposed burning surfaces are assumed to contribute gases to the volume of interest. Assume that we know the geometry of the burn surface so we can determine A_b (m²) and $\dot{m}_{in}$ (kg/s). We know from our definition of characteristic velocity that the mass flow exiting the nozzle may be written as

$$\dot{m}_{out} = \frac{p_c A_t}{c^*} \tag{6.26}$$

where $\dot{m}_{out}$ = mass-flow rate out of control volume (kg/s)
p_c = chamber pressure (Pa)
A_t = throat area (m²)
c^* = characteristic velocity (m/s), see Eqs. (3.130–3.133)

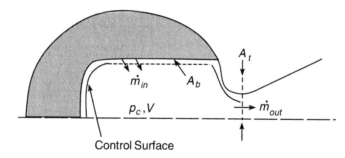

Fig. 6.25. Control Volume for a Ballistics Analysis Using Lumped Parameters. This figure defines several parameters used in Eqs. (6.25) to (6.29).

The conservation of mass states that the rate of change of mass (within the chamber) is equal to the difference between the mass entering the chamber and the mass leaving through the throat:

$$\frac{dm}{dt} = \dot{m}_{in} - \dot{m}_{out} = r_b \rho_p A_b - p_c \frac{A_t}{c^*} \tag{6.27}$$

If the motor is operating under *steady-state* conditions, the two mass flows balance exactly, giving $dm/dt = 0$. In essence, this assumption implies that the chamber pressure adjusts instantaneously to any changes in burning rate, burning surface area, or other parameters. For this case, Eq. (6.27) becomes

$$r_b \rho_p A_b = p_c \frac{A_t}{c^*} \tag{6.28}$$

which can be solved for the motor-chamber pressure by inserting Eq. (6.18) for the burn rate:

$$p_c = \left(\frac{a \rho_p A_b c^*}{A_t} \right)^{1/(1-n)} \tag{6.29}$$

where p_c = chamber pressure (Pa)
 a = propellant regression rate coefficient
 n = propellant regression rate parameter
 ρ_p = solid propellant density (kg/m³)
 A_b = burn surface area (m²)
 c^* = characteristic velocity (m/s)
 A_t = throat area (m²)

This relation defines the motor-chamber pressure for given motor geometry and propellant characteristics, assuming steady-state operation. Note that the pressure essentially follows the trend of the burn surface history $A_b(t)$. Because thrust is proportional to p_c, the thrust also mimics this trend.

We must carefully define units for Eq. (6.29). Because the burn rate constant (a) is in units of cm/s/MPa^n, we must ensure that the group $\rho_p A_b c^*/A_t$ is in MPa/cm/s, so the group in brackets in Eq. (6.29) is in units of MPa$^{(1-n)}$.

To illustrate the use of Eq. (6.29), consider the motor/grain design shown in Fig. 6.21 with r_i, r_o, and L equal to 10, 20, and 200 cm, respectively. Further, we assume a throat radius of 5 cm, a propellant characteristic velocity of 1500 m/s, a propellant density of 0.0015 kg/cm³, and a burning rate law $r_b = 0.5 p_c^n$ (p_c in MPa, r_b in cm/s). Using these values and the fact that a cylindrical burning surface area can be written as $A_b = 2\pi L (r_i + W)$, we can reduce Eq. (6.29) to

$$p_c = [0.18(10 + W)]^{1/(1-n)} \tag{6.30}$$

Here the *web distance* (W, or propellant web) is defined as the linear amount of propellant consumed as measured normal to the local burn surface (see Fig. 6.21). We can solve for the pressure history by remembering that the web distance must be the integral of the burning rate history:

$$W = \int_0^{t_b} r_b(t) \, dt \tag{6.31}$$

Here we note that burning rate is a function of time because, in general, pressure is a function of time. This nonlinearity leads us to solve the governing equations by integrating numerically. We can use the above two relations to construct a predicted pressure history using a method of Euler numerical integration:

---

1. Calculate the initial pressure ($W = 0$).
2. Assume this pressure is valid over some small time increment Δt. With the pressure assumed constant over this interval, r_b is also constant, and Eq. (6.31) gives $\Delta W = r_b \Delta t$ (Euler integration).
3. Add this incremental web distance burned to the previous web distance to obtain the new location of the burning front, i.e., new *web* value.
4. Use Eq. (6.30) to calculate the pressure at the new web location.
5. Repeat the process in Steps 2–4 until the web distance is equal to the total web distance (40 cm in our example).

Fig. 6.26. **Algorithm of Performance History for a Cylindrical Port Using Euler Numerical Integration.** This algorithm allows us to integrate Eq. (6.31) no matter how complex parameters such as A_b become.

Note that we also could have taken steps in web and calculated the resulting time difference to consume the incremental web, i.e., $\Delta t = \Delta W / r_b$. This modified procedure produces identical results provided we choose small enough web increments to keep our assumption of constant pressure over that interval numerically valid. Figure 6.27 shows the result of our calculation if we assume values for the burning rate exponent of 0.3, 0.4, and 0.5. Note the increasing exponent effectively reduces burning time and increases pressure by increasing the burning rate; however, the total pressure integral remains constant.

Also note from Eq. (6.29) that the chamber pressure gets large as n approaches unity. In fact, we can show that we must have $n < 1$ to attain the equilibrium condition. If $n > 1$, a perturbation in the chamber pressure causes this quantity to grow without bound. This behavior is consistent with *explosives*. Most propellants have n values in the range $0.2 < n < 0.6$, so they remain stable even during large disturbances.

How can we "solve" the steady-state problem without considering conservation of energy? If we assume an adiabatic situation and a calorically perfect gas, the conservation of energy states that stagnation enthalpy must be constant. We may write this condition as

$$\dot{m}_{in} c_p T_c - \dot{m}_{out} c_p T_{out} = 0 \qquad (6.32)$$

where $\dot{m}_{in}$ = mass-flow rate into control volume (kg/s)

$\dot{m}_{out}$ = mass-flow rate out of control volume (kg/s)
c_p = constant pressure specific heat of burned propellant gas (J/kg·K)
T_c = chamber temperature (K)
T_{out} = temperature of gases exiting the control volume (K)

Under steady-state operation $\dot{m}_{in} = \dot{m}_{out}$ so Eq. (6.32) degenerates to $T_c = T_{out}$, which is the standard result stating that stagnation temperature is fixed in an adiabatic flow (with no work interaction).

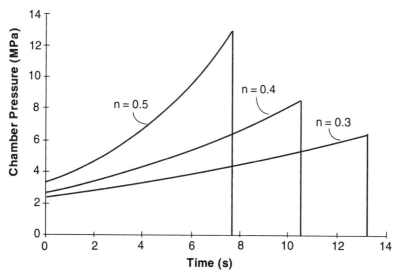

Fig. 6.27. History of Calculated Chamber Pressure for Example Problem: n = 0.3, 0.4, and 0.5. Note that increasing n increases the sensitivity of chamber pressure to changes in burning surface area. (Note that total impulse—represented by the area beneath the curve—is identical for each case.)

Unsteady Lumped-Parameter Method

How do we know when we can assume steady-state operation of the motor? When we are operating under unsteady conditions, **mass accumulates** within the chamber. To understand the accumulation concept, consider the analogy of pouring some liquid into a funnel. If you pour the liquid at a certain rate, you can keep the fluid at a constant level in the funnel (steady-state case). However, if you increase the rate at which you pour the fluid, the level in the funnel increases and the vessel accumulates mass. The same thing can happen in a combustion chamber if the mass flows do not balance exactly. In this case, $dm/dt \neq 0$, and mass accumulation will cause the chamber pressure to rise above the steady-state value. The opposite effect can also occur; i.e., if outflow exceeds inflow, the chamber pressure will fall below the steady-state value.

To address these issues, we can write the mass of gas in the chamber from the perfect gas law (see Sec. 3.2.3):

$$m = \frac{p_c V M}{\mathcal{R}_u T_c} \qquad (6.33)$$

where m = mass of gas in chamber (kg)
 p_c = chamber pressure (Pa)
 V = volume available for gases within the chamber (m³)
 M = molecular mass of gas (kg/kmol)
 $\mathcal{R}_u$ = universal gas constant (8314 J/kmol·K)
 T_c = chamber temperature (K)

To find the rate of change of m, we just need to differentiate Eq. (6.33). Before we do this, let us reason as to which terms in this differential might be important (i.e., in general, p_c, T_c, M, and V_c can vary with time). We are trying to get an equation for changes in chamber pressure in terms of grain (and throat) geometry and propellant properties. In Chapter 4, we show that propellant properties (T_c, M, γ) are relatively insensitive to changes in pressure under equilibrium conditions. In this case, pressure is also varying, so it is not obvious that variations in these parameters are negligible. Heister and Landsbaum [1991] show that we can safely neglect variations of temperature and molecular mass unless a very rapid and large pressure disturbance is present. Physically speaking, the energy of adiabatic compression (due to an increase in pressure) is balanced by an increase in energy flux through the throat.

By taking the log of both sides of Eq. (6.33) and differentiating the result, we obtain

$$\left(\frac{1}{m}\right)\left(\frac{dm}{dt}\right) = \left(\frac{1}{p_c}\right)\left(\frac{dp_c}{dt}\right) + \left(\frac{1}{V}\right)\left(\frac{dV}{dt}\right) \qquad (6.34)$$

The rate of change of chamber volume is proportional to the current burning surface area and burning rate:

$$\frac{dV}{dt} = r_b A_b \qquad (6.35)$$

We can combine Eqs. (6.27), (6.34), (6.35) to give

$$\frac{1}{p_c}\frac{dp_c}{dt} = \frac{1}{m}\left[r_b A_b \left(\rho_p - \rho_g\right) - p_c \frac{A_t}{c^*}\right] \qquad (6.36)$$

where we have substituted for $\dot{m}$ from Eq. (6.33) and defined the gas density as $\rho_g = m/V$. If the first term on the right-hand side of Eq. (6.36) exceeds the second term, the pressure increases with time because of the accumulation effect. The ρ_g contribution results from the chamber's increase in volume with time because of consumption of the propellant. Typically, $\rho_p/\rho_g \approx 300$, so the ρ_g term is usually neglected. Under this assumption, we can write the dimensionless form of Eq. (6.36) as [see Heister and Landsbaum (1991) for derivation]

$$\tau \frac{dp'}{dt} = k_b p'^n - k_t p' \qquad (6.37)$$

where k_b = dimensionless burn surface = A_b/A_{b_i}

k_t = dimensionless throat area = A_t/A_{t_i}

p' = dimensionless pressure = p_c/p_{c_i}

τ = motor time constant = $\rho_g V / \dot{m}_i$

subscript "i" refers to the initial condition

The motor time constant is an important parameter that represents the length of time required to eject the mass of gas in the chamber, prior to the disturbance, under the equilibrium flowrate $\left(\dot{m}_i = \left(p_{c_i} A_{t_i}\right)/c^* = a p_{c_i}^n A_{b_i} \rho_p\right)$. For most SRMs, the time constant lies in the range $0.01 < \tau < 0.5$ seconds, implying that steady-state conditions exist unless the chamber pressure is changing appreciably within this time scale. We can integrate Eq. (6.37) to determine the unsteady response of chamber pressure to imposed changes in burning surface or throat area. In general, this relation is valid for $\tau\, dp'/dt < 1.0$, for if the disturbance exceeds (or approaches) this limit, stagnation temperature variations become important as well.

First, let us investigate the behavior of Eq. (6.37) for the **blowdown process** which occurs at the end of the firing, when the propellant is completely consumed. The specifics of the grain design usually give a nonzero pressure at this time, so the combustion gases remaining in the chamber at this instant give us residual thrust. This problem is important in operating space motors because we must have negligible thrust when the SRM separates from the payload to avoid a collision between the spent motor and the payload. Under this condition, $k_b = 0$ because no burn surface remains and we assume negligible change in throat area over the blowdown process, $k_t = 1$. Therefore, Eq. (6.37) becomes

$$\frac{dp'}{p'} = -\frac{dt}{\tau} \qquad (6.38)$$

with the initial condition that $p' = 1$ at $t = 0$. Integrating Eq. (6.38) for this condition gives the simple result:

$$\frac{p_c}{p_{c_i}} = e^{-t/\tau} \tag{6.39}$$

where p_c = chamber pressure (Pa)
 p_{c_i} = initial chamber pressure (Pa)
 t = time (s)
 τ = time constant (s)

which implies the pressure decays exponentially to zero under our assumptions. Because $e^{-4} = 0.018$, the pressure decays to only 2% of its original value after four time constants. Once again, we see the importance of this parameter in determining the motor's transient behavior.

Finally, we note that Eq. (6.39) is derived assuming negligible temperature variations. This "isothermal blowdown" is reasonable if the motor burns out at moderate to low pressures because heat transfer from hot surfaces of the chamber wall keep chamber gases at high temperatures. We cannot make this assumption for the blowdown of a large high-pressure tank because, in this case, an adiabatic situation might be more reasonable. For an adiabatic, reversible process, we can use $T \propto p^{(\gamma-1)/\gamma}$ to change Eq. (6.37) for this situation.

Changes in Burn Surface or Throat Area. For k_b and k_t constant, we can integrate Eq. (6.37) to give

$$p' = \frac{p_c}{p_{c_i}} = \left[\frac{k_b}{k_t} - \frac{k_b - k_t}{k_t} e^{(-k_t(1-n)t/\tau)} \right]^{1/(1-n)} \tag{6.40}$$

We can use this relation to predict pressure changes from instantaneous changes in throat area or burning surface area. For example, the motor's response to the instantaneous exposure of a crack or void in the propellant (which tends to increase exposed burn surface) can be predicted by assuming negligible throat area variation ($k_t = 1$) and a fixed $k_b > 1$. Under these conditions, we can write Eq. (6.40) as

$$p' = \frac{p_c}{p_{c_i}} = \left[k_b - (k_b - 1) e^{(-(1-n)t/\tau)} \right]^{1/(1-n)} \tag{6.41}$$

By comparing the exponential term in this relation with Eq. (6.39), we see that the effective time constant governing the decay is $\tau/(1-n)$, so propellants with high exponents respond more slowly to the imposed disturbance. Also note that if $n > 1$, the exponential is positive, so the disturbance grows without bound, as an explosive does. Figure 6.28 shows the time response for an assumed 10% increase in burning surface (k_b = 1.1) for two different burn-rate exponents. Note that the new equilibrium pressure, $p' = k_b^{(1-n)^{-1}}$, is essentially reached in 3–4 time constants.

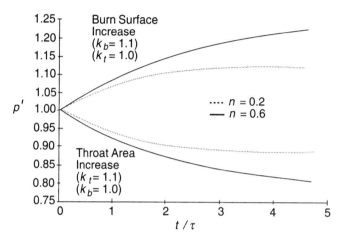

Fig. 6.28. **Chamber-Pressure Response to Changes in Throat or Burning Surface Area.** Note: propellants with higher exponents show larger pressure changes for a given disturbance.

Equation (6.41) can also be used to evaluate instantaneous changes in throat area. This situation has arisen in development and flight tests for several programs from the loss of part of the throat or ejection of internal motor components which temporarily block the throat. We can investigate these cases by assuming negligible change in burning surface over the brief history of the event. In this case, Eq. (6.41) gives

$$p' = (1/k_t)^{1/(1-n)}[1 - (1 - k_t)e^{(-(1-n)k_t\, t/\tau)}] \tag{6.42}$$

indicating an effective time constant governing the decay of $\tau/k_t(1 - n)$. Figure 6.28 also shows the behavior for an assumed instantaneous 10% increase in throat area. In this case, Eq. (6.42) reveals that the new equilibrium pressure is $k_t^{-1/(1-n)}$ showing the obvious notion that pressure is inversely proportional to throat area.

Remember, we are assuming that the motor time constant is indeed a constant. Because τ depends on the volume of the open chamber, we are restricted to looking at time intervals in which the volume changes a negligible amount. If we need to integrate over much of the motor burning time, we must use numerical integration to account for changes in chamber volume during this time.

6.5.2 Ballistics with Variations in Spatial Pressure

Single-Element Model

If the motor is long (as for most boosters), the static pressure drops considerably as the gases accelerate down the length of the bore. In this case, lumped-parameter methods are inappropriate, and we must consider spatial variations for reasonable accuracy. To illustrate this case, consider the simple circular-port motor shown in Fig. 6.29. Let us assume adiabatic flow from the end of the grain to the throat. In this case, the Mach number at the aft-end of the grain (M_a) must be related to the port-to-throat area ratio (A_p/A_t) through the implicit isentropic flow relation (derived in Section 3.2):

$$\frac{A_p}{A_t} = \frac{1}{M_a}\left[\frac{2+(\gamma-1)M_a^2}{1+\gamma}\right]^{(\gamma+1)/2(\gamma-1)} \tag{6.43}$$

We can determine the pressure variation down the length of the bore by considering a momentum balance on the control volume shown in Fig. 6.29. For a steady flow, the change in momentum must be equal to the net force on the control volume. Assuming a port area is constant and only pressure forces act on the boundaries of the control volume gives the following relation:

$$\dot{m}v + pA - (\dot{m}+d\dot{m})(v+dv) - (p+dp)A = 0 \tag{6.44}$$

Ignoring second-order terms and simplifying Eq. (6.44), we find

$$d(\dot{m}v) = d(\rho v^2 A) = -Adp \tag{6.45}$$

Now, because $\rho v^2 = \gamma p M^2$ (see Sec. 3.2.5 for discussion of Mach number), we can write

$$\frac{dp}{p} = -\frac{\gamma dM^2}{1+\gamma M^2} \tag{6.46}$$

We can integrate this equation along the length of the bore if we assume a constant port area. At the head-end of the grain, we assume a stagnation condition ($M_h = 0$, $p_h = p_{o_h}$, where p_h is static pressure at the head end, and p_{o_h} is total pressure at the head end). Integration from this point to the aft-end of the grain gives

$$\frac{p_a}{p_h} = \frac{1}{1+\gamma M_a^2} \tag{6.47}$$

where p_a = static pressure at aft end (Pa)
       p_h = static pressure at head end (Pa)

6.5 Performance Prediction

γ = ratio of specific heats of burned propellant gas
M_a = Mach number of gas at aft end

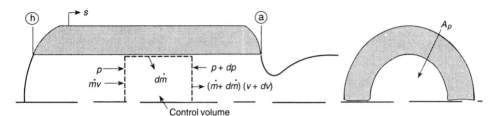

Fig. 6.29. **Effects of Added Mass on the Flow in a Circular-Port Motor.** Locations "h" and "a" refer to head- and aft-ends of propellant grain. The relations shown are used to derive Eqs. (6.44) through (6.48).

Therefore, the static pressure drop is a function of the Mach number at the end of the grain. We are usually more interested in the change in stagnation pressure because the motor's thrust can be related to this quantity. Using the isentropic relation between static and stagnation pressure, we can write

$$\frac{p_{o_a}}{p_{o_h}} = \phi = \frac{\left\{1+[(\gamma-1)/2]M_a^2\right\}^{\gamma/(\gamma-1)}}{1+\gamma M_a^2} \qquad (6.48)$$

where p_{o_a} = stagnation pressure at aft end (Pa)
p_{o_h} = stagnation pressure at head end (Pa)

where the common notation $p_{o_a}/p_{o_h} = \phi$ has been introduced to describe the stagnation pressure ratio.

Figure 6.30 shows the behavior of M_a, as well as the pressure ratios derived above, as a function of the port-to-throat area ratio. Because we want to place the most propellant possible within a given motor-case diameter, we want to make A_p/A_t as small as possible. As indicated in the figure, stagnation pressure losses and high Mach numbers within the bore are the prices we must pay to achieve this goal. Because large pressure changes imply large loads on the grain (and increased MEOP values), we are limited as to how small a port-to-throat area ratio we can use. In addition, as port Mach numbers increase, erosive burning of the propellant can take place. This phenomenon, described in Section 6.5, is undesirable and limits port Mach numbers. For these reasons, current motors usually have initial port-to-throat ratios in the 2.0–3.0 range.

It is interesting to note that we have investigated the effect of mass addition on the bore flowfield without actually solving for the amount of mass we add. The actual amount of added mass sets the level of the pressure field in the motor. In

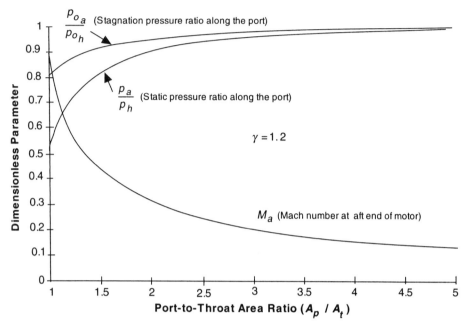

Fig. 6.30. How Added Mass Affects Losses in Static and Stagnation Pressure. See text for additional discussion.

other words, the more mass flow, the higher the pressure for a given A_p/A_t ratio. Because the burning rate depends on the local static pressure, we expect this quantity to drop as we progress down the bore. Assuming this effect is negligible (for the time being), we estimate the burning rate using the head-end pressure so the mass flow into our simple motor in Fig. 6.29 is

$$\dot{m}_{in} = a p_{0_h}^n \rho_p A_b \qquad (6.49)$$

where a = burning rate coefficient
 n = burning rate exponent
 p_{0_h} = stagnation pressure at head end (Pa)
 ρ_p = density of solid propellant (kg/m³)
 A_b = burn surface area (m²)

The mass flow through the throat is related to the stagnation pressure at the aft-end of the grain:

$$\dot{m}_{out} = \frac{p_{0_a} A_t}{c^*} = \frac{\phi p_{0_h} A_t}{c^*} \qquad (6.50)$$

6.5 Performance Prediction

Equating these two mass flows for our assumed steady-state condition gives

$$p_{o_h} = p_h = \left(\frac{a \rho_p A_b c^*}{\phi A_t} \right)^{1/(1-n)} \tag{6.51}$$

which is the maximum pressure level in the motor. Note that, because $A_b = 2L\sqrt{A_p \pi}$ (see Fig. 6.24), lengthening the grain increases the pressure level in the motor. Also note that ϕ in Eq. (6.51) raises the pressure level in the head-end of the motor because by definition $\phi \leq 1$.

Ballistic Element Methods. Actual motors usually have port areas which change as a function of the axial distance along the grain. These variations can lead to large variations in pressure (and hence burning rate). In this case, our simple solution discussed in the previous section is no longer valid, so we must consider another approach. We consider the bore to be comprised of a series of ballistic elements—each having its own pressure, Mach number, burning rate, geometry, and mass flow—and then create a general model for the flow in the bore. This technique, which computer codes use for state-of-the-art analysis, is illustrated in Fig. 6.31. While this figure shows a circular port geometry, the method is not restricted to this case. We can consider arbitrary grain geometries.

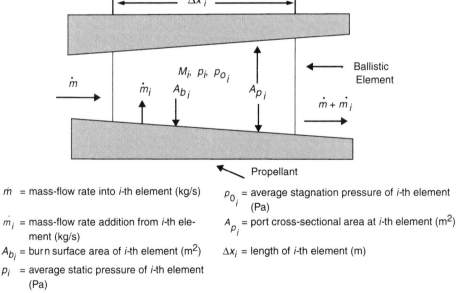

$\dot{m}$ = mass-flow rate into i-th element (kg/s)

$\dot{m}_i$ = mass-flow rate addition from i-th element (kg/s)

A_{b_i} = burn surface area of i-th element (m²)

p_i = average static pressure of i-th element (Pa)

p_{o_i} = average stagnation pressure of i-th element (Pa)

A_{p_i} = port cross-sectional area at i-th element (m²)

Δx_i = length of i-th element (m)

Fig. 6.31. Control Volume for Mass/Momentum Balance on a Ballistic Element. The main flow must provide momentum to accelerate gases entering the control volume.

To begin, we assume a perfect gas is in adiabatic flow along one dimension and know all geometry associated with the element. In other words, the element length (Δx_i), port area $\left(A_{p_i}\right)$, and burning surface area $\left(A_{b_i}\right)$, must be known functions of the local web distance (W_i). Element lengths are prescribed with a small enough variation in port area across the element so we can use a suitable average of parameters in calculations. Flow variables are defined at the center of the element where we assume local mass flow is $\dot{m} + \frac{1}{2}\dot{m}_i$. Finally, because port-divergence angles are typically small, we assume the mass enters normal to the main flow.

We also assume the burning rate in each element can be expressed in terms of the static pressure and Mach number in the element, $r_i = r_i(p_i, M_i)$. In this case, the mass flow contributed by element i can be written as

$$\dot{m}_i = r_{b_i} \rho_p A_{b_i} \tag{6.52}$$

Using mass and momentum balances, the equation of state, and definitions of Mach number and stagnation pressure, we can obtain the following relationship (see Zucrow and Hoffman [1976] for derivation):

$$p_{0_i} = \frac{p_{0_{(i-1)}}}{1 + \gamma M_i^2 \frac{\dot{m}_i}{\dot{m}}} \tag{6.53}$$

where

$$\dot{m} = \sum_{j=1}^{i-1} \dot{m}_j \tag{6.54}$$

is the mass flow entering element i as noted in Fig. 6.31. Static and stagnation pressure within the element are related by the familiar isentropic relation [see Eq. (3.95)]

$$\frac{p_{0_i}}{p_i} = \left(1 + \frac{\gamma-1}{2} M_i^2\right)^{\frac{\gamma}{\gamma-1}} \tag{6.55}$$

Finally, we can relate the mass flow at the center of the element $(\dot{m} + \dot{m}_i/2)$ to the flow variables through a continuity equation of the form (development similar to discussion in Sec. 3.3 where $M_i \neq 1$):

$$\dot{m} + \frac{\dot{m}_i}{2} = \frac{M_i p_{0_i} A_{p_i}}{\sqrt{RT_c/\gamma}} \left(1 + \frac{\gamma-1}{2} M_i^2\right)^{\frac{-(\gamma+1)}{2(\gamma-1)}} \tag{6.56}$$

6.5   Performance Prediction

Equations (6.52–6.56) can be used to solve for M_i, p_i, p_{o_i}, and $\dot{m}_i$ by applying a stepping procedure down the length of the bore. We start by guessing a pressure in the first element and assuming the Mach number in this element is zero. This approach provides initial values to permit stepping through the rest of the elements $i = 2, N$.

We can take mass accumulation into account if we assume that all accumulated mass is "lumped" at the end of the chamber. In this case, from Eq. (6.34), we can express the rate of accumulation as

$$\frac{dm}{dt} = m\left(\frac{1}{p_{oN}}\frac{dp_{oN}}{dt} + \frac{1}{V_N}\frac{dV_N}{dt}\right) \tag{6.57}$$

where p_{oN} represents the stagnation pressure in the last element, and the volume derivative results from the fact that the port area is increasing with time. This relation assumes all the accumulated mass is lumped into the last element in the motor. This assumption makes sense because accumulation effects are usually second order at most. We can determine the rate of change of stagnation pressure feeding the nozzle from the finite-difference approximation:

$$\frac{dp_{oN}}{dt} = \frac{p_{oN}^j - p_{oN}^{j-1}}{\Delta t} \tag{6.58}$$

where j = the time level
 i = the element
 Δt = the timestep for the calculation

In addition, we can calculate the rate of change volume in the last element by applying Eq. (6.35) to element "N."

The rate of accumulation is also the difference between the total inflow and outflow for the motor:

$$\frac{dm}{dt} = \sum_{i=1}^{N} \dot{m}_i - \frac{g p_{oN} A_t}{c^*} \tag{6.59}$$

In general, the dm/dt values obtained from Eqs. (6.57) and (6.59) do not agree unless we are incredibly talented at guessing the head-end pressure. Therefore, we must repeat the entire process until the postulated head-end pressure causes these relations to agree (within tolerance). At this point, we have obtained the proper mass flows and pressure level at the given time step, and we can "march" in time to a new step and repeat the same process. If the motor operates at a steady-state condition, we end up iterating the head-end pressure until $dm/dt = 0$ in Eqs. (6.57) and (6.59). Note that we have a new geometry at a new time level because the web in each element has moved by a distance $r_{bi} \Delta t$. Fig. 6.32 illustrates the ballistic-element algorithm.

Treatment of Slotted Grains. Radial slots are often included in the grain to ease manufacturing for large boosters or to provide more surface area for proper shaping of thrust over time. The radial mass flow caused by propellant burning in the slot contracts the main flow as noted in Fig. 6.33. This contraction can lead to a large pressure drop in the main flow, which has contributed to the failure of some slotted motors with high velocity flow in the bore. The constriction in the flow drives up the head-end chamber pressure and causes the motor case to burst. Also, the lower pressure near the slot can cause the propellant to deflect inward, reducing the bore area and increasing the effect.

The detailed two-dimensional flow near a slot is best described using CFD techniques. However, we have simple approximations for incompressible flow and corrections for compressibility to help us predict the loss in stagnation pressure across the slot. The methods are based on Kays' analytical formula [1950]:

$$\Delta p_o = \frac{1}{2}\rho V_1^2 \left(1 - \frac{A_1}{A_2}\right)^2 \qquad (6.60)$$

and

$$\Delta p_o = 0.4 f(M_2) \frac{1}{2}\rho V_2^2 \left(1 - \frac{A_2}{A_1}\right)^2 \qquad (6.61)$$

Here A_1 and A_2 are the port areas immediately upstream and downstream from the slot, respectively.

Equation (6.60) is valid for a sudden expansion ($A_2 > A_1$), whereas Eq. (6.61) applies to the more common case of a sudden contraction ($A_2 < A_1$). The term $f(M_2)$ in Eq. (6.61) is a compressibility correction correlated from the experimental results of Perry [1949]:

$$f(M_2) = 1 + 3.05 M_2^4 \qquad (6.62)$$

These relations can be used with balances between mass and momentum to predict conditions on the downstream side of the slot, as long as we know the upstream conditions and amount of mass flow contributed from the slot itself. The ratios of mass flows and momentums between the slot flow and main flow are important in determining the loss in pressure at these junctions. We use the treatment described above only for preliminary calculations; for more detailed calculations we use the approach of Roessler and Landsbaum [1991] (or CFD analysis).

6.5.3 Calculating Specific Impulse, Mass Flow, and Thrust

Assuming we have calculated the internal ballistics, we know the chamber-pressure history for the motor. We next need to use this information to predict the

6.5 Performance Prediction

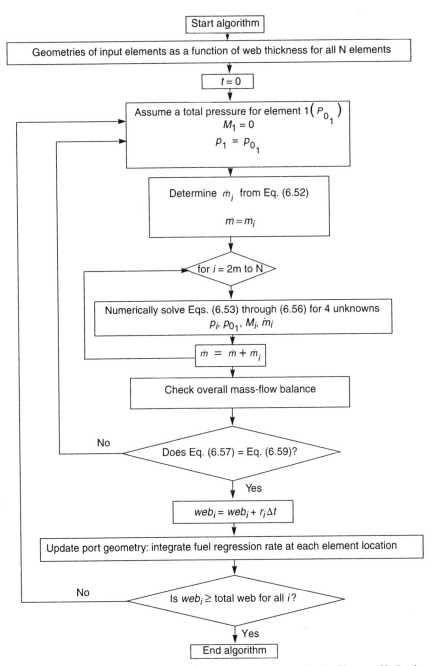

Fig. 6.32. Algorithm for Calculating Performance Based on Ballistic-Element Method.

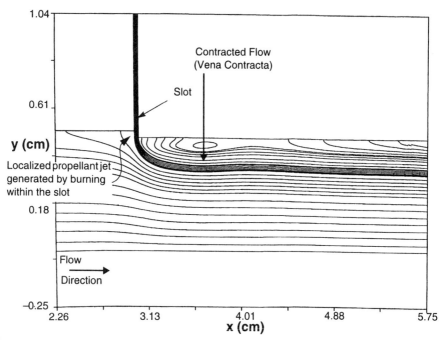

Fig. 6.33. Flowfield in the Region Near a Radial Slot in the Grain. The presence of the slot flow causes a contracted flow downstream from the injection point.

motor's other performance parameters. Engineers routinely use test data to determine the motor's efficiency compared to ideal analyses. Because the mass of propellant loaded into a test motor is carefully measured, we can introduce a combustion *efficiency*,

$$\eta_{c*} = \frac{1}{m_{prop}} \int_0^{t_b} \frac{p_c A_t}{c^*} dt \qquad (6.63)$$

to give the proper expended mass, where c^* is the theoretical characteristic-velocity value calculated from flow relations. Of course the integration in Eq. (6.63) is for the actual pressure history measured in the test (with an appropriate prediction for the throat area as a function of time).

Typical frozen-flow η_{c*} (see Chap. 4) values lie between 0.98 and 1.01, indicating a very high combustion efficiency in solid motors. Because this efficiency acts like a discharge coefficient, heat transfer in the throat can cool the gas, increase the density, and can actually give efficiencies slightly greater than unity. This result occurs because the ideal mass flow is based on an adiabatic situation at all points. The actual, or delivered c_d^* is then

6.5 Performance Prediction

$$c_d^* = \eta_{c^*} \cdot c^* \tag{6.64}$$

which we can use to predict the pressure for future motors.

The average delivered vacuum I_{sp} can be determined from the test data by dividing the total impulse by the loaded propellant mass:

$$I_{sp_v} = \frac{1}{g_0 m_{prop}} \int_0^{t_b} (F + p_e A_e)\, dt \tag{6.65}$$

where I_{sp_v} = vacuum theoretical specific impulse (s)
g_0 = 9.81 m/s²
m_{prop} = mass of propellant (kg)
t_b = total burn duration (s)
F = time varying thrust magnitude (N)
p_e = exhaust nozzle exit pressure (Pa)
A_e = exhaust nozzle exit area (m²)
t = time (s)

Based on this measurement, we can define the motor's overall efficiency as

$$\eta_o = \frac{I_{sp_v}}{I_{sp_{vth}}} \tag{6.66}$$

where $I_{sp_{vth}}$ is the vacuum theoretical I_{sp} for shifting equilibrium (see Chap. 4).

Typical η_o values are 80–87% for small motors and 88–96% for large motors. Small motors have lower efficiencies mainly because the two-phase flow losses are larger in these motors. In other words, particles have less time to accelerate in small nozzles than in larger nozzles, so performance suffers more. In addition, a larger fraction of the flow in small motors is subject to viscous drag in the boundary layers along the nozzle wall.

Using the overall efficiency, the vacuum thrust of the motor is simply

$$F_v = \eta_o I_{sp_{vth}} \dot{m} g_0 \tag{6.67}$$

where the predicted mass flow contains c^* efficiency as defined in Eq. (6.63). Predictions of chamber pressure for new motors include this factor as well, as evidenced by the c^* dependence in Eq. (6.29). By predicting chamber pressure using current data on the burn rate for the motor under consideration, we can immediately determine mass flow and thrust using the notions discussed in this subsection. As discussed in Chap. 2, propulsion engineers typically report vacuum thrust and I_{sp} because they can readily correct these conditions to an arbitrary ambient pressure.

6.6 Case Study

To illustrate some of the concepts discussed in this chapter, we do a preliminary design of an upper-stage motor for an orbital-transfer vehicle. We want to define a motor that can place a 2270-kg (5000-lb) satellite in a circular geosynchronous orbit. We presume the launch vehicle has placed the stage in a geosynchronous transfer orbit using the first stage of a Hohmann transfer, as discussed in Chap 2. We also assume the perigee of this elliptic orbit is at 185 km (100 nm) and the apogee is at the geosynchronous altitude of 35,800 km (19,350 nm).

The propulsion system we are designing for this mission circularizes a highly elliptic orbit to place the satellite in a geosynchronous state. By firing the motor at the apogee of the orbit, we can achieve the circular orbit. Motors used for this application are often called *apogee kick motors*, or *AKMs*.

To begin the design, we must first specify the technology level for various components as well as what each component must do. To evaluate the effects of changing technology, we can repeat the entire process outlined below for each candidate material. In addition, we often assume values for critical performance parameters such as average chamber pressure and burning duration as discussed in Sec. 6.3. Table 6.10 summarizes the assumed performance and design characteristics of our AKM. This table shows our design uses technologies similar to the Thiokol Star Motors, using an aft-end ignition system and a submerged nozzle with a Flexseal to provide thrust-vector control. We also use a state-of-the-art, graphite-composite motor case as well as an industry-standard EPDM insulation with silica loading.

An assumed I_{sp} value of 290 seconds is a reasonable estimate based on the data in Table 6.2. Of course as we refine the design, we can improve on this estimate using the ideas discussed in Chapter 3. By its very nature, design must be iterative because we must reasonably estimate critical parameters to get started. As a result, we select parameters for Table 6.10 by assessing the characteristics of other motors of this type. We normally obtain the best chamber pressure, burning time, and motor case L/D by repeating the design process for a range of values. The factor $p_{c_{max}}/p_{c_{avg}}$ is a function of the particular propellant's grain design and the motor's sensitivity to temperature.

6.6.1 Requirements for Velocity and Propellant Mass

We employ the Hohmann transfer equations (Chapter 2) to estimate the velocity increment (Δv) required for this mission. The elliptic orbit is specified by the radii r_1 and r_2 at the perigee and apogee, whereas the apogee burn Δv is specified in terms of the circular velocity at the perigee. For our 185 × 35,800-km orbit, these quantities are

$$r_1 = 6.507 \times 10^6 \text{ m} \; ; \; r_2 = 4.221 \times 10^7 \text{ m} \; ; \; v_{c1} = 7791 \text{ m/s} \quad (6.68)$$

Table 6.10. Assumed Motor Performance and Design Characteristics. Assumed characteristics match state-of-the-art space motors.

Performance Parameters	Motor Case
Avg. chamber pressure = 5.17 MPa Burning time = 80 s Vacuum specific impulse = 290 s	Graphite composite construction Overall L/D = 1.5 Volumetric loading = 92% $p_{c_{max}}/p_{c_{avg}}$ = 1.4
Propellant 89% solids, HTPB formulation Density = 1800 kg/m^3 Characteristic velocity = 1527 m/s Head end web (with slot) grain design	Safety factor = 1.25 F_{tu} = 1.34 GPa **Insulation** Silica-loaded EPDM formulation Density = 1100 kg/m^3
Nozzle State-of-the-art materials Flexseal TVC 20% submergence 18° effective divergence angle	**Igniter** Aft-end flame stick design (Similar to Fig. 6.7)

and the increment of burn velocity at apogee is

$$\Delta v = \left[v_{c1} \sqrt{\frac{r_1}{r_2}} - \sqrt{\frac{2r_1}{r_2\left(1 + \frac{r_2}{r_1}\right)}} \right] = 1478 \text{ m/s} \qquad (6.69)$$

The mass ratio we must attain for this Δv is

$$R = e^{\frac{\Delta v}{g_0 I_{sp}}} = e^{\left(\frac{1478}{9.81 \times 290}\right)} = 1.681 \qquad (6.70)$$

We can see that we require several thousand kilograms of propellant because the burnout mass must certainly be greater than the 2270-kg payload we have assumed. From our plots of mass fraction in Figs. 6.9 and 6.10, we can expect the propellant mass fraction of our motor to exceed 0.9 for a propellant mass in this range. For this reason, we arbitrarily assume that $f_{prop} = 0.9$ so we can proceed with our design. We must assume some value that allows us to determine the propellant mass. Later, we will see how to improve on our initial guess.

Given that $f_{prop} = 0.9$, we can immediately solve for the required propellant mass using Eq. (6.7):

$$m_{prop} = m_{pay} \left(\frac{f_{prop}(R-1)}{1 - R(1 - f_{prop})} \right) = 1670 \text{ kg} \qquad (6.71)$$

We must estimate the required amount of propellant early in the design because almost all other system masses depend on it. Some preliminary design tools require propellant mass to size other components properly. Iteration is also required using this procedure because we cannot know in advance whether the assumed propellant mass is enough to provide the required Δv.

6.6.2 Preliminary Design of the Motor Case

Having an estimate of the propellant mass and case volumetric loading, we can design the case. For our required propellant mass, the internal volume of the case must be

$$V_{cs} = \frac{m_{prop}}{\eta_v \rho_p} = \frac{1670}{0.92 \times 1800} = 1.007 \text{ m}^3 = 1.007 \times 10^6 \text{ cm}^3 \quad (6.72)$$

We can determine the diameter of the case required to hold this volume by

$$V_{cs} = D^3 \left[\frac{\pi}{6} + \frac{\pi}{4}((L/D) - 1) \right] \quad (6.73)$$

where D is the internal diameter of the case. This expression assumes the domes of the case are spherical, so the first term on the right-hand side of Eq. (6.72) is the dome volume, and the second term is the volume of the cylindrical section. Given an assumed L/D of 1.5, we combine Eqs. (6.71) and (6.72) to give

$$D = 103 \text{ cm} \; ; \; L = 155 \text{ cm} \; ; \; L_{cy} = 51.6 \text{ cm} \quad (6.74)$$

where L_{cy} is the length of the cylindrical section.

Now that the case dimensions are known, we can determine the case thickness from strength considerations. The predicted burst pressure (p_b) for our assumptions in Table 6.10 becomes

$$p_b = p_{c_{avg}} \frac{p_{c_{max}}}{p_{c_{avg}}} f_s = 5.17 \times 1.4 \times 1.25 = 9.05 \text{ MPa} \quad (6.75)$$

Table 6.4 gives an ultimate tensile strength of 2.34 GPa for the graphite composite material used in our motor case. Using this value in Eq. (6.72), we can find the case thickness,

$$t_{cs} = \frac{p_b D}{2 F_{tu}} = \frac{9.05 \times 103}{2 \times 2340} = 0.348 \text{ cm} \quad (6.76)$$

Here, we assume that both domes and the cylinder section of the case are the same thickness [given by Eq. (6.76)], even though pressure theory for thin walls states that dome stresses are only half that in the cylinders. We presume the additional material is used as local buildup in polar boss and skirt regions.

At this point, we have defined the entire case geometry and we can estimate the mass from the volume of material present. Because the thickness is much less than the case radius, we can estimate the enclosed volume as the surface area times the thickness. Using this approach combined with the density of the graphite composite material (1550 kg/m³) gives a pressure vessel weight of:

$$m_{pv} = \rho_{cs} t_{cs} D^2 \pi (1 + L_{cy}/D) = 27.0 \text{ kg} \tag{6.77}$$

where the first and second terms on the right-hand side of this equation represent contributions from the domes and cylindrical sections, respectively. A calculation of the pV/W for this case gives a value of 35 km, which is a bit high even for the high-strength graphite material. Simple techniques such as these often underestimate masses because complex loadings require more material due to bending stresses and stiffness considerations.

6.6.3 Preliminary Design of the Thrust Skirt and Polar Boss

For this analysis, we assume both forward and aft skirts are equal in length to the forward and aft domes. While most skirts are not this long, the additional length amounts to what would normally be interstage structure. We also assume the skirts are as thick as the case itself. Note that because the skirts withstand compressive loadings and the case must withstand mainly tensile loadings, this assumption may be poor. We could analyze the compressive load and buckling; see Huzel and Huang [1971] for a relatively simple sizing procedure. But we assume the 0.348-cm thickness calculated above is adequate.

We can calculate the skirt masses much like we did the mass of the pressure vessel:

$$m_{sk} = \rho_{cs} t_{cs} \pi D^2 = 18.0 \text{ kg} \tag{6.78}$$

To account for the aft polar boss, we arbitrarily increase by 10% our calculated masses for the pressure vessel and skirt. Under this assumption, the motor case mass becomes

$$m_{cs} = 1.1(m_{pv} + m_{sk}) = 49.5 \text{ kg} \tag{6.79}$$

Of course, our analysis of the motor case is highly simplified in this preliminary design. Manufacturers of solid motors have created much more sophisticated design codes, but overall sizing proceeds along the lines we have detailed in this section.

6.6.4 Preliminary Design of the Insulation

In preliminary design, we simply estimate the insulation's mass. As discussed in Section 6.3.6, actual design thicknesses depend not only on the grain design and burning time but also on the type of insulation used and the detailed contour of the motor's aft wall. We use the correlation [Eq. (6.16)] developed for space motors using the EPDM insulation assumed for our design. The exposed area of the wall surface in our case is

$$A_w = \pi D^2 + \pi D L_{cy} = 50{,}130 \text{ cm}^2 \tag{6.80}$$

and for 20% nozzle submergence, Eq. (6.16) gives

$$m_{insul} = 876 \times 10^{-9} 1670^{-1.33} 80^{0.965} 1.5^{0.144} 20^{0.058} 50{,}130^{2.69} = 35.2 \text{ kg} \tag{6.81}$$

Because the density of this type of insulation is 1108 kg/m³, we can calculate an average thickness distributed over the interior of the case:

$$t_{insul} = \frac{m_{insul}}{\rho_{insul} A_w} = 0.635 \text{ cm} \tag{6.82}$$

Assuming an average exposure time equal to half the burning time gives us an average erosion rate of more than 0.16 mm/s, which is quite reasonable for a space motor design.

6.6.5 Designing the Nozzle and Igniter

As mentioned previously in this chapter, the nozzle often contains most of the motor's parts. Even preliminary designs of this component can be quite involved and depend on the amount of submergence, the type of thrust-vector-control system, and the types of materials for components. As we noted in Fig. 6.6, performance trade-offs may show that we should include an extendible exit cone as well. For top-level design, we need a more comprehensive design tool to address detailed issues such as these.

Because we have selected a space motor for our design, we can use the correlation in Eq. (6.17). To begin, we need to estimate the expansion ratio for our design. First, calculate the throat diameter from Eq. (6.43):

$$D_t = \sqrt{\frac{4 c^* m_{prop}}{\pi t_b P_c}} = 8.86 \text{ cm} \tag{6.83}$$

We want to make the exit diameter less than the diameter of the motor case for good packaging efficiency. Assuming that the exit diameter is 80% of the case diameter, we may determine the expansion ratio,

$$\varepsilon = \frac{(0.8 \times 103)^2}{8.86^2} = 86 \tag{6.84}$$

In addition, we can determine the length of the nozzle from the throat to the exit (L_{noz}) by

$$L_{noz} = \frac{D_e - D_t}{2\tan(\theta_{cn})} = 113 \text{ cm} \tag{6.85}$$

Using this result in Eq. (6.17), the predicted nozzle mass becomes

$$m_{noz} = 0.256 e^{-4} \left[\frac{16 \prime 0 \times 1527^{1.2} 86^{0.3}}{9.05^{0.8} 80^{0.6} (\tan 18°)^{0.4}} \right]^{0.917} = 39.7 \text{ kg} \tag{6.86}$$

Now we have yet to consider the mass of the igniter and the thrust-vector control system. Because we are using an aft-end igniter, the correlation in Eq. (6.85) is not appropriate for this configuration. Use of a Flexseal TVC system requires shims and pads for the Flexseal (not a major mass factor) as well as a pair of hydraulic actuators (with associated attachments and power systems) to push the nozzle in pitch and yaw directions. To account for these systems, we apply a historical factor of 1.5. Thus, the assumed mass of the nozzle/TVC/ignition system is

$$m_{noz-sys} = 1.5(39.7) = 59.6 \text{ kg} \tag{6.87}$$

6.6.6 Preliminary Ballistics Analysis

Finally, we can check to see what propellant burning rate will be required for our design. Here we must assume a size for the port of the motor. We can do this by assuming a *web fraction*, which is the fraction of the motor radius comprised of propellant. For most space motors, the web fraction lies in the range of 70–85%. Assuming a web fraction of 80% gives us an initial port radius of 10.4 cm. To keep the pressure history reasonably neutral, we include a 5-cm radial slot running from the bore to the outer case radius. In practice, one would normally stop the slot before the wall, but this situation is much easier to analyze.

Under these assumptions, the grain geometry is as shown in Fig. 6.34. The internal radius of the case and domes is simply the external radius minus case and insulation thicknesses (50.8 cm). Using this grain geometry, we need to determine A_b as a function of web distance. In Fig. 6.34, we have divided the grain into five regions to help us calculate A_b. The initial burn surface area is simply the sum of the contributions from these five surfaces:

$$A_{bo} = 2\pi 10.4^2 + 2\pi 10.4 \times 25.8 + \pi(50.8^2 - 10.4^2)$$
$$+ \pi(50.8^2 - 10.4^2) + 2\pi 10.4 \times 75.5 \tag{6.88}$$

We leave to you constructing a general equation for A_b as a function of web distance from the specified geometry. Note from Fig. 6.34 that at $w = w_1$ surface "2" vanishes and the axial surfaces "3" and "4" enter the dome region. At $w = w_2$, propellant in the aft dome burns out and surfaces "4" and "5" vanish. Finally, burning is complete when the forward dome burns out.

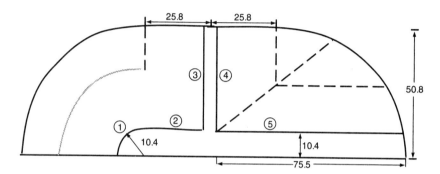

Fig. 6.34. **Preliminary Design of the Grain.** All dimensions are in cm. Numbers in circles refer to particular geometric regions.

We can construct a ballistic prediction, based on steady-state lumped parameters (as in Section 6.5.1) to predict the pressure history for our motor. We assume a burning rate exponent of 0.3 and we vary the burning rate coefficient until we get the assumed burning time of 80 seconds. The resulting pressure prediction is shown in Fig. 6.35. The average pressure derived from this curve is 4.95 MPa, which is slightly lower than our assumed value. The ratio $p_{c_{mas}}/p_{c_{avg}}$ from Fig. 6.35 is 1.3, which implies a 10% "margin" to account for temperature effects. The resulting expression for the burning rate is

$$r_b = 0.316 p_c^{0.3} \tag{6.89}$$

Comparing this result with formulation B in Table 6.9 indicates that we need to reduce the burning rate coefficient (presumably through minor changes in the formulation) by 20%. We must explore this issue in more detail and investigate the implications of such a change on propellant formulation. Of course, the easiest way to reduce the required burning rate for future iterations is to increase the input burning time.

Because we have defined a grain geometry in Fig. 6.34, we can also calculate the volumetric loading and propellant mass resulting from this configuration. We can determine the propellant volume to show that

$$m_{prop} = 1607 \text{ kg} \; ; \; \eta_v = 0.95 \tag{6.90}$$

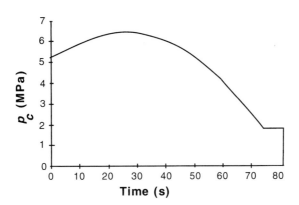

Fig. 6.35. Preliminary Prediction of Chamber Pressure using Lumped-Parameter Method. See Section 6.5.1.

which shows we actually can achieve a higher volumetric loading but lower propellant mass than we originally assumed. The primary reason for the drop in propellant mass is that we ignored the volume of the case and insulation in defining the case geometry. We can incorporate the updated volumetric loading in future design iterations, but we must remember we have used a simplified grain geometry. For instance, we probably would need to remove some propellant at the end of the grain to allow ample clearance for the nozzle. We could roughly estimate this effect for use in future design iterations.

A sketch of our motor is in Fig. 6.36. We have taken the liberty of removing some of the propellant in the aft dome to accommodate the nozzle. We also assumed an area contraction ratio of 3.0 in deciding where to place the nosecap tangency point, which is the forward-most point of the nozzle (shown in Fig. 6.16). Also, we assumed that the throat was about ten centimeters aft of this tangency point. All other dimensions are from our analysis.

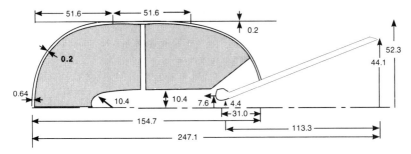

Fig. 6.36. Sketch of the Preliminary Design for an Apogee Kick Motor. All dimensions are in cm.

6.6.7 Preliminary Prediction of Performance

Now that we have an idea what our motor looks like, we can update the performance. For example, we can predict the specific impulse more precisely using a thermochemistry code for the actual nozzle design obtained in the previous subsection.

We can also update the motor mass fraction based on the masses calculated in the previous section. The total predicted inert mass is

$$m_{inert} = m_{cs} + m_{insul} + m_{noz-sys} = 144.3 \text{ kg} \quad (6.91)$$

so the new mass fraction becomes

$$f_{prop} = \frac{1607}{1607 + 144.3} = 0.918 \quad (6.92)$$

and our vehicle mass ratio is

$$\frac{m_i}{m_f} = \frac{2270 + 1607 + 144.3}{2270 + 144.3} = 1.666 \quad (6.93)$$

Next we can compute the actual velocity gain using our updated information:

$$\Delta v = 9.81 \times 290 \ln(1.666) = 1452 \text{ m/s} \quad (6.94)$$

which implies that we have about 2% less than the required velocity for the payload. We need to repeat our analysis by first determining the amount of propellant required to attain our 1478 m/s Δv using our new mass fraction. Then, we can update the motor design.

6.6.8 Conclusions

By presenting this case study, we have tried to illustrate the methods for designing a solid rocket motor. As we mentioned, the simplistic approaches used for component designs are often replaced with detailed computer models (using roughly the same parameters), which can make the results much more accurate. Even in more advanced models, iteration is required to obtain design parameters that meet mission requirements. We have included many references to allow access to more detailed design approaches.

The procedure discussed above can be repeated often (presumably using a computer code) to find the best burning time, chamber pressure, motor case length-to-diameter(L/D), and nozzle expansion ratio. For burning time, we must trade off insulation mass and nozzle mass; we can have a smaller nozzle throat for a long burning time (average mass flow is smaller) but we must "pay" for this luxury in terms of increased insulation. For the nozzle expansion ratio, we must trade off between the increased performance (I_{sp}) of a larger nozzle and increased mass, usually requiring a detailed nozzle mass model. All of these trades may be performed under constraints such as overall length (or diameter), allowable range of burning ratio, acceleration loads on the payload, and so forth.

The chamber pressure affects the masses of nearly all components, so we must optimize it in any reasonable design study. Using advanced materials for motor cases (such as the high-strength graphite composite in this study) inevitably drives us to higher optimal chamber pressures compared to materials with lower strengths. Optimizing the motor case L/D is more challenging. Changing this parameter affects the propellant grain design. In multistage systems (such as the Pegasus), we often want to have the same diameter on adjacent stages which constrains the L/D on the upper stage. Drag becomes an important factor in this trade-off for airborne systems because higher L/D motors have lower diameters (and drag). For upper-stage systems, volumetric packing efficiency is a major factor in deciding on the best value.

Acknowledgments

For help in developing this chapter, I gratefully acknowledge: Dr. E. Landsbaum and Dr. M. Adams (The Aerospace Corporation), Dr. C. Chase (Chemical Systems Division), Mr. W. Daines (Hercules Aerospace), Mr. S. Stein and R. Hamke (Thiokol Corporation), Mr. J. Horton and Mr. G. Andrews (Aerojet Corporation), Dr. T. Hawkins (Phillips Laboratory), Lt. P. Whittaker (USAF Space Systems Division), Dr. V. Yang (Penn State University), and Mr. J. Hilbing and A. Anderson (Purdue University).

References and Bibliography

A. J. McDonald, R. R. Bennett, J. C. Hinshaw, and M. W. Barnes. 1991. Chemical Rockets and the Environment. *Aerospace America.* 25(5): 33.

American Institute of Aeronautics and Astronautics. 1991. *Atmospheric Effects of Chemical Rocket Propulsion.* Report of an AIAA Workshop in Sacramento, California. Oct. 1991.

Anderson, J. D. Jr. 1982. *Modern Compressible Flow with Historical Perspective.* New York: McGraw Hill Book Company.

Anon. 1987. *Design Methods in Solid Rocket Motors.* AGARD LS-150, 1987.

Anon. 1979. *Solid Rocket Motor Technology.* AGARD CP-259.

Anon. 1991. Combustion of Solid Propellants. AGARD LS-180.

Barrere, M., A. Jaumotte, B. Frais de Veubeke, and J. Vandenkerckhove. *Rocket Propulsion.* Elsevier Publishing Company.

Bartz, D. R. 1957. A Simple Equation for Rapid Estimation of Rocket Nozzle Convective Heat Transfer Coefficients. *Jet Propulsion.* 1:49–51.

Bate, R.R., D. D. Mueller, and J. E. White. 1971. Fundamentals of Astrodynamics. New York: Dover Publishing Company.

Cornelisse, J. W., H. F. R. Schoyer, and K. F. Wakker. 1979. *Rocket Propulsion and Spaceflight Dynamics*. London: Pitman Publishing Co.

Feodosiev, V. I., and G. B. Siniarev. 1959. *Introduction to Rocket Technology*. Orlando, FL: Academic Press.

Gordon, S., and B. J. McBride. 1971. *Computer Program for Calculation of Complex Chemical Equilibrium Composition, Rocket Performance, Incident and Reflected Shocks, and Chapman-Jouget Detonations*. NASA SP-273, NASA Lewis Research Center.

Green, L. 1954. Erosive Burning of Some Composite Solid Propellants. *Jet Propulsion*. 24:9–15.

Heister, S. D. 1990. *A Computer Code for Solid Rocket Motor Design (SRMDES Version 2.0) Technical Description and User's Manual*. Technical Report TOR-0090(9975)-3. El Segundo, CA: The Aerospace Corporation.

Heister, S. D., and E. M. Landsbaum. 1991. Analysis of Ballistic Anomalies in Solid Rocket Motors. *AIAA Journal of Propulsion and Power*. 7(6):887–893.

Heister, S. D., and R. J. Davis. 1992. Predicting Burning Time Variations in Solid Rocket Motors. *Journal of Propulsion and Power*. 8(1):564–569.

Hilbing, J. H., and S. D. Heister. 1993. Radial Slot Flows in Solid Rocket Motors. Paper No. AIAA-93-2309 presented at the 29th AIAA Propulsion Conference.

Hill, P. G. and C. R. Peterson. 1970. *Mechanics and Thermodynamics of Propulsion*. Reading, MA: Addison-Wesley Publishing Co.

Incropera, F. P., and Dewitt, D. P. 1985. *Fundamentals of Heat and Mass Transfer*. 2nd ed. New York: John Wiley and Sons.

Kays, W. M. 1950. Loss Coefficients for Abrupt Changes in Flow Cross Section with Low Reynold's Number Flow in Single and Multiple Tube Systems. *Transactions of the ASME*. 72:1067-1074.

Kays, W. M., and M. E. Crawford. 1980. *Convective Heat and Mass Transfer*. New York: McGraw-Hill Book Company.

Kordig, J. W., and G. H. Fuller. 1967. Correlation of Nozzle Submergence Losses in Solid Rocket Motors. *AIAA Journal*. 5:175–177.

Kuentz, Craig. 1964. *Understanding Rockets and Their Propulsion*. John F. Ryder Publisher.

Kuo, K. K. 1986. *Principles of Combustion*. New York: Wiley Interscience.

Kuo, K. K., and M. Summerfield. (eds.) 1984. *Fundamentals of Solid Propellant Combustion*. Progress in Astronautics and Aeronautics. Vol. 90. Washington, DC: AIAA.

Landsbaum, E. M., M. P. Salinas, and J. P. Leary. 1980. Specific Impulse Predictions of Solid-Propellant Motors. *J. Spacecraft and Rockets*. 17:400–406.

Lenoir, J. M., and G. Robillard. "A Mathematical Method to Predict Effects of Erosive Burning in Solid Propellant Rockets." in *Proceedings of 6th International Symposium on Combustion*. New York: Reinhold Publishing Company.

McAdams, W. H. 1954, *Heat Transmission*. 3rd ed. New York: McGraw-Hill Book Company.

National Aeronautics and Space Administration. 1970. *Solid Rocket Motor Metal Cases.* NASA SP-8025.

National Aeronautics and Space Administration. 1971. *Solid Rocket Igniters.* NASA SP-8051.

National Aeronautics and Space Administration. 1971. *Solid Rocket Motor Performance Analysis and Prediction.* NASA SP-8039.

National Aeronautics and Space Administration. 1971. *Solid Propellant Selection and Characterization.* NASA SP-8064.

National Aeronautics and Space Administration. 1972. *Solid Propellant Grain Design and Internal Ballistics.* NASA SP-8076.

National Aeronautics and Space Administration. 1975. *Solid Rocket Motor Nozzles.* NASA SP-8115.

National Aeronautics and Space Administration. 1974. *Solid Rocket Thrust Vector Control.* NASA SP-8114.

National Center for Advanced Technologies, Astronautics Laboratory (AFSC). 1990. *National Rocket Propulsion Strategic Plan.* Edwards AFB, California. August.

Oates, G. C. 1984. *Aerothermodynamics of Gas Turbine and Rocket Propulsion.* Washington, DC: AIAA Education Series.

Ordway, F. I., III, ed. 1989.*History of Rocketry and Astronautics.* American Astronautical Society History Series, Vol. 9. New York: American Astronautical Society.

Pellet, G. L., D. I. Sebacher, R. J. Bendura, and D. E. Wornom. 1983. HCL in Rocket Exhaust Clouds: Atmospheric Dispersion, Acid Aerosol Characteristics, and Acid Rain Deposition. *Journal of the Air Pollution Control Association.* 33(4): 304.

Penner, S. S. 1957. *Chemical Problems in Jet Propulsion.* New York: Pergamon Press.

Perry, J. A., Jr. 1949. Critical Flow Through Sharp Edged Orifices. *Transactions of the ASME.* 71(7) 757.

Roessler, W. U., and E. M. Landsbaum. 1991. Stagnations Pressure Losses at Abrupt Contractions With Transverse Flow. *Astronautica Acta.* 18:349–358.

Seifert, H. S., ed. 1959. *Space Technology.* New York: John Wiley and Sons.

Siegel, B., and L. Schieler. 1964. *Energetics of Propellant Chemistry.* New York: John Wiley and Sons.

Sutton, G. P., and D. M. Ross. 1976. *Rocket Propulsion Elements.* New York: John Wiley and Sons.

Van Wylen, G. J. and R. E. Sonntag. 1973. *Fundamentals of Classical Thermodynamics.* 2nded. New York: John Wiley and Sons.

Von Braun, Wernher, and Frederick I. Ordway. 1967. *Space Travel: A History.* New York: Harper and Row.

Wood, K.D. 1964. *Spacecraft Design.* Johnson Publishing Co.

Zucrow, M. J. and Hoffman. 1976. *Gas Dynamics.* Vol. I. New York: John Wiley and Sons.

CHAPTER 7
HYBRID ROCKET PROPULSION SYSTEMS

David Altman, *Sr. V.P. Chemical Systems Div/UTC (ret.)*
Consultant
Ronald Humble, *U.S. Air Force Academy*

> 7.1 History
> 7.2 Hybrid-Motor Ballistics
> 7.3 Design Process
> 7.4 Preliminary Design Decisions
> 7.5 Performance Estimate
> 7.6 Preliminary Component Design
> 7.7 Case Study

A *hybrid rocket* stores propellant in two different states—liquid and solid. In a typical hybrid, the fuel is a solid and the oxidizer is a liquid (see Fig 7.1). However, it is possible to use "reverse hybrids"—solid oxidizers with liquid fuels. Unless we specify otherwise, our discussion in this chapter assumes the classical hybrid with a solid fuel grain. Because of this separation of oxidizer and fuel into two different states, combustion differs from that of either the solid or liquid rocket. In those cases, a small element of volume in the combustor contains an essentially uniform mixture of oxidizer and fuel. However, the hybrid burns as a macroscopic diffusion flame, in which the oxidizer-to-fuel ratio (O/F) varies down the length of the fuel port (see Sec. 7.2). This distinguishing characteristic gives rise to several interesting features. Some of the advantages include:

- **Safety**—An intimate mixture (as in solid motor grains where oxidizer and fuel crystals touch each other) of oxidizer and fuel is not possible. Thus, explosive mixtures that can occur in liquid or solid rockets are virtually impossible in hybrids. Also, because the fuel is inert, storage and handling are much simpler.

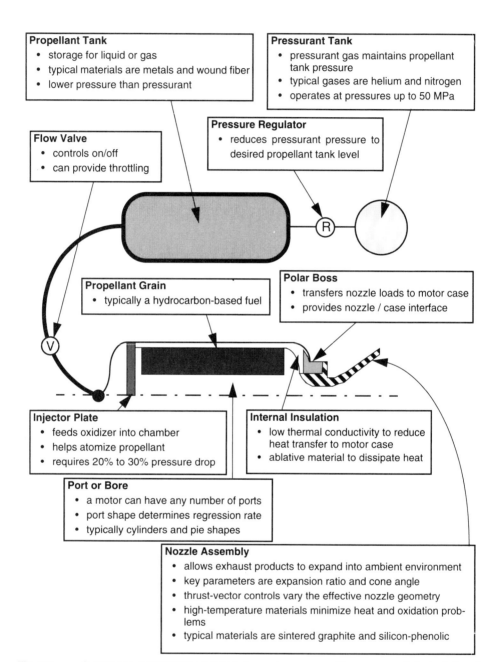

Fig. 7.1. **Schematic of Hybrid Rocket.** This figure shows many of the components typically found in a hybrid rocket system and some of their key features.

- **Throttling**—The engine can be throttled by simply modulating the liquid flow rate, whereas liquid rockets require two flow rates to be synchronized while being modulated. They must also satisfy certain momentum requirements for stable combustion. This throttling feature is useful for trajectory shaping of the booster and energy management in many tactical missile applications.
- **Shutdown and restart**—We can shut down and restart the system. This feature is an important safety consideration because it allows for an abort procedure.
- **Propellant versatility**—The hybrid's liquid oxidizer provides a higher energy level than any of the conventional solid oxidizers. But the hybrid's solid fuel permits the simple addition of many other ingredients, such as energetic metals, whereas a liquid system requires slurry mixtures.
- **Environmental cleanliness**—The large range of available propellant ingredients permits the selection of many combinations which do not produce hydrochloric acid, aluminum oxide, or other undesirable chemicals in the exhaust.
- **Grain robustness**—Unlike solid rockets, fuel grain cracks are not catastrophic because burning occurs only in the port where it encounters the oxygen flow.
- **Low temperature sensitivity**—Because the temperature effect on burn rate is negligible (as in liquids), we need not apply a margin to the thrust chamber weight to account for the variation of maximum expected operating pressure (MEOP) with ambient temperature.
- **Low cost**—Lower system cost results from greatly reduced failure modes, which permit the use of commercial-grade, instead of Mil-Spec, ingredients. Furthermore, the system can tolerate larger design margins, resulting in a lower fabrication cost. Because the fuel does not explode, storage and handling requirements are simpler.

But hybrids also have some disadvantages:

- **Low regression rate**—A small fuel web means that most combustion chambers over 30 cm in diameter require multiple ports. However, this characteristic may be an advantage for long-duration applications such as target drones, hovering vehicles, and gas generators.
- **Low bulk density**—The volumetric fuel-loading density in the combustion chamber is low because we must inject the total oxidizer at the head end and allow for a mixing volume aft of the grain. This results in a lower mass fraction than in liquids or solids, so the hybrid exhibits a larger envelope for a given mission.
- **Combustion efficiency**—The nature of the large diffusion flame results in a lower degree of mixing and hence lower impulse efficiency. This loss is

usually 1–2% more than in liquids or solids. Compared to solids, however, the delivered specific impulse is greater because of higher-performance propellants.

- **O/F shift**—Opening the port during burning causes an *O/F* shift with burning time, which can lower theoretical performance. But proper design minimizes this loss.

7.1 History

Although hybrid rockets have not enjoyed the same extensive development as solid and liquid motors, hybrid combustion involving a solid and a fluid has been nature's way of burning fuels and oxidizers. Examples are:

- The wax candle or oil lamp, burning in the presence of atmospheric oxygen, with a wick as the flame-holding device
- The fireplace, in which the bellows increases the "oxidizer mass flux" and, therefore, the burning rate
- On a grander scale, the forest fire, involving turbulent mixing of air and the vaporized fuel from the trees. Here, again, wind velocity augments combustion.

The earliest work on hybrid rockets was conducted in the late 1930s at I. G. Farben in Germany and at the California Rocket Society in the United States. Leonid Andrussow, an ex-Russian World War I cavalry officer working for I. G. Farben, conceived of a fluid-solid propellant rocket. In 1937, with O. Lutz and W. Noeggerath, he tested a 10-kN hybrid using coal and gaseous nitrous oxide. During the same period, Oberth did some work in Germany on the more energetic LOx-graphite propellant combination. Neither of these last two efforts was successful because carbon's very high heat of sublimation results in a negligible burning rate. That is why the early ICBM programs in the U.S. selected this material for reentry nose cones!

In the early 1940s, the California Pacific Rocket Society conducted a more successful effort, which employed LOx and several fuels such as wood, wax, and finally rubber. Of these combinations, the LOx-rubber combination was the most successful: a rocket using these propellants was flown in June 1951 to an altitude of about 9 km. Although the Society did not report any ballistic analyses, they did have an accurate concept of the fundamentals of hybrid burning as evidenced by the following statement: "The chamber pressure of a solid-liquid rocket engine is proportional to the oxidizer flow and not to the internal surface area exposed to the flame. Thus, there is no danger of explosions due to cracks and fissures in the charge as with solid propellant rockets."

In the mid-1950s, two significant hybrid efforts occurred. One was by G. Moore and K. Berman at General Electric, involving the use of 90% hydrogen peroxide and polyethylene in a rod-and-tube grain design. This effort was successful in that combustion was very smooth, resulting in a high combustion efficiency. These authors drew several very significant conclusions:

- The longitudinal uniformity of burning was remarkable
- Grain cracks had no effect on combustion
- Hard starts* were never observed
- Combustion was stable because the fuel surface acted as its own flame holder
- Throttling was easy, using a single valve
- A high liquid-to-solid ratio ($O/F = 7$) was desirable to simplify combustion

Moore and Berman observed, however, that the burning rate was low and nearly invariable. They admitted the inherent thermal instability of peroxide could be a practical problem. The second significant effort was by William Avery at the Applied Physics Laboratory (US Patent [1964]). He investigated a "reverse hybrid" composed of a liquid fuel (JP) and a solid oxidizer (ammonium nitrate)—selected for their low cost. Technically the program was not successful because of rough combustion and poor performance. Interestingly, the liquid-to-solid ratio of this combination is in the range of 0.035, which is about 200 times smaller than that used by Moore and Berman!

During the 1960s, organizations in two European countries engaged in hybrid studies leading to flight tests of sounding rockets. These organizations were ONERA (with SNECMA and SEP) in France and Volvo-Flygmotor in Sweden. The ONERA development used a hypergolic propellant based on nitric acid and an amine fuel. The first flight of this vehicle occurred in April 1964, followed by three flights in June 1965 and four flights in 1967. All eight flights were successful, reaching altitudes of 100 km. The Volvo-Flygmotor rocket was based on a hypergolic combination using nitric acid and Tagaform (polybutadiene (PB) plus an aromatic amine). It was flown successfully in 1969 carrying a 20-kg payload to an altitude of 80 km.

United Technologies Center (Chemical Systems Division) and Beech Aircraft developed a high-altitude, supersonic target drone in the late 1960s. Called the Sandpiper, it used MON-25 (25% NO, 75% N_2O_4) and polymethyl methacrylate (PMM)-Mg for fuel. The first of six flights occurred in January 1968, and flew for more than 300 seconds and 160 km. The HAST, a second version, carried a heavier payload and was based on an IRFNA-PB/PMM propellant combination. This 33-cm diameter motor was throttleable over a 10/1 range. A later version of this vehicle, the Firebolt, was developed by Chemical Systems Division (CSD) and Teledyne Aircraft, using the same propulsion configuration as the HAST. This successful program ended in the mid-1980s.

---

* *Hard starts* refer to the tendency in liquid bipropellant engines to get very high pressure spikes during ignition.

In the mid-1960s, CSD investigated high-energy hybrid propellants based on a fuel containing lithium and FLOx (F_2+O_2) as the oxidizer. This experiment led to a hypergolic propellant system that was throttleable and demonstrated a vacuum specific impulse of 380 seconds at 93% combustion efficiency, assuming equilibrium flow (LaForce and Wolff [1970]). Figure 7.2 shows a firing of this 107-cm diameter motor in 1970.

Fig. 7.2. **Test Firing of a FLOx / Lithium Hybrid Motor.** A liquid fluorine / liquid oxygen mixture is injected into a solid lithium fuel grain. (Courtesy of Chemical Systems Division of United Technologies)

The largest hybrid rockets built to date—designed for a space booster—were made by AMROC in the late 1980s and the early 1990s. Their H-500 engine developed 312,000 N of thrust for 70 seconds using a LOx/HTPB (Hydroxyl Terminated PolyButadiene) propellant. Figure 7.3 shows a firing of this engine at the Air Force Phillips Laboratory. Their later version, H-250F, using the same propellant, has been tested at a thrust level of more than 1,000,000 N.

In January 1994, the U.S. Air Force Academy flew a 6.4-m long hybrid sounding rocket. This vehicle used HTPB/LOx for propellant. Peak thrust was approximately 4400 N with a thrust duration of 17 seconds. The vehicle reached an altitude of about 5 km.

Fig. 7.3. **Test Firing of the AMROC H-500 Hybrid Motor (LOx/HTPB).** This engine was static tested at the U.S. Air Force's Phillips Laboratory. It developed 312,000 N of thrust and burned for 70 seconds. (Courtesy of AMROC)

7.2 Hybrid-Motor Ballistics

At this point in the other technology chapters in this book, we talk about the design process. However, designing Hybrid Rocket Propulsion Systems (HRPSs) depends on the unique nature of hybrid combustion or internal ballistics. Therefore, we start the design discussion with ballistics.

The question we must answer is: How does the solid fuel vaporize and enter the port for combustion, and more specifically, how fast does the fuel vaporize or regress? Intuitively, we can postulate that the quantity of liquid oxidizer flowing into the combustion chamber affects the thrust produced. Experiments have verified this postulate. Thus, the solid-fuel burn rate (regression rate) is also a function of oxidizer flow rate. The following discussion theoretically verifies these assertions and shows that the fuel regression rate is given by the relatively simple expression:

$$\dot{r} = aG^n x^m \tag{7.1}$$

where $\dot{r}$ = fuel regression rate (m/s)
G = total propellant (oxidizer and fuel) mass flux (kg/m^2·s)
x = distance down the port (m)
a,n,m = regression rate constants, characteristic of the propellants

From this equation we can see that regression rate varies with mass flux and also varies along the length of the port. Notice that the mass flux (G) is the total mass flux, including both the oxidizer injected and the fuel that has vaporized from the surface of the fuel wall, thus causing G to increase continuously down the port.

The combustion process in the hybrid is distinctly different from that in the liquid or solid motor. We might instinctively assume that we can design the combustion chamber in a hybrid by combining the technologies of liquid and solid combustion, but this is a fallacy. Although many of the hardware components in the feed system and insulation are common to chemical rockets, the grain design, interior ballistics, and performance predictions are different. These characteristics vary because the hybrid combustion occurs as a macroscopic diffusion flame,[*] so the composition (O/F) varies down the length of the motor. In contrast, the O/F in liquid and solid motors is essentially independent of axial location. In liquid motors, all of the propellant is injected at the head end, so we simply need enough volume in the chamber (volume-to-throat area ratio, L^*) to allow for vaporization, local small-scale diffusion, and combustion kinetics. For a solid rocket, the "injector" is uniformly distributed down the bore of the grain, and every small element of gas has a constant O/F. Although this fundamental distinction may make ballistic design more complex, separation of the oxidizer and fuel keeps the hybrid from exploding and makes it safer to handle and operate.

7.2.1 Interior Ballistics Model

In the hybrid combustion chamber, an atomized or vaporized liquid flows down the port and reacts near the surface of the solid fuel. Experiments have shown that the controlling factors in combustion are the rate of heat transfer to the solid surface and the heat of decomposition of the solid-phase fuel. The mass flux, which is regulated by the rate of flow of the liquid phase, determines the rate of heat generated in the combustion zone and, hence, the heat transfer and thrust magnitude.

The combustion phenomenon is similar to that of a turbulent diffusion flame, for which the flame zone is established within the boundary layer[†] (see Fig. 7.4). We can conveniently represent this process by an idealized model which treats the flame zone as a point of discontinuity in temperature gradient and composition. In the real case of finite-combustion kinetics, the flame zone is thickened with continuous gradients in both temperature and composition. Oxidizer enters the flame zone from the port free-stream core by diffusion, while the fuel enters the boundary layer as a result of vaporization at the wall surface. The combustion zone is

---

[*] A diffusion flame occurs in a region where the variable fuel and oxidizer concentrations allow combustion. In this case, the fuel is more concentrated near the solid fuel wall, and the oxidizer is more concentrated near the center of the port. Diffusion allows the chemicals to mix.

[†] The *boundary layer* is the region of the flow near a surface that has reduced flow velocity due to viscous shear forces between the flow and the surface. Velocity is zero at the surface and must increase to a maximum at the boundary layer edge.

established at that point where an approximate stoichiometric mixture ratio has been achieved. This model shows that the combustion zone occurs within the turbulent boundary layer at a distance from the solid wall, which is less than the thickness of the momentum boundary layer. The axial velocity at the flame (v_b) is also less than that at the outer edge of the boundary layer (v_e). Experimental measurements of gaseous oxygen reacting with Plexiglass fuel have confirmed this basic model, as is shown in the Schlieren photograph[*] in Fig 7.5 (Muzzy [1963]). We use this basic model to develop the mathematics of combustion, as discussed in the following sections.

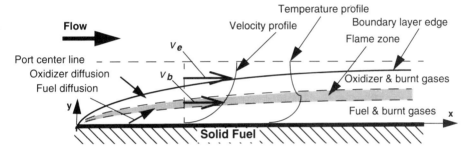

Fig. 7.4. **Schematic of Combustion Zone Above Hybrid Fuel.** This basic model is used to develop the theoretical expressions for regression rate.

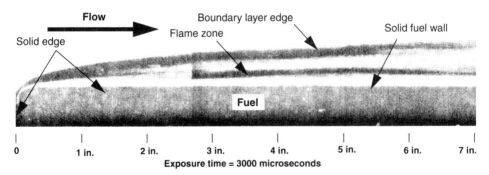

Fig. 7.5. **Schlieren Photograph of a Boundary Layer for Hybrid Combustion.** This photograph shows that our basic model (Fig. 7.4) is based on empirical observation. The "flame zone" and boundary layer features are shown [Muzzy, 1963].

---

[*] A Schlieren photograph is a technique that photographs variations in gas density. In this figure, the boundary layer and the flame zone show up because they have different densities.

7.2.2 The Burning-Rate Equation

At the steady state, the heat transferred from the flame to the surface is exactly balanced by the mass flow rate of vaporized fuel times the total heat of gasification. This equation is

$$\dot{Q}_w = \dot{M}_f h_v \qquad (7.2)$$

where $\dot{Q}_w$ = heat flux rate transferred from the flame to the solid surface (J/m²·s)
$\dot{M}_f$ = mass flux rate of vaporized fuel, perpendicular to surface (kg/m²·s)
h_v = heat content of a unit mass of gasified fuel at the surface minus the heat content of the solid at ambient temperature (J/kg)

The h_v parameter includes three terms:

1) the heat to warm the solid to the surface temperature
2) thermal changes prior to gasification (such as depolymerization)
3) the heat of vaporization

In complex polymer systems (cross-linked), we may combine steps 2 and 3 listed above because cracking of the polymer may occur and directly generate gas fragments, such as H_2 or CH_2, in a nonequilibrium process. Therefore, this variable (h_v) is the amount of heat required to decompose the solid fuel into vapors and fragments that can move into the flame zone. Because of this complexity, the evaluation of h_v is best made by pyrolysis experiments in the laboratory. In these experiments we can directly measure the heat of gasification, or analyze the composition of gas products, thus allowing us to calculate h_v from heats of formation.

The heat transfer through the boundary layer occurs by conduction, which is proportional to the thermal gradient:

$$\dot{Q}_w = -k \frac{\delta T}{\delta y} = -\frac{k}{c_p} \frac{\delta h}{\delta y} \qquad (7.3)$$

where k = the molecular/turbulent boundary layer gas conductivity (J/m·s·K)
T = temperature of the gas at any point y (K)
y = the distance from the solid fuel wall (m)
h = specific enthalpy of the gas at any point y (J/kg)
c_p = specific heat of the gas at constant pressure (J/kg·K)

The second part of the equation, in terms of the enthalpy (h), is an alternate form useful when chemical recombination occurs in the boundary layer.[*] Equation (7.3)

---

[*] An additional heat term can be considered, resulting from the reaction of active oxidizer species that have escaped through the combustion zone on the fuel surface. Altman and Wise [1956] have treated the special case of chemical recombination in the boundary layer and surface reaction.

can also be conveniently expressed in terms of the Stanton number C_H, as in

$$\dot{Q}_w = C_H \rho_b v_b \Delta h \tag{7.4}$$

where C_H = the Stanton number (dimensionless)
$\rho_b v_b$ = the axial mass flux in the flame zone (kg/m²·s)
Δh = the total specific enthalpy difference between the flame and the wall (J/kg)
ρ_b = the gas density at the flame (kg/m³)
v_b = the gas velocity at the flame (m/s)

To evaluate the Stanton number, we invoke the Reynolds analogy[*] and determine C_H in terms of the friction coefficient (C_f), for which experimental data are available for turbulent flow over flat plates. Because friction is a shear stress (τ) between adjacent layers, and if we assume the Lewis[†] and Prandtl[‡] numbers are unity, the Reynolds analogy across the boundary layer can take the form:

$$\frac{\dot{Q}}{\partial h/\partial y} = \frac{\tau}{\partial v/\partial y} \tag{7.5}$$

where $\dot{Q}$ = heat flux rate at any point y (J/m²·s)
τ = the shear stress (force per unit area) between adjacent layers of gas at y (Pa)
v = the axial velocity of the gas at any point y (m/s)

To integrate Eq. (7.5), we note from thermal conductivity arguments that[**]

$$\frac{\dot{Q}}{\dot{Q}_w} = \frac{\tau}{\tau_w} \tag{7.6}$$

where $\dot{Q}_w$ = heat flux at the solid fuel wall (J/m² ·s)
τ_w = shear stress at the solid fuel wall (Pa)

---

[*] The *Reynolds analogy* means that the temperature/enthalpy profile (dh/dy) through a boundary layer is proportional to the velocity profile (dv/dy). Experiments have validated this analogy.
[†] Lewis number being unity implies that the diffusivity of heat and molecular species are equal.
[‡] Most of the empirical data relating C_H and C_f have been normalized for Prandtl number $Pr = 1$. This number relates several gas constants [conductivity (k), viscosity (μ), and specific heat (c_p): $Pr = c_p \mu / k$]. For particular gases, $Pr \neq 1$ and the following equations require an additional factor. However, Eq. (7.13) introduces an empirical constant (C) which takes the simplification, $Pr = 1$, into account for all gases.
[**] Heat flux is constant through the boundary layer in the steady state.

which permits integration of Eq. (7.5) between the burning zone and the surface to yield

$$\frac{\dot{Q}_w}{\Delta h} = \frac{\tau_w}{v_b} \qquad (7.7)$$

Dividing both sides of this equation by the mass flux rate $(\rho_b v_b)$ and comparing with Eq. (7.4), we obtain

$$C_H = \frac{\dot{Q}_w}{\rho_b v_b \Delta h} = \frac{\tau_w}{\rho_b v_b^2} \qquad (7.8)$$

From the definition of the friction coefficient (C_f):

$$\tau_w = \frac{1}{2}\rho_e v_e^2 C_f \qquad (7.9)$$

where ρ_e = free-stream gas density (kg/m²)
 v_e = free-stream gas velocity (m/s)
 C_f = flow friction coefficient

and from Eq. (7.8) we get

$$C_H = \frac{C_f}{2}\frac{\rho_e v_e^2}{\rho_b v_b^2} \qquad (7.10)$$

Compare this result to the conventional noncombusting (and, therefore, nonblowing[*]) boundary layer, for which

$$C_{H_0} = \frac{C_{f_0}}{2} \qquad (7.11)$$

The subscripts 0 on the Stanton number and friction coefficient imply that no fuel mass is being added to the flow from the solid fuel wall and therefore no blowing is taking place.

We need one more assumption before identifying the friction coefficient in Eq. (7.10) with the well-known empirical law on flat plates. In the current model, the boundary layer is extended as a result of blowing due to the fuel vaporizing at the surface. Let us denote C_{H_0} and C_{f_0} as the appropriate values for nonblowing. Then,

---

[*] As the solid fuel vaporizes, the gaseous fuel ejects from the solid, perpendicular to the surface. Physically, this is similar to *blowing* gas through holes in a solid surface, hence, the name. Blowing substantially reduces the friction shear stress. In this equation, no combustion is taking place, so heat transfers between the surface and the free stream. Thus, the density $(\rho_e = \rho_b)$ and velocity $(v_e = v_b)$ are equal.

because we expect blowing to have similar effects on the heat and momentum transfers, we may write

$$\frac{C_H}{C_{H_0}} = \frac{C_f}{C_{f_0}} \tag{7.12}$$

We may now evaluate the friction coefficient from the well-known turbulent empirical law with Prandtl number equal to one ($Pr = 1$):*

$$\frac{C_{f_0}}{2} = CR_e^{-0.2} \tag{7.13}$$

where
- C = 0.03 (an empirically derived constant)
- R_e = the Reynolds number [see Eq. (7.15)]

By combining Eqs. (7.2), (7.4), (7.12) and (7.13), we finally arrive at the basic expression for burning rate in the hybrid rocket:

$$\dot{M}_f = \dot{r}\rho_f = 0.03(\rho_e v_e)R_e^{-0.2}\left(\frac{C_H}{C_{H_0}}\right)\left(\frac{v_e}{v_b}\right)\left(\frac{\Delta h}{h_v}\right) \tag{7.14}$$

Assuming the boundary layer does not severally restrict the port flow, we can identify $\rho_e v_e$ with G, the total mass flux rate through the port, and note that

$$R_e = \frac{Gx}{\mu} \tag{7.15}$$

$$\dot{M}_f = \dot{r}\rho_f = \left(\frac{0.03\mu^{0.2}}{x^{0.2}}\right)G^{0.8}\left(\frac{C_H}{C_{H_0}}\right)\left(\frac{v_e}{v_b}\right)\left(\frac{\Delta h}{h_v}\right) \tag{7.16}$$

where
- $\dot{M}_f$ = mass flux rate of the fuel from the solid surface (kg/m²·s)
- $\dot{r}$ = regression rate (m/s) of the solid fuel
- ρ_f = density (kg/m³) of the solid fuel
- μ = viscosity (kg/m·s) of the burned propellant gas
- x = distance along port (m)
- G = total propellant mass flux down port (kg/m²·s)
- Δh = enthalpy difference between flame zone and wall (J/kg)
- h_v = total heat of vaporization of the solid fuel (J/kg)
- C_H = Stanton number with blowing

---

*No accuracy is lost here by letting $Pr = 1$.

C_{H_0} = Stanton number without blowing
v_e = velocity of the gas down the port at boundary layer edge (m/s)
v_b = velocity of the propellant gas down the port at the flame (m/s)

The question at this point is—how do we evaluate the constants while taking blowing into account?

Blowing is an effect that has been studied for various problems other than our fuel-addition problem. For example, reentry heat shields can ablate, thereby changing thermal conditions. Blowing through or along aerodynamic surfaces can affect lift and drag on vehicles. Generalized studies of the blowing problem have provided us an empirical approach to evaluate the more difficult constants in Eq. (7.16). Using the experimental data of Mickley and Davis [1957], Marxman [1964] evaluated blowing and showed its effect on the skin-friction coefficient (C_f). This is given in terms of the blowing parameter (mass-transfer number, B), defined as

$$B = \frac{\dot{M}_f}{\rho_e v_e \frac{C_f}{2}} \tag{7.17}$$

By combining Eqs. (7.2), (7.4), and (7.10), we can also express B as

$$B = \left(\frac{v_e}{v_b}\right)\left(\frac{\Delta h}{h_v}\right) \tag{7.18}$$

which is a thermochemical parameter determined basically from propellant properties.* Based on our combustion model, Marxman [1964] derived an empirical expression for C_f/C_{f_0}:

$$\frac{C_f}{C_{f_0}} = \left[\frac{\ln(1+B)}{B}\right]^{0.8}\left[\frac{1 + 1.3B + 0.364B^2}{\left(1 + \frac{B}{2}\right)^2(1+B)}\right]^{0.2} \tag{7.19}$$

which he represented by the simple formula:

---

* The velocity ratio (v_e / v_b) is a function of where the flame is in the boundary layer, which in turn is a function of the combustion properties of the propellants. The enthalpy difference between the flame and the wall (Δh), is strictly based on thermochemistry (see Chap. 4). We designate the specific vaporization enthalpy of the solid by h_v (a propellant property), and p_c is the chamber pressure.

$$\frac{C_f}{C_{f_0}} = 1.2B^{-0.77} \tag{7.20}$$

However, in the range $5 < B < 20$, which is of interest in most hybrid systems, the formula,

$$\frac{C_f}{C_{f_0}} = B^{-0.68} \tag{7.21}$$

fits Marxman's expression much better. Figure 7.6 shows a fit of three simplified equations that have been used to represent C_f/C_{f_0} as given by Marxman's formula [Eq. (7.19)]. Netzer [1972] discussed two of the formulas and compared them in the B range of interest for hybrids. The superiority of the simple representation in Eq. (7.21) is obvious. This apparently small change in the exponent is significant because it indicates a greater sensitivity of regression rate to h_v, (about 40%) than suspected from Eq. (7.20). Because increasing the low regression rates in hybrids has been a challenge to researchers, the greater significance of h_v is an important idea.

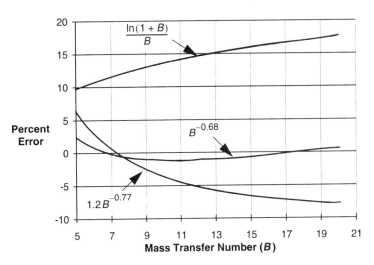

Fig. 7.6. **Comparing Simplified Expressions and a Precise Expression for the Friction Coefficient Factor.** The graph shows the percent difference between the "exact" empirical expression and several simpler approximations.

We get the final equation for regression rate by incorporating Eqs. (7.18) and (7.21) into (7.16). (Recall that $C_H/C_{H_0} = C_f/C_{f_0}$):

$$\dot{M}_f = \dot{r}\rho = 0.03\left(\frac{\mu}{x}\right)^{0.2} G^{0.8} B^{0.32} \qquad (7.22)$$

In contrast to the empirical expression for a burning solid (as presented in Sec. 6.4.5):*

$$r = a p_c^n \qquad (7.23)$$

Equation (7.22) allows a fairly good estimate of the regression rate from theory. The general form of Eq. (7.22) can be written as

$$\dot{r} = a G^n x^m \qquad (7.24)$$

where a = regression-rate coefficient
 G = total propellant (includes oxidizer & fuel) mass flux rate (kg/m²·s)
 x = distance down the port (m)
 n = regression-rate exponent
 m = regression-rate exponent

The coefficient a contains the properties of the propellants and the combustion gas. In particular, it contains the thermodynamic term $(\Delta h/h_v)^{0.32}$. It is important to note that the propellant and gas properties—μ, Δh, and h_v—appear as fractional powers and therefore only slightly influence the regression rate. Indeed, most hybrid fuels tested to date show that the burning rates fall within a rather narrow range regardless of fuel composition. The B exponent of 0.32 shown here, however, is about 40% larger than the Marxman value of 0.23. Although this change appears small, it is significant when we try to increase the burn rate by adding a more volatile component. In such an event, the percent change in $\dot{r}$ improves by 40% over that calculated using the 0.23 exponent. The exponents n and m depend on the fluid dynamics and are empirically determined. These values usually deviate somewhat from the theoretical (0.8 and –0.2 respectively) because the model represents an idealized combustion system. Section 7.2.5 presents corrections to the regression rate due to radiation and combustion kinetics.

Effect of Axial Length. The final burning rate, Eq. (7.24), is given at a specific axial position x in the fuel port. From a practical point of view, we would like to have a burn rate that is averaged along the grain because engineering tests typically measure loss of motor mass over time. Using an average burn rate for space is valid, however, only if the instantaneous rate down the port is approximately constant. Examination of Eq. (7.24) shows the terms that depend on the axial posi-

---

* Notice that the nomenclature between the two technologies is slightly different. Solids (Chap. 6) use "r" and hybrids use "$\dot{r}$" to represent regression rate. This is consistent with actual practice.

tion are x^m and, to a lesser extent, oxidizer mass flux (G_o)—through its dependence on the cross-sectional area. We can write

$$G(x) = G_o + G_f(x) \tag{7.25}$$

where $G(x)$ = total propellant mass flux in port at x (kg/m²·s)
G_o = oxidizer mass flux in port (kg/m·s)
$G_f(x)$ = the fuel flux added to port in front of point x (kg/m²·s)

The regression rate explicitly showing the x dependence is

$$\dot{r}(x) = ax^m [G_o + G_f(x)]^n \tag{7.26}$$

where $G_f(x)$ is determined by integrating the fuel-mass addition along the port:

$$G_f(x) = 4\rho \int_0^x \frac{\dot{r}(x)}{D_H} dx \tag{7.27}$$

where D_H = hydraulic diameter (m) = $4A/P$
A = port cross section area (m²)
P = port perimeter (m)

Because of the implicit nature of Eqs. (7.26) and (7.27), we can best solve them with stepwise integration by computer (we discuss this algorithm in Sec. 7.5). Table 7.1 is a detailed computer solution. It shows the variation of several parameters with respect to both axial length and time in a cylindrical grain. It applies to an experimental hydrocarbon system with exponents slightly different from the theoretical one. Note the relatively small variations of the burn rate and port diameter as a function of axial distance, except near the origin where the Blasius effect[*] occurs. This leading-edge effect is not as pronounced in practice, probably because chemical kinetics delay establishing the combustion zone in the boundary layer. The downstream displacement of the combustion zone would tend to decrease the heat transfer at the leading edge.

Figure 7.7 shows these data graphically. The significant conclusion from these calculations is that hybrid combustion down the fuel port provides fairly constant burn rates. This relative uniformity results from two compensating factors:

1. The boundary layer growth represented by x^m, which causes a decrease in heat flux, is approximately balanced by the increase in total mass flux from adding fuel.

---

[*] The *Blasius effect* refers to the initial buildup of the boundary layer. At the start, the boundary layer is thin, so heat transfer, and therefore fuel-vaporization rate, is high.

2. A self-regulating feature, by which a spurious, higher, local cross section causes a reduced local mass flux, thereby tending to level the contour.

Table 7.1. **Burning-Rate Contours in a Circular Port.** ($n = 0.75$, $m = -0.15$, oxidizer mass flow = 7.95 kg/s, $a = 2.066 \times 10^{-5}$, fuel density = 1000 kg/m^3)

x (m)	t = 0.1 sec			t = 20 sec			t = 60 sec		
	Port Dia. (m)	$\dot{r}$ (cm/s)	O/F	Port Dia. (m)	$\dot{r}$ (cm/s)	O/F	Port Dia. (m)	$\dot{r}$ (cm/s)	O/F
0.381	0.152	0.263	17.19	0.227	0.143	21.34	0.315	0.087	24.11
0.762	0.152	0.231	9.102	0.220	0.131	11.40	0.301	0.081	13.20
1.143	0.152	0.221	6.316	0.217	0.127	7.983	0.296	0.078	9.640
1.524	0.152	0.217	4.835	0.216	0.125	6.139	0.293	0.077	7.249
1.905	0.152	0.215	3.917	0.215	0.123	4.974	0.292	0.076	6.006
2.286	0.152	0.215	3.310	0.215	0.123	4.177	0.291	0.076	4.972
2.667	0.152	0.215	2.852	0.215	0.123	3.594	0.291	0.076	4.342
3.048	0.152	0.216	2.506	0.215	0.123	3.115	0.291	0.075	3.757
3.429	0.152	0.218	2.234	0.216	0.123	2.771	0.291	0.075	3.375
3.810	0.152	0.219	2.016	0.216	0.123	2.495	0.292	0.075	3.008
4.191	0.152	0.221	1.831	0.216	0.124	2.263	0.292	0.075	2.747
4.572	0.152	0.224	1.677	0.217	0.124	2.073	0.293	0.076	2.492

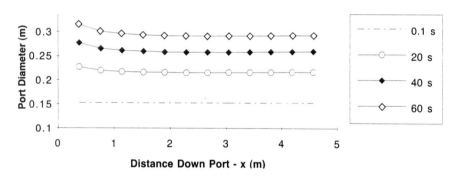

Fig. 7.7. **Port Contour for the Data in Table 7.1.** Regression is highest at the head end ($x = 0$) and decreases down the port length.

Because of these considerations, some investigators have used the classic burning-rate expression [Eq. (7.24)], in which mass flux rate is an average value* at any instant, and x refers to the grain length:

---

* To calculate average mass flux, divide the total mass flow rate by the average port cross-sectional area.

$$\dot{r}_{avg} = a G_{avg}^n L_p^m \tag{7.28}$$

where $\dot{r}_{avg}$ = average fuel regression rate along the port (m/s)
G_{avg} = average propellant flux rate along port (kg/m²·s)
L_p = port length (m)

7.2.3 Alternate Expressions for Burning Rate

A practical disadvantage in using Eq. (7.24) is that the total mass flux (G), at the end of the grain is not an input parameter. Instead, we implicitly determine it after knowing the fuel flux (G_f), which in turn depends on G. This approach can be awkward, especially when studying many experimental design configurations. It would be more convenient to have an equivalent expression based on oxidizer mass flux (G_o) which is an explicit input parameter. Because of the small variation of port diameter with x, we can derive the following integrated burning rate. From Eq. (7.27),

$$G_f(x) = \frac{4\rho}{D_{Havg}} \int_0^x \dot{r}(x)\, dx \tag{7.29}$$

where D_{Havg} = average port hydraulic diameter (m)
ρ = density of the solid fuel (kg/m³)
$\dot{r}(x)$ = fuel regression rate at x (m/s)
x = position measured from head end of port (m)
$G_f(x)$ = fuel mass flux down the port at x (kg/m²·s)

If we define an auxiliary variable to help in integrating this equation,

$$q = \int_0^x \dot{r}(x)\, dx \tag{7.30}$$

Then, combining this expression with Eqs. (7.26) and (7.29) gives

$$\dot{r} = \frac{dq}{dx} = a G_o^n x^m \left(1 + \frac{4\rho q}{D_{Havg} G_o}\right)^n \tag{7.31}$$

Because the variables are now separated in this first-order differential equation, integration leads to

$$q = \frac{D_{Havg} G_o}{4\rho} \left[\left(1 + \frac{4(1-n)a\rho x^{1+m}}{(1+m)D_{Havg} G_o^{1-n}}\right)^{\frac{1}{1-n}} - 1 \right] \tag{7.32}$$

The average value of regression rate $\dot{r}_{avg}$ along the port (from $0 \to x$) is

$$\dot{r}_{avg} = \frac{1}{x}\int_0^x \dot{r}\,dx = \frac{q}{x} \tag{7.33}$$

As a further simplification, the second term in parentheses in Eq. (7.32) is much less than 1, permitting a Taylor series expansion. Combining them, we get

$$\dot{r}_{avg} = \left(\frac{a}{1+m}\right) G_o^n x^m \left(1 + \frac{2n(a/1+m)\rho x^{1+m}}{D_{Havg} G_o^{1-n}}\right) \tag{7.34}$$

Because a is an empirical constant obtained by using Eq. (7.24), we can define a new coefficient $a' = a/(1+m)$, resulting in

$$\dot{r}_{avg} = a' G_o^n x^m \left(1 + \frac{2n a' \rho_f x^{1+m}}{D_{Havg} G_o^{1-n}}\right) \tag{7.35}$$

Equations (7.24) and (7.35) are essentially equivalent expressions for the burning rate. Although $\dot{r} = aG^n x^m$ is derived as a point value, the theory (Fig. 7.4) and experiments show that an average burn rate is a reasonable practical assumption. Because of this view, some investigators have commonly used Eq. (7.24) as an integrated average. In a sense, then, this is no more presumptuous than using Eq. (7.35), which involves the same assumption (average hydraulic diameter at any instant) in its derivation. Which is the preferred equation? The decision is based on a balance between ease of use and accuracy in correlating and predicting data.

Figure 7.8 shows the agreement between these two expressions for the theoretical set of parameters. Equation (7.24) is shown as exact and Eq. (7.35) is shown as average. The average deviation between the two equations employing the same burning-rate constants is less than 5% over a wide range of mass fluxes. The practical advantage of Eq. (7.35) is that it uses G_o, which is explicit, instead of G, which typically requires an iterative solution.

7.2.4 Variations of Ballistic Parameters during a Burn

Because of the dependence of the internal ballistics on mass flux, parameters such as pressure, thrust, and O/F vary during the motor burn at a constant oxidizer flow rate ($\dot{m}_{ox}$). Table 7.1 shows the detailed variation for several of the parameters at any position in the grain at a given time. It is often useful, and certainly more convenient, to have simple analytical expressions that can predict the behavior of the various ballistic parameters with reasonable accuracy. For motors of a given length, investigators have found suitable for this purpose the simplified burning expression,

7.2 Hybrid-Motor Ballistics

$$\dot{r} = a_o G_o^n \tag{7.36}$$

where $\dot{r}$ = regression rate (m/s)
 a_o = regression-rate coefficient incorporating grain length term (L_p^m)
 n = regression-rate exponent
 G_o = oxidizer mass flux rate (kg/m²·s)

This equation condenses Eq. (7.34), using a fixed grain length and an average constant value for the term in parentheses, which varies much slower than G_o. To get the best accuracy from the simplified Eq. (7.36), we should determine the constants a_o and n directly from experimental data. Typical experimental values of n range from 0.5 to 0.8.

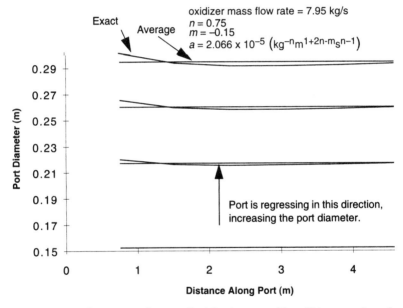

Fig. 7.8. **Exact Port Contour vs. Average Port Contour over Time.** This comparison shows that the regression rate is higher at the head end of the port and decreases.

The following formulas are given for both circular and noncircular ports. The table at the end of this section summarizes the general expressions. For noncircular cross sections, we use the hydraulic diameter,

$$D_H = 4\frac{A}{P} \tag{7.37}$$

where D_H = hydraulic diameter (m)
       A = port cross-sectional area (m^2)
       P = port perimeter (m)

Note that A and P must be calculated from the selected configuration.

Variation of O/F with Time and Diameter—During burning, the oxidizer-to-fuel ratio (O/F, defined as G_o / G_f) increases because the port diameter increases over time. This shift to higher O/F values is common with all classical hybrids at a fixed oxidizer flow rate. We must understand this behavior in motor design to get the best average specific impulse during the motor burn. The following formula illustrates this effect. Assuming a fixed oxidizer mass flow rate during the run, fuel mass flux is given by

$$G_f = \frac{\dot{r} \rho PL}{A} = \frac{a_o G_o^n \rho PL}{A} \qquad (7.38)$$

where G_f = fuel mass flux rate (kg/m^2·s)
       ρ = density of the solid fuel (kg/m^3)
       L = length of the port where G_f is measured (m)

The expression for O/F is

$$O/F = G_o/G_f = \frac{\dot{m}_{ox}^{1-n} A^n}{\rho a_o LP} \qquad \text{(noncircular)} \qquad (7.39)$$

$$O/F = G_o/G_f = \frac{\dot{m}_{ox}^{1-n} D^{2n-1}}{KL} \qquad \text{(circular)} \qquad (7.40)$$

where $K = 4^n \rho \pi^{1-n} a_o$ and $\dot{m}_{ox} = G_o A$ (mass-flow rate of oxidizer).

For a typical grain design at a volumetric loading efficiency of 60% (volume of grain divided by volume of chamber holding the grain), $D_{final} / D_{initial} \approx 1.6$. This value translates, at $n = 0.75$, to a ratio $(O/F)_{final} / (O/F)_{initial}$ of about 1.26. Notice that at $n = 0.5$, O/F does not vary with diameter and burning time. Figure 7.9 illustrates this behavior for various n exponents.

O/F Shift with Throttling—Eq. (7.40) also shows the effect of throttling the oxidizer flow rate on O/F. This effect is important in booster applications for which throttling is required during passage through the atmosphere to satisfy maximum aerodynamic-pressure limits (max Q) on the vehicle. Figure 7.10 illustrates this variation based on Eq. (7.40) at a constant port diameter. As you would expect, at $n = 1$, there is no O/F shift. In some applications, such as space-vehicle boosters, the trajectory program calls for throttling down during the burn, as mentioned above. The combined effect of the increase in diameter with time and the decrease in oxidizer mass flow partially neutralizes the O/F shift, as the term in the numerator of Eq. (7.40) shows.

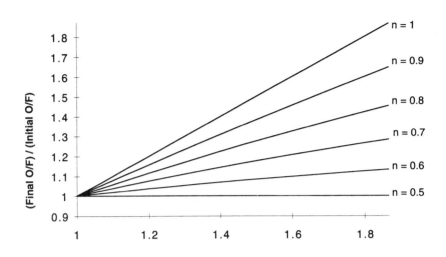

Fig. 7.9. *O/F* **Shift vs. Change in Port Diameter for Several Regression-Rate Exponents (n).**
As the port area increases during a burn, the *O/F* increases.

Stoichiometric Length. The *stoichiometric length** is that position in the grain where the integrated fuel burned satisfies the required O/F. In chemical terms the stoichiometric point [$(O/F)_{st}$] occurs at complete oxidation, when all carbons and hydrogens react. In a moderate-energy system, optimum specific impulse occurs near complete oxidation; in a higher-energy system, the optimum specific impulse occurs on the fuel-rich side. This is true for hybrid propellants employing liquid oxygen as the oxidizer and even more true when the fuel contains aluminum. The relation between the length-to-diameter ratio (L/D) and O/F is obtained from Eq. (7.38) to yield

$$(L/D)_{st} = \frac{G_0^{1-n}}{4a_0 \rho (O/F)_{st}} \quad \text{(circular and noncircular)} \quad (7.41)$$

where $(L/D)_{st}$ = port length-to-diameter ratio at the stoichiometric point
  O/F_{st} = oxidizer-to-fuel ratio at the stoichiometric point
  G_0 = oxidizer mass flux rate (kg/m²·s)
  n = regression-rate exponent
  a_0 = regression-rate coefficient incorporating term for grain length

---

* In general, *stoichiometry* refers to chemical reactions that occur at the "correct" mixture ratio. In common usage, this usually means the mixture for which all of the carbon, hydrogen, etc. is consumed in the reaction. However, in general, the stoichiometric ratio can refer to ANY desired mixture ratio.

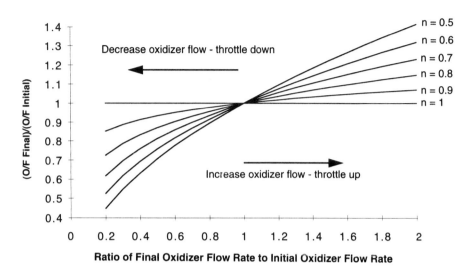

Fig. 7.10. *O/F* **Shift with Throttling of Oxidizer Flow.** As the oxidizer flow increases, the *O/F* also increases.

This equation is important in the preliminary design of a hybrid motor because it determines the envelope for a given propellant weight. The equation tells us if (and how many) multiple ports are required to satisfy the envelope constraint. A typical value of L/D is about 25 with initial values of G_o = 352 kg/m²·s, $\dot{r}$ = 0.178 cm/s, ρ = 1000 kg/m³ and O/F = 2. The L/D range for most hybrids lies between 20 and 30.

During the burn, L_{st} [the length at which the desired stoichiometry $(O/F)_{st}$ occurs] moves downstream. Its position at any time t is

$$L_{st}(t) = \frac{G_o^{1-n} D_H}{4 a_o \rho (O/F)_{st}} = \frac{\dot{m}_{ox}^{1-n} A^n(t)}{a_o \rho P(t) (O/F)_{st}} \quad \text{(noncircular)} \tag{7.42}$$

$$L_{st}(t) = \frac{G_o^{1-n} D(t)}{4 a_o \rho (O/F)_{st}} = \frac{\dot{m}_{ox}^{1-n} D(t)^{2n-1}}{K (O/F)_{st}} \quad \text{(circular)} \tag{7.43}$$

In designing the hybrid motor, therefore, we would define the O/F ratio to correspond to the optimum specific impulse and select that L/D to occur roughly in the middle of the burn. For larger rockets, a typical L/D is 25 to 30, whereas smaller rockets tend to have smaller L/Ds. We discuss this design consideration in Sec. 7.4.

Thrust Variation—Assuming a constant specific impulse, the thrust varies only with the total mass flow. Because the fuel flow varies with time, the thrust varies accordingly. Noting that $G_o/G_f = \dot{m}_{ox}/\dot{m}_{fuel}$, we can use Eq. (7.40) to give the following result:

$$\dot{m}_{prop} = \dot{m}_{ox}\left(1 + \frac{a_o\rho LP}{\dot{m}_{ox}^{1-n}A^n}\right) = \dot{m}_{ox}\left(1 + \frac{1}{O/F}\right) \quad \text{(noncircular)} \quad (7.44)$$

$$\dot{m}_{prop} = \dot{m}_{ox}\left(1 + \frac{KL}{\dot{m}_{ox}^{1-n}D^{2n-1}}\right) = \dot{m}_{ox}\left(1 + \frac{1}{O/F}\right) \text{ (circular)} \quad (7.45)$$

This equation implies that mass-flow rate ($\dot{m}_{prop}$), and therefore thrust magnitude, decrease as port diameter increases for a constant oxidizer flow rate. The second term in the parenthesis is typically less than 1 (about 0.5 for many hydrocarbon systems), so the thrust decrement with burning time at a constant oxidizer flow rate is diluted. For the example in Table 7.1, the mass-flow decrement is only 13% for a 93% increase in D.

Summary—Table 7.2 compares various ballistic parameters using the simplified expression instead of the more "exact" expression used for Table 7.1. The a_o and n exponents in $\dot{r} = a_o G_o^n$ are assigned to give a best fit with the burning-rate expression $\dot{r} = a\, G^n L^m$. The behavior agrees well with variations in O/F, stoichiometric length, and throttling. But note that the simplified expression is less accurate in scaling when dimensions change.

Table 7.2. Comparison of Ballistic Parameters Based on Different Models.

(Fuel density = 1000 kg/m^3, oxidizer mass flow rate = 7.95 kg/s; L_p = 4.57 m
exact equation: $\dot{r} = 2.066 \times 10^{-5} G^{0.75} L^{-0.15}$ in m/s; $G = G_o\left(1 + \frac{1}{O/F}\right)$
approximate equation: $\dot{r} = 1.451 \times 10^{-5} G_o^{0.83}$ (in m/s).

Port Dia. (m)	G_o kg/m^2·s	Regression Rate (cm/s)		O/F		L at O/F = 2.5 (m)		Mass Flow Rate (kg/s)	
		Exact	Approx	Exact	Approx	Exact	Approx	Exact	Approx
0.1524	435.19	0.224	0.225	1.68	1.62	3.048	2.972	12.745	12.836
0.2172	214.43	0.124	0.125	2.08	2.05	3.810	3.733	11.793	11.838
0.2598	149.75	0.092	0.092	2.32	2.30	4.267	4.216	11.385	11.385
0.2933	117.41	0.076	0.075	2.49	2.50	4.546	4.572	11.112	11.112

Table 7.3 summarizes the equations for circular and noncircular ports, where

$\dot{m}_{ox}$ = oxidizer mass flow rate (kg/s)
A = port cross-sectional area (m^2)

ρ = density of the solid fuel (kg/m³)
a_0 = regression-rate coefficient using just G_o, ($a_0 = a\,L^m$)
n = regression-rate exponent
L = length down the port (m)
L_p = port length = 4.57 m, where L = axial length at O/F = 2.5
L_{st} = port length at the stoichiometric point (m)
P = port perimeter (m)
D = diameter of the circular port (m)
K = parameter ($4^n \pi^{1-n} a_0 \rho$)
G_o = oxidizer mass flux rate (kg/m²·s)
D_H = hydraulic diameter of the port = $4A/P$ (m)
O/F_{st} = oxidizer-to-fuel ratio at the stoichiometric point
L/D = ratio of the port's length to its diameter
L/D_H = ratio of the port's length to its hydraulic diameter

Table 7.3. **Formulas for Ballistic-Parameters.** This table summarizes the important design formulas we have developed.

Ballistic Parameters	Geometry of Ballistic Parameters			
	Noncircular		Circular	
O/F	$\dfrac{\dot{m}_{ox}^{1-n} A^n}{\rho a_0 LP}$	Eq. (7.39)	$\dfrac{\dot{m}_{ox}^{1-n} D^{2n-1}}{KL}$	Eq. (7.40)
L_{st}	$\dfrac{G_o^{1-n} D_H}{4 a_0 \rho (O/F)_{st}}$ $\dfrac{\dot{m}_{ox}^{1-n} A^n}{a_0 \rho (O/F)_{st} P}$	Eq. (7.42)	$\dfrac{G_o^{1-n} D}{4 a_0 \rho (O/F)_{st}}$ $\dfrac{\dot{m}_{ox}^{1-n} D^{2n-1}}{K(O/F)_{st}}$	Eq. (7.43)
$\dot{m}$	$\dot{m}_{ox} + \dfrac{a_0 \rho \dot{m}_{ox}^n LP}{A^n} = \dot{m}_{ox}\left(1 + \dfrac{1}{O/F}\right)$	Eq. (7.44)	$\dot{m}_{ox} + \dfrac{\dot{m}_{ox}^n KL}{D^{2n-1}} = \dot{m}_{ox}\left(1 + \dfrac{1}{O/F}\right)$	Eq. (7.45)
L/D or L/D_H	$\dfrac{G_o^{1-n}}{4 a_0 \rho (O/F)_{st}}$	Eq. (7.41)	$\dfrac{G_o^{1-n}}{4 a_0 \rho (O/F)_{st}}$	Eq. (7.41)

Temperature Sensitivity

The temperature sensitivity refers to the increase in burning rate and, therefore, chamber pressure with ambient temperature. This quantity is especially important in solid rockets because of its impact on Maximum Expected Operating Pressure (MEOP). In typical hybrid fuel compositions, the latent heat of vaporization (or decomposition) of the solid is large compared to the variation in heat content due to the expected extremes of ambient temperature. So the initial temperature of the solid material only slightly affects the rate of regression. Characteristic data showing this effect in a plexiglass/oxygen (PMM-O_2) system are in Fig. 7.11, as taken from Ordahl and Altman [1962]. The observed temperature dependence qualitatively and quantitatively agrees with theory. From Eqs. (7.18) and (7.22), we note that the term most sensitive to fuel temperature is h_v, so

$$\dot{r} \propto [h_v - c(T - T_0)]^{-0.32} \qquad (7.46)$$

where h_v = heat of gasification at the reference temperature – T_0 (J/kg)
  c = specific heat of the solid (J/kg·K)
  T = temperature of the solid fuel (K)
  T_0 = reference temperature (K)
  $\dot{r}$ = fuel regression rate (m/s)

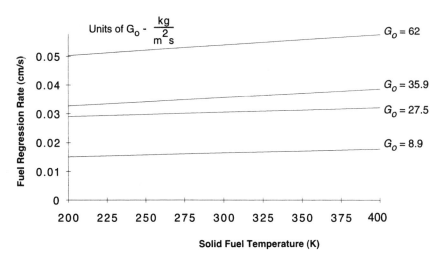

Fig. 7.11. Effect of Fuel Temperature on the Regression Rate of Plexiglass and Oxygen. As the solid fuel temperature goes up, so does the fuel regression rate. The effect is fairly linear [Ordahl and Altman, 1962].

The temperature sensitivity of burn rate (σ_p) is defined as

$$\sigma_p = \frac{1}{\dot{r}}\frac{\partial \dot{r}}{\partial T} = \frac{0.32c}{h_v - c(T - T_0)} \tag{7.47}$$

Because the main influence of ambient temperature on the flow rate of combustion gases is through the burning-rate equation, the temperature sensitivity of pressure for hybrids (π_K) is simply

$$\pi_K = \frac{1}{p}\frac{\partial p}{\partial T} = \frac{\sigma_p}{1 + O/F} \tag{7.48}$$

For hybrids with pressure-sensitive burn rates (see Sec. 7.2.5), where the pressure exponent is n' (we include the "prime" here to distinguish from the exponent without a pressure effect), the equation becomes

$$\pi_K = \frac{\sigma_p}{(1-n')(1+O/F)} \tag{7.49}$$

In solids, the term $(1 + O/F)$ is omitted because the entire propellant is temperature sensitive. A typical value for π_K for solids is $\approx 0.2\%/$ K. By contrast, if we take for a typical hybrid: $c = 1676$ J/K·kg, $h_{vo} = 1.257 \times 10^6$ J/kg at $T_0 = 273$ K, and $O/F = 2$, the corresponding value of π_K at 298 K = 0.015% / K. This value is an order of magnitude less than for solids, so in designing the thrust chamber, we need not assign a weight margin (penalty) for variations in ambient operating temperatures.

7.2.5 Effect of Pressure on Radiation and Kinetics

Our burning-rate expressions are based on assuming that convection is the only source of heat transfer and that the combustion kinetics are fast relative to diffusional processes. As a consequence, the burning-rate equation is independent of pressure. Although most hybrid data agree well with this approach, the regression rate depends on pressure for certain fuels and regimes of operation. These regimes typically are at both low and high oxidizer flow rates.

Radiation

In the presence of radiation, we must change the energy equation to include radiative heat transfer. This equation now reads:

$$\dot{M}_f h_v = \dot{r}\rho h_v = \dot{Q}_c + \dot{Q}_r \tag{7.50}$$

where $\dot{M}_f$ = mass flux from the fuel surface (kg/m²·s)

h_v = heat of fuel vaporization (J/kg)
$\dot{r}$ = fuel regression rate (m/s)
ρ = fuel density (kg/m³)
$\dot{Q}_c$ = heat flux into the fuel due to conduction (J/m²·s)
$\dot{Q}_r$ = heat flux into the fuel due to radiation (J/m²·s)

From Eq. (7.22), $\dot{Q}_c$ is

$$\dot{Q}_c = 0.03\left(\frac{\mu}{x}\right)^{0.2} G^{0.8} B^{0.32} h_v \qquad (7.51)$$

where G = mass flux down the port (kg/m²·s)
B = blowing parameter
μ = viscosity of the combustion gases (kg/m·s)
x = distance down the port (m)

The radiation term is given by

$$\dot{Q}_r = \sigma \varepsilon_w \left(\varepsilon_g T_r^4 - T_w^4\right) \qquad (7.52)$$

where σ = the Stefan-Boltzmann constant (W/m²·K⁴)
ε_w = emissivity of the wall (fuel surface)
ε_g = emissivity of the gas
T_r = average effective radiation temperature of the combustion gases (K)
T_w = temperature at fuel surface (K)

We can now incorporate Eqs. (7.51) and (7.52) into (7.50) to yield

$$\dot{r}\rho h_v = (0.03)\left(\frac{\mu}{x}\right)^{0.2} G^{0.8} B^{0.32} + \sigma \varepsilon_w \varepsilon_g T_r^4 \qquad (7.53)$$

Note that the term T_w^4 is dropped because it is usually negligible relative to the T_r^4.

Before talking about these equations, we need to comment on some limitations. For example, these equations assume a uniform radiation environment, which is certainly not the case whenever a turbulent boundary layer contains a thin combustion zone relative to the diameter of the port. To take this feature into account, it is customary to assign a typical average temperature in the combustion region. Although the equation does not provide an accurate quantitative prediction of the radiation transfer, it shows a reasonable mathematical form of the correction, which can then be useful for assigning empirical constants from experimental data.

Note also that, because radiation increases the heat transfer and hence the blowing, convection contributes less to the heat transfer. This indicates that the two terms in Eq. (7.53) are coupled. Marxman, et al. [1969], show that the ratio of the blowing parameter with radiation (B_r) to the unaltered parameter (B) is determined from the equation:

$$\frac{B_r}{B} = 1 + \frac{\dot{Q}_r}{\dot{Q}_c}\left(\frac{B_r}{B}\right)^{0.68} \tag{7.54}$$

where we have replaced their exponent of 0.77 with 0.68, as justified previously. An expression which provides a good fit over the range $0 < \dot{Q}r/\dot{Q}c < 3$ is

$$\frac{B_r}{B} = e^{1.1\frac{\dot{Q}_r}{\dot{Q}_c}} \tag{7.55}$$

Incorporating this expression into Eqs. (7.50)–(7.54), we can now show that

$$\dot{r} = \dot{r}_c \left(e^{\frac{-0.75\dot{Q}_r}{\dot{Q}_c}} + \frac{\dot{Q}_r}{\dot{Q}_c} \right) \tag{7.56}$$

where

$$\dot{r}_c = \frac{\dot{Q}_c}{\rho h_v} \tag{7.57}$$

is the burn rate due to conduction only. Table 7.4 shows the magnification in $\dot{r}$ due to radiation:

Table 7.4. Comparison of Regression Rate with Radiation Included. Regression rate increases as the fraction of radiative heat added to the fuel increases.

$\dot{Q}_r / \dot{Q}_c$	0	0.2	0.5	1.0	2.0
$\dot{r} / \dot{r}_c$	1.00	1.06	1.19	1.47	2.22

The correction, therefore, is small when $\dot{Q}_r < 20\%$ of $\dot{Q}_c$. The table shows the blocking effect caused by the blowing parameter (B) because, even when $\dot{Q}_r = \dot{Q}_c$, the resulting heat transfer to the surface does not double but increases by only 47%.*

The dominant term for radiation in Eq. (7.53) lies in the gas emissivity (ε_g). For transparent gases, this term is small because of the low optical density. In a metallized system, however, the radiating particles considerably increase the emissivity. The equation for emissivity is

$$\varepsilon_g = 1 - e^{-\alpha N z} \qquad (7.58)$$

where α = empirically derived constant (m²/particle)
 N = number density of particles (particles/m³)
 z = optical path length, or distance from flame to wall (m)

The product αN basically relates to the radiating surface area per unit volume of radiating gas. Therefore, it is a function of the metal concentration in the fuel, the O/F, the density of the particle, its optical average size, and the pressure. In studying a given system over a range of oxidizer mass flows and pressures, we find the terms that vary most are the pressure, which determines αN, and O/F, which determines both αN and the radiating temperature. Wherever the experimental measurements are made over a restricted O/F range, the dominant parameter is the pressure, and we can represent Eq. (7.58) by

$$\varepsilon_g \cong 1 - e^{-bpz} \qquad (7.59)$$

where b = an empirical constant which depends on the percent metal and O/F
 p = the pressure (Pa)

Figure 7.12 shows the relationship of the burn rate with the oxidizer mass flux rate for various pressures using Eqs. (7.53), (7.56), and (7.59). At lower pressures, for which the radiation effect is negligible, regression rates behave according to the classical convective model. This is true even at low oxidizer mass flux. At moderate to high pressures, however, the burn rates tend to level off and are determined mainly by radiative transfer.

Kinetics

The diffusion-controlled model we have presented assumed that reaction rates are fast relative to diffusion rates and, therefore, the combustion zone is thin relative to the boundary layer. As the pressure is reduced, the reaction rates decrease and the combustion zone thickens. Further, as the mass flow increases, the diffusion rates increase and the combustion zone again thickens because the transport

---

* Marxman [1969] has shown that if the B exponent is 0.77 in Eq. (7.54) instead of 0.68, the increase in $\dot{r}$ is only 35%.

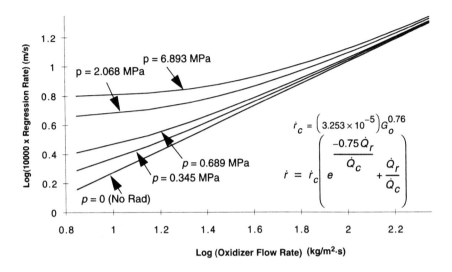

Fig. 7.12. Radiation Effect on Fuel Regression Rate. Increases in pressure tend to increase regression rate at the lower oxidizer flux.

of combustibles through the reaction zone is fast relative to its consumption. In this region of either high mass flux or low pressure, one can expect the burning rate to be kinetically limited and depend on pressure. Qualitatively, therefore, the important parameter is G^n/p, where either a high mass flux or a low pressure creates a kinetic limitation.

The physical parameter that determines when the burning rate starts depending on pressure is the ratio of the kinetic reaction time (t) to diffusion time (T). Wooldridge and Marxman [1969] show the ratio of these two times as

$$\theta_t = \frac{t}{T} = C \frac{l_1 (\mu/x)^{0.2} G^{0.8} B^{0.32}}{l_2 p^{\frac{n}{2}} T_f^{1+\frac{n}{2}} e^{\frac{-E}{RT_f}}} \tag{7.60}$$

where θ_t = time ratio
 t = kinetic reaction time (s)
 T = chemical diffusion time (s)
 C = empirical constant
 l_1 = characteristic flame length
 l_2 = characteristic mixing length
 μ = gas viscosity (kg/m·s)
 x = axial distance down the port (m)
 p = chamber pressure (Pa)

n = global order of the chemical reaction
T_f = flame temperature (K)
E = activation energy of the combustion reaction (J/mol)
R = gas constant (J/kmol)

For a given propellant combination, we can assume that l_1/l_2 is roughly constant, the T_f variation is small, and the order of the global reaction is typically 2. With these assumptions, we can condense Eq. (7.60) to

$$\theta = \frac{kG^{0.8}}{p} \qquad (7.61)$$

where k is an empirically derived constant. Note that the form of the equation is essentially correct because $G^{0.8}$ determines the diffusion time and pressure determines the reaction time. Thus, it satisfies the definition of θ as the ratio of these two times.

These authors have further shown that we can write the regression equation as

$$\dot{r} = \dot{r}_0 \left(\frac{2}{\theta_t}\right)^{0.5} \left[1 - \frac{1}{\theta_t}\left(1 - e^{-\theta_t}\right)\right]^{0.5} \qquad (7.62)$$

where $\dot{r}_0$ is the nonkinetic regression rate. Figure 7.13 shows a graph of the burning rate ratio given by Eq. (7.62) in terms of the parameter θ_t. The figure shows that as θ_t increases, caused either by an increase in mass flux or a decrease in pressure, the ratio decreases. This behavior satisfies our physical intuition of the process, as discussed earlier. Figure 7.14 shows a set of experimental data for a PMM/O_2 system taken at constant mass flux and varying pressure (private communication from Chemical Systems Division of United Technologies, circa 1965). The theoretical curve uses Eqs. (7.61) and (7.62) with one adjustable constant (k). The value of $\dot{r}_0$ is obtained at large values of the pressure (where diffusion is controlling). Note that the regression rate prediction at low pressures is very good over the entire range.

7.2.6 Correlating Experimental Data

With a given set of experimental data, we must determine the type of burning-rate equation which provides the best correlation. Two situations are relevant here. In the first case, we have a given set of data over a fixed range of variables and we want to obtain a best fit so we can interpolate accurately. We are also interested in verifying the form of the equation and, therefore, the model. This occurs when the ballistic constants are established for a small size, but development leads to a large motor, and we want to predict the best correlating equation for the large motor. Such extrapolations are more exacting and provide an acid test for the best analytical model.

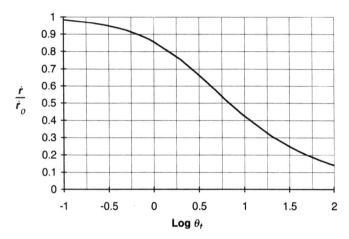

Fig. 7.13. Regression Rate Ratio versus Log Kinetic Time Ratio. As the time ratio increases, the regression ratio decreases.

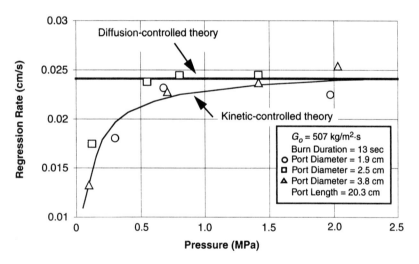

Fig. 7.14. Experimental Data on the Kinetic-Controlled Regression Rate for a Plexiglass and Oxygen System. At higher pressures, kinetics approaches the idealized solutions. At lower pressures, kinetics reduces the actual regression rate [United Technologies, Chemical Systems Division, circa 1965].

For convenience, we can establish the burning-rate law as equal to the theoretical rate times a correction term which, in general, is a function of the additional parameters, the port diameter, and the pressure. The theoretical burning rate can be based on total mass flux (G), as shown in Eq. (7.24), or on the oxidizer mass flux (G_o), as shown in Eq. (7.36). In either case, the general form for the burning-rate equation is

$$\dot{r} = \dot{r}_o f(p, D) \tag{7.63}$$

For the ideal case, in which combustion occurs as shown in the classical model of boundary layer diffusion (see Fig. 7.4), the major corrections result from radiation and kinetics. But keep in mind that these corrections are only suggestive because the actual combustion process is much more complex than in the idealized model.

In the case of radiation, Eqs. (7.51), (7.52), (7.56) and (7.59) combine to show the nature of the correction. Thus, we see that the form of the correction for lower values of emissivity is given by

$$\dot{Q}_r / \dot{Q}_c \sim \frac{\left(1 - e^{bpz}\right)}{G^n} \sim \frac{p}{G^n} \tag{7.64}$$

For small $\dot{Q}_r / \dot{Q}_c$, we can expand the exponentials in Eqs. (7.56) and (7.59) to give, along with Eq. (7.51), the following form of correction:

$$\dot{r} \approx \dot{r}_o \left(1 + 0.25 \frac{\dot{Q}_r}{\dot{Q}_c}\right) \approx \dot{r}_o \left(1 + \frac{\text{Const } p}{G^n}\right) \tag{7.65}$$

When the radiation term becomes dominant (typically in metallized systems at high pressures and low oxidizer flux), the regression rate becomes largely dependent on $\dot{Q}_r$ and independent of mass flux rate (G). In other words,

$$\dot{r} \cong \text{Const } \dot{Q}_r \tag{7.66}$$

The left side of Fig. 7.15 clearly shows this behavior. The kinetic effect on the right side is calculated from Eqs. (7.61) and (7.62). This result is more clear if we expand Eq. (7.62) into two limiting cases. For small θ_t, (where the kinetic effect is just starting to be observed),

$$\frac{\dot{r}}{\dot{r}_o} \to 1 - \frac{\theta_t}{6}$$

resulting in

$$\frac{\dot{r}}{\dot{r}_0} \cong \left(1 - \frac{C_1 G^{0.8}}{p}\right) \tag{7.67}$$

When θ_t is large (the far right portion of Fig. 7.15):

$$\frac{\dot{r}}{\dot{r}_0} \cong \sqrt{\frac{2}{\theta_t}} \cong C_2 \frac{\sqrt{p}}{G^{0.4}} \tag{7.68}$$

where C_1 and C_2 are empirical constants. Equation (7.67) shows how decreasing pressure causes a drop in regression rate, and Eq. (7.68) shows how further changes in mass flux and pressure can extinguish burning.

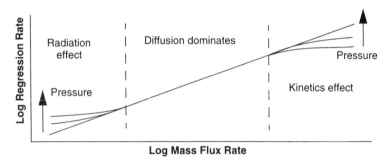

Fig. 7.15. Schematic of the Effects of Pressure on Regression Rate. This figure summarizes the results discussed in the text.

The question now arises: If a pressure effect is observed, how do we determine whether it arises from radiation or kinetics? Figure 7.15 provides the clue. It shows how pressure affects the regression rate over the entire range of practical mass fluxes (G) resulting from the effects of radiation and kinetics. The point at which either radiation or kinetics becomes important is a unique characteristic of each propellant system and motor scale. If radiation is dominant and the mass flux decreases at constant pressure, the regression rate tends to level out and exceeds the extrapolated value for the diffusion-controlled case. If kinetics is dominant, decreasing the pressure at a constant high mass flux causes the rate to fall off from the extrapolated diffusion-controlled case, as shown in Fig. 7.14. In this event, an increase in pressure can bring the burning rate up only to the extrapolated curve.

Because of the awkwardness in handling theoretical pressure-correction equations, it is convenient to employ either of the following corrections whenever we observe a pressure effect:

$$f(p, D) = p^c D^d \tag{7.69}$$

$$f(p, D) = \left(1 - e^{-p/p_o}\right)\left(1 - e^{-D/D_o}\right) \tag{7.70}$$

where c, d = empirically derived exponents
p_o, D_o = empirically derived reference pressure and diameter (Pa, m)

Equation (7.69) is a familiar form used extensively in the chemical-engineering literature. As Altman [1991] pointed out, however, using it to extrapolate to large sizes is hazardous, especially when moderate experimental error exists. Equation (7.70) is better behaved in that the asymptotic behavior is proper at both high and low values of the parameters p and D.

Fitting Experimental Data

A study has been made of three sets of experimental motor measurements which cover three different fuels running with liquid oxygen (LOx) as the oxidizer. These data were analyzed for the excellence of fit with respect to several burning-rate equations and also with regard to the accuracy of predicting data for motors of larger or smaller size.

Table 7.5 represents an analysis of 35 tests with a polybutadiene fuel and LOx. Six regression-rate equations were studied. The first three columns of data show the analysis for all 35 motors. Note that all of the average absolute-percent errors are quite small and all of the n exponents are close to the theoretical value of 0.8. This propellant combination, which does not contain aluminum, depends very little on pressure. The next three columns show the regression-rate constants and the average errors derived from the 24 small motors. The average percent errors for the extrapolated data for the large motors in the last column show generally larger values. But system 6 shows gross errors, indicating that the form of the pressure and diameter terms is invalid.

Table 7.6 shows the same type of analysis made for an aluminized polybutadiene fuel with LOx (courtesy of CSD/NASA [1993]). These two sets of data differ in significant ways. First, because the fuel is metallized, there is a pressure dependence, as is evidenced from the constants in systems 5 and 6. Second, the presence of aluminum decreases the combustion efficiency, which probably accounts for the greater scatter of the data, as evidenced by the average percent error. Notice also that the extrapolated results are poor, except for the case of Eq. 5.

Summary Observations

For the fuel system without aluminum (AMROC tests), which depends very little on pressure, equations 1 through 5 are all acceptable, with some preference for system 5. For the propellant with aluminum for which radiation causes pressure dependence, system 5 is clearly preferable because it allows for a pressure term.

Table 7.5. Summary of AMROC Test Results for 24 Small Motors and 11 Large Motors. Oxidizer mass flux ranged from 85 to 352 kg/m²·s. Port diameters in small motors ranged from 4.3 cm to 5.6 cm, and length ranged from 61 cm to 173 cm. Dimensions of large motors ranged from 19 cm to 29 cm in diameter and 3.05 m to 7.37 m in length (data provided courtesy of AMROC).

	Equation	All Motors			Small Motors			Large* Motors
		n	m	Avg. Error	n	m	Avg. Error	Avg. Error
1	$aG^n L^m$	0.800	−0.200	5.5%	0.800	−0.200	3.9%	9.5%
2	$aG^n L^m$	0.763	−0.148	4.7%	0.829	−0.256	5.7%	13.9%
3	$aG_o^n L^m$	0.756	−0.165	6.4%	0.668	0.028	4.7%	38.5%
4	$aG_o^n L^m \left(1 + \dfrac{2 a n \rho L^{1+m}}{D G_o^{1+n}}\right)$	0.765	−0.162	5.1%	0.740	−0.103	4.9%	11.5%
5	$aG^n L^m \left(1 - \exp\left(\dfrac{-D}{1.06}\right)\right)$	0.767	−0.254	3.8%	0.757	−0.242	3.6%	3.5%
6	$aG_o^n L^m D^d p^{\phi_p}$	0.722	0.034	4.6%	0.633	0.076	4.3%	108.3%
		$\phi_p = -0.05$			$\phi_p = -0.046$			

*The large-motor errors are calculated using small-motor constants, indicating the scaling effects.

Although system 6 is acceptable for interpolation, it is clearly inferior for extrapolation.

Significantly, the n exponents fall in the range 0.72–0.76 range (near the theoretical 0.8 value), regardless of the equation. For the propellant with aluminum, the scatter is somewhat larger, and the average n values are lower (0.53–0.68). This reduction in n can be expected in systems where radiation contributes to the heat transfer because it does not depend on mass flux, so the mass-flux dependence is diluted. Indeed, as $\dot{Q}_r$ increases, n decreases and, in the limit, approaches zero. The reduction in n also occurs for the kinetically limited case, as the limiting form Eqs. (7.67) and (7.68) show.

In the final observation, all n exponents vary significantly among the models selected for the same set of data. That is probably because in the mathematical regression analysis, which optimizes the fit, we vary the n value to minimize the impact of data scatter. This effect is particularly severe when the database is relatively small.

Table 7.6. Summary of Test Results for an Aluminized Regression Rate. Thirty-five small motors and 9 large motors were tested. Oxidizer mass flux ranged from 15 to 297 kg/m²·s. Port diameters in small motors raged from 4.3 cm to 5.1 cm, and length ranged from 30.5 cm to 91.4 cm. Dimensions of small motors ranged from 10.4 cm to 20.1 cm in diameter and 1.13 m to 1.22 m in length. Port pressures ranged from 1.688 to 5.721 MPa [CSD /NASA, 1993].

	Equation	All Motors			Small Motors			Large* Motors
		n	m	Avg. Error	n	m	Avg. Error	Avg. Error
1	$aG^n L^m$	0.800	−0.200	19.0%	0.800	−0.200	15.2%	36.1%
2	$aG^n L^m$	0.676	−0.063	15.9%	0.749	−0.185	15.1%	27.9%
3	$aG_o^n L^m$	0.597	0.113	17.4%	0.618	0.142	17.8%	16.6%
4	$aG_o^n L^m \left(1 + \dfrac{2an\rho L^{(1+m)}}{DG_o^{(1+n)}}\right)$	0.645	0.025	18.8%	0.677	−0.016	19.0%	21.3%
5	$aG_o^n L^m \left(1 - e^{\frac{-D}{1.4}}\right)\left(1 - e^{\frac{-p}{625}}\right)$	0.535	−0.052	6.5%	0.547	−0.048	5.8%	10.1%
6	$aG_o^n L^m D^d p^{\phi_p}$	0.532	0.145	11.3%	0.565	0.027	6.7%	27.9%
		$\phi_p = 0.574$			$\phi_p = 0.677$			

*The large-motor errors are calculated using small-motor constants, indicating the scaling effects.

7.3 Design Process

Table 7.7 represents a typical preliminary design process for a Hybrid Rocket Propulsion System (HRPS). The HRPS requirements and design outputs are similar to those seen in previous chapters.

As with other propulsion systems, the design process starts by establishing the mission requirements—given in terms of performance and allowable envelope. The performance is typically described in terms of maneuver Δv for the stage, range of specific impulse, thrust versus time, or the total impulse. The dimensional envelope usually specifies the maximum length and diameter permitted. We then use these requirements to select the associated masses that satisfy the mission. From this point forward, preliminary design has three major parts:

1. **Make Preliminary Design Decisions**—based on the propulsion-

system requirements, we must make some basic design decisions, such as the propellants to be used, the number of ports, and the port configuration.
2. **Estimate Performance**—once we have determined the basic grain configuration, we must simulate the engine performance to ensure the design meets the system requirements.
3. **Size and Configure Components**—we need a preliminary design of the major components to estimate their configurations and masses.

Of course, iteration paths exist between these major steps, as well as within the steps themselves. We discuss each of these steps in the following sections. Section 7.4 discusses preliminary design decisions, Sec. 7.5 discusses performance evaluation, Sec. 7.6 discusses component design, and Sec. 7.7 presents a design example.

7.4 Preliminary Design Decisions

This section deals with the design decisions that are unique to the hybrid rocket (many decisions are common among chemical systems). For the propellant candidates, we are concerned not only with the general and thermodynamic properties but also with the regression rate behavior and O/F shift during a burn.

7.4.1 Choose Propellants

As previously mentioned, the usual combination of propellants for hybrid rockets is a solid fuel with a liquid oxidizer. However, we may use a solid oxidizer, such as ammonium perchlorate, and a liquid fuel, such as kerosene, hydrazine, or liquefied hydrogen. This latter combination is sometimes called a reverse hybrid. The solid-fueled hybrid, as the most popular approach, gets most of the attention here.

Fuels for hybrid rockets, with a few exceptions, are carbon-based polymers in the form of plastic or rubber. Examples of typical fuels include plexiglass (polymethyl-methacrylate or PMM), polyethylene (PE), and polybutadiene (PB). Each of these polymers has chain terminators such as the hydroxyl or carboxyl ion. Figure 7.16 shows the molecular configuration of these examples. The total list of possibilities is almost endless and could include virtually every known polymer. Originally, most of the work on combustion research used PMM because of its low cost, easy accessibility, and transparency, which makes it good for demonstration models. At present, the most popular fuel is hydroxyl-terminated polybutadiene (HTPB)—a rubber compound that is quite energetic and extremely safe to handle. Studies show that HTPB soaked in liquid oxygen is not explosive.

Typically, the polymer fuels do not allow for fuel densities as high as in solid rockets. Some hybrid-rocket applications mix powdered metal, such as aluminum, with the polymer to enhance propellant density and therefore reduce motor volume.

Table 7.7. Preliminary Design Process for a Hybrid Rocket Propulsion System.

Step	Action	Comments
1.) Summarize requirements	Summarize	• Performance requirements • Envelope constraints • Mass requirements • Thrust history • The "ilities"
2.) Make preliminary design decisions Section 7.4	Choose propellants	• Evaluate thermochemistry (c, c_f, I_{sp}) • Choose design O/F and allowable O/F range • Environmental requirements • Performance requirements
	Determine pressure levels for the engine and feed system	• Pump or pressure feed system • Combustion-chamber pressure & nozzle expansion • Injector pressure drop • Profile of feed-system pressure • Dynamic pressure • Tank pressure • Pressurant system
	Determine requirements for the initial propellant flow	• Estimate initial specific impulse • Determine required flow rates based on required thrust level
	Size system	• Choose the inert mass fraction • Use the ideal rocket equation • Estimate fuel and oxidizer masses • Estimate inert mass
	Configure combustion ports	• Choose number of ports • Port cross section, length, and web thickness
3.) Estimate performance Section 7.5	Simulate the burn	• Estimate operating parameters • Predict the grain and nozzle configuration over the burn duration
4.) Size and configure components Section 7.6	Nozzle	• Characteristic velocity sizes the throat (Eq. 3.133) • Expansion ratio sizes the exit • Required efficiency determines length • See Secs. 5.4.1 and 6.3.7
	Combustion chamber	• Include fore and aft sections with grain length • Perform hoop-stress analysis • Include an injector mass • See Sec. 6.3.3
	Oxidizer tank	• Use hoop-stress or structural mass-factor approach • See Sec. 5.4.4
	Pressurant system	• See Sec. 5.4.5
	Support structure and ancillary parts	• 10% of mass • See Sec. 5.4.7
5.) Iterate as required		• Iteration can occur from any point to any other point in the process • Iteration ends when an adequate design emerges

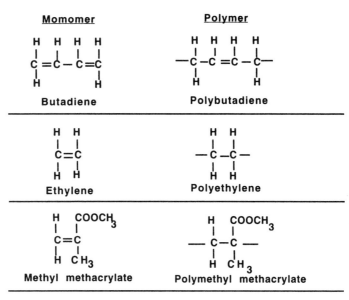

Fig. 7.16. **Chemical Structure of Typical Fuels.** This figure shows three of the more common polymer fuels. The left column shows the monomer structure. The right column shows the basic polymer element with empty bond positions where chain linking can occur.

Oxidizers include the entire list from Appendix B and many more. In this list are gaseous or liquid oxygen (O_2), hydrogen peroxide (H_2O_2), nitrogen tetroxide (N_2O_4), as well as the fluorines.

Appendix B shows specific impulses versus ratios of oxidizer-to-fuel mass flow rates (O/F) for various propellant combinations at a specific design condition, as well as frozen-flow thermochemical data for several fuel/oxidizer combinations. The data, as a function of O/F, includes flame temperature (T_f), the isentropic parameter (γ), and molecular mass of the products ($\mathcal{M}$).

Sometimes, carbon black is also used in hybrid rockets. This is a light black powder with a consistency similar to icing sugar. It is mixed into the solid fuel to enhance the fuel's radiative heat absorption. This mixing improves the fuel-regression rate by concentrating heat absorption near the fuel surface and inhibits radiative heat transfer through the fuel to the motor case. Typically, the amount added is less than 1%, so it has little effect on thermochemical parameters such as the flame temperature.

In many hybrid rocket systems, we have a problem getting the injected liquid propellant to vaporize properly. This leads to problems of fuel usage at the head of the fuel grain as liquid streams strike the solid fuel. It also leads to combustion inefficiencies. We would like to vaporize the liquid before it enters the fuel port if we can. To do so, we can

- Inject a hypergolic fluid into the oxidizer stream: the chemical reaction of the hypergolic fluid with the oxidizer creates enough heat to vaporize the nonreacted oxidizer. A typical hypergolic liquid for liquid oxygen is triethyl aluminum (TEA). Hydrazine works with nitrogen tetroxide.
- Inject hot gases from a gas generator: hot gases can be generated from a separate gas generator and then fed into the pre-combustion region to vaporize the liquid oxidizer. One successful system has used a hydrogen/oxygen gas generator [CSD/NASA, 1993].
- Use a monopropellant to generate hot oxidizer. The best example of this approach is the catalytic decomposition of liquid hydrogen peroxide into hot oxygen and steam (hydrazine can also be used for reverse hybrids). This system resembles the gas generator described above.

Choose the propellant combination—by weighing two major considerations:

- Quantitative—The vehicle-level design establishes specific performance requirements, usually in terms of propellant mass and specific impulse. An allowable lower limit on specific impulse definitely eliminates some of the possibilities. For example, if we need a vacuum I_{sp} greater than 340 seconds, we are limited to using O_2 or perhaps the more exotic fluorines. Appendix B shows vacuum I_{sp} vs. O/F for various propellants at a specific design condition. For preliminary design, we often use the characteristic velocity (c^*) equation from Sec. 3.3.3. The parameters necessary to determine c^* are also in Appendix B. The advantage of using c^* is that it varies only slightly with chamber pressure and is independent of all of the expansion considerations.
- Qualitative—Once we have established which of the possible combinations meet the performance requirements, we select from these possibilities based on handling, storability, transportability, ignition, toxicity, and other qualities. Table B.1 lists some of the considerations for many of the more widely used oxidizers. The fuels are all basically benign and nontoxic. Choice of fuel usually depends on cost and the process required to cast the fuel grain. For example, we often use HTPB because we have a lot of experience in handling this chemical as a binder in solid rockets. Polyethylene can be purchased in a pellet form, heated in an oven, and allowed to resolidify in its desired configuration. PMM is not widely used in larger engines because it has a tendency to crack while under heat load.

Choose design O/F and allowable range—The main question raised here is: "What is the performance penalty from a shifting O/F when the average value makes the vehicle perform best?" To clarify the problem, consider an example of a typical LOx/HTPB propellant run in a cylindrical grain, as shown in Table 7.8.

The spread in O/F is from 1.72 to 2.53, varying about 20% from the mean value of 2.13, which corresponds to the maximum I_{sp}. However, the average value of I_{sp}

Table 7.8. Variation of Oxidizer-to-Fuel Ratio and Specific Impulse over Time. As the fuel burns, the port diameter increases, reducing the fuel-regression rate. At a constant oxidizer flow rate, the reduced fuel flow increases the O/F. This is a simulation of a HRPS with a single cylindrical grain with an initial diameter = 7 cm. The initial oxidizer flux = 350 kg/m²·s. Chamber pressure = 3.45 MPa, exit pressure = 0.101 MPa, and the I_{sp} efficiency is 92% of equilibrium flow and n = 0.75.

Time (seconds)	0	5	10	20	30	40	50
O/F	1.72	1.89	2.01	2.20	2.33	2.44	2.53
I_{sp} (seconds)	249.4	253.3	254.7	255.1	254.3	253.2	252.1

is 253.9 seconds. This is only 1.2 seconds (0.5%) less than the maximum I_{sp} value of 255.1 seconds. The penalty, in this case, is nearly zero and depends, in general, on the propellant, the grain design, and operating conditions, but in most practical cases, it is less than 2%. A simulation is required to predict the actual number more accurately (see Sec. 7.5). We usually choose an initial O/F and then iterate with simulations to maximize the average I_{sp}.

Evaluate thermochemistry—To determine the engine performance, we must look at the thermochemistry by running one of the available computer codes or by using the frozen-flow data from Appendix B. For hybrid rockets, we need to evaluate the chemistry over the desired O/F range, but the initial and average design points are particularly important. The most important parameter is the characteristic velocity (c^*). We can determine the isentropic parameter (γ), the molecular mass ($\mathcal{M}$), and flame temperature (T_f) from the appendix. These all combine to give us

$$c^* = \frac{\eta_{c^*}\sqrt{\gamma R T_f}}{\gamma\left(\frac{2}{\gamma+1}\right)^{\frac{\gamma+1}{2\gamma-2}}} \tag{7.71}$$

$$R = \frac{8314}{\mathcal{M}} \quad (\text{J/kg·K}) \tag{7.72}$$

The combustion efficiency (η_{c^*}) is somewhat lower for hybrids than for liquids or solids (see Fig. 4.3). If we assume equilibrium flow, we can get up to 0.95. However, if we do not allow for adequate mixing, efficiency can drop below 0.9. If we assume frozen-flow chemistry, the efficiency can approach 1.0.

7.4.2 Determine Pressure Levels for the Engine and Feed System

We determine the pressure levels through the HRPS system using an approach similar to that discussed for liquid rockets in Chap. 5. We start by determining the combustion-chamber and exhaust-nozzle pressures (nozzle pressures define expansion ratio) to get the performance required and then move back through the feed system and finish the job with an estimate of propellant-tank and pressurant-tank pressures. See Sec. 5.3.5 for the details of this process.

7.4.3 Size System

The next step is to roughly size our overall propulsion system, based on the ideal rocket equation (Secs. 1.1.4 and 1.1.5). To use this equation, we need to know

- Payload mass (m_{pay})—This includes our payload and any additional stages that operate after our present stage
- Specific impulse (I_{sp})—The specific impulse to use is the average specific impulse over the burn. We estimate this by reducing the maximum I_{sp} by 1% to 2% as discussed in Sec. 7.4.1.
- Δv—Mission analysis determines this value.
- Inert-mass fraction (f_{inert})—Inert-mass fractions for hybrid rockets are usually somewhat poorer than for solid rockets because the loading of the solid fuel is less efficient for hybrids than for solids and solid rocket propellant is usually more dense. An exception arises if the hybrid rocket oxidizer is pumped, which lowers the structural mass of the oxidizer tank. We can determine the inert-mass fraction ($1 - f_{prop}$) from the mass-fraction data in Table 6.2. Inert-mass fractions for solid rockets vary from 0.07 up to 0.19. Assuming inert-mass fractions at the upper end of this range is appropriate (0.16–0.20).

Once we have determined these values, we can estimate the required propellant mass from Eq. (1.27):

$$m_{prop} = \frac{m_{pay}\left[e^{\left(\frac{\Delta v}{I_{sp}g_0}\right)} - 1\right](1 - f_{inert})}{1 - f_{inert} e^{\left(\frac{\Delta v}{I_{sp}g_0}\right)}} \quad (7.73)$$

The inert mass (m_{inert}) can then be estimated from Eq. (1.24):

$$m_{inert} = \frac{f_{inert}}{1 - f_{inert}} m_{prop} \quad (7.74)$$

This allows us to determine the fuel mass (m_{fuel}) from

$$m_{fuel} = \frac{m_{prop}}{1 + O/F} \quad (7.75)$$

and the oxidizer mass (m_{ox}) from

$$m_{ox} = m_{prop}\frac{O/F}{O/F + 1} = m_{prop} - m_{fuel} \quad (7.76)$$

The O/F used here is the average O/F over the burn. We usually try to hit the optimal O/F as an average, or we can use a simulation result to estimate an actual average. The total volume (V) required to hold this mass is simply the sum of the two volumes:

$$V = V_{ox} + V_{fuel} = \frac{m_{ox}}{\rho_{ox}} + \frac{m_{fuel}}{\rho_{fuel}} = \frac{m_{prop}}{\rho} \quad (7.77)$$

where m_{fuel} = fuel mass (kg)
m_{ox} = oxidizer mass (kg)
O/F = oxidizer-to-fuel ratio
m_{prop} = total mass of propellant (kg)
V = total volume required to hold propellant (m^3)
V_{ox} = volume of oxidizer (m^3)
V_{fuel} = volume of fuel (m^3)
ρ = propellant bulk density (kg/m^3)
ρ_{fuel} = fuel density (kg/m^3)
ρ_{ox} = oxidizer density (kg/m^3)

7.4.4 Determine Requirements for the Initial Propellant Flow

Before we can configure the solid-fuel grain, we need to determine what the initial propellant flow rates are and how much propellant is required. To determine the initial port configuration, we need to know the initial thrust required, our initial O/F target, and the initial expected specific impulse. To estimate the total propellant requirements, we need to know the values of these same parameters at some average O/F.

Estimate initial specific impulse—Our thermochemical analysis (Sec. 7.4.1) and our nozzle analysis (Sec. 7.4.2) give us a good estimate of the initial and average specific impulses to expect. The initial specific impulse, used to size the initial port configuration, is adequate as is. However, we must reduce the average specific impulse by a couple percent to account for performance losses over the O/F shift.

Determine required flow rates—We now have enough data to estimate the propellant flow rates to meet our requirements. Given the initial thrust level (F) and an average thrust level (determined by the given thrust-history requirement), as well as the initial and average specific impulses (I_{sp}), we can determine the propellant flow rate ($\dot{m}_{prop}$):

$$\dot{m}_{prop} = \frac{F}{I_{sp}\, g_0} \tag{7.78}$$

Knowing the O/F, we can determine the fuel and oxidizer flow rates:

$$\dot{m}_{fuel} = \frac{\dot{m}_{prop}}{1 + O/F} \tag{7.79}$$

$$\dot{m}_{ox} = \dot{m}_{prop}\frac{O/F}{O/F + 1} = \dot{m}_{prop} - \dot{m}_{fuel} \tag{7.80}$$

7.4.5 Configure Combustion Ports

At this point in the design we want to decide on the number and configuration of ports in the combustion chamber. The simplest design is a single, cylindrical port. When multiple ports are required, the circular cross section is not efficient because a uniform web cannot be constructed between adjacent ports. When motors become large, the single cylindrical grain loses efficiency, largely because the web thickness (which typically does not vary much with size) becomes a smaller fraction of the diameter. This situation requires a longer motor, creating control problems in flight. Of the various multi-port designs with circular cross section, the seven-cylinder cluster is one of the most effective because it is conveniently packaged in a larger cylinder, with minimal volume loss between the cylinders (see Fig. 7.18). Fig. 7.17 shows an example of how the volumetric fuel efficiency (volume of fuel / volume of chamber) and the grain length-to-diameter ratio (L/D) varies with the total mass of the fuel grain. The single cylinder is more volumetrically efficient for fuel masses under 700 kg, but the port L/D becomes quite large. On the other hand, a multiple-port configuration shows a fairly constant volumetric efficiency in the modest range of 50% to 58%, but it has the great advantage of being compact, with a small L/D in the range of 2 to 6.

Figure 7.18 shows typical grain configurations. Note that for two ports, a suitable design is the double-D grain. The most common multiple-port configuration is the wagon wheel. If the number of ports is large (more than seven) the port cross section can be approximated by a triangle.* For a more accurate description of the burning contour, we should base the geometry on a "pie-sector" (the base section follows the curve of the outer grain case, making the geometry a bit more complicated than the simple triangle).

The multiple-port wagon wheel typically has a center hole. We have two options for handling this center hole—allow burning in it or block it to make it inert. We improve volumetric fuel efficiency somewhat if we allow burning in the

---

* For fewer ports (see the 4-port configuration in Fig. 7.18), the curvature of the triangle base becomes significant.

center hole, but the circular-hole geometry differs from that of the main ports affecting ballistic performance. If the center-hole web is improperly designed, an unburned fuel sliver can negate any beneficial result.

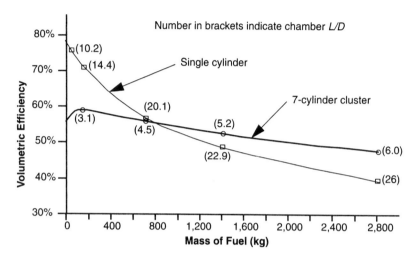

Fig. 7.17. Volumetric Efficiency and Chamber Length-to-Diameter Ratio (L/D) as a Function of Required Fuel Mass. For this particular example, a single cylindrical port has better volumetric efficiency than the seven-port configuration for fuel mass below 700 kg. Above 700 kg, the 7-cylinder multiple port cluster is better.

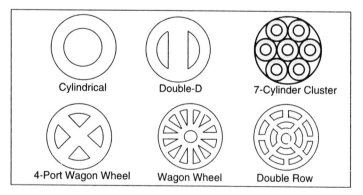

Fig. 7.18. Typical Port Configurations for Hybrid Rockets. The trick is to choose a configuration that optimizes volumetric loading efficiency of the solid fuel. Such a configuration requires the fuel web to be constant around the perimeter of each port.

In the mid-1960s, the Chemical Systems Division of United Technologies (CSD) tested a motor with a diameter of 96.5 cm using burning in the center hole. Figure 7.19 shows the port configuration before and after the motor was fired. This motor had 12 triangular ports and the additional center port. The uniformity of burning is shown in the figure and supports the theory discussed in Sec. 7.2.

Fig. 7.19. **Section of a Twelve-Port Combustion Chamber Before and After Firing.** The initial triangle shape burns very efficiently and leaves very small fuel residuals [Feemster, 1968].

The last example in Fig. 7.18 is a double row of ports suitable for very large motors (diameters greater than 3.5 m). As of 1995, motors of this design have not yet been tested. The concerns with this design are fabrication, potential burning dissimilarity in the two different port shapes, and excessively large fuel slivers at the end of the burn.

Section 7.2.6 discusses the problem of scaling. This situation arises when we develop a large multiport grain using one of two available concepts. The conventional approach is going directly to the full-size configuration. Because testing a full-size motor is expensive, however, an alternate approach involves testing just two full-size ports, which can be done in a motor with a smaller diameter (see Fig. 7.20). We use two ports instead of one because we want to check the interactions between adjacent ports.

One other consideration for multiple ports is placing web supports or spacers midway in the web. The supports typically are a slow-burning, stiff, plastic material which helps support the shape of the port toward the end of the firing, when the fuel web gets thin. It also helps to avoid the possibility of pieces of fuel breaking off the grain near the end of the burn. Figure 7.21 shows a typical design of a web support.

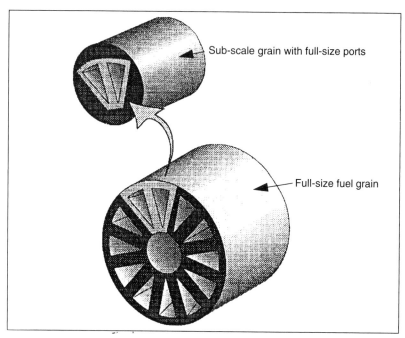

Fig. 7.20. Concept for Testing Two Full-Scale Ports Instead of Scaling from Smaller Motors. We can test large motor ports without testing the full-size motor. (Courtesy AMROC)

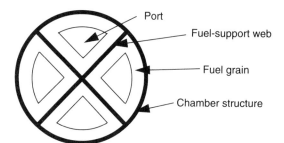

Fig. 7.21. Schematic of a Fuel-Web Support. As the motor burns, the fuel grain becomes too thin to support itself. We insert structural webs to hold the fuel in place.

Port Configuration

From previous sections, we have chosen the propellants and each propellant's initial mass flow rate. We must now determine the number of ports, their arrangement in the overall grain, their length, and their cross-sectional configuration (including web thickness). Section 7.2 introduces us to much of the mathematics required to make these decisions.

How do we determine the number of ports? As a rule of thumb, we expect to achieve a volumetric efficiency (η_{V_g}) above 50% for the port grain:

$$\eta_{V_g} = \frac{volume\ of\ solid\ fuel}{total\ volume\ of\ the\ grain} \tag{7.81}$$

The number of ports, web thickness, and oxidizer mass flux usually influence volumetric efficiency. Figure 7.22 shows a typical relationship between volumetric efficiency and the number of ports for various web thicknesses in a wagon-wheel design.

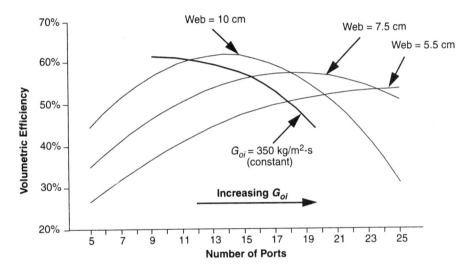

Fig. 7.22. Schematic of Volumetric Efficiency Versus the Number of Ports. Three curves show how the grain efficiency changes as the number of ports changes. For these curves, the oxidizer flow rate is constant. The fourth (heavier line) curve shows the variation of port geometry if the initial oxidizer flux (G_{oi}) is constant. These curves are for a wagon-wheel configuration with a nonburning center hole.

For a web of a given size in a given chamber diameter, we want to be as near to the maximum volumetric efficiency as possible because this gives us the smallest engine. It is important to understand why there is a maximum in these curves for web thickness. For a small number of ports (N), the corresponding number of webs limits the total web cross section. As N increases, so does the number of webs. However, from geometry, the center hole also increases. As a result, for each web thickness and chamber diameter, the maximum cross-sectional fuel area has a unique N value. Usually, we cannot get this maximum because

- Mass flux has a practical upper limit. With liquid oxidizers, *flooding* can occur, extinguishing the flame. Gaseous oxidizers also have this limit, termed *blow-off*. This phenomenon is analogous to extinguishing a candle by vigorously blowing on it. Typically, the upper limit on oxidizer mass flux is between 350 and 700 kg/m²·s. The lower limit is a conservative initial estimate for liquids; the higher limit is more appropriate for gases.
- Figure 7.22 also shows the benefit of a thicker web (the 10-cm web has a maximum efficiency higher than the 7.5-cm web). However, a thicker web gives a longer burn time and a greater O/F shift, thus reducing the integrated specific impulse over the burn duration. We must balance the decrease in average I_{sp} with increase in mass fraction to arrive at the best web thickness.

Determine the initial port configuration—The first step is to select the number of ports (N). As discussed, we get this number by maximizing the volumetric efficiency, subject to envelope constraints. We do not know the best number of ports a priori; we must look at several individual cases to determine this number.

The next step is to choose the initial oxidizer flux (G_{oi}). As previously mentioned, the maximum upper limit ranges between 350 and 700 kg/m²·s, depending on the propellants used. Using this limit and knowing the initial oxidizer mass flow rate ($\dot{m}_{oi}$—from Sec. 7.4.3), we can determine the initial port area,

$$A_{pi} = \frac{\dot{m}_{oi}}{N G_{oi}} \tag{7.82}$$

$$G_{fi} = \frac{\dot{m}_{fi}}{N A_{pi}} = \frac{G_{oi}}{O/F_i} \tag{7.83}$$

where A_{pi} = initial port area (m²)
$\dot{m}_{oi}$ = initial oxidizer flow rate (kg/s)
$\dot{m}_{fi}$ = initial fuel mass flow rate (kg/s)
O/F_i = initial oxidizer-to-fuel ratio
G_{fi} = initial fuel flux (kg/m²·s)
G_{oi} = initial oxidizer flux (kg/m²·s)
N = number of ports

The next step is to determine the port length. We know that the required mass flow rate of the fuel at any time is (we have dropped the subscript i here because these equations apply all of the time)

$$\dot{m}_{fuel} = \dot{r} \rho S \tag{7.84}$$

where $\dot{m}_{fuel}$ = fuel mass flow rate (kg/s)
  ρ = solid fuel density (kg/m³)
  S = total fuel surface area (m²)
  $\dot{r}$ = fuel regression rate (m/s)

If we assume a regression rate of the form,

$$\dot{r} = aG^n L_p^m \tag{7.85}$$

where a = regression-rate coefficient
  G = total propellant mass flux (kg/m²·s)
  n = mass-flux exponent
  m = port-length exponent
  L_p = port length (m)

Combining these two equations gives

$$\dot{m}_{fuel} = N\rho S_p a (G_o + G_f)^n L_p^m \tag{7.86}$$

for the total grain with N ports. We need to solve this equation for the port length, but the surface area of a single port (S_p) is a function of length as well as the shape of the port cross section:

$$S_p = P_p L_p \tag{7.87}$$

where P_p is the port perimeter (circumference). Then we can solve for the port length,

$$L_p = \left(\frac{\dot{m}_{fi}}{N\rho a (G_{oi} + G_{fi})^n P_{pi}} \right)^{\frac{1}{m+1}} \tag{7.88}$$

Notice that we have included the i subscript to indicate initial conditions.

If we assume a circular port configuration, the port surface area and cross-sectional area become, respectively,

$$S_p = \pi D_p L_p \tag{7.89}$$

$$A_{pi} = \pi \left(\frac{D_{pi}}{2} \right)^2 \tag{7.90}$$

where D_p = port diameter (m)

Then, we relate the port's length to its diameter by

$$L_p = \left(\left[\frac{\dot{m}_{fi}\pi^{n-1}}{a(4\dot{m}_i)^n \rho}\right] D_{pi}^{2n-1}\right)^{\frac{1}{m+1}} \tag{7.91}$$

where $\dot{m}_i$ = total initial propellant flow rate (kg/s)

Notice that the term in square brackets has parameters we know from previous calculations and decisions.

Determine the final port configuration—At the end of the burn, we have consumed the entire volume of fuel:

$$V_{fuel} = \frac{m_{fuel}}{\rho} \tag{7.92}$$

Assuming the port's length does not change (which is usually the case), we know its final cross-sectional area:

$$A_{pf} = \frac{V_{fuel}}{NL_p} + A_{pi} \tag{7.93}$$

From this final port area, we can determine the port's final dimensions and its fuel web, depending on the port geometry. If we are using triangular ports, we can determine the web thickness (w) from Fig. 7.19, by solving the quadratic:

$$w^2 + \left(\frac{2l+b}{\pi}\right)w + \frac{1}{\pi}\left(\frac{hb}{2} - A_{pf}\right) = 0 \tag{7.94}$$

where all of these parameters are defined in Fig. 7.23.

For circular ports:

$$D_{pf} = \sqrt{\frac{4m_{fuel}}{\pi L_p \rho N} + D_{pi}^2} \tag{7.95}$$

$$w = \frac{D_{pf} - D_{pi}}{2} \tag{7.96}$$

where m_{fuel} = total fuel mass (kg)
N = number of ports
D_{pi} = initial port diameter (m)
D_{pf} = final port diameter (m)
w = thickness of the fuel web (m)

We can estimate the O/F shift by comparing the O/Fs at the beginning and end of the burn. We know the initial O/F from the required flow rates

$$O/F_{initial} = O/F_i = \frac{\dot{m}_{oi}}{\dot{m}_{fi}} \quad (7.97)$$

We determine the final O/F by iteratively solving the following implicit equation [remember that regression rate is a function of itself, see Eq. (7.85)]:

$$O/F_f = \frac{\dot{m}_{oi}}{N\dot{r}_f \rho S_{pf}} \quad (7.98)$$

where $\dot{r}_f$ = final rate of fuel regression (m/s)

S_{pf} = final surface area of the port (m^2)

Figure 7.23 shows how we can configure a wagon-wheel grain, assuming triangular port geometry.

7.5 Performance Estimate

At this point in the design, we have some idea of what the rocket system should look like and how it should perform. However, we need to check the validity of our simplifying assumptions, in terms of performance, before looking at the individual components. To realistically estimate performance, we must numerically simulate the engine system's operation over time. This confirms whether we have successfully configured our engine system. If not, we can iteratively "tweak" the design to get what we want based on results from the simulation.

7.5.1 Analytical Solution for Circular Ports with Constant Oxidizer Flow Rate

Suppose we assume the port has a circular cross section and we can accurately describe the fuel-regression rate by (see Sec. 7.2.3)

$$\dot{r} = aG_o^n L_p^m \quad (7.99)$$

If so, we can develop a relatively simple expression or simulation for the change in port diameter over time, which is just twice the regression rate:

$$\dot{r} = \frac{1}{2}\frac{dD_p}{dt} = a\left(\frac{4\dot{m}_{ox}}{\pi D_p^2}\right)^n L_p^m = a\left(\frac{4\dot{m}_{ox}}{\pi}\right)^n L_p^m D_p^{-2n} \quad (7.100)$$

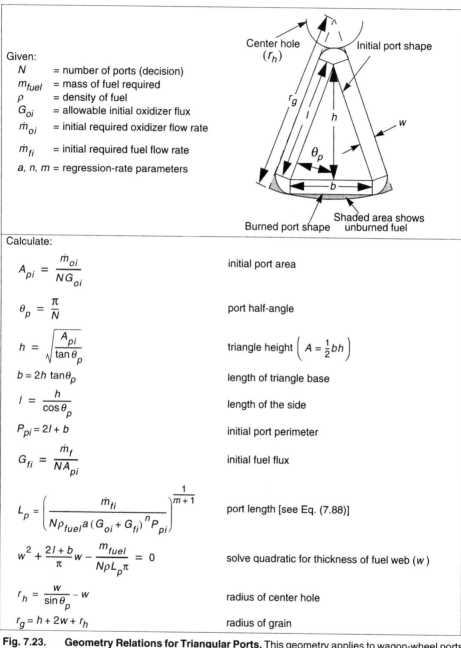

Given:
- N = number of ports (decision)
- m_{fuel} = mass of fuel required
- ρ = density of fuel
- G_{oi} = allowable initial oxidizer flux
- $\dot{m}_{oi}$ = initial required oxidizer flow rate
- $\dot{m}_{fi}$ = initial required fuel flow rate
- a, n, m = regression-rate parameters

Calculate:

$A_{pi} = \dfrac{\dot{m}_{oi}}{N G_{oi}}$ initial port area

$\theta_p = \dfrac{\pi}{N}$ port half-angle

$h = \sqrt{\dfrac{A_{pi}}{\tan\theta_p}}$ triangle height $\left(A = \tfrac{1}{2}bh\right)$

$b = 2h\tan\theta_p$ length of triangle base

$l = \dfrac{h}{\cos\theta_p}$ length of the side

$P_{pi} = 2l + b$ initial port perimeter

$G_{fi} = \dfrac{\dot{m}_f}{N A_{pi}}$ initial fuel flux

$L_p = \left(\dfrac{\dot{m}_{fi}}{N\rho_{fuel}\, a\,(G_{oi}+G_{fi})^n P_{pi}}\right)^{\frac{1}{m+1}}$ port length [see Eq. (7.88)]

$w^2 + \dfrac{2l+b}{\pi}w - \dfrac{m_{fuel}}{N\rho L_p \pi} = 0$ solve quadratic for thickness of fuel web (w)

$r_h = \dfrac{w}{\sin\theta_p} - w$ radius of center hole

$r_g = h + 2w + r_h$ radius of grain

Fig. 7.23. Geometry Relations for Triangular Ports. This geometry applies to wagon-wheel ports in which the total number of ports is greater than seven. For fewer ports, we must take into account the curvature of the triangle base.

where all of these parameters, such as oxidizer flow rate, are for an individual port. This equation can be integrated because the variable parameters (D and t) are separable (assuming oxidizer mass flow rate and port length are constant):

$$\int_{D_{pi}}^{D_p} D_p^{2n} dD_p = 2a\left(\frac{4\dot{m}_{ox}}{\pi}\right)^n L_p^m \int_0^t dt \qquad (7.101)$$

Integrating gives us the port diameter at any time:

$$D_p = \left(a(4n+2)\left(\frac{4\dot{m}_{ox}}{\pi}\right)^n L_p^m t + D_{pi}^{2n+1}\right)^{\frac{1}{2n+1}} \qquad (7.102)$$

where D_p = port diameter at any time (t) (m)
a = regression-rate coefficient based on g_0
n = regression-rate exponent
m = regression-rate exponent
$\dot{m}_{ox}$ = oxidizer mass flow rate through a single port (kg/s)
L_p = port length (m)
D_{pi} = initial diameter of the circular port (m)
t = time since the burn start (s)

This solution applies to single- and multiple-port grains, so long as the port cross section is circular.

7.5.2 Generic Simulation Algorithm

To numerically simulate the system operation, we start at an initial time and condition. We assume the parameters of this condition are constant over a very small time period (a scheme that allows Euler numerical integration). At the end of that period, we modify the parameters based on things such as fuel regression changing the port configuration, erosion of the nozzle throat, and atmospheric conditions. We then progressively step along in time, changing things as we go.

Table 7.9 shows the generic algorithm, which is fairly self-explanatory. The biggest problem with algorithms of this sort, including solid-rocket algorithms, is estimating the effect on port geometry. In step 6, we can easily update the fuel-web thickness through the burn, but relating it to the port cross section requires knowing characteristics of the port geometry (such as those shown in Fig. 7.23 for triangular ports).

We have also introduced throat erosion. For ablatively cooled exhaust nozzles, the throat tends to erode over time. This erosion increases the throat area and affects the engine's performance. Section 6.3.7 discusses this problem, and Table 6.5 gives estimates of typical erosion rates.

Table 7.9. Simulation Algorithm. This algorithm estimates the performance of a hybrid rocket over time. We assume that regression rate is constant down the port length.

Action	Inputs	Outputs
1. Initialize simulation $t = 0$	• Grain geometry • History of oxidizer flow • Nozzle configuration • Regression-rate parameters • Throat-erosion parameters • History of ambient pressure • Simulation parameters (time step size - Δt) • Thermochemistry data	
2. Determine fuel-regression rate	• $\dot{m}_{ox}$ - oxidizer mass flow rate (kg/s) • A_p - cross-section area of port (m²) • ρ - fuel density (kg/m³) • N - number of ports	• G_{ox} - oxidizer mass flux in each port (kg/m²·s) • G - total mass flux (kg/m²·s) • $\dot{r}$ - fuel regression rate (m/s)
3. Determine total mass flow rate	• $\dot{r}$ • S_p - port surface area (m²) • ρ • N • $\dot{m}_{ox}$	• $\dot{m}_{fuel} = \dot{r} \rho S_p N$ - fuel flow rate (kg/s) • $\dot{m}_{prop} = \dot{m}_{fuel} + \dot{m}_{ox}$ - total flow rate (kg/s)
4. Determine thermo-chemistry	• $\dot{m}_{ox}$ • $\dot{m}_{fuel}$ • Thermochemistry data (tables or curve-fit expressions, Appendix B)	• $\dfrac{O}{F} = \dfrac{\dot{m}_{ox}}{\dot{m}_{fuel}}$ - oxidizer-to-fuel ratio • T_c - flame temperature (K) • γ - ratio of specific heats • $\mathcal{M}$ - molecular mass (kg/kmol) • c^* - characteristic velocity [Eq. (3.133)]
5. Determine chamber, nozzle, and engine performance	• $\dot{m}_{prop}$ • A_t - nozzle throat area (m²) • ε - nozzle expansion ratio • p_a - ambient pressure (Pa) • λ - nozzle efficiency • c^* • γ	• $p_c = \dfrac{\dot{m}_{prop} c^*}{A_t}$ - chamber pressure (Pa) • M_e - exit Mach number [Eq. (3.100)] • p_e - nozzle exit pressure [Eq. (3.95)] • v_e - nozzle exit velocity [Eq. (3.112)] • $A_e = \varepsilon A_t$ [Eq. (3.100)] • $F = \lambda \left[\dot{m}_{prop} v_e + \left(p_e - p_a \right) A_e \right]$ [Eq. (1.6)] • $I_{sp} = \dfrac{F}{\dot{m}_{prop} g_0}$

7.5 Performance Estimate

Table 7.9. Simulation Algorithm. (Continued)

Action	Inputs	Outputs
6. Upgrade grain and nozzle geometry	• w - port web size (m) • A_t • Port geometry • $\dot{r}$ • $\dot{e}$ - throat erosion rate (m/s)	• $w = w - \dot{r}\Delta t$ • $r_t = r_t + \dot{e}\Delta t$ - throat radius (m) • new port geometry (A_p, S_p) • A_t • $t = t + \Delta t$
7. Output desired parameter values to storage	• As required	• As required
8. If burn duration is NOT complete, go to Step 2		• $w = w - \dot{r}\Delta t$ • $r_t = r_t + \dot{e}\Delta t$ - throat radius (m) • new port geometry (A_p, S_p) • A_t • $t = t + \Delta t$

Another problem with this algorithm is determining the total mass flux so we can determine the fuel-regression rate. If we have chosen the regression-rate expression based only on oxidizer flow, flux is a given and there is no problem. However, if we have chosen to use the regression expression involving total mass flux, we have a more difficult problem. We need to know fuel-mass flow to determine fuel regression, which is required to estimate fuel flow—a circular problem. One process for estimating total mass flux is as follows:

1. Start by assuming the total flux is just the oxidizer flux
2. Determine fuel regression and flow rate based on this flux
3. Add the fuel flow to oxidizer flow to get a new port mass flux
4. If the old guess at flux is equal (or nearly equal) to the new guess, stop
5. If not, go back to step (2) and iterate

This simple algorithm is very robust and converges to the "correct" estimate of the total mass flux.

How do we choose the time-step size? Answer: choose the time-step size that gives a result close to a previous simulation's time-step size. First, choose a duration (say 1 second). When the simulation is complete, choose a smaller size (say 0.5 seconds) and compare the results of the two simulations. If they are similar, we have chosen a good time-step size. If not, choose another (say 0.25 seconds) and compare with the 0.5-second result. Continue until the results of consecutive simulations agree to some defined degree of accuracy.

7.6 Preliminary Component Design

The entire propulsion system consists of two main parts—the liquid tank, with the feed-system components, and the thrust chamber, which houses the fuel and provides the thrust. The liquid tankage and feed system are similar to that of a liquid system (Chap. 5), and the thrust chamber is similar to that in solid rockets (Chap. 6). We do not discuss here how to design the storage and feed system for a liquid propellant. Our discussion of the combustion chamber considers only the differences between solids and hybrids because we can analyze hybrid combustion chambers the way we analyze solids.

The thrust chamber is divided into five or six parts—the injector assembly, the pre-combustion chamber (optional), the grain area (combustion chamber), the aft mixing chamber, the nozzle, and the thrust-vector control (which typically is part of the nozzle). The major differences here from solid rockets involve the injector and the aft mixing area.

7.6.1 Combustion-Chamber Design

The size (diameter and length) of the combustion chamber has already been determined by the grain configuration. The combustion chamber must be large enough to contain the fuel. Length is also usually added to enhance propellant mixing and combustion (see Sec. 7.6.3). The combustion chamber's mass depends on the strength considerations presented in Sec. 6.3.3.

7.6.2 Injector Assembly

We can use any of the injector types discussed in the chapter on liquid propellants, plus several unique to hybrids. There are two basic injector design philosophies: one directly injects the oxidizer down the port and the second injects it into a precombustion chamber, where it is largely gasified and heated before flowing down the port. Direct injection was used in early work involving small motors with single ports, and it seemed the obvious choice. With larger-diameter motors requiring multiple ports, we want a precombustion chamber to vaporize the oxidizer and provide a uniform entrance condition to the ports.

In the early work with multiple ports, individual injectors sprayed individual streams down each port. In this case, all injectors must have identical flow characteristics to guarantee uniform burning down each of the ports. This arrangement has a further disadvantage: if a hypergolic fuel is required for ignition or for vaporization, each injector needs its own means of injecting fuel. Despite these caveats, individual injectors do work well, giving relatively high combustion efficiency (91–93% of equilibrium flow prediction).

When we use a precombustor, we employ one of the more conventional liquid injectors to provide a uniform atomized spray. A hypergolic liquid is injected in much the same fashion as the fuel in a liquid rocket except, of course, in much smaller quantities. The injector types that can be used include the showerhead, impinging jets (doublets and triplets), splash plates, and swirl sprays involving both hollow and full cone sprays. See Sec. 5.4.1 and other texts on liquid rockets for a detailed discussion of these injectors. In the early French and Swedish programs [Altman, 1991, Calabro, 1991], for which nitric acid was the oxidizer, the main solid fuel was hypergolic with the oxidizer, so no separate ignition fuel was necessary. If we use a precombustor, we select the length-to-diameter ratio (L/D) so there is enough residence time to vaporize the oxidizer. For typical arrangements, the L/D of the precombustor area is about 0.5. For details on sizing and estimating the mass of injectors, see Chap. 5.

7.6.3 Aft Mixing Chamber

As the boundary layer builds down the length of a fuel port, concentrations vary perpendicular to the port wall. If we exhaust the propellant immediately at the end of the port, all possible combustion is not complete. To get the best performance from propellants, we must mix the gasified propellants to make sure the combustion process is complete before the gases enter the exhaust nozzle.

At the end of the grain, there is usually insufficient length to give good mixing, so we must provide more mixing. To do so, we can add length to the motor, provide a physical mixing surface (vortex generators), use a submerged nozzle, or inject a gas or some of the oxidizer into the aft end. The most practical approach has been to provide more length and a submerged nozzle. Typically, the aft mixing section has an L/D of 0.5 to 1.0. To account for this mass, we simply add the length to the combustion chamber. Figure 7.24 schematically shows an aft mixing area with a submerged nozzle.

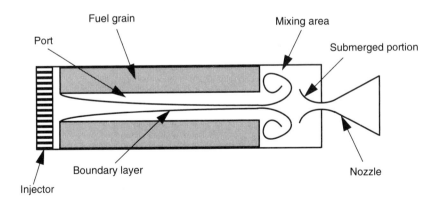

Fig. 7.24.   Schematic of Aft Mixing Area with a Submerged Nozzle. We need enough volume at the end of the propellant grain so the propellants can completely mix and react.

7.6.4 Nozzle Design

Nozzle design closely parallels that of the solid rocket, with one major difference. In the solid rocket, the oxygen-to fuel ratio (O/F) is constant. As discussed in previous sections, for hybrid rockets, O/F typically increases with time and becomes oxygen-rich in the latter stages of the burn. Hence, the nozzle material must withstand an oxygen-rich atmosphere. Graphite, good for solids, is not good for hybrids that go oxygen-rich, because graphite oxidizes easily. Phenolic-based materials are better in this case.

Equation (7.103) gives an empirically derived expression for nozzle mass:

$$m_{noz} = 125 \left(\frac{m_{prop}}{5400} \right)^{\frac{2}{3}} \left(\frac{\varepsilon}{10} \right)^{\frac{1}{4}} \qquad (7.103)$$

where m_{noz} = nozzle mass (kg)
m_{prop} = total mass of propellant (kg)
ε = nozzle expansion ratio

Regeneratively cooled nozzles (Chap. 5) are possible for hybrids but are more complex than the standard ablative nozzles discussed in Chap. 6. If we use the oxidizer as the coolant, the nozzle may oxidize (burn), so we must carefully analyze this approach before using it.

7.6.5 Thrust-Vector Control

Techniques for controlling the thrust vector are the same as for solids with one exception. Liquid Injection Thrust Vector Control (LITVC) is easier to use in hybrids because the system already has a liquid. The mass penalty for additional liquid is small and is easily accounted for in the design.

7.6.6 Summary of Relationships between Mass Estimates for Components

Because there is virtually no published database concerning mass estimates for hybrid rockets, we estimate system masses using numbers for liquid and solid rockets, as appropriate. Table 7.10 summarizes the equation or section in which the component is discussed.

Of course, these relationships derive from empirical data on existing systems, so they may or may not apply to hybrids, depending on vehicle design. For example, we typically do not use separate internal insulation in hybrid rockets. Instead, we add a bit of excess fuel. Another example is nozzle design, which is very specific for solid rockets. It must account for particulate erosion and other solid-rocket problems. However, without a good database for hybrids, we rely on these numbers for estimates.

Table 7.10. Summary of Relationships between Mass Estimates for Components.

Component	Relationship
Motor case	• Section 6.3.3 (m_{mc})
Thrust skirt and polar boss	• Section 6.3.4 • $m_{tspb} = 0.1 m_{mc}$ (kg)
Nozzle	• Section 7.6.4 [Eq. (7.103)] • $m_{noz} = 125 \left(\dfrac{m_{prop}}{5400} \right)^{\frac{2}{3}} \left(\dfrac{\varepsilon}{10} \right)^{\frac{1}{4}}$ (kg) • m_{prop} = total propellant mass (kg) • ε = nozzle expansion ratio
Thrust-vector control	• Section 6.3.8 (m_{tvc})
Injector	• Section 5.4.1 (m_{inj})
Oxidizer tank	• Section 5.4.4 (m_{tank})
Pressurant system	• Section 5.4.5 (m_{press})
Turbo-pumps	• Section 5.4.3 (m_{pump})
Feed system	• Section 5.4.2 (m_{fs})
Structural supports	• $m_{ss} = 0.1 [m_{mc} + m_{noz} + m_{tvc} + m_{inj} + m_{tank} + m_{press} + m_{pump} + m_{fs}]$ • 10% of the other inert masses

7.7 Case Study

In Chap. 10, we derive propulsion-system requirements for a liquid-rocket system assuming that the propellants are RP-1 (kerosene) and liquid oxygen (LOx). The performance of these propellants is very close to the performance of a polybutadiene / LOx hybrid system, so we believe we can design a hybrid propulsion system based on these similar propellants.

7.7.1 Summarize Requirements

We intend to design the Option 4 system outlined in Chap. 10. As in the design case study for liquid rockets (Sec. 5.5), we need to add margin to the numbers to increase the likelihood that our design is successful. Remember that this option is a two-stage system in which the first stage is a high-thrust chemical rocket providing a transfer from low-Earth orbit to a 5000-km circular orbit. From there, a low-thrust ion rocket is used to spiral up to geostationary altitude. Table 7.11 lists the key requirements and suggests reasonable design margins.

Table 7.11. Summary of Propulsion-System Requirements. The middle column gives the requirement from Chap. 10, and the far right column discusses the approach we use to ensure the requirement is met.

Name	Requirement	Modification
Maximum initial vehicle mass	12,000 kg	We iteratively select the structural mass fraction to get this number. Hopefully, the resulting mass fraction is conservative.
Payload mass	4914 kg	Includes 3387-kg payload and an electric stage
Δv	1721 m/s	1893 m/s (10% margin)
Minimum thrust-to-weight ratio	0.3	We choose 1.5 to decrease the burn duration.
Envelope limits	3 m long x 3 m diameter	Diameter is not a problem, but length may be difficult to meet.

In any design, some iterating and optimizing are necessary. Usually, we can iterate to develop a "better" design until the end of time. In the following design, we make several decisions necessary to the design so we can illustrate the process. These decisions give us a workable design (ignoring the length restriction) but are not "optimal." We could then vary the decision parameters to derive a better design.

7.7.2 Make Preliminary Design Decisions

Basic Decisions

As mentioned above, we need to make several decisions that allow us to design the system. These decisions generate a solution but are not necessarily optimal. To find an optimal solution, we would iterate on these decisions to improve the result.

- Number of ports = 8 (wagon wheel with no center hole burning)
- Initial O/F = 1.2
- Initial thrust-to-weight ratio = 1.5
- Initial chamber pressure = 1.4 MPa
- Helium gas pressurant stored in a titanium pressure vessel
- Conventional metallic structure for thrust chamber and oxidizer tank

Choose Propellants

The mission-level analysis (Chap. 10) suggests that RP-1 and LOx propellants are a good choice to meet the requirements. Review of Appendix B indicates that the performance of these propellants is very similar to that of HTPB / LOx, so we choose them for our preliminary design. The optimal O/F is 2.1, based on the frozen-flow thermochemical analysis presented in Appendix B. Remember that the O/F shifts over the burn duration. We want to restrict the O/F shift to the smallest range possible. Thus, we choose an initial O/F of 1.2.

Determine Engine and Feed System Pressure Levels

We start by looking at how chamber pressure affects system performance. Assuming the optimal O/F of 2.1 as an analysis point, we can determine the thermochemical parameters for combustion from Appendix B (assume a frozen-flow combustion efficiency of 1.0):

- Combustion temperature = 3593 K
- $\gamma = 1.231$
- Molecular mass = 22.84 kg/kmol
- $c^* = 1747.4$ m/s   [Eq. (7.71)]

Using the process outlined in Table 5.5, we can calculate the variation in vacuum specific impulse as a function of nozzle expansion ratio. Fig 7.25 shows this relationship, assuming a nozzle efficiency of 0.98. A reasonable expansion ratio is 70, using the argument of diminishing returns. This assumption gives us a specific impulse of 330 s.

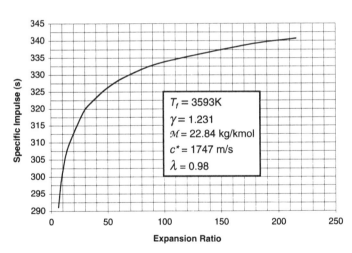

Fig. 7.25. Performance of Vacuum Specific Impulse for Polybutadiene and Liquid Oxygen. The chart shows the performance at an O/F of 2.1 and a nozzle efficiency of 98%.

Because we are designing a space engine (expanding to a vacuum) the specific-impulse performance is a function of expansion ratio alone (independent of chamber pressure). This characteristic allows us to choose a low chamber pressure which, in turn, lowers the mass of the overall system. The only negative impact that low chamber pressure has on our system is that the nozzle can get large. A reasonable lower limit for combustion-chamber pressure is the 1.4 MPa mentioned above.

- Chamber pressure = 1,400,000 Pa (203 PSI) (decision)
- Injector pressure drop = 0.2(1,400,000) = 280,000 Pa [Eq. (5.20)]
- Feed line pressure drop = 50,000 Pa [Eq. (5.18)]
- Oxidizer dynamic pressure = 0.5(1142)(10)^2 = 57,100 Pa [Eq. (5.16)]
- Oxidizer tank pressure = 1,400,000+280,000+50,000+57,100 = 1,787,100 Pa
- Pressurant tank pressure = 21,000,000 Pa (3046 psi) (decision)

Notice we have excluded a pump to meet the simplicity requirement (see Chap. 10). From the nozzle analysis used to generate Fig. 7.25, we get the following for an expansion ratio of 70:

- Nozzle exit Mach number = 4.86 [Eq. (5.15)]
- Nozzle exit pressure = 1255 Pa [Eq. (5.14)]

Determine Initial Propellant Flow Requirements

To size our ports, we need to determine the initial thrust magnitude and determine the initial mass flow rates required to achieve this thrust level. Our requirement, as discussed in Sec. 7.7.1, is for an initial vehicle mass of 12,000 kg and a minimum thrust-to-weight ratio (F/W) of 0.3. We want to keep the initial F/W as low as possible to keep the nozzle size as small as we can. To avoid excessive burn time, however, we have selected an initial F/W of 1.5, as mentioned above.

We also need to evaluate the thermochemistry at the initial design point. Assuming an initial O/F of 1.2 (also mentioned above), so we can get a reasonable O/F shift, the thermochemical parameters are:

- $\gamma = 1.304$
- Combustion temperature = 2203 K
- Molecular mass = 17.2 kg/kmol
- $c^* = 1545$ m/s

These parameters give us an initial specific impulse of 280 s (Eq. 5.11). The initial mass flow rate then becomes

$$\dot{m}_{prop} = \frac{F}{I_{sp}\, g_0} = \frac{1.5\,(12,000)\,(9.807)}{280\,(9.807)} = 64.2 \text{ kg/s}$$

We can now determine the initial fuel and oxidizer flow rates from the initial O/F:

$$\dot{m}_{fuel} = \frac{\dot{m}_{prop}}{1+O/F} = \frac{64.2}{1+1.2} = 29.2 \text{ kg/s} \quad [\text{Eq. (5.29)}]$$

$$\dot{m}_{ox} = \dot{m}_{prop}\frac{O/F}{1+O/F} = 35.0 \text{ kg/s} \quad [\text{Eq. (5.30)}]$$

Size System

We can determine the total mass of propellant required from the ideal rocket equation if we assume an average specific impulse. Typically, we get specific impulse losses from the O/F shift that vary between 1% and 2% of the maximum. We take the maximum I_{sp} (330 s) and reduce it by 7 seconds to account for the O/F shift loss. As we shall see in the following performance evaluation, this number is a bit conservative, as we would like.

$$m_f = m_i \, e^{\frac{-\Delta v}{I_{sp}\, g_0}} = 12,000 \, e^{\frac{-1893}{323\,(9.807)}} = 6601 \text{ kg}$$

Then the propellant mass is just the difference between the initial and final masses (m_i and m_f):

$$m_{prop} = 12,000 - 6601 = 5399 \text{ kg}$$

Assuming an average O/F of 2.3 (an assumed parameter verified by the simulation in Sec. 7.7.3), we get the individual propellant masses:

- Total fuel mass = 1636 kg
- Total oxidizer mass = 3763 kg
- Total allowable inert mass = 12,000 − 4914 − 1636 − 3763 = 1736 kg

Configure Combustion Ports

To uniquely configure the ports, we must decide on the initial oxidizer flux and the number of ports. We choose the maximum recommended flux number and 8 ports (decision made above) for this design point. The maximum flux number minimizes the O/F shift.

- Initial oxidizer mass flux = 350 kg/m^2·s
- Number of ports = 8

The oxidizer flow per port ($\dot{m}_{op}$) is

$$\dot{m}_{op} = \frac{\dot{m}_{ox}}{N} = \frac{35.0}{8} = 4.38 \text{ kg/s}$$

This number is held constant over the burn duration. The initial port area and fuel flux become

$$A_{pi} = \frac{4.38}{350} = 0.0125 \text{ m}^2 \qquad \text{[Eq. (7.82)]}$$

$$G_{fi} = \frac{G_o}{O/F_i} = \frac{350}{1.2} = 292 \text{ kg/m}^2\cdot\text{s} \qquad \text{[Eq. (7.83)]}$$

For this many ports, we choose a wagon-wheel configuration with no burning in the center hole. For the 8 ports we have specified, we can completely define the initial port configuration (see Fig. 7.23):

$$\theta_p = \frac{\pi}{8} = 0.3927 \text{ rad} \qquad \text{half angle of the port triangle}$$

$$h = \sqrt{\frac{A_{pi}}{\tan \theta_p}} = \sqrt{\frac{0.0125}{\tan (0.3927)}} = 0.1738 \text{ m} \qquad \text{triangle height}$$

$$b = 2h \tan \theta_p = 0.1440 \text{ m} \qquad \text{triangle base}$$

$$P_i = \frac{2h}{\cos \theta_p} + b = 0.5201 \text{ m} \qquad \text{triangle perimeter}$$

We choose the following regression-rate parameters with the help of Table 7.5:
- $n = 0.75$ (flux exponent)
- $m = -0.15$ (length exponent)
- $a = 0.000\,02$ (coefficient)

We can determine the port length,

$$L = \left(\frac{\dot{m}_f}{N\rho a G^n P}\right)^{\frac{1}{m+1}} \qquad \text{[Eq. (7.88)]}$$

$$L = \left(\frac{29.2}{8\,(950)\,(0.000\,02)\,(350+292)^{0.75}\,(0.5201)}\right)^{\frac{1}{1-0.15}} = 3.493 \text{ m}$$

We know that each port must contain one-eighth of the total fuel. Remembering that the port half angle does not change over the burn, we can determine the final port dimensions using the equations in Fig. 7.23:

$$V_{fp} = \frac{m_{fuel}}{N\rho} = \frac{1636}{8(950)} = 0.2153 \text{ m}^3 \qquad \text{fuel volume per port}$$

$$A_{pf} = \frac{V_f}{L_p} - A_{pi} = 0.0741 \text{ m}^2 \qquad \text{final area of the port}$$

$$w = 0.0799 \text{ m} \qquad \text{fuel-web thickness}$$

$$P_f = 1.0222 \text{ m} \qquad \text{final perimeter of the port}$$

$$S_f = P_f L = 3.5706 \text{ m}^2 \qquad \text{final surface area of the port}$$

This geometry allows us to solve numerically for the port regression rate so we can determine the final O/F:

$$O/F_f = \frac{G_{of}}{G_{ff}} = 2.933$$

We determine the final grain geometry by using the last two equations in Fig. 7.23:
- Grain radius = 0.4675 m
- Center hole radius = 0.1289 m

Figure 7.26 shows a top view of the grain configuration. Notice we have not allowed any extra space for structural support of the grain.

7.7.3 Estimate Performance

Now that we have defined the grain and nozzle dimensions, we have enough information to simulate the system performance over time. Table 7.9 gives the simulation algorithm. To evaluate the thermochemistry changes as the O/F changes, we use "curve fit" expressions from Appendix B.

Basically, we take the initial grain configuration and a constant oxidizer flow rate. As the port area increases because of fuel regression, all of the thermodynamic properties, including thrust and specific impulse, also change. For this preliminary design, we assume that we have no throat erosion. This assumption is a bit optimistic, as throat erosion tends to decrease the expansion ratio and thus the overall specific impulse.

Using an Euler integration scheme and a time-step size of 1 second gives good results. Table 7.12 lists selected parameters. Average values for the run are
- O/F = 2.260
- Specific impulse = 323.7 seconds (better than assumed)

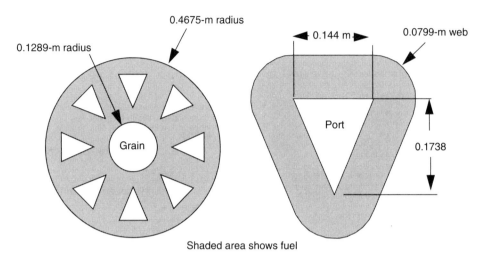

Fig. 7.26. Grain Configuration. This is a top-view drawing of the grain layout, looking down the ports. Eight ports are shown in their initial configuration with a nonburning center hole. The additional figure shows a single port.

Table 7.12. Selected Simulation Results. The table shows selected results of the simulation analysis for the nominal design point.

Time (s)	Web (cm)	Port Area (m^2)	O/F	I_{sp} (s)	Thrust F (N)	F/W
0	0	0.01251	1.200	280.3	176,467	1.50
10	1.621	0.02176	1.690	324.2	177,200	1.58
20	2.731	0.02905	1.965	330.0	171,042	1.61
30	3.626	0.03550	2.161	329.9	165,711	1.63
40	4.393	0.04142	2.317	328.3	161,437	1.67
50	5.074	0.04699	2.446	326.5	157,992	1.73
60	6.691	0.05228	2.557	324.8	155,168	1.79
70	6.258	0.05736	2.655	323.2	152,809	1.86
80	6.786	0.06227	2.743	321.8	150,804	1.95
90	7.280	0.06702	2.822	320.5	149,071	2.05
100	7.746	0.07165	2.895	319.3	147,552	2.17
106	8.014	0.07437	2.936	318.7	146,724	2.25

7.7 Case Study

Figures 7.27, 7.28, and 7.29 show the O/F, the thrust level and the vehicle's overall thrust-to-weight ratio as a function of time. As expected, the O/F increases and the thrust decreases over the burn. Notice that the vehicle's F/W increases even though the thrust level is dropping.

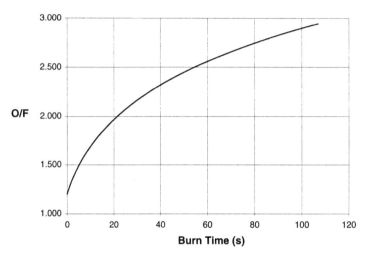

Fig. 7.27. O/F Ratio Shift Versus Burn Time. The O/F increases through the burn as expected.

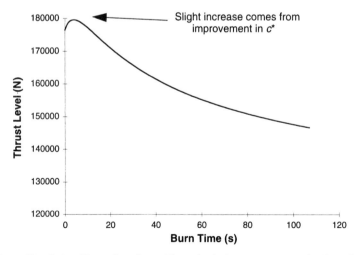

Fig. 7.28. Simulation Thrust Level over Time. As the burn progresses, the thrust level drops. The slight increase at the start is because of the rapid increase in O/F and the corresponding increase in characteristic velocity.

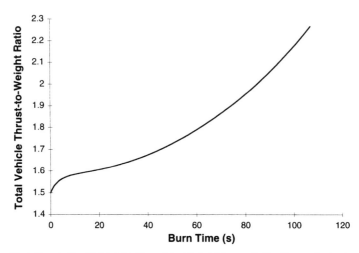

Fig. 7.29. Variation in the Vehicle's Thrust-to-Weight through the Burn. As the burn progresses, the F/W increases despite the decrease in thrust level.

7.7.4 Size and Configure Components

Now that we are fairly confident our system meets the basic performance requirements, we must configure the various systems. From the ideal-rocket analysis, we know

- Total mass = 12,000 kg
- Payload mass = 4914 kg
- Propellant mass = 5399 kg
- Inert mass of the stage = 12,000 − 5399 − 4914 = 1687 kg
- Inert-mass fraction = $\dfrac{1687}{12,000 - 4914}$ = 0.238

The sum of the individual component masses must be less than the inert mass number by some reasonable margin.

Nozzle—We start the component analysis at the nozzle and work back through the system. We need to design our nozzle so we get the correct initial mass flow rate (to get the specified initial thrust level). From the characteristic velocity equation, we can determine the throat area (A_t). We can then determine the nozzle exit area (A_e) from the expansion ratio (ε). We choose the initial condition to specify the thrust level [Eq. (3.133)].

$$A_t = \frac{\dot{m} c^*}{p_c} = \frac{64.2\,(1545)}{1,400,000} = 0.0708 \text{ m}^2$$

7.7 Case Study

$$A_e = A_t \varepsilon = 0.0708\,(70) = 4.959 \text{ m}^2$$

These translate into:
- Throat diameter = 30.03 cm
- Exit diameter = 251.3 cm

If we were to use a conical nozzle with a 15° half angle, the length of this nozzle would be:

$$\frac{2.513 - 0.300}{2\tan 15} = 4.128 \text{ m} \qquad [\text{Eq. (5.41)}]$$

Using the bell nozzle figures in Chap. 5 (Fig. 5.25) and requiring a 98% nozzle efficiency, we can determine the bell nozzle's length:
- Bell-nozzle fraction = 0.675
- Nozzle length = 0.675 × 4.128 = 2.787 m

We then use the empirical mass equation for phenolic nozzles [Eq. (7.103)] to determine the nozzle mass:

$$125\left(\frac{5399}{5400}\right)^{\frac{2}{3}}\left(\frac{70}{10}\right)^{\frac{1}{4}} = 203 \text{ kg}$$

Combustion chamber—The next step is to evaluate the thrust chamber. The port length is 3.493 meters, but we must allow for an aft mixing section and an injector section. The typical rule of thumb for the aft mixing section is 0.5–1.0 grain diameter. We choose the total grain diameter:
- Grain length = 3.493 m
- Fore and aft mixing section length = 2 × 0.4675 = 0.9350 m
- Chamber length = 3.493 + 0.9350 = 4.4280 m

The injector section uses a hemispherical dome with a radius equal to the case radius (0.4675 m). Figure 7.30 shows the engine configuration.

The process for estimating the thrust-chamber mass is similar to that for the cases of solid rocket motors (Sec. 6.3.3). To be conservative, we assume an aluminum case material with material properties taken from Table 6.4:
- Ultimate allowable stress (F_{tu}) = 414,000,000 Pa
- Density = 2800 kg/m^3

We assume that burst pressure is twice the maximum expected operating pressure:
- Burst pressure = 2 × 1,400,000 = 2,800,000 Pa

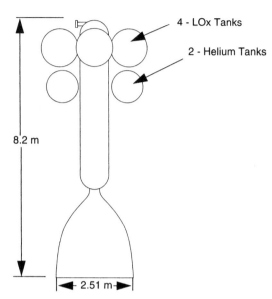

Fig. 7.30. Configuration of the Thrust Chamber. This is the design-point configuration showing the main dimensions.

From Eq. (6.11), we know the required wall thickness is (assuming a thickness factor of 1.5 to allow for stress concentrations):

$$t_{cs} = \frac{2\,(1.5)\,(2,800,000)\,(0.4675)}{414,000,000} = 0.0048 \text{ m}$$

The approximate material volume (surface area × thickness) is 0.0754 m³, giving a

- Mass of the combustion chamber case = 209 kg

If we estimate the interior volume of the tank assuming a cylinder capped by two hemispherical ends, we calculate a volume of 3.47 m³. Multiplying this by the chamber pressure and dividing by the case weight, we end up with a structural mass factor of 2366 m. This result agrees with Chap. 5 for typical numbers (2500 m).

We estimate the injector mass by calculating the mass of a 2.5-cm thick aluminum plate spanning the chamber radius:

- Injector mass = $2800 \times \pi(0.4675)^2(0.025) = 48$ kg

The total thrust-chamber mass is:

- Thrust-chamber mass = 209 + 48 + 203 = 460 kg

Oxidizer tank—The required volume of oxidizer is 3.295 m³ (allowing for 10% ullage), and the pressure is 1,787,100 Pa from Sec. 7.7.2. Assuming a structural tank factor of 2500 m (see Sec. 5.4.4) for an aluminum tank and assuming a 2.0 burst-pressure factor, we can determine the oxidizer tank's mass:

- Oxidizer tank's mass = 528 kg.

Pressurant system—Using the process discussed in Sec. 5.4.5 for designing a regulated pressure system for isentropic blowdown with a helium pressurant, we need

- Pressurant volume = 1.132 m³
- Pressurant mass = 42 kg (assuming 21,000,000-Pa initial pressure)
- Tank mass = 191 kg (assuming 12,700 as the pressurant tank factor)

Support structure and ancillary parts—As in the other sections, we allow for mass of the support structure by assuming it is 10% of the inert mass:

- Mass of the support structure = 0.1 × (460 + 528 + 191) = 118 kg

Summary—Adding up the masses of all components gives

- Total inert mass = 1339 kg
- Inert mass margin = 1687 − 1339 = 348 kg

These values give us a 27% margin in inert mass. Figure 7.30 shows the layout of the overall system.

7.7.5   Iterate

The analysis of this case study in the previous sections gives us a single point in the "design space." Varying the decision parameters (Sec. 7.7.2) could allow us to improve our design somewhat. For example, decreasing the initial thrust-to-weight ratio or increasing the chamber pressure decreases the nozzle size. Changing the number of ports changes the volumetric efficiency, which in turn changes the size of the thrust chamber.

Notice that we have violated the length constraint. Long, narrow, thrust chambers are typical for hybrid rockets. It is unlikely that we can design a practical hybrid rocket system to meet the length requirement. Reducing the nozzle expansion ratio, increasing the chamber pressure, and using a segmented and deployable nozzle skirt can substantially decrease the nozzle size. Decreasing the fore and aft mixing area and increasing the number of ports can decrease the combustion-chamber length. However, extreme changes in these design parameters compromise the utility of this system, making the liquid system much more attractive.

References

Altman, D. 1991. Hybrid Rocket Development History. Paper presented at the AIAA/SAE/ASME/ASEE 27th Joint Propulsion Conference.

Altman, D. 1991. Scaling of Hybrid Motors. Paper presented at the AIAA 29th Aerospace Sciences Meeting. Hybrid Propulsion Lecture Series. Reno, NV.

Altman, D. and H. Wise. 1956. Effect of Chemical Reaction in the Boundary Layer on Convective Heat Transfer. *Jet Propulsion*. April: 256–269.

Calabro, M. 1991. (Jan. 10). European Hybrid Propulsion History. AIAA Hybrid Propulsion Lecture Series.

Chemical Systems Division, United Technologies. c. 1965. Private communication.

CSD/NASA. 1993. Data acquired through private communication with Ben Goldberg at the Marshall Space Flight Center. Work done under NASA contract by Hercules/United Technologies, Chemical Systems Division and by the JIRAD group.

Feemster, J.R. 1968. Demonstration of a High Thrust Hybrid Thrust Chamber Assembly. USAF AFRDL-TR-68-56.

LaForce, A. D. and H. Wolff. 1970. Demonstration of an Upper Stage High Energy Hybrid Rocket. Paper presented at the 12th JANNAF Liquid Propulsion Meeting.

Lees, L. 1958. "Combustion and Propulsion." *Third AGARD Colloquium*. New York: Pergamon Press.

Marxman, G. A. 1966. "Boundary-Layer Combustion in Propulsion" Eleventh Symposium (Int'l) on Combustion. p. 269. (Comments by attendees discussed an alternate kinetic effect due to surface reaction (paper by Price and Smoot). The equation is: $r = CDG^nP/(CP+DGPTnPT)$

Marxman, G. A. 1964. "Combustion in the Turbulent Boundary Layer on a Vaporizing Surface" Tenth Symposium (Int'l) on Combustion. p. 1337. (Provides the derivation for C_H/C_{H_0} in terms of B. The expression $C_f/C_{f_0} = 1.2\ B^{-.77}$ is derived from skin-friction coefficient with blowing from the data of Mickley and Davis)

Marxman, G. A., C. E. Wooldridge, and R. J. Muzzy. 1964. "Fundamentals of Hybrid Boundary-Layer Combustion." *Progress in Astronautics and Aeronautics*. p. 485. NY: Academic Press. (Treats r with radiation and transient r. Also gives equation for flame height.

Mickley, H.S. and R. S. Davis. 1957. *Momentum Transfer for Flow over Flat Plate with Blowing*. NACA TN-4017. Washington, DC: National Advisory Committee on Aeronautics.

Moore, G. E. and K. Berman. 1956. A Solid-Liquid Rocket Propellant System. *Jet Propulsion*. Vol. 26: 965–968.

Muzzy, R. J. 1963. Schlieren and Shadowgraph Studies of Hybrid B-L-Combustion. *AIAA Journal*. 1:2159.

Ordahl, D. D. and D. Altman. 1962. Hybrid Propellant Combustion Characteristics and Engine Design. Paper presented at the Third Symposium on Advanced Propulsion Concepts. U.S. Air Force/OSR.

Smoot, L.D., C. F. Price, and C. M. Mihlfeith. 1966. The Pressure Dependence of Hybrid Fuel Regression Rates. AIAA Preprint No. 66113, 3rd Aerospace Sciences Mtg.

Wooldridge, C.E., G. A. Marxman. and R. J. Kier. 1969. *Investigation of Combustion Instability in Hybrid Rockets.* Final report NASA CR-66812, Contract NAS 1-7310.

Bibliography

Barrere, M. and A. Moutet. 1963. La Propulsion par Fusées Hybrides. Paper presented at the International Astronautical Congress, September.

Green, L., Jr. 1964. Heterogeneous Combustion. *Progress in Astronautics and Aeronautics* 15: 45.

Marxman, G. A. and M. Gilbert. 1962. "Turbulent Boundary Combustion in the Hybrid Rocket" Ninth Symposium (Int'l) on Combustion. p. 371. (Combustion zone in the boundary layer occurs at about 1/2 $(O/F)_{st}$ due to diffusion limitation, also observed by Rocketdyne.

Muzzy, R. J. 1972. Applied Hybrid Combustion Theory. AIAA Paper 72-1143, AIAA/SAE 8th Joint Propulsion Specialist Conference. New Orleans, LA.

Netzer, D. W. 1972. "Hybrid Rocket Internal Ballistics" CPIA Publication 222. Baltimore, MD: Johns Hopkins University, Applied Physics Laboratory.

Ramohalli, K. and J. Yi. 1990. Hybrids Revisited. AIAA Paper 90-1962. AIAA/SAE/ASME/ASEE 26th Joint Propulsion Conf. Orlando, FL.

Rocketdyne Division of Rockwell International. 1962. *Research on Hybrid Combustion. Summary Report January 30, 1962.* Contract No. 3016(00).

Smoot, L. D. and C. F. Price. 1965. Regression Rates of Metallized Hybrid Fuel Systems. Paper presented at the AIAA 6th Solid Propellant Rocket Conference. Washington D.C.

Smoot, L.D., C. F. Price, and L E. Taylor. 1965. Regression Rate Mechanisms of Metallized Hybrid Fuel Systems. Paper presented at the AIAA 6th Solid Propellant Combustion Conference. February 1–3. [Uses Spalding thin film model to demo: $C_f/C_{fo} = \ln(1+B)/B$]

Wooldridge, C. E. and R. J. Muzzy. 1964. "Measurements in a Turbulent Boundary with Porous Wall Injection and Combustion" Tenth Symposium (Int'l) on Combustion. p. 1351.

Wooldridge, C.E. and R. M. Muzzy. 1966. Internal Ballistics Considerations in Hybrid Rocket Design. AIAA Paper No. 66-628, Colorado Springs. Treats burning rate corrections for metallized propellants and radiation. Gives calculated h_v for PMM as 350 cal/gm. Calculations of typical B values for hybrids in the range 5-20. Gives design equations for cartwheel grains. Discusses O/F shift, impact on average specific impulse and optimizing specific impulse. Also design equations for cartwheel grain.

CHAPTER 8
NUCLEAR ROCKET PROPULSION SYSTEMS

Timothy J. Lawrence, *U.S. Air Force Academy*
Jonathan K. Witter, *U.S. Nuclear Regulatory Commission*
Ronald W. Humble, *U.S. Air Force Academy*

8.1 Introduction
8.2 Design Process
8.3 Preliminary Design Decisions
8.4 Size the Reactor
8.5 Size the Radiation Shield
8.6 Evaluate Vehicle Operation
8.7 Case Study

Ideas for using nuclear energy for space propulsion began shortly after the first controlled nuclear chain reaction in 1942. Starting in the late 1940s, several development programs were pursued by the United States Air Force, the Atomic Energy Commission (now the Department of Energy), and the National Aeronautics and Space Administration. Some of the systems developed from these programs are still in use today. To use nuclear power for space propulsion, a propellant is heated in a suitable nuclear reactor to create hot, high-pressure gas which is expanded through a nozzle. The resultant high thrust and high specific impulse enhance or enable missions which may not be feasible using conventional chemical rocket engines. Nuclear reactors, at the simplest level, are heat sources; they can heat a propellant directly (nuclear thermal) or create electricity (nuclear electric). Although we can use this electric power to develop thrust (see Chap. 9), in this chapter we discuss only nuclear thermal rockets.

We intend to introduce how to design and analyze nuclear rocket propulsion systems. The design process applies basic theory to show how and why we configure nuclear systems as we do. Using additional data from current and past programs allows us to use requirements for the propulsion system to define a preliminary design.

8.1 Introduction

Most rockets are thermally driven gas devices in which energy is added in the form of heat. This heat energy ejects propellant from the engine, giving us the required momentum exchange or thrust. Energy can come from any number of sources. In chemical propulsion, the propellant releases energy through combustion. In a nuclear rocket, the propellant heats up when energy releases from the controlled fission of uranium or other fissionable material.

Fission involves the absorption of neutrons in a fuel material such as uranium (Fig. 8.1). This absorption excites the uranium atom until it splits into fragments and releases, on average, two new nuclei and one to three free neutrons. The fission fragments have high kinetic energy from the release of nuclear binding energy (see Sec. 8.4.1). This energy becomes thermal energy through collisions and interactions with other atoms. The neutrons also give up kinetic energy and slow down so they can be absorbed into the other fuel material. This process occurs more readily in lighter materials such as carbon, hydrogen, and beryllium because of their cross sections (Sec. 8.4.2). If each fission results in one other fission event, the core is said to be critical. Neutrons can either be absorbed by other engine materials or can leak from the reactor. The neutrons that leak out are lost from the cycle. Two or three neutrons are usually released in each fission event to ensure that at least one is absorbed by the fuel and causes another fission event.

The thermal energy produced from fission transfers to the coolant or propellant. For nuclear rockets, we refer to the solid uranium as the fuel and to the gas, such as hydrogen or ammonia, as the coolant or propellant. Conduction across the fuel material and convection into the coolant can heat the coolant gas to high temperatures (3000 K), limited only by the requirement to keep the fuel system below the fuel's melting point. The following sections explain these concepts of fission and heat transfer in more detail.

We can envision the nuclear rocket as a simple cold-gas thruster (discussed in Chap. 3) with a heat source added. As the propulsion system "fires" to generate thrust, acceleration, and velocity change, it consumes large quantities of propellant. To compare the efficiency of these different systems, we use specific impulse (Sec. 1.1.2). To increase specific impulse, the gas must have either a higher exit temperature or lower propellant molecular mass. Nuclear propulsion offers an advantage over chemical systems because we can choose the propellant with the lowest molecular mass. We can still impart large quantities of thermal energy to get high exit velocity without worrying about the combustion properties. Nuclear-

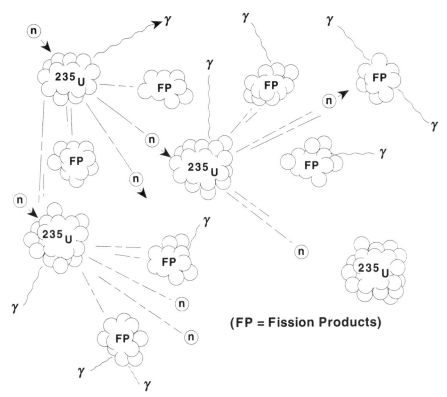

Fig. 8.1. **The Fission Chain Reaction.** The reactor materials are configured to support the controlled nuclear-fission chain reaction, so the number of neutrons generated equals the number of neutrons lost. Uranium (^{235}U) splits to give two products of the nuclear fission (FP), several neutrons (n), and gamma rays (γ) [Angelo & Buden, 1985].

fission rockets can have a specific impulse double that of chemical rockets. To get this advantage, we usually choose a lightweight gas, such as hydrogen, as the reactor coolant/propellant, but we can use higher-density propellants, such as methane, whenever storage volume is limited. Figure 8.2 shows fission rockets can produce high thrust levels (low specific mass) with good specific impulse. Having high specific impulse, high thrust, and high thrust-to-weight ratio is a tremendous advantage for a propulsion system. Systems with high specific impulse but high specific mass and low thrust, such as those using electric propulsion, require trip times of hundreds of days to go from low-Earth orbit (LEO) to geosynchronous-Earth orbit (GEO). But nuclear-propulsion systems need only hours.

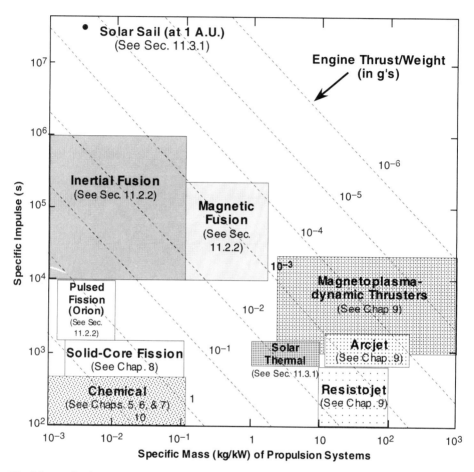

Fig. 8.2. Performance of Propulsion Systems. Fission rockets have better specific impulse than chemical rockets at equivalent thrust-to-weight and specific mass ratios [NASA, 1990].

Assuming we can build a nuclear system to achieve a specific impulse of 1000 s, we can compare Δvs (see Sec. 1.1.4 and Chap. 2) as follows:

1. **Launch vehicles**—require up to 10 kilometers per second Δv capability to place payloads into LEO—roughly from 250–1000 km altitude (see Sec. 2.6). Nuclear propulsion systems as a single stage, or as an upper stage on a chemical launch vehicle's existing first stage, can carry 4–6 times more payload than chemical systems because of their higher specific impulse.

2. **Orbital-transfer vehicles**—are designed to carry payloads (spacecraft) between different orbits. The Δv requirement for a mission from low-Earth orbit (LEO) to geosynchronous orbit is approximately 4–5 km/s (depending on the plane change requirement). A nuclear system can carry two times more payload than a comparable chemical system.
3. **Mars mission**—A manned mission to Mars has various Δv requirements. A long mission (hundreds of days) requires a Δv of approximately 3.5 km/s from LEO to Mars. A 40-day transfer from LEO to Mars requires 85 km/s. For a longer-duration mission, galactic radiation makes such space travel hazardous. In addition, humans suffer physical and mental difficulties in a constant free-fall environment (microgravity), so we must achieve the shortest possible trip time. Table 8.1 assesses total radiation exposure in relation to trip time for hypothetical Mars missions using chemical and nuclear propulsion. For both cases, the stay time on Mars's surface is 30 days. But because of the shorter length of the mission, we actually get *reduced* **radiation exposure by using nuclear propulsion**, as compared with a mission using a conventional chemical rocket.

Table 8.1. Comparison of Radiation Exposure for Nuclear and Chemical Systems. Nuclear systems can reduce overall radiation exposure by reducing the trip duration. This shorter trip time occurs because the high specific impulse of nuclear rockets allows higher Δvs for a given mission [Sager, 1993]. A *rem* (from roentgen equivalent man) is a measure of radiation dosage based on the type of radiation an individual receives (Sec. 8.5).

Radiation Source	433-Day Mars Mission Using a Chemical Rocket	316-Day Mars Mission Using a Nuclear Rocket
Van Allen belts	2 rem	2 rem
Mars surface	1	1
Galactic radiation	31	22
Solar flares	26	15
Reactor	--	5
Mission total	60 rem	45 rem

4. **Other applications**—A Δv of 12–20 km/s allows use of small, high-velocity projectiles as interceptors. A Δv of 15 km/s also could place payloads into an intercept trajectory with the Sun (1/10 perihelion of Mercury), which would allow the safe disposal of hazardous waste. Nuclear propulsion systems can use any working fluid as a propellant and reactor coolant. Hydrogen, ammonia, methane, octane, carbon dioxide, water, and nitrogen have been considered as propellants.

Because of their higher molecular mass, specific impulse is lower than for hydrogen. However, these working fluids offer advantages for long missions, in which storability is an issue, or for interplanetary missions, in which propellants could be acquired from a planet, moon, or asteroid.

The above discussion shows the tremendous payoffs from using nuclear propulsion. The payload savings alone offers huge economic benefits which justify the development costs. In many cases the high performance enables missions, such as interplanetary missions with a human payload, that are otherwise impossible.

8.1.1 System Configuration and Operation

A nuclear rocket's configuration is similar to that of a chemical system, except that it requires a nuclear reactor as a heat source. Figure 8.3 shows a typical nuclear propulsion system, which consists of a propellant tank, shielding, feed system (configured depending on type of engine cycle), reactor, and nozzle. The tank and feed system are very similar to those in chemical systems discussed in Chap. 5 and work in the same manner. The chemical rocket requires an oxidizer and fuel (each with its own feed system) for combustion to a hot gas. The nuclear rocket works the same way, except that only one propellant feeds through the core, where the reactor heats it to produce thrust. The main difference is engine control because the gas's thermodynamic conditions are coupled to the reactor and its unique requirements.

Figure 8.4 shows a schematic of a reactor for nuclear propulsion. The reactor is complicated by various components needed to keep the fission reaction under control. Let us discuss the major reactor components that differ from those of a chemical system.

- **Radial reflector**—On the outside of the core is a radial reflector. To have a controlled chain reaction (neutrons produced = neutrons used, see Sec. 8.3) and to reduce the size of the core, a reflector reflects neutrons produced in the chain reaction back into the core. We must prevent them from leaking out of the system in a large enough quantity to destroy the neutron balance and cause the reactor to shut itself down. The reflector is usually made of beryllium.
- **Reactor pressure vessel**—The reactor vessel is needed to maintain reactor pressure (3 MPa–8 MPa). It is made of aluminum or composite material to withstand the high radiation, heat flux, and pressures from the reactor. The vessel may require cooling to support the heat flux in some reactor designs.
- **Moderator**—Reactors are said to be either thermal or fast, depending on the neutron energy with which most of the fissions take place. In a thermal reactor, most of the fissions are caused by neutrons having an

8.1 Introduction

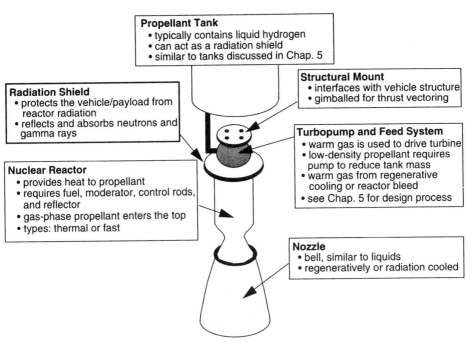

Fig. 8.3. Schematic of a Nuclear Rocket. A nuclear rocket operates as a monopropellant liquid system, with the nuclear reactor as a heat source.

energy less than 1 eV (an electron volt is a unit of energy equal to the energy acquired by an electron falling through a potential difference of one volt = 1.602×10^{-19} joules). Most neutrons produced from a nuclear-fission reaction have energies much higher than 1 eV. To slow the neutrons down, we use a moderator assembly made of a material with a low atomic mass (beryllium, plastics, lithium hydride, graphite). In a fast reactor, the energy range in which most of the fissions take place is much wider, extending from 100 keV to the top range of the fission spectrum (15 MeV). If we wish to build a fast reactor, we avoid light elements (moderating materials) and have no moderator. Section 8.4 discusses these concepts in more detail. Some reactors effectively mix moderating material with fuel material to limit the system's size and mass.

- **Fuel-element assembly**—The fuel-element assembly contains the uranium fuel and propellant/coolant flow channels. The fuel produces the heat to be transferred to the propellant flowing past the fuel. Different configurations of a fuel element can take advantage of surface area to better transfer heat and to make sure some kind of

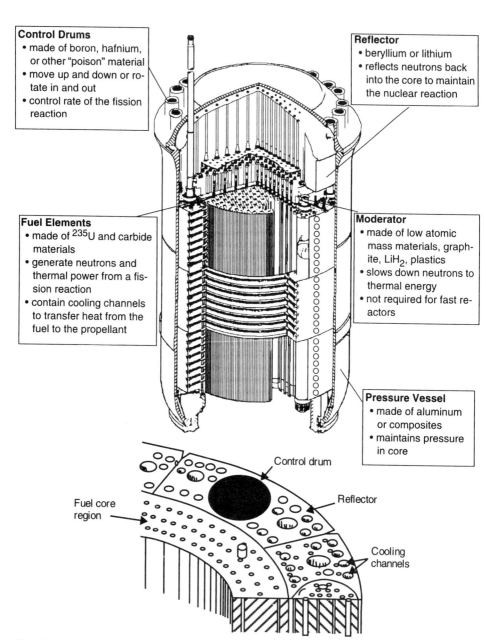

Fig. 8.4. Schematic of a Reactor in a Nuclear Propulsion System. All elements work together to control the neutron population and the energy level of individual neutrons. This control makes sure we have a reliable source of heat energy (NASA, [1990]).

barrier contains the fission products. The reflector, control rods, and moderator are placed around the fuel to maintain the proper flow and control of neutrons. Geometry of this configuration is important and is discussed in Sec. 8.4.

- **Control rods or drums**—The control rods or drums contain materials (usually boron) that absorb neutrons to decrease the neutron population. The rods control the reaction rate and can shut the reactor down. This material is known as a "poison" because it lowers the number of fission reactions when inserted in the core. The rods are dispersed around the core to ensure the neutron population can be properly controlled and adjusted to meet engine power level requirements. The control rods can be inserted into the reactor axially or rotated. For the axial insertion, the depth of the rods controls the amount of neutrons captured. For the rotation insertion, one side of the rod contains boron, whereas the other side contains beryllium. When the boron side is rotated into place, neutrons are absorbed. When the beryllium side is rotated, neutrons are reflected back into the core.
- **Coolant flow paths**—Coolant piping cools components and provides the propellant gas needed to generate thrust from the reactor. We usually want propellant to be completely vaporized before it enters the reactor core.

Now that we basically understand most of the components associated with the reactor, let us look at the operation of the core—discussed further in Sec. 8.6. The core is placed on the launch pad with the control poison fully inserted. With the control poison in this position, the core produces no power and has negligible radioactivity (only the natural radioactivity of the fuel). This condition allows workers to handle the reactor with no protective shielding. Once the mission requires thrust, the control poison is withdrawn (either lifted vertically or rotated outward) and a neutron source is put into the reactor to provide the initial neutrons for fissioning. With the control poison withdrawn, the fissioning causes the thermal power to increase exponentially to the desired power level. When the poison is removed, the feed system immediately supplies gas to the core for cooling and thrust production. When the reactor reaches the desired full-power level, the control poison's position is adjusted to keep the power at a steady-state (number of neutrons produced = number of neutrons used). This is a delicate balance.

At mission's end, the control poisons are inserted back into the core and the power decays exponentially. But the reactor needs to cool down for some time, depending on how long the core was at full power, and it may or may not need active cooling. This cooling may be necessary because of delayed neutrons and residual heat production that result from radioactive decay of by-products from the fission process ("fission products").

8.1.2 Concepts

The following section discusses nuclear systems that have been developed or proposed for propulsion applications in space. Many other possibilities exist (18 concepts proposed by NASA [1990]), but we believe these are the most likely candidates for near-term missions.

NERVA Derivative or Enabler

The NERVA (Nuclear Engine for Rocket Vehicle Applications) is a modified design of the reactor used in the NERVA program (NERVA-1). The NERVA program started in 1947 under the U.S. Air Force to design a reactor that could propel intercontinental ballistic missiles (ICBMs). In 1958, NASA took control of NERVA as part of their space-exploration program. NASA ran the program until 1972, achieving

- 28 full-power tests with restarts
- Up to 30-minute test duration
- Reactor sizes ranging from 300 MW to 200,000 MW
- Design of a hydrogen flow system
- Development of a way to contain affluents (propellant exhaust gases) for safe ground testing (used only in later tests)
- Development of a high-temperature fuel
- Solution to the problem of fuel erosion from hot hydrogen
- Specific impulse levels as high as 835 seconds
- Thrust levels as high as 890,000 N

The reactor developed for the NERVA program contains approximately 300 hexagonally shaped, graphite fuel elements in which uranium-carbide fuel particles coated with pyrocarbon are disbursed (see Fig. 8.5). The fuel particles provide the heat source while the graphite matrix serves as the moderator and structural component of the fuel elements. Niobium-carbide or zirconium-carbide coatings protect the surfaces of the fuel elements from the hydrogen propellant. Twelve rotary drums in the radial reflector control the core. The drums have boron plates which rotate in toward the core or out toward the perimeter, as required, to absorb and control the neutron population. A shield containing boron carbide, aluminum, and titanium hydride is at the top of the reactor. This shield limits nuclear-radiation heating of the engine assembly's nonnuclear components, including the propellant in the storage tank.

Particle-bed Reactor

The particle-bed reactor (PBR) (see Fig. 8.6) is designed to provide a core with a high power density (by increasing the propellant's temperature and the fuel's surface area) and a nuclear rocket engine with high thrust-to-weight. The core consists of a number of fuel particles (Fig. 8.6a) packed in a bed and surrounded by

8.1 Introduction

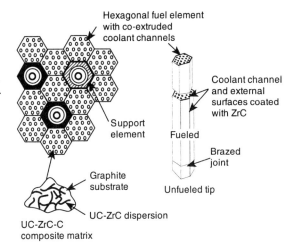

Fig. 8.5. **NERVA Fuel Element.** This figure shows how individual fuel elements are configured and nested together with additional elements for structural support [NASA, 1990]. Refer to Fig. 8.4 to see how this element fits into the core.

hexagonal moderator blocks arrayed in a cylindrical assembly (Fig. 8.6b). Its distinguishing feature is that the hydrogen propellant directly cools small (200–500-µm diameter), coated, particulate fuel spheres. The uranium-carbide fuel (coated with graphite buffer layers and an outer layer of zirconium hydride) is packed between two concentric, porous cylinders called frits, which confine the fuel but allow coolant flow. These small, annular fuel elements rest in a cylindrical moderator block. Candidate materials for the moderator block are beryllium or lithium hydride. Coolant flow moves radially inward through the cold frit, the packed bed, and the hot frit. At the same time, it moves axially out through the inner annular channel and then expands through the nozzle to produce thrust.

The PBR's advantages come from its high specific impulse, thrust, and thrust-to-weight ratio (~40:1). This performance enables missions that the NERVA, with its lower thrust-to-weight (~5:1) cannot do. The PBR has undergone several small proof-of-concept tests but has not had a full-scale engine test. Because the PBR and NERVA have had the most money invested in them, they represent near-term options. Most of the PBR development was done as part of the Air Force Space Nuclear Thermal Propulsion program (SNTP). This program lasted until 1993, with a budget of about $40 million per year. It intended to design a particle-bed reactor for various Air Force missions, but it ended having achieved

- Fuel tests to temperatures as high as 3000 K (NERVA only reached 2650 K)
- Nuclear tests of single fuel elements

- Criticality experiments for a prototype 1000-MW core
- Tests of thermal hydraulics in multiple fuel elements at high power densities (40 MW/liter as compared to about 2 MW/liter for NERVA)
- Various mission designs (detailed analysis of a stage for a mission to Mars)
- Verification of computer codes for simulating reactor physics

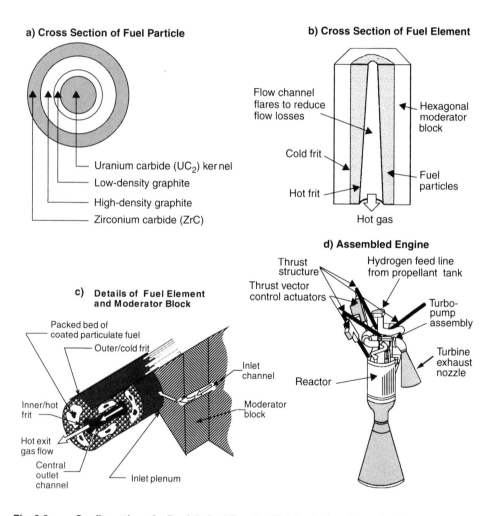

Fig. 8.6. **Configuration of a Particle-bed Reactor.** This illustration shows the details of a) a fuel particle; b) a fuel element; c) flow through the moderator block to the fuel bed; d) an assembled engine [NASA, 1990 and Maise, 1995].

CERMET

The CERMET-core nuclear rocket (see Fig. 8.7) uses a fast fissioning spectrum (greater than 1 MeV) compared to thermal reactors that slow down neutrons to fission energies of less than 1eV. Therefore, it does not need a moderator. The CERMET has a lower thrust-to-weight than the particle bed and has not been tested as extensively as the NERVA-type engines. Fuel tests show the CERMET-type fuel is much more robust than that for either NERVA or PBR. This feature makes CERMET attractive for applications such as a reusable orbital-transfer vehicle (OTV), for which we may need as many as 50 burns.

The reactor consists of hexagonal fuel elements similar to those in the NERVA design, except that the fuel is made of uranium-dioxide particles imbedded in a tungsten or tungsten/rhenium matrix. The advantages of CERMET fuel are its potential for very long operating life (more than 40 hours), ability to restart, handling of temperature cycling at high temperature, and greater compatibility between the fuel and hydrogen coolant (resulting in high fuel integrity and retention of fission products). However, the metal matrix can increase system mass because of competition for neutron absorption in uranium and tungsten. Using an axial, two-zone fuel element reduces system mass. In this two-zone concept, the fuel loading in the upper (low-temperature) half of the core uses a molybdenum/urania matrix configuration. The lower (high-temperature) part of the core uses a tungsten-rhenium/urania matrix configuration. This concept reduces the system's mass and gives it a thrust-to-weight ratio of 5.3:1—slightly better than NERVA. The system, shown in Fig. 8.7, consists of a CERMET core surrounded by a cooled pressure-containment shell. A neutron reflector and reactivity-control assembly mounts to the outside of the pressure vessel. So far, tests of the CERMET have checked only the fuel—up to 1900 K for 10,000 hours.

Comparing Reactors Depends on Mission

Of the reactors discussed here, NERVA-1 (the flight engine developed in the early 1970s) offers the lowest performance in terms of specific impulse. But it has had the most money invested in it and was ready for a flight before the program ended. The PBR offers the highest performance for a solid-core design. The CERMET may be a good design to pursue if reusability becomes an issue because its fuel lasts longer than other types of cores investigated. Table 8.2 shows the characteristics of the three engine concepts. Table 8.3 lists the reactors best suited for particular missions, with number 1 being the best.

8.2 Design Process

Figure 8.8 shows a preliminary-design flow chart for nuclear thermal propulsion. If we compare this figure with Fig. 5.7, we see it is very similar to liquid-rocket design. The key difference, of course, is the reactor design. In this chapter

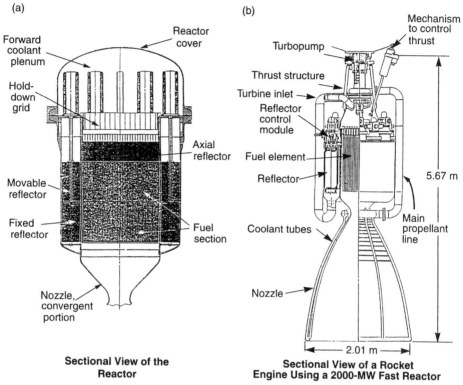

Fig. 8.7. Configuration of the CERMET Engine. (a) shows details of the reactor (b) shows a split view of the engine configuration, with interior details on the left side and exterior details on the right side [Bhattacharyya, et al., 1988].

we do not repeat the discussions of Chap. 5; instead, we concentrate on the aspects specific to designing nuclear thermal rockets.

Preliminary Design Decisions

In this step, we must evaluate the propellant chemistry, trade the nozzle's expansion ratio to determine specific impulse, and size the overall system using the rocket equation. But the major emphasis in this section (different from liquid rockets) is on determining the propellant-heating requirements. From the specific-impulse estimate and the required thrust level, we can determine the necessary mass flow rate. Knowing this rate, we must determine the amount of heat to deliver to the propellant. This delivered-heat number is directly proportional to the amount of power required from the reactor.

Table 8.2. Comparison Of Possible Near-Term Concepts for Reactors. The engine using a particle-bed reactor has higher performance and is lighter than the NERVA, but NERVA is more developed [Clark, 1991]. CERMET is a possible fast-reactor concept that we can also reuse.

	NERVA	Particle-Bed	CERMET
Power (MW)	1570	1945	2000
Thrust (N)	334,061	333,617	445,267
Propellant	H_2	H_2	H_2
Fuel element	solid rod	porous particle bed	solid rod
Maximum propellant temperature (K)	2361	3200	2507
I_{sp} (s)	825	971	930
Chamber pressure (MPa)	3.102	6.893	4.136
Nozzle expansion ratio	100	125	120
Engine mass (kg)	10,138	1705	9091
Total shield mass (kg)	1590	1590	1590
Engine thrust / weight (no shield)	3.4	20.0	5.0

Table 8.3. Applying Different Technologies to Missions. Numbers rank order the concept based on mission: 1—most applicable; 3—least applicable. We assume a near-term Mars mission that does not allow for the significant development required for other concepts.

Reactor	Mars	OTV	Military	Upper Stage	High Δv
NERVA	1	3	2	F/W too low	3
Particle-bed	2	2	1	1	1
CERMET	3	1	3	N/A	2

Determining the system pressure levels is similar to the method used for liquid rockets, except for the often-significant pressure drop through the reactor core.

Size the Reactor

Once we have the power requirement, we can size the reactor core. We look at three possible design approaches: NERVA-type, particle-bed reactor (PBR), or CERMET-type. The trick is to have a reactor large enough to produce the power and configured correctly to ensure criticality.

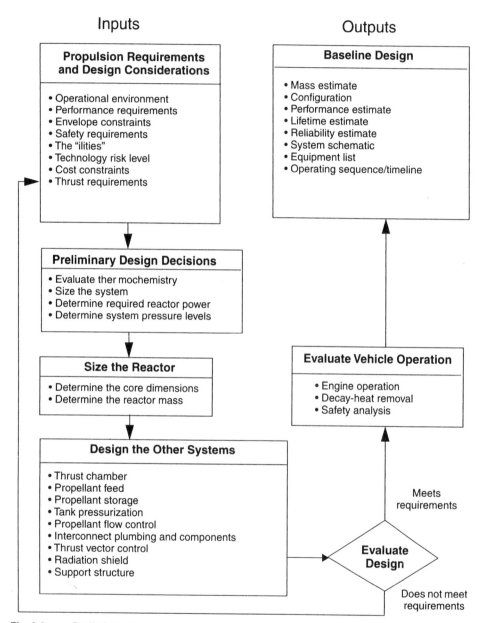

Fig. 8.8. **Preliminary Design for a System Using Nuclear Rocket Propulsion.** We can break this process into four distinct steps. They are actually coupled, but we treat them separately for preliminary design. Many possible iteration pathways are not shown.

We start the discussion with a review of basic nuclear physics. We then look at each of the three different reactors, going through theory appropriate for the NERVA-type cores to produce design curves. For the PBR and CERMET, we cannot rely on simplified theory, so we must present a family of curves derived from extensive computer simulations to do our preliminary sizing.

Design the Other Systems

Other systems on a nuclear engine are very similar to those on liquid rockets, except for the radiation shielding. Fission results in the release of energetic particles and electromagnetic radiation. We must protect the payload and other vehicle systems from this radiation.

Evaluate Vehicle Operation

Operating the engine, removing decay heat from the reactor, and analyzing safety are unique aspects of nuclear propulsion that affect our vehicle and its operation.

8.3 Preliminary Design Decisions

8.3.1 Evaluate Thermochemistry

Propellant Temperature

Evaluating the thermochemistry in nuclear rockets is simpler than in liquids because no combustion takes place. We are simply heating a working fluid from its storage temperature to the maximum chamber temperature, which depends not on the propellant but on the core materials and the configuration of the fuel elements. The maximum allowable temperature of the reactor core is our major concern. We can drive the temperature of the solid core materials as high as 3500 K, but the heat-transfer properties of the materials usually limit us to lower numbers.

The NERVA-type core restricts us to gas temperatures of about 2361 K (Table 8.2 gives typical numbers) because of its configuration. The periphery of the fuel rod is near the 2650-K limit, but there is a thermal gradient from the periphery to the coolant-channel wall near the center of the rod. Increasing the coolant (propellant) temperature raises the peripheral core temperature beyond the allowables, so melting results.

The CERMET-type core has a similar restriction. Fuel tests show that we can drive the propellant temperature up to 2507 K. The PBR reactors can go to a much higher temperature (3200 K) because the temperature change through the individual particles is much less than for the other reactor types.

Gas Properties

We evaluate gas properties (isentropic parameter and molecular mass) so we can evaluate the characteristic velocity (c^*). Both depend on our propellant choice. Molecular mass ($\mathcal{M}$) for the more common propellants (others are in Table 3.1) are

- Hydrogen (H_2) — 2.016 grams/mole
- Methane (CH_4) — 16.043 grams/mole
- Carbon dioxide (CO_2) — 44.01 grams/mole
- Water (H_2O) — 18.015 grams/mole

We can determine the isentropic parameter (γ) from (see Sec. 3.2.2)

$$\gamma = \frac{c_p}{c_p - \frac{8.314}{\mathcal{M}}} \tag{8.1}$$

taking the specific heat (c_p) from the following empirical equations, as a function of temperature (T in K). These relations are from VanWylen and Sonntag [1976] and are accurate from 300 K to 3500 K.

$$c_{pH_2} = \frac{1}{\mathcal{M}}\left[56.505 - 702.74\,\phi_T^{-0.75} + \frac{1165}{\phi_T} - 560.7\,\phi_T^{-1.5}\right] \tag{8.2}$$

$$c_{pCH_4} = \frac{1}{\mathcal{M}}\left[-672.87 + 439.74\,\phi_T^{0.25} - 24.875\,\phi_T^{0.75} + 323.88\,\phi_T^{-0.5}\right] \tag{8.3}$$

$$c_{pCO_2} = \frac{1}{\mathcal{M}}\left[-3.7357 + 30.529\,\phi_T^{0.5} - 4.1034\,\phi_T + 0.024198\,\phi_T^{2}\right] \tag{8.4}$$

$$c_{pH_2O} = \frac{1}{\mathcal{M}}\left[143.05 - 183.54\,\phi_T^{0.25} + 82.751\,\phi_T^{0.5} - 3.6989\,\phi_T\right] \tag{8.5}$$

$$\phi_T = \frac{T}{100} \tag{8.6}$$

Some of these values are plotted in Fig. 3.5 for the typical temperature range found in nuclear rockets.

Nozzle Expansion Ratio and Specific Impulse

Once we have determined the basic thermochemistry, we can use this data to trade the expansion ratio of the nozzle against specific impulse. We follow the procedure discussed in Table 5.5, but we find that the specific impulse level possible for nuclear rockets is usually much higher than that from chemical systems. We must accurately estimate specific impulse before proceeding to the next step: system-level sizing.

8.3.2 Sizing the System

Inert-Mass Fraction

Chapters 1 and 5 discuss the process for sizing the various systems, based on the ideal rocket equation. The major difference for nuclear rockets is in choosing the inert-mass fraction. Typically, nuclear engines are heavy, compared to liquid rockets, because of the large nuclear reactor mass. We also usually have a radiation shield which, in some cases, is as heavy as the engine. Finally, we usually choose hydrogen as our propellant—with density an order of magnitude lower than typical propellant combinations in chemical systems. All these effects combine to give us inert-mass fractions that can range from 0.5 to 0.7.

Unfortunately, we do not have a good database of flight engines to give us inert-mass fractions for nuclear systems. So we must estimate the inert-mass fraction and iterate back from our analysis of components.

Propellant Flow Rate

Given the specific impulse (I_{sp} from Sec. 8.3.1) and the required thrust level, we can determine the propellant mass flow rate from

$$\dot{m}_{prop} = \frac{F}{I_{sp}g_0} \tag{8.7}$$

where g_0 = 9.81 m/s^2
F = thrust (N)
I_{sp} = specific impulse (s)
$\dot{m}_{prop}$ = propellant mass flow rate (kg/s)

8.3.3 Determine the Required Reactor Power

To determine the reactor power needed to heat the mass flow to the desired temperature, we use an equation based on the first law of thermodynamics (from Secs. 3.2.2 and 3.2.3). Basically the heat added to the propellant flow, which is equal to the reactor power generated at steady-state, is equal to the increase in enthalpy of the flow. If we assume low flow velocity, the governing equation is

$$P_{core} = \dot{m}_{prop}\left(h_v + \int_{T_1}^{T_2} c_p dT\right) = \dot{m}_{prop} P \tag{8.8}$$

where P_{core} = reactor power generated (W)
$\dot{m}_{prop}$ = mass flow rate of propellant through the reactor (kg/s)
h_v = enthalpy required to vaporize the liquid propellant (J/kg)
T_1 = temperature of the propellant entering the reactor (K)
T_2 = temperature of the propellant leaving the reactor (K)

c_p = specific heat of the propellant (J/kg·K)
T = temperature of the propellant (K)
P = specific reactor power (W·s/kg)

Figure 8.9 shows the relationship between required reactor power and the desired maximum propellant-gas temperature, assuming the propellant is initially a liquid. We must use this chart carefully because the gas vaporizes at different temperatures and pressures, with a corresponding change in h_v, depending on our system. However, these results are accurate to a few percent, which is usually good enough for preliminary design. To use this curve, we choose the required propellant temperature, based on the reactor type and the typical numbers given in Table 8.2. Given this temperature, we can use Fig. 8.9 to determine the reactor power required to heat up 1 kg/s of the propellant flow. To determine the total reactor power required, we multiply the power number by the total propellant flow rate.

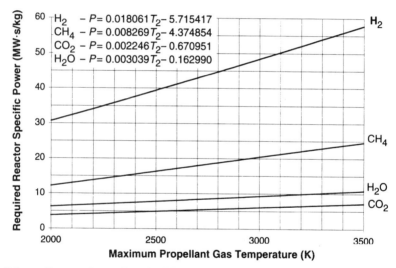

Fig. 8.9. **Reactor Power Required For a Given Propellant Temperature.** This chart shows the power required to heat a 1 kg/s flow to a particular temperature (T_2), assuming the gas is initially a liquid. To get power required for different flow rates (P_{core}), simply multiply the result (P) from the chart by the flow rate ($\dot{m}_{prop}$). The relations in the upper left corner are linear approximations of these complicated relationships.

8.3.4 Determine System Pressure Levels

The propellant feed system for a nuclear thermal rocket is similar to that for the chemical combustion of liquids. But nuclear rockets have some differences:

- There is no significant injector pressure drop. We want to keep the flow rates relatively low through the reactor core. Low-speed flow enhances heat transfer to the propellant and avoids flow-induced vibration. We do not need to accelerate the flow to aid mixing, vaporizing, or burning (Mach number is about 0.2) because the propellant is usually injected in a gaseous state.
- There is almost always a pressure drop in a regenerative cooling jacket. A typical range for this drop is 20% to 30%. The reactor vessel and exhaust nozzle are usually regeneratively cooled, with enough heat added to the propellant to completely vaporize it. We want the propellant to be a gas when it enters the reactor core.
- There is a pressure drop through the reactor core because of flow losses. In the NERVA-1 core, pressure dropped 27.5%, but proposed improved versions of this core show drops of only 10%. Design studies for particle-bed and CERMET-type reactors predict drops of 5% and 30%, respectively. The actual numbers (from Clark [1991]) are in Table 8.4. Pressure drop calculations and resultant sizing are important for detailed design of core fuel elements. Flow instabilities from viscous effects may exist and are a key consideration for this type of analysis.

Table 8.4. **Typical Pressure Numbers for Several Engine Designs.** We can use these indicative numbers to make initial estimates for various parameters requiring decisions.

	NERVA-1	Enhanced NERVA	Particle Bed	CERMET
Chamber pressure (p_c in Pa)	3,101,786	6,892,857	6,176,000	4,135,714
Pressure drop through core (Δp in Pa)	1,178,679	732,021	310,179	2,219,500
% of chamber pressure	38.0	10.6	5.0	53.7
% of reactor head pressure	27.5	9.6	4.8	34.9

8.4 Size the Reactor

Designing the reactor is the key step in designing a nuclear rocket. Essentially, once we have designed the reactor, we can treat it as the "heater" and size most of the other components (feed system, tanks, nozzle) just as for a chemical system. To design the reactor, we need to draw from three disciplines: reactor physics, thermal hydraulics, and materials. Reactor physics defines the system geometry for a controlled fission at a required power level. Thermal hydraulics is the study of energy removal from the reactor and the transfer of this energy to the propellant. A good thermal balance allows us to achieve high performance (high gas temperature) without overheating the reactor. Finally, through iteration, we must analyze materials to build a reactor that can withstand high temperatures, radiation, and structural loads.

8.4.1 Nuclear Physics

To understand reactor physics and design, we need to know the basics of *nuclear physics*—the study of the nucleus of atoms. Our intent is to present a picture of the nucleus, discuss the forces and energies involved in fission, and present the concept of nuclear cross sections. Once we understand these concepts, we can jump into reactor physics.

Atomic Structure

The atom is made up of protons, neutrons, and electrons. We can picture these particles as small spheres (billiard balls) with a heavy central nucleus (protons and neutrons) surrounded by orbiting electrons (see Table 8.5 and Fig. 8.10).

Table 8.5. Characteristics of Atomic Particles. This table shows the size, mass, and charge of particles to give us an initial feel for an atom's configuration. A = atomic mass number (the total number of protons and neutrons), and Z = number of protons or electrons [Henry, 1986].

Object	Mass (kg)	Charge (coulomb)	Radius (m)
Electron	9.11×10^{-31}	-1.6×10^{-19}	1.88×10^{-15}
Proton	1.67×10^{-27}	1.6×10^{-19}	10^{-18}
Neutron	1.67×10^{-27}	0	10^{-18}
Nucleus	$A \times 1.67 \times 10^{-27}$	$Z \times 1.6 \times 10^{-19}$	10^{-15}
Atom	$A \times 1.67 \times 10^{-27}$	0	10^{-10}

A complete or uncharged atom has the same number (Z) of protons and electrons. The atomic mass number (A) defines the total number of protons and neutrons. These account for most of the atom's mass. An uncharged atom is electrically neutral. But the number and resulting configuration of the orbital electrons uniquely determine the atom's chemical properties and hence, its identity as an element. (The atomic mass number governs the elements' alignment in the periodic table). With two exceptions, all naturally occurring atoms have nuclei with the number of neutrons present equal to or greater than the number of protons present. The nucleus of ordinary hydrogen is the proton itself, and the nucleus of a very rare form of helium-3 contains two protons and one neutron. Atoms having nuclei containing the same number of protons but different numbers of neutrons are called isotopes. For example, three different isotopes of uranium (^{233}U, ^{235}U, ^{238}U) all have 92 protons with 141, 143, and 146 neutrons, respectively. All naturally occurring elements have an atomic mass number greater than two times the number of protons. The isotopes with atomic numbers from 2 Z to 2.6 Z are stable, and all isotopes outside that range (except hydrogen and helium-3) are unstable and decay.

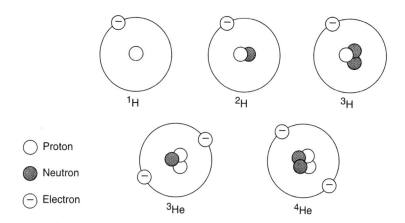

Fig. 8.10. **The Nucleus.** The nucleus is made up of protons and neutrons. This illustration shows possible compositions of the first two elements on the periodic table.

The collection of neutrons and protons is held together by a very strong, short-range, attractive force, called the *strong nuclear force*. This force is thought to be the same between two neutrons, two protons, or a neutron and a proton. The electrons stay in their orbits because of Coulomb attraction with the protons. (This electromagnetic force causes like charges to repel and opposite-charged particles to attract.) The nuclear force overrides the Coulomb repulsive force to bind the positively charged protons* (along with the uncharged neutrons) into the compact nucleus. The range of the strong nuclear force is 10^{-15} m, roughly the size of the nucleus.

From the viewpoint of a free neutron (not bound to a nucleus), most of the space through which it is traveling is "unoccupied." Let us use an analogy with distances more common to our experience. If the nucleus were 0.1 m in radius, the outer electrons of the atom would be about 10,000 m away, and the next nearest nucleus would be about 20,000 m away. Even in a material as dense as uranium, the neutron is in a virtual vacuum.

If a neutron is moving through a medium, it does not feel the presence of a nucleus until it gets into the 10^{-15}-meter range. Therefore, it continues traveling in a straight line until it gets into this range. On the other hand, if a proton or nucleus

---

* Particles in the nucleus—protons and neutrons—are usually called *nucleons*.

starts to move through a medium, it is repelled by the positive charges of neighboring nuclei as soon as it penetrates the electron clouds surrounding those nuclei. Long-range Coulomb forces cause this reaction. Therefore, because the neutron has a neutral charge, it is the best particle for fission.[*]

Now that we know the atom's basic structure, we need to quantify the forces involved with the atom. By understanding these forces, we can understand how energy is harnessed from the atom to propel our nuclear rocket.

Binding Energy of the Nucleus

The binding energy of the nucleus is directly related to the amount of energy released in a nuclear-fission reaction. *Binding energy* is the energy required to separate all of the nucleons to a distance at which they no longer exert the strong nuclear force on each other.[†] We may also think of it in reverse as the amount of energy released when the free nucleons coalesce to form the nucleus.

The mass of an atom is less than the sum of the masses of the individual constituents. When all of the particles are assembled, the product atom has "missing mass." This *defect mass* is given by

$$\Delta = [Z(m_p + m_e) + (A - Z)m_n] - m_{atom} \tag{8.9}$$

where Δ = defect mass (kg)
  m_p = mass of the proton (kg)
  m_e = mass of the electron (kg)
  m_n = mass of the neutron (kg)
  m_{atom} = mass of the atom (kg)
  Z = number of protons
  A = number of protons and neutrons = atomic mass number

The defect mass converts into energy when the nucleus forms. The conversion of defect mass to energy is described by Einstein's famous equation ($E = mc^2$)

$$E = \Delta c^2 \tag{8.10}$$

where E = energy (J, where 1 eV = 1.602×10^{-19} J),
  c = the speed of light in a vacuum = 3×10^8 m/s

---

[*] The neutron can "strike" a nucleus, causing fission, without being deflected by the Coulomb force. In the early 1900s, Rutherford looked at using protons to break apart nuclei. Although this approach is possible, the energy required for a proton to overcome the Coulomb forces makes it impractical. Rutherford further concluded from this experiment that nuclear fission is impractical.

[†] The "binding energy" should not be confused with chemical "bond energy." Bond energy is what holds different atoms together to form a molecule. Binding energy is what holds the nucleus of an atom together.

The energy associated with the mass defect is the binding energy. In Fig. 8.11, the binding energy is plotted as a function of atomic mass number. This figure shows that, on average, the nuclides in the center of the range are more tightly bound than those at the very low or very high masses.

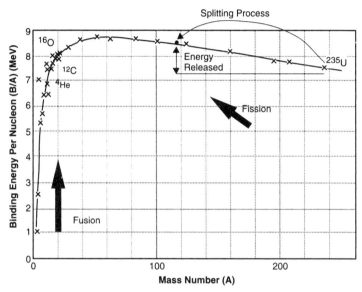

Fig. 8.11. The Average Binding Energy Per Nucleon for Stable Atoms. ^{235}Uranium splits, yielding two new molecules with average masses of 117, releasing the energy shown (200 MeV/2 × 117 = 0.855 MeV/nucleon) [Knief, 1992].

Figure 8.11 also shows the existence of fission and fusion. We can increase the binding energy per nucleon in a nucleus with A = 240 by nearly 1 MeV if we split it into two fragments of A = 120. Because 240 nucleons are involved, a total of about 240 MeV of energy is available to cause such a transition to take place. Compared to nuclei of half its mass, the ^{235}U nucleus is bound relatively lightly. Energy must be released to split the loosely bound ^{235}U into two tightly bound fragments. We can get a reasonably good estimate of the energy released in fission by using data from Fig. 8.11 (~200 MeV of energy is actually released in fission, as we discuss later). The low A side (left) of this curve shows that if an $^{4}_{2}$He is made by putting together two deuterons $^{2}_{1}$H, there is an increase of 6 MeV per nucleon. ($^{4}_{2}$He represents a helium atom with mass number of 4 and atomic number of 2.) Reactions of this type, known as *fusion reactions*, are the source of energy for the Sun. To clarify this point:

- *Nuclear fission* is splitting a nucleus
- *Nuclear fusion* is the assembling of two nuclei into a single nucleus

Natural Radioactivity

Before we discuss the products of fission in more detail, we need to know more about radioactivity. Many radioactive nuclides, or radionuclides, exist in nature (and did so before the Manhattan Project). Naturally occurring radioactive decay may produce any of the three radiations—alpha, beta, and gamma. Whenever a nucleus can attain a more stable (more tightly bound) configuration by emitting these radiations, spontaneous disintegration (radioactive decay) occurs.

Alpha radiation is a helium nucleus, which may be represented as either ^4_2He or α. An important alpha-decay process with ^{235}U takes the form:

$$^{235}_{92}\text{U} \rightarrow \,^{231}_{90}\text{Th} + \alpha$$

where ^{231}thorium is the decay product of the ^{235}U nucleus. This decay (and all decay processes) obeys the conservation of mass (mass number A) and charge (atomic number Z) on both sides of the equation.

Beta radiation is an electron of nuclear, rather than orbital, origin (chemical reactions occur when electrons are exchanged from the outer orbits). Here, an electron is emitted from the nucleus. Remember, the electron has a negative charge equal in magnitude to that of the proton and a mass number of zero, shown by $^{\;\;0}_{-1}\text{e}$ or $^{\;\;0}_{-1}\beta$. An example of a beta-decay reaction of ^{239}neptunium to ^{239}plutonium is

$$^{239}_{93}\text{Np} \rightarrow \,^{239}_{94}\text{Pu} + \,^{\;\;0}_{-1}\beta + \,^0_0\nu$$

where ν is an uncharged, massless particle called an antineutrino. It does not contribute significantly to nuclear energy.

Why does the nucleus emit an electron? The nuclear basis for beta decay is:

$$^1_0\text{n} \rightarrow \,^1_1\text{p} + \,^{\;\;0}_{-1}\text{e} + \,^0_0\nu$$

where the uncharged free neutron (n) emits an electron (e) and an antineutrino (ν) while leaving a proton (p) and a net positive charge in the nucleus. The slight mass difference between the neutron and proton is enough to allow for electron emission as well as a small amount of kinetic energy.

Gamma radiation (γ) is high-energy electromagnetic radiation that originates in the nucleus. It is emitted in the form of photons, which are discrete bundles of energy. Gamma radiation is emitted by excited or metastable nuclei. (These nuclei have a slight mass excess—usually a result of a previous alpha or beta transition of less than maximum energy during a transition to a lower, more stable energy level). An example of one gamma-decay reaction for metastable cobalt-60 (the * indicates a metastable state) is:

$$^{60*}_{27}\text{Co} \rightarrow \,^{60}_{27}\text{Co} + \gamma$$

The gamma-ray process increases the binding energy but does not affect the nucleus's charge or mass number. The gamma-ray energies represent the transition between discrete energy levels in a nucleus, so these distinctive gamma energies often identify or differentiate nuclides from each other.

Products of Fission

Fission occurs most easily in materials with extra neutrons or high mass (see Fig. 8.11). Besides producing neutrons (on average, 2.5), fission results in two heavy fragments which carry most of the energy released in the form of kinetic energy. Nuclear stability usually causes one of the fission fragments to be larger than the other. Figure 8.12 shows the characteristic asymmetrical, or "double-humped," distribution for thermal fission of ^{235}U. Equal-mass fragments (A = 117) are produced in only about 0.01 percent of the fissions (shown by the "valley" in the middle of Fig. 8.12). Fragment nuclides in the ranges of roughly 90–100 and 135–145 occur in as many as 7% of the fissions (indicated by the "hills" on both sides of Fig. 8.12). The fission fragments tend to be neutron-rich with respect to stable nuclides of the same mass number. The related energy imbalance is generally rectified by successive beta emissions, each of which converts a neutron to a proton. Figure 8.13 shows two beta-decay "chains" from different fissions. The gamma rays are emitted whenever a beta particle leaves the nucleus in an excited state. Again, we do not show the antineutrinos because they have little effect.

Because each fission results in two such decay-chains, the process produces a lot of beta and gamma radioactivity and toxic fission products. This is the source of most of the radioactive fallout from nuclear weapons. These radioactive fission products also cause serious problems in nuclear reactors because they make the spent fuel extremely radioactive after the reactor has been run at power.* Because of the delayed nature of radioactive decay, energy (*decay heat*) continues to radiate even after the reactor has been shut down. A major design problem for reactors is to make sure this radioactive material cannot escape.

The two fragments of the split nucleus quickly re-form into roughly spherical shapes. Because they are separated by a distance greater than the range of the nuclear force and are charged, a very large Coulomb repulsive force pushes them apart. They pick up speed and go careening through the medium, repelling (again by Coulomb interaction) other charged nuclei and transmitting kinetic energy to them. These charged nuclei, in turn, push their neighbors. As a result, the initial kinetic energy of the fission fragments (~168 MeV) cascades, through Coulomb interaction, to the nuclei near the site where the original fission took place. Thus, the average kinetic energy (temperature) of these neighboring nuclei increases from these collisions (or interactions) in a time which, for reactor calculations, is instantaneous. About 82% of the energy released in fission converts to a local increase in temperature in this manner. Table 8.6 summarizes the rest of the energy

---

* This is the source of a common misconception. The radiation level of reactors is very low <u>before</u> operation. Reactors become dangerous only after operation creates a large number of highly radioactive fission products, such as those shown in Fig. 8.13.

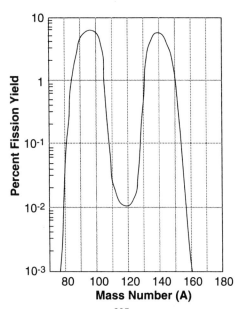

Fig. 8.12. **Mass Yield Curve for Fission of ^{235}U by Thermal Neutrons.** This curve shows that fission does not often produce two fission products of equal mass. Getting two fission products with masses of 97 and 139 is much more likely. The sum of these numbers is 233, showing that two neutrons are released to bring the total mass to 235 (mass conservation) [Benedict, 1957].

released from fission. The total kinetic energy of neutrons emitted in a fission is approximately 5 MeV.

All of these energy sources contribute to the approximate total 200 MeV of energy released per fission. Fission converts mass energy into kinetic energy of the fission fragments, prompt neutrons,* and the radiation energy of gamma rays. The fission fragments subsequently decay, further converting mass energy into gamma rays and into kinetic energy of released beta particles and delayed neutrons. The neutrons transmit their energy to other charged particles by direct collisions with nuclei, through gamma rays, and by electromagnetic interactions. Therefore, all the energy released from fission transmits to charged particles. Because these particles interact with nuclei making up the material medium, they transmit their energy to other charged nuclei through Coulomb interaction in a cascade effect, raising the average temperature of the material comprising the medium.

---

* *Prompt particles* (neutrons, electrons, gamma rays) are released immediately during fission (within nanoseconds); *delayed particles* are emitted after a several-second delay, depending on the particular nuclide's stability.

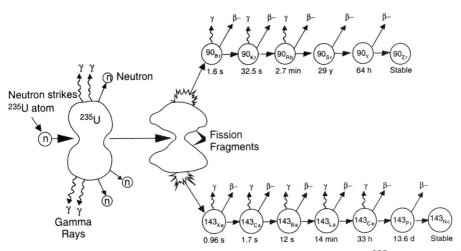

Fig. 8.13. Nuclear Fission and Subsequent Decay. A neutron strikes the ^{235}U atom, giving it enough energy to split. Fission releases neutrons, gamma rays, and two fission products (FP). The masses of the FPs depend on a probability distribution (Fig. 8.12). The resulting FPs decay radioactively as shown, producing more energy [Knief, 1992].

Table 8.6. Representative Distribution of Fission Energy. Although most of the energy is in fission fragments, other particles take up some additional energy.

Energy Source	Fission Energy (MeV)
Fission fragments	168
Neutrons	5
Prompt gamma rays	7

Delayed radiation
  Beta particles
  Gamma rays
  Radiative capture gammas
	
	8
	7
	5

Total	200

The prompt and delayed gamma rays emitted at the time of fission and later from the fission products each contribute about 7 MeV of energy. The amount of gamma energy released from neutron capture depends on the material in which the capture takes place but is usually 3–12 MeV. Because gamma rays are not charged particles, they are not subject to Coulomb interaction. Instead, they lose their energy by other kinds of electromagnetic interactions with charged particles:

- *Photoelectric-effect* photons of a certain wavelength incident on the surface of a material emit electrons (low-energy gamma rays, less than 0.3 MeV)
- *Compton scattering* of photons from charged particles (gamma-ray energy of 0.3–10 MeV) occurs when photons transfer energy to charged particles
- *Pair production* occurs when a gamma ray is completely absorbed and an electron-positron pair is produced for gamma-ray energies greater than 10 MeV

All of these mechanisms result in energy transfer to a charged particle. This particle then moves only a short distance before its energy converts to heat through the cascade effect. Therefore, once a gamma ray interacts with matter, the energy it transfers is deposited locally. But the gamma ray itself may move a number of centimeters, or occasionally meters, before losing all of its energy. This range of movement makes shielding important. In this respect, the gamma ray is like the neutron, which also moves a number of centimeters on average before being absorbed. Because space reactors are small, gamma-ray and neutron leakage is higher than the few-percent loss in larger, terrestrial reactors.

The beta particles released from radioactive decay of the fission fragments contribute an energy of about 8 MeV per fission. Because these beta particles are charged, this energy immediately converts to local heating.

Nuclear Cross Sections

A neutron interacting with a nucleus may produce

- Scattering: the neutron strikes the nucleus, imparts some of its momentum and kinetic energy to the nucleus, and then moves off in a direction different from its original path
- Absorption: the nucleus absorbs a neutron and de-excites itself by emitting a charged particle or gamma ray
- Fission: the neutron splits the nucleus and releases, on average, 2.5 neutrons

We must define numbers which specify the probability that a neutron, having a given kinetic energy, moves through a medium containing a given material and interacts in one of these three ways. Tracking these interactions is important in designing our reactor.

We can describe the probability of a neutron interaction with a material nucleus in terms of a conceptual quantity known as a *cross section*. For conceptual understanding,[*] consider a thin target as shown in Fig. 8.14. The interaction rate (R)

---

[*] We use the "analogy" of a nuclear cross section to help us understand neutron interactions, but the analogy quickly breaks down if we try to take it too far. For example, the cross section changes as a function of the energy of the incident neutrons. This does not make much physical sense, based on our simple geometric model.

of neutrons with the target atoms is proportional to the beam intensity, slab area, slab thickness, and target concentration. We can represent this relation as

$$R = \sigma_{cs} \Phi \, n \, dA \, dx \tag{8.11}$$

where R = rate of interactions of incident neutrons with atoms in the slab (1/s)
  σ_{cs} = nuclear cross-section parameter (cm^2)*
  Φ = neutron flux on the slab (neutrons/[cm$^2 \cdot$s])
  n = concentration of target atoms (atoms/cm^3)
  dA = area of the slab (cm^2)
  dx = thickness of the slab (cm)

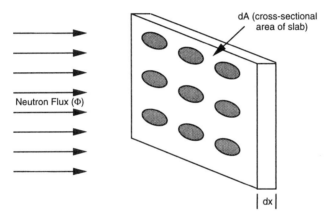

Fig. 8.14. **Neutrons Incident on a Target.** This model helps us predict the probability of a given neutron interaction with a nucleus. *dA* is the physical cross-sectional area of a target slab of material with a thickness of *dx*.

Rearranging terms, we can define the cross section as

$$\sigma_{cs} = \frac{R}{\Phi \, (n \, dA \, dx)} \tag{8.12}$$

and, because $(n \, dA \, dx)$ is the total number of target nuclei in the volume, σ_{cs} is the interaction rate per atom in the target per unit intensity of the incident beam. By

---

* We use a subscript on "σ" to distinguish this parameter from the Stefan-Boltzmann constant. There are different cross sections, depending on whether we are talking about absorption, fission, scattering, or removal. We eliminate the "cs" subscript when appropriate.

the definition of beam intensity, the number of neutrons that strike the target per second is ΦdA. The relative probability of interaction is the ratio of interaction rate divided by the number of neutrons incident on the target, or

$$\frac{R}{\Phi dA} = \frac{\sigma_{cs} \Phi n\, dA\, dx}{\Phi dA} = \frac{\sigma_{cs}}{dA}(n\, dA\, dx) \tag{8.13}$$

The volume sets the number of target nuclei. Then, because area is fixed, the nuclear cross section (σ_{cs}) alone determines the probability of interaction.

Any particular interaction has some probability of occurring. This probability depends on the relative energy of the incident neutron and the type of target nucleus. Although the units of the cross section are cm^2, the cross section can be much larger than the geometrical cross section of the nucleus because of quantum-mechanical effects. For example, the absorption cross section for $^{135}_{54}Xe$ is almost a million times larger than its geometrical cross section.

We can generalize the concept of interaction probability to any target isotope j (such as ^{235}U or ^{238}U) and interaction of type i (scattering, absorption, and fission) with neutrons of energy (E) as

$$P_i = n^j \sigma_i^j(E)\, dx = \Sigma_i^j(E)\, dx \tag{8.14}$$

where P_i = probability of any type of interaction i (scattering, fission, absorption)
n^j = concentration of isotope j in the material (atoms/cm^3)
$\sigma_i^j(E)$ = microscopic cross section of isotope j for interaction i (cm^2), as a function of neutron energy (E in eV).
dx = distance neutron travels (cm)
Σ_i^j = macroscopic cross section of isotope j for interaction i (cm^{-1})

Equation (8.14) defines the difference between a microscopic and macroscopic cross section. The microscopic cross section (σ_i) is an "effective area" used to characterize a single nucleus. The macroscopic cross section (Σ_i) is the probability that a neutron interacts in traveling a unit distance through a (macroscopic) sample of material.

The microscopic cross sections for absorption, scattering, or fission between isotopes and neutrons derive from experiments (see Table 8.7 for microscopic cross sections from absorption and fission). Brookhaven National Laboratory tabulates the data [Drake, 1970]. This job includes creating computer programs for manipulating this data and making data tapes available. The tapes contain a current set of recommended numbers for all possible interactions between a neutron and a given isotope.

Table 8.7. Characteristics of Some Isotopes Important for Reactors. This table shows the cross section for absorption of a thermal neutron (0.25 eV ≈ 2200 m/s). Units are barns (10^{-24} cm^2). The final two rows show what happens when a thermal neutron strikes ^{235}U—it can be absorbed or it can cause fission. The larger number in the last row indicates that fission is more likely than absorption. Notice that boron (^{10}B) has a very large absorption cross section, which is why it is used as a control poison.

Atomic Number	Isotope	Absorption Cross Section σ_a for Thermal Neutrons (barns)
1	^{1}H	3.32×10^{-1}
1	^{2}H	5.30×10^{-4}
3	^{6}Li	9.40×10^{2}
3	^{7}Li	3.7×10^{-2}
5	^{10}B	3.84×10^{3}
5	^{11}B	5.50×10^{-3}
6	^{12}C	3.40×10^{-3}
6	^{13}C	9.00×10^{-4}
92	^{235}U (absorption)	106
92	^{235}U (fission)	577

Neutron Energy

We have described the fission process and discussed the energy released from each fission to produce the heat we need. Now we need to discuss the neutrons produced from fission.

If neutrons with energy ranging from 0.25 eV to 10 MeV strike a nucleus, there is enough energy to split the nucleus. Neutron energies higher than this do not interact with the nucleus, and lower energies are not sufficient to cause fission. However, nuclear cross sections are also a function of neutron energy. Uranium cross sections are large for neutron energies close to 0.25 eV and 1.0 MeV. Cross sections are very small for other neutron energies. For this reason, we design a reactor to moderate the neutron energies to one of these values. *Thermal reactors* are designed to operate at 0.25 eV neutron energies (*thermal neutron*), and *fast reactors* operate at 1.0 MeV neutron energies (*fast neutrons*).

The energy distribution, or spectrum, for neutrons emitted by fission [$\chi(E)$] is relatively independent of the energy of the neutron causing the fission. For many purposes, the expression

$$\chi(E) = 0.453e^{-1.036E} \sinh\sqrt{2.29E} \tag{8.15}$$

adequately approximates the neutron density for ^{235}U shown in Fig. 8.15. This figure shows that most of the neutrons produced from fission have energies between 0.1 and 5 MeV. The highest fraction of neutrons are at 0.75 MeV (appropriate for fast reactors). Reactor concepts based on using thermal neutrons with energy less than 1 eV must moderate or slow down the fission neutrons by a factor of one million or more. This is why our thermal space reactors have moderators.

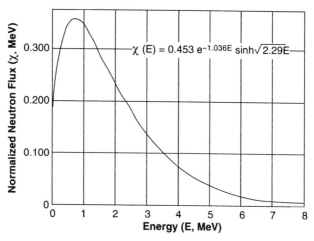

Fig. 8.15. Empirical Energy Spectrum for Fission Neutrons from the Thermal Fission of ^{235}U. A thermal neutron (kinetic energy about 0.25 eV) creates a nuclear fission. Fission releases more neutrons with an energy distribution as shown. Most of the neutrons have energies much greater than the thermal level and do not support fission. Moderators must slow them down so they are useful for future fission reactions [Knief, 1992].

8.4.2 Size the NERVA-Type Reactor

Now we take the basic physical concepts and apply them to reactor design. Several effects—such as thermal hydraulics of the coolant and detailed reactor physics, including control rod effects, Doppler feedback, dimensional effects, and other nasty stuff—complicate detailed design. Here, we want to roughly size the core, taking simplified reactor physics into account. To do this, we use a *lumped-parameter* approach. By assuming the reactor is homogenous, with no variation in performance through the core, we get the basic dimensions needed to estimate the core's mass (Sec. 8.4.5).

Estimate the Core Volume

The requirement on our core is a given power level, as discussed in Sec. 8.3.2. Each technology can produce a certain power density (P_D) or range of power densities. NERVA-type reactors have a fairly consistent core power density of 1,570 MW/m³. Various studies have shown this number can be improved. However, this is the core power density of the actual flight hardware tested (but never flown), so we know we can do this much! PBR and CERMET systems have highly variable ranges of power density—one reason these two reactor types appear to be more useful for space reactors, mainly at the lower power levels.

Given the required power level (P_{core}), we can estimate the volume of our reactor core as

$$V_{core} = \frac{P_{core}}{P_D} \quad (8.16)$$

This equation gives us the core volume if the burn duration is relatively short (minutes). However, to maintain the required power level over the entire burn, we need to factor in more fuel to account for the uranium lost from fission. The energy produced by the core [E_{core} in MW-hours or W·s (joules)] is the product of power and burn duration (t_b):

$$E_{core} = P_{core} t_b \quad (8.17)$$

The amount of *core burn-up* is related to the energy produced. Knowing the fission of a single uranium atom produces 200 MeV (3.206×10^{-11} joules) of energy, we can determine the number of uranium atoms consumed by

$$N_{cons} = \frac{E_{core}}{3.206 (10)^{-11}} \quad (8.18)$$

We can estimate the mass of uranium consumed (m_{cons}, in kg) by knowing the molecular mass (235 g/mol = 0.235 kg/mol):

$$m_{cons} = \frac{0.235 N_{cons}}{0.6023 (10)^{24}} \quad (8.19)$$

To get the additional ^{235}U volume needed for burn-up (V_{cons}), we then divide the mass by the density of ^{235}U (19,100 kg/m³):

$$V_{cons} = \frac{m_{cons}}{19,100} \quad (8.20)$$

Combining Eqs. (8.16) through (8.20) gives us the total core volume, including burn-up:

$$V_{core} = P_{core}\left[6.4\,(10)^{-19} t_b + \frac{1}{P_D}\right] \qquad (8.21)$$

where V_{core} = volume of the reactor core (m³)
P_{core} = power output of the core (W)
t_b = burn duration (s)
P_D = power density (W/m³)

Notice that we have reverted to a pure version of SI units to avoid error. This volume is higher than what is really needed at the start of the burn. We use rods, inserted into the reactor core, to vary the number of neutrons causing fission.* At the beginning of the burn, the control rods are placed to restrict the number of fissions. As the burn progresses, we reposition the rods (by rotating them or pulling them out of the core (see Sec. 8.1.2). As the neutron production drops because of fuel burn-up, the repositioned rods absorb fewer neutrons, allowing a constant reaction rate.

Ensuring Criticality Using a Modified Four-Factor Formula †

Our next step is to create a critical reactor. When the number of neutrons produced is steady and not equal to zero, a fission chain reaction is self-sustaining, and the system is said to be *critical*. When neutron production exceeds losses, the system is said to be *supercritical*, and the reactor power increases (start up). *Subcritical* systems are in a state for which neutron losses exceed production (shutdown). Therefore, to have a critical reactor, we must track the neutron population.

In a *nuclear reactor* (any collection of materials in which nuclear fission can occur), we can describe the neutron population with the following equation:

rate of increase in the number of neutrons = rate of production of neutrons
− rate of absorption of neutrons − rate of leakage of neutrons (8.22)

A system produces a neutron either from fission or scattering. The absorption term represents a loss of neutrons to the system. However, if a uranium nucleus absorbs a neutron, the production term increases immediately because of subsequent fis-

---

* The number of fission reactions taking place in a given time period defines the level of reactor power. Thus, a constant number of neutrons are available for fissioning. With the higher volume of fuel, the equilibrium condition would give us more fissions than we want, so we use control rods to reduce the available neutron count to our desired level.

† This preliminary design approach is based on diffusion theory, an appropriate simplification that allows for quick, rough sizing of the core. See Henry [1986] for this and other more complex approaches (Monte Carlo and discrete ordinate).

sion. Finally, the leakage term equals the number of neutrons that leave the system (leak out) without interacting with anything.

We define the effective multiplication factor (k_{eff}) as

$$k_{eff} = \frac{\text{neutron production}}{\text{neutron losses}} = \frac{\text{production}}{\text{absorption + leakage}} \qquad (8.23)$$

For a critical reactor, the effective multiplication factor is one. For supercritical reactors, it is greater than one. We estimate k_{eff} of a thermal reactor core[*] by defining k_{eff} as a combination of four factors, for a homogenous infinite reactor, with a fifth factor that accounts for the neutron leakage of a finite-size reactor.[†] Expressing this combination as an equation, we get:

$$k_{eff} = p f \eta \varepsilon P_{nl} \qquad (8.24)$$

where p = resonance escape probability, which is defined as the ratio of the rate of absorption at thermal energies throughout the fuel cell to the rate of absorption at all energies. Hence, it is the probability that a neutron is not captured while at energies above the thermal level. In the following equation, N_t is the total absorption of neutrons into the fuel (such as ^{235}U). The parameter "I" (resonance integral) is an effective neutron absorption/loss term. It takes into account all of the physics that occurs as the neutron decreases energy. The lethargy (ξ) represents the number of collisions a neutron must incur to slow down to the fission energy. The scattering cross section (Σ_s) represents the neutron-scattering cross section for each material in the core. This factor represents the likelihood that the neutron will collide with an atom and scatter. The combination of these parameters is integrated over the energy region as the neutron energy decreases to the fission level. The result is an exponential effect:

$$p = e^{\left(\frac{-N_t^{235}U \cdot I}{\xi\Sigma s_H + \xi\Sigma s_C + \xi\Sigma s_{238}U}\right)} \qquad (8.25)$$

f = thermal utilization factor, defined as the ratio of the rate at which thermal neutrons are absorbed in fuel (Σ_f) to the total rate at which thermal neutrons are absorbed in the entire core ($\Sigma_{acore}+\Sigma_f$).

---

[*] This analysis approach is appropriate only for thermal reactors.
[†] This approach has a historical precedent, based on the work of Enrico Fermi in the 1940s for the first controlled fission reaction. Fermi used only the first four factors, and assumed a reactor of infinite dimensions, which does not allow for neutron leakage. By adding the final P_{nl} term, we can modify the four-factor approach to account for all effects.

$$f = \frac{\Sigma_{t\,^{235}U}}{\left(\Sigma_{acore} + \Sigma_{t\,^{235}U}\right)} \qquad (8.26)$$

η = neutron production effectiveness factor, defined as the ratio of the rate at which neutrons are created by thermal fission to the rate at which thermal neutrons are absorbed in the fuel. In the following equation we take the average total number of neutrons produced by a fission event (v) and multiply it by the ratio of fission cross section (σ_f) over absorption cross section (σ_a). This ratio gives the probability that a neutron is usable for a future fission event.

$$\eta = v\frac{\sigma_f}{\sigma_a} \qquad (8.27)$$

ε = fast-fission factor (for thermal reactors, this number is close to 1.00). This factor is defined as the ratio of the total rate of production of neutrons from fissions in the fuel occurring at all energies to the rate of production of neutrons from fissions occurring in the thermal range. We can determine this factor through experiment by irradiating a bare foil with fast neutrons at different flux levels for a fixed time. Then, we measure the level of radioactivity (β's or γ's emitted per second) caused by decay of the radioactive fragments which fissions in the irradiation period have created.

P_{nl} = nonleakage probability, defined as the fraction of neutrons that remain in the reactor-core region, including any reflector (or one minus the fraction of neutrons that leak out of the system). All of the other factors in the modified four-factor formula are fixed, depending on our choice of materials. The leakage probability defines the reactor geometry needed for criticality. We choose P_{nl} to get a critical reactor and then use this number to define the geometry:

$$P_{nl} = \frac{1}{p\eta\varepsilon f} \qquad (8.28)$$

These factors seem disjointed at first glance, but the following description puts them into context. Considering that fission produces a large population of neutrons,

- A fraction (p) of neutrons reach thermal energies after escaping capture from materials in the reactor while slowing down.
- A fraction (f) of the thermal neutrons are absorbed in the fuel (while the rest are absorbed elsewhere).

8.4 Size the Reactor

- For each thermal neutron absorbed in the fuel, fission produces an average of η fast neutrons.
- The fast neutrons are augmented by a factor ε from those resulting from the fissions caused directly by fast neutrons (actually by some of the "$1-p$" fraction that does not reach thermal energies).
- A fraction of the neutrons are lost from the system because of its finite size, and the remaining fraction of the neutron population are available for the interactions modeled by the other four factors.

Reflectors can reduce neutron leakage. Thermal reactors, by definition, rely mainly on thermal neutrons to cause fissions. At these low energies, the fission cross section for fissile nuclides is quite large (see Table 8.7). Fission produces fast neutrons (as shown in Fig. 8.15), so some form of moderation and reflection is a crucial part of the process.

Size a NERVA-Type Reactor without a Reflector

As a first step, we want to design a reactor using 90% enriched* ^{235}U as the fuel, hydrogen (H_2) as the propellant,† and no reflector. For other propellants, the cross sections change by a few percent (CH_4 or CO_2 make the reactor a bit larger), which is negligible for preliminary design. If we use other fuels (different enrichment levels or ^{233}U), we need to reevaluate this design-point sizing with different cross sections.

For the neutron population to remain steady, or critical, the product of the five factors must equal one. To get this result, we must assemble our materials, define the cross sections, and determine the values of the factors in the formula. We start with the nuclear physical constants shown in Table 8.8.

$$N = \frac{0.6023 \times 10^{24} \frac{\text{atoms}}{\text{mole}} \cdot \rho \cdot 10^{-24} \text{cm}^2/\text{barn}}{A} \qquad (8.29)$$

where N = number density (atoms/cm^3)
 ρ = density of material (kg/cm^3)
 A = atomic number of material

The following analysis comes from the neutron-transport equation, which takes into account all neutrons gained and lost through the reactor core and all microscopic interactions. Solving it is very difficult and is inappropriate for preliminary design. To get something a bit simpler, we have derived the equations in

---

*Enrichment level refers to the fraction, by mass, of the fuel that is ^{235}U. Naturally occurring uranium is made up of both ^{238}U and ^{235}U. We need to process the natural ore to remove excess ^{238}U and get our desired enrichment level.

†The H_2 propellant also acts as a moderator.

Table 8.8. Nuclear Physical Constants and Design Process for Nuclear Thermal Rockets. The microscopic cross sections are physical properties of the material. The number density (N) is set by the density of the material as a fraction of theoretical density.

Isotope	N	σ_a	Σ_a	σ_s	Σ_s	σ_t	Σ_t	σ_f	Σ_f	A	ξ
^{235}U	9.74E-04	106	1.032E-01	0.00	0.000E+00	683	6.652E-01	577	0.562	235	0.008
^{238}U	7.33E-05	7.6	5.571E-04	8.30	6.084E-04	16	1.173E-03	--	--	238	0.008
C	0.057	0.004	2.280E-04	4.80	2.736E-01	4.8	2.736E-01	--	--	12	0.158
H	3.74E-06	0.33	1234E-06	38.00	1.421E-04	38	1.421E-04	--	--	1	1.000
Be	0.099	0.01	9.900E-04	7.00	6.930E-01	7.01	6.940E-01	--	--	9.01	0.207

σ_a = microscopic absorption cross section (cm^2)
Σ_a = macroscopic absorption cross section (cm^{-1})
σ_s = microscopic scattering cross section (cm^2)
Σ_s = macroscopic scattering cross section (cm^{-1})
σ_t = microscopic total neutron removal cross section (cm^2)
Σ_t = macroscopic total neutron removal cross section (cm^{-1})
σ_f = microscopic fission cross section (cm^2)
Σ_f = macroscopic fission cross section (cm^{-1})
ξ = lethargy $\approx$ (2/(atomic weight + 2/3))

the rest of this section by making simplifying assumptions for diffusion theory. In many cases, they come from an intuitive feel for the microscopic physics. They are accurate enough for preliminary design.

We first need to find the total macroscopic absorption cross section (Σ_{acore}) for the core:

$$\Sigma_{acore} = \Sigma_a\,^{235}\text{U} + \Sigma_a\,^{238}\text{U} + \Sigma_a\,\text{C} + \Sigma_a\,\text{H} = 0.103 \text{ cm}^{-1} \tag{8.30}$$

Because we are at the molecular level, this cross section is a simple summation.

Next, we find the approximate transport length (D_{core}), which equals the mean distance a thermal neutron travels through the core:

$$D_{core} \approx \frac{1}{3\left(\Sigma_t\,^{235}\text{U} + \Sigma_t\,^{238}\text{U} + \Sigma_t\,\text{C} + \Sigma_t\,\text{H}\right)} = 0.354 \text{ cm} \tag{8.31}$$

We then need to find the thermal-diffusion length in the core (L_{core}), or the mean square distance a thermal neutron travels before it is absorbed in the core:

$$L_{core} = \sqrt{\frac{D_{core}}{\Sigma_{acore}}} = 1.846 \text{ cm} \tag{8.32}$$

Given these values, we can calculate the five factors of our modified four-factor formula [see Eqs. (8.25) through (8.28)]:

$$\eta = v\frac{\sigma_f}{\sigma_a} = 2.0529 \tag{8.33}$$

$$f = \frac{\Sigma_{t\,^{235}U}}{\left(\Sigma_{acore} + \Sigma_{t\,^{235}U}\right)} = 0.9988 \tag{8.34}$$

$$p = \exp\left(\frac{-N_{t\,^{235}U} \cdot I}{\xi\Sigma s_H + \xi\Sigma s_C + \xi\Sigma s_{^{238}U}}\right) = 0.7139 \tag{8.35}$$

$$\varepsilon \approx 1.0 \tag{8.36}$$

$$P_{nl} = \frac{k_{eff}}{p\eta\varepsilon f} \tag{8.37}$$

where I = 15—the resonance integration factor
       v = 2.43—the average number of neutrons produced per fission event based on empirical evidence [Henry, 1986]

Notice we have not calculated an exact value for P_{nl}. For a critical reactor (k_{eff} = 1.0), but we usually design our reactor to be a bit supercritical. A typical margin is 10% (k_{eff} = 1.1). We leave the choice of k_{eff} as a design decision.

To determine the height (H_{core}) and the radius (R_{core}) of our cylindrical core,[*] we have our volume requirement but need a second equation (two unknowns requires two equations). We get this equation from the criticality requirement. (Remember, we need to define the geometry to get the correct nonleakage factor—P_{nl}.) We start by defining the core *material buckling parameter* (B_m):

$$B_m = \sqrt{\frac{1 - P_{nl}}{P_{nl}\left(L^2_{core} + \tau\right)}} \tag{8.38}$$

where τ is the neutron age (a term that represents the average lifetime of neutrons in a core), with a magnitude of 100 cm². Because of the geometric distribution of

---

[*] A cylinder is the usual configuration because of the high pressure in the core and the structural requirements for containment. Spheres have also been looked at but not built.

the neutron population, the dimensions of the critical reactor adhere to certain constraints. By defining a *geometric buckling* (B_G), which must equal the material buckling in a critical reactor, we can determine an equation that defines the dimensions of the core. For a cylindrical shape, the expression is

$$B_m^2 = B_G^2 = \left(\frac{\pi}{H_{core}}\right)^2 + \left(\frac{2.405}{R_{core}}\right)^2 \tag{8.39}$$

where R_{core} and H_{core} are the core radius and height, respectively. This equation tells us that, as long as H_{core} and R_{core} satisfy this equality, we have a critical or supercritical reactor, depending on our choice of k_{eff}.

The question remains: what dimensions give us the desired power output and the required criticality? Assuming a cylindrical core, we can relate the required core volume to the core dimensions:

$$V_{core} = \pi R_{core}^2 H_{core} \tag{8.40}$$

With this equation and our criticality equation [Eq. (8.39)], we have two equations we can solve for the two unknowns—core radius and height. Solving each equation for core height and equating them gives us a single equation containing only the core radius term:

$$\left(R_{core}^2\right)^3 - \left(\frac{BV_{core}}{\pi^2}\right)^2 \left(R_{core}^2\right) + \left(\frac{2.405 V_{core}}{\pi^2}\right)^2 = 0 \tag{8.41}$$

As shown, we can analytically solve this 6th-order polynomial equation by recognizing that it contains only even orders of R_{core}, giving us a cubic equation in terms of the square of R_{core}. This equation gives us three solutions for the square of R_{core}, but one is not real. Assuming no core burn-up, the solution to this equation for a NERVA power density is shown in Fig. 8.16. This figure shows two curves—one for reactor height and the other for reactor radius. The dashed parts of height and radius correspond to each other, and the solid parts also correspond to each other. The shaded region is restricted by thermal hydraulics; the core gets too hot at the aft end if it is much longer than 130 cm. Notice that we have two choices for powers below 1400 MW. Either works, so we can choose the appropriate solution for our application. But we try to avoid "pencils and pancakes," or excessive dimensions in either direction. A larger core radius also gives us a larger radiation shield. We cannot get a critical reactor core for powers below 1290 MW because the neutron leakage for these smaller cores becomes excessive. To overcome this leakage, we must include a radial reflector.

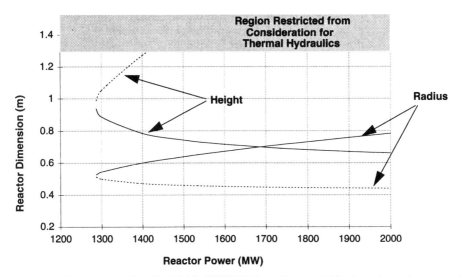

Fig. 8.16. Dimensions of an Unshielded NERVA-Type Reactor. This chart shows the two real solutions to Eq. (8.41). We have solved the four-factor formula assuming a 10% margin on criticality (k_{eff} = 1.1). The two dashed lines are one solution, and the two solid lines are another. This type of reactor cannot be made critical below 1290 MW because of excessive neutron leakage.

Size a NERVA-Type Reactor with a Reflector

Adding a neutron reflector to our NERVA-type core allows us to reduce the radius. To accommodate the reflector, we change the buckling equation to include a multiplication factor (f_{refl}) on the core-radius parameter:

$$B^2 = \left(\frac{\pi}{H_{core}}\right)^2 + \left(\frac{2.405}{f_{refl}R_{core}}\right)^2 \quad (8.42)$$

A practical range for f_{refl} is between 1.1 and 1.2. Combining this equation with the core volume [Eq. (8.40)] gives us

$$\left(R_{core}^2\right)^3 - \left(\frac{BV_{core}}{\pi^2}\right)^2\left(R_{core}^2\right) + \left(\frac{2.405 V_{core}}{f_{refl}\pi^2}\right)^2 = 0 \quad (8.43)$$

Figure 8.17 shows the solutions to this equation. The core radii have decreased, but we are still limited in the minimum core power. We can further reduce this minimum power by doing more creative reflecting and moderating for different geometries, but we typically do not use these techniques for space applications. We can go higher than 2000 MW, but ground test limitations usually keep us from using these large cores. Remember, we can always use multiple cores or engines.

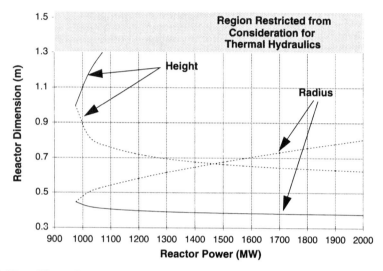

Fig. 8.17. **Dimensions of a NERVA-Type, Solid-Core Reactor with a Reflector.** This figure shows the reactor dimensions required to give us a critical reactor assuming a 10% margin on criticality ($k_{eff} = 1.1$) and a radial neutron reflector. The dashed lines are for one possible core; the solid lines are for another.

8.4.3 Size the Particle-Bed Reactor Core

Preliminary sizing of a particle-bed reactor (PBR) from basic principles is a bit more difficult than for the NERVA-type cores. Diffusion theory and assumption of a homogenous core do not give us a very accurate description of what is going on. The major problem is ensuring criticality in the core. As an alternative, Fig. 8.18 shows typical PBR core sizes (R_{core} and H_{core}) as a function of required reactor power.

The three sets of curves shown in Fig. 8.18 represent the results of a parametric analysis using the design code for the Los Alamos reactor [Los Alamos National Laboratory, 1986]. This code evaluates criticality and power-level requirements in a Monte Carlo fashion. We have chosen three possible fuel-element configurations (7, 19, or 37 fuel elements) to span the range of possible requirements for core power. The particular numbers represent the configurations that work based on geometry requirements for fuel-element packing (see Fig. 8.5).

The basic approach is to pick the core size directly off this curve, given the power requirement. In some cases, there are two possible solutions. We choose the

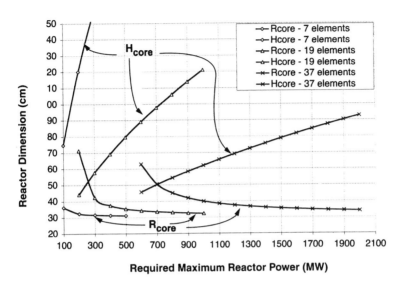

Fig. 8.18. Dimensions of a Particle-Bed Reactor. This figure shows three sets of sizing curves for a PBR. The curves descending to the right are for the core radius, and the ascending curves are for core height. Each family of curves represents a design with a fixed number of fuel elements. The particular numbers (7, 19, 37) are chosen for geometric and criticality reasons (all elements must nest together). Given the required reactor power, we pick the reactor dimensions directly from these curves. If two solutions are possible, we choose the one that best meets our requirements. The best solution is usually the core with the smallest radius, so we can reduce the shield size.

solution that best meets our particular requirements. For most situations, we choose the configuration with the smallest core radius so our radiation shield is smaller. However, if length is constrained, a "fatter" core may be desirable.

Curve fits for the core radius are

$$R_{core} = 9.0958\,(10)^{-10} P_{core}^4 - 1.3261\,(10)^{-6} P_{core}^3 + 7.1665\,(10)^{-4} P_{core}^2 \quad (8.44)$$
$$- 0.1735 P_{core} + 47.625 \quad \text{(for 7 elements)}$$

$$R_{core} = -2.655\,(10)^{-12} P_{core}^5 + 8.946\,(10)^{-9} P_{core}^4 - 1.1703\,(10)^{-5} P_{core}^3 \quad (8.45)$$
$$+ 7.427\,(10)^{-3} P_{core}^2 - 2.2955 P_{core} + 313.34 \quad \text{(for 19 elements)}$$

$$R_{core} = 4.905\,(10)^{-11}P_{core}^4 - 2.881\,(10)^{-7}P_{core}^3 + 6.2522\,(10)^{-4}P_{core}^2 \quad (8.46)$$
$$- 0.5992 P_{core} + 252.28 \text{ (for 37 elements)}$$

Similar curve fits for the core height are:

$$H_{core} = -0.000283 P_{core}^2 + 0.5203 P_{core} + 26.06 \text{ (for 7 elements)} \quad (8.47)$$

$$H_{core} = -4.027\,(10)^{-5} P_{core}^2 + 0.1427 P_{core} + 17.9883 \text{ (for 19 elements)} \quad (8.48)$$

$$H_{core} = -6.502\,(10)^{-6} P_{core}^2 + 0.05009 P_{core} + 18.335 \text{ (for 37 elements)} \quad (8.49)$$

where R_{core} = the reactor core radius (cm)
H_{core} = the reactor core height (cm)
P_{core} = required core power (MW)

We choose the configuration that gives us the lowest mass. Section 8.4.5 makes this choice clear.

8.4.4 Size the CERMET / Fast-Reactor Core

We have the same problem with this type of reactor as with the PBRs—simple diffusion theory does not work well (diffusion theory is not applicable to fast reactors). So we take the same approach to preliminary design. Figure 8.19 shows the sizing data for CERMET. We simply pick the reactor radius and height directly from this curve, given the required power level.

8.4.5 Estimate the Reactor-Core Mass

Once we have determined the basic reactor-core dimensions, we must estimate the core mass. This is not a simple task because reactor cores have complexities such as the number and size of the coolant channels, the "void space" between the fuel particles and fuel rods, and the distribution of the various materials through the core. To determine the core mass, we divide the overall core density by the core volume. For preliminary design, we assume a constant reactor-core density which should make our mass estimate at least 90% accurate.

We have assumed a cylindrical core, so the core volume is given by

$$V_{core} = \pi R_{core}^2 H_{core} \quad (8.50)$$

Based on the dimensional data for the different reactor types, Fig. 8.20 shows the core volume as a function of power level.

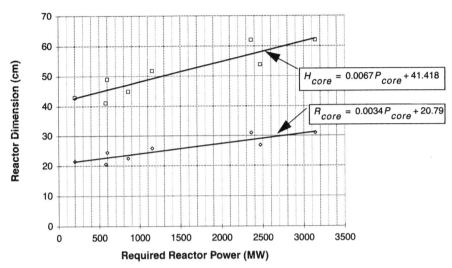

Fig. 8.19. **Dimensions of a CERMET-Type Reactor Core.** Given the required power level, we can directly pick off the reactor dimensions. The two equations show linear, least-squares curve fits for the data. P_{core} is the reactor power in MW; R_{core} and H_{core} are the reactor radius and height in cm. The radius dimension includes the reflector.

$$H_{core} = 0.0067 P_{core} + 41.418$$

$$R_{core} = 0.0034 P_{core} + 20.79$$

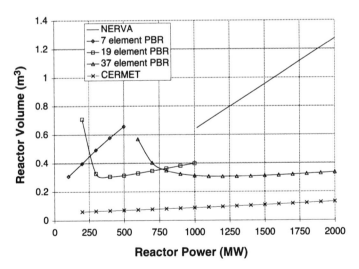

Fig. 8.20. **Reactor Volume versus Power Available.** CERMET reactors have the highest density because they have no moderator. Power density for PBRs lies between that for the CERMET and NERVA.

Then, knowing the reactor core density (ρ_{core}), we can estimate the core mass:

$$m_{core} = \rho_{core} V_{core} \qquad (8.51)$$

Typical reactor densities are:
- NERVA type—2300 kg/m^3
- Particle-bed reactor—1600 kg/m^3
- CERMET type—8500 kg/m^3

Note that these numbers are for the density of the entire reactor, including coolant passages, moderator, reflector, and all other components except the pressure vessel. In comparison, the density of uranium is approximately 19,100 kg/m^3. Figure 8.21 plots the reactor core mass as a function of required power. At the higher power levels, the CERMET cores are two to three times better than the NERVA, the PBR cores are five to six times better than the NERVA cores, and the PBRs are four times better than the CERMET cores. This result is consistent with the data in Table 8.2 and with other reference data.

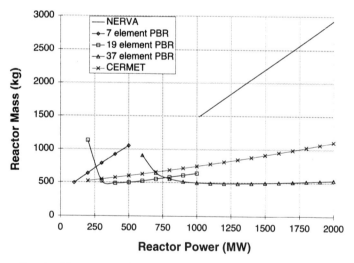

Fig. 8.21. **Reactor Mass versus Reactor Power.** Particle-bed reactors are better than both of the other reactor technologies except at very low power levels. NERVA has the lowest performance.

Figure 8.21 also helps us choose a PBR configuration. For powers less than 250 MW, the 7-element numbers give the lowest mass. Between 250 MW and 750 MW, the 19-element configuration is best. For power levels greater than 750 MW, the 37-

element configuration is best. Notice that the CERMET reactor has the lowest mass at power levels below 300 MW.

8.4.6 Thermal Hydraulics of the Reactor

Now that we understand fission, we need to see how fission energy transfers to the propellant so we can generate thrust. To make sure that the reactor or parts of the reactor do not overheat, we must extract all of the heat generated by the reactor once it reaches steady state. Compared to chemical systems and terrestrial-based nuclear systems, energy removal in space reactors is more complex because

- Some designs incorporate very high power densities (i.e., MW/kg)
- Fuel is not consumed in the usual sense; instead, it must maintain a fixed geometry for the mission duration
- Inherent nuclear-radiation fields limit selection of fuel and structural materials
- Fission power decay provides a heat source long after the neutron chain reaction shuts down

Detailed analysis of the thermal hydraulics in a nuclear reactor is beyond the scope of our discussion here (although thermal hydraulics is a major consideration in the development of Figs. 8.18 and 8.19). However, let us look at some numbers for a particle-bed reactor. A typical power density for a particle-bed reactor is 40 MW/litre. Most uranium fuel pellets are spherical, with diameters on the order of 500 μm, giving us about 13,459,304 pellets per liter. We can determine the amount of power generated per pellet from

$$\frac{\frac{\text{power}}{l}}{\frac{\text{pellets}}{l}} = \frac{\text{power}}{\text{pellet}} = \frac{40,000,000}{13,459,304} = 2.972 \frac{\text{watts}}{\text{pellet}} \tag{8.52}$$

Thus, each pellet in a particle-bed reactor produces 2.972 joules of energy per second and must transfer this energy to the propellant. In a typical pellet, the coating material surrounding the pellet is about 100 μm thick. Let us assume

- Flow rate is very low, making the temperature change across the flow boundary layer small
- The highest static (and total temperature for low flow rates) temperature of the propellant is 3000 K, which occurs at the position in the reactor where the propellant is fully heated
- The temperature profile through the coating surrounding the uranium fuel is modeled by

$$power = -kA\left(\frac{\Delta T}{\Delta x}\right) \tag{8.53}$$

where k = coefficient of thermal conductivity (w/m·K)
 ≈ 1.9 average for the coating material
 A = outer surface area of the pellet (m²)
 = 7.854 (10)$^{-7}$ m²
 ΔT = temperature change through the coating (K)
 Δx = coating thickness (m)
 = 100 μm is typical

By rearranging Eq. (8.53), we can find the temperature at the center of the fuel particle (T_f):

$$T_f = \frac{2.927\,(0.0001)}{1.9\,(7.854)\,(10)^{-7}} + 3000 = 3196\,\text{K} \tag{8.54}$$

8.4.7 Reactor Materials

For space reactors, the fuel is highly enriched (% ratio of ^{235}U to the total amount of uranium present). A typical range of enrichment level is 90–99%. These higher enrichment levels are needed to decrease the reactor size so we can package it better for space (also applies to navy vessels). ^{238}U does not fission as easily as ^{235}U, so we want to increase the fraction of ^{235}U in our fuel. Typical uranium ore samples contain different fractions of ^{235}U and ^{238}U, as well as some impurities. Enrichment is possible using a centrifuge, gaseous diffusion, or laser separation of isotopes.

The main selection criteria for other reactor materials are

- Nuclear properties (cross sections, particle behavior, burn up)
- Physical properties, such as density change and phase transformation (melting is an extreme example)
- Thermal properties, including thermal expansion, heat transfer, and heat removal
- Mechanical or structural properties, such as strength, ductility, and cracking from stress fatigue
- Chemical effects, such as impurities, corrosion cracking, hot-hydrogen erosion, and stability under irradiation

The reactor must also have a safety system to allow safe operation or shutdown if a primary component fails.

It is tough to select materials that are available at a reasonable cost and resist irradiation. One serious effect on these materials is *embrittlement*. As neutrons constantly leak out of the core and hit the outer vessel, their energy affects the crystal lattice of the vessel material by gradually compressing the molecules in the lattice.

This microscopic compression makes the material more brittle and can eventually lead to failure. It is more of a problem in commercial, terrestrial reactors because of their longer lifetimes. Material swelling, irradiation creep, and the need to compensate for high temperatures and high pressures also make material selection, testing, and qualification important.

8.5 Size the Radiation Shield

A nuclear reactor in space often requires a radiation shield to protect the crew, payload, or other spacecraft equipment sensitive to radiation. The unit that describes the damaging effects of radiation is the *rem*:

$$\text{rem} = \text{absorbed radiation dose} \times \text{quality factor} \tag{8.55}$$

The absorbed radiation dose is traditionally defined in terms of the rad (radiation absorbed dose). One *rad* equals the amount of radiation required to cause the absorption of 100 ergs (1 joule = 10^7 ergs) per gram of material. Therefore, the "rad" is the amount of energy imparted to a component or person.* Experimental data has shown that different types of radiation produce different effects. So we add the *quality factor* to account for the effect of a particular type of radiation. Table 8.9 shows the biological consequences of acute, short-term radiation effects from whole-body exposure to gamma radiation.

To put these numbers into context, we can list some typical exposures:
- Natural radioactive material in the bones—0.034 rem/year
- Flight in an aircraft—0.001 rem/hr at a 9-km altitude
- Chest x-ray (lung dose)—0.01 rem
- Living one year in Houston, Texas—0.25 rem
- Living one year in Denver, Colorado—0.35 rem (higher elevation allows more cosmic radiation)
- Working in a nuclear power plant for 1 year—less than 0.5 rem
- Skylab astronauts on 84-day mission—18 rem
- 90-day space station mission—16 rem (NASA estimate)
- Van Allen belts—16 rem/year
- Cosmic radiation outside of the Van Allen belts—19 rem/year
- A single solar flare with moderate energy protons—303 rem/year
- A properly shielded nuclear space engine—10 rem/year

---

* To determine the radiation dose from a reactor, 1 rad = 2×10^9 MeVs/cm^2. From Table 8.6, about 14 MeVs of gamma rays are produced per fission and a bit less than 1 MeV of neutrons escape. We can roughly predict the dosage if we know the reactor power, the distance from the reactor to the target, and the quality factor of gammas (1), fast neutrons (10), and thermal neutrons (2).

Table 8.9. Acute Radiation Effects From Whole-Body Exposure to Gamma Radiation. As the radiation dose increases, the biological effect increases correspondingly. These somatic effects appear after exposure to acute doses over short times rather than over longer periods, during which the cells can repair some damage.

Acute Irradiation Level (rem)	Acute Somatic Effect
15–25	Subtle reductions in white-blood-cell counts; not generally apparent from exposure for one person unless a blood sample was taken before the exposure
50	Reduction in white-blood-cell count after exposure; the count returns to normal in a few weeks
75	10% chance of nausea
100	10% chance of temporary hair loss
200	90% chance of radiation sickness; moderate depression of white-blood-cell fractions
400–500	50% chance of death within 30 days without extensive medical treatment
>600	Lethal to most people in 3 to 30 days; even with extensive medical treatment, death is likely within a few months from infection and hemorrhage
>10,000	Lethal within 24 hours from damage to central nervous system

Table 8.10 shows allowable skin radiation dosages, as given by the National Committee on Radiation Protection and Measurement.

Many factors influence the geometry, composition, and mass of the radiation shield, including:

- Size and nature of the power source
- Type of radiation
- Configuration of the spacecraft or platform and its payload (radiation flux level decreases by a factor of $1/\text{distance}^2$ from the radiation source)
- Generic operational procedures and requirements for the mission
- Length of the mission
- Total permitted levels of radiation dosage

To design the shield correctly, we need to compare the radiation flux from the engine with the allowable dose. For example, if the engine is releasing 10^7 rem per year and the payload is allowed a dose of only 10 rem per year, we need to attenuate the radiation by a factor of 10^6. For shield design in a nuclear rocket, we usually attenuate the radiation flux to avoid excess propellant heating while the

Table 8.10. Allowable Limits for Skin Dosage. The National Committee on Radiation Protection and Measurement has agreed to these allowables.

Criteria	General Public	Occupational Workers	Astronauts
30-day limit	0.04 rem	0.4 rem	150 rem
annual limit	0.50 rem	5 rem	300 rem
career limit	N/A	5 × (age in years − 18) rem	600 rem
accident limit	25 rem	100 rem	N/A
acute limit	N/A	N/A	10 rem

propellant is still in the tank. The structure of the tank, the propellant, and the additional distance of the payload from the radiation source then reduce the radiation further (see Fig. 8.22).

By combining different shielding materials, we can tailor the shield to a particular form of radiation. For example, usually a shadow shield (shields the reactor from the tank) is directly under the tank. Figure 8.23 shows a typical cross section of a shadow shield. In this example, the radiation first sees a neutron reflector material such as beryllium (Be). Next, it encounters a thin layer of heavy material used to shield gamma rays. Tungsten (W) is a good candidate for this part of the shield because it has a high neutron-absorption cross section plus high gamma attenuation. Finally, a lighter-weight material finishes attenuating the neutron flux. We often suggest using lithium hydride (LiH_2) because it has good neutron-slowing properties from the hydrogen component and a high neutron-absorption cross section from the lithium component.

Table 8.11 summarizes the properties and resultant radiation attenuation across each thickness and for the overall shield. Because the size and mass of shields are significant, we keep the layers as thin as possible for an effective shield. But the shield shown in Fig. 8.23 can reduce the gamma ray flux by a multiplication factor of 0.001 05 and the neutron flux by 4.0 $(10)^{-9}$, which is adequate for most applications. We can use the data in Table 8.12 to increase or decrease the attenuation. For example, adding 1.872 cm of tungsten further decreases the gamma-ray flux by a factor of 10.

For preliminary design, we use the shield cross section shown in Fig. 8.23 as our baseline. This shield configuration has a mass of 3,500 kg/m². As a first cut in sizing our shield, we assume it has a radius equal to that of the reactor core (R_{core}). We can change this baseline shield configuration a bit for lower or higher reactor power, burn duration, and reactor type. However, scaling this shield is highly nonlinear and requires complicated computer analysis. The simplified approach we discuss here gives us an adequate estimate.

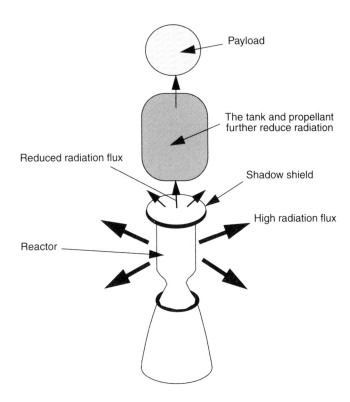

Fig. 8.22. **Radiation Attenuation with a Shadow Shield.** Nuclear rockets typically use a shadow shield to reduce the radiation levels at the payload. The shadow shield reduces the radiation flux to a level that prevents heating and boiling of the propellant. The propellant, the tank, and distance further reduce the radiation level to the payload's allowable level.

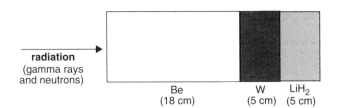

Fig. 8.23. **Typical Shield—Materials and Thicknesses.** This figure shows a typical shadow-shield cross section for a space reactor. The radiation attenuation is given in Table 8.11. If we need to modify the radiation attenuation, we can use the data given in Table 8.12.

8.5 Size the Radiation Shield

Table 8.11. Physical Properties of Shielding Material and Effectiveness for Sample Shield Layout. The information in the last two lines is most important. As gamma rays pass through the complete shield (Fig. 8.23), they are attenuated by a factor of 1000 (0.001 05). Similarly, neutrons are attenuated by 9 orders of magnitude (4×10^{-9}).

Parameter/Shield Material	Be	W	Li-H$_2$
Density (ρ – kg/m^3)	1850	19,300	500
Molecular weight (gm/mol)	9.01	183.86	Li—6.94; H—1.01
σ_a - absorption cross section (barn)	0.009	19.2	Li—71; H—0.33
Attenuation factor for 1MeV gamma rays (μ – cm^{-1})	0.104	1.235	0.0444
Shield thickness (cm)	18	2	5
Gamma attenuation of incident beam	0.1538	0.0854	0.8011
Number density (atom/(barn·cm))	0.124	0.063	0.047
Σ_a - absorption cross section (cm^{-1})	0.001	1.214	3.373
Attenuation of incident neutron beam	0.9802	0.0882	4.7×10^{-8}
Integrated gamma-flux reduction	0.1538	0.00131	0.00105
Integrated neutron-flux reduction	0.9802	0.0864	4.0×10^{-9}

Table 8.12. Radiation Attenuation for Shielding Example. To reduce gamma rays by 50%, we would need a tungsten shield 0.584-cm thick.

Reduction Factor	Thickness Required for Neutron Attenuation (cm)			Thickness Required for Gamma Attenuation (cm)		
$\Phi(x)/\Phi(0)$	Be	W	LiH$_2$	Be	W	LiH$_2$
0.5	622.7	0.571	0.205	6.665	0.564	15.611
0.2	1,446	1.326	0.477	15.475	1.308	36.249
0.1	2,069	1.897	0.683	22.141	1.872	51.861
0.01	4,138	3.794	1.365	44.281	3.744	103.721
0.001	6,206	5.691	2.048	66.421	5.616	155.581

8.6 Evaluate Vehicle Operation

The operation cycle for a single-thrust burn for a nuclear engine (based on a NERVA-type reactor) includes the following steps:

1. Make the reactor critical and raise the power to a level between 0.15% and 1% of full power by rotating the control drums.

2. *Condition* the engine by bleeding propellant flow through the system to chill the pump assembly for a few seconds while maintaining power at less than 1%. This conditioning helps avoid pump cavitation at the higher power levels.

3. "Bootstrap" the turbopump: rapidly increase the pressure and temperature to allow turbopump operation from the heat energy of the engine system. This phase usually occurs over a few seconds. Increase the temperature by increasing the reactor power output without increasing the propellant flow rate by an equivalent amount.

4. Place the engine on a thrust-buildup ramp, so the pressure increases slowly and the temperature increases more rapidly. This phase lasts 15 to 30 seconds.

5. Once the system reaches the rated temperature, rapidly increase the pressure to its rated condition. This phase takes about 5–10 seconds. Increase pressure by increasing the propellant flow rate. Also increase reactor power to maintain the constant temperature for an increased flow rate.

6. Hold rated conditions for the required thrust period. Typical thrust duration is from 600 to 3600 seconds.

7. After the thrust period, ramp the pressure back while maintaining the temperature at rated conditions. Keep the ramp at the ramp rate for maximum allowable pressure and then hold the power constant for a period to allow some reduction in the decay heat load.

8. Start a *scram* (shut down) and ramp the temperature back in a controlled manner at its maximum cooldown rate by manipulating the pressure. Stop the ramp once you reach the desired cooldown mode temperature. This phase usually lasts for 15–30 seconds.

9. Maintain a period of engine cooldown at some set temperature as long as necessary to keep components below their temperature limitations while the fission-product heat power decays. Because radioactive decay can be slow, the energy of beta and gamma rays from fission fragments and from captured gamma rays transforms into heat relatively slowly.

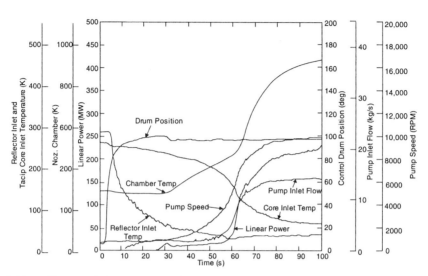

Fig. 8.24. **Characteristics of Engine Start-up for Nuclear Rockets.** This chart shows the transient start-up characteristics of several engine parameters. Things are changing drastically for the first 80 seconds before stabilizing [Angelo and Buden, 1985].

8.7 Case Study

In this section, we examine an example of a nuclear rocket design, based on the mission case study from Chap. 10—in particular, the first-stage system for "Option 4." We have already concluded that this approach is not feasible for technical reasons. Our case study has extremely restrictive envelope constraints. This means that any system using a low-density propellant, such as H_2, is impractical. If it were not for the restrictive envelope, a nuclear system would be a good technology candidate for this mission. But we use the requirements to develop a system we can compare to chemical designs. We choose the particle-bed reactor technology as our baseline because of its high performance. We also choose hydrogen as our propellant because it gives the highest specific impulse and because it is the baseline propellant for most nuclear-propulsion applications. Choosing a more dense propellant may allow us to meet the case-study requirements from Chap. 10, but the result would not represent a typical nuclear-rocket design.

8.7.1 Requirements and Design Considerations

For this mission, we use a high-thrust propulsion system to take us from an initial 400-km circular orbit to a 5000-km altitude circular orbit. From there we use a propulsion system based on an electric-ion engine to spiral up to the final orbit. The basic, propulsion-system-level requirements for this "stage" are

- Δv = 1721 m/s (we give ourselves a 10% margin for a design Δv of 1893 m/s)
- Payload mass = 4914 kg (includes 2600-kg payload plus the ion-stage mass)
- Continuous thrust with a minimum thrust-to-weight ratio of 0.3
- Initial vehicle mass < 12,000 kg
- Envelope limitations = 3-m diameter by 3-m long cylinder

8.7.2 Make Preliminary Design Decisions

Evaluate Thermochemistry

Thermal considerations allow us to drive the propellant temperature up to 3200 K. From the specific heat equation [Eq. (8.2)] and Eqs. (5.16) through (5.19), we get the following propellant properties:

- Chamber temperature = 3200 K
- c_p = 18,641.8 J/kg·K
- γ = 1.284
- c^* = 5468.1 m/s

Applying these numbers to the isentropic flow equations for a nozzle gives the relationship between specific impulse and nozzle expansion ratio shown in Fig. 8.25. We choose the following:

- Specific impulse = 1000 s
- Nozzle expansion ratio = 67

System-Level Sizing

How much propellant do we need and what kind of system masses can we expect? We answer these questions by using the ideal rocket equation discussed in Chap. 1. We already know all parameters needed to solve this equation except the structural mass fraction (f_{struc}). For this low-Δv mission, the inert mass of nuclear rockets tends to be much higher than for chemical systems, for two main reasons: first, the thrust-to-weight ratio for the engine alone is very low for nuclear systems because of the heavy reactor and the additional mass of the radiation shield. Second, using pure hydrogen requires a very large propellant tank and pressurant system. We choose a mass fraction number of 0.703 that gives us the 12,000-kg maximum allowable mass. The resulting ideal rocket analysis gives us

- Δv = 1893 m/s
- Payload mass = 4914 kg
- Structural mass fraction = 0.703

- Specific impulse = 1000 s
- Propellant mass = 2106.6 kg
- Structural mass = 4979.4 kg
- Initial vehicle mass = 12,000 kg
- Initial thrust level = 0.3 × 9.81 × 12,000 = 35,305.2 N
- Mass flow rate = 35,305.2 / (9.81 × 1000) = 3.600 kg/s
- Thrust duration = 585.2 s

Notice that we have selected an initial vehicle thrust-to-weight ratio of 0.3. This choice reduces gravity loss as discussed in Chap. 10.

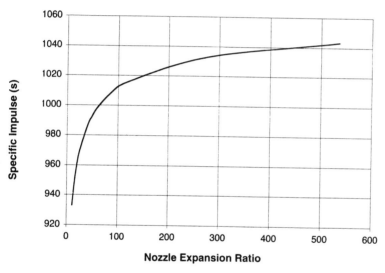

Fig. 8.25. **Specific Impulse versus Nozzle Expansion Ratio.** Marginal performance increases diminish for increases in expansion ratios above about 100. We choose an I_{sp} of 1000, which corresponds to an expansion ratio of 67. We assume a nozzle efficiency of 0.98.

Determine the Required Reactor Power

From Fig. 8.9 we know the required reactor power is

$$P_{core} = 3.6\,[0.018061\,(3200) - 5.715417] = 187.5 \text{ MW}$$

For this preliminary design, we have assumed that core burn-up is negligible. This logic is based on the relatively short burn duration.

Determine System Pressure Levels

Typical reactor-chamber pressures vary from 3 MPa to 10 MPa (some proposals have a chamber pressure as low as 0.1 MPa). We are not severely limited in engine length in this design, so we choose 3.5 MPa. From Fig. 5.19, we choose a propellant tank pressure of 250,000 Pa. Now we need to look at the feed system that will give us the required flow rate for these pressures at opposing ends of the feed system. Assuming an expander cycle that uses all of the propellant flow to drive the pump, we get:

- Chamber pressure = 3,500,000 Pa
- Reactor pressure drop = $0.05 \times 3{,}500{,}000$ = 175,000 Pa
- Turbine discharge pressure = 3,500,000 + 175,000 = 3,675,000 Pa
- Turbine pressure ratio = 1.1008 (this number is based on an iterative turbopump power balance as discussed below)
- Turbine inlet pressure = 4,045,340 Pa
- Pressure drop at the cooling jacket = $0.2 \times 4{,}045{,}340$ = 809,068 Pa
- Pressure in the propellant tank = 250,000 Pa
- Pressure drop over the feed line = 50,000 Pa
- Propellant dynamic pressure = $0.5 \times 71 \times 10^2$ = 3550 Pa
- Rise in pump pressure = 4,045,340 + 809,068 − 250,000 + 50,000 + 3550
  = 4,657,958 Pa

We determine the turbopump parameters from the power-balance Eq. (5.29). If we assume we can design our regenerative cooling system to raise the propellant temperature to 300 K (a temperature allowing the use of almost any material for the pressure vessel), we get the following turbine drive gas numbers:

- c_p = 14,209 J/kg·K   [Eq. (8.2)]
- γ = 1.409   [Eq. (8.1)]

Next, if we assume standard pump and turbine efficiencies of 0.8 and 0.7, respectively, we know the required numbers are

- Pump-head rise = 6687.6 m   [Eq. (5.53)]
- Pump power = 295,223 W   [Eq. (5.26)]
- Pump speed = 6568.4 rad/s   [Eq. (5.55)]
- Pump torque = 44.95 N·m   [Eq. (5.65)]

Then, solving the power-balance Eq. (5.29) for the turbine pressure ratio, we get 1.1008. Using the other equations in Sec. 5.3.5, we find our turbopump spins at 62,724 RPM and all our design parameters are reasonable.

8.7.3 Size the Reactor

To size our reactor, we take our required reactor power and use Eqs. (8.44) through (8.49). Because our reactor power is so low, we must use Eqs. (8.44) and (8.47) for the 7-element configuration. We then get

- Core radius = 0.327 m
- Core height = 1.136 m
- Core volume = 0.382 m^2

Assuming a core density of 1600 kg/m^3, we can then determine the core mass:

- Core mass = 0.382 × 1600 = 610 kg

8.7.4 Size the Remaining Systems

Thrust Chamber

Nozzle—From the required mass flow rate, chamber pressure, and c^*, we can determine the critical nozzle dimensions:

- Throat area = 0.005 62 m^2 [Eq. (3.130)]
- Exit area = 67 × 0.005 62 = 0.3768 m^2
- Throat diameter = 8.46 cm
- Exit diameter = 69.27 cm

Then, using charts for a bell-nozzle design in Fig. 5.25, we find that, to get a 98% efficient bell nozzle, we need

- 15° nozzle length = 1.1346 m
- Bell fraction = 0.675
- Diverging nozzle length = 0.7659 m
- Converging nozzle length = 0.0635 m
- Total nozzle length = 0.7665 m

We estimate the nozzle mass by using the equations for a tapered nozzle from Table 5.6. To be conservative, we assume the nozzle material is a high-temperature nickel alloy (such as columbium) with an assumed

- Ultimate tensile strength = 310,000,000 Pa
- Density = 8500 kg/m^3

We use the hoop-stress approach to estimate the nozzle thickness. For analysis, we assume the throat pressure is equal to the chamber pressure. (This assumption is conservative, because the throat pressure is about 50% of the chamber pressure).

We also assume a stress multiplication factor of 3.0 to account for stress concentrations and thermal effects. This gives us a 1.5-mm thickness at the throat:

$$\frac{3\,(3,500,000)\left(\frac{0.0846}{2}\right)}{310,000,000} = 0.0015 \text{ m} \quad [\text{Eq. (5.44)}]$$

If we assume the thickness tapers to zero at the exit, the nozzle length is 0.8293 m, and the conical nozzle approximation is accurate enough:

- Nozzle mass = 4.6 kg

Reactor Containment Vessel—The shape of the reactor pressure vessel is more like a propellant tank than the combustion chamber in a conventional chemical-rocket. Thus, we use the pV/W approach (see Sec. 5.4.5) to estimate the reactor vessel mass. We assume the volume of the vessel equals the reactor volume plus the volume of a hemisphere at either end of the cylinder. We assume the tank mass factor is 2500 m and the burst pressure is 2 times the storage pressure. This gives us

- Reactor vessel volume = 0.386 + 0.147 = 0.533 m³
- Vessel mass = $\dfrac{2\,(3,500,000)\,(0.533)}{9.81\,(2500)}$ = 152 kg

Hardware for the Cooling and Feed System—We know from Table 5.5 that the nozzle and combustion chamber of liquid rockets make up about 40% of the total system mass. The cooling system takes up about 35.1%, and the injector/feed system makes up the last 24.9%. (Remember, we assume the ablative cooling numbers to be a conservative estimate of a regeneratively cooled system.) Nuclear systems are somewhat different, but we use these numbers in the absence of better empirical ones. These percentages give us

- Mass of the thrust chamber structure = 152 + 4.6 = 156.6 kg
- Mass of the feed system = 391.5 × 0.249 = 97.5 kg
- Mass of the cooling system = 391.5 × 0.351 = 137.4 kg
- Total mass of the thrust chamber = 156.6 / 0.4 = 391.5 kg

Turbopump—From the engine balance analysis above, we know our pump requires a torque of 44.95 N·m. Using the empirical relation for estimating turbopump mass [Eq. (5.64)], we can get:

- Turbopump mass = $1.5\,(44.95)^{0.6}$ = 14.7 kg

Propellant-Storage System

From the system-level sizing (Sec. 8.7.2), we require 2106.6 kg of hydrogen. At a density of 71 kg/m³, we need a tank volume of 32.64 m³, allowing ourselves a

10% margin in volume. These requirements translate to a single spherical tank with a 2-m radius, or four tanks, each with a 1.24-m radius (but these do not fit into our volume envelope, as we already suspected). Using the pV/W approach for estimating the propellant tank mass and assuming a standard metallic tank, we find

- Tank mass = $\dfrac{2\,(250,000)\,(32.64)}{9.81\,(2500)}$ = 665.6 kg

Pressurant System

Using the isentropic blowdown approach to sizing our pressurant system, as discussed in Chap. 5, we get the following results with a helium pressurant:
- Initial tank pressure = 21,000,000 Pa
- Final tank pressure = 250,000 Pa
- Initial temperature = 273 K
- Final isentropic expansion temperature = 46.9 K
- $\gamma = 1.66$
- Molecular mass = 4.003 kg/kmol
- Pressurant mass = 94.9 kg (with a 5% margin)
- Pressurant volume = 2.562 m^3
- Tank mass factor = 6350 kg
- Pressurant tank mass = 863.8 kg

Notice that the mass of the pressurant system is excessive. In detailed design we would want to come up with a more elegant approach—perhaps a pressure tap from the exhaust nozzle or a heater in the H_2 tank to increase the vapor pressure.

Radiation Shield

We use the standard shield configuration discussed in Sec. 8.8.5. This shield has a mass of 3500 kg/m^2, and we assume the shield radius is equal to the core radius. This analysis gives us a
- Shield radius = 0.327 m
- Shield area = 0.3353 m^2
- Shield mass = 1173.7 kg

Support Structure

We estimate the mass of the support structure by assuming it is 10% of the system's inert mass (stage mass without propellant). Adding up all masses gives us

  Thrust chamber = 610 + 391.5 = 1001.5
  Propellant tank = 665.6 kg

Pressurant tank	= 863.8 kg
Pressurant	= 94.87 kg
Shield	= 1173.7 kg
Turbopump	= 14.7 kg
Total	**= 3725.0 kg**
Support structure	= 0.1 × 3725 = 372.5 kg

8.7.5 Baseline Design

Figure 8.26 shows the baseline configuration based on our analysis of the preliminary design. Adding up all the system masses gives a total stage mass of 6204 kg with an inert (structural) mass of 4098 kg. Remember that we originally assumed an inert-mass fraction of 0.703, which gave us an inert mass of 4979 kg. The result gives us a margin of 882 kg, which is about 22% of the estimated mass. This margin seems conservative enough to ensure that we can meet the mass requirements. But, as previously mentioned, our system does not meet the envelope constraints.

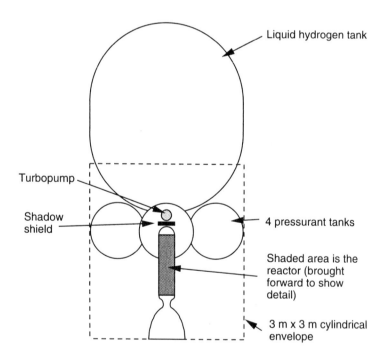

Fig. 8.26. Nuclear Rocket System Configuration. This figure shows the basic layout of our system; unfortunately, it does not fit into our envelope constraints because of the low propellant density.

To get the best advantage from nuclear rockets, we need to look at high-Δv missions (>4 km/s) that require shorter trip times. At these higher Δvs, the core criticality requirement becomes less prohibitive and the higher specific impulse gives substantially higher benefits than chemical systems (remember, the rocket equation shows an exponential improvement with specific impulse). Nuclear propulsion is a critical technology for human exploration of the solar system. Without this high thrust and high-specific-impulse option, it is highly unlikely that any significant exploration will be done by humans.

References

Angelo, Joseph and Buden, David. 1985. *Space Nuclear Power*. Malabar, Florida: Orbit Book Company, a division of Krieger Publishing Company.

Benedict, M. and Pigford, T.H. 1957. *Nuclear Chemical Engineering*. New York, NY: McGraw-Hill Book Company.

Bhattacharyya, Samit et al. 1988. *CERMET Reactor Orbit Transfer Vehicle Concept*. AFAL-TR-88-033. Edwards AFB, CA: U.S. Air Force Astronautics Laboratory.

Clark, John S. 1991. *An Historical Collection of Papers on Nuclear Thermal Propulsion*. Washington D.C.: American Institute of Aeronautics and Astronautics.

Drake, M.K. 1970. *Data Formats and Procedures for the ENDF Neutron Cross Section Library*. BNL-50274 (T-601). Brookhaven National Laboratory.

Henry, Allan F. 1986. *Nuclear Reactor Analysis*. Boston: MIT Press.

Knief, Rolland Allen. 1992. *Nuclear Engineering: Theory and Technology of Nuclear Power*. Washington D.C.: Hemisphere Publishing.

Los Alamos National Laboratory. 1986. *Monte Carlo Neutron/Photon Transport Code*. Los Alamos, New Mexico.

Ludewig, Hans. 1993. *Summary of Particle Bed Reactor Designs for the Space Nuclear Thermal Propulsion Program*. BNL-52408. Brookhaven National Laboratory.

Maise, George, and Hans Ludewig. 1995. Brookhaven National Laboratory. Private communication.

Ma, I. 1983. *Materials for Nuclear Applications*. New York: McGraw Hill.

NASA. 1990. *NASA/DOD/DOE Nuclear Thermal Propulsion Workshop Notebook*. Cleveland Ohio: NASA Lewis Research Center.

Witter, Jonathan Keay. April,1993. Modeling for the Simulation and Control of Nuclear Reactor Rocket Systems. Thesis, Massachusetts Institute of Technology.

CHAPTER 9
ELECTRIC ROCKET PROPULSION SYSTEMS

Peter J. Turchi, *The Ohio State University*

9.1	History and Status
9.2	Design Process
9.3	Specify the Mission
9.4	Select an Electric Thruster
9.5	Select Space Power
9.6	Assess System Performance
9.7	Evaluate the System
9.8	Case Study

Of the various methods for generating high speed reaction-mass, electromagnetic techniques offer the only way that, in principle, is not limited by the bond strengths of matter. In chemical systems, the amount of energy added to the propellant flow is equal to the energy released by chemical reactions. In nuclear-thermal devices, bond-strengths of materials limit structural integrity at high temperatures. In electric propulsion systems, electromagnetic forces directly accelerate the reaction-mass, so we are theoretically limited only by our ability to apply these forces at the desired total power levels. Table 9.1 compares performance values for chemical and nuclear thrusters, and for the three general categories of electric propulsion.

In chemical propulsion, exhaust speed depends on the quotient of reaction energies and reaction-product molecular masses. With nonchemical sources of energy, molecular masses can be arbitrarily reduced to hydrogen values, so higher exhaust speeds are possible (See Table 9.1). *Electrothermal propulsion* uses electrical energy to heat a propellant flow, by heat transfer from electrically heated surfaces, or by direct heat deposition in the flow itself. The former technique is known as a *resistojet*; if the latter approach involves an arc discharge in the flow, it is called an *arcjet*.

Table 9.1. Comparison of Space Thruster Values. This table compares some attributes of chemical and nuclear propulsion with those for the three basic types of electric propulsion.

Type	Specific Impulse (s)	Thrust/Weight	Thrust Duration
Chemical	200–465	1–10	minutes
Nuclear (thermal)	750–1500	1–5	hours
Electrothermal	300–1500	$< 10^{-3}$	years (intermittent) months (steady)
Electromagnetic	1000–10,000	$< 10^{-4}$	years (intermittent) months (steady)
Electrostatic	2000–100,000+	$< 10^{-4}$–10^{-6}	months – years (steady)

Electrical energy can create high exhaust speeds in other ways. Electrostatic forces applied to charged particles can accelerate matter directly for *electrostatic propulsion*. If the charged particles are ionized atoms or molecules, the device is known as an *ion engine*. (Electrostatic thrusters can use other types of charged particles, such as fine sprays of liquid droplets employed in *colloid thrusters*.) Magnetic forces can also accelerate matter to high speed. In many arrangements, electrical currents through the propellant interact with magnetic fields to provide the accelerating (Lorentz) force. Such techniques often employ a charge-neutral, ionized gas called *plasma*, so the technology is broadly known as *plasma propulsion*. The more general category is *electromagnetic propulsion*. It comprises gaseous, liquid, and solid electrically conducting materials and both steady-state and time-varying electrical power.

The basic elements of an electric propulsion system are indicated by the block diagram of Fig. 9.1. A conventional chemical rocket would have equivalent blocks for the *propellant handling system* and for the *thruster system* itself. The electric propulsion system adds blocks for functions that a chemical system leaves on Earth: components for accumulating and processing energy. The *electric power system* consists of the original power source, devices for converting the original form of power into electricity, and conditioning subsystems that shape the initial electrical power into the form required by the electric thruster. If the power source is the Sun or another radiator, such as a microwave transmitter, it does not accompany the electric propulsion system. The *thermal management system* comprises sensors, controls, coolers and radiators. The radiators dispose of heat generated by onboard components not otherwise carried away by the exhaust flow or radiated locally.

The added mass demanded by electric propulsion for these additional blocks—the so-called "power supply penalty"—comes at the expense of the payload mass and thereby diminishes some of the promise of electrical thruster techniques. *Technical* limitations thus exist in terms of minimum masses on board the

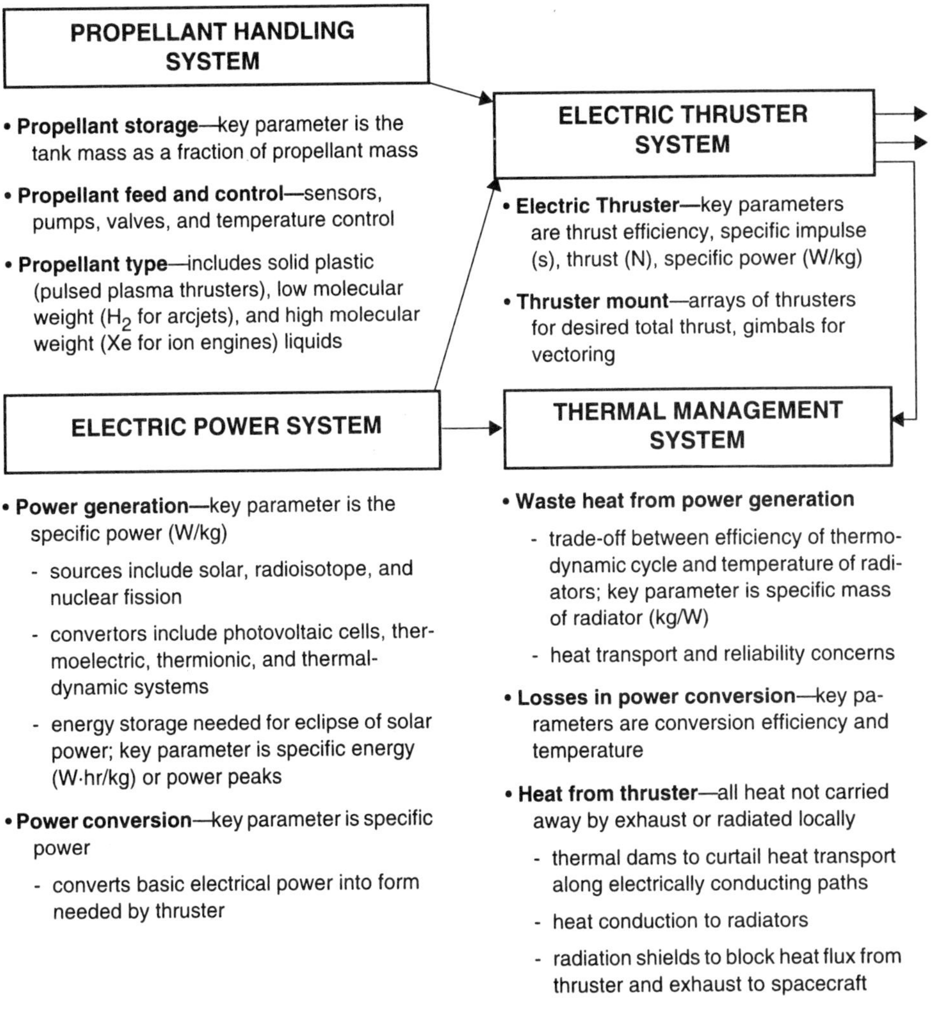

PROPELLANT HANDLING SYSTEM

- **Propellant storage**—key parameter is the tank mass as a fraction of propellant mass
- **Propellant feed and control**—sensors, pumps, valves, and temperature control
- **Propellant type**—includes solid plastic (pulsed plasma thrusters), low molecular weight (H_2 for arcjets), and high molecular weight (Xe for ion engines) liquids

ELECTRIC THRUSTER SYSTEM

- **Electric Thruster**—key parameters are thrust efficiency, specific impulse (s), thrust (N), specific power (W/kg)
- **Thruster mount**—arrays of thrusters for desired total thrust, gimbals for vectoring

ELECTRIC POWER SYSTEM

- **Power generation**—key parameter is the specific power (W/kg)
  - sources include solar, radioisotope, and nuclear fission
  - convertors include photovoltaic cells, thermoelectric, thermionic, and thermaldynamic systems
  - energy storage needed for eclipse of solar power; key parameter is specific energy (W·hr/kg) or power peaks
- **Power conversion**—key parameter is specific power
  - converts basic electrical power into form needed by thruster

THERMAL MANAGEMENT SYSTEM

- **Waste heat from power generation**
  - trade-off between efficiency of thermodynamic cycle and temperature of radiators; key parameter is specific mass of radiator (kg/W)
  - heat transport and reliability concerns
- **Losses in power conversion**—key parameters are conversion efficiency and temperature
- **Heat from thruster**—all heat not carried away by exhaust or radiated locally
  - thermal dams to curtail heat transport along electrically conducting paths
  - heat conduction to radiators
  - radiation shields to block heat flux from thruster and exhaust to spacecraft

Fig. 9.1. Basic Elements of an Electric Propulsion System. There are four basic elements in a typical system, shown along with some of the key considerations for each element.

spacecraft needed to generate and process electrical power, and to reject heat. In addition, the total electrical power available in space is usually insufficient to permit rapid expulsion of a significant fraction of the total spacecraft and propellant mass. We must therefore use spiral or multiple-burn trajectories for orbital-transfer missions. Spiral trajectories introduce inefficiencies in terms of lifting propellant mass that is later expelled at higher gravitational potential ("gravity losses").

Cyclic operation and extended exposure to radiation (e.g., Van Allen belts) require de-rating of component performance values. Increased radiation dose may be unacceptable to nonpropulsive portions of the spacecraft, such as electronics or crew.

Programmatic limitations are both obvious and subtle, involving the difficulty of displacing (or even supplementing) an existing technology, such as chemical rockets, with new technology. Missions that depend on unproven technology do not compete well against those that use standard techniques, so a mission requirement for new technology never develops, and neither does the technology. Historically, without an urgent requirement for progress on high-power electrical sources in space, competition from alternatives precluded opportunities for electric propulsion. For example, microcircuitry allowed high-speed computers to precisely calculate and control gravity-turn maneuvers. Chemical rockets launched from the Earth's surface thereby achieved interplanetary missions, such as the Neptune fly-by, previously reserved for electric propulsion.

Although electric propulsion missed most of the early successes of the space age, these successes have laid the foundation for commercializing and exploring space. In the former category, the substantial growth of Earth-orbiting satellites for communication and surveillance has led to renewed interest in electric propulsion for stationkeeping, orbit-raising and repositioning. (Stationkeeping does not involve gravity penalties relative to impulsive thruster operation.) In these cases, electric propulsion may reduce the launch-vehicle class (e.g., Pegasus, Delta, Titan) or increase the spacecraft's mission capability by providing more time on station for a given propellant load or by providing faster repositioning. Future interplanetary missions can realize the traditional benefit of electric propulsion—higher spacecraft mass fraction to higher speeds—depending on the national will and power-system development.

9.1 History and Status

The earliest reference to applying electromagnetic forces for space travel is by Edmund Rostand's character, Cyrano de Bergerac, who suggested he could reach the Moon by casting a magnet into the air while standing on an iron plate. Robert H. Goddard and Hermann Oberth independently mention the possibility of electrical propulsion in studies that focused on chemical propulsion. E. Stuhlinger extensively discusses electric-propulsion analyses through 1964. His book, *Ion Propulsion for Space Flight*, details the technical foundations for electric propulsion, particularly ion engines [Stuhlinger, 1964]. A slightly later book by R. G. Jahn, *Physics of Electric Propulsion* [1968], covers the operation of a wider variety of electric thrusters, focusing on the physical processes that must be understood and mastered to develop efficient devices. Other publications discuss Russia's efforts on various electric thrusters [Grishin and Leskov, 1989].

Much of the modern history of electric propulsion in the West has involved laboratory exploration and development of concepts and components. In 1964, the first U.S. test of electric propulsion in space occurred with SERT I (Space Electric Rocket Test), in which a chemical rocket carried an ion thruster onto a sub-orbital trajectory. A subsequent test, SERT II, involved an actual Earth-orbiting testbed dedicated to operating an ion engine. Launched in 1970, it continued (with restart in 1979) through 1990, using a solar-cell power supply. SERT II successfully demonstrated operation of a mercury-fueled ion engine in space, including charge-neutralizing the exhaust beam [Kerslake and Ignaczak, 1993].

During the early 1970s, electric propulsion achieved operational use for east-west stationkeeping on Lincoln Laboratories' LES satellites [Vondra, et al., 1970]. The thruster discharged a small, high-voltage capacitor to create a spark periodically across a block of Teflon. A small amount of the Teflon surface was ablated by the electrical discharge and then accelerated by electromagnetic forces in the discharge, achieving peak exhaust speeds of 20 km/s. To compensate for losing the ablated mass, a simple negator spring fed the Teflon block into the electrode region. This pulsed-plasma thruster (PPT) created impulse bits corresponding to micropounds of thrust. To date, millions of bits have been generated on orbit in stationkeeping applications for several satellites. Efforts through the 1980s included developing PPTs in the millipound range, but the program was cancelled when funds were shifted to nonpropulsion areas [Vondra, 1984].

Electric propulsion also progressed during the 1980s in the form of small, electrically heated, hydrazine engines for attitude control and stationkeeping. A modest gain in performance was achieved by increasing the enthalpy of the hydrazine-product flow using resistively heated, solid surfaces. These devices, now known as "augmented-hydrazine thrusters," can extend specific impulse values from the usual 240 s up to 330 s. Even the relatively small gain in specific impulse provides enough competitive advantage for satellite operations to justify the slight increase in system complexity. Most recently, kilowatt-level arcjets (specific impulse = 502 s) have been employed for stationkeeping on commercial communication satellites (Telstar IV) [Smith, et al., 1991]. Hydrazine gas generators provide the propellant source, thereby connecting electric propulsion to a developed space technology. As with the PPT, electric propulsion has become practical not by demanding a role in primary propulsion but by working within the available system limits, such as onboard power, for existing missions.

9.2 Design Process

The design process for electric propulsion has two different forms, depending on the general nature of the mission. The traditional approach presumes the mission is important enough to demand a power supply developed specifically for an electric-propulsion system. Such missions may include interplanetary flights for which electric propulsion's high specific impulse justifies multi-megawatt power sources, such as nuclear reactors, for primary propulsion. Because the power supply is dedicated largely to electric propulsion, the designer must choose from various subsystem possibilities, based mainly on the subsystem mass required to handle the necessary power, with due regard for reliability at the total powers and lifetimes the mission requires.

The "modern" design approach, which led to flight application of electric propulsion, starts with the power source already determined by the spacecraft and its mission. We then design the rest of the electric-propulsion system to match the power level and availability of this source, along with the spacecraft's other operating constraints. For example, we may need to specify propellant type based on sharing with other systems or prior experience. Propulsion system design choices involve the type of electric thruster, power-conditioning components, and trade-offs between transfer times and propellant mass. Figure 9.2 displays the basic elements of electric-propulsion design for both approaches. The accompanying flowchart (Table 9.2), steps through these elements. It also provides the location of discussion and data within this chapter.

9.3 Specify the Mission

As Fig. 9.2 shows, the first step summarizes mission requirements developed previously. Typically, these include Δv, allowable thrust time, mass requirements, environmental factors, launch-vehicle constraints, reliability requirements, and, if possible, cost requirements. We now discuss how some of these requirements are related to each other and to later design steps.

By the simple rocket equation, high exhaust speed is important for achieving a desired speed change (Δv) with the highest final mass fraction [Eq. (1.17)]:

$$\frac{m_f}{m_i} = e^{\frac{-\Delta v}{v_e}} \qquad (9.1)$$

where m_f = the total mass obtaining the Δv (kg)
m_i = the initial mass (kg)
Δv = effective vehicle velocity change (m/s)
v_e = exhaust velocity (m/s)

All of the electrical approaches to advanced rocket propulsion require equipment for providing electrical energy to the exhaust flow. In chemical propulsion, with relatively low exhaust speed and therefore relatively high propellant mass, the mass of the energy source is largely the propellant mass itself; corrections for the tankage and pump mass are relatively small. On the other hand, systems with high exhaust speed need less propellant mass, and external equipment may dominate the mass of the energy source. Thus, the final mass (m_f) includes the mass of the external power source and the desired mission payload. Typically, the mass of

9.3 Specify the Mission

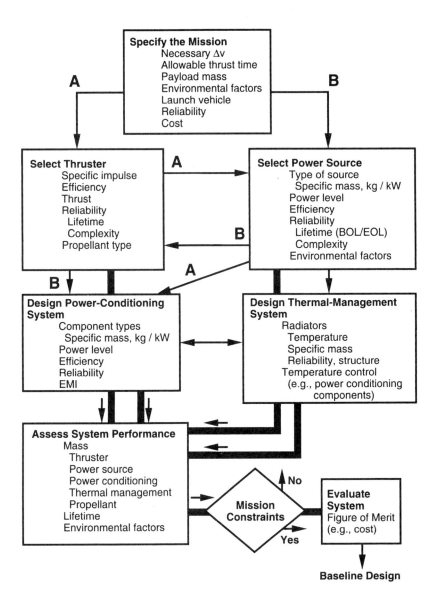

Fig. 9.2. **Designing Electric Propulsion.** We take path (A) when we base the power source on the chosen thruster and mission. We follow path (B) when we base the power source on what is available from the spacecraft. We must complete all of the activities indicated by the boxes, but the path depends on the initial requirements and constraints.

Table 9.2. Flowchart for Designing the Electric-Propulsion System. This table goes through each of the process steps in Fig. 9.2, showing where it is discussed and giving an example calculation.

Design Process Element	Where to Look
1. Specify the Mission • Define the mission in terms of the necessary maneuver, time of accomplishment, and mass delivered or allowed initially.	Sec. 9.3
2. Select Thruster (Design Path A) • Choose an initial candidate for analysis from the menu of thruster and propellant types. • For specific impulse and thrust efficiency of candidate thruster, find the necessary specific power to optimize the deliverable payload fraction for the selected exhaust speed. • Obtain the allowable specific mass of the power supply by multiplying the jet specific mass (β_J) by thruster and power processor efficiencies.	Sec. 9.4.1; Table 9.3 Sec. 9.6; Table 9.11; Eq. (9.15) Sec. 9.6; Table 9.11
3. Select Power Source (Design Path A) • Select a candidate power source technology that satisfies the allowable value of specific mass; divide specific power of source by converter efficiency for thermal sources.	Sec. 9.5; Fig. 9.42 Table 9.10 for solar arrays
4. Select Power Source and Thruster (Path B) • If the spacecraft or mission specifies the power supply, we know the value of β_s. Base the candidate thruster and propellant on the optimum exhaust speed (v_{eo}).	Eq. (9.15)
5. Design Power Conditioning System • The thruster, propellant, and power supply determine the design needed for power conditioning, which is quite complex in detail. Preliminary design specifies only the efficiency and specific mass of the power processor.	Sec. 9.5.2, 9.6; Table 9.11; Fig. 9.42
6. Design Thermal Management System • Thermal management of the thruster and power processor are included in respective design values for specific mass. • Thermal sources and convertors need to radiate substantial powers but radiator specific mass is relatively low in most cases.	Sec. 9.6 Fig. 9.42
7. Assess System Performance • Compute the system masses based on the mass of either the initial package or the desired payload. • The propellant mass (m_{prop}) depends only on the necessary Δv, and the selected v_e. • Calculate the propellant tank's dry mass, based on the known mass and type of propellant. • Compute the jet power and source power using the average mass flow rate (known because the propellant mass is expended over the thrust time). • Calculate the dry mass of the total electric-propulsion system, using known values for the specific mass and efficiency of various parts of the system. • Obtain the payload mass, not counting the power system, by subtracting the dry mass, m_{ts}, from the final mass delivered with the specific Δv.	Sec. 9.6 Sec. 9.3 Table 9.12 or Eq. (9.88) Eq. (9.4); Eq. (9.6) Sec. 9.8.6; Eq. (9.87) Sec. 9.3; Eq. (9.1); Sec. 9.6

9.3 Specify the Mission

Table 9.2. Flowchart for Designing the Electric-Propulsion System. (Continued)

Design Process Element	Where to Look
8. Decide on Mission Constraints • Our calculations so far should have satisfied the initial parameter values used to specify the mission. Now, consider other factors, such as the reliability of unfolding the solar array from the allowed launch vehicle housing. If the design is acceptable for these other conditions, retain and compare it with other candidate designs obtained by repeating this design process with different technologies.	Sec. 9.7; Sec. 9.8.8
9. Evaluate System • Compare all candidate designs based on a common figure-of-merit, such as total cost, to select the final (preliminary) design for our system.	Sec. 9.7; Sec. 9.8.8

the external source is proportional to the processed power, which is proportional to the thrust power (through a concatenation of processes, each with associated efficiency values). For a kinetic energy per unit mass ($v_e^2/2$) of the axially-directed exhaust, the thrust or *jet power* is

$$P_J = \frac{\dot{m}_{prop} v_e^2}{2} \quad (9.2)$$

where $\dot{m}_{prop}$ = mass flow rate of propellant from the thruster (kg/s)

The propellant mass (m_{prop}) is the difference between the initial and final masses of the spacecraft:

$$m_{prop} = m_i \left(1 - e^{\frac{-\Delta v}{v_e}}\right) \quad (9.3)$$

If the total efficiency of connecting source power to jet power is η_T, the required electric power is

$$P_s = \frac{\frac{1}{2} \dot{m}_{prop} v_e^2}{\eta_T} \quad (9.4)$$

The system's mass (m_{inert}) is related to the power by[*]

---

[*] We assume here that the inert mass of our entire propulsion system is a function of the power level. Further, this assumption implies that the masses of the propellant tanks and additional structure are also a function of power level, which is not, in general, correct. However, as a "first cut" we assume that these masses are negligible. This approach is at odds with that taken for chemical rockets, which are dominated by propellant. Here, our electric system mass is dominated by power level.

$$m_{inert} = \beta P_s = \frac{P_s}{\alpha} \tag{9.5}$$

where β = *specific mass* of the propulsion system (kg/W)
α = *specific power* of the propulsion system (W/kg)

Note that confusion may occur because some authors use α to denote specific mass. The advantage of working with specific mass is that usually the system design involves many components, all of which process powers proportional (through efficiency factors) to the jet power. We can then compute the total specific mass by summing the component values (with the weighting factors due to efficiencies), rather than dealing with the inverses of specific powers.

For constant thrust and exhaust speed over the thrust duration (τ), the propellant mass flow rate is

$$\dot{m}_{prop} = \frac{m_{prop}}{\tau} \tag{9.6}$$

The mass of the power system is then obtained by combining this rate with Eqs. (9.4) and (9.5):

$$m_{inert} = \frac{\beta v_e^2 m_{prop}}{2\eta_T \tau} \tag{9.7}$$

If the principal concern is the mass of the power-related components, the payload mass is

$$m_{pay} = m_f - m_{inert} \tag{9.8}$$

where m_{inert} = inert mass of the propulsion system (kg)
m_f = final mass achieving Δv (kg)

and substitution from Eq. (9.1) then gives the payload mass fraction,

$$\frac{m_{pay}}{m_i} = e^{\frac{-\Delta v}{v_e}} - \left(1 - e^{\frac{-\Delta v}{v_e}}\right) \frac{\beta v_e^2}{2\eta_T \tau} \tag{9.9}$$

where m_{pay} = payload mass (kg)
m_i = initial vehicle mass (kg)
Δv = characteristic velocity change for mission (m/s)
v_e = exhaust velocity (m/s)
β = power system specific mass (kg/W)
η_T = efficiency of converting source electric power to jet power (P_j)
τ = thrust duration (s)

9.3 Specify the Mission

A design goal is to maximize the payload mass fraction. Figure 9.3 shows that, for any desired maneuver characterized by a speed change Δv, an optimum exhaust speed (v_{eo}) maximizes the payload fraction (m_{pay}/m_i). This optimum speed depends on the total thrust time (τ) allowed (or specified) for the thrust maneuver, and the choice of propulsion system, as represented by β and η_T. Figure 9.3a shows example curves for particular numbers. In Fig. 9.3b, all speeds have been normalized by

$$v_0 = \sqrt{\frac{\eta_T \tau}{\beta}} \tag{9.10}$$

to dimensionless speeds:

$$\Delta v^* = \frac{\Delta v}{v_0} \tag{9.11}$$

and

$$v_e^* = \frac{v_e}{v_0} \tag{9.12}$$

Eq. (9.9) can then be converted to a dimensionless form:

$$\frac{m_{pay}}{m_i} = e^{\frac{-\Delta v^*}{v_e^*}} - \frac{1}{2}\left(1 - e^{\frac{-\Delta v^*}{v_e^*}}\right) v_e^{*2} \tag{9.13}$$

Note, in addition to an optimum exhaust speed, Fig. 9.3b also indicates a maximum-attainable, normalized characteristic speed (Δv^*) for a payload mass fraction greater than zero. Any speed (Δv) can be achieved, but more time (τ) is required to provide higher values of v_0.

The propulsion system mass per unit of jet power, or a *jet-specific mass* (β_J) is

$$\beta_J = \frac{\beta}{\eta_T} = \frac{1}{\alpha \eta_T} \tag{9.14}$$

The optimal exhaust speed is then

$$v_{eo} = k\sqrt{\frac{\tau}{\beta_J}} \tag{9.15}$$

where k is a number near unity, depending on the normalized characteristic speed (Δv^*). For high values of payload mass fraction, the maximum is rather broad, so the exhaust speed can differ considerably from the optimum value without

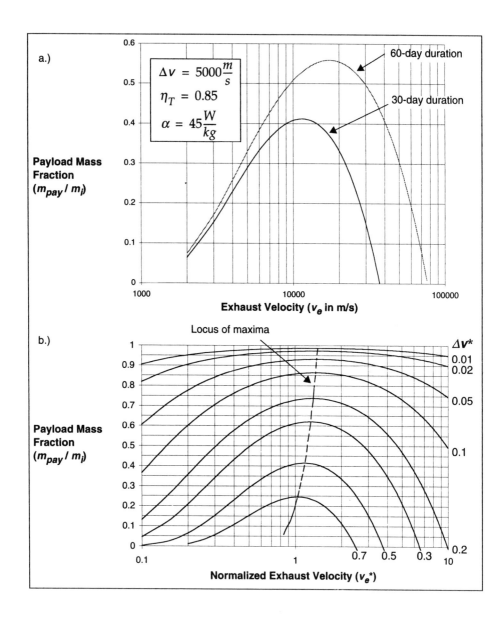

Fig. 9.3. Payload Mass Fraction versus Exhaust Velocity. The top figure represents a particular example and the lower figure shows the normalized curves, which can represent all speed possibilities. To use the normalized curves, evaluate Δv^* from Eq. (9.11), choose the v_e^* and determine the exhaust velocity from Eq. (9.12). Note the maximum payload fraction is obtained for v_e^* between 0.7 and 1.4.

9.3 Specify the Mission

significantly affecting the result.* We can also optimize exhaust speed (v_e), mission speed (Δv), and thrust time (τ) in various other ways. For examples, see the extensive discussions in Chap. 4 of Stuhlinger [1964] and Chap. 2 of this book.

Recognize that Eq. (9.15) provides an immediate, though broad, gauge to determine whether we should consider electric propulsion for a mission. If the jet-specific mass (β_J) is too high or the allowable thrust time (which is less than mission time) is too low, the calculated optimum exhaust speed may be less than that available from chemical propulsion. The deliverable payload fraction is thus less than that from chemical rockets, so there is no special point in using electrical thrusters. We may still wish to use electric propulsion for missions involving an electrical power supply, part of which could be used by the electrical thruster. This application effectively reduces the jet-specific mass by recognizing the mission's need for power, power conditioning, and heat rejection. If the fraction of the power source mass devoted only to the propulsion system is f, the effective jet-specific mass is $f\beta_J$, and the optimum exhaust speed increases by $f^{-1/2}$. For such an increase to be reasonable, we need requirements for propulsion power that match the mission needs. For auxiliary propulsion such as stationkeeping, this means we should design electric thrusters to match the mission's need for power. Primary electric propulsion will not benefit from power-system sharing until we specify large-scale missions requiring upwards of many megawatts of power.

Electric propulsion may also benefit some missions, such as lifting a large space structure from LEO to GEO, because thrust applied continually at a low level rather than as a succession of short, high-thrust events would allow greater control of structural dynamics. Similarly, distributing many electric thrusters over a large, flexible structure may avoid difficulties of stress concentrations and undesirable dynamics during spacecraft maneuvering and orbit-raising.

To optimize the payload mass fraction using Fig. 9.3, we assume the value of Δv does not depend on the exhaust speed. However, the Δv required to achieve particular orbital conditions depends on the distribution of impulse over the mission trajectory. For example, if the speed increment for escape from orbit is Δv_i when we apply thrust in a short burst (impulsively), the necessary characteristic speed (Δv) is about 2.3 Δv_i for thrust applied continually over a gentle spiral [Hill and Peterson, 1970]. Thus, the relative payload fraction that can escape orbit is higher for the low-thrust systems only if its exhaust speed exceeds that of the high-thrust system by more than a factor of 2.3. The exhaust speed must be even higher to account for the power supply mass. Based on deliverable mass fraction to escape conditions, electric rockets for primary propulsion would need I_{sp} values of at least

---

* We have assumed that η_T is a constant, which is not the case, in general. Instead, η_T is usually a function of exhaust velocity. If η_T increases with an increase in exhaust velocity, we tend to choose an exhaust velocity a bit to the right of the optimal value shown in Fig. 9.3. Conversely, if efficiency decreases, we choose a number to the left.

1000 s to compete with state-of-the-art chemical rockets, such as Centaur, which has an I_{sp} above 450 s.

Some trajectories, of course, have the same Δv for high-thrust and low-thrust systems. In particular, stationkeeping involves small impulses delivered at the same gravitational potential during the mission, so there is no "gravity penalty," and the advantage of higher exhaust speed depends only on the power supply and other system concerns. Similarly, for large orbital changes, multiple firings can accumulate the needed total velocity change in short increments at approximately the same gravitational positions as the end points of a Hohmann transfer used for impulsive operation. Thus, multiple apogee burns can follow multiple perigee burns to transfer between two circular orbits at lower thrust but with nearly the same net Δv. We need more time to complete the maneuver, but we avoid the penalty for relatively low thrust. We actually need <u>less</u> time using electric propulsion for some missions because higher specific impulse provides higher Δv with the same propellant mass. Higher midcourse altitudes are accessible for satellite repositioning, and faster passage to the outer planets is possible.

Usually, the mission specifies the allowable or desired time for thruster performance and the availability of onboard electrical power. Choice of power technology provides the various values of specific mass for components that combine to total β, and selecting a particular type of electric thruster gives us η_T, thereby allowing us to calculate β_J and the best exhaust speed (v_{eo}). If this speed is not high enough to warrant electric propulsion, we can return to chemical or nuclear-thermal propulsion. If it is within range of the candidate electric-propulsion technology, we can calculate the deliverable mass fraction, allowing for the propulsion system's dryweight (thruster, tankage, structure) not already counted in the specific mass values. This allowance should include the need for multiple engines (either to achieve the required thrust or for reliability), as well as the associated structures and control mass.

In evaluating and selecting a thruster, we must consider power-supply degradation and thrust efficiency vs. exhaust speed. Design of components of the electric-propulsion system, including the thruster itself, can influence the available thrust duration and the allowable values of power and thrust for reliable operation. If mission requirements do not limit the thrust duration, degradation of the power supply with time can constrain the allowable thrust time or reduce operating power values. For example, LEO-to-GEO transfer through the Van Allen radiation belts can degrade solar-cell performance. Similarly, long-duration missions with nuclear power sources (e.g., a voyage to Neptune) can involve problems with materials at high temperature, such as thermal creep. For constant inert mass and exhaust velocity, the time required to expel the necessary amount of propellant mass is proportional to β_J. Thus, if the spacecraft accelerates too slowly, relative to the time for β_J to increase because of power supply or thruster degradation, it may never attain the desired Δv for the mission. An obvious and extreme example of increased β_J is complete failure of the power supply or thruster.

The relationship of efficiency to exhaust speed is a critical element in selecting technology for electric propulsion. For example, the efficiency of converting electrical energy into thrust energy can be quite high for electrostatic devices because the basic acceleration mechanism (electrostatic fields operating on single particles) within the thruster is well-controlled. The efficiency largely depends on the energy cost to create the charged particle, such as an ion, compared to the particle's kinetic energy. The cost per ion (W_i) is dictated by atomic physics (and efficiency of technique) and is essentially independent of the desired exhaust speed. The kinetic energy (W_k) depends on the square of the exhaust speed, so the elementary thrust efficiency,

$$\eta_k = \frac{W_k}{W_k + W_i} \qquad (9.16)$$

decreases significantly at lower exhaust speeds. Similar considerations are involved in arcjets and electromagnetic thrusters, in which the energy involved in creating the high-temperature working fluid may be lost in the high-speed, low-density exhaust (*frozen-flow* loss).

Propellant choice strongly affects the efficiency (in terms of relative cost per ion or frozen flow), but issues of physics and chemistry are not critical. Instead, we choose propellants based on availability of proven technology and compatibility with other systems on board the spacecraft. For example, hydrazine has been selected for use with low power (~1.8 kW) arcjets for satellite stationkeeping in part because of established space-based technology for gas generators using hydrazine decomposition. For lengthy missions, propellants must be storable. So selecting hydrogen as a propellant depends on the availability of cryogenic technology, and not merely on the atomic properties of hydrogen in an arc discharge.

9.4 Select an Electric Thruster

This section includes three levels of discussion. Section 9.4.1 surveys the types of electric propulsion at a level sufficient for selecting candidate electric thrusters in the basic, systems-level design of Fig. 9.2. Section 9.4.2 provides the propulsion scientist or engineer with specialized material needed to understand the detailed behavior and design of electric thrusters. It assumes a background in gas dynamics and some familiarity with electricity and magnetism. Section 9.4.3 uses concepts from Sec. 9.4.2 to describe the operation of several types of electric thruster. This section also presents detailed data on thruster performance.

9.4.1 Overview of Electric Thrusters

Within the three categories of electric propulsion (electrothermal, electromagnetic, and electrostatic) given in Table 9.1, very different arrangements can share common problems and physical processes. Some devices have characteristics from more than one category. In particular, electrothermal and electromagnetic devices involving direct-current arc discharges behave similarly. Figure 9.4 shows the types of electric thrusters.

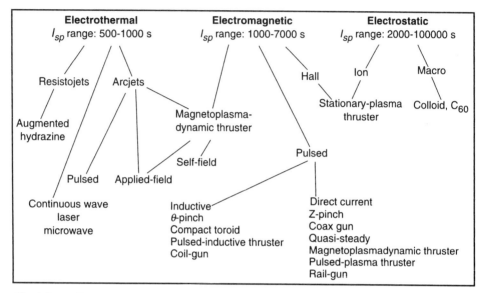

Fig. 9.4.   Types of Electric Thrusters. Different thrusters share common problems and physical processes. As shown, some devices have characteristics from more than one category.

Electrothermal Propulsion

In electrothermal propulsion, electrical energy heats the working fluid. Actual flight devices within this category include a special class of resistojet, in which electrical energy merely augments the much larger chemical energy of the hydrazine monopropellant [Olin Aerospace Co., 1993a], and kilowatt-class arcjets [Olin Aerospace Co., 1993b]. The latter now provide stationkeeping for communication satellites. The behavior of the flow in an electrothermal thruster is basically the same as for standard rocket engines: hot gas expands through a converging-diverging nozzle to achieve high exhaust speeds. The same concerns exist for the relative energy in kinetic vs. internal energy, as well as for the loss of energy to heat transfer, radiation, and viscosity. High exhaust speeds in electrothermal rockets imply high stagnation temperatures and the possibility of accessing high-energy regions of phase space. Higher speed and smaller diameter (because of limited electrical power availability) mean shorter time for expansion of the flow relative to the time ions and atoms need to de-excite or recombine. This can result in greater retention of internal energy in the exhaust, often called *frozen-flow loss*.

In resistojets (Fig. 9.5), the electrically heated channel wall has a higher temperature than the flow, as opposed to the situation in a chemically heated or arc-heated flow. Thus, the flow-stagnation temperature attained electrically is less than that available from advanced chemical systems, such as O_2/H_2, and also less than that attained in arc-heated, electrothermal devices. In arcjets (Fig. 9.6), conditions for operating the arc discharge involve sharp gradients in flow temperature between the arc core (at 10,000–40,000 K) and the thruster walls (1000–3000 K), which are cooled. Flow heated by electromagnetic waves, such as microwaves, provides similar gradients (Fig 9.7) [Balaam and Micci, 1991].

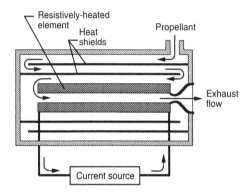

Fig. 9.5. **Schematic Diagram of a Resistojet** (configured with re-entrant flow). The propellant passes over concentric, tubular heating surfaces before exhausting through a conventional nozzle.

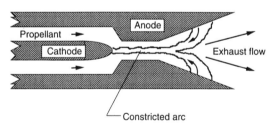

Fig. 9.6. **Schematic Diagram of a Constricted Arcjet.** Cool propellant flows into the throat where it is heated by an electric arc passing through the propellant from the anode to the cathode.

For arcjets (direct-current discharge) and electromagnetic wave-heated devices (induced-current discharge), the distribution of heating depends on the distribution of electrical conductivity, which in turn depends on the details of local electrical heating and flow cooling. Power delivery into the propellant flow requires heat transfer from the electrically heated parts of the flow to the colder input flow. Without such heat transfer, most of the flow may bypass the electrical discharge, and the discharge itself does not absorb power effectively from the

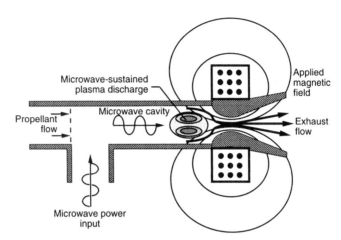

Fig. 9.7. Schematic of a Microwave-Heated Thruster. Microwaves sustain a plasma discharge that heats the propellant flow. A magnetic field helps to stabilize the plasma discharge in the microwave cavity upstream of a conventional nozzle.

electrical source. Conduction of electrical current in direct-current devices (arcjets) demands power deposition in and near the electrodes (see Electrical Discharge Behavior in Sec. 9.4.2). Nonuniform distribution of current in the arcjet flow, as well as high voltage drops at the electrodes relative to the voltage across the high-temperature discharge, can concentrate power deposition in small spots on the electrode's surface. This condition can severely decrease electrode life.

To improve the azimuthal symmetry of the discharge, high-power (100 kW–10 MW) arcjets usually employ solenoidal magnetic fields, as shown in Fig. 9.8. Such fields rotate the discharge flow, thereby depositing energy more uniformly within the plasma and on surfaces such as electrodes. In higher-power versions, the applied magnetic field also improves performance, resulting in a device that is sometimes referred to as a *magnetoplasmadynamic* (MPD) arcjet to emphasize the importance of electromagnetic processes.

Electromagnetic Propulsion

Electrothermal thrusters resemble conventional rockets discussed in earlier chapters. The source of heat is the only difference. Electromagnetic forces provide new ways to accelerate propellant flow. Section 9.4.2 discusses these forces and properties of the propellant flow. For the thrusters described here, we need to know a few basic concepts. In particular, the equation describing the electromagnetic force per unit volume acting on a gas that carries current in a magnetic field is

$$\vec{F}_m = \vec{j} \times \vec{B} \qquad (9.17)$$

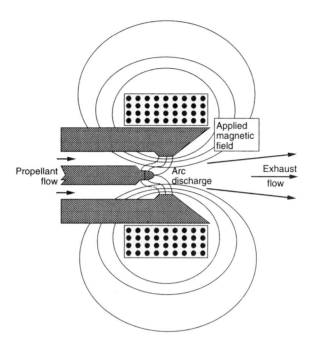

Fig. 9.8. **Schematic of an Arcjet with an Applied-(Solenoidal) Magnetic Field.** The magnetic field rotates the plasma discharge to improve azimuthal symmetry of heating.

where $\vec{F}_m$ = electromagnetic force per unit volume of gas (N/m³)

$\vec{j}$ = electric current density passing through the gas (A/m²)

$\vec{B}$ = magnetic (induction) field in gas (Tesla or T)

Note that the electromagnetic force density is perpendicular to the local current density and to the local magnetic field. In Fig. 9.9, current flows through the propellant gas from the anode to the cathode of a simplified thruster. Poles of a permanent magnet provide a magnetic field perpendicular to the current and propellant flow directions. The resulting electromagnetic force accelerates the propellant downstream.

Instead of the permanent magnet, an electromagnet or solenoid can provide the magnetic field. Current flow through the gas also induces a magnetic field. For a long, cylindrical current path, like a straight wire, the magnetic-field lines are circular loops surrounding the axis of symmetry. In a plane containing this axis, the magnetic field is represented by small circles with crosses or dots, indicating a field directed into or out of the plane, as shown in Fig. 9.10. The magnetic field at a point induced by current flow is the vector sum of the magnetic fields surrounding each part of the current path through the propellant and the electrodes. The magnetic fields applied by external sources add to the field induced by the current itself. The local current interacts with the total magnetic field at a point to create the electromagnetic force.

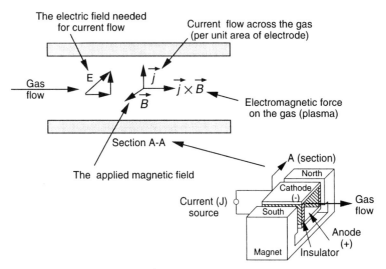

Fig. 9.9. **The Electromagnetic Force on a Plasma.** Current through a gas interacts with an applied magnetic field to give a body force on the gas. The electric field needed for current flow has components parallel and perpendicular to the current. The voltage across the gas depends on the parallel part. Slight charge separation in the gas causes the perpendicular part.

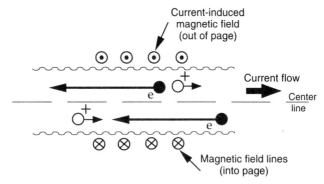

Fig. 9.10. **Magnetic Field Lines around a Current Flow.** The motion of electrons and positive charges are an electric current which is encircled by magnetic self-field lines as shown.

Electromagnetic propulsion includes various devices that divide along two separate axes: unsteady vs. steady, and self-field vs. applied field. The latter division is based on discharge currents whose own magnetic fields (self-fields) are high enough for efficient thruster performance without needing external (applied) magnetic fields. This situation demands high power (> megawatts). Such power is readily available in a short pulse from a capacitor bank, which historically links self-field arrangements with unsteady operation. Devices include *z-pinch* engines, in which the current flow has components parallel to the axis of symmetry (Fig. 9.11), and *θ-pinch* engines, in which the current is in the azimuthal direction. In both arrangements, currents and self-fields combine to implode or "pinch" plasma at high speed (10–40 km/s). In the *pulsed inductive thruster* [Dailey and Lovberg, 1979, 1993] (Fig. 9.12), coil and plasma currents are azimuthal, and the intervening magnetic field is basically radial, so plasma accelerates parallel to the axis of symmetry. Pulsed-plasma thrusters may operate without ionizing all the propellant, and thereby reduce frozen-flow loss. Conditioning power for such repetitively pulsed devices and providing transient pulses of propellant to match the electromagnetic pulses significantly complicate the propulsion system. Ablation-supplied thrusters avoid this last problem but require ingenuity in arranging the propellant feed.

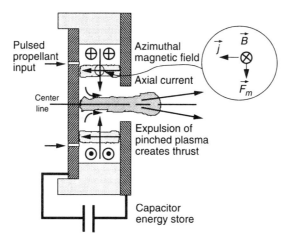

Fig. 9.11. **Schematic of a Z-Pinch Discharge Thruster.** The axial flow of current through the propellant gas interacts with the associated azimuthal magnetic field to create a radial, inward, electromagnetic force that pinches the plasma.

Pulsed-arc discharges have long been used for satellite stationkeeping, for which simple, reliable operation is more important than sophisticated physical concepts. The device, known as a *pulsed-plasma micro-thruster* [Vondra, et al., 1970], usually abbreviated as PPT, involves a spark discharge across a spring-driven block of Teflon propellant (Fig. 9.13). Some researchers have suggested extending the pulsed-plasma thruster's simplicity to electrothermal and electromagnetic devices for applications needing higher total energy and using quasi-steady operation [Turchi, 1982].

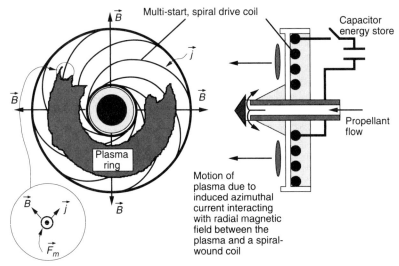

Fig. 9.12. **Schematic of Pulsed Inductive Thruster.** Spiral-wound drive coil induces azimuthal current in plasma formed from gas "puffed" back toward an insulated coil (see Dailey and Lovberg [1979, 1993]).

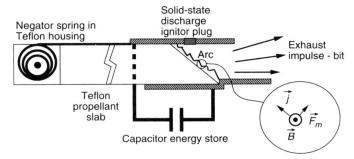

Fig. 9.13. **Schematic of a Pulsed-Plasma Thruster (PPT).** A negator spring advances the slab of Teflon propellant into the discharge chamber. The arc discharge interacts with its associated magnetic field to accelerate ablated plasma away from the Teflon surface (see Vondra, et al. [1970]).

Steady-state electromagnetic thrusters can include self-field and applied-field arrangements. Self-field devices (Fig 9.14) closely resemble electrothermal arcjets, differing only in the use of lower flow densities to attain higher exhaust speeds. Indeed, an empirical survey by Ducati discovered the magnetoplasmadynamic arcjet when the mass flow to an electrothermal arcjet was reduced to low levels [Jahn, 1968]. Instead of destroying the electrodes by overheating, the discharge achieved a diffuse operation with lower rates of electrode erosion. At lower particle densities, the flow becomes more susceptible to electromagnetic processes. In particular, the electromagnetic force density in the flow can greatly exceed pressure gradients in the gas.

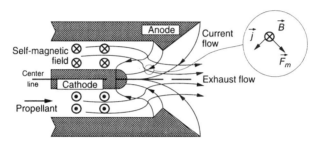

Fig. 9.14.   Basic Arrangement of Self-Field Magnetoplasmadynamic Thruster. The discharge current interacts with its associated (self) magnetic field to accelerate the flow axially and radially.

Applying solenoidal magnetic fields (Fig. 9.15) in magnetoplasmadynamic thrusters allows higher voltages at lower discharge currents, but results in a very complex thruster flow field [Myers, 1991]. Such field application is typical of moderate power operation (~100 kW), in which the direct arc current is relatively low (1 kA vs. 20 kA). Without applied fields, electrode fall voltages can dominate the voltage across the thruster at low current, which decreases the thrust efficiency. Also, at low currents, self-magnetic fields are too low for significant electromagnetic acceleration. Applied magnetic fields increase electromagnetic forces in the plasma. They also force the current to flow in spiral paths, increasing the total voltage relative to the electrode falls. This operation represents the *Hall effect*, discussed in Secs. 9.4.2 and 9.4.3.

The Hall effect is particularly evident in electromagnetic thrusters at low particle densities. Thus, many such devices are known as *Hall thrusters*. A special coaxial arrangement, developed in Russia [Bugrova, et al., 1991], uses a basically radial magnetic field and axial current flow from an upstream anode (Fig. 9.16). It is called a *stationary-plasma thruster (SPT)*. This device, operated at powers of 5–20 kW, has provided xenon plasma flows at specific impulses of 1500–2000 s with relatively high efficiency. Axial current flow across the radial magnetic field generates an azimuthal electron flow. An internal, Hall electric field in the axial direction

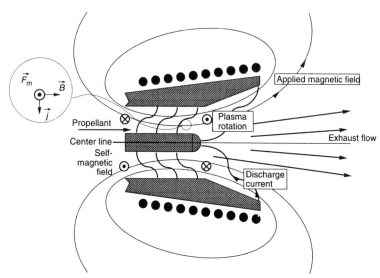

Fig. 9.15.   Basic Arrangement of Applied-Field Magnetoplasmadynamic Thruster. Applying a solenoidal magnetic field can improve thrust performance, particularly when the discharge current is too low to create a satisfactory self-magnetic field.

transmits the axial electromagnetic force on this electron flow to the plasma ions. Some have characterized the SPT as a gridless ion thruster because this axial electric field accelerates ions. The plasma is essentially charge-neutral, however, so the device is actually distinct from the vacuum-diode arrangement of electrostatic thrusters.

Electrostatic Propulsion

The simplest notion of an electric rocket engine is based on the attraction (or repulsion) of electric charges. The basic arrangement is shown schematically in Fig. 9.17. A source supplies charged particles of a single sign to the region of an electrostatic field. The charges are accelerated by the field and then pass out to a region in which the overall flow is neutralized to prevent a net charge buildup on the spacecraft. Such a buildup would return ions to the spacecraft, causing damage and cancelling the thrust. The usual form of an electrostatic thruster is the *ion engine* [Beattie, et al., 1990], in which positively charged ions are supplied to an accelerating region comprising a set of grids to control the potential distribution and ion motion. Adding an equal current of electrons from thermionic emitters near the thruster exit neutralizes the high-speed flow of positive ions. The ion kinetic energy simply depends on the net accelerating voltage. Accumulation of charge within the accelerating region limits the ion current when this space-charge offsets the applied charges on the grids (see Sec. 9.4.3).

9.4 Select an Electric Thruster

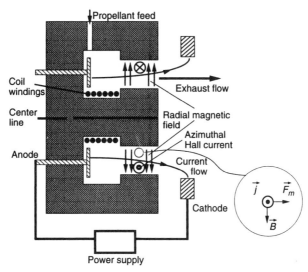

Fig. 9.16. Schematic Arrangement of a Hall Thruster. A radial magnetic field inhibits electron transport from the cathode to the anode. This field creates an azimuthal Hall current that interacts with it to give an axial thrust. The Hall current is associated with an axial Hall electric field that accelerates ions created near the anode.

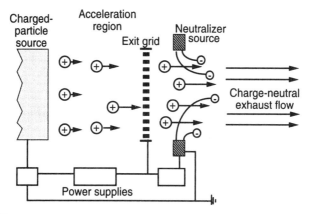

Fig. 9.17. Basic Elements of an Electrostatic Thruster. Ions created at the source are accelerated by the net electric field created by the grid and intervening space charge. Electrons are added after the exit grid to neutralize the charge on the exhaust flow.

Although several types of ion source have been developed, the principal technique presently used is an electron-bombardment plasma source (see Sec. 9.4.3). The efficiency of generating ions (the cost-per-ion in volts) compared to the net accelerating voltage is the chief limitation in applying ion engines to missions needing lower specific impulses (< 2000 s), at least for simple atomic ions, such as xenon. Using charged droplet sprays (and other macroparticle sources) has been explored to achieve particles with higher mass per unit charge, but for reasons that are not entirely technical, this technique has not been applied to space propulsion [Zafran, 1984]. The relatively low thrust density of ion engines, together with the need to control carefully the various voltage stages, has challenged their application in near-term missions, but they will soon be used for satellite stationkeeping. The potential for almost unlimited specific impulse certainly recommends them for deep-space missions, especially to the outer planets and beyond.

Table 9.3 lists various electric-thruster arrangements with associated specific impulse values and selected references.

Table 9.3. **Some Details of Electric Thrusters.** We can use these numbers to perform the systems-level thruster calculations discussed in Secs. 9.2 and 9.3. See Sec. 9.4.3, Fig. 9.41, and Table 9.11 for more details.

Type	Specific Impulse (s)	Thrust (N)	Efficiency	References
Resistojet	300	0.5	N/A	Olin Aerospace Company, 1993a
Kilowatt-level arcjet	500	0.2	0.3	Olin Aerospace Company, 1993b
10–100 kW arcjet	1000	5.0	0.3	Sovey and Mantenieks, 1988
Microwave thrusters*	600	0.3	0.5	Balaam and Micci, 1991
Pulsed electrothermal thruster (PET)	1500	0.13	0.4	Burton, et al., 1990
MW-level, self-field MPD (quasi-steady)	5000	10–100	0.5	Myers, Mantenieks, and LaPointe, 1991
100 kW-level, steady, applied-field MPD	1300	4.0	0.2	Myers, 1991
Pulsed-plasma thruster (PPT)	1500	10^{-5}–10^{-3}	0.2	Vondra, Thomassen, and Solbes, 1970
Pulsed-inductive thruster (PIT)	4000	0.1–1.0	0.5	Dailey and Lovberg, 1979, 1993
Compact toroid*	10,000	10–100	0.3	Bourque, et al., 1992
Stationary-plasma thruster (SPT)	2000	0.1–1.0	0.6	Bugrova, et al., 1991
Ion engines	2000–10,000	0.1–1.0	0.9	Beattie, Matossian, and Robson, 1990

* Indicates a concept for which values are based on theory or preliminary laboratory development (as opposed to actual thrust-stand data).

9.4.2 Background Physics for Electric Thrusters*

Aerospace-engineering education typically does not include the specialized aspects of electromagnetism, physics of ionized gases, and electrical-discharge behavior commonly encountered in electric thrusters. Propulsion scientists and engineers making decisions to select or extend electric propulsion for their missions need some exposure to these specialties. Fundamental understanding also helps in evaluating new thruster concepts, especially where common limitations are obscured by mere novelty of arrangement. For more complete discussions, please refer to the classic text, *Physics of Electric Propulsion* [Jahn, 1968], and to standard treatises in various specialty areas, such as statistical mechanics [Morse, 1964] and electromagnetic theory [Reitz and Milford, 1960]. Here we briefly discuss three specialized topics of particular importance: electrical forces and currents, ionization, and gas-discharge behavior.

Electrical Forces and Currents

To understand the forces and currents within electric-propulsion devices, we need to develop equations for the dynamics of single particles and collections of particles (fluids). We must combine these dynamical equations consistently with equations for the electric and magnetic fields to determine important features, such as thermal loadings and forces.

Single particle dynamics—For a single charged particle, the dynamical equation in the presence of both electric and magnetic fields is

$$m_i \frac{d\vec{u}_i}{dt} = q_i\left(\vec{E} + \vec{u}_i \times \vec{B}\right) + \sum_k \vec{P}_{ik} \qquad (9.18)$$

where m_i = particle mass (kg)
q_i = particle charge (coulombs or C)
$\vec{u}_i$ = particle velocity (m/s)
$\vec{E}$ = electric field (volts/meter or V/m)
$\vec{B}$ = magnetic (induction) field (teslas or T)
$\vec{P}_{ik}$ = collisional force per particle (N)
$q_i\left(\vec{u}_i \times \vec{B}\right)$ = *Lorentz force* (N)

---

* Editor's note: This section provides the basic theory of electric propulsion and is included for completeness. It is not necessary to perform basic design and sizing of electric propulsion systems. We assume readers of this section are already familiar with electric theory.

The sum on the right includes all momentum-changing encounters between the particle and other particles, (including particles of the same type, $k = i$). Note that the Lorentz force is perpendicular to the particle's velocity and therefore does not change the particle's energy but merely redirects its motion.

In the absence of encounters, the particle can gyrate in a magnetic field based on a balance between centrifugal effects and the Lorentz force:

$$m_i \frac{u_i^2}{r_g} = q_i |\vec{u}_i \times \vec{B}| \tag{9.19}$$

where the radius of gyration is

$$r_g = \frac{m_i |\vec{u}_i \times \vec{B}|}{q_i B^2} \tag{9.20}$$

and the (angular) frequency of gyration is

$$\omega_g = \frac{q_i B}{m_i} \tag{9.21}$$

Applying an electric field to a gyrating particle causes the center-of-gyration to drift perpendicular to the magnetic field with a drift velocity

$$\vec{u}_d = \frac{\vec{E} \times \vec{B}}{B^2} \tag{9.22}$$

while accelerating freely along the magnetic field. Figure 9.18 displays charged-particle motions in the presence of magnetic fields.

Dynamics of collections of particles—If we add up the dynamical equations for all particles of the same type present with number density (n_i), we obtain the fluid-momentum equation for the i-th fluid. This equation has various levels of sophistication. A simple Euler representation serves our needs:

$$n_i m_i \frac{D\vec{v}_i}{Dt} = n_i q_i (\vec{E} + \vec{v}_i \times \vec{B}) - \nabla p_i + n_i m_i \sum_k \frac{(\vec{v}_k - \vec{v}_i)}{\tau_{ik}} \tag{9.23}$$

where $\frac{D}{Dt}$ = $\partial/\partial t + \nabla$, the convective derivative operator
n_i = number density of particles of i-th fluid (m^{-3})
p_i = the (isotropic) pressure of the i-th fluid (Pa)
$\vec{v}_i$ = mass-averaged velocity of the i-th fluid (m/s)
τ_{ik} = characteristic time (s)

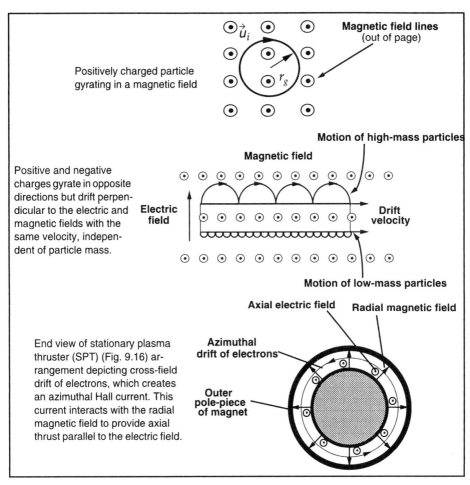

Fig. 9.18. Charged-Particle Motions within a Magnetic Field. Gyrating motion couples with translational motion in a magnetic field as shown.

We take the sum on the right over all the other types of particles, with respective fluid velocities ($\vec{v}_k$). The characteristic time for the i-th and k-th fluids to achieve a common velocity (and thereby alter the velocity of the i-th fluid) is denoted by τ_{ik}. Adding up the momentum equations for all the fluids and imposing the quasi-neutrality condition that the net electrical charge in the fluid is approximately zero,

$$\sum_i n_i q_i = 0 \tag{9.24}$$

we can obtain the momentum equation for the total, charge-neutral fluid, called *plasma*:

$$\rho \frac{D\vec{v}}{Dt} = -\nabla p + \vec{j} \times \vec{B} \tag{9.25}$$

where ρ = the plasma mass density (kg/m³)
p = the plasma pressure (Pa)
$\vec{j}$ = electrical current density (A/m²)

We have introduced the current density, which may be written in terms of the particle densities and velocities of each species of charged fluid:

$$\vec{j} = \sum_i n_i q_i \vec{v}_i - n_e e \vec{v}_{el} = n_e e (\vec{v} - \vec{v}_{el}) \tag{9.26}$$

In this equation, we have explicitly separated the electron contribution (subscript "*el*") to the current density from the motions of the other charges in the plasma; the summation ("*i*") is therefore over the charged, heavy particles in the plasma, with all velocities taken as that of the fluid, $\vec{v}_i = \vec{v}$. Eq. (9.24) tells us that the product of particle density and charge for the electrons equals the summation of such products for the heavy particles.

Note that the effects of particle collisions among the constituents of the plasma cancel out, along with the effect of the electric field. For meaningful, local, simply related, averaged quantities, such as fluid velocity and pressure, the continuum approximation is valid. Our momentum equation for the plasma then shows that the only ways to accelerate a plasma are pressure gradients and current flow perpendicular to magnetic fields. The former situation is the basis for electrothermal thrusters, while the latter is, of course, the mechanism for various types of electromagnetic thrusters. The complexity in electromagnetic thrusters derives from the fact that we do not directly control the pressure and current density, but must self-consistently solve the fluid equations and Maxwell's laws for the electric and magnetic fields.

Calculating the Electric Field in a Plasma—For many situations, electrons in a plasma can accelerate to accommodate the various forces exerted on them over time, as well as the distance scales that are much smaller than similar scales for heavy-particle dynamics. This condition is represented mathematically by setting the electron inertial term equal to zero in the momentum equation, [Eq. (9.23)], specialized for the electron fluid:

$$\rho_e \frac{D\vec{v}_{el}}{Dt} = 0 = -n_e m_e (\vec{E} + \vec{v}_i \times \vec{B}) - \nabla p + n_e m_e \sum_k \frac{(\vec{v}_k - \vec{v}_{el})}{\tau_{ek}} \tag{9.27}$$

where ρ_e = mass density of the electron fluid (kg/m³)
       e = magnitude of the electron charge (1.6×10^{-19} C)
       $\vec{V}_{el}$ = electron fluid velocity (m/s)

We may then derive a relationship for the electric field in the plasma consistent with electron motion relative to the other particles, in the presence of a magnetic field and an electron pressure gradient. To find this relationship, we must first evaluate the last term in Eq. (9.27). The electrons (e) collide with other particles (k) with a collision frequency for momentum transfer (v_{ek}^p), that is proportional to the number density of the other species:

$$v_{ek}^p = \frac{1}{\tau_{ek}} = n_k < Q_{ek}^p |\vec{c}_e - \vec{c}_k| > \tag{9.28}$$

where the symbol $<>$ indicates averaging the product of the collision cross section for momentum transfer (Q_{ek}^p in m²) and the relative encounter speed over the distributions of particle velocities, denoted by $\vec{c}_e$ and $\vec{c}_k$ respectively. If collisions with charged particles predominate (because of the relatively long range of the Coulomb electrostatic force compared to atomic forces), the velocity difference in the collision term involves only charged particles, and we can represent it by the current density from Eq. (9.26):

$$\vec{V} - \vec{V}_{el} = \frac{\vec{j}}{n_e e} \tag{9.29}$$

The electric field is then

$$\vec{E} = \eta \vec{j} - \vec{V}_{el} \times \vec{B} - \left(\frac{\nabla p_e}{n_e e}\right) \tag{9.30}$$

where we have combined several terms into the quantity:

$$\eta = \frac{m_e v_{ce}}{n_e e^2} \tag{9.31}$$

By comparison with Ohm's law, we can identify this quantity as the plasma's electrical resistivity (ohm·m). The total collision frequency for momentum transfer of electrons with heavy particles is denoted by v_{ce}, which is just the sum over the individual collision frequencies. The resulting equation for $\vec{E}$ is often referred to as a *generalized Ohm's law*.

Typically, the electron-fluid velocity is eliminated in favor of the plasma (positively charged-particle) velocity ($\vec{V}$) and the current density ($\vec{j}$) to obtain the form

$$\vec{E} = \eta\vec{j} - \vec{v}\times\vec{B} + \vec{j}\times\frac{\vec{B}}{n_e e} - \frac{\nabla p_e}{n_e e} \qquad (9.32)$$

We recognize the first and second terms on the right as the contributions of resistance and back electromotive force, respectively, to the electric field. The third term represents the *Hall effect*. This latter term operates on the charged heavy-particles of the flow to create useful momentum in a charge-neutral plasma, providing the basis for propulsion. To express the importance of this term, typically in situations of low particle density, we refer to some plasma thrusters as *Hall accelerators*. Note that the acceleration of charged-particles by the Hall field is in the direction of the $\vec{j}\times\vec{B}$ force, which agrees with the overall momentum equation for the plasma.

The scalar product of the electric field with the current density provides the power per unit volume deposited in the plasma:

$$\vec{E}\cdot\vec{j} = \eta j^2 - \vec{j}\cdot(\vec{v}\times\vec{B}) = \eta j^2 + \vec{v}\cdot(\vec{j}\times\vec{B}) \qquad (9.33)$$

where for simplicity, we have neglected the pressure-gradient term. The first term on the right is the resistive heating, used in arcjets, while the last term is the rate of work performed on the flow by the electromagnetic force, used in electromagnetic thrusters. (In magnetohydrodynamic generators, the force opposes the flow, so the flow does work to create electromagnetic energy.)

From the generalized Ohm's law, we observe that the electric field in a plasma thruster is not simply due to a voltage applied across two electrodes; it arises in a much more complicated way to satisfy the requirements for current conduction in the presence of magnetic fields and plasma flow. In the very-low-density, nonneutral flows of electrostatic thrusters, we add up the electric fields from individual charges. In plasma flows, we use Faraday's law:

$$\nabla\times\vec{E} = -\frac{\partial\vec{B}}{\partial t} \qquad (9.34)$$

Substituting the electric field from the generalized Ohm's law into Faraday's law allows us to calculate the distribution of magnetic field within the plasma. From this we may obtain the local current density (via Ampere's law) and thereby compute how much the electromagnetic processes heat and accelerate the flow. Even for many apparently simple situations, numerical computation must replace analytical calculation. Some features of magnetized plasma flows, however, allow easy estimates of conditions inside electromagnetic thrusters, such as thermal and dynamic loadings.

Convection of Magnetic Flux—By combining Faraday's law, the generalized Ohm's law, and particle conservation (i.e., continuity equation), we can derive an equation linking the magnetic field in a plasma to the flow density. In simplified form appropriate for a plasma of relatively high density and uniform electrical resistivity, the ratio of magnetic field to plasma density is given by the following convective derivative:

$$\frac{D(\vec{B}/\rho)}{Dt} = \left(\frac{\eta}{\mu\rho}\right)\nabla^2\vec{B} + \frac{(\vec{B}\cdot\nabla)\vec{v}}{\rho} \qquad (9.35)$$

where μ is the permeability ($= 4\pi \times 10^{-7}$ henries/m, for nonmagnetic materials).

If the electrical resistivity of the plasma flow is very low, and if we neglect the field-curvature term so the right side of the equation can be set equal to zero, the ratio of magnetic field to density is preserved as a plasma fluid element convects. This is a statement of Faraday's law, by which the magnetic flux within a closed loop (of zero resistance) is constant as the loop area, enclosing a fixed amount of mass, changes inversely with the mass density. The condition of very "low" resistivity, permitting this so-called *frozen-field* approximation, is quantified more exactly by a dimensionless number:

$$R_m = \sigma\mu v L_c \qquad (9.36)$$

where σ = electrical conductivity of the plasma, (Ω^{-1} m^{-1}, the inverse of the resistivity)
L_c = a characteristic length scale of the flow (m)

In analogy with the comparison of convective-to-viscous diffusive effects in a flow, R_m is called the *magnetic Reynolds number*. For R_m much greater than one, convection dominates, so we have the frozen-field (or more generally, frozen-flux) approximation. If R_m is much less than one, magnetic flux diffuses readily across the plasma, and electrical current distributes in the manner of resistors in parallel. We may estimate the characteristic thickness through which current conducts in a plasma flow by calculating the distance for which $R_m = 1$. Thus, current conduction across high-speed plasma flows in electromagnetic thrusters may use only a fraction of the available electrode area (greatly increasing thermal loads), if the streamwise dimensions of the electrodes are large compared to $(\sigma \mu v)^{-1}$.

Magnetic Pressure and Tension—To calculate the electromagnetic thrust, we could add up the electromagnetic force per unit volume ($\vec{j}\times\vec{B}$) acting in the charge-neutral plasma. The current density, however, is related to the magnetic field by Ampere's law (without Maxwell's addition of the displacement current):

$$\mu\vec{j} = \nabla\times\vec{B} \qquad (9.37)$$

By vector identity, we obtain

$$\vec{j} \times \vec{B} = -\nabla\left(\frac{B^2}{2\mu}\right) + \vec{B} \cdot \frac{\nabla \vec{B}}{\mu} \qquad (9.38)$$

The first term on the right shows that the electromagnetic force density can act as the gradient of a scalar ($B^2/2\mu$), identified as the *magnetic pressure*. The second term depends on the shape of the magnetic field. In the case of circular magnetic-field lines, we may write this term as $-B_\theta^2/\mu r$, where B_θ is the magnitude of the field, and r is the radius of the field line. The force acts radially inward and is called *magnetic tension*. [Note that, without the second term, the ($1/r$) variation in strength of a magnetic field surrounding a current-carrying wire would appear to provide a force due to the gradient of magnetic pressure even in vacuum.]

From electromagnetic theory (Poynting's theorem), the energy per unit volume of the magnetic field in most plasma situations is $B^2/2\mu$, which equals the magnetic pressure. A highly magnetized plasma flow, with high electrical conductivity (for which $B^2/2\mu \gg p$ and $R_m \gg 1$), behaves as a gas with specific heat ratio $\gamma = 2$, as long as we can also ignore field-curvature effects. The distribution of magnetic field and current within the flow then follows simple, constant-γ, gas dynamics. For example, the current density distribution and thermal loading at the exit of a plasma thruster are derived from the magnetic field variation, which is proportional to the variation in mass density. We find these variations by using the Prandtl-Meyer function [Liepmann and Roshko, 1956], with $\gamma = 2$, for expansion into field-free vacuum.

Forces on Conductors—Because metal conductors are high-density plasmas, our equations for electromagnetic force density and magnetic pressure apply to the electrodes and circuit-conductors delivering current to the flow. For simple situations, such as straight, coaxial transmission lines, we can readily calculate the magnetic field at any position from the circuit currents. Direct use of the magnetic field and its distribution is then convenient. In many arrangements, however, the relationship of local magnetic field to the circuit currents is difficult to calculate. In such circumstances, we could use numerical tools, such as general-purpose field-solving programs. A readily available alternative approach employs existing formulas for the inductance of various conductor arrangements. Electrical-engineering manuals and the well-known *Handbook of Chemistry and Physics* contain these formulas.

The *inductance* (L) of an electrical circuit is defined as the flux (ϕ) divided by the circuit current (J), where the flux is determined from Faraday's law in terms of the voltage (V):

$$\frac{\partial \phi}{\partial t} = V \qquad (9.39)$$

For a single-turn circuit (vs. a multi-turn coil, for example), we can also compute the flux from the magnetic field through an area A:

$$\phi = \int \vec{B} \cdot d\vec{A} \tag{9.40}$$

In the absence of magnetic materials, B is proportional to the circuit current (J) in terms of the conductor geometry. We can therefore write the flux as

$$\phi = LJ \tag{9.41}$$

where L = inductance (henries or h)
J = current (A)

The inductance is a function only of the size and arrangement of the circuit conductors. It is independent of the instantaneous circuit current. Inductance formulas capture the complexity of the circuit arrangement and replace the results of a complicated magnetic-field calculation. Combining Eqs. (9.39) and (9.41), we have

$$\frac{dLJ}{dt} = V \tag{9.42}$$

For a coil consisting of N turns, the magnetic field generated by the circuit current (J) is increased by N. The circuit voltage (V) is the sum of the voltages around the N turns. The inductance of a multi-turn circuit is therefore greater than a single-turn circuit with the same geometry by the square of the number of turns.

The power delivered to a circuit from a voltage source is the product VJ. For a constant inductance (fixed circuit dimensions and arrangement), the integral of the power with time allows us to identify the magnetic energy in the circuit as

$$W_m = \frac{LJ^2}{2} \tag{9.43}$$

For the general case of a circuit that changes size or arrangement with time, the power equation gives

$$V_J = \frac{d}{dt}\left(LJ^2/2\right) + \left(J^2/2\right)\frac{dL}{dt} \tag{9.44}$$

where the first term on the left is the rate of increase in circuit (magnetic) energy, and the second term is the rate the circuit does work while increasing its inductance. This term corresponds to the rate of increase of the propellant's kinetic energy in pulsed, unsteady thrusters, such as the pulsed-plasma microthruster and railgun schemes. Because the inductance is a function of geometry, we can write inductance's rate of change over time as the product of the derivative of the inductance function with respect to the changing value of circuit dimension and the rate at which this dimension is changing with time. As a scalar product of vectors, this is

$$\frac{dL}{dt} = \vec{v} \cdot \nabla L \tag{9.45}$$

where $\vec{v}$ is the velocity at which the circuit dimensions change. The rate at which work is done, however, is the scalar product of velocity and force, so we can calculate the electromagnetic force ($\vec{F}_m$) from the gradient of inductance with respect to its dimensional parameters:

$$\vec{F}_m = \left(\frac{J^2}{2}\right) \nabla L \tag{9.46}$$

For example, suppose we have a coaxial circuit of length z, with inner and outer radii (r_1 and r_2, respectively). The current travels axially in opposite directions in the inner and outer conductors, crossing radially at the each end to complete the circuit. The inductance of this arrangement is

$$L = \left(\frac{\mu}{2\pi}\right) z \ln\left(\frac{r_2}{r_1}\right) \tag{9.47}$$

The axial force on the circuit is then

$$F_z = \left(\frac{\mu}{4\pi}\right) J^2 \ln\left(\frac{r_2}{r_1}\right) \tag{9.48}$$

By the same kind of analysis used for single circuits, the magnetic energy of two coupled circuits, carrying currents J_1 and J_2, is $M_{12} J_1 J_2$. Here, M_{12} is the mutual inductance between the two circuits. The associated force between the circuits is then

$$\vec{F}_{12} = J_1 J_2 \nabla M_{12} \tag{9.49}$$

where the component of force on any particular circuit element is evaluated in terms of the derivative of the mutual inductance with respect to that element's change in position. This force acts along the direction of the derivative.

To use standard inductance formulas, we must know the circuit geometry in terms of specific values of length, width, or radius. Distributions of current can cause problems. Within a current-carrying region, we might overlook mutual inductances coupling each part of the current to the rest of the distribution. In these cases, we should compute an effective inductance (L_e) by equating the integral of the magnetic energy density ($B^2/2\mu$) over the conducting volume (V) to the circuit energy for the total current (J) carried by the distribution:

$$\frac{1}{2}L_e J^2 = \int \frac{B^2}{2\mu} dV \qquad (9.50)$$

For example, a plasma column carrying a total current (J) along its length, with uniform current density, has an azimuthal magnetic field that increases linearly with radius within the column. By Eq. (9.50), the inductance per unit length associated with the magnetic energy within the column is simply $\mu/8\pi$, which is half the value obtained from the magnetic flux calculations of Eqs. (9.40) and (9.41). With the correct inductance formula, we can then compute the electromagnetic force on the end face of an electrode:

$$F = \left(\frac{J^2}{2}\right)\frac{dL}{dz} = \frac{\mu J^2}{16\pi} \qquad (9.51)$$

If J is the current carried at the cathode tip in a coaxial plasma thruster, this force adds to the electromagnetic portion of the thrust. It also contributes to motion of molten cathode material.

Ionization

Except for certain arrangements using solid or liquid propellants (e.g., electromagnetic railguns or ferromagnetic accelerators), electric-propulsion devices create electrically active working fluids by ionizing atoms or molecules. Different processes cause ionization, but all result in the removal of an electron from a normally charge-neutral system (atom or collection of atoms). Applying strong electrostatic fields from electrode structures of common dimensions (> millimeters) usually cannot remove electrons directly from their atomic orbits. Asperities, such as microscopic whiskers at metal surfaces or needles less than about 0.1 mm in diameter, can ionize atoms directly, if voltages are greater than about 10^5 V. Charged-droplet sprays for colloid thrusters are obtained by applying lower voltages to the tips of fine hypodermic needles emitting the liquid propellant flow [Krohn, 1963]. Locally high electric fields are also used in field-emission ion sources and play a role in contact ionization. Neutral atoms of low ionization potential (few eV) can ionize by contact with a heated metal surface. Cesium passing through a porous plug of heated tungsten was an early source for ion engines [Stuhlinger, 1964].

More commonly, ionization requires electrons or photons colliding with electrons bound in atoms or molecules. Single collisions can cause ionization if the relative energies exceed the ionization potential. Multiple collisions at lower relative energies can also combine to create ionization, with the bound electron first excited to higher states and then finally given enough energy to escape. Figure 9.19 shows the dependence of ionization cross section on the energy of a bombarding electron for several noble gases. Note that a threshold energy must be surpassed to obtain ionization (by a single collision). Figure 9.19 also shows Maxwellian distributions of electron energies, indicating that the high-energy tails of these distributions (above the threshold energy) are responsible for ionization. Thus, an equilibrium collection of atoms, ions, and electrons can often be nearly fully ionized even though the ionization potential is five to ten times the electron temperature.

Recall from basic mechanics that collisions of light particles with heavy particles (e.g., ping-pong ball on a bowling ball) are inefficient in elastically transferring kinetic energy. But they are very effective in converting light-particle energy to heavy-particle internal energy (e.g., bullet into a block of wood), if energies are above the threshold for inelastic changes. Thus, ionizing (and exciting) a gas depends primarily on the electron temperature, rather than the heavy-particle (atom or ion) temperature. Note that we must use the term "temperature" with care for a nonequilibrium, multi-component gas, especially when instabilities or particle beams might alter the population of high-energy electrons. Table 9.4 lists some ionization potentials and cross-section (maximum) values for ionization from collisions of single electrons.

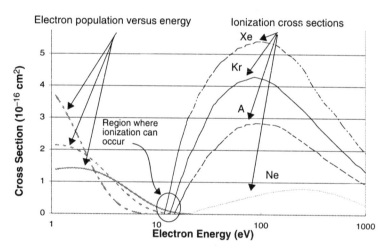

Fig. 9.19. **Dependence of Cross Section for Ionization from Collisions of Single Electrons vs. Electron Energy for Various Noble Gases**, [Mitchner and Kruger, 1973]. Also shown is a sketch of relative populations of electrons vs. electron energy for different electron temperatures. (1, 2, 3 eV, respectively). Only electrons in the high-energy tail of the population have energies that exceed the threshold values for ionization.

Although we can try to calculate the ionization of a gas by integrating or comparing the rate equations for all ionization and deionization processes, some situations allow us to substitute statistical-mechanical formulas. Examples are the Saha equation at high densities and coronal models at low densities [McWhirter, 1965].

Table 9.4. Sample Data for Ionization Processes. The cross section for ionization when single electrons collide can be characterized by the threshold energy (ionization potential) and a nearly linear rise to a maximum cross-section value at a higher-energy location.

Material	(First) Ionization Potential (eV)	Maximum Cross Section (10^{-20} m²) and Location of Ionization from Single-Electron Collisions (eV)
H	13.6	0.70/75
H$_2$	15.4	1.0/70
N$_2$	15.6	3.0/100
He	24.6	0.38/90
Ne	21.6	0.80/200
Ar	15.8	2.75/75
Kr.	13.9	4.1/80
Xe	12.1	5.3/100
Cs	3.9	9.4/20
Li	5.4	5.2/15

Formulas based on particle temperatures may not be useful for thruster design because of so-called plasma *microinstabilities* [Krall and Trivelpiece, 1973]. Typically, these occur when the drift speed of charged particles (v_d) necessary to support the local current density is no longer small compared to other characteristic speeds (v_c), such as the electron or ion thermal speeds:

$$v_d = \frac{j}{n_e e} > v_c = \sqrt{\frac{2kT_{e,i}}{m_{e,i}}} \tag{9.52}$$

where the subscripts e and i refer to electrons and ions, respectively (in various combinations for different instabilities). These instabilities are caused by high-frequency electromagnetic waves that scatter the directed motion of the charged particles. Such scattering can behave, in part, like particle collisions, increasing the electrical resistivity and the ability of charged particles to cross magnetic fields. Particles can also scatter to higher energies. Thus, the population of electrons above the threshold for ionization and excitation can be much higher than would be calculated from a Maxwellian distribution at the temperature associated with simple electron transport, such as current conduction. This means that the ionization levels in a plasma discharge with high current density and low particle density can be quite different from those found by conventional models, including approaches that directly employ reaction rates based on local temperatures.

Instead, we need to return to equations that determine the distribution of electron velocity itself (e.g., Boltzmann equation) to compute the actual populations of electrons in response to electromagnetic wave scattering and inelastic events. We do not usually perform these calculations for electric thrusters at this time because they strain computer resources. The behavior of a device or subsystem may depend critically on microinstability phenomena [Choueiri, et al., 1991]. Modern codes therefore incorporate "phenomenological" models for increases in transport properties (electrical resistivity and thermal conductivity) caused by microinstabilities [Frese, 1986].

We must also consider the size of the device relative to the physical scale for various important processes. For example, neutral atoms in a plasma thruster can travel distances comparable to, or larger than, the flow dimension before ionization by (thermal) electron impact. Calculations based on local values of density and particle energies can be faulty in many regimes of interest. In general, ionization in plasma systems, including ion sources and plasma thrusters, is so complex that we must use great caution when we extrapolate from empirical results to other sizes, powers, and gas types, even when validated numerical modeling is available.

Electrical Discharge Behavior

Electrical discharges can provide ions for electrostatic engines and momentum and energy for electrothermal and electromagnetic thrusters. The type of electric discharge created by depositing electrical energy in a gas depends on the rate and intensity of the energy input, as well as on the thermodynamic and chemical properties of the gas and surrounding materials.

The main elements of electrical-discharge geography are shown in Fig. 9.20. This geography includes the anode and cathode *fall* regions, in which significant departures from charge-neutrality can occur, and the intervening, quasi-neutral, plasma region, often called the *positive column*. The anode and cathode are the positive and negative electrodes, respectively, of the discharge. Even though the anode is positive, the anode fall may actually repel electrons back into the plasma in some situations. The properties of each region depend on power balances needed to conduct current at the required level. Near the electrodes, these balances include interactions with the solid-state surfaces that can conduct electricity and heat, and also can absorb heat and introduce material to the discharge by ablation. Along the plasma discharge, interactions can occur with insulating surfaces, which may also ablate, and with surrounding gas flows.

In the plasma region, the rate at which electrical energy supplies heat must balance the rate of loss of local plasma energy to heat transfer, including convection of internal energy (e.g., ionization) and radiation. In general, all of these rates depend on the particle temperatures and densities. The behavior of a discharge needed to carry a desired current can, therefore, vary widely depending on the gas and its environment. For example, in the simplest case of resistive heating, for

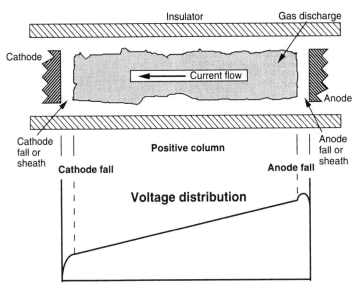

Fig. 9.20.   Basic Geography of a Gas Discharge. Electrode fall regions and intervening positive column are displayed, along with a sketch of voltage distribution.

which, $E = \eta j$, the electrical resistivity (η) of the plasma is often a strong function only of electron temperature. The thermal conductivity is also a strong function of temperature, whereas the radiation rate (per unit volume) and the degree of ionization depend significantly on the particle densities. Typically, the total current and total mass flow rate, not the current or particle densities, are the fixed quantities in a design problem. The self-consistent values of temperature and current density needed to compute the local electric field are not well-controlled. Thus, we cannot calculate even the total voltage drop across the plasma without consulting very closely related experiments. Nonlinear interactions of the discharge with the electrodes makes predictions of the general behavior of electrical discharges even more difficult.

Next to the electrode surfaces, within so-called sheath regions, the particle density of electrons does not closely equal the product of the average ion charge number and the ion particle density. The net charge densities create local electrical fields which can alter particle trajectories. If we characterize the distribution of electron velocities in the plasma by a temperature (T_e), the distance over which the plasma can depart from electrical neutrality is scaled by the Debye length [Spitzer, 1964]:

$$\lambda_D = \sqrt{\frac{\varepsilon_0 k T_e}{n_e e^2}} \tag{9.53}$$

where λ_D = Debye length (m)
ε_0 = permittivity of free space 8.885×10^{-12} A·s/V·m
k = Boltzmann's constant (1.38×10^{-23} J/K)
T_e = electron temperature (K)
n_e = electron density (m^{-3})
e = electron charge, 1.6×10^{-19} (C)

Over distances longer than a Debye length, the intervening plasma electrons can move to cancel the effect of charge-density perturbations. Surfaces immersed in a thermal plasma can provide such perturbations to satisfy boundary conditions on current collection by the surface. Sheath regions scaled by the Debye length cover these surfaces. In many plasma thrusters, the Debye length is less than 10^{-6} m. For devices operating at high voltages and low densities, the non-neutral region can be much larger. For example, in ion engines, it is larger than the electrode spacing, and the entire flow field is non-neutral.

In plasma discharges, charged particles will drift into the sheath region surrounding an electrode. For sheaths in which the electric field tends to repel particles of charge q, the flux of particles that reaches the surface is reduced by a so-called Boltzmann factor:

$$f_B = e^{\frac{-q \Delta V}{kT}} \tag{9.54}$$

where ΔV = the voltage across the sheath (V)
T = the particle temperature (K)

On the other hand, positively charged ions near the (negative) cathode experience rapid acceleration by the electric field in the sheath and bombard the cathode's surface with much more energy than the average kinetic energy of a particle in the thermal plasma. For example, a singly charged ion that drifts into the sheath near the cathode, with a thermal energy of 0.3–3 eV, can "fall" through the potential gradient in the sheath and strike the cathode with a kinetic energy of 10–50 eV ($kT = 1$ eV $= 1.6 \times 10^{-19}$ J, corresponding to a temperature of 11,605 K). This kinetic energy can liberate electrons from the cathode surface which then accelerate in the opposite direction because of the same voltage gradient. The kinetic energy (in eV) of each of these electrons equals the cathode-fall voltage and helps to heat the local plasma. The combination of ion and electron motion through the sheath represents the current density at the cathode, which in steady state must equal the current density of the local plasma discharge. Typically, electrons are so mobile compared to positive charges that the sheath fields repel them to limit or prevent current collection by the surface. In plasma discharges at relatively high density, this may be true even at the (positive) anode.

For discharges of low current density ($\sim 10^{-3}$ A/cm^2), at low pressure (~ 1 torr), the ion flux to the sheath combines with the flux of electrons liberated when ions bombard the cathode surface to account for the total current. The liberated electrons arrive in the local plasma with enough energy to re-ionize the flux of heavy particles returning from the cathode (after ions recombine at the surface). The modest heavy-particle flux required at low current density implies low heat flux to the walls surrounding the discharge. The electron temperature, however, must remain high enough to support some ionization so current can conduct in the plasma. Thus, heavy-particle temperatures can be much lower than electron temperatures in discharges at low current density. This nonequilibrium, low-intensity operation is often referred to as a *glow discharge*. Fluorescent lighting is a common example of a discharge in which hot electrons ionize and excite gas inside a tube that is relatively cool.

At higher current densities (> 0.1 A/cm^2), electron release by direct ion impact (secondary emission) cannot support the density of the discharge current over the cathode's entire surface. Instead, the current concentrates into a more intense arrangement, known as an *arc*. Ion bombardment heats the local surface of the cathode enough to cause thermionic emission of electrons. The higher current density also dissipates more power in the plasma near the cathode, resulting in higher degrees of ionization and higher heavy-particle temperatures. The flux of ions to the cathode sheath and the subsequent heating by ion bombardment increase as well, enhancing thermionic emission.

Typically, thermionic emission of electrons from the cathode surface is characterized by the *Richardson equation* for the current density emitted by a surface at temperature T:

$$j_{th} = A_R T^2 e^{\frac{-e\phi_w}{kT}} \tag{9.55}$$

where $A_R = 1.2 \times 10^6$ A/m$^2\cdot$K^2 is *Richardson's constant*, and ϕ_w is the so-called *work function* (in electron volts, or eV) associated with the electrode material. Pure metals typically have work functions of a few volts, but additives such as thorium, barium oxide, and various multi-species arrangements can lower values to less than two volts. The operating temperature limits of the cathode material are less than a few tenths of an electron volt, so small differences in work function can have large effects. The electric field of the sheath next to the cathode surface also helps electrons leave the surface by lowering the apparent value of the work function. This is known as the *Schottky effect* [Cobine, 1958], and the effective work function (ϕ_{eff}) is given in terms of the field E_0 at the cathode surface:

$$\phi_{eff} = \phi_w - e\frac{|E_0|}{4\pi\varepsilon_0} \tag{9.56}$$

To find the electric field at the cathode surface, we must solve equations for particle motions and densities in the non-neutral sheath that connects the surface to the local plasma. But conditions in the local plasma, such as ion density and electron temperature are determined by the fluxes of particles and power from the sheath (e.g., high-energy electrons), as well as the free-stream plasma (positive column). We require balances of power and particle fluxes involving all three elements (cathode surface, sheath, free-stream), including radiation transport, heat conduction in the cathode, and evaporated or sputtered cathode material.

The cross-sectional area of the discharge near the cathode surface depends on nonlinear processes in the surface and the local plasma. Surfaces at locally higher temperatures emit electrons more easily. Higher current densities in the plasma increase the local electron temperature and lower the electrical resistivity. As the total discharge current increases, it tends to concentrate in a single intense column, often called a *spoke*. Local plasma (heavy-particle) temperatures above 10,000 K and local surface temperatures above 1000 K can occur, severely damaging the surface (e.g., erosion).

To overcome the tendency to spoke, we need to distribute the heat generated in the electrodes and plasma. Magnetic fields can move the discharge attachment around on the electrode surfaces and also swirl the discharge through the available gas volume. Operation at lower particle densities within the plasma increases the mean free path between particle collisions, which improves heat conduction. Increased heat transfer because of hydrodynamic turbulence or radiative transport also helps in this manner. Furthermore, use of cathode materials with low work functions can increase the possibility of thermionic emission at moderate temperatures and thus allow more of the cathode surface to participate in carrying the discharge current.

Regimes exist between situations of glow and arc discharge (*glow-to-arc transitions*). A graph of voltage vs. current often summarizes the types and behaviors of electrical discharge in gases [see Cobine, 1958]. The actual point of operation of the discharge depends on the interplay of this characteristic graph with the voltage-current behavior of the power source. For example, if the source has a constant internal impedance (Z) and a zero-current (open-circuit) voltage (V_0), the voltage-current characteristic of the source is simply a straight line:

$$V = V_0 - ZJ \qquad (9.57)$$

The intersections of this line with the discharge characteristic provide points of possible operation. The actual operating point depends on how we start (and later manipulate) the discharge, and on how stable the discharge operation is near points of intersection. Thus, for an idealized arc operating at constant power ($P = VJ$), which corresponds to current conduction at a constant cooling rate, intersection typically occurs at two points, but the arc operates at the lower voltage point (see Fig. 9.21). At this point, a slight excursion to higher current lowers the available voltage relative to that needed by the arc, so the current returns to a lower value. A perturbation to a lower current, however, raises the available voltage and increases the current. But at the higher point of intersection, excursions to a higher current <u>increase</u> voltage, and perturbations to a lower current <u>decrease</u> it, so the power arc never stabilizes there. We can analyze other combinations of discharge and power-supply characteristics in the same way to select or design propulsion-system components and to specify operating procedures, such as start-ups, shut-downs, and thrust variations.

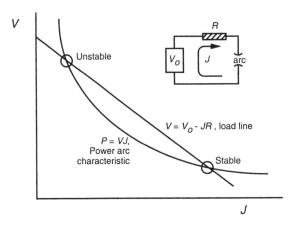

Fig. 9.21. **Sketch of the Voltage-Current Characteristic for a Power Arc and Its Circuit Load-line.** The graph shows two possible solutions for discharge operation, only one of which is stable.

9.4.3 Details of Selected Thrusters

We have chosen for discussion several thrusters that represent their respective sub-categories of electric propulsion and that may apply to near-Earth space missions. In most cases, we can scale up the same thruster configuration (or operate them as a cluster) to accomplish higher-energy missions, such as interplanetary exploration. For the electrothermal thrusters, however, bond strengths limit the exhaust speed to values for stationkeeping and low-energy maneuvers. Even this limitation does not restrict their use to mundane applications because spacecraft may need these capabilities around distant planets. Using electric propulsion at the destination reduces the propellant load. This increases the delivered payload or decreases the launch mass and initial mass in Earth orbit.

Electrothermal Propulsion—Arcjets

Electrothermal propulsion has seen considerable flight use as resistojets (augmented hydrazine thrusters). Table 9.5 summarizes typical values for these thrusters. Figure 9.22 displays thrust-to-power ratio vs. I_{sp} for resistojets operating with several types of propellant. Most of their engineering involves details of heat transfer, thermal management, and material selection found in other sources. But we can explore arcjet principles here to understand the complexities of electrically active flows accelerated to speeds well above those chemical thrusters can achieve.

Table 9.5. Typical Values for Resistojets. Pure resistojets derive their power entirely from the electrical source versus the largely chemical power of electrically heated thrusters using chemically decomposed hydrazine.

	Input Power (kW)	Thrust (N)	I_{sp} (s)
"Pure" Resistojet			
Hydrogen (low power)	3.0	0.53	838
Hydrogen (high power)	30.0	6.0	846
Ammonia (NH_3)	12.3	2.9	423
Augmented Hydrazine N_2H_2 products	0.7	0.5	300

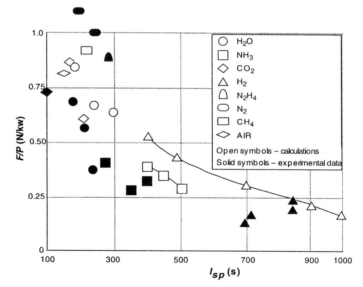

Fig. 9.22. Thrust-to-Power Ratio for Resistojets as a Function of Specific Impulse for Various Propellants [Mantenieks, 1990]. The thrust-to-power ratio decreases with exhaust speed and because of increased frozen-flow loss. Note, however, the improved (experimental) performance of H_2. (Courtesy of NASA-Lewis)

The arcjet's high values of specific impulse, compared to chemical-combustion rockets, provide commercial advantages for new communication satellites. For example, the recent Telstar IV satellite uses 1.4-kW, hydrazine-fed arcjets for stationkeeping. The schematic diagram in Fig. 9.23 shows details of the arcjet system, including the hydrazine-product gas generator. Developing this class of thruster and applying it to commercial communication satellites required even more than the already challenging task of combining a high-temperature, arc-discharge flow with power sources and suitable ignition circuitry. User acceptance was critical, so extensive testing and analysis addressed concerns such as the possibility that free electrons remaining in the exhaust plume would interfere with the satellite's communication beam, or that the exhaust would chemically or thermally contaminate the spacecraft.

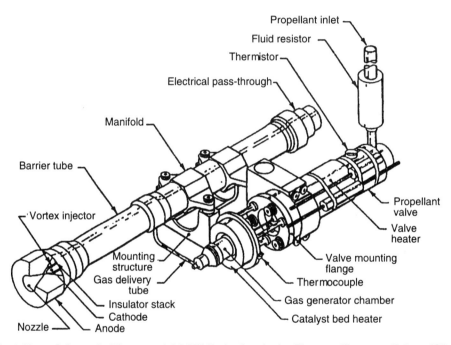

Fig. 9.23. Schematic Diagram of 1.8-kW Hydrazine Arcjet Thruster Flown on Telstar [Olin, 1993b]. Note that much of the system is the hydrazine gas generator which provides the propellant flow.

Figure 9.24 shows the *constricted-arc* configuration used in present arcjet thrusters. In this form, the arc attaches to a central cathode that usually starts its life as a sharp cone. The arc extends through the throat of the exhaust nozzle and attaches to the anode, which forms part of the nozzle's divergent section. This attachment should be axisymmetrical and cover a large area of the anode, as opposed to a localized "spoke," to reduce eroding of anode material. Heat transfer to the propellant flow occurs because this flow is forced close to the arc in the throat. A typical throat diameter for a kilowatt-class arcjet is 0.5 mm. The walls of the throat are cooler than most of the propellant flow, which in turn is cooler than the propellant that participates in the arc discharge. The basic thermal profile in an arcjet resembles that in chemical engines, rather than that of resistojets (or solid-core nuclear-thermal rockets), in which the walls are hotter than the flow. Except for some heat transfer by plasma radiation, the propellant flow protects the nozzle walls from the arc.

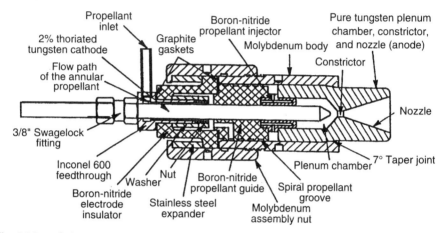

Fig. 9.24. Schematic Drawing of Constricted Arcjet Thruster Rated at 30 kW and Using Ammonia as Fuel [Pivirotto and Goodfellow, 1991]. This device has also been operated in steady-state life tests at 10 kW, and has basic features of other constricted arcjets. (Courtesy of the Jet Propulsion Laboratory)

The arc discharge, however, must attach to solid surfaces at the electrodes. The central cathode operates by thermionic emission, so heat transfer to the cathode is countered by electron emission, which represents a heat flux equal to the product of electron current density and the energy cost of emitting an electron, $\sim j_e \phi_w$ (see Sec. 9.4.2). The anode does not have such a cooling mechanism. It must rely instead on heat transfer by radiation (and conduction). Adding swirl to the inlet flow improves heat transfer to the propellant. In principle, this swirl persists in the nozzle's diverging section and may help to improve axisymmetry on the anode surface.

For the rather small dimensions and relatively high temperatures of the flow field, the Reynolds number is quite low (~10–100), which implies significant friction and heat transfer. Therefore, improvements in specific impulse associated with increased temperature incur lower efficiencies at fixed power because of lower flow densities. Also, the Knudsen number (= mean free path/flow dimension) may not be low enough for valid design calculations using the fluid equations. Instead, we may require computational techniques for noncontinuum flow, such as Direct Simulation Monte Carlo methods [Bird, 1994].

9.4 Select an Electric Thruster

At a fixed mass flow rate, the propellant's ability to absorb heat limits power dissipation in the arc discharge. Thus, the arc voltage decreases with higher current. The voltage-current characteristic (see Sec. 9.4.2) admits two solutions for a power source with a simple resistive load-line, the lower voltage of which is stable. But ignition of the arc requires a high-voltage pulse, so we must dynamically control the power supply. Arcjets have recently become operational for space in part because of new circuitry that allows repetitive ignition, followed by ramp-up of power to steady levels.

Table 9.6 provides typical performance data for kilowatt-level arcjet thrusters [Olin Aerospace Co., 1993b]. Figure 9.25 displays thrust, specific impulse, and thrust efficiency varying with power, and mass flow rate.

Table 9.6. Typical Values for Kilowatt-Level Arcjets [Olin Aerospace Co., 1993b]. These arcjets are used for stationkeeping on communication satellites such as Telstar.

Propellant	Hydrazine decomposition products
Input power	1.8 kW
Voltage	96 V
Thrust	0.2 N
Specific impulse	502 s
Throat diameter	0.5 mm
Overall diameter (arcjet)	4.8 cm
Arcjet mass (including connection)	1.3 kg

Higher-power arcjets have been proposed for orbit-raising missions and other applications of primary propulsion [DiVincenzi, et al, 1990.]. The basic thermodynamics and chemistry of the arcjet flow remain about the same as we increase power. The overall flow dynamics are also similar until magnetic forces become important. Therefore, values of specific impulse improve over kilowatt-level thrusters only by factors of about two, attaining values (for hydrogen) of 1000 to 1500 s. Such values are well within the best range for orbit-raising missions, including LEO-to-GEO transfer. Arcjets in the 10–30-kW range have been based on laboratory-tested versions, shown schematically in Figs. 9.24 and 9.26. Figure 9.27 shows specific impulse and thrust efficiency varying with specific energy supplied to hydrogen propellant flows. Ammonia often replaces hydrogen as the propellant to achieve the benefits of highly hydrogenic (low molecular mass) operation without requiring cryogenic storage and handling. Planners select ammonia instead of hydrazine because most laboratories avoid hydrazine.

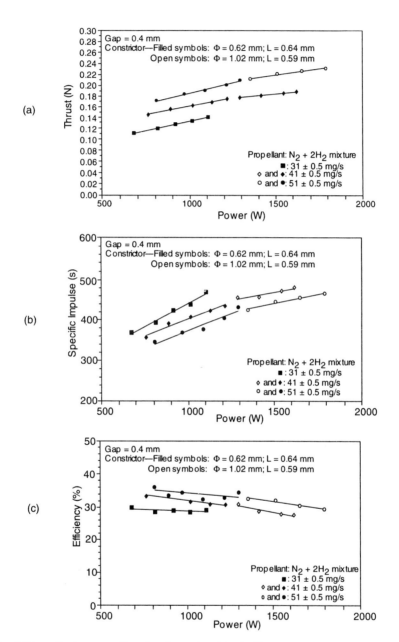

Fig. 9.25. Performance Data for Constricted Arcjet Thruster Using Hydrazine-like Propellant.
[Cruciani, et al., 1991] (a) Thrust vs. Power. (b) Specific Impulse vs. Power. (c) Efficiency vs. Power. (Courtesy of BPD Difesa a Spazio)

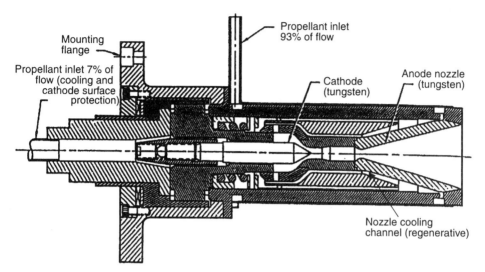

Fig. 9.26. Schematic Drawing of a Laboratory-Version Arcjet Thruster Using a Constricted Arc and Hydrogen Propellant (10–30 kW level) [Sankovic, et al., 1991]. This arrangement uses regenerative cooling of the constrictor, nozzle, and anode. (Courtesy of NASA-Lewis)

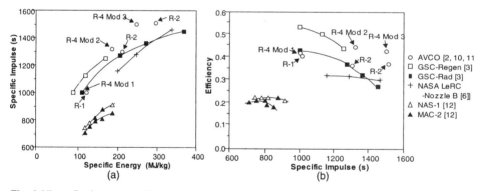

Fig. 9.27. Performance of Laboratory Version Constricted Arcjets. Propellant is H_2 [Sankovic, et al., 1991]. Slight decline of efficiency with higher specific impulse is associated with increased frozen-flow loss. Arcjets are similar to that in Fig. 9.26. (Courtesy of NASA-Lewis)

Electromagnetic Propulsion— Pulsed-Plasma Thrusters

The interaction of current in a plasma discharge with the magnetic field that this current creates is the basis for many plasma thrusters. The pulsed-plasma thruster (PPT) is perhaps the simplest electromagnetic device, which may account for its early acceptance and use in space missions [Vondra, et al., 1970]. Previously shown in Fig. 9.13, the PPT uses an electrical discharge across the flat, rectangular endface of a block of Teflon propellant. The Teflon surface initially supports a high voltage between two electrodes in vacuum. Two semiconductor sources in one electrode of the PPT provide sparks that ignite the discharge. A single, high-voltage capacitor supplies the initial voltage and powers the discharge. As heat transfer from the discharge causes the Teflon surface to ablate, a negator spring pushes the block into the discharge chamber. Because heat transfer to the propellant is an important feature of PPT operation, some electrothermal component of thrust is always present. But the PPT does not attempt to convert efficiently the plasma's thermal energy into directed kinetic-energy. Instead, plasma accelerates mainly because of the $\vec{j} \times \vec{B}$ force density created during the short, high-current discharge.

We can estimate the electromagnetic force acting in the plasma by idealizing the discharge geometry as a simple flat sheet of width (w) and length (h). If the distance between this sheet and an equal sheet serving as the return conductor for the discharge current is z, this part of the PPT circuit has an inductance:

$$L = \frac{\mu z h}{w} \tag{9.58}$$

so the component of the inductance gradient in the z-direction is

$$L' = \frac{\mu h}{w} \tag{9.59}$$

As discussed in Sec. 9.4.2, the force on the plasma discharge is then

$$F_z = \frac{\mu J^2 h}{2w} \tag{9.60}$$

where F_z = thrust force in the z-direction (N)
h, w = dimensions of the plasma discharge in the plane normal to the z-direction, parallel and perpendicular to the current flow, respectively (m)
J = discharge current (A)
μ = permeability of free space, $4\pi \times 10^{-7}$ henries/m

We can integrate this force over the pulse duration simply in terms of the current waveform $J(t)$, to compute the impulse per discharge event:

$$I = \mu \left(\frac{h}{2w}\right) \int J^2(t) \, dt \tag{9.61}$$

Table 9.7 provides typical values for PPT operation; Figure 9.28 shows typical records of discharge current and capacitor voltage vs. time.

Table 9.7. **Typical Values for Pulsed-Plasma Thruster** [Myers, 1993]. Various arrangements provide different performance values.

Propellant	Teflon (CF$_2$)
Input power	0.12 kW (average at 0.2-Hz rep-rate) 20 MW (peak during pulse)
Voltage	28 V from spacecraft bus 1.2 kV initial capacitor charge
Thrust	4.5×10^{-3} N at 0.2-Hz rep-rate
Specific impulse	2000 s
Exit dimensions (propellant endface)	2 cm × 2 cm
Overall dimensions (box with two thrusters and PCU)	25 × 35 × 50 cm
Total thruster mass	7.0 kg

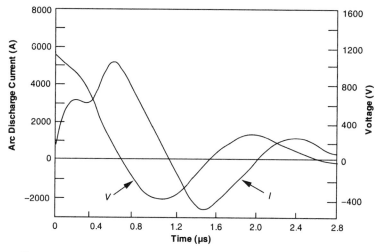

Fig. 9.28. **Typical Current and Voltage History for a Pulse from a Pulsed-Plasma Thruster** [Vondra, et al., 1970]. The oscillatory wave forms indicate inefficiency from mismatch of the power circuit and the time-varying impedance of plasma discharge.

For ablation-fed systems like the PPT, we have to calculate the propellant mass given the impulse (I). Experimental data (Fig. 9.29) for the PPT indicate the mass ablated is simply proportional to the total energy of the discharge event (~1–3 µg/J).

For a discharge near the propellant's surface, we expect some fraction of the energy of resistive heating in the plasma to transfer to the surface, thereby ablating propellant mass. The resistive heating rate is proportional to the square of the discharge current, so the total mass ablated can scale with the same integral of J^2 over time used in Eq. (9.61). Thus, the specific impulse, obtained by dividing the total impulse by the total mass, could be constant over at least reasonable ranges of systems operation. However, the time that the discharge remains near the propellant's surface clearly depends on the rate at which it accelerates vs. the rate at which heat transfer to the surface increases the discharge mass. As a result, we should expect differences in behavior if we change circuit power (e.g., charging voltage), propellant area, and propellant type.

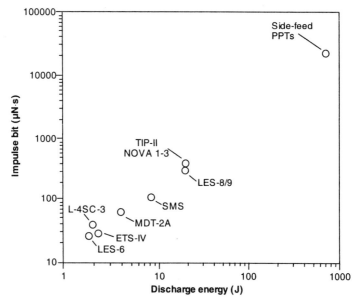

Fig. 9.29. **Variation of Impulse Bit with Discharge Energy for Several Pulsed-Plasma Thrusters Operated in Space.** Note roughly linear dependence. The acronyms indicate particular satellites. [Myers, 1993]. (Courtesy of NASA-Lewis)

Attempts to scale up the PPT's total impulse and thrust have ranged from larger-energy versions of the same microthruster arrangement (for millipound performance) [Vondra, 1982] to using a PPT as an ignitor for a second stage of ablation-fed plasma acceleration [Turchi, 1982]. In the former approach, the Teflon fuel is wrapped in a spiral slab, which suggests the difficulty of keeping PPT design and operation simple in higher-energy missions that require larger propellant mass. The latter approach requires a separate power source for the main plasma discharge but permits this discharge to be created over larger propellant areas, thereby shortening the propellant slab. Experiments [Turchi, et al., 1984] indicate the discharge can achieve quasi-steady operation when driven by a source of constant current, such as a pulseline. Efficiency of transferring electrical energy to plasma energy improves by matching the impedance of the ablation-fed arc and the source impedance.

Electromagnetic Propulsion—Magnetoplasmadynamic Thrusters

As the current and magnetic field in an arcjet increase, electromagnetic forces may dominate pressure forces. We can accomplish this transition from electrothermal to electromagnetic propulsion by increasing the discharge current and its associated (self) magnetic field, or by applying a magnetic field from an external source. This external source can be a set of permanent magnets or current-carrying coils, either superconducting or normal. In this section, we discuss self-field and applied-field magnetoplasmadynamic (MPD) thrusters in coaxial electrode geometries that derive in part from earlier experiences with arcjets.

Extensive laboratory studies of MPD thrusters have been largely directed toward goals for primary propulsion, especially those requiring exhaust speeds higher than are available with arcjets, such as ferrying cargo to the Moon and exploring nearby planets. Because electric power available in space is still too low for such applications, no MPD thrusters are operating. Thus, the data provided do not represent completed results for space-ready hardware. In fact, we cannot claim the largely empirical surveys of a few thruster arrangements and operating regimes have captured limiting values for efficiency, lifetime, and specific impulse. Instead, we simply try here to show the basic principles of operation. Future decisions on developing and using MPD thrusters will depend on understanding these principles.

Self-Field MPD Thrusters—We can derive the basic equation for the thrust from electromagnetic forces in a self-field MPD thruster by using the inductance-gradient analysis of Sec. 9.4.2, Eqs. (9.48) and (9.51). We need to add a term, however, to account for a pressure force on the downstream end of the center electrode due to the electromagnetic pinch effect. This axial force results from the pressure distribution required to balance the $\vec{j} \times \vec{B}$ "pinch" force density within a current-carrying column extending from the endface of the electrode. The pressure gradient in the radial direction is

$$-\frac{\partial p}{\partial r} = |j_z B_\theta| \qquad (9.62)$$

where p = plasma pressure (Pa)
  r = radial position (m)
  j_z = axial current density (A/m^2)
  B_θ = azimuthal magnetic field (T)

For a uniform current density in the column, the azimuthal magnetic field is

$$B_\theta = \frac{\mu j_z r}{2} \quad (9.63)$$

where the axial current density is the total current (J_1) carried by the column divided by its cross-sectional area:

$$j_z = \frac{J_1}{\pi r_1^2} \quad (9.64)$$

The pressure inside the column rises from its value (p_1) at the column edge. The total force due the pressure increase above p_1, as a result of the magnetic pinch effect, is

$$F_p = \frac{\mu J_1^2}{8\pi} \quad (9.65)$$

If the column carries the same discharge current used for the previous thrust terms (i.e., $J_1 = J$), we can assemble a thrust equation for the total effect of electromagnetic forces:

$$F = \left(\frac{\mu}{4\pi}\right) J^2 \left[\ln\left(\frac{r_2}{r_1}\right) + \frac{3}{4} \right] \quad (9.66)$$

We have used a simplified representation of the current distribution, in which the all the current flows within radius r_1 at the center electrode and attaches at radius r_2 at the outer electrode. The quantity in brackets depends on the actual distribution of current. Equation (9.66) displays the dependence of thrust on total current and on a factor that involves the thruster geometry. Dividing the thrust by the propellant mass flow rate ($\dot{m}_{prop}$) provides the exhaust speed:

$$v_e = \left(\frac{\mu}{4\pi}\right)\left(\frac{\Gamma J^2}{\dot{m}_{prop}}\right) \quad (9.67)$$

where the factor Γ represents the various geometric effects.

This equation suggests that we can obtain any specific impulse value by increasing the current or decreasing the mass flow rate. In practice, however, at some level of the quotient $J^2/\dot{m}_{prop}$, thrusters experience the onset of oscillations of the terminal voltage and increased erosion. Several very different physical models claim to explain this change in behavior [Turchi, 1986], known by researchers in the field simply as "onset." It is sometimes related to *Alfven critical speed*:

$$V_{crit} = \sqrt{\frac{2W_i}{m_{pi}}} \qquad (9.68)$$

where V_{crit} = Alfven critical speed (m/s)
  W_i = energy required to ionize a propellant molecule (J)
  m_{pi} = mass of propellant ion (kg)

This is the speed a propellant ion would have if its kinetic energy were to equal the energy needed to create the ion.

For self-field plasma thrusters, at high magnetic Reynolds number (Sec. 9.4.2), the current tends to concentrate in two regions:

- At the thruster entrance, where magnetic flux is added to the electrically conducting flow
- At the exit of the thruster, where flux separates from the flow in a two-dimensional expansion to the field-free vacuum

The upstream current region has a streamwise extent based on the balance of convection and diffusion of magnetic flux. We can therefore estimate the size scale (Δ) for this region by the dimension for which the magnetic Reynolds number is unity:

$$\Delta = \frac{\eta}{\mu v} \qquad (9.69)$$

where we may equate the characteristic flow speed (v) to the exhaust speed. The resistance of the discharge region at the entrance of the thruster is then

$$\begin{aligned} R_i &= \frac{\eta h}{2\pi r \Delta} \\ &= \left(\frac{\mu h}{2\pi r}\right) v \end{aligned} \qquad (9.70)$$

where h and r, respectively, are the length and mean radius of the current path. The resistive voltage across the discharge at the entrance to the thruster is thus proportional to the characteristic back electromotive-force (EMF):

$$\begin{aligned} V_i &= R_i J_i \\ &\sim vBh \end{aligned} \qquad (9.71)$$

where the current near the entrance scales as the total discharge current ($J_i \sim J$) and the characteristic magnetic field is

$$B = \frac{\mu J}{2\pi r} \qquad (9.72)$$

The resistive dissipation in the thruster therefore matches the thrust power, so we can obtain flow kinetic energy only to the extent that we can generate heat within the discharge. This heat may be absorbed and convected by the flow as energy required for ionization, until the mass flow is fully (singly) ionized, which would correspond approximately to Alfven critical speed. Beyond this speed, the additional energy dissipated in the flow must be absorbed or transferred by other processes. These processes can include the development of higher-energy particles, with the possibility of higher heat transfer to solid surfaces. This may increase erosion, so the total mass flow rate would be higher, tending to limit the exhaust speed. Higher-energy electrons can also create higher ionization and excitation states, which would represent increased frozen-flow loss in the exhaust. These states would also mark a greater chance to transport energy to solid surfaces by radiation or by direct bombardment of the surfaces with more energetically active particles.

The mechanisms for generating high-energy electrons include plasma microinstabilities. These instabilities typically develop when the number density of electrons becomes so low that their drift speed, needed to carry the local current density, exceeds other characteristic speeds, such as the electron or ion thermal speed (see Sec. 9.4.2). When we try to increase the exhaust speed by decreasing the mass flow rate and increasing the current, the particle density may become too low to match the current density, particularly near the edges of electrodes. Design of MPD thrusters can avoid difficulties by limiting operation to exhaust speeds below some value proportional to $J^2/\dot{m}_{prop}$. We may estimate this value from experimental behavior, but we need to know the distributions of current and mass density in each device.

Figure 9.30 shows experimental results for operating a self-field MPD thruster. Note that the voltage-current characteristic has a positive slope, as opposed to the negative slope for operation of an electrothermal arcjet. The slope increases rapidly for exhaust speeds exceeding Alfven critical speed. Apart from electrode fall voltages (see Sec. 9.4.2), the voltage across the discharge is proportional to the product of exhaust speed and magnetic field. The shape of the characteristic changes from a linear to cubic dependence on J because the exhaust speed increases with J^2, when electromagnetic forces predominate (compared to axial pressure gradients). This change also implies the electromagnetic force accelerates all the propellant mass. Coupling of ions and neutral atoms by collision can be too weak at low (ion) density to accelerate the total mass flow. Ions are then accelerated by electromagnetic forces until the associated resistive dissipation ionizes more atoms, increasing the mass flow of charged particles. This situation resembles boiling, effectively at the ionization potential. The speed of the charged-particle flow is then constant at approximately Alfven critical speed, so the back EMF across the discharge is only linear in J, through the magnetic field.

Self-field arrangements are the simplest to create and (in principle) to analyze, but they require high currents for significant electromagnetic forces. High currents

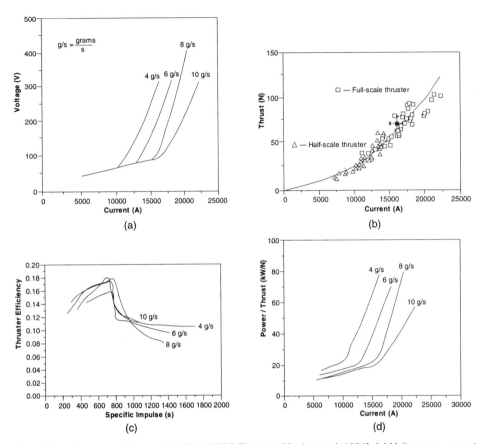

Fig. 9.30. Data for a Typical Self-Field MPD Thruster [Andrenucci, 1991]. (a) Voltage vs. current at several mass flow rates. (b) Thrust vs. current (independent of mass flow rate) for two geometrically scaled thrusters. (c) Efficiency vs. specific impulse at several mass flow rates. (d) Power-to-thrust ratio vs. current at several mass flow rates. Note that the flow rate units are grams per second. (Courtesy of Centrospazio)

typically imply megawatt powers, which are not yet available in space for steady-state operation. Using intermittent pulses of high power, however, permits self-field behavior at much-reduced average power levels, as in the PPT. If the current and voltage are approximately constant during the pulse, this operation is referred to as *quasi-steady*. With the times of start-up and shut-down limited to small portions at the beginning and end of the pulse, additional losses incurred during these times can be relatively small. Factors of over a thousand between erosion rates for cold- vs. hot-cathode operation at high current densities (1–100 vs. 10^{-3} μg per coulomb of charge transfer) suggest that maintaining the cathode temperature at a high value between pulses would be useful. The simplicity of self-field, electromagnetic propulsion can be lost as we try to operate at average powers below a megawatt.

Applied-Field MPD Thrusters—If external means (e.g., multi-turn coils, permanent magnets) supply magnetic fields, total electrical power can be lower even for steady operation. The overall arrangement is simpler than repetitively pulsed, self-field systems for sub-megawatt applications. But physical interactions within the thruster flow field are much more complex. By Ampere's law, an axisymmetrical magnetic field provided by currents external to the thruster has components only in a plane containing the axis of the thruster (i.e., the rz-plane in cylindrical polar coordinates). Multipole fields can also be applied, with components in the plane normal to the axis, but these fields are not axisymmetrical. (Such fields are sometimes used in experiments for controlled nuclear fusion to contain plasma away from solid boundaries or for increased stability). The discharge current and the rz-components of an applied field interact to create a torque on the plasma electrons and on the plasma itself. The former is manifested by the generation of a Hall current circulating around the axis of the thruster. We can calculate this current in steady-state operation by setting the azimuthal electric field equal to zero in the generalized Ohm's law [Eq. (9.30)]. The vector products of this current with the radial and axial components of magnetic field produce, respectively, an axial thrust force and a radial pinching force in the plasma. The latter force opposes the centrifugal effect resulting from rotating the plasma due to the vector product of the discharge current and the magnetic field in the rz-plane $\left(\vec{j} \times \vec{B}\right)_\theta$. Thus, besides a relatively weak contribution from the self-magnetic field, the total thrust of an applied-field MPD thruster has three main sources:

- Increased resistive heating because of azimuthal Hall current
- Higher pressure and temperature in the thrust chamber because of plasma rotation
- Direct electromagnetic acceleration by interaction of the azimuthal Hall current and the radial part of the applied field

The first two components depend on expansion of the plasma to convert thermal and rotational energy into the kinetic energy of axial exhaust.

The relative magnitudes of pinch forces vs. centrifugal effects and azimuthal Hall current vs. circuit current depend on the particle density and its distribution in the thruster flow field. These properties in turn respond to the distributions of force density and resistive heating. Thus, the behavior of applied-field thrusters can be quite nonlinear in terms of the physical interactions that govern performance. Curiously, however, experiments indicate simple relationships within particular operating regimes [Fradkin, 1973; Myers, 1991]. For example, as shown in Fig. 9.31, the voltage across the thruster is linear in the value of the applied magnetic field, in the low-voltage mode of operation. In the high-voltage mode, the voltage scales as the square of the applied-field value.

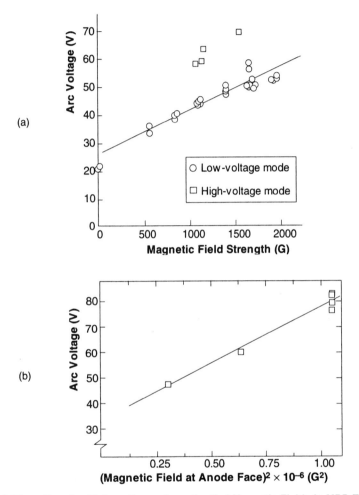

Fig. 9.31. How Arc Voltage Depends on Applied-Magnetic Fields in MPD Thrusters [Fradkin, 1973]. (a) Low-voltage mode (linear in field). (b) High-voltage mode (linear in field-squared).

We can explain the latter variation readily in terms of the back EMF ($-\vec{v} \times \vec{B}$) component of the electric field in the swirling plasma flow. At fixed mass flow rate and discharge current, the electromagnetic torque and subsequent rotational speed should increase linearly with magnetic field. The voltage therefore scales as the square of the applied field. The linear dependence of the low-voltage mode may indicate that the rotational speed of the plasma is limited to a constant value. Indeed, observed correlations of exhaust speed with Alfven critical speed have led some researchers to suggest this limitation. Recent calculations [Mikellides, et al. 1995], however, show the rotational speed becoming constant because viscous drag in the thrust chamber increases with applied field. Centrifugal effects press the plasma against the inside surface of the outer conductor, and viscosity increases with ion temperature.

The complexity of interactions within the applied-field MPD thruster demands modeling by numerical computation using an appropriately sophisticated code. For example, the MACH2 code [Frese, 1986] allows us to calculate axisymmetrical, magnetohydrodynamic flows, including the effects of applied-magnetic fields, internal plasma currents, anomalous resistivity, and Hall effect. Figure 9.32 displays sample results obtained with MACH2 for a steady-state, applied-field, MPD thruster at the NASA Lewis Research Center [Myers, 1991].

These results indicate many of the trends in the experimental data, but do not include the effects of electrode fall voltages (Sec. 9.4.2) on the terminal voltage and overall thruster performance. Such falls are a major problem with steady-state, low-power operation of plasma thrusters because much of the total voltage across the thruster is associated with the electrode falls. The falls can add heat to the flow for electrothermal thrust but mostly represent power lost in heating and eroding the electrodes.

Experimental data for operation in the 100-kW regime indicate that electrode falls are major sources of inefficiency in applied-field MPD thrusters. Figure 9.33 displays the efficiency and specific impulse for various MPD thrusters operating from 10 kW to megawatt levels [Sovey and Mantenieks, 1988]. By increasing the value of the applied field and lowering the plasma density, we can increase the voltage across the plasma flow relative to the electrode falls. Also, the Hall current will be much higher than the direct current through these falls, so relatively less power is dissipated in the electrode regions. Operation in this limit represents the regime of *Hall thrusters*.

Electromagnetic Propulsion—Stationary-Plasma Thrusters

Various arrangements have been proposed for using the Hall effect to achieve an effective space thruster. After two decades of successful development in Russia, a particular form has recently become the subject of considerable interest in the West. The *stationary-plasma thruster* (SPT), shown in Figs. 9.16 and 9.34, has an axisymmetrical geometry, in which axial current flows from a cathode next to the plasma exhaust plume to the anode at the upstream end of the thrust chamber across a region of radially directed magnetic field.

Current flow across a magnetic field often requires a Hall electric field within the plasma. In the SPT, however, $E_\theta = 0$ and the axial current generates an azimuthal Hall current. We can compute this current from the generalized Ohm's law [Eq. (9.30)]. The azimuthal current must cross the radial magnetic field, which

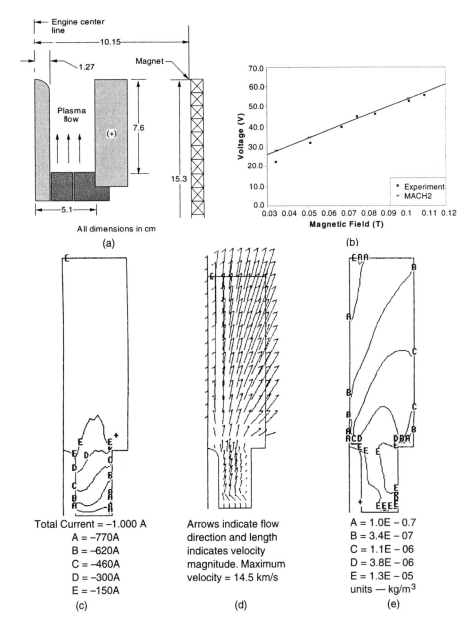

Fig. 9.32. Numerical Code Calculations for an Applied-Field MPD Thruster [Mikellides, et al., 1995]. (a) Schematic drawing of MPD thruster modeled in calculations. (b) Calculated (using MACH2 code) and experimental dependence of (plasma) voltage on applied field. (c) Contours of the discharge current. (d) Flow velocity vectors. (e) Isodensity contours.

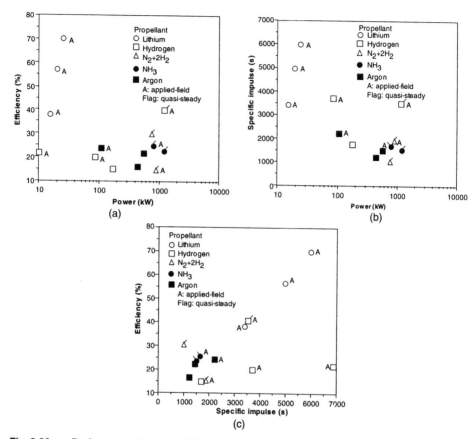

Fig. 9.33. Performance Data for MPD Thrusters Under Various Conditions of Self- and Applied-Field for Several Propellants [Sovey and Mantenieks, 1988]. (a) Efficiency vs. power. (b) Specific impulse vs. power. (c) Efficiency vs. specific impulse.

requires a Hall electric field in the axial direction. This electric field accelerates the plasma ions axially. Again, from the generalized Ohm's law, the total axial electric field is

$$E_z = \eta j_z \left(1 + \Omega_e^2\right) \tag{9.73}$$

where η is the effective resistivity (including the effects of electron scattering by electromagnetic waves due to microinstabilities), j_z is the axial current density, and Ω_e is the electron Hall parameter, as in the earlier discussion of applied-field MPD thrusters. For a given regime of particle density, the voltage across the thruster will, therefore, scale as the discharge current and the square of the magnetic field. Total resistance to electron motion in the axial direction includes the effect of collisions with ions, which represents a drag force on the ions in the axial direction. The Hall electric field provides the net acceleration of the ions.

With $\Omega_e^2 \gg 1$, the kinetic energy of the ions approximately equals the voltage across the thruster. This is also true for electrostatic thrusters, in which ions are accelerated by the voltage applied to the thruster electrodes. Indeed, in much of the literature, the stationary-plasma thruster is considered a form of "gridless" ion thruster. This picture ignores the fact that the Debye length (Sec. 9.4.2) in the SPT is much smaller than the dimensions for flow acceleration, so the plasma is quasi-neutral, as in most other plasma thrusters. The electric field is supported by the local dynamic conditions for current flow, as embodied by the generalized Ohm's law, not by the charge distributions on the electrodes calculated from the electrostatic potential. These charge distributions must, of course, be consistent with the voltages across the plasma, including electrode falls. The point here is that the SPT operates with a load line based on conduction in the plasma, and not merely in response to a voltage applied to its electrodes.

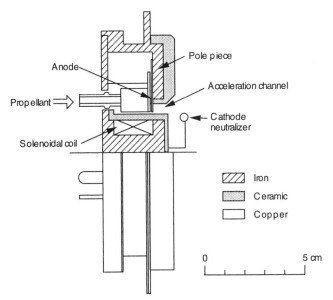

Fig. 9.34. System Schematic for an Experimental (Japanese) Stationary-Plasma Thruster [Komurasaki, et al., 1991]. Data for this Hall thruster are given in Fig. 9.35. (Courtesy of Department of Aeronautics, University of Tokyo)

The actual behavior of the SPT involves many features that are much subtler than even the rather complex interactions discussed here. For the Japanese thruster experiment [Komurasaki, et al., 1991] of Fig. 9.34, the efficiency increases with specific impulse, as shown in Fig. 9.35. Many features have been explored during the lengthy development of the SPT in Russia [Bober, et al., 1991]. They include proper variation of the radial magnetic field with axial position and the effects of layers near the insulating walls of the thrust chamber. Table 9.8 summarizes the results of this development. Figure 9.36 displays the thrust efficiency and thrust-to-power ratio vs. exhaust speed. It is particularly interesting that xenon can produce specific impulses of 1500 to 2000 s at efficiencies above 50%. Alfven critical speed for xenon is 4200 m/s; yet, the SPT does not experience the deleterious effects found in other electromagnetic thrusters at exhaust speeds well above Alfven critical speed. Because the net voltage across the flow is in the same direction as the ion acceleration, the back EMF caused by plasma motion in a magnetic field is not limited by the resistive drop across low-speed flow at the entrance to the thrust chamber, as it is in the self-field MPD thruster. The propellant's ability to absorb resistive dissipation therefore does not constrain the power delivered to flow kinetic energy.

SPT performance may be restricted to relatively low particle densities to operate with high values of electron Hall parameter. This restriction implies relatively low thrust density. However, the small number of parts compared to ion engines, as well as their rugged construction, suggests we can achieve satisfactory total thrust levels by scaling up thruster dimensions or by simply using arrays of smaller devices.

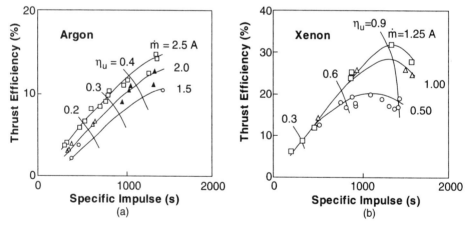

Fig. 9.35. **Performance Data for (Japanese) Hall Thruster Experiment of Fig. 9.34** [Komurasaki, et al., 1991]. These data correspond to the thruster in Fig. 9.34 and have not achieved values for Russian devices. (Courtesy of Department of Aeronautics, University of Tokyo)

Table 9.8. Typical Values for Stationary-Plasma Thruster [Bober, et al., 1991]. Twenty years of development, including many subtle features, were necessary to achieve these results.

Propellant	Xenon					
Input power (kW)	0.7	1.5	3.0	6.0	12	25
Thrust (N)	0.01–0.05	0.02–0.1	0.04–0.2	0.08–0.4	0.15–0.6	0.3–1.0
Specific impulse (s)	1200	1600	1600	2000	2500	3000
Efficiencies	0.4	0.5	0.5	0.6	0.6–0.7	0.7–0.75
Diameter of acceleration channel (cm)	5	7	10	14	20	28
	(Production models)		(Prototypes)		(Experimental models)	

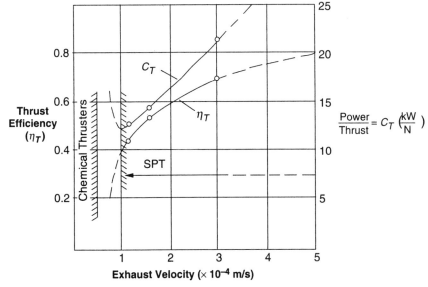

Fig. 9.36. Thrust Efficiency and Power-to-Thrust Ratio (C_T) vs. Exhaust Speed for (Russian) Hall Accelerator. Stationary-Plasma Thruster (SPT) using xenon propellant [Bober, et al., 1991].

Electrostatic Propulsion—Ion Thrusters

Direct acceleration of electrically charged matter by an electric field is conceptually the simplest approach to electric propulsion. While physically straightforward, efficient devices with adequate values of total thrust and lifetime required several decades of development. Matching capabilities to mission needs is the main challenge today. For tasks near the Earth, such as satellite stationkeeping and orbit-raising, we must not only satisfy the needs for efficiency, lifetime, and specific impulse but also displace competing technologies. Thus, issues of reliability, parts count, and propellant availability can become critical. Applying electric propulsion in space has favored systems that can use the on-board infrastructure of propellant handling (e.g., hydrazine) and electrical power. But electrostatic propulsion requires xenon as a propellant to achieve values of specific impulse needed for orbit-raising (< 2000 s) with reasonable thrust efficiency. Also, the high voltages associated with electrostatic propulsion demand considerable and precise power conditioning of electrical power available from the spacecraft, including individual power supplies for separate tasks. Both requirements are disadvantages compared to other techniques for orbit-raising missions near the Earth. For primary propulsion to the outer planets, on the other hand, electrostatic techniques may have no rivals.

The basic elements of an electrostatic thruster are a source of charged particles, an acceleration region, and a way to neutralize the exhaust flow so a net charge does not develop on the spacecraft that would prevent further thruster operation. For ion thrusters, proposed ion sources include:

- Contact ionization, in which material with low-ionization potential (e.g., cesium) is ionized by passage through a hot, porous plug
- Field-emission, in which high electric fields near the edge of an electrode create ions from a liquid metal (e.g., cesium)
- Electron bombardment, which uses magnetic fields in a low-density gas discharge to retain electrons while extracting ions for acceleration
- Radio frequency-induced gas discharge (RIT), which uses an RF discharge vs. direct current to provide a plasma from which ions are extracted

The main techniques now used are the electron-bombardment thruster (Fig. 9.37) and the RIT-system [Groh, et al., 1991]. Accumulation of charge within the acceleration region, rather than the type of ion source, limits the thrust density.

Current Flow Limited by Space-charge—Child-Langmuir Equation—Acceleration of a solitary particle of charge q and mass m by a uniform electric field E is quite easy to calculate. When a large number of such particles are assembled as a charged flow, however, we must account for the effect of the charge density on the electric field. In particular, we expect the field in the charged-particle flow will be significantly altered, if the total charge in the flow between the electrodes is comparable to the charge on the accelerating electrodes. With particles of fixed mass-to-charge value, the total voltage across the electrodes is determined by the necessary particle kinetic energy to achieve a desired specific impulse. The thrust density (thrust per unit area of electrode) depends merely on the current density of charged particles and their mass per unit charge value. The higher the current density, however, the higher the particle density at fixed exhaust speed and thruster (electrode) area, and thus the stronger the charge in the inter-electrode space. This

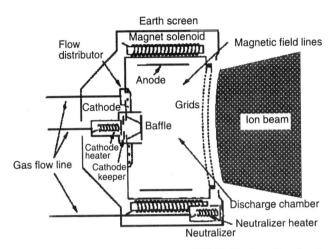

Fig. 9.37. Schematic Drawing of Ion Engine, (UK-10 Electron-Bombardment) [Courtesy of Culham Laboratory]. Note that the accelerating grids occupy only a small portion of the thruster volume compared to the electron-bombardment discharge chamber.

space-charge limits the thrust density in electrostatic thrusters. We want to maximize the engine's thrust density to minimize the area of the accelerating-electrode structure.

The variation of the voltage (V) is given by Poisson's equation in terms of the charge density (per unit volume), (ρ_q, in C/m³) and the permittivity of free space ($\varepsilon_0 = 8.885 \times 10^{-12}$ farad/meter):

$$\nabla^2 V = \frac{\rho_q}{\varepsilon_0} \tag{9.74}$$

For a one-dimensional problem, with the charge density expressed as the product of particle density and charge (nq), we obtain the voltage from the ordinary differential equation:

$$\frac{d^2 V}{dx^2} = \frac{nq}{\varepsilon_0} \tag{9.75}$$

The particle density, however, depends on the voltage in terms of the speed attained in traveling through a difference in potential ($V - V_0$):

$$u = \sqrt{2\left(\frac{q}{m}\right)(V - V_0)} \tag{9.76}$$

(where V_0 is the potential at the entrance to the acceleration region) and the current density:

$$j = nqu \qquad (9.77)$$

Thus, the governing equation for the voltage distribution involves the current density, the mass-to-charge ratio, and the initial value of the voltage, which may be set equal to zero:

$$\frac{d^2V}{dx^2} = \frac{j}{\varepsilon_0 \sqrt{2\left(\frac{q}{m}\right)V}} \qquad (9.78)$$

This equation can be integrated by multiplying both sides by dV/dx:

$$\left(\frac{dV}{dx}\right)\frac{d^2V}{dx^2} = \frac{d}{dx}\left[\frac{1}{2}\left(\frac{dV}{dx}\right)^2\right] \qquad (9.79)$$

and

$$\left[\frac{j\sqrt{\frac{m}{2q}}}{\varepsilon_0}\right]\frac{\left(\frac{dV}{dx}\right)}{\sqrt{V}} = \left[\frac{j\sqrt{\frac{m}{2q}}}{\varepsilon_0}\right]2\frac{d}{dx}\left(\sqrt{V}\right) \qquad (9.80)$$

For boundary conditions, we set $V_0 = 0$ and recognize that the maximum space-charge for particles to enter the acceleration region with negligible speed corresponds to an electric field at the entrance:

$$E_0 = \left(\frac{dV}{dx}\right)\bigg|_0 = 0 \qquad (9.81)$$

We then obtain the *Child-Langmuir equation* for current density limited by the space-charge:

$$j = \left[\frac{4}{9}\varepsilon_0\sqrt{\frac{2q}{m}}\right]\frac{V^{3/2}}{x^2} \qquad (9.82)$$

where the value of the voltage (V) is specified at the end of the acceleration region of length x. To achieve high current density, we need high voltages applied across small gaps. The value of charge-to-mass should also be high, if we want high current density (which is the case for systems heated by particle beams). High thrust density, however, requires high mass-to-charge ratios at fixed specific impulse, because the thrust per unit area is

$$\frac{F}{A} = nmu^2$$

$$= j\left(\frac{m}{q}\right)u \qquad (9.83)$$

$$\sim \sqrt{\frac{m}{q}}$$

Thus, high-mass particles are favored, such as ions of cesium, mercury, and xenon. Still higher mass-to-charge ratios are possible with colloidal droplets and, more recently [Leifer and Saunders, 1991], with buckminsterfullerenes (e.g., C_{60}).

The specific impulse is determined by the total voltage difference the particles experience, whereas the conditions for flow limited by space-charge depend mainly on local values of the electric field. We can therefore accelerate particles with a voltage that obtains high current density and then decelerate the particles with an opposing electric field to achieve a lower net difference in voltage. The total power thereby decreases for a given thrust, which reduces the size of the electric power system. Apart from detailed problems of maintaining a stable flow, there are limits to the efficacy of such an *accel-decel* arrangement: the cost per charged-particle (in volts) may become too high a fraction of the net difference in voltage, causing the thrust-power efficiency to suffer.

Electron-bombardment Source of Ions—Space-charge limits the current density through the acceleration region, but we still need to create ions with a dense enough source current. The *electron-bombardment* source of ions matches the acceleration region exactly. As displayed in Fig. 9.37, the source occupies most of the thruster's volume, with ions extracted directly to a thin pair of grids that form the end surface of the thruster and serve as the accelerating electrodes. All of the preceding discussion of space-charge limited flow is confined to the relatively thin region between these grids.

The electron-bombardment source consists of a central cathode that provides electrons, a propellant feed that provides neutral atoms ionized by collisions with electrons, and a magnetic field that causes electrons to circulate in the plasma as ions are extracted through an exit grid. Typically, it has a *hollow cathode*. This device, shown schematically in Fig. 9.38, consists of a cylindrical insert with low work function surrounding a plasma of relatively high density ($\sim 10^{15}$ cm^{-3}) and low temperature (~ 1 eV). The insert and plasma are held in a cylindrical chamber ($\sim$ few mm dia.) that is heated (at least initially) to temperatures (~ 1000°C) allowing thermionic emission from the inner surface of the insert. Propellant fed into this chamber is ionized by the interior discharge and exits, along with electron flow, through a very small orifice (< 1 mm dia.) in the chamber endplate. The potential distribution just outside of the orifice is adjusted by means of an intermediate electrode ("keeper") to maintain the discharge as the particle density decreases to a relatively low value for the rest of the source volume. The low particle density allows electrons from the hollow cathode to attain higher kinetic energies (several eV), so ionization of additional propellant gas occurs efficiently within the source volume (see Sec. 9.4.2). The magnetic field inhibits loss of electrons to positive electrode surfaces before ionizing collisions can occur in the low-density gas.

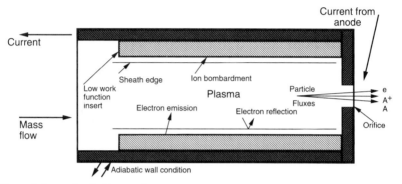

Fig. 9.38. **Schematic of a Hollow Cathode** [Salhi and Turchi, 1993]. Ions from the plasma heat the cathode surface, causing thermionic emission of electrons which ionize the mass flow and the neutral atoms returning from the cathode surface. Electrons, ions, and atoms are extracted through the orifice for use in the discharge chamber.

The most critical task in designing and developing any ion source is to concentrate the population of heavy particles in a single species and charge state, at least near the exit grid. Obviously, if we create significant numbers of ions of different charge-to-mass ratio than the desired value, their subsequent acceleration could follow trajectories that impinge on electrode structures, resulting in reduced thruster life. Even neutral atoms can be harmful because collisions with ions would also scatter particles onto trajectories that strike electrodes. At the relatively high kinetic energies associated with the voltage needed to achieve desired specific impulse values, ion bombardment of surfaces can sputter off additional material, which then may further interfere with free passage of ions through the thruster grids.

The electrode grids must be able to provide the necessary potential over the exit area while allowing ions to be extracted from the source, accelerated to high speed, and expelled into space. For the particle energies of interest to electric propulsion, the grids must have open spaces, such as holes. These holes spread over the area of the grids and align from grid to grid to form a large number of focusing electrode pairs. Focusing results from the potential distribution near a hole in a plate. Equipotential surfaces along the plate's surface dip into the hole, effectively providing a locally concave electrode. This focusing helps to direct the flow through the local pair of electrode holes. The effective area of the thruster for mass flow is reduced from the total exit area by the metal required between holes and by the fraction of the hole area that cannot be used for a focused flow without impinging on the sides of the holes. To avoid electrical breakdown along short insulator surfaces, the grid separation is maintained by support at the periphery of the grid area, rather than by many internal, electrically insulated supports. This approach limits the total area of the grids because we must precisely maintain relatively small gaps (~ mm) over grid diameters of up to 30 cm. For larger thruster areas, we must therefore use arrays of thrusters, each with the largest proven area.

Performance of the Ion Engine—Ion engine design has progressed for much of the twentieth-century, so most practical trade-offs are complete. These trade-offs include choosing and operating an ion source to achieve satisfactory energy cost-per-ion, and reducing spurious particles (e.g., neutrals, multiply-charged ions, electrode impurities) to low enough levels for mission life. Indeed, life-testing of ion engines in ground-based laboratories becomes an issue because the background neutral gas within the vacuum tank penetrates into the engine and causes collisions that limit electrode life. The size and pumping capacity of the vacuum facility sets a maximum value for thrust power during the life test. This value keeps us from developing and validating larger ion engines, so we must use arrays of smaller thrusters for high-energy missions. Figure 9.39 and Table 9.9 provide typical values for ion-engine performance.

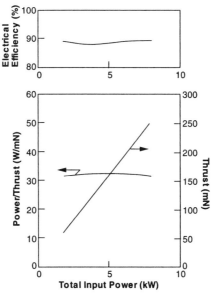

Fig. 9.39. **Performance Data for a UK-25 Ion Engine,** [Fearn, 1987]. (a) Electrical efficiency vs. input power. (b) Power-to-thrust and thrust vs. input power.

Table 9.9. Typical Performance Values for Ion Engines. For other values of specific impulse, see Fig. 9.42. Xenon engines also operate at 10 kW per engine.

Propellant		Mercury	Xenon
Input power (kW)	1.0	25	0.44
Thrust (N)	0.03	0.5	0.017
Specific impulse (s)	5000	9000	2585
Diameter (cm)	15	30	13

9.5 Select Space Power

People who hear about electric propulsion of space vehicles are often concerned about very long extension cords. As discussed in Sec. 9.3, however, we can more seriously discuss electric power in space by calculating the specific mass of systems for electric power and thermal management. Here, we just briefly mention the types of power components available for electric propulsion in space.

9.5.1 Matrix of Power Technology for Space

Figure 9.40 shows the basic choices for electric-power systems in space, including efficiencies of power conversion and values for specific mass. Energy is available to the spacecraft in two forms: onboard and radiated. The first form means *nuclear energy*. If we use batteries, fuel-cells, and other techniques to obtain electrical energy from chemical energy, the system's effective specific impulse reduces to that of chemical propulsion. For missions requiring very high energy, such as interstellar exploration, even nuclear energy starts to have this problem. The second form of energy includes a particularly available type of nuclear energy, namely radiation from the Sun, usually called *solar power*. Besides using photons with the various wavelengths of the solar spectrum, we can, in principle, use microwave transmitters and laser beams to radiate energy to a spacecraft. A major difference between onboard and radiated power is that onboard systems must reject waste heat from the former, whereas receiving power from a remote source requires rejection of heat only from power conversion devices (and parts of the propulsion system).

Several types of power-conversion equipment can work with either source of energy. Two broad categories are photovoltaic (including large-scale photon receivers, such as antennas), and thermal. The latter category includes devices that are thermoelectric, thermionic, and thermal-dynamic, any of which can process heat into electricity. These devices absorb heat in a working fluid, such as electrons or liquid metal. A thermal-dynamic converter involves the working fluid's pressure or momentum, either in turbine-driven, rotating machinery or the linear flow of a magnetohydrodynamic generator. Photovoltaic systems combine only with

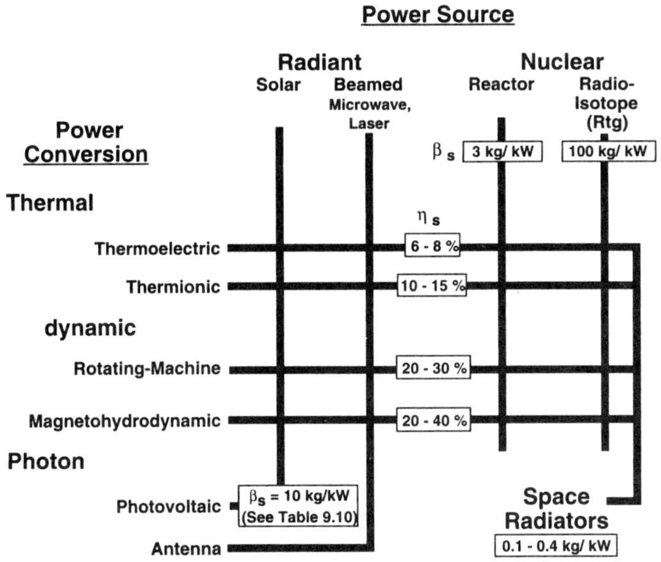

Fig. 9.40. Matrix of Space-Power Technologies. The power source type is shown on the horizontal axis, and conversion type is shown on the vertical. Intersections mark possible combinations. Efficiencies (η_s) of converting power from the source to electricity and specific masses of the power sources (β_s) are indicated.

radiated-power sources, except in some schemes in which plasma radiation transfers power from nuclear fission or fusion reactors on board the spacecraft.

The *specific mass of a power source* (β_s in kg/W) may be quoted in terms of raw output power, which in the case of nuclear sources would be the thermal power. We then divide this specific mass value by the efficiency (η_s), for the converter that changes raw power into electrical power. For photovoltaic array systems, the output is already in the form of electricity, so $\eta_s = 1$.

Recently, selecting the best combination of source and converter technology has become much simpler: nuclear-reactor technology for space power has been postponed into the indefinite future. Our only real option is solar power. Table 9.10 provides data on several types of solar arrays. Furthermore, for near-term applications, the mission (e.g., communication satellite) will already have determined the electrical-power system. We must therefore adapt the thruster to the total power available for the propulsion system during the mission.

9.5.2 Power Conditioning

We must convert the basic electrical power available from the spacecraft into forms needed by the thruster. Typically, this power conditioning begins with the standard power for the spacecraft bus. In principle, for missions that also need

Table 9.10. Values for Solar-Photovoltaic Systems (Beginning-of-Life) [Pollard, et al., 1993].*
Knowing the required beginning-of-life power, we can determine the array mass and size.

Array Technology	Panel/Cell Type	Power/Mass (α_s in W/kg)	Power/Area (W/m^2)	Mass/Power (β_s in kg/W)
Planar	Conventional Rigid / Si	40	100–140	0.0250
Planar	Advanced Rigid / GaAs	60	220	0.0167
Planar	Flexible Blanket / Si	138	140	0.0072
Concentrator (15:1)	Survivable Low Aperture/ GaAs	55	250	0.0182
Concentrator	Mini-dome / multi-junction	96	290	0.0104

*Editor's note: These numbers are a bit optimistic (particularly power per area), see Larson and Wertz [1992].

power conditioning, we might tap into power at higher than normal voltages or AC vs. DC. The need for extra coordination and testing, however, makes such integration unlikely. In general, the performance of the power-processing subsystems, in terms of specific mass (kg/ W) and efficiency, will depend on design details that are beyond the scope of this chapter. Typical values for the power processor's efficiency (η_{pp}) are 0.92 for ion-engine systems and 0.98 for steady, arcjet systems. The design of power processing for unsteady, pulsed, and quasi-steady plasma thrusters is particularly complex. For example, we must obtain capacitors of low mass that can store energy at the needed voltage, deliver current at the desired value, tolerate voltage reversal, and fire reliably for ~10^7 shots. If capacitor mass dominates the thruster system, the product of specific energy (J/kg) and shot-life replaces the ratio of thrust time to specific mass (τ / β) used in Sec. 9.3 to evaluate performance. In a recent effort involving a pulsed-plasma thruster, we required a totally new circuit design to achieve adequate energy per unit mass (~45 J/kg) at the relatively low voltages (~100 V) and high currents (~10 kA) needed by an ablation-arc. Limiting values for such a circuit are η_{pp} = 0.9 and specific mass β_{pp} = 0.02 kg/W at a repetition rate of 2 Hz.

Specific mass values for power processors are combined in some modeling (e.g., Table 9.11) with the specific mass of the thruster itself [Mantenieks, 1990]. Power processors for steady, kilowatt-level, arcjets have achieved a specific mass value of 0.2 kg/ kW [Cassady, et al., 1987].

9.5.3 Thermal Management

All of the thermally based systems require heat rejection as part of their operating cycle. Radiators typically add mass with specific mass values (β_H) in the range of 0.1–0.4 kg/ kW for the output of thermal-dynamic systems. Even with rather low converter efficiency, the mass of the radiator may be relatively low compared to that of the thermal power source itself [Sercel, 1993]. For the heat that must be rejected due to inefficiencies in the power processor, the radiator mass is also only a minor part of the overall system. Inefficiency of the thruster itself includes energy lost to ionization and electrode heating. The former loss is retained (or radiated) in the exhaust flow, whereas the latter may be handled by local radiative-cooling, rather than burdening the spacecraft's radiators. Detailed design of the thruster must address this cooling, along with features (e.g., thermal dams, radiation baffles) that shield the spacecraft and thruster structure from heat generated in and near the thruster.

9.6 Assess System Performance

From the data curves and formulas, we can relate thrust efficiency and exhaust speed. Variation of thrust efficiency with exhaust speed, at low payload fraction, can strongly determine the proper exhaust speed and possibly, the selection of thruster type, as well. Figure 9.41 summarizes the dependence of thrust efficiency on specific impulse for several types of thrusters.

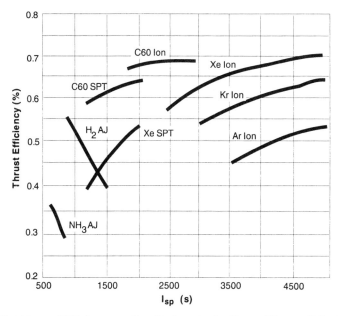

Fig. 9.41. **Summary of Efficiency vs. Specific Impulse for Several Types of Electric Thrusters** [Miller and Seaworth, 1993]. Although we assume constant efficiencies in Sec. 9.3, we can see here that the efficiency varies with exhaust speed. (Reproduced with permission from the American Institute of Aeronautics and Astronautics)

To calculate the jet-specific mass for a particular electric-propulsion system, we must compute the jet power and the total mass of the propulsion system. We can write the former quantity in terms of the source power and a product of efficiency factors:

$$P_J = \eta_{th}\, \eta_{pp}\, \eta_s\, P_s = \eta_T P_s \tag{9.84}$$

where P_J = jet power (W)
P_s = the raw power from the power source (W)
η_s = the efficiency of converting raw power to electric power (see Fig. 9.40)
η_{pp} = the efficiency of power conditioning for the thruster (see Sec. 9.5.2)
η_{th} = the efficiency of converting electric power to thrust power [see Eq. (9.4)]
η_T = total efficiency, source power to jet power [see Eq. (9.4)]

We can find the thrust efficiency from experimental data or associated empirical formulas. For example [Gilland, et al., 1990], we may write the efficiency of a MPD thruster as

$$\eta_{th} = \frac{b\, v_e}{(v_e + d)} \tag{9.85}$$

where v_e is exhaust velocity (m/s) and the constants b and d depend on propellant: for hydrogen, $b = 0.858$ and $d = 2.59 \times 10^4$ m/s. Similarly [Gilland, et al., 1990], for ion engines,

$$\eta_{th} = \frac{b\, v_e^2}{\left(v_e^2 + d^2\right)} \tag{9.86}$$

with $b = 0.835$ and $d = 2.2 \times 10^4$ m/s, for argon. We can create other formulas simply by fitting curves to available data (see Sec. 9.4.3). Table 9.11 provides another formula that models thruster efficiency, with constants appropriate to various thruster types and propellants [Mantenieks, 1990]. It also gives a model and associated constants for the specific mass of the thruster system (not including the propellant system, power source, and radiator). This model is based on the electrical power delivered to the power processor vs. the jet power. The literature differs on the forms and values for modeling thruster systems, so we must base final designs on specific and complete engineering analyses.

The mass of the electric-propulsion system includes the thruster, the propellant storage and feed system, and those parts of the power processor, electrical power system, and heat rejection system not needed by the mission. We may reasonably write the sum of subsystem masses in terms of products of their respective specific masses and power levels, except for the propellant subsystem:

9.6 Assess System Performance

$$m_{inert} = \frac{\beta_{th} P_J}{\eta_{th}} + \beta_{pp} P_{pp} + \beta_s P_s + \beta_H P_H + m_{pr} \tag{9.87}$$

where m_{inert} = mass of the propulsion system without propellant (kg)
m_{pr} = dry mass of the propellant-handling subsystem (kg)
pp (subscript) indicates power-processor system
H (subscript) indicates heat-rejection system

Table 9.11. Models for Characteristics of Electric Thrusters.* System Efficiency Model: $\eta = \eta_t \eta_{PP} = A + B \ln I_{sp}$. System Specific Mass Model: $\beta_{pp} + \beta_{th} \eta_{pp} = C I_{sp}^D$ [kg / kW]. These numbers indicate efficiency and specific mass for a complete thruster system. This system includes the thruster (η_{th}, β_{th}) and the power processor (η_{pp}, β_{pp}).

Thruster System	Constants for Models			
	A	B	C	D
H_2 Arcjet	--	--	5.0	0
NH_3 Arcjet	--	--	1.8	0
Ar Ion	-2.024	0.307	4490	-0.781
Ar Ion (MW_e design)	--	--	0.49	0
Xe Ion	-1.776	0.307	123,100	-1.198
Hg Ion	-0.765	0.181	82,870	-1.136
Ar MPD (Lab version)	-0.591	0.126	7	0
H_2 MPD (2.5 MW_e design)	--	--	0.17	0
Ar PIT	-1.99	0.32	7	0

*Adapted in part from Mantenieks, 1990. *Electric Propulsion System Design Notes*, NASA Lewis Research Center, Cleveland, OH

We may estimate the mass of the propellant tank using the formulas in Table 9.12 [Palaszewski, 1989]. A different equation has been used in studies of arcjets for orbit-raising to estimate the mass of the propellant tank and feed system [Deininger and Vondra, 1987]:

$$m_{pr} = 100 + 0.2 \, m_{prop} \tag{9.88}$$

where m_{prop} is the initial mass of propellant (kg), and the constant portion of this mass estimate results from minimum structural needs and the feed system for operating an arcjet at 30-kW.

In Fig. 9.42, the payload mass fraction is displayed vs. the exhaust speed for several values of Δv but now with m_{pr}/m_i, based on Eq. (9.88), also subtracted from the final delivered mass fraction:

$$\frac{m_{pay}}{m_i} = e^{\frac{-\Delta v}{v_e}} - \left(1 - e^{\frac{-\Delta v}{v_e}}\right)\left\{\beta_J \frac{v_e^2}{2\tau} + 0.2\right\} - \frac{100}{m_i} \qquad (9.89)$$

where m_{pay} = payload mass (kg)
m_i = initial mass of the entire vehicle (kg)
Δv = velocity change (m/s)
v_e = exhaust velocity (m/s)
β_J = jet-specific mass (kg/W)
τ = burn duration (s)

Table 9.12. Formulas for Propellant Tank Mass. These empirical tank mass equations can be used to estimate tank mass, given the propellant mass (m_{prop}). The equations are applicable over the indicated range.

Propellant Type	Propellant Mass, m_{prop} (kg)	Tank Mass (m_{pr}) (kg)
NH_3 (small systems)	5000–18,300	$120 + 0.173\, m_{prop} + 2.28\, m_{prop}^{2/3}$
NH_3 (larger systems)	18,300–22,000	$1020 + 0.198\, m_{prop}$
H_2	5000–13,000	$610 + 0.493\, m_{prop}$
Xe	5000–22,000	$52 + 0.075\, m_{prop} + 0.154\, m_{prop}^{2/3}$

For the curves in Fig. 9.42, the last term in Eq. (9.89) is set equal to 0.1, corresponding to $m_i = 1000$ kg. All speeds are normalized by $(\tau / \beta_J)^{1/2}$, as in Fig. 9.3b. Note that the optimum exhaust speed is still approximately $v_{eo} = (\tau / \beta_J)^{1/2}$.

We can relate the parts of the system mass that depend on power to the source power P_s:

$$P_{pp} = \eta_s P_s, \qquad P_J = \eta_{th}\, \eta_{pp}\, \eta_s\, P_s \qquad (9.90)$$

including the heat that must be rejected:

$$P_H = (1 - \eta_s)\, P_s + (1 - \eta_{pp})\, P_{pp} + \frac{K(1 - \eta_{th})\, P_J}{\eta_{th}} \qquad (9.91)$$

A factor $K (\leq 1)$ in the last term recognizes that only a fraction of power loss in the thruster needs to be handled by the heat rejection system. The rest either leaves with the exhaust flow or radiates locally; K may reasonably approach zero. We then obtain the jet-specific mass for the electric-propulsion system by multiplying the individual component powers by their respective specific mass values and dividing the sum of the resulting mass values by the jet power:

$$\beta_J = \left\{ (\beta_{pp} + \beta_{th}\eta_{pp})\,\eta_s + \beta_s + \frac{\beta_H[K(1-\eta_{th})\,\eta_{pp}\eta_s]}{\eta_{th}\eta_{pp}\eta_s} \right\} \qquad (9.92)$$

Note that β_J depends only on the choice of technologies. It is independent of the actual mission (payload mass, thrust duration, Δv requirement). But the mission parameters then combine with the jet-specific mass to determine the optimum exhaust speed, the propellant mass, the initial or payload mass [through the payload mass fraction of Eq. (9.89) or Fig. 9.42], the masses of various components, and the operating power levels. The source power level then specifies the array area in the case of solar-photovoltaic systems. (If the array-handling requirements are incompatible with other aspects of the mission, such as the launch-vehicle shroud, the candidate design would fail the decision diamond for mission constraints in Fig. 9.2). The thrust efficiency depends on the exhaust speed, so the calculation is nonlinear, allowing system improvement by moving off the exact peaks in Fig. 9.42.

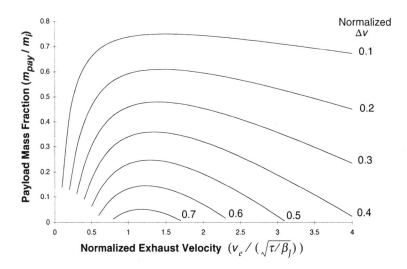

Fig. 9.42. Payload Mass Fraction Versus Normalized Exhaust Speed. Normalized by $(\tau/\beta_J)^{1/2}$. Each curve represents a normalized Δv. The relative dry mass of the propellant system, estimated as $m_{pr}/m_i = 0.1 + 0.2\,m_{prop}/m_i$, is subtracted here from the total delivered mass fraction.

9.7 Evaluate the System

The design equations used in our sample calculation are simplified versions of modern analysis techniques in which subsystem masses and efficiencies are treated in greater detail [Meserole, 1993; Miller and Seaworth, 1993]. These techniques combine analyses of mission trajectories that include the effects of optimal thrust vectoring, solar-array degradation and shadowing, atmospheric drag, and a host of other, often important, effects. [Some factors we ignored in Eqs. (9.87) and (9.91) include the mass of gimbal systems and separate specific masses for heat transfer from different subsystems.] At this time, computer programs such as OASIS (Optimized Advanced System Integration and Simulation), an expert system-based software package, have been applied to analyze solar-electric orbital-transfer vehicles [Miller and Seaworth, 1993]. With these programs, we can consider very complex design problems, including figure-of-merit analysis based on costs. Such analysis allows us to compare the costs of technology options for complete missions from launch through operations on orbit. Figure 9.43, for example, shows the money saved by systems using smaller launch vehicles with electric propulsion for on-orbit maneuvering.

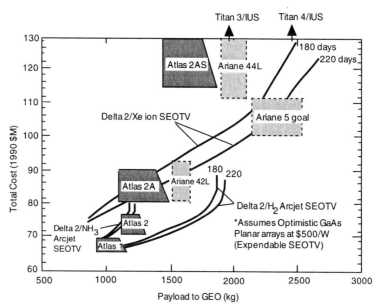

Fig. 9.43. Comparing Total Costs of Various Propulsion Systems for Earth-to-GEO Mission, Including Solar-Electric Orbital-Transfer Vehicle's Upper Stage [Miller and Seaworth, 1993, reprinted with permission of the American Institute of Aeronautics and Astronautics].

In Chapter 10, mission analysis of propulsion options indicates electric propulsion would be a leading candidate for orbit-raising of large spacecraft. Furthermore, recent work suggests electric propulsion can play an important role even after orbit has been achieved by repositioning spacecraft needed for Earth-surveillance of critical areas [Stocky and Vondra, 1994]. Such repositioning needs can occur with some urgency: a loss of a weather satellite in hurricane season, or in military missions, such as Desert Storm. Analyses show, for example, that the higher exhaust speeds provided by ion propulsion offer higher midcourse altitudes than chemical propulsion (580 vs. 110 km). This difference permits 90° repositioning of a 10,000-lb, geosynchronous satellite in only 41% of the time needed by chemical propulsion (25 days vs. 60 days). Alternatively, we can use the propellant savings to increase a satellite's ability to reposition repeatedly (see Fig. 9.44). These predicted advantages, and the experience now developing on commercial satellites, encourage continued application of electric propulsion near the Earth, while we await the next initiative toward human exploration of space.

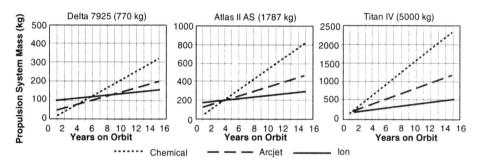

Fig. 9.44. Comparing Masses (kg) of (Wet) Propulsion Systems Required for GEO Stationkeeping and Repositioning vs. Time on Orbit [Stocky and Vondra, 1994]. As the time on orbit increases, electric propulsion becomes more attractive.

9.8 Case Study

9.8.1 Specify the Mission

This design example looks at the ion-engine design specified in Option 4 from the case study in Chap. 10. For this option, we use a chemical system to take us from low-Earth orbit to a 5000-km circular orbit. The plan is then to use an ion engine to get us the rest of the way to geostationary orbit. The requirements for the ion stage are:

- Payload mass = 3387 kg
- Δv = 4358 m/s
- Maximum allowable initial mass = 4914 kg
- Burn duration = 14,100,073 s = 163.2 days

To allow some margin, before beginning our systems-level design, we increase the Δv by 5%:

$$\Delta v = 1.05 \times 4358 = 4575.9 \text{ m/s}$$

9.8.2 Select the Thruster

We have already selected a xenon-ion engine as the candidate thruster. We still need to specify the exhaust velocity and number of engines. We can use Fig. 9.3 to choose a normalized exhaust speed. At high mass fractions, the maximum is rather broad, and Fig. 9.43 suggests that including the effects of propellant system (inert) mass further broadens these curves and shifts the optimum exhaust speed to higher values. We therefore select $v_e^* = 1.3$ and manipulate Eq. (9.13) to calculate the Δv^* that allows the payload mass fraction. We then have

- $\dfrac{m_{pay}}{m_i} = \dfrac{3387}{4914} = 0.689$
- $v_e^* = 1.3$
- $\Delta v^* = 0.240$

For the assumed, higher value of Δv, Eq. (9.11) gives

$$v_0 = \frac{4575.9}{0.240} = 19,085 \text{ m/s}$$

Equation (9.12) then provides the exhaust speed

$$v_e = 1.3(19,085) = 24,810 \text{ m/s}$$

which corresponds to a specific impulse of 2530 s. From Table 9.11, we use this value to determine the efficiency of converting solar-array power (with $\eta_s = 1$) to thrust power

$$\eta_T = \eta_s \eta_{th} \eta_{pp} = (1.0)(A + B \ln I_{sp}) = 0.630$$

where, for xenon-ion engines, we have $A = -1.776$ and $B = 0.307$.

For the specified initial mass, Δv, and derived exhaust velocity, Eq. (9.3) gives us the propellant mass

$$m_{prop} = 4914 e^{\frac{-4575.9}{24,810}} = 828 \text{ kg}$$

We divide this mass by the specified thrust duration, $\tau = 14{,}100{,}073$ s, to find the mass flow rate $\dot{m}_{prop} = 5.87 \times 10^{-5}$ kg/s. The system power, Eq. (9.4), is then

$$P_s = \frac{0.5 \, (5.87)\,(10)^{-5}\,(24,810)^2}{0.630} = 28,693 \text{ W}$$

Three engines in the 10-kW class can satisfy this mission's thrust requirement (see Table 9.3).

9.8.3 Select the Power Source

Equation (9.10) gives us the allowable (maximum) value of specific mass to achieve the velocity change in the specified thrust duration, $t = 14{,}100{,}073$ s:

$$\beta = \frac{\eta_T \tau}{v_0^2} = \frac{0.630\,(14{,}100{,}073)}{(19{,}085)^2} = 0.0244 \text{ kg/W}$$

corresponding to a specific power of 41.0 W/kg. This number is well within the range of solar arrays (Table 9.10), if the other masses of the system are not too large. We defer selection of the particular form of solar-array technology until the mass assessment is complete.

9.8.4 Power Conditioning Mass

From Table 9.11, the specific masses of the thruster and power processor combine to give us

$$\beta_{pp} + \beta_{th}\eta_{pp} = CI_{sp}^D = 0.0103 \text{ kg/W}$$

We obtain the mass of the thruster and power processor by multiplying this number by the power output of the solar array:

$$m_{th} + m_{pp} = (0.0103)(28{,}693) = 296 \text{ kg}$$

9.8.5 Thermal Management System Mass

The thermal-management system must gather and radiate all the heat associated with inefficiencies in the electric-propulsion system. We can estimate this heat by subtracting the power leaving in the thruster exhaust from the system's total power. The power in the exhaust is the thrust power plus the energy of ionization (which is the ionization potential shown in Table 9.4, not the total energy cost-per-ion). The kinetic energy of a xenon ion at the specified exhaust velocity is

$$W_k = \frac{m_{Xe} v_e^2}{2}$$
$$= \frac{\left(2.19 \times 10^{-25} \text{ kg}\right)(24{,}810 \text{ m/s})^2}{2}$$
$$= 6.74 \times 10^{-17} = 421 \text{ eV}$$

To convert to electron volts from joules, we divide by 1.602×10^{-19}. The ionization potential of xenon is 12.1 eV (see Table 9.4), so the exhaust power is 0.97% higher than the thrust power. The total thermal power that must be radiated is then

$$P_H = 28{,}693 \text{ W} - (0.630)(28{,}693 \text{ W})(1 + 0.0097)$$
$$= 10{,}108 \text{ W}$$

To be conservative, we increase the specific mass of the radiators quoted in Fig. 9.40 from 0.4 kg/kW to 4 kg/kW. The radiator mass is then

- $m_H = (4 \text{ kg/kW})(10.108 \text{ kW}) = 40 \text{ kg}$

9.8.6 Propellant Tank and Power System Masses

Before finally selecting the solar-array technology, we find the inert mass of the propellant system (m_{pr}) using the formula from Table 9.12 (even though we are below its explicit range of validity). With $m_{prop} = 828$ kg,

- $m_{pr} = 52 + 0.075(828) + 0.154(828)^{2/3} = 128$ kg

From Table 9.10, the lowest mass of solar array results from using planar, flexible-blanket technology. With a beginning-of-life value of specific power of 138 W/kg, degraded by 3% by radiation damage in the Van Allen belts, the end-of-life value for specific mass is 0.00747 kg/W. The mass of the solar array is then

- *mass of solar array* = $(0.00747 \text{ kg/W})(28{,}693 \text{ W}) = 214$ kg

9.8.7 Assess System Performance

The propulsion system inert mass is

$$m_{inert} = (m_{th} + m_{pp}) + m_{pr} + m_H + m_s$$
$$= (296 \text{ kg}) + 128 \text{ kg} + 40 \text{ kg} + 214 \text{ kg}$$
$$= 678 \text{ kg}$$

The mass of the mission payload is 3387 kg, so the total inert mass of components is 4065 kg. The total initial mass, including propellant, is then 4893 kg. This is less than the specified initial mass of 4914 kg, which allows us about 3% (inert mass) margin.

9.8.8 Mission Constraints

Although the present calculation is successful within the initially specified parameters of the mission, the solar array is rather large. The flexible-blanket technology selected has a beginning-of-life power density (Table 9.10) of 140 W/m^2. For the calculated power of 28,693 W, and assuming a 3% degradation to end-of-life, the necessary area is 205 m^2. Deployed as two "wings," each 3 meters wide, the array extends almost 35 m, which is rather long. Still, for a mission to create a high-power microwave transmitter, the present design might be quite attractive.

To reduce the array area and account for possible deficiencies in component performance, we can repeat the design calculation with higher values of v_e^* and τ. Higher exhaust speed means higher ion-thruster efficiency and low inert mass, so the payload mass fraction maximizes at higher values of v_e^* than indicated by the curves in Fig. 9.3b, which assume constant efficiency. Longer thrust times reduce the mission power and directly reduce the size of the solar array.

9.8.9 Evaluate the System

The trade-off between thrust duration and array area demands an economic analysis of the entire mission. Indeed, we could reconsider the cost of each subsystem expressed in terms of mass in the preceding discussion. We must then combine these unit costs with operating costs (in the appropriate budget period) to achieve the best system within budget constraints. Using the same economic ground rules and critical technical specifications, we can then select the best propulsion system.

References

Andrenucci, M., et al. 1991. Scale Effects on the Performance of MPD Thrusters. Paper 91-123 presented at the 22nd International Electric Propulsion Conference, Viareggio.

Balaam, P. and M.M. Micci. 1991. Performance Measurements of a Resonant Cavity Electrothermal Thruster. Paper 91-031 presented at the 22nd International Electric Propulsion Conference, Viareggio.

Beattie, J.R., J.N. Matossian, and R.R. Robson. 1990. Status of Xenon Ion Propulsion Technology. *J. Propulsion and Power.* 6:145.

Bird, G.A. 1994. *Molecular Gas Dynamics and the Direct Simulation of Gas Flows.* Oxford: Clarendon Press.

Bober, A.S. et al. 1991. State of Works on Electrical Thrusters in the USSR. Paper 91-003 presented at the 22nd International Electric Propulsion Conference, Viareggio.

Bourque, R.F., et al. 1992. "Pulsed Plasmoid Thruster." in *Proc. of the Nuclear Electric Propulsion Workshop.* JPL D-9512, Vol. 1. Pasadena, CA: Jet Propulsion Laboratory.

Bugrova, A.I., et al. 1991. Physical Processes and Characteristics of Stationary Plasma Thrusters with Closed Electron Drift. Paper 91-079 presented at the 22nd International Electric Propulsion Conference, Viareggio.

Burton, R.L., et al. 1990. Experiments on a Repetitively Pulsed Electrothermal Thruster. *J. Propulsion and Power.* 6 (2): 139.

Cassady, R.J., E.J. Britt, and R.D. Meya. 1987. Performance Testing of a Lightweight 30 kW Arcjet Power Conditioning Unit. AIAA Preprint 87-1085. Washington, DC: American Institute of Aeronautics and Astronautics.

Choueiri, E.Y., A.J. Kelly, and R.G. Jahn. 1991. Current-Driven Plasma Acceleration Versus Current-Driven Dissipation. Part II. Electromagnetic Wave Stability Theory and Experiments. Paper 91-100 presented at the 22nd International Electric Propulsion Conference, Viareggio.

Cobine, J.D. 1958. *Gaseous Conductors.* New York: Dover.

Cruciani, G. et al. 1991. Advanced Laboratory Model, 1 kW-Class Arcjet Engine Testing. Paper 91-044 presented at the 22nd International Electric Propulsion Conference, Viareggio.

Dailey, C.L. and R.H. Lovberg. 1979. "Large Diameter Inductive Plasma Thrusters." AIAA Preprint 79-2093. Washington, DC: American Institute of Aeronautics and Astronautics. Also in 1993. Pulsed Inductive Thruster Performance Data Base for Megawatt-Class Engine Applications. AIP Proc. No. 271 presented at the *10th Symposium on Space Nuclear Power and Propulsion.* New York: American Institute of Physics.

Deininger, W.D. and R.J. Vondra. 1987. Design and Performance of an Arcjet Nuclear Electric Propulsion System for a Mid-1990s Reference Mission. AIAA Preprint 87-1037. Washington, DC: American Institute of Aeronautics and Astronautics.

DiVincenzi, D.L., et al. 1990. Elite Systems Analysis. AIAA Preprint 90-2530. Washington, DC: American Institute of Aeronautics and Astronautics.

Fearn, D.G., A.R. Martin, and A. Bond. "The Development Status of the UK-10 and UK-25 Ion Thruster Systems." IAF-87-293, October 1987.

Fradkin, D.B. 1973. Analysis of Acceleration Mechanisms and Performance of an Applied Field MPD Arcjet. Ph.D. diss. Work done under Government contract: Report No. 1088-T. Dept. of Aerospace and Mechanical Sciences, Princeton University, Princeton, NJ.

Frese, M.H. 1986. *MACH2: A Two-Dimensional Magnetohydrodynamics Simulation Code for Complex Experimental Configurations.* Report AMRC-R-874. Albuquerque, NM: Mission Research Corporation.

Gilland, J.H., R.M. Myers, and M.J. Patterson. 1990. *Multimegawatt Electric Propulsion System Design Considerations.* AIAA Preprint 90-2552. Washington, DC: American Institute of Aeronautics and Astronautics. Also, NASA Technical Memorandum 105152. Washington, DC: National Aeronautics and Space Administration.

Grishin, S.D., and L.V. Leskov. 1989. *Electrical Rocket Engines of Space Vehicles.* (Elektricheskiye Raketnyye Dvigateli Kosmicheskikh Apparatov). Moscow: Mashinostroyeniye.

Groh, K.H., et al. Recent Performance Results of the RIT 15 Auxiliary Propulsion Engine. Paper 91-082 presented at the 22nd International Electric Propulsion Conference, Viareggio.

Hill, P.G. and C.R. Peterson. 1970. *Mechanics and Thermodynamics of Propulsion.* Reading, MA: Addison-Wesley.

Jahn, R.G. 1968. *Physics of Electric Propulsion.* New York: McGraw-Hill, Inc.

Kerslake, W.R. and L.R. Ignaczak. 1993. Development and Flight History of the SERT II Spacecraft. *J. Spacecraft and Rockets.* 30:258.

Komurasaki, K., M. Hirakawa, and Y. Arakawa. 1991. Plasma Acceleration Process in a Hall-Current Thruster. Paper 91-078 presented at the 22nd International Electric Propulsion Conference, Viareggio.

Krall, N.A. and A.W. Trivelpiece. 1973. *Principles of Plasma Physics.* New York: McGraw-Hill.

Krohn, V.E., Jr. 1963. "Glycerol Droplets for Electrostatic Propulsion." in *Electric Propulsion Development*. E. Stuhlinger, ed. Progress in Astronautics and Rocketry. 9:435. New York: Academic Press.

Larson, Wiley J. and James R. Wertz, eds. 1992. *Space Mission Analysis and Design*. 2nd ed. Torrance, CA and Norwell, MA: Microcosm, Inc. and Kluwer Academic Publishers.

Leifer, S.D. and W. A. Saunders. 1991. Electrostatic Propulsion Using C_{60} Molecules. Paper 91-154 presented at the 22nd International Electric Propulsion Conference, Viareggio.

Liepmann, H.W. and A. Roshko. 1956. *Elements of Gasdynamics*. New York: Wiley.

Mantenieks, M. 1990. NASA / University Electric Propulsion Design Notes. Cleveland, OH: NASA Lewis Research Center. Personal communication.

McWhirter, R.W.P. 1965. "Spectral Intensities." in *Plasma Diagnostic Techniques*. R.H. Huddlestone and S.L. Leonard, eds. New York: Academic Press.

Meserole, J.S. 1993. Launch Costs to GEO Using Solar-Powered Orbit Transfer Vehicles. AIAA Preprint 93-2219. Washington, DC: American Institute of Aeronautics and Astronautics.

Mikellides, P.G., P.J. Turchi, and N.F. Roderick. 1995. Theoretical Model of an Applied-Field MPD Thruster. AIAA Preprint 95-2646. Washington, DC: American Institute of Aeronautics and Astronautics

Miller, T.M. and G.B. Seaworth. 1993. An Approach to System Optimization for Solar Electric Orbital Transfer Vehicles. AIAA Preprint 93-2222. Washington, DC: American Institute of Aeronautics and Astronautics.

Mitchner, M. and C.H. Kruger. 1973. *Partially Ionized Gases*. New York: Wiley.

Morse, P.M. 1964. *Thermal Physics*. New York: W.A. Benjamin.

Myers, R.M. 1991b. *Applied-Field MPD Thruster Geometry Effects*. NASA Contractor Report 187163. AIAA Preprint 91-2342. Washington, DC: American Institute of Aeronautics and Astronautics.

Myers, R.M. 1993. Electromagnetic Propulsion for Spacecraft. AIAA Preprint 93-1086. Washington, DC: American Institute of Aeronautics and Astronautics.

Myers, R.M., M.A. Mantenieks, and M.R. LaPointe. 1991. MPD Thruster Technology. AIAA Preprint 91-3568. Washington, DC: American Institute of Aeronautics and Astronautics.

Olin Aerospace Company. 1993a. *Electrothermal Hydrazine Thruster*. MR-501. Redmond, WA: Olin Aerospace Company.

Olin Aerospace Company. 1993b. *Low Power Arcjet System*. MR-507, 508. Redmond, WA: Olin Aerospace Company.

Palaszewski, B. 1989. Lunar Transfer Vehicle Design Issues with Electric Propulsion Systems. AIAA Preprint 89-2375. Washington, DC: American Institute of Aeronautics and Astronautics.

Pivirotto, T.J. and K.D. Goodfellow. 1991. An Experimental and Numerical Investigation of an Applied-Field Magnetoplasmadynamic Space Propulsion Engine. IEPC91-074 presented at the 22nd International Electric Propulsion Conference. Viareggio, Italy.

Pollard, J.E. et al. 1993. Electric Propulsion Flight Experience and Technology Readiness.

AIAA Preprint 93-2221. Washington, DC: American Institute of Aeronautics and Astronautics.

Reitz, J.R. and F.J. Milford. 1960. *Foundations of Electromagnetic Theory.* Reading, MA: Addison-Wesley.

Salhi, A. and P.J. Turchi. 1993. Theoretical Modeling of Orificed, Hollow Cathode Discharges. Paper 93-024 presented at the 23rd International Electric Propulsion Conference, Seattle.

Sankovic, J.M., et al. 1991. Hydrogen Arcjet Technology. Paper 91-018 presented at the 22nd International Electric Propulsion Conference, Viareggio.

Sercel, J.C. 1993. Multimegawatt Nuclear Electric Propulsion: First Order System Design and Performance Evaluation. JPL D-3898. Pasadena, CA: Jet Propulsion Laboratory.

Smith, W.W., et al. 1991. Low Power Hydrazine Arcjet Flight Qualification. Paper 91-148 presented at the 22nd International Electric Propulsion Conference, Viareggio.

Solbes, A. and R.J. Vondra. 1973. Performance Study of a Solid Fuel-Pulsed Electric Microthruster. *J. Spacecraft and Rockets.* 10 (6): 406.

Sovey, J.S. and M.A. Mantenieks. 1988. *Performance and Lifetime Assessment of MPD Arc Thruster Technology.* NASA Technical Memorandum 101293. Washington, DC: National Aeronautics and Space Administration. Also AIAA Preprint 88-3211. Washington, DC: American Institute of Aeronautics and Astronautics

Spitzer, L., Jr. 1964. *Physics of Fully Ionized Gases.* New York: Wiley.

Stocky, J.F. and R.J. Vondra. 1994. Personal communications.

Stuhlinger, E. 1964. *Ion Propulsion for Space Flight.* New York: McGraw-Hill, Inc.

Turchi, P.J. 1982. An Electric Propulsion Development Strategy Based on the Pulsed Plasma Microthruster. AIAA Preprint 82-1901. Washington, DC: American Institute of Aeronautics and Astronautics.

Turchi, P.J. 1986. Critical Speed and Voltage-Current Characteristics in Self-Field Plasma Thrusters. *J. Propulsion and Power.* 2:398.

Turchi, P.J., C.N. Boyer, and J.F. Davis. 1984. "Multi-Stage Plasma Propulsion." Paper IEPC 84-51 in *Proc. of 17th International Electric Propulsion Conference,* Tokyo.

Vondra, R.J. 1982. One Millipound Pulsed Plasma Thruster Development. AIAA Preprint 82-1877. Washington, DC: American Institute of Aeronautics and Astronautics.

Vondra, R.J. 1984. Personal communication.

Vondra, R.J., K. Thomassen, and A. Solbes. 1970. Analysis of Solid Teflon Pulsed Plasma Thruster. *J. Spacecraft and Rockets.* 7:1402.

CHAPTER 10
MISSION DESIGN CASE STUDY

David Baker, *DAB Engineering, Inc.*

10.1 Define Mission Requirements
10.2 Develop Criteria to Evaluate and Select a System
10.3 Develop Alternative Mission Concepts
10.4 Define the Vehicle System and Select Potential Technologies
10.5 Develop Preliminary Designs for the Propulsion System
10.6 Assess Designs and Configurations
10.7 Compare Designs and Choose the Best Option

This chapter illustrates how to design a propulsion system for a realistic mission. We follow the steps listed in Table 1.3 to provide a systematic approach for preliminary design of propulsion systems.

Designing is interesting because there are many different ways to achieve a workable solution. Here, we describe four approaches to design. The first approach specifies what must be done (defines requirements) and brings together existing systems to get the job done (satisfies the requirements). This is called top-down or spiral design. Sometimes, we need to develop new systems in order to complete a top-down design. The second approach starts with a past solution to the problem and improves the design incrementally through small changes. This is the evolutionary approach to design. The third approach advocates using a particular technology to solve other designers' problems. This bottom-up approach succeeds only when technologists convince designers to use their technology. The fourth approach randomly mixes and matches different technologies without a focused mission or goal. This approach is usually not intended to solve a particular problem. Designers must apply useful solutions from basic research to their particular design problems. In this chapter we employ the top-down approach to design introduced in Sec. 1.2.

Design should not be isolated. Many decision criteria and constraints are not engineering issues; rather, they involve business development and marketing disciplines. Even if our organization is not commercially based, we must always understand the issues of the agency paying for the device we are charged with designing or developing. Following this process requires discipline on our part, especially in defining and sticking with the figure-of-merit criteria.

Design Solutions

Designing a propulsion system (or any other system) has many similarities to the mathematical process of constrained optimization. We can use the concepts and nomenclature of constrained optimization to clarify our thoughts regarding what is good about a system and what is mandatory. In constrained optimization, we define a few terms common to our propulsion-design process. A *feasible* solution is one which meets all the constraints and limitations (performance, cost, reliability). In other words, a feasible solution is one that can work—it may not be the best solution, but it meets all the criteria which *must* be met. A "best" design is called an *optimal* solution—one that scores highest with our figure of merit. But an optimal solution may not meet all the constraints, so it may not be feasible. Obviously, we are searching for the *optimal feasible* solution—the one that meets all the constraints and limitations and does it better than any other feasible solution. We must find a solution feasible before assessing how optimal it is. Working with a design or grading its performance does not make sense when the solution is not feasible.

We must temper our search for optimal designs because it uses up the very resources (usually money) we are trying to save. How much sense does it make to have a team spend one month trying to add a few kilograms of payload to an existing rocket system if the value of that payload is less than the team's labor expenses? Figure 10.1 shows this point graphically. The asymptote of a design which is 100% optimal is shown across the top. As we spend time and effort finding better solutions (moving from left to right), we are consuming value. In Fig. 10.1, the heavy line is the value added by optimizing less its cost. We should stop optimizing when we reach the maximum customer value—the point at which we are consuming and adding value at the same rate. Unfortunately, we can never reach the ultimate design. The more it costs to search for an optimum, the less optimal the final system can afford to be.

All too often designers want to look impressive by using their full-blown analysis tools at every opportunity. This inclination slows down initial design work and confuses the bigger issues. The basis of design is resolving all of the issues at a certain level of fidelity before doing more work. A formal review at the end of each cycle reinforces this approach by guaranteeing that all issues of equal importance are satisfactorily resolved before allowing the team to go the next level. For example, NASA has the following reviews for a typical campaign: preliminary

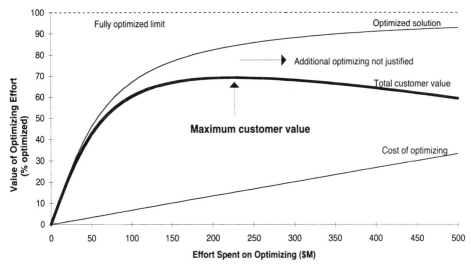

Fig. 10.1. Optimization Model with Resource Consumption. As the optimizing effort increases, the marginal return rate drops, but the cost does not. At some point (maximum value), the cost does not justify further optimization.

requirements review (PRR), conceptual design review (CoDR), preliminary design review (PDR), critical design review (CDR), and flight readiness review (FRR). Each of these reviews ends a cycle of design or build, and ensures all aspects of the design are ready for the next cycle.

When designing a system, we strive to create models of real systems in our heads, in computer simulations, and when testing physical models. Any design problem, in reality, is infinitely complex. Think of Newton's falling apple. Newton's breakthrough identified the most relevant force on the apple—gravity. But additional forces act on falling apples—aerodynamic and buoyancy forces caused by the air, centripetal forces from Earth's spinning, gravitational perturbations from the Moon, Sun, and even Newton himself! Ultimately, we must model the entire universe to understand the motion of any given part of it—obviously, an impossible task. Hence, we must draw a line somewhere and resign ourselves to understanding only the more significant characteristics of the system.

Each cycle of the design process increases its level of detail. But complexity from one cycle to the next increases geometrically. At first, a designer may choose to model velocity only in terms of the ideal rocket equation, which consists of mass fraction and specific impulse. To move beyond this level of analysis multiplies by several times the number of things we must look at. Issues such as gravity losses, staging (or shedding of inert mass), propellant storage, trajectory alternatives, and engine jet power all have similar effects on the velocity capability of the vehicle. One can see that each of these seven issues represents a dimension in a constrained-optimization problem. Adding up these independent variables means our search for feasible and optimal velocity capability has seven dimensions. If the average number of options in each dimension is five, we are faced with 5^7 or 78,125 configurations to study!

When we are searching for the best solution in many dimensions, a comprehensive understanding of the design space is impossible. We are resigned to searching for better solutions by making limited changes within one dimension, such as a change from a single-stage to a two-stage design. We do not know where the optimum is because we cannot evaluate all of the potential configurations. All we know is the slope of the terrain under our feet. This approach is formally called sensitivity analysis. System designers demand sensitivity information from their team so they can get a feel for the local slope of the terrain. In optimization theory this technique is called generating the local gradient. By moving in the direction of the gradient, we can incrementally improve the design. A good systems engineer strives to keep the slope at the same angle in all directions. This is just another way of saying all issues in the design are in balance with each other and no "big swingers" remain to be optimized. We must keep in mind that following the gradient is inherently myopic, so it can lead us to a false summit, without our ever knowing of higher summits.

10.1 Define Mission Requirements

Let us move through the design process with "real" mission requirements. In step one, we must identify the objective of this system. Often it comes from an outside source such as the customer or the project engineer. Usually, we must help finish the mission-definition statement because our customer or project engineer may not know everything needed to begin. Consider a customer who simply says: "Create the propulsion system for a GSO (geostationary orbit) satellite mission." At first blush, this seems straightforward enough. After some thought, however, we recognize that we do not know if we must also design the propulsion systems for stationkeeping and attitude control. Further, we know some launch vehicles place their payloads into low-Earth orbit (LEO) and others into geosynchronous transfer orbit (GTO). Which launch vehicle do we use? How much time can the transfer take? These are the sorts of things that can get a project into trouble.

Mission Requirements

We need to define what is to be done (the objective) along with the associated constraints. After much discussion with the project engineer, our requirements become clearer:

> *Create a propulsion system to transfer a 2.6 t* dry mass) spacecraft from LEO (400 km circular at 28.5°) to GSO (geostationary orbit: 35,786 km circular altitude at*

---

* Here, t refers to a metric ton (or tonne), which equals 1000 kg, or about 2200 lb.

zero inclination) at a longitude of 165° W; to provide the satellite with an attitude-control system that dumps its accumulated angular momentum (at 1200 N·m·s per year); and to provide stationkeeping for at least eight years. Adding up to four years of on-orbit life is worth $50M per year to the project. The launch vehicle can lift a total of 12 t into LEO, so the allowable system mass is 9.4 t (12 – 2.6). The diameter of the launch-vehicle fairing is 3 m, and its length is 8 m. The satellite takes up 5 m of that length, leaving a 3-m length for the propulsion system.

Delineate the Political, Economic, and Institutional Considerations

This commercial project is quite straightforward in terms of political and institutional considerations. We need not concern ourselves with alternative agendas or sensitive issues. For government-sponsored projects, this step becomes significant. If a government agency is sponsoring advanced technology development, the objective is focused less on economic return and more on maturing a technology or technologies. If the mission is for national security, we use the latest technology to create a system with more capability than ever before and, hopefully, with more capability than the adversary has. Many government projects have turf issues regarding who does what. In selecting our figure of merit in the coming steps, we should take into account whose funding stream we may be threatening or supporting.

10.2 Develop Criteria to Evaluate and Select a System

In working with the customer, as well as the marketing and financial-planning teams, we learn that cost is the most important characteristic of the propulsion design. But they have told us a system failure represents huge cost in the form of lost income, so reliability is critical, too. Once the satellite is launched, the mission-operations costs, lost revenue, and cost of money mean that each hour added to the time before revenue can be generated costs the project $5,000. Finally, every year of operation, beyond the first eight, is worth $50M per year in increased revenue. Yet, from a financial perspective, once the first eight years have earned revenue, additional years of operation have a diminished value because of the time value of money and because of improving capability in replacement satellite designs and reduced launch costs.

With this information we can build an equation for our preliminary figure of merit. Because this is a commercial project, we use a money-based figure of merit:

Figure of Merit (*FOM*) = Flight-system cost
   + Reflight cost × Probability of failure
   + Cost of delayed operation × Delivery time
   − Extended lifetime profits × (Lifetime − 8 years).

We want to minimize the figure of merit in this case. The first term is the flight-system cost (C_{fs}). This amount represents the cost of the propulsion system and is made up of design, testing, integration into the spacecraft, launch support, and flight support. The second term is the cost to the customer if the propulsion system fails, which, in most cases, means a complete loss of the mission from that point forward. We determine the cost of losing the satellite and then multiply it by the propulsion system's probability of failure (*POF*). For example, if the cost of losing the satellite is $1000M in lost revenue and the propulsion system's probability of failure is one chance in one hundred (0.01), this term gives a cost of failure of $10M. The cost to deliver the satellite to its operational orbit is simply the time it takes (T_d) times $5,000 per hour. Finally, extended on-orbit life (*Life*) is worth $50M per year—up to a total of 12 years. Using the actual numbers, we restate the figure of merit (*FOM*):

$$FOM = C_{fs} + \$1B \times POF + \$5000/\text{hr} \times T_d - \$50M/\text{yr} \times (Life - 8 \text{ yrs}) \quad (10.1)$$

Note that we have a single figure of merit which combines four different issues. Four separate figures of merit work badly because what might improve one would hurt the others, and we would not have a clear means of comparing them. This figure of merit combines terms so all comparisons are "apples-to-apples," or, more specifically, "dollars-to-dollars." Creating equivalency between figures of merit is extremely difficult. We are often tempted to track separately all the different attributes of each system. This approach can be effective, but we must eventually pick just one system. We must synthesize all of these issues, perhaps subconsciously, to create a "gut feeling" about the best overall system.

10.3 Develop Alternative Mission Concepts

Examine All Past Solutions to Similar Objective(s) and Constraints

We examine all existing upper stages to see if any can meet our requirements. Table 10.1 lists the world's current and planned inventory of upper stages of the major launch vehicles. (To date we have no space-based vehicles to consider for orbit transfer.)

Decide Best Previous Solution for Our Mission

As we can see from Table 10.1, none of the existing upper stages is able to fit into our tight space and still provide the impulse needed. Hence, we cannot pick a "best" previous solution because none is even feasible for our problem. Studying the data of Table 10.1 gives us some useful information. Aside from noticing we have very demanding requirements, we see our biggest issues are volumetrics and Δv for the given masses. TOS comes very close to fitting into our space and weighs only slightly more than we can allow, but it does not have quite enough energy in its solid propellant to give us our required Δv. The third stage of the CZ-3 has just the right performance, but it grossly exceeds our volumetric limitations due to the low-density hydrogen fuel.

10.3 Develop Alternative Mission Concepts

Table 10.1. Data for Existing Propulsion Options. This table lists existing upper stages that could apply, along with their key parameters. No existing systems completely meet the requirements. Shaded boxes show the parameters that do not meet the requirements.

Name	Propulsion Type	Launcher	Propulsion System Mass (t)	Propulsion System Length × Diameter (m)	Δv Available to Transfer a 2.6 t Payload (km/s)
PAM-DII	solid (Star-63D)	Shuttle/Delta	3.5	2.0 × 1.6	2.1
Transtage	bi-prop	Titan	13.5	4.5 × 3.0	3.1
TOS	solid (Orbus-21)	Shuttle/Titan	10.8	3.3 × 3.4	3.7
IUS	solid (2-stage)	Shuttle/Titan	14.86	5.2 × 2.9	4.0
Centaur	H_2/O_2	Titan-3, 4	23.8	9 × 4.3	6.3
H-10	H_2/O_2	Ariane-4	12.1	9.9 × 2.6	5.8
L-7/EPS	bi-prop	Ariane-5	9.3	4.5 × 5.4	2.9
H-2 Stg-2	H_2/O_2	H-2	16.8	10 × 4	5.6
TDK-DM	kerosene/O_2	Proton/Zenit	19.95	5.8 × 3.7	5.0
CZ-3	H_2/O_2	3rd stage	10.5	7.5 × 2.25	4.3
Requirement			9.4	3 × 3	4.2

At this point, we have at least three options:
- Develop a new system
- Go back and challenge the requirements
- Modify an existing system

To investigate the impact of a new system, we want to look at this option. However, once we have completed this analysis (Secs. 10.3 through 10.6), we need to remember that we consciously chose the new technology route. It is then appropriate to ask whether one of the other options would be a better choice (Sec. 10.7).

We have been assuming that a single stage can do the perigee and apogee burns. If we break these burns into two stages, we may be able to improve our system mass and performance. We explore this option further in the next section.

Develop Alternative Trajectory Concepts

To get from LEO to GSO, we can use four classes of trajectory transfer:

1. One or more perigee burns followed by one or more apogee burns, which we refer to as segmented Hohmann transfers.

2. Three-burn, "super-GSO" transfer, in which the plane change is performed higher than GSO altitude to reduce the Δv needed for the plane change.

3. A continuously thrusting spiral trajectory for low-thrust rocket technology.

4. A combination of options 1 and 3, in which we use a high-thrust chemical propulsion system and a Hohmann transfer to boost us out of the deepest part of Earth's gravity well and then switch to a low-thrust/high-I_{sp} propulsion system and spiral to our destination orbit. The high-thrust Hohmann transfer takes the vehicle to a higher altitude, reducing eclipse time, drag, and atmospheric effects. The minimum altitude of the transfer from high-thrust to low-thrust propulsion is approximately 5000 km.

Space maneuvers always trade off flight time and propellant usage. For a transfer from LEO to GSO, we can transfer faster or slower than a Hohmann transfer. To speed up the transfer, we can increase the semimajor axis of the transfer ellipse over that of the Hohmann ellipse. To slow down the transfer, we can put in multiple burns. Multiple burns reduce Δv only if the total perigee burn time exceeds about 10 minutes or the apogee burn time exceeds about 45 minutes.*

For design, we choose four reference trajectories:

1. A two-burn Hohmann transfer with all the plane change performed at apogee (the standard approach). Figure 10.2 shows this trajectory from two different views.

2. "Segmented Hohmann" transfer using two perigee burns and a single apogee burn, with some plane change in each perigee burn and the rest in the apogee burn. Figure 10.3 shows this trajectory from two different perspectives. The plane change in the transfer orbit is barely noticeable because it is only about 2° different from the parking orbit.

3. A spiral transfer using continuous thrust, with all the plane change occurring after reaching GSO altitude.

4. A two-stage transfer. The first stage is a high-thrust Hohmann transfer, and the second stage is a low-thrust spiral. Staging occurs at a 5000-km altitude. All of the plane change is performed after reaching GSO altitude.

---

* These "rules of thumb" are based on "gravity losses," as discussed in Chap. 2. Impulsive (burn time = 0) maneuvers show negligible gravity losses. However, real, finite burns always incur losses but do not become excessive if they are short. Figure 10.6 shows that, for our requirements, a thrust-to-weight ratio greater than 0.3 allows us to assume impulsive burns with accuracy.

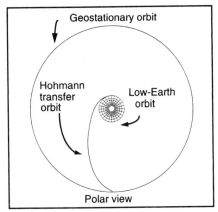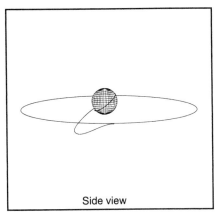

Fig. 10.2. **Trajectory Option Using a Two-Burn Hohmann Transfer.** These two views of the same trajectory show transfer from LEO to GSO. [Figures 10.2–10.4 created using DAB Orbit (Version 2.0) software.]

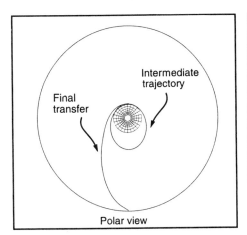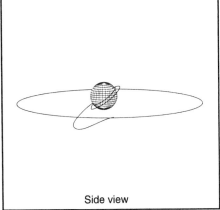

Fig. 10.3. **Trajectory Option Using a Segmented Hohmann Transfer.** These two views show configurations for the intermediate and final transfer trajectories.

We do not include a three-burn, super-GSO transfer, as shown in Fig. 10.4 below, because the relatively low inclination of 28.5° for the parking orbit causes the Δv savings in apogee plane change to be less than the additional Δv needed for the higher-energy transfer orbits. We do not know this intuitively; we must calculate it for several super-apogee altitudes to see if we can find any savings. For launches from latitude above 45°, this approach does have a benefit.

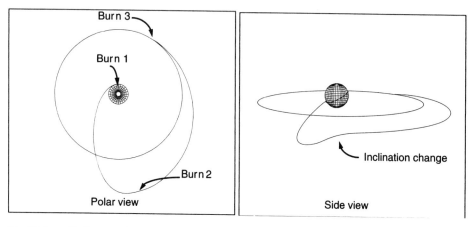

Fig. 10.4. Trajectory Option Using a Super-GEO, Three-Burn Transfer. To reduce the Δv requirement, it is sometimes best to transfer to a much higher altitude than desired to take advantage of the low apogee velocity for the plane-change maneuver. In our particular problem, this does NOT give an advantage.

Calculate Δv for Option 1

Hohmann transfers are ideal for high-thrust propulsion systems such as solid motors or liquid systems. We assume impulsive burns to compute the Δvs. For the Hohmann transfer with apogee plane change, we need to compute the velocities at different points in the three orbits before we can compute the Δvs (see Chap. 2):

Parking orbit:

$$V_{park} = \sqrt{\frac{\mu}{r_p}} = \sqrt{\frac{398,600}{400+6378}} = 7.669 \text{ km/s} \tag{10.2}$$

Geosynchronous transfer orbit (GTO):

$$a = \frac{h_a + h_p}{2} + R_e = \frac{400 + 35,786}{2} + 6378 = 24,471 \text{ km} \tag{10.3}$$

$$V_{peri} = \sqrt{\frac{2\mu}{r_p} - \frac{\mu}{a}} = \sqrt{\frac{2(398,600)}{6378+400} - \frac{398,600}{24,471}} = 10.066 \text{ km/s} \tag{10.4}$$

$$V_{apo} = \sqrt{\frac{2\mu}{r_a} - \frac{\mu}{a}} = \sqrt{\frac{2(398,600)}{6378+35,786} - \frac{398,600}{24,471}} = 1.618 \text{ km/s} \tag{10.5}$$

Geostationary orbit (GSO):

$$V_{GSO} = \sqrt{\frac{\mu}{r_f}} = \sqrt{\frac{398,600}{35,786+6378}} = 3.075 \text{ km/s} \quad (10.6)$$

The next step is to calculate the difference in velocities. For the apogee burn, we cannot simply subtract the velocities because they do not lie in the same plane. Remember, our transfer orbit is inclined at 28.5° and our final orbit is equatorial, so the apogee maneuver must carry out the plane change while it increases velocity.

$$\Delta V_2 = \sqrt{\left(V_{GSO}^2 + V_{apo}^2\right) - 2V_{GSO}V_{apo}\cos(28.5)} = 1.824 \text{ km/s} \quad (10.7)$$

Notice from Fig. 10.5 how efficient it is to do the 28.5° plane change when it combines with the velocity increase during the apogee burn. If we simply subtract the GTO apogee speed from the GSO circular speed, we require a Δv of 1.457 km/s. Adding in the plane change increases the Δv by only 367 m/s. Because all of the plane change is done at apogee, the perigee burn is planar and is simply

$$\Delta V_1 = V_{peri} - V_{park} = 2.397 \text{ km/s} \quad (10.8)$$

$$\Delta V_{opt1} = \Delta V_1 + \Delta V_2 = 4.221 \text{ km/s} \quad (10.9)$$

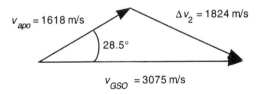

Fig. 10.5. Geometry for the Out-Of-Plane Apogee Burn. This figure shows the velocity-vector diagram for a plane-angle change done with the apogee burn.

Calculate Δv for Option 2

Performing two perigee burns with some plane change in each and a single apogee burn with the remaining plane change is ideal for low-thrust chemical propulsion systems with initial thrust-to-weights from 0.1 to 0.3 (see Fig. 10.7). To calculate the Δvs, we assume impulsive maneuvers as we did with option 1. To determine the optimal amount of plane-change for each burn, we assume the optimal inclination of the ultimate GTO ellipse is the same whether we use one or two perigee burns. This assumption allows us to optimize Δv as a function of only one parameter—the GTO transfer ellipse's inclination. We do this by building a spreadsheet to compute the Δvs at perigee and apogee, sum both Δvs, and vary GTO inclination until we find the smallest possible combined Δv. This analysis gives an optimal transfer inclination of 26.3° and reduces the overall Δv by 24 m/s. We compute the apogee Δv in the same manner as option 1 except for using 26.3° instead of 28.5°.

$$\Delta V_3 = \sqrt{\left(v_{GSO}^2 + v_{apo}^2\right) - 2v_{GSO}v_{apo}\cos(26.3)} = 1.776 \text{ km/s} \qquad (10.10)$$

To compute the perigee burn Δv, we first assume a single burn.

$$\Delta V_{1\&2} = \sqrt{\left(v_{park}^2 - v_{peri}^2\right) - 2v_{park}v_{peri}\cos(2.2)} = 2.421 \text{ km/s} \qquad (10.11)$$

$$\Delta V_{opt2} = \Delta V_{1\&2} + \Delta V_3 = 4.197 \text{ km/s} \qquad (10.12)$$

To get the Δvs for the two-perigee burn, we simply break $\Delta V_{1\&2}$ into two equal impulses of 1.211 km/s, which creates an intermediate orbit with a period of 2.9 hours and an apogee of 7400-km altitude. Keep in mind that the velocity savings of option 2 over option 1 comes from the 2.2° plane change at perigee, which we could just as easily have used in option 1 as well. The real savings of option 2 comes from a more involved analysis of gravity losses incurred by the vehicle during the perigee burns (see Fig. 10.6). If a vehicle has high thrust, it is more reliable to execute a single perigee burn.

Calculate Δv for Option 3

For propulsion systems with very low thrust, such as electric thrusters, impulsive computation of the Δv does not correctly represent the mission. We must use the appropriate method to estimate the Δv for spiral trajectories. From Sec. 2.7.2 we find the planar Δv to change from one circular orbit to another is

$$\Delta V_{spiral} = |\, V_{park} - V_{GSO}\,| = 7.669 - 3.075 = 4.594 \text{ km/s} \qquad (10.13)$$

Using Eq. (2.36), we must add 1.514 km/s to the planar transfer Δv for the plane-change Δv we do after reaching 35,786-km altitude. Hence, the total Δv for low-thrust transfers from LEO to GSO is

$$\Delta V_{opt3} = \Delta V_{spiral} + \Delta V_{plane\ change} = 6.108 \text{ km/s} \qquad (10.14)$$

Just as with the impulsive case, we can reduce this Δv by combining the tasks of raising the orbit and changing inclination. One can imagine an optimal transfer that would have the satellite begin sinusoidal yawing (one cycle per orbit) as the orbit nears GSO altitude to take out the inclination as it continues to climb. This analysis is beyond our scope, so we use the 6.108 km/s worst-case value.

10.3 Develop Alternative Mission Concepts

Calculate Δv for Option 4

The first step in the combined-trajectory transfer is to compute the Hohmann transfer Δv from 400 km to 5000-km altitude with no plane change.

Park orbit:
$$V_{park} = 7.669 \text{ km/s from option 1} \tag{10.15}$$

High-thrust transfer orbit:
$$a = \frac{h_a + h_p}{2} + R_e = \frac{(400 + 5000)}{2} + 6378 = 9078 \text{ km} \tag{10.16}$$

$$V_{peri} = \sqrt{\frac{2\mu}{r_p} - \frac{\mu}{a}} = \sqrt{\frac{2(398,600)}{6378 + 400} - \frac{398,600}{9078}} = 8.585 \text{ km/s} \tag{10.17}$$

$$V_{apo} = \sqrt{\frac{2\mu}{r_a} - \frac{\mu}{a}} = \sqrt{\frac{2(398,600)}{6378 + 5000} - \frac{398,600}{9078}} = 5.114 \text{ km/s} \tag{10.18}$$

$$V_{final} = \sqrt{\frac{\mu}{r_f}} = \sqrt{\frac{398,600}{5000 + 6378}} = 5.919 \text{ km/s} \tag{10.19}$$

$$\Delta v_1 = V_{peri} - V_{park} = 0.916 \text{ km/s} \tag{10.20}$$

$$\Delta v_2 = V_{final} - V_{apo} = 0.805 \text{ km/s} \tag{10.21}$$

$$\Delta v_{total} = \Delta v_1 + \Delta v_2 = 1.721 \text{ km/s} \tag{10.22}$$

Now we calculate the Δv to spiral from 5000 km up to GSO radius at 35,786 km. As for the previous option, it is simply the difference between the circular-orbit velocities:

$$V_{start} = \sqrt{\frac{\mu}{r_{start}}} = 5.919 \text{ km/s} \tag{10.23}$$

$$V_{GSO} = 3.075 \text{ km/s} \tag{10.24}$$

$$\Delta v_{total} = V_{start} - V_{GSO} = 2.844 \text{ km/s} \tag{10.25}$$

Finally, we must add in the plane-change, as we did in option 3. We already calculated the Δv as 1.514 km/s, so our total low-thrust Δv for a spiral trajectory is

$$\Delta v_{opt4} = 4.358 \text{ km/s}$$

Calculate Δv for Stationkeeping

From Sec. 2.5 we know the east-west Δv is about 37 m/s per year and the north-south Δv is about 14.5 m/s per year. Given our longitude of 165° W, we can also determine the J_{22} resonance Δv using Eq. (2.51) at 1.5 m/s per year. Hence, the stationkeeping Δv is 53 m/s per year.

Calculate Δv for Removing Accumulated Angular Momentum

The amount of this Δv depends on the locations of the thrusters. The farther the thruster is from the center of mass, the greater its torque impulse for a given amount of propellant usage. We can determine an upper bound for the lever-arm's length of the thrusters from the center of mass based on the payload fairing envelope. The radius is 1.5 m, and the longitudinal arm's length can be about 2.5 m. Thruster arms are usually not designed to deploy to increase the moment-arm because of the unreliability of a joint in the propellant feed line, so we have to compute the Δvs based on the fixed lengths. Because we do not know the direction of the accumulated momentum vector, we must assume the worst case and use the shortest arm, which is 1.5 m. The 1200 N·m·s (see Sec. 10.1) can be removed with a total impulse of 1200 N·m·s/1.5 m = 800 N·s. Because the total impulse is equal to the propellant mass times the specific impulse, we can determine the annual propellant mass consumed only after we have selected a propulsion technology and know its I_{sp}. Table 10.2 summarizes all of our propulsion requirements.

Evaluate Gravity Loss

We have assumed that impulsive burns for the high-thrust maneuvers give us accurate Δv estimates. However, even high-thrust burns require a finite duration. What magnitude of thrust allows us to minimize gravity losses and validate our assumption?

We can use a numerical-integration analysis that allows us to compare finite-duration burns to the impulsive-burn approximation. Figure 10.6 shows the result of this type of analysis. If we require an initial thrust-to-weight ratio greater than 0.3, the gravity losses are small.

10.4 Define the Vehicle System and Select Potential Technologies

We must now select several possible technologies and see how these choices affect the size of our potential systems. We start by looking at the spacecraft's onboard propulsion system. Once we have sized the completely loaded spacecraft, we can look at the stage or stages for the propulsion system for orbit transfer.

10.4 Define the Vehicle System and Select Potential Technologies

Table 10.2. Summary of Requirements for Delivery and Stationkeeping of a LEO-GSO Satellite. This table summarizes the trajectory requirements for the four trajectory concepts analyzed. It also lists the technical requirements from Sec. 10.1.

Trajectory Option	Δv Requirements
Option 1 (Hohmann)	4.221 km/s
Perigee Δv	2.397 km/s
Apogee Δv	1.824 km/s
Option 2 (segmented Hohmann)	4.197 km/s
Perigee-1 Δv	1.2105 km/s
Perigee-2 Δv	1.2105 km/s
Apogee Δv	1.776 km/s
Option 3 low thrust (spiral)	6.108 km/s
Altitude raising Δv	4.594 km/s
Plane change Δv	1.514 km/s
Option 4 Hohmann/spiral	6.079 km/s
Hohmann Δv to 5000 km	1.721 km/s
Spiral to GEO altitude	2.844 km/s
Plane change Δv	1.514 km/s
Stationkeeping Δv	53 m/s per year
Momentum control impulse	800 N·s per year
Maximum propulsion mass (Sec. 10.1.1)	9400 kg
Payload dry mass (Sec. 10.1.1)	2600 kg
Maximum length and diameter (Sec. 10.1.1)	3 m diameter and 3 m length
Maximum volume (Sec. 10.1.1)	21.2 m^3
Required lifetime (Sec. 10.1.1)	8 years
Maximum lifetime (Sec. 10.1.1)	12 years

10.4.1 Size the On-board Propulsion System

We are given the mass of our payload (2600 kg) without a propulsion system for attitude control and stationkeeping. So, the first step in our design is to size a propulsion system to meet these requirements (see Table 10.2 for a summary of requirements). The typical approach is to design a single system that can handle both the attitude control and stationkeeping tasks. We choose to use a monopropellant hydrazine (N$_2$H$_4$) system which, again, is a typical choice. Specific impulse for these thruster types can be as high as 235 s, but, to be conservative, we reduce this value by a fairly arbitrary amount down to 220 s. In a pulsed-mode operation,

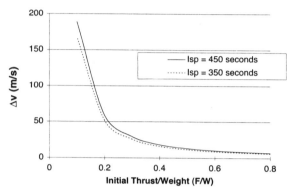

Fig. 10.6. Gravity Loss for Perigee Burn as a Function of Initial Thrust-to-Weight Ratio. This figure, developed by numerically integrating the motion equations over a finite thrust duration, shows the error (in Δv) introduced by assuming an impulsive maneuver. For F/W's greater than 0.3, the effect is negligible.

such as we see for attitude control, the specific impulse is further reduced. A conservative choice is 210 s. Table 10.3 shows the calculation for the propulsion-system sizing. All of the numbers are justified as shown in the right-hand column. We have added a 5% mass margin to the total propellant mass.

10.4.2 Size Potential Orbit-Transfer Stages

Now that we know the mass of our payload (3387 kg), we can design the propulsion system (or systems) to get the payload from low-Earth orbit to its operational orbit. We know from Table 10.1 that our toughest requirement is the envelope constraint. So our first step is to see how this constraint affects our choice of different technologies and propellants.

Our 12,000-kg initial mass and 3387-kg payload mass provide a limiting I_{sp} (from the ideal rocket equation), and we have a limiting density for the propellant based on the volume available and the necessary amount of propellant. Figures 10.7 and 10.8 graphically illustrate the feasible domain of solutions for Hohmann (high-thrust) and spiral (low-thrust) transfers, respectively. These figures relate inert-mass fraction, specific impulse, and propellant density.

We can generate the mass-fraction curves as follows:

- Choose an inert-mass fraction (we choose 0.05, 0.10, 0.15, 0.20, 0.25, and 0.30 for Fig. 10.7 and 0.25, 0.30, 0.35, 0.40, 0.45, and 0.50 for Fig. 10.8)

- Given a payload mass (3387 kg from Table 10.3) and a Δv (4.2 km/s for Fig. 10.7 and 6.2 km/s for Fig. 10.8—approximate values from Table 10.2), vary the specific impulse (I_{sp}) over an appropriate range and calculate propellant mass with Eq. (1.27) at various discrete points.

10.4 Define the Vehicle System and Select Potential Technologies

Table 10.3. Calculations of Mass for the Attitude-Control System. This table summarizes the calculations for sizing the attitude-control and stationkeeping systems on our payload.

Parameter	Result	Calculation or Source
Payload mass	2600 kg	Requirement: Table 10.2
Operational life	8 years	Requirement: Table 10.2
Stationkeeping Δv	53 m/s/yr	Table 10.2
Attitude-control impulse	800 N·s/yr	Table 10.2
Inert-mass fraction	0.2	Fig. 5.22
Stationkeeping specific impulse	220 s	Table 1.2 shows upper value of 235 s, degrade to 220 s for conservatism
Attitude-control specific impulse	210 s	Same thrusters as for stationkeeping, but pulsed mode has reduced performance
Total stationkeeping Δv	424 m/s	8×53
Total attitude control impulse	6400 N·s	8×800
Propellant mass for stationkeeping	597 kg	Eq. (1.27)
Propellant mass for attitude control	3 kg	Eq. (1.33)
Total propellant mass (includes a 5% margin)	630 kg	$1.05 \times (597 + 3)$
Propulsion system inert mass	157 kg	Eq. (1.24)
Total payload mass	**3387 kg**	$2600 + 630 + 157$

- From the allowable volume (75% of the 3 m × 3 m cylinder gives 0.75 × 21.2 m³ = 15.9 m³) and the propellant mass, determine the allowable propellant density (minimum-average propellant density) at each discrete point on the particular mass-fraction curve
- Plot the discrete points and connect them with a curve (anything above the curve is feasible, subject to other limits)

To construct the "initial-mass limit" curve, we do the following:

- Given the initial mass (12,000-kg requirement) and the payload mass (3387 kg from Table 10.3), calculate the propellant mass for a range of specific impulse (I_{sp}—on horizontal axis) using Eq. (1.19)
- Given the propellant mass and allowable volume (15.9 m³), calculate the minimum-average propellant density (on vertical axis) allowed
- Plot discrete points and connect them with a curve (anything below this curve is feasible, subject to other limits).

Notice that we assume only 75% of the volume is usable for propellant storage. The remaining 25% is assumed to be wasted space or other components such as thrusters. To use the figures, select a propulsion technology from Table 10.4 and trace lines along the correct values of I_{sp} and average propellant density. If the intersection of the two orthogonal lines is in the unshaded region, a vehicle with that inert-mass fraction is feasible.

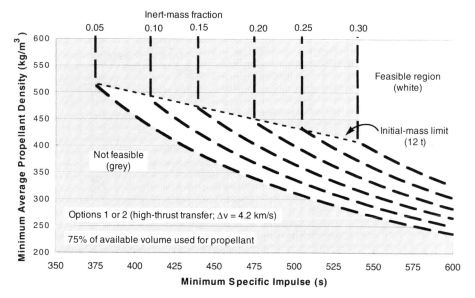

Fig. 10.7. Boundary of Solutions for High-Thrust Propulsion (Options 1 and 2). Having propellant densities greater than a given limit allows a feasible solution. The ideal rocket equation is used to generate these parametric curves. Reasonable inert-mass fractions range from 0.05 to 0.30. The initial-mass limit is based on the rocket equation with initial mass of 12 t and the Δvs as shown. The shaded region is not feasible.

Figure 10.7 shows that using a single-stage chemical rocket is probably not feasible. For example, from Table 10.4, the average density of a H_2/O_2 system is about 384 kg/m^3. If we take this number to Fig. 10.7, we find that a mass fraction of 0.05 requires a specific impulse of greater than 440 seconds. This level of technology is unlikely. Similar observations can be made with the other chemical propellant combinations. Using the numbers for electric thrusters from Table 10.4, we see in Fig. 10.8 that the single-stage ion system seems feasible but that the hydrogen-arc-jet approach seems unlikely.

The next step is to choose particular technologies and see how these choices affect our vehicle size. We use the propellants and technologies listed in Table 10.4 and look at the following concepts for propulsion systems.

10.4 Define the Vehicle System and Select Potential Technologies

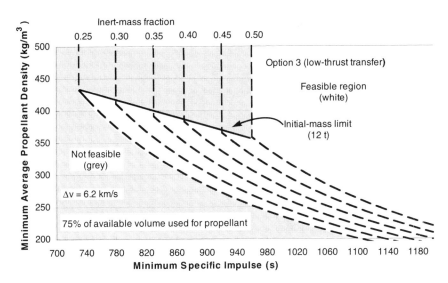

Fig. 10.8. Boundary of Solutions for Low-Thrust Propulsion (Option 3). Having propellant densities greater than a given limit allows a feasible solution. The ideal rocket equation is used to generate these parametric curves. Reasonable inert-mass fractions range from 0.25 to 0.50. The initial-mass limit is based on the rocket equation with initial mass of 12 t and the Δvs as shown. The shaded region is not feasible.

Table 10.4. Key Properties for Candidate Propellants. This table lists the specific impulse and propellant densities for feasible options (see Appendix B and Chap. 9).

Propellants (see Appendix B)	O/F	I_{sp} (s)	Fuel Density (kg/m³)	Oxidizer Density (kg/m³)	Average Density (kg/m³)
Xenon (ion)	monoprop	2000	N/A	N/A	2730
Hydrogen (arcjet)	monoprop	1000	N/A	N/A	71
N_2H_4	monoprop	220	N/A	N/A	1010
H_2/O_2	6	435	77	1140	384
CH_4/O_2	3	335	445	1140	820
RP-1/O_2	2.25	325	807	1140	1012
MMH/N_2O_4	1.85	320	1023	1447	1263
Solid (TP-H-1202)	N/A	283	N/A	N/A	1843

- An electric single stage using xenon-ion thrusters (at 25-kW and 100-kW power levels)
- An electric single stage using hydrogen arcjet thrusters (at 25-kW and 100 kW power levels)
- A chemical single stage using H_2 and O_2
- A chemical single stage using CH_4 and O_2
- A two-stage system with a CH_4 and O_2 first stage and a 25-kW xenon-ion second stage
- A two-stage system with an RP-1 and O_2 first stage and a 25-kW xenon-ion second stage
- A two-stage system with a CH_4 and O_2 first stage and a 25-kW H_2 arcjet second stage
- A two-stage system with an RP-1 and O_2 first stage and a 25-kW H_2 arcjet second stage

Notice that we are still considering the single-stage chemical systems, despite the indication that they are not feasible. The purpose here is to size these systems for comparison with the other possibilities.

Recognizing that chemical rocket mass is driven primarily by propellant mass and that electric systems are driven primarily by power levels, we need to look at two separate sizing algorithms.

Sizing Algorithm for Electric Stages

To size each of the proposed electric stages, we use a "specified-power" or "power-limited" algorithm. With this approach, we choose an allowable power level and size the system accordingly (see Chap. 9). Table 10.5 gives this algorithm.

To determine the specific mass, we use the data given in Fig. 9.41 and Table 9.10 along with much of the intelligence in Secs. 9.2 and 9.6. If we assume flexible-blanket technology for our solar arrays (specific power = 138 W/kg), a specific power of 50 W/kg seems a reasonable choice for the entire system (specific mass = 0.02 kg/W).

Table 9.3 indicates that reasonable thruster performance for hydrogen arcjets is an efficiency of 0.3 at a specific impulse of 1000 s. The discussion in Sec. 9.4 confirms that these numbers are accurate. Table 9.3 indicates that I_{sp}s over the range of 2000 s to 10,000 s are possible for xenon-ion thrusters. In addition, Table 9.11 shows us that system efficiency for ion thrusters changes as a function of specific impulse. So how do we choose a specific impulse in step 2 of Table 10.5?

To gain some insight into this question, we look at an example using the single-stage, xenon-ion concept. We can vary the choice of specific impulse over the 2000-s to 10,000-s range and plot the initial mass of the vehicle and the flight duration as a function of I_{sp}. We do this by repeating the algorithm in Table 10.5 for several points. The givens are

10.4 Define the Vehicle System and Select Potential Technologies

- $\Delta v = 1.05 \times 6108 = 6413.4$ m/s (includes a 5% margin)
- Payload mass = 3387 kg
- Source-power level = 25,000 W

Table 10.5. Electric-Stage Sizing. This algorithm allows us to size an electric-propulsion stage, if our allowable power is specified.

Step	Comments
Given	• Δv (m/s) • Payload mass (m_{pay} in kg) • Power level (P_s in W)
1. Assume a system technology	• System specific mass (β in kg/W) from Chap. 9 • This fixes stage inert and final masses (m_{inert} and m_f in kg)
2. Assume thruster technology, choose propellant, and choose I_{sp}	• I_{sp} (s) fixes propellant and initial masses (see Table 9.3) • I_{sp} and power level fix the flight or burn duration (τ)
3. Determine exhaust velocity (v_e) and system efficiency (η_T)	• $v_e = I_{sp} \times 9.81$ (m/s) • Efficiency is a function of I_{sp} (see Table 9.3 or Table 9.11)
4. Determine the stage's inert mass and final mass	• $m_{inert} = \beta \times P_s$ [Eq. (9.5)] • $m_f = m_{pay} + m_{inert}$ [Eq. (1.25)]
5. Determine the stage's initial mass (m_i) and the propellant's mass (m_{prop})	• $m_i = m_f e^{\frac{\Delta v}{v_e}}$ [Eq. (1.17)] • $m_{prop} = m_i - m_f$ [Eq. (1.18)]
6. Determine jet power (P_J), propellant mass flow rate ($\dot{m}_{prop}$), and thrust level (F)	• $P_J = \eta_T \times P_s$ [Eq. (9.4)] • $\dot{m}_{prop} = \dfrac{2 P_J}{v_e^2}$ [Eq. (9.2)] • $F = \dot{m}_{prop} \times v_e$ [Eq. (1.4)]
7. Determine burn duration (τ)	• $\tau = \dfrac{m_{prop}}{\dot{m}_{prop}}$ [Eq. (9.6)]

The results of an individual calculation are shown in the results summary (Table 10.7). Initial mass and burn duration, as a function of our choice of specific impulse, are shown in Figs. 10.9 and 10.10, respectively. We use the system efficiency model from Table 9.11. However, this model lets efficiency get above 1.0 at the higher I_{sp} levels. To avoid this problem, we use an efficiency of 0.9 (see Table 9.3) whenever the model predicts efficiencies above 0.9. This decision explains the small kink in the curve on Fig. 10.10 at $I_{sp} = 6200$ s.

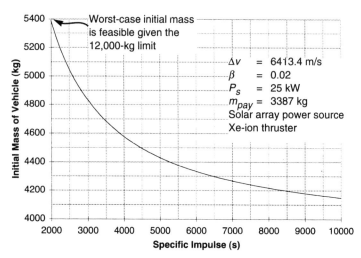

Fig. 10.9. Vehicle Initial Mass vs. Specific Impulse. As we increase I_{sp}, the initial mass decreases.

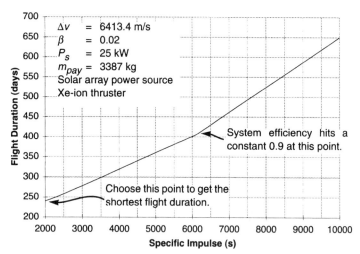

Fig. 10.10. Mission Duration vs. Specific Impulse. Although an increase of specific impulse decreases the vehicle mass (Fig. 10.9), the required burn duration increases.

From Figs. 10.9 and 10.10, we can see the initial mass of the vehicle decreases and the burn duration increases as I_{sp} increases. We also see that the minimum flight time is 240.22 days, corresponding to an initial vehicle mass of 5390 kg. The initial mass is well within our limit of 12,000 kg, so we choose the lowest specific impulse possible. This choice gives us the shortest flight time for a feasible system, so we choose the lowest possible I_{sp} for all of our cases listed above.

Figure 10.11 shows the same results as our example but directly compares the vehicle's initial mass with the flight time. If we were to use an argument for diminishing marginal returns, and were not constrained by burn time, we would probably choose a flight time of about 390 days.

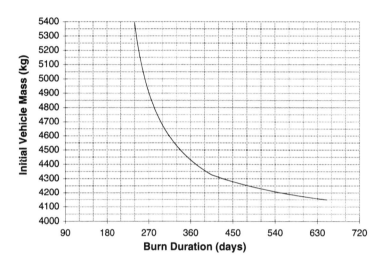

Fig. 10.11.   Initial Vehicle Mass vs. Burn Duration. As we increase the allowable burn duration, the initial mass of the vehicle decreases quickly at first and then levels out. If we are not constrained by burn duration, we would choose a flight time in the flatter section of the curve.

Algorithm for Sizing Chemical Systems

Because chemical systems have lower specific impulses than electric rockets do, they are dominated by propellant mass. For this reason, we use the concept of the inert-mass fraction, as discussed in Sec. 1.1.5, to size our chemical stages. Table 10.6 shows the sizing process.

Summary of Sizing Results

Using the algorithms described in Tables 10.5 and 10.6, we size the various vehicle concepts. Table 10.7 summarizes results for the electric system; Table 10.8 summarizes results for chemical systems. From these results, we see that the single-stage systems using liquid rockets violate the initial-mass limit. The single-stage electric systems are all feasible from a mass standpoint, but the ion rocket is substantially lighter than the arcjet system. The two-stage systems using an ion-engine second stage are feasible, but the arcjet-based systems violate the initial-

Table 10.6. Algorithm for Sizing Chemical Systems. We use the equations developed in Sec. 1.1.5.

Step	Comments
Given	• Δv (m/s) • Payload mass (m_{pay} in kg)
1. Assume propellants and specific impulse	• Use numbers from Table 10.4 • Data comes from Appendix B
2. Assume an inert-mass fraction (f_{inert})	• Use Table 5.14 • Assume 0.15 for the smaller (less propellant) first stages • Assume 0.12 for the single stages using denser propellants • Assume 0.13 for the H_2/O_2 system (lower-density propellants mean larger tanks)
3. Calculate propellant mass	• Use Eq. (1.27)
4. Calculate inert mass	• Use Eq. (1.24)
5. Calculate initial and final vehicle masses	• Use Eq. (1.26) for initial mass • Use Eq. (1.25) for final mass

mass limit. Results for the methane-based two-stage system are similar to the RP-1 results. However, because methane is cryogenic and RP-1 is not, we toss out the methane system. The following is a list of "down-selected" options:

- Single-stage—25 kW ion system (the 100-kW system is substantially larger)
- Two-stage—RP-1/O_2 first stage and a 25-kW second stage

10.5 Develop Preliminary Designs for the Propulsion System

We have chosen two possibilities that may meet our mission requirements. The first exclusively uses an electric, low-thrust trajectory. The second concept uses a two-stage combination with chemical and electric propulsion. We send the basic requirements discussed above (Tables 10.2 and 10.3) back to the individual chapters for more detailed analysis.

In Chaps. 5, 7, and 8, we look at nonelectric approaches to the first stage of the two-stage system. Chapter 5 looks at the liquid-rocket (RP-1/LOx) solution to the first stage. The result is a feasible solution. Chapters 7 and 8 look at hybrid- and nuclear-rocket applications respectively, for the first stage. In both of these studies (hybrid and nuclear), the resulting stage exceeds the length restriction, making them infeasible. The Chap. 9 case study for the ion-thruster second stage gives a

10.5 Develop Preliminary Designs for the Propulsion System

Table 10.7. Summary of Analysis Results for the Electric Concepts. These results are generated using the algorithm from Table 10.5.

	1 Stage Ion 25 kW	1 Stage Ion 100 kW	1 Stage Arcjet 25 kW	1 Stage Arcjet 100 kW	2nd Stage Ion 25 kW	2nd Stage Arcjet 25 kW
Δv + 5% (m/s)	6413.4	6413.4	6413.4	6413.4	4575.9	4575.9
Payload mass (kg)	3387	3387	3387	3387	3387	3387
Source power (W)	25,000	100,000	25,000	100,000	25,000	25,000
Specific mass (kg/W)	0.02	0.02	0.02	0.02	0.02	0.02
Specific impulse (s)	2000	2000	1000	1000	2000	1000
Exhaust velocity (m/s)	19,620	19,620	9810	9810	19,620	9810
System efficiency	0.557	0.557	0.3	0.3	0.557	0.3
Inert mass (kg)	500	2000	500	2000	500	500
Final mass (kg)	3887	5387	3887	5387	3887	3887
Initial mass (kg)	5389.9	7469.8	7473.8	10,357.9	4914.0	6197.2
Propellant mass (kg)	1502.9	2082.8	3586.8	4970.9	1027.0	2310.2
Jet power (W)	13,936.9	55,747.7	7500.0	30,000.0	13,936.9	7500.0
Propellant flow rate (kg/s)	7.24^{-5}	2.90^{-4}	1.56^{-4}	6.23^{-4}	7.24^{-5}	1.56^{-4}
Thrust (N)	1.421	5.683	1.529	6.116	1.421	1.529
Duration (s)	20,754,770	7,191,018	2.3^7	7,973,010	14,100,073	14,821,351
Duration (days)	240.2	83.2	266.3	92.3	163.2	171.5

feasible design. The net result is that Sec. 5.5 contains details of a feasible first-stage design, and Sec. 9.8 contains similar details for a feasible second-stage design.

We use an approach identical to that specified in Sec. 9.8 for our electric-system design with a single stage, but we leave out the analysis details. Table 10.9 summarizes the results of the more detailed preliminary design for both of our concepts. Notice that in both of the electric designs, the detailed analysis gives us a larger inert mass than we originally estimated. To make the system work, we arbitrarily fix burn duration and the initial mass and allow the power level to "float." This decision increases our power requirement and specific impulse by a small amount.

Table 10.8. Summary of Analysis Results for Chemical-System Concepts. These numbers are generated by using the algorithm in Table 10.6. The shaded boxes show systems that violate the constraints.

	1 Stage H_2/O_2	1 Stage CH_4/O_2	Stage 1 CH_4/O_2 Ion	Stage 1 CH_4/O_2 Arc	Stage 1 $RP-1/O_2$ Ion	Stage 1 $RP-1/O_2$ Arc
Δv + 5% (m/s)	4432	4432	1807	1807	1807	1807
Payload mass (kg)	3387	3387	4914	6197	4914	6197
Specific impulse (s)	435	335	335	335	325	325
Inert-mass fraction	0.13	0.12	0.15	0.15	0.15	0.15
Propellant mass (kg)	8500.5	15,808.4	4137.2	5217.4	4330.1	5460.6
Inert mass (kg)	1270.2	2155.7	730.1	920.7	764.1	963.6
Final mass (kg)	4657.2	5542.7	5644.1	7117.7	5678.1	7160.6
Initial mass (kg)	**13,157.7**	**21,351.1**	**9781.3**	**12,335.2**	**10,008.2**	**12,621.3**
Initial stage mass (kg)	9770.7	17,964.1	4867.3	6138.2	5094.2	6424.3
Payload mass/initial mass	25.7%	15.7%	50.2%	50.2%	49.1%	49.1%

Table 10.9. Summary of More Detailed Analysis Results for Our Remaining Concepts. The list shows the salient numbers for our single-stage electric concept and for our two-stage concept.

	Single Stage: Electric-Ion	First Stage: Chemical	Second Stage: Electric-Ion
Initial vehicle mass (kg)	5389.9	11,180	4914
Initial stage mass (kg)	1966.8	6266	1505.8
Stage inert mass (kg)	718.9	1034	678.1
Stage propellant mass (kg)	1247.9	5232	827.7
Thrust level (N)	1.464	35,305	1.456
Specific impulse (s)	2483	337	2530
Thrust duration	240.22 days	489.8 s	163.20 days
Electric power (W)	28,577	N/A	28,693

10.6 Assess Designs and Configurations

Having completed the systems analysis, we come down to the bottom line: how do we choose the best system? Section 10.1 specifies that our assessment criterion is cost, with the cost figure of merit:

$$FOM = C_{fs} + \$1B \times POF + \$5000/hr \times T_d - \$50M/yr \times (Life - 8 \text{ yrs})$$

We now run into the classic problem in our industry: how do we accurately predict the flight-system cost and how do we predict the system reliability? The answers to these questions are very difficult, and some people would argue that they are impossible during preliminary design. The answer invariably contains a caveat phrase such as, "It depends!" Historically, we have tried to shy away from making system-design decisions based on cost, mainly because it is so subjective. In many cases, changing our model can drastically change the result.

Having said all this and keeping in mind the limits of any costing approach, we still try to answer the question: what is the best system based on cost? We assume

- All systems operate for the same lifetime. This means the last term of the figure of merit is not a discriminator, so we ignore it.
- All research, development, testing, and evaluation (RDT&E) costs are lumped into the first unit.
- Each propulsion system has a reliability of 97%. This means each propulsion system has a 0.03 probability of failure, so our two-stage system has twice this probability of failure, or 0.06.
- We use the cost-modeling algorithms documented by R. Wong, Chap. 20 in Larson and Wertz [1992].
- The only difference between the chemical and electrical systems is the power system. The chemical system includes only the cost of the propulsion system. The electrical system includes the cost of the propulsion system itself and the additional cost of the extensive power system. All other systems are relatively independent of the propulsion-system type, should be similar on each system, and are ignored for comparison.

The parametric cost algorithms in Larson and Wertz are based on systems developed for the U.S. Air Force. Using these algorithms poses at least two problems. First, the Air Force typically has requirements that tend to drive the cost higher than a similar commercial system. This means the development cost of our system should be lower than the cost predicted by these numbers. Wong suggests that the development costs for commercial systems are 80% of the numbers predicted by the Cost Estimating Relationships (CERs). The second problem concerns extrapolation. Wong cautions against using these CERs for parameters more than

25% above the applicable range. However, because we lack a better algorithm, we use these techniques.

For three-axis-stabilized propulsion systems (extrapolated from apogee-kick motors), the RDT&E CER of the propulsion system is

RDT&E Cost in 1992 U.S. dollars = 1.56 × total impulse in N·s

where the applicable impulse range is 45,000 N·s to 2,440,000 N·s with a standard (1 standard deviation) error of $1M. For power systems, the CER is

RDT&E Cost in 1992 U.S. dollars = 5,303,000 + 108 × (mass × power)$^{0.97}$

where the applicable range is 104 to 415,000 kg·W (mass is in kilograms, and power is in watts). The standard (1 standard deviation) error is $5.7M.

The costs to produce the first flight unit, above and beyond the RDT&E cost, are given by the following CERs:

Propulsion System First Unit Cost in 1992 U.S. dollars =
0.52 × total impulse in N·s

Power System First Unit Cost in 1992 U.S. dollars =
183,000 × (mass × power)$^{0.29}$

The applicable range for the propulsion is 240,000 N·s to 2,440,000 N·s, with a cost standard deviation of $500,000. The applicable range for the power system is 104 kg·W to 414,920 kg·W, with a standard deviation in cost of $2.254 million.

Tables 10.10 and 10.11 show the results of our cost analysis. The single-stage system based on an ion thruster has a better (lower) figure of merit than the two-stage system does.

The figures of merit for the single-stage ion system and the second-stage ion system are virtually identical. This is not surprising! As expected, the difference between the single-stage system and the two-stage system is the cost of the liquid first stage ($65 million).

The numbers for the power system's RDT&E seem unreasonably high. If we look at the CER, we find it is essentially a linear relationship. This relationship is accurate for smaller systems, but we would expect diminishing marginal costs as the system gets larger. This argument makes the absolute cost numbers for the power system suspect. However, recognizing that the single-stage system and the second-stage system are virtually identical, we can argue that the two-stage approach is always more expensive than a single-stage system.

A key cost issue we have avoided to this point is propellant cost. The cost of RP-1 and LOx is negligible, but the cost of xenon is extremely high. In addition, the mass of xenon required for this one mission probably exceeds the world's supply. An important consideration for the next iteration is to look at this cost. A better propellant choice may be mercury, but we have environmental concerns with this propellant.

Table 10.10. Evaluating Cost and Figure of Merit for the Single-Stage Electric System. Cost numbers are reduced by 20% to account for the difference between commercial and government systems.

Power (W)	28,577
Thrust (N)	1.464
Burn duration (hrs)	5765
Probability of failure	0.03
Power system mass (kg)	204
Power system factor (W·kg)	5,829,708
Impulse (N·s)	30,384,983
Power RDT&E CER	$319,881,750
Propulsion RDT&E CER	$37,920,459
Power first unit CER	$13,414,628
Propulsion first unit CER	$12,640,153
Flight system cost	$383,856,991
Flight time cost	$28,826,069
Figure of merit	**$442,683,060**

10.7 Compare Designs and Choose the Best Option

From the analysis in Sec. 10.6, we can conclude that the single-stage system is our best choice because the figure of merit is 13% better than the two-stage system. However, it is appropriate at this point in the design to go back and examine the requirements. If we can relax some of them, we may be able to use one of the existing systems in Table 10.1. Alternatively, we may be able to modify one of these existing systems to meet the given requirements at a much lower cost than the new system.

As we can see, the all-electric system has the lowest overall figure of merit (cost) at $443 million. We can see that the two-stage design reduces costs for delivery, but increases development cost because of the additional chemical stage. Hence, we have selected the single-stage ion system as our best design from the set of feasible designs. Figure 10.12 shows a conceptual design drawing of the all-electric ion stage. It is dominated by the size of the solar panels which provide the power. Our design must allow folding solar-power arrays to fit within the 3 × 3-meter launch-vehicle dimensions. The stage also includes the on-orbit attitude-control system which accounts for the two smaller hydrazine tanks. Five ion thrusters are placed on the stage to protect against thruster burnout or other unexpected failures. Radiators are placed at right angles to the solar panels to keep solar

Table 10.11. Evaluating Cost and Figure of Merit for the Two-Stage System. Cost numbers are reduced by 20% to account for the difference between commercial and government systems.

	Stage 1	Stage 2
Power (W)	500	28,693
Specific power (W/kg)	45	
Thrust (N)	35,305	1.456
Burn duration (hrs)	2	3,917
Probability of failure	0.03	0.03
Power system mass (kg)	11	214
Power system factor (W·kg)	5,556	6,140,302
Impulse (N·s)	17,292,389	25,659,405
Power RDT&E CER	$4,612,995	$336,181,019
Propulsion RDT&E CER	$21,580,901	$25,621,073
Power first unit CER	$1,784,482	$13,618,087
Propulsion first unit CER	$7,193,634	$8,540,358
Flight system cost	$35,172,012	$383,960,537
Flight time cost	$10,680	$19,583,435
Figure of merit	**$65,182,692**	**$433,543,971**
Composite figure of merit		**$498,726,663**

energy off them. Finally, avionics (flight computer, sensors, batteries) are located away from the hot PCUs.

Notice the tremendous cost we have paid because of the small volume and limited maximum stage mass. If we could afford to increase these margins, we could use an all-chemical system, which costs only about $65 million. The money saved may be better spent on buying a larger launch vehicle. Considering that the most expensive U.S. rocket—the Titan IV—costs less than $300 million, we see that our upper-stage development cost is extremely high compared to other options. Only when a payload is pushing the upper limits of the largest launch vehicle does a new stage design, such as our all-electric ion design, make sense.

The next step in the design of our stage is to build a sound design-management structure to keep track of all the subsystem's key parameters, such as cost, power, mass, and interfaces with other subsystems. Spreadsheets are very useful tools for tracking the configuration and design changes. A system-level spreadsheet (see Fig. 10.12) contains all the requirements and assumptions, high-level mass budgets, power budgets, and so forth. Subsystem engineers develop their systems to

10.7 Compare Designs and Choose the Best Option

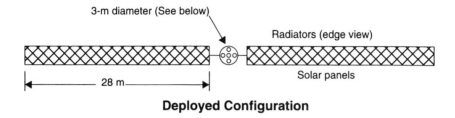

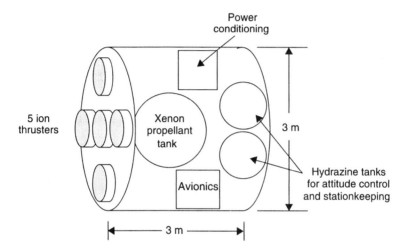

Fig. 10.12. Preliminary Layout for the Ion-Propulsion Stage. This figure shows the basic configuration of the single-stage electric system using only ion propulsion. The upper figure shows the deployed array to scale. The bottom figure shows how we can arrange the different subsystems into our envelope.

higher fidelity and enter their results in their summary spreadsheets. The system-level spreadsheet is then updated with the new subsystem data via the links. The system engineer can then apply margin to subsystems which have underestimated their numbers or reduce other subsystems' allocations. Even with poor guesses for subsystem specifications, a summary table is critical for tracking the design. The system engineer must always be able to respond to a request for the current design numbers so management (and investors) can track the design's progress.

An interesting way to model a system is to link the subsystem engineer's design spreadsheet into the system requirements and summary table, so any changes update instantly and automatically on the system resources table. The subsystem's design equations are built into the subsystem spreadsheet so it can instantly create a new design if the input requirements change. Links can be added from the system-level requirements down to the subsystem spreadsheets. These

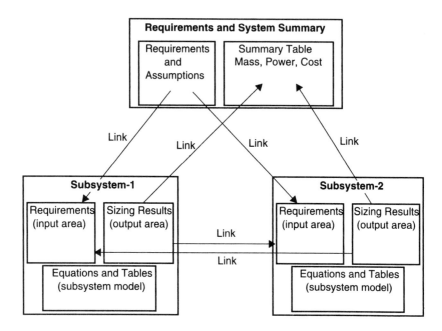

Fig. 10.13. **Linked Spreadsheets Can Create a Dynamic System-Model.** By linking various subsystem models to a system model, we can quickly evaluate the integrated effects of these subsystems on the overall design.

links allow us to do system-wide trade studies by simply changing design assumptions or system-level requirements. Links back to the system spreadsheet (and between subsystems) create a linked network that responds to any changes originating in the systems or subsystem. Complex linked spreadsheets such as this can become numerically unstable. This instability parallels the real design process, which can also become unstable if changing requirements from subsystem to subsystem keep the overall design in flux. Hunting down these instabilities and studying them closely are valuable exercises, as these are the design's sensitive issues and can cause significant overruns in cost and scheduling.

References

Larson, Wiley J. and James R. Wertz, eds. 1992. *Space Mission Analysis and Design.* 2nd edition. Norwell, MA: Kluwer Academic Publishers and Torrance, CA: Microcosm, Inc.

CHAPTER 11
ADVANCED PROPULSION SYSTEMS

Robert L. Forward, *Forward Unlimited*

    11.1   Air-Augmented Rockets
    11.2   Rocket Advancements
    11.3   Nonrocket Advancements
    11.4   Interstellar Flight

    Let us shift from design details to discussing some novel approaches to propulsion. Except for air-augmented rockets, little technical experience supports some of the more likely possibilities for future propulsion systems. We discuss the basic concepts and indicate the present state of the technology. For each topic, we have listed one or two references for those interested in learning more.
    If any of these concepts work technically, they could provide a breakthrough in space propulsion by lowering the mass ratios needed. Thus, they would decrease the costs of carrying out missions or decrease mission times by increasing the obtainable Δv. Of the propulsion systems without rockets, only the solar thermal thruster (Sec. 11.3.1) or tether propulsion (Sec. 11.3.4) is likely to transition soon from far term to near term.

11.1 Air-Augmented Rockets

    Strictly speaking, air-augmented rockets are not advanced propulsion systems. For 40 years they have been used mostly in missiles and Remotely Piloted Vehicles (RPVs), such as the US Navy's Firebrand target vehicle, which need a mixture of high speeds and good endurance.
    Considering performance only, airbreathing rockets are not all that different from ramjets or supersonic-combustion ramjets (scramjets). The main difference is

that airbreathing rockets, or ramrockets, can provide static thrust and thrust in a vacuum when operating as a pure rocket. Figures 11.1 through 11.4 represent types of ramrockets used as airbreathers. As pure rockets, they would need an additional inlet cover or closeable inlet.

The *solid-propellant, gas-generator ramrocket* shown in Fig. 11.1 is probably the most popular because it is simple. In this ramrocket, the fuel and part of the oxidizer burn in a conventional solid rocket motor. The fuel-rich propellant gases and atmospheric air feed into a secondary combustion chamber. Additional combustion takes place here, and the combined flows exhaust through a single nozzle.

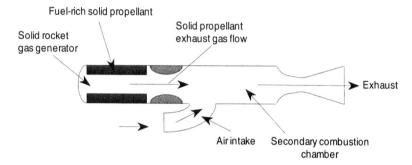

Fig. 11.1.   Solid-Propellant, Gas-Generator Ramrocket. The generator creates fuel-rich gases which exhaust from the air intake into the airflow. The gases mix with the air and burn until they exhaust to the atmosphere.

Fig. 11.2 shows the *liquid-fuel ramrocket*, in which high-speed air slows down and compresses as it enters the combustion chamber. Fuel and possibly more oxidizer feed into the combustion chamber, where these materials mix, burn, and move out through a conventional converging/diverging nozzle.

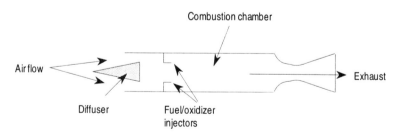

Fig. 11.2.   Liquid-Fuel Ramrocket. Fuel and oxidizer are injected into the internal airstream where they mix and burn. In a vacuum, the inlet is sealed off and this system works as a pure rocket.

11.1 Air-Augmented Rockets

Figure 11.3 shows a *solid-fuel ramrocket*, which operates like the solid fuel hybrid rockets discussed in Chap. 7. Atmospheric air slows and compresses in the inlet to the combustion chamber. The hot gases in the combustion chamber vaporize the solid fuel, and combustion occurs in the axial flow field. The exhaust then expands through the nozzle. Modulating the flow mass flux rate of the air and oxidizer controls the thrust level.

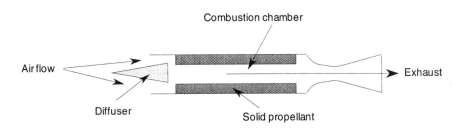

Fig. 11.3. Solid-Fuel Ramrocket. Air flows down the solid-propellant port and mixes with a fuel-rich combustion gas burning from the solid grain.

Figure 11.4 shows the most "advanced" version of the ramrocket, which operates in two stages. This is called the *dual-mode ramrocket*. The first stage is a conventional solid- or liquid-fuel ramrocket (we show the solid-fuel version). This stage operates fuel rich and exhausts into a secondary combustion chamber. The secondary chamber operates in the supersonic regime, though combustion is still subsonic. This means that the secondary chamber must be long enough to allow a complete burn. The dual-mode ramrocket has advantages over the solid-fuel ramrocket at velocities above Mach 5 because the ambient air burns at higher speeds and therefore lower static temperatures, allowing more efficient combustion. Recent research also suggests this dual-mode combustor will reduce ozone-destroying emissions of nitrogen oxide by burning away from the stoichiometric mixture.

The controllable inlet common to these four types of ramrockets is a key to operating in the pure rocket mode because the inlet must close to provide static thrust or thrust in a vacuum. As ram air pressure becomes greater than the combustion chamber pressure, the inlet can open if we use a

- *removable cover*—essentially a door which opens and closes as required
- *consumable cover*—a cover that is destroyed when we want to open the inlet
- *variable inlet*—an inlet we can open, close, and adjust to control flow rates of the air mass at the inlet

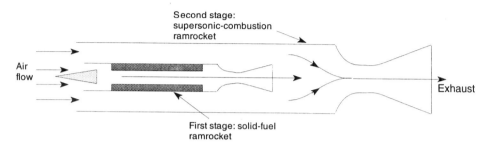

Fig. 11.4. Dual-Mode Ramrocket. This is a combination of a solid-fuel ramrocket (Fig. 11.3) and the gas-generator ramrocket (Fig. 11.1).

Performance—The simple one-dimensional equation for momentum thrust (F) is similar to the equation for a rocket except that we must take the air's initial momentum into account:

$$F = \dot{m}_{air}(V_e - V_{intake}) + \dot{m}_{prop} V_e \qquad (11.1)$$

where F = thrust (N)
$\dot{m}_{air}$ = mass flow rate of the air through the engine (kg/s)
V_e = velocity of the exhaust gas at the nozzle exit (m/s)
V_{intake} = velocity of the freestream air (unaffected by the vehicle in m/s)
$\dot{m}_{prop}$ = mass flow rate of the propellant stored on the vehicle; could include auxiliary oxidizer (m/s)

When the propulsion system is operating as a pure rocket, only the last term is significant. When it is operating as an airbreather, the last term usually has little effect. Oates [1988] shows that the rocket works best when the mass flow rate of the air is as high as possible and the difference between the exit and intake velocities is as low as possible. Increasing the mass flow through the engine is a significant advantage because it reduces the fuel needed to heat this mass.

Figure 11.5 shows the typical specific impulse for airbreathing propulsion as a function of the freestream air Mach number. In this figure, only the fuel mass flow rate matters because the stored oxidizer mass flow is negligible in the air-breathing phase.

$$I_{sp_f} = \frac{F}{g_0 \dot{m}_{fuel}} \qquad (11.2)$$

where g_0 = standard acceleration of Earth gravity (9.81 m/s²)
$\dot{m}_{fuel}$ = mass flow rate of the fuel (kg/s)
I_{sp_f} = fuel specific impulse (s)

11.1 Air-Augmented Rockets

There are two curves, depending upon the fuel used. As expected, hydrogen performs much better than hydrocarbon fuel because the exhaust gases are lightweight. We have included a curve for turbojets for comparison. It provides good specific impulse at the low Mach numbers but quickly tapers off.

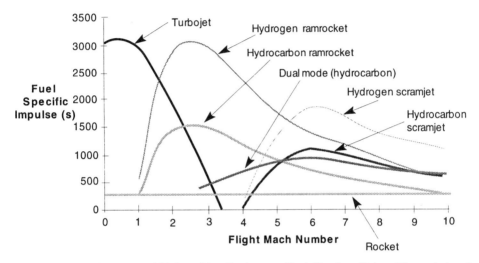

Fig. 11.5. Performance of Air-breathing Engines vs. Mach Number. Air-breathing rocket performance is a function of vehicle speed. At lower speeds, an air-breathing engine is better than a pure rocket. At higher speeds, air temperature reduces the performance of air-breathing systems.

As we would expect, at low Mach numbers, the "ram" effect is insignificant, reducing ramrocket performance to that of a rocket. As the Mach number increases above one, the ramrocket becomes much better than a rocket. As the Mach number goes above three, the inlet air slows down or "diffuses" in the inlet, thus increasing its static temperature. This high temperature reduces the combustion efficiency. We start seeing much of the combustion energy going to produce ions rather than to provide thrust. At Mach numbers greater than five, the scramjet starts to perform better than a ramjet because the air velocity is higher at the inlet, keeping the inlet temperature much lower.

Knowing the specific impulse does not give the whole picture. Figure 11.6 gives estimates of thrust-to-engine weight (F/W) for the different technologies. We have plotted ramjets because we do not have a good database for ramrockets, but ramrockets should not be more than 10% higher than a comparable ramjet. We must treat the data in Figs. 11.5 and 11.6 with caution because we have used several simplifying assumptions to keep the presentation simple.

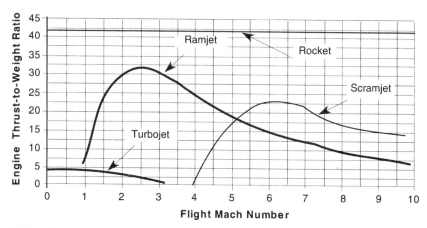

Fig. 11.6. Thrust-to-Weight Ratios vs. Mach Number for Air-breathing Engines. The figure shows that, although air-breathing engines have better performance (Fig. 11.5), they can be quite a bit heavier than a pure rocket.

11.2 Rocket Advancements

Nearly all present space-propulsion systems use the same technology as the first rockets—producing thrust by using a nozzle to direct the expanding hot gases resulting when a fuel chemically reacts with an oxidizer. A few other propulsion systems have flown, such as *electric rockets*, where onboard electrical energy heats an expellant, or electromagnetic forces expel ionized particles or plasmas at high exhaust velocities. This section briefly discusses additional concepts under study for advanced rockets.

11.2.1 Chemical Fuels with High-Energy Density

All of the 107 chemical elements, and nearly all possible combinations of those elements into molecular compounds, are known. So far, the most energetic combination for fuel we have (that is not highly poisonous) is liquid-oxygen/liquid-hydrogen (LOX/hydrogen). It has an energy release of 12.56 MJ/kg, an ideal specific impulse of 511 s, and a realistic specific impulse of 457 s or an exhaust velocity of 4.5 km/s. Because a rocket's mass ratio decreases exponentially with an increase in exhaust velocity, a small improvement in exhaust velocity can greatly increase payload size for a given vehicle. The search continues for a more energetic propulsion fuel with a higher exhaust velocity. The surveys to date [U.S. Air Force, 1992] have found some candidates worth further research but no promising near-term candidates. Table 11.1 [Stwalley, 1991] compares candidate high-energy density fuels with LOX/hydrogen; we discuss some of them in the following sections.

Table 11.1. Examples of Potential High-Energy Density Fuels. The table shows that certain chemical reactions can dramatically improve specific impulse.

Reaction (l) = liquid (g) = gas (s) = solid	Energy Release (MJ/kg)	Specific Impulse (s)	
		Ideal	Realistic
$H_2(l) + \frac{1}{2} O_2(l) \rightarrow H_2O(g)$	12.56	511	457
$H_2(l) + \frac{1}{3} O_3(s) \rightarrow H_2O(g)$	14.65	555	494
$H_2(l) + \frac{1}{3}$ metastable $O_3(s) \rightarrow H_2O(g)$	17	600	535
$H_2(l) + \frac{1}{4}$ cyclic $O_4(s) \rightarrow H_2O(g)$	19	635	565
$H_2(l)$ + free radical $O(g) \rightarrow H_2O(g)$	29	775	690
2 free radical $H(g) + \frac{1}{2} O_2 (l) \rightarrow H_2O(g)$	36	875	780
2 metallic $H(l,s) \rightarrow H_2(g)$	138	1700	1500
2 free radical $H(g) \rightarrow H_2(g)$	210	2100	1870
metastable $He(g) \rightarrow He(g)$	480	3150	2800
$H^+(g) + H^-(g) \rightarrow H_2(g)$	835	4190	3730

Compounds Using High-Energy Molecular Bonds

One of the approaches to storing more energy in a fuel is to form molecules that have a lot of energy stored in their interatomic bonds. This usually means forcing the formation of a molecule which does not usually form in nature because a similar or smaller molecule exists with a lower energy of formation. One example (Table 11.1) is the ozone molecule O_3, which has more stored energy per atom than O_2, the normal molecular form of oxygen. Laboratory quantities of frozen amorphous ozone (a dark blue ice), which have been made and kept for long periods at cryogenic temperature, are being tested [U.S. Air Force, 1992]. Amorphous ozone ice seems to be less sensitive to shock-initiated decomposition than the crystalline form of the ice. Ring forms of oxygen, such as cyclic O_4, O_6, O_8, and O_{12}, have also been studied. Cyclic O_4 is the most energetic of the oxygen-ring molecules (Table 11.1), probably because of the higher strain in the bonds. Other strained-bond molecules studied are those in the shape of a tetrahedron—tetrahydrogen, H_4, and tetranitrogen, N_4—plus cubane, C_6, in the shape of a cube. The bond angles in these molecules are much more acute than the normal bond angles between adjacent atoms, so the bonds store considerable strain energy.

Another approach is to apply high pressure to produce a solid form of an element with a crystalline lattice that has more stored energy than the normal ground-state lattice. We expect that under pressure a light element in the first or second rows of the periodic table assumes the crystalline structure of the heavier elements in their periodic column. A well-known example is the element carbon.

Its normal ground-state lattice structure is graphite, whereas the heavier element below it in the periodic table, silicon, has a dense crystalline lattice structure. Under high pressure graphite turns into diamond, a solid phase of carbon with a crystalline structure like silicon. This high-energy form of carbon turns out to be metastable, with an essentially infinite lifetime if kept from getting too hot.

Working from this example, researchers are looking at using high pressure to turn normal, transparent, nonconducting, molecular-hydrogen ice with a specific density of 0.088 into a silvery, conducting (possibly superconducting) atomic metal like lithium, but with a specific density of 1.15 and an energy release of 138 MJ/kg (Table 11.1). Other candidates are polymeric nitrogen with a structure like phosphorous, and metallic boron with a metallic lattice structure like aluminum. We still do not know whether hydrogen, nitrogen, or boron can be forced to assume these high-energy forms, or whether they remain stable after the pressure is released.

Metastable Excited Molecules

Any atom or molecule can be excited by electrons or photons to raise one of the orbital electrons of the molecule from its ground state to a higher energy state. The amount of energy stored in the molecule in this manner often exceeds the energy released when that molecule is "burned" in a combustion reaction. Thus, excited molecules would be excellent candidates for a high-energy propulsion fuel. Unfortunately, most molecules rapidly return to their ground state, giving off the stored energy in the process. But some molecules can form in a *metastable* excited state with a long lifetime.

The best known candidate is metastable helium, the electronically excited triplet state of a helium atom. When the helium atom decays into the ground state it releases 19.8 eV/atom, or 480 MJ/kg of energy, almost 40 times that of LOX/hydrogen (Table 11.1). Metastable helium atoms are easy to produce by running electrical current through helium gas. (Every helium-neon laser produces a lot of metastable helium inside at electrical efficiencies greater than 10%). The theoretical lifetime of the metastable state of helium is limited by spin-orbit coupling to 2.5 hours in a vacuum. This lifetime is marginal for use as a propellant. If we could suppress this effect, the next mode of decay is by double-photon emission, with a predicted lifetime of 8 years. In a typical container, however, interaction with the walls or other atoms gives a lifetime for metastable helium of less than 0.1 s.

We need new ideas for storage and improved lifetime, or we must find other molecules having similar high-energy metastable states with long lifetimes. One such candidate is the predicted metastable excited state of ozone. If burned stoichiometrically with liquid hydrogen to produce water, it has a predicted energy release of 17 MJ/kg, or 35% better than LOX/hydrogen (Table 11.1).

Energetic Free Radicals

Many elements used in propulsion are gases that tend to form diatomic molecules, in which two atoms share their electrons to form closed electronic shells. Well-known examples are H_2, O_2, N_2, and F_2. Other elements prefer to form solid crystalline lattices, like Li, Be, B, and C. Removing one of these atoms from its diatomic partner or from its crystalline lattice requires considerable energy. These isolated atoms, called *free radicals* or *radicals*, thus have a higher energy than the bound atoms. If we could store these free radicals for long periods at high densities, we could use them as a fuel with high energy density. Electromagnetic traps [Cline, 1980] have held spin-polarized hydrogen radicals for a few hours at extremely low temperatures (less than 1 K) and high magnetic fields (greater than 10 T). Enough has been stored [Sprik, 1983] at densities of 2×10^{18} atoms/cm^3 (10^{-5} that of normal matter densities) to explode the cryogenic glassware when containment failed, but the lifetime was found to be inversely proportional to the storage density.

Present research efforts involve storing the free radicals in a cryogenic solid matrix such as hydrogen, argon, and xenon. For propulsion, the preferred cryogenic solid matrices would be hydrogen, oxygen, or nitrogen. Table 11.2 gives the propulsion performance of the neutral free radicals of the first eight elements reacted with a stoichiometric quantity of solid H_2 used as a cryogenic storage matrix [Stwalley, 1991]. All of the reaction products are gaseous at the reaction temperatures expected. If the actual percentage of the free radical in the solid H_2 matrix is less than the stoichiometric percentage, we must adjust the energy release appropriately.

Table 11.2. Reactions of Neutral Free Radicals with Liquid Hydrogen. The table shows the performance of free radicals that are stored in a solid matrix and then react with the matrix material.

Reaction	Energy Release (MJ/kg)	Specific Impulse (s)	
		Ideal	Realistic
$\frac{1}{2} H_2$ + free radical H → H_2	104.0	1480	1320
$\frac{1}{2} H_2$ + free radical Li → LiH	2.5	230	205
H_2 + free radical Be → BeH_2	18.4	620	555
$3H_2$ + 2 free radical B → B_2H_6	37.3	855	790
$2H_2$ + free radical C → CH_4	49.4	1020	905
$1\frac{1}{2} H_2$ + free radical N → NH_3	30.5	800	710
H_2 + free radical O → H_2O	28.5	775	690
$\frac{1}{2} H_2$ + free radical F → HF	17.6	610	540

Considerable research has been done on the storage of H radicals in cryogenic solid H_2. The solid H_2 is first doped with radioactive tritium. Each energetic beta particle emitted by the tritium creates hundreds of H radicals as it passes through the solid, producing up to 0.5% atomic H in solid H_2. So far, no one has been able to get higher concentration levels. Free radical Li has also been stored for days in solid H_2 at 3 K, but this material has low energy release compared to standard rocket fuels (Table 11.2).

A free radical does not have to be neutral. It can be charged, in which case the energy obtainable is much higher than from a comparable neutral free radical. The most spectacular example is shown in the last column of Table 11.1, where H^+ (a proton) combines with H^- (a hydrogen atom with an extra electron). Although we can easily make positively and negatively charged hydrogen ions, and then store small quantities at very low densities in electromagnetic traps, no good ideas exist yet for storing them in large numbers at densities suitable for use as a propellant.

11.2.2 Unconventional Propulsion Systems Using Nuclear Rockets

The conventional nuclear rockets described in Chap. 8 use a self-sustained, controlled-nuclear-fission reactor either to heat an expellant fluid (usually hydrogen) for thermal propulsion or to produce electricity for electric propulsion. The nuclear rockets discussed here have received considerable study, but their development involves high technological risk.

Propulsion Systems Using Nuclear-Fusion Rockets

In nuclear fusion, under conditions of high temperature and pressure, the nuclei of two light atoms "fuse" together to make a heavier nucleus, plus some particles and a large amount of energy. If properly contained, the released energy can maintain high temperature and pressure, causing a chain reaction in the rest of the fusion fuel. Magnetic fields can contain and direct the charged particles resulting from the fusion reactions to provide rocket thrust. Ideally, fusion can produce specific impulses of hundreds of thousands of seconds. In reality, the specific impulse depends on thermal limits in the reaction chamber but still can be many thousands of seconds. Haloulakos and Bourque [1989] include a review and bibliography of fusion propulsion.

The fusion fuels usually considered are the nuclei of atoms such as hydrogen with a nucleus consisting of a proton (p^1); deuterium (D^2), with one proton and one neutron; tritium (T^3), with one proton and two neutrons; or helium-3 (He^3), with two protons and one neutron. The major reaction products are the normal helium-4 nucleus (He^4), neutrons (n^1), and protons. The primary fusion reactions, in order of ease of ignition, are

$$D^2 + T^3 \rightarrow He^4 (3.5 \text{ MeV}) + n^1 (14.1 \text{ MeV}) \tag{11.3}$$

$$D^2 + D^2 \rightarrow T^3 \,(1.01 \text{ MeV}) + p^1 \,(3.02 \text{ MeV})$$
$$50\%$$
$$\rightarrow He^3 \,(0.82 \text{ MeV}) + n^1 \,(2.45 \text{ MeV})$$
$$50\% \tag{11.4}$$

$$D^2 + He^3 \rightarrow He^4 \,(3.67 \text{ MeV}) + p^1 \,(14.7 \text{ MeV}) \tag{11.5}$$

where 1 MeV = 1.60×10^{-13} J. The energy released in these reactions divides inversely according to the mass of the product. The total energy released is the sum of the individual parts. But in Eqs. (11.3) and (11.4), for example, the energy in the neutron product (n^1) is hard to use because the particle is not charged and is difficult to control, so the energy comes out as heat instead of directed kinetic energy.

We can produce nuclear fusion through magnetic or inertial magnetic confinement. In *magnetic-confinement fusion*, the ionized plasma of fusion atoms collects in a magnetic "bottle," where it heats and compresses until the temperature and pressure are sufficient to start fusion. Deuterium and tritium mixtures have produced fusion, but the fusion energy output has been less than the energy input required to create the fusion conditions. All of the magnetic-confinement systems built to date are very massive compared to their energy output, with a projected energy density more suitable for ground-based power plants than flight-weight rocket engines.

In *inertial-confinement fusion*, a small pellet of frozen fusion fuel is confined within a hollow sphere (typically gold or uranium) and then compressed and heated by forces applied from all directions by beams of laser photons, elementary particles, antimatter, or hypervelocity pellets. Again, fusion has occurred, but the energy out is much less than the energy in. Except for antimatter-induced implosion, the mass of the beam-generating apparatus and the power plant needed to close the loop by converting part of the fusion energy into the electricity needed to run the beam apparatus makes these systems more suitable for power plants than rocket engines. Because inertial-confinement fusion produces "microexplosions," versions which do not completely confine the resulting explosion would also fall under propulsion systems using nuclear-pulse rockets, which we discuss below.

Another major problem with nearly all fusion-propulsion concepts is that either the primary reaction or secondary reactions produce uncharged neutrons, which not only carry away energy without producing thrust but also create a large hazard from operational and residual radiation. Unless we can fuse deuterium with helium-3 [Eq. (11.5)] without activating deuterium-deuterium fusion, fusion rockets, like fission rockets, will be copious neutron emitters.

Propulsion Systems Using Nuclear-Pulse Rockets

In these systems, nuclear explosions (either fission or fusion) impart thrust to a space vehicle. The total system consists of the space vehicle coupled through a momentum-conditioning unit (shock absorber) to a momentum absorber (pusher plate). In operation, a nuclear bomblet explodes behind the pusher plate as it reaches the end of its stroke at maximum velocity. A fraction of the plasma debris from the nuclear explosion strikes the strong, ablative plate, reversing its velocity. As the plate approaches the vehicle, the shock absorber decelerates it, finally bringing it to a halt with respect to the vehicle. Part of the energy collected by the pusher plate is now in the vehicle, which has accelerated forward when the spring compresses, while part of the energy remains in the spring. The springs in the shock absorber return the plate to the original state, ready for the next explosion, and in the process maintain the acceleration on the vehicle.

Enough work has been done on the concept, especially the survivability of the pusher plate under repetitive shocks from the nuclear blasts, to determine that nuclear-pulse propulsion will work. A review and bibliography are in Martin and Bond [1979]. A recently proposed alternate design has a "parachute" out in front of the main vehicle replacing the "pusher plate" and elastic "shrouds" acting as the "shock absorbers" [Solem, 1992]. Development of nuclear-pulse rockets has essentially stopped because the nuclear test ban treaty excludes testing in the atmosphere or space. This system would emit a lot of radioactive waste during operation, and rockets using thermal propulsion from a solid-core nuclear reactor (see Chap. 8) provide strong competition.

11.2.3 Propulsion by Annihilating Antimatter

The concept of using antimatter as the energy source for space propulsion has been in the literature for decades. The fuel is conceptually simple to use; antimatter particles spontaneously annihilate with normal matter particles to obtain a total conversion of mass to energy, with an energy release per unit mass of antimatter (m) of

$$E = 2c^2 m = 1.8 \times 10^{17} m \quad \text{(J)} \tag{11.6}$$

where $c = 3 \times 10^8$ m/s (the speed of light).

The interaction of positrons (antielectrons) with normal-matter electrons always generates two energetic gamma rays with an energy of 0.511 MeV. Because the gamma rays cannot be directed to produce thrust, the only way their energy can be used for propulsion is to absorb the energy in a refractory metal heat exchanger and use that to heat an expellant. The antiproton is more suitable for propulsion [Martin, 1982; Forward, 1988]. As is shown in Fig. 11.7, when an antiproton (p^-) annihilates with a normal-matter proton (p^+), it does not produce prompt gamma rays. Instead, it produces between 3 and 7 elementary particles called pions.

$$p^+ + p^- \rightarrow i\,\pi^- + j\,\pi^0 \tag{11.7}$$

where $i = 1$ or 2 and $j = 1, 2,$ or 3. The amount of energy released is 1876 MeV minus the rest-mass energy of the pions, which is 140 MeV for the charged pions (π^+ and π^-) and 135 MeV for the neutral pions (π^0).

For the case of five pions consisting of one neutral pion and two charged pion pairs, as shown in Fig. 11.7, the kinetic energy distributed among the five reaction products is 1876 MeV $- 4 \times 140$ MeV $- 135$ MeV $= 1181$ MeV. This distribution gives us an average of 236 MeV per pion, with a spread from 0 to 500 MeV. Because the average kinetic energy of the pions is nearly twice their rest-mass energy, the pions travel typically at 94% of the speed of light. As illustrated in Fig. 11.7, the charged pions can be contained and directed by a suitably designed, high-strength, "magnetic nozzle," whereas the neutral pions escape and are lost. If the magnetic fields are strong enough, the charged pions move in tight spirals that follow the magnetic field lines, as shown. Figure 11.7 does not show that charged pions which start out in the wrong direction inside the nozzle move closer to the throat of the nozzle, where the magnetic field gets stronger. At this point, their spirals rapidly reverse direction, so they move back up the nozzle and out of the exit plane along with the charged pions that started in the proper direction. Thus, a simple magnetic nozzle can convert much (60–80%) of the charged pions' directed kinetic energy into directed thrust.

The neutral pions, which have a lifetime of only 9×10^{-19} s, almost instantly decay into two gamma rays. These rays divide equally an energy ranging from 270 to 700 MeV, depending upon the amount of kinetic energy in the neutral pion:

$$\pi^0 = 2\gamma \tag{11.8}$$

Thus, about one third of the initial 1876 MeV of annihilation energy eventually converts into gamma rays. The other two-thirds of the annihilation energy stays in the charged pions. The charged pion pairs have a normal half-life of 26 ns, but they are moving at speeds averaging 94% of the speed of light, so their lives lengthen to 76 ns [Eq. (11.34)]. At an average speed of 0.94 c, the charged pions travel an average of 21 m in their tight spirals before they decay. As Fig. 11.7 shows, if the antiproton-proton annihilation were arranged to take place in a reaction chamber containing a magnetic field, the field could either direct the pions to provide thrust or contain them long enough to transfer their kinetic energy to a working fluid, such as hydrogen or water, introduced into the chamber. After an average of 76 ns, the charged pions decay. As shown in Fig. 11.7, the positive pion decays into a positive muon (μ^+) plus a muon neutrino (v_μ):

$$\pi^+ \rightarrow \mu^+ + v_\mu \tag{11.9}$$

whereas the negative pion decays into a negative muon (μ^-) plus a muon antineutrino ($\bar{v}_\mu$):

$$\pi^- \rightarrow \mu^- + \bar{v}_\mu \tag{11.10}$$

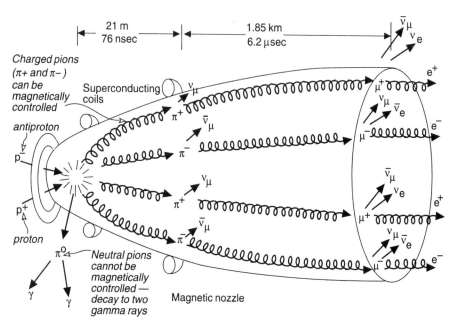

Fig. 11.7. Using Magnetic Fields to Obtain Thrust from Antiproton Annihilation. A proton and antiproton annihilate each other, creating various energetic particles that can be directed by a magnetic nozzle to provide thrust. See the text for a complete explanation of the decay process and symbols.

The charged muons have a normal lifetime of 2.2 μs, but because of their speed they last 6.2 μs, long enough to travel 1.82 km. This period is long enough for them to deposit all of their energy into a working fluid before decaying. The positive muon decays into a positive electron (e^+) (positron), an electron neutrino (v_e), and a muon antineutrino ($\bar{v}_\mu$):

$$\mu^+ \rightarrow e^+ + v_e + \bar{v}_\mu \tag{11.11}$$

whereas the negative muon decays into an electron (e^-), an electron antineutrino ($\bar{v}_e$), and a muon neutrino (v_μ):

$$\mu^- \rightarrow e^- + \bar{v}_e + v_\mu \tag{11.12}$$

The only source of low-energy antiprotons is at CERN, in Switzerland. But the Fermi National Accelerator Laboratory or the Brookhaven National Laboratory could be modified for less than $25 million to produce low-energy antiprotons. Present experiments are concentrating on the capture and storage problem. At

11.2 Rocket Advancements

CERN more than 100,000 antiprotons have been captured [Gabrielse, 1993] in a cryogenic electromagnetic trap using an ultrahigh vacuum. Similar traps have kept antiprotons for more than two months with zero loss.

Storing large quantities of antiprotons is difficult. One method is to add a positron to make antihydrogen. The antihydrogen is frozen to create "ice," which can then be levitated by a magnetic or electrostatic field. The feasibility of using magnetic fields to levitate milligram-sized balls of antihydrogen ice has been demonstrated [Paine, 1991] using normal hydrogen—the magnetic and mechanical properties of hydrogen and antihydrogen are the same.

Two Antiproton Science and Technology Workshops [Augenstein, 1988] found no technical barriers to antimatter propulsion but determined that producing the desired grams per year requires first a pilot plant to prove economic feasibility and then, a large production plant. This approach would take 30 years and 100 billion dollars but could save hundreds of billions from the cost of future space programs.

Antimatter Rocket Equation

To those rocket engineers inured to the inevitable rise in vehicle mass ratio with increasing mission difficulty, antimatter rockets provide relief. The mass ratio of an antimatter rocket for any mission is always less than 4.9:1 [Shepherd, 1952], and cost-optimized mass ratios are as low as 2:1 [Forward, 1985]. In an antimatter rocket, the source of the propulsion energy is separate from the reaction fluid. Thus, the rocket's total initial mass consists of the vehicle's empty mass, the reaction fluid's mass, and the energy source's mass, half of which is the mass of the antimatter. According to the standard rocket equation, the mass ratio is now (assuming $m_r \gg m_e$)

$$\frac{m_i}{m_f} = e^{\Delta v/v_e} = \frac{m_v + m_r + m_e}{m_v} \approx \frac{m_v + m_r}{m_v} \quad (11.13)$$

where Δv = change in vehicle velocity (m/s)
  v_e = rocket exhaust velocity (m/s)
  m_i = initial mass of the vehicle (kg)
  m_f = final mass of the vehicle (kg)
  m_v = empty mass of the vehicle (kg)
  m_r = mass of the reaction fluid (kg)
  m_e = mass of the energy source (kg)

The kinetic energy (K.E.) in the expellant at exhaust velocity (v_e) comes from converting the fuel's rest-mass energy into thrust with an energy efficiency (η_e):

$$K.E. = \frac{1}{2} m_r v_e^2 = \eta_e \, m_e c^2 \quad (11.14)$$

where K.E. = kinetic energy (kg·m²/s²)
    c = speed of light (3×10^8 m/s)

Solving Eq. (11.14) for the reaction mass (m_r), substituting into Eq. (11.13), and solving for the energy source's mass (m_e) produces

$$m_e = \frac{m_v \, v_e^2}{2 \eta_e \, c^2} \left(e^{\Delta v/v_e} - 1 \right) \qquad (11.15)$$

We can find the minimum antimatter required to do a mission with a given Δv. We set the derivative of Eq. (11.15) with respect to the exhaust velocity v_e equal to zero, and solving (numerically) for the exhaust velocity:

$$v_e = 0.63 \, \Delta v \qquad (11.16)$$

Substituting Eq. (11.16) into Eq. (11.13), we find that, because the optimal exhaust velocity is proportional to the mission Δv, the vehicle mass ratio is a constant:

$$\frac{m_i}{m_f} = e^{\Delta v/v_e} = e^{1.59} = 4.9 = \frac{m_v + m_r}{m_v} \qquad (11.17)$$

The reaction mass (m_r) is 3.9 times the vehicle mass (m_v), while the antimatter fuel mass is negligible. Amazingly enough, this constant mass ratio is independent of the efficiency (η_e) with which the antimatter energy is converted into kinetic energy of the exhaust. (If the antimatter engine has low efficiency, we will need more antimatter to heat the reaction mass to the best exhaust velocity. The amount of reaction mass needed remains constant.) If we can develop antimatter engines that can handle jets with the very high exhaust velocities Eq. (11.16) implies, this constant mass ratio holds for all conceivable missions in the solar system. It starts to deviate significantly only for interstellar missions in which the mission Δv approaches the speed of light [Cassenti, 1984].

We can obtain the amount of antimatter needed for a specific mission by substituting Eq. (11.16) into Eq. (11.15) to get the mass of the energy source (m_e). The antimatter needed is just half of this mass. We find it to be a function of the square of the mission velocity (Δv) (essentially the mission energy), the empty vehicle's mass (m_v), and the conversion efficiency (η_e):

$$m_a = \frac{1}{2} m_e = \frac{0.39 \Delta v^2}{\eta_e \, c^2} m_v \qquad (11.18)$$

The amount of antimatter calculated from Eq. (11.18) is typically measured in milligrams. Thus, no matter what the mission, the vehicle uses 3.9 tons of reaction mass for every ton of vehicle and an insignificant amount (by mass, not cost) of antimatter. Depending on the relative cost of antimatter and reaction mass after they have been boosted into space, missions trying to lower costs may use more antimatter than that given by Eq. (11.18) to heat the reaction mass to a higher exhaust velocity. If so, they would need less reaction mass to reach the same mission velocity. Such cost-optimized vehicles could have mass ratios closer to 2 than 4.9 [Forward, 1985].

The low mass ratio of antimatter rockets enables missions which are impossible using any other propulsion technique. For example, a reusable antimatter-powered vehicle using a single-stage-to-orbit has been designed [Pecchioli, 1988] with a dry mass of 11.3 tons, payload of 2.2 tons, and 22.5 tons of propellant, for a lift-off mass of 36 tons (mass ratio 2.7:1). This vehicle can put 2.2 tons of payload into GEO and bring back a similar 2.2 tons while using 10 milligrams of antimatter. Moving 5 tons of payload from low-Earth orbit to low Martian orbit with an 18-ton vehicle (mass ratio 3.6:1) requires only 4 milligrams of antimatter.

Antimatter rockets are a form of nuclear rocket. Although they do not emit many neutrons, they do emit large numbers of gamma rays and so require precautions concerning proper shielding and stand-off distance.

11.2.4 Magnetic Thrust Chambers

All advanced concepts have the same problem when they try to produce high thrust at high specific impulse (greater than 1500 s) in a compact engine. The problem is independent of the type of engine, whether it involves high-power electromagnetic thrusters; atomic hydrogen, metallic hydrogen, or metastable atom fuels; solar, laser, or microwave-heated thrusters; and fission-, fusion-, or antimatter-powered reaction chambers. The high-energy exhaust from any of these processes produces a blazing plasma that melts or evaporates any reaction chamber and nozzle made of ordinary materials. One solution is to make or shield the reaction chamber and nozzle with magnetic fields. Research in this field is still in its preliminary phases [Gerwin, 1990] and we are not sure a good design is possible.

11.3 Nonrocket Advancements

A generic rocket consists of payload, structure, energy source, expellant, and engine. In a chemical rocket, the energy source and expellant combine in the "fuel," where there is a compromise between energy content and molecular mass. In nuclear thermal rockets, the energy source is the nuclear material, picked for its high energy content, whereas the expellant is hydrogen, picked for its low molecular mass. All rockets must obey the rocket equation [see Eq. (1.15) in Chap. 1] which predicts an exponential increase in mass ratio for the vehicle when the mission Δv greatly exceeds the exhaust velocity of the propulsion system. This exponential increase in the mass of the expellant is a major limitation of rockets because it is "impossible" to build a space vehicle that can hold more than 1000 times its mass in expellant. Staging can help, but the exponential growth of the rocket equation applied to the whole stack soon dominates. This means that, depending on the exhaust velocity, some missions are "impossible" for a given rocket-propulsion system.

The "rocketless rocketry" concepts discussed in this section are not limited by the exponential growth of the system mass implied by the rocket equation. These propulsion systems are missing one or more of the "parts" of a generic rocket, such as the energy source, or the expellant mass, or the engine, and sometimes all three.

11.3.1 Systems That Collect Solar Power

The light from the Sun can provide energy and reaction mass to an advanced space-propulsion system designed to use that light. The power available from sunlight in space is considerable: at the Earth, it is 1.36 kW/m^2 or 1.36 GW/km^2 [see Eq. (11.19)].

Solar-Thermal Propulsion

The first nonrocket propulsion systems were solar-electric propulsion, described in more detail in Chap. 9. In *solar-thermal* propulsion, large inflated concentrating mirrors gather and focus sunlight onto a light-absorber, which converts the solar energy into thermal energy. The thermal energy can then operate a heat engine to produce electricity for electric propulsion, or to heat an expellant (typically hydrogen) we can exhaust to produce thrust.

A diagram of one version of a solar-thermal thruster is in Fig. 11.8. An Air Force program [Laug, 1989] has conducted ground tests in which sunlight from a 25-kW solar collector was focused into a small thruster. The thruster consisted of a cylindrical absorption chamber lined with rhenium tubing through which hydrogen gas flowed. This solar-heated rocket produced 4.45 N of thrust at a specific impulse of 600 s. New absorption-chamber designs with volumetric absorption are under study. Work on inflatable concentrator mirrors produced an inflatable reflector concentration ratio of 10,000:1. This off-axis reflector was 7 m by 9 m. Larger mirrors are under study. Nonimaging secondary optics are being studied as a means of achieving higher concentrations than those from imaging optics. Work is proceeding on a flight test of a system which uses solar-thermal thrusters for orbital transfers. It will have two off-axis mirrors of 30-m projected diameter, delivering a combined 3.6 MW of thermal power at a concentration ratio of 10,000:1, and two thrusters operating at a specific impulse of 900 s and 445 N (100 lb) combined thrust. The dry mass of the vehicle is 1800 kg, and it uses 9100 kg of hydrogen expellant to deliver 11,800 kg of payload from LEO to GEO in 10 to 30 days.

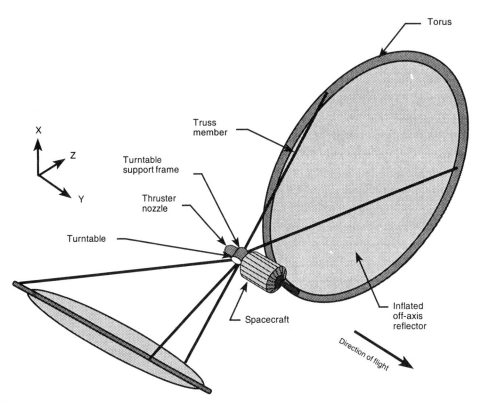

Fig. 11.8. Solar Thermal Thruster. Sunlight is collected by mirrors and focused into a thrust chamber. The light heats a working fluid that expands to provide thrust.

Solar-Lightsail Propulsion

Solar lightsails are large, lightweight reflectors attached to a spacecraft that use light pressure from solar photons to obtain thrust. Although the thrust available from sunlight is small, the solar lightsail never runs out of fuel. Over a long enough time, the small thrust can build up to extremely high Δvs, allowing solar lightsails to take on missions that cannot be attempted by vehicles limited by the exponential growth of the rocket equation. A solar lightsail is ideal for shuttling of interplanetary cargo or rendezvous missions to several asteroids because it does not require refueling. The thrust energy and the reaction mass for a solar-lightsail propulsion system come from sunlight. The solar-light flux (S) in space at the distance (D) from the Sun is

$$S = S_o\left(\frac{D_o}{D}\right)^2 \tag{11.19}$$

where S = solar-light flux (W/m²)
 S_o = 1.36 kW/m² = solar-light flux at 1 Astronomical Unit (AU)
 D_o = 1 AU = 150 × 10⁹ m
 D = distance from the Sun (AUs or m)

If this solar flux (S) is intercepted by a flat solar lightsail of area (A) that is tilted at an angle (θ) with respect to the Sun-sail line [Fig. 11.9], the effective interception area (A_i) of the lightsail is

$$A_i = A \sin\theta \tag{11.20}$$

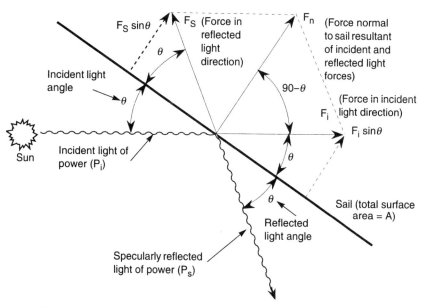

Fig. 11.9. **Radiation Pressure Forces on a Flat Solar Sail.** Light from the Sun is reflected or absorbed by the sail and exchanges momentum. This figure shows the basic geometry.

The incident-light power (P_i) in the solar-light flux intercepted by the lightsail is

$$P_i = SA_i = SA \sin\theta \tag{11.21}$$

This incident-light power (P_i) produces an incident-radiation force (F_i) through pressure on the lightsail [Forward, 1990] of

$$F_i = \frac{P_i}{c} = \frac{SA\sin\theta}{c} \quad (11.22)$$

The direction of this force vector is in the direction of the incident light.

Realistic lightsails are "grey" [Forward, 1990], in that part of the light transmits through the lightsail, part is absorbed, and part is specularly reflected. Let us reasonably assume the lightsail is flat and highly reflecting. If so, the reflected light power is equal to the incident-light power, and the magnitude of the specularly reflected force vector (F_s) arising from radiation pressure is equal to the incident-force magnitude given by Eq. (11.22) ($F_s = F_i$). The direction of the specularly reflected force vector, however, is opposite to the direction of travel of the reflected light. The components of the incident force vector and the reflected force vector in the plane of the lightsail are equal and opposite, so they cancel. But the components normal to the plane of the lightsail (see Fig. 11.9) sum to

$$F_n = F_s \sin\theta + F_i \sin\theta = 2F_i \sin\theta = \frac{2SA}{c} \sin^2\theta \quad (11.23)$$

This force is normal to the back of the flat lightsail. By tilting the lightsail to change the force direction, the light pressure can increase the spacecraft's orbital speed, sending it outward from the Sun, or decrease its orbital speed, allowing it to fall inward toward the Sun.

Over the decades, lightsail designs and missions have undergone considerable study. Detailed engineering studies [Friedman, 1988; Wright, 1992] have been carried out on a square lightsail and a 12-blade "heliogyro" lightsail designed to rendezvous with Halley's Comet, not just fly by at high relative speed. The square lightsail is 820 m on a side, is made of 2µm aluminum-coated Kapton™ film, and has a 3400-kg mass without payload.

One major problem with solar lightsails is that, at the low-Earth orbital altitudes the Space Shuttle can reach, the atmospheric drag force on the lightsail is greater than the light-pressure force. One solution is to deploy the lightsail from the Space Shuttle or the Space Station at the end of an upward-going tether that is 100 km long, with the lightsail kept edge-on to the orbital motion to minimize drag. When the lightsail is released, it rises upward in an elliptical orbit [see Sec. 11.3.4], where it can turn to the Sun and fly into space on its own power.

If a solar lightsail is light enough, it can "hover" without orbiting—the light pressure from the solar photons balancing the attraction of the Sun's or Earth's gravity. Solar-lightsail orbits around the Sun [McInnes, 1989] can produce nearly any desired orbital period at nearly any desired orbital distance, in or out of the ecliptic plane. Examples include zero—hovering anywhere over the Sun, moving heliosynchronously with features on the solar surface, or matching a planet's orbital period. As shown in Fig. 11.10, a solar lightsail can also continuously hover over the Earth's polar regions, thus functioning as a *statite* or stationary spacecraft [Forward, 1991] instead of a satellite, which by definition must orbit the Earth.

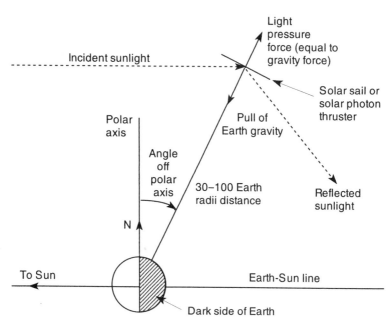

Fig. 11.10. Statite Concept. Light forces are used to allow a spacecraft to hover over the rotating Earth, without actually orbiting the Earth.

Propulsion Using the Solar Wind and a Magnetic Loop Sail

The Sun emits not only photons but also particles, mostly charged protons, called the *solar wind*. These particles cause the aurora borealis and keep the Van Allen radiation belts populated. The energy flux in the solar wind varies considerably depending on the sunspot activity, but it is always much less than the energy available from the much larger and relatively constant photon flux.

It is possible to extract propulsion thrust from the solar wind by using a magnetic loop sail, or Magsail [Zubrin and Andrews, 1991]. As shown in Fig. 11.11, a *magsail* is a simple loop of high-temperature, superconducting wire carrying a persistent current. The charged particles in the solar wind are deflected by the magnetic field, producing thrust. The thrust density in the ion flux from the solar wind is 5000 times less than the thrust density in the solar-photon flux. But the mass of a solar sail goes directly as the area of the sail, whereas the mass of the magsail goes as the perimeter of the area enclosed. In addition, the effective cross-sectional area of the magnetic fields around the magsail is about 100 times the physical area of the loop. As a result, calculations show that the thrust-to-weight of a magsail can be 20 times better than that of a solar sail. Thermal analyses of the loop's radiation cooling against the 2.7 K background temperature of space indicate that a properly Sun-shielded cable can be passively maintained at a temperature of 65 K in space, well below the superconducting transition point for many of the new high-temperature superconductors.

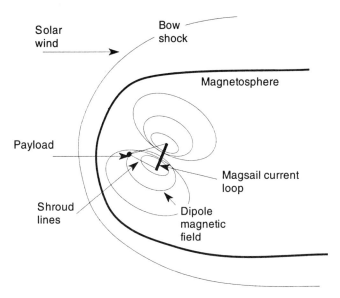

Fig. 11.11. Propulsion System Using a Magnetic Loop in the Solar Wind. A current loop creates a magnetic field, which interacts with charged particles in the solar wind to produce thrust [Zubrin, 1993].

When the magsail is inside the Earth's or another planet's magnetosphere with a magnetic field, the solar wind is deflected by the magnetosphere and does not interact with the magsail. In this situation, the magsail can interact usefully with the planet's magnetic poles to generate large amounts of thrust. It switches the current in the magsail on and off, synchronized with the orbital motion, to push and pull against the planetary poles and raise the orbit apogee to escape. About 200 perigee kicks would drive a magsail with a payload three times its mass into interplanetary space within two months. Once out of the magnetosphere, it could use the solar wind for interplanetary travel [Zubrin, 1993].

11.3.2 Propulsion Systems Using External Beamed Power

An alternate method of collecting power in space is to beam the power to the using vehicle in the form of coherent electromagnetic radiation. A transmitting antenna of electromagnetic radiation of wavelength (λ) and diameter (D) is able to create a spot of diameter (d) at a distance (s) given by [Forward, 1984]

$$Dd = 2.44s\lambda \tag{11.24}$$

where the 2.44 factor shows that the spot-size diameter (d) is measured not at the half-power point but at the first null in the Bessel function for a circular aperture, where 84% of the transmitted power falls. If the collecting aperture on the receiving vehicle equals or exceeds this spot size, it collects essentially all of the transmitted electromagnetic power. If the receiving antenna has a smaller diameter, the power collected is proportional to the relative areas.

For example, a 1-km diameter transmitter at 3-cm wavelength (10 GHz) can fill a 30-m diameter collector at 400 km (LEO). A 10-km diameter transmitter antenna at 1-cm wavelength (30 GHz) can fill a 100-m diameter collector at 36,000 km (GEO). A 30-m mirror focusing 1 µm infrared laser wavelength can fill a 30-m collector at the Moon and a 10-km collector at Mars. Larger space-based lenses or mirrors can even transmit laser power over interstellar distances (see Sec. 11.4.4). Although the transmitting apertures must be carefully constructed to transmit a nearly diffraction-limited beam, the receiving apertures can be low-tolerance "rectennas," solar-cell arrays, "light buckets," or reflective sails.

Conceptually, beamed microwave or laser radiation could be reflected off a conducting sail to provide thrust from the radiation pressure, as is done with sunlight reflecting off a solar sail, discussed in Sec. 11.3.1. This process is not energy-efficient, however, because the reflected radiation still retains most of its energy (the infinite specific impulse of the radiation being poorly matched to the mission Δv). Only when the reflecting sail is moving at speeds approaching the speed of light (see Sec. 11.4.4), where the reflected radiation is Doppler shifted to lower frequencies, can reflective-sail systems driven by beamed power become efficient.

Beamed microwave and laser power received at a vehicle can be converted to electrical power through rectennas or solar-cell arrays, which can then operate any sort of electric-propulsion system. If the electric-propulsion system happens to use microwave power, we may coherently collect the microwaves with a receiving antenna and send them directly to the propulsion system. The major experimental effort to date concerning beamed power has been for laser-thermal propulsion.

Beamed-Laser Thermal Propulsion

Most of this effort has focused on the nozzle-less planar thruster [Kantrowitz, 1972]. As Fig. 11.12 shows, the payload sits on a solid block of ablative propellant such as plastic or water ice. An "evaporation" laser pulse ablates a few-micrometer-thick layer of propellant, forming a thin layer of gas. A second laser pulse then "explodes" this gas layer, producing thrust on the plate of propellant. The process takes a few microseconds and is repeated at 100–1000-Hz rates. An important feature is that we do not need a nozzle because the explosive expansion takes place so close to the plate of propellant.

The resultant thrust is normal to the plate and independent of the direction of the incident laser light, allowing the vehicle to fly at an angle to the laser beam. The vehicle can therefore transition into a near-circular orbit without requiring an apogee kick motor. The vehicle is steered from the ground by moving the laser beam

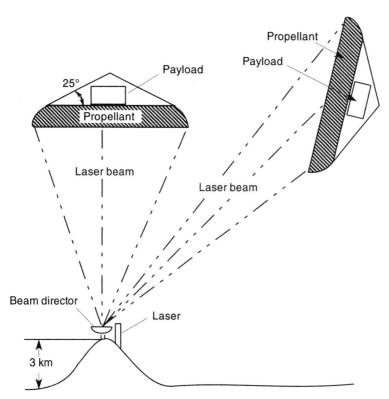

Fig. 11.12. Schematic of a Generic Rocket Using a Flat-Plate Laser. The figure shows a laser rocket in two possible flight attitudes. A high-energy laser pulse ablates a solid propellant. A second pulse "explodes" the vapor, producing a thrust pulse [Kare, 1989].

off the center of the base plate, so the vehicle does not need an onboard guidance system. The payload size depends on the laser power; 20 MW can launch a 150-kg vehicle carrying a 20-kg payload in a single stage. Higher powers can launch proportionately larger payloads. Peak accelerations compare well to those of chemical rockets. Experiments [Kare, 1989] at several industry and government laboratories show the double-pulse thruster concept works, producing high thrust efficiency and specific impulses up to 800 seconds. The actual thrust efficiencies obtained to date are only about 10%.

Another approach to laser propulsion using beamed power is to absorb the laser light in a plasma "flame" sustained by laser light focused in the center of a flowing stream of propellant gas. Thrust levels as high as 10,000 N with specific impulses of 1000 seconds appear achievable using hydrogen as the propellant gas. Experimenters working with a 1-kW CW CO_2 laser have reported an absorption efficiency of 86% and a thermal efficiency of 38% in an argon plasma at 2.5 atmospheres. The experiments were repeated using an rf-linac free-electron laser that produces a 0.1-ms burst of 10-ps pulses separated by 46 ns. These short pulses ignited the argon gas and formed a plasma that absorbed 92% of the laser power [Keefer, 1990].

Laser powers as low as 1 MW would be useful for raising vehicles from LEO to GEO orbit without relay optics. 10–100-MW lasers can launch small payloads from the ground. With up to 100 launches a day, a launcher of 20-kg payloads, using 20 MW of power, could place several hundred tons in orbit per year. Beamed-power laser rockets will have better payload fractions than chemical rockets because the heavy power plant remains on the ground and the higher specific impulse results in lower propellant fractions.

11.3.3 Catapult-Propulsion Systems

Catapult-propulsion systems use a fixed energy source powering a catapult "engine" with a long aspect ratio to apply external propulsive forces to a payload-carrying vehicle as it moves along the catapult. A catapult can also be used as a type of electromagnetic rocket in which the catapult is carried aboard the vehicle and fires small pellets as the expellant mass. The discussion in this section, however, concerns fixed catapults, either on the surface of a planetoid or orbiting a planetoid, that apply propulsion to a payload-carrying vehicle.

The basic performance characteristics of all catapults depend on simple mechanics equations. Because the payloads to be catapulted can stand only a certain maximum acceleration, we assume the catapult has a constant acceleration at that maximum or below. At a constant acceleration (a), the velocity of the projectile v as a function of time (t) is

$$v = at \tag{11.25}$$

The distance, s, that the projectile has traveled along the catapult at any time (t) is

$$s = \frac{1}{2}at^2 = \frac{v^2}{2a} \tag{11.26}$$

The energy (E) in the projectile as a function of time is

$$E = \frac{1}{2}mv^2 = \frac{ma^2t^2}{2} \tag{11.27}$$

which means the power (P) that the catapult must supply as a function of time is

$$P = \frac{dE}{dt} = ma^2t = mav \tag{11.28}$$

which increases linearly with time. Because most driving sources have limited power, this means that after the projectile attains the power-limited speed, the catapult must switch over from constant acceleration to constant power. The acceleration then drops with time, so we must use other equations.

Let us put some general numbers into Eqs. (11.25) to (11.28). We need to consider terminal velocities for lunar escape (3 km/s), Earth orbit (8 km/s), Earth escape (12 km/s), and a planetary mission (30 km/s). Our range of accelerations must include the Earth's acceleration (1 g = 10 m/s²), nominal crewed-mission accelerations (3 g's = 30 m/s²), maximum crewed-mission accelerations (10 g's = 100 m/s²), and unmanned-payload accelerations of 1000 and 10,000 m/s². Table 11.3 shows the lengths of the catapults needed for these terminal velocities and accelerations, as calculated from Eq. (11.26).

Table 11.3. Lengths of Catapults for Various Velocities and Accelerations. As the allowable acceleration increases, the required cable length decreases. Cable length increases for a higher velocity requirement.

Velocity (km/s)	Acceleration (m/s²) (g's)	Catapult Length (km)				
		10 / 1	30 / 3	100 / 10	1000 / 100	10,000 / 1,000
3		450	150	45	4.5	0.5
8		3200	1100	320	32	3
12		7200	2400	720	72	7
30		45,000	15,000	4500	450	45

This table shows we need catapults hundreds to thousands of kilometers long to launch crewed payloads from the Earth into orbit or escape. This length seems extreme, but remember that when the Space Shuttle takes off from Earth, its engines fire for many minutes, supplying three or more g's of acceleration to the vehicle over the long down-range trajectory that extends for thousands of kilometers over the ocean. Because of the difficulty of finding a site for such a long catapult on Earth, and the amount of mass that must be placed into orbit to construct a long catapult in space, catapults usually apply only to uncrewed payloads that can take high accelerations. At high accelerations, the lengths of the catapults drop to a more manageable hundreds of meters to tens of kilometers. Many types of catapults have been proposed and studied. Those that may be able to achieve at least Earth-orbital speed, with payloads of several kilograms to one metric ton, appear in the following sections.

Light-Gas Guns

A light-gas gun typically has two stages [Seigel, 1979], as Fig. 11.13 illustrates. The first, or *driver*, stage is a long, large-bore, gas-filled tube that has a drive piston at one end. This tube is connected by a tapered section to the second, or *launcher*, stage, which is a very long, smaller-bore, evacuated tube. A diaphragm designed to fail at a given pressure separates the two tubes, and the projectile to be launched is placed in the second tube near the diaphragm. The "light" gas used is typically a mixture of hydrogen and helium. The first stage is similar to a conventional gun in that an explosive drives the massive piston down the tube, compressing the gas ahead of it until the gas reaches a high temperature and high pressure. At a pressure determined by the break point of the diaphragm, the gas releases into the launcher stage, where it pushes the projectile down the bore at nearly constant acceleration and out the muzzle at high velocity. The piston (with a deformable plastic front end) stops in the tapered section. Muzzle velocities up to 10 km/s are possible for small projectiles. Present research is concentrated on attaining higher velocities with more massive projectiles, so we can throw multi-kilogram to metric-ton payloads of fuel and structural materials off the Earth at high accelerations.

Ram Accelerators

The *ram accelerator* [Hertzberg, 1991; Bruckner, 1991] is a ramjet-in-tube concept for accelerating projectiles to very high velocities. Its propulsive cycle is similar to the aerothermodynamic cycle that generates the thrust in a conventional, supersonic, airbreathing ramjet; however, the device operates differently. There is no propellant on board the projectile. Instead, the projectile travels through a stationary tube filled with a premixed gaseous fuel and oxidizer mixture. The projectile acts as the centerbody of a conventional ramjet, while the tube acts as the outer cowling of the ramjet. The combustion process travels with the projectile, generating a pressure distribution which produces forward thrust on the projectile. Several modes of operation are possible for ram accelerators, distinguished by the operating velocity range and the way the combustion process starts and stabilizes. In subsonic, *thermally choked* combustion, a subsonic combustion zone stabilizes behind the bluff base of the projectile (see Fig. 11.14), allowing the projectile to be accelerated to the detonation (supersonic combustion) speed of the gaseous propellant mixture.

A transdetonative mode follows to smoothly transition from 85% to 115% of the detonation speed. We do not understand completely how combustion works, but we believe it occurs partially on the projectile body. Next is the *superdetonative* mode, during which the projectile accelerates while always traveling faster than the detonation speed. In this mode, oblique shock waves ignite supersonic combustion as shown in Fig. 11.15.

These various operational modes should allow the ram accelerator to span the velocity range of 0.7 to 12 km/s. Experiments to date have used a ram accelerator

11.3 Nonrocket Advancements

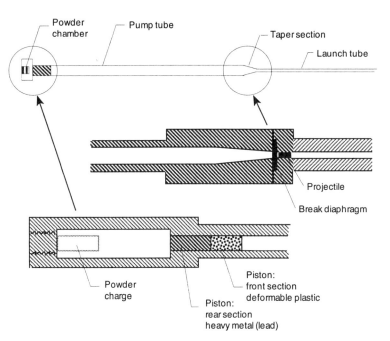

Fig. 11.13. Concept of the Light-Gas Gun. The powder charge propels a piston down the pump tube, compressing the gas contained there. At a given pressure, a diaphragm bursts, releasing the high-pressure gas into the launch tube and accelerating the projectile [Seigel, 1979].

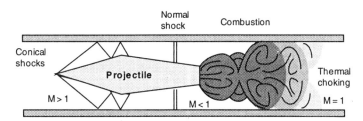

Fig. 11.14. Ram Accelerator with Subsonic Combustion and Thermal Choking. A projectile travels down a tube containing a combustible gas. Combustion behind the projectile provides the pressure to accelerate the projectile. This shows what goes on at slower speeds.

which is 16 m long, has a bore 38 mm in diameter, and employs 45 to 90-g projectiles. They have attained velocities up to 2.7 km/s [Hertzberg, 1991]. Research has started on ram accelerators with larger calibers, including a 90-mm system at the Institut Saint-Louis in France and a 120-mm system at the U.S. Army Aberdeen Proving Ground in Maryland. We expect the ultimate limit to the attainable veloc-

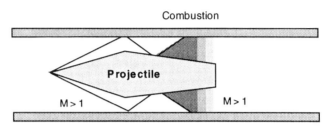

Fig. 11.15. Ram Accelerator during Supersonic Combustion. The figure shows the shock and combustion configuration for projectiles moving faster than the detonation speed of the propellant mixture.

ities to depend on the heating and drag produced as the projectile moves through the stationary fuel mixture.

Magnetic-Coil Accelerators

Magnetic-coil accelerators or *mass drivers* use a series of sequentially energized magnetic coils to push or pull a conductive "bucket" carrying a payload. Magnetic-coil accelerators use electrical power instead of chemical "fuel" as the light-gas guns and ram accelerators do. As a result, they have been the preferred approach for catapult-propulsion systems in space-application studies involving lunar mining, asteroid mining, and asteroid moving. They can accelerate much higher than 10,000 m/s^2 or 1000 g's and can potentially attain lunar-escape velocity (greater than 2.3 km/s) with catapult lengths of only a few hundred meters. Work on mass drivers tapered off in the late 1970s and early 1980s after experimenters constructed several models with lengths up to 2.5 m, payloads of tens of grams, and accelerations up to 18,000 m/s^2 (1800 g's) [Snively, 1983]. Because these models were short, velocities were only 20 m/s. The mass of the coils, capacitor banks, and power supplies makes these magnetic-coil accelerators massive and complex for space, but they require no expendable mass, such as fuel, to operate.

Cable Catapults

Another type of electromagnetic accelerator is the *cable catapult* [Forward, 1991], which uses a long space tether with many separated, conductive strands as a launch rail. The cable points in the desired direction of travel (Fig. 11.16), and a payload attaches to a linear electromagnetic motor capable of traveling along the tether inside the conductive strands. A power supply generates radio-frequency power which travels down the tether's interior, functioning as an rf transmission line. The rf power is absorbed by the motor, as a moving load matched to the transmission line. The linear motor accelerates along the cable until the payload reaches the desired launch velocity, at which point the payload is released. The linear motor then decelerates to a halt to await the arrival of an incoming payload.

Because the cable "track" is always under tension, this concept can have an extremely low mass for a space catapult. The "mass ratio" of the cable mass (M) to the mass of the motor plus payload (m) has been shown [Forward, 1991] to be

$$\frac{M}{m} = \frac{v^2}{2u^2} \qquad (11.29)$$

where we assume the mass of the power supply is negligible compared to the mass of the cable, and v is the final velocity reached by the payload. The "characteristic velocity" of the cable material (u) is given by

$$u = \sqrt{\frac{F_{ro}}{\rho}} \qquad (11.30)$$

where F_{ro} is the design tensile strength and ρ is the density of the tether material.

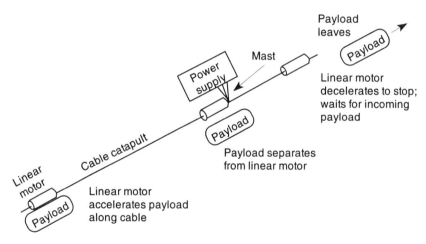

Fig. 11.16. **Schematic of a Cable Catapult and Payload.** Power is transmitted to a linear motor through a cable. The motor accelerates along the cable until it reaches the desired speed. The payload is then released, and the motor decelerates.

The "mass ratio" is independent of the acceleration, which means that a very long cable operating at low accelerations has the same mass as a shorter cable operating at higher accelerations. Also, the "mass ratio" for a cable catapult rises only as v^2, whereas a rocket's mass ratio rises as the exponential of v, and the mass ratio of a rotating tether rises as the exponential of v^2 (Eq. 11.32). Although linear motors exist, there has been no design work on an rf-powered motor capable of moving along flexible strands at high accelerations, so the practical limitations of the cable catapult are unknown.

11.3.4 Tether-Propulsion Systems

Tether propulsion is an advanced space-propulsion technology [Bekey, 1986] that has already been flight-tested. The first space test of a 2.5-mm diameter, 20-km long, electrodynamic tether using the Space Shuttle was attempted in 1992. Unfortunately, the test had to be aborted early because of a minor mechanical problem with the deployment reel. Future Shuttle experiments involve a repeat of the 20-km, upward-going, electrodynamic tether, then a 120-km downward-going tether trolling an aerodynamic experiment through the Earth's upper atmosphere. A Small Expendable-tether Deployment System (SEDS) with a 20-km long, 0.75-mm diameter tether successfully deorbited a small payload from an expendable second stage in early 1993. More SEDS experiments are planned. The following sections describe some ways in which a long tether can propel a vehicle. Specific tether programs and detailed design data are in the *Tethers in Space Handbook* [Penzo, 1989], which also has an extensive bibliography.

Vertically Stabilized Tethers

The simplest way to use a tether for propulsion in space is merely to pay out a payload-carrying vehicle from the main vehicle on the end of a long cable or tether. The Earth's gravity-gradient field combines with the orbital-motion gradients to cause the payload to go up or down with respect to the orbital altitude of the system's center of mass. After the transients have died out (or are damped out by judiciously using the drag and reel speed controls), the tether is vertically stabilized. If the main vehicle is in a circular orbit, the payload is along the radius vector between the center of the Earth and the main vehicle. It moves in a circular orbit at a higher or lower orbital altitude than the main vehicle, but moves at the same angular rate as the main vehicle. This means that if the payload is above the main vehicle, it is traveling faster (forced-overspeed circular orbit) than it normally would if it were in a pure circular gravity orbit at that altitude. If the tether is cut, the payload switches from its forced-overspeed, circular orbit to an elliptical orbit, with the cut point being the perigee of the elliptical orbit. Analysis has shown that a half-orbit later, the apogee of the elliptical orbit has an altitude that is higher than the original circular orbit by a distance that is 7 times the tether length [Section 4.4.5 of Penzo, 1989]. Thus a 120-km tether can transfer a payload from a nominal Space Shuttle altitude of 400 km to a higher orbit of 1240 km. There it could either fire a rocket to circularize its orbit or rendezvous with another vehicle at that altitude. A 5000-km long tether would suffice to toss the payload from low-Earth orbit to the 35,800-km altitude of geosynchronous Earth orbit. Downward-going tethers can be used to deorbit payloads from space without using rockets. For example, an SEDS with a 100-km long tether having a mass of 200 kg can deorbit 2000 kg.

11.3 Nonrocket Advancements

More ambitious vertically stabilized tethers that would attach to the two moons of Mars (Fig. 11.17) and range in length up to 6100 km have been proposed [Penzo, 1986]. They would provide a space-transportation system that can send payloads from low Martian orbit to escape or back to Earth without using rockets (needs power supplies on the moons). Like all tether-transport systems, it operates equally well in both directions, and once set up, it requires no more fuel to keep it operating.

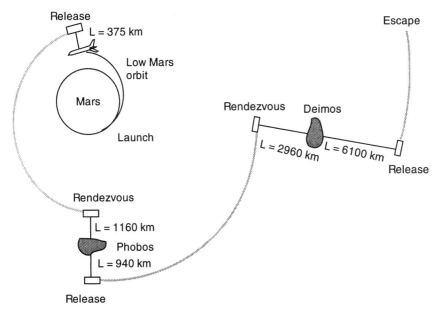

Fig. 11.17. Tether-Transport System for Mars. By judiciously using the orbital energy of Mars' moons, tethers can progressively "lift" a payload to an escape orbit [Penzo, 1986, courtesy of the Jet Propulsion Laboratory].

All vertically stabilized tethers are actually rotating tethers because they are inertially rotating once per orbit. If the tethers are deliberately set into angular rotation, the amount of propulsion available increases dramatically.

Librating and Rotating Tethers

A *librating* or swinging tether can change the orbit-transfer abilities of the vertically stabilized tether. Because the gradients in space act as a pseudo-gravity field, the payload can swing back and forth about its nominal vertical position. Depending on when the payload is cut loose from the tether, the nominal change in altitude of 7 times the tether length can vary from 1 to 13 times the tether length [Section 4.4.5 of Penzo, 1989]

A rapidly rotating tether can act like a sling to toss payloads at a velocity equal to the center-of-mass velocity plus the tip velocity. The same system can "catch" payloads passing by the moving tether (either at the tip or somewhere along the tether arm) at low relative velocity. For a rotating tether throwing a payload-carrying vehicle, the required ratio of the mass (M) of the tether to the mass (m) of the vehicle is a function of the ratio of the maximum tip speed (v) of the rotating tether to the "characteristic velocity" (u) of the tether material given by Eq. (11.30). From [Moravec, 1977],

$$\frac{M}{m} = \sqrt{2\pi}\, erf\left(\frac{v}{2u}\right)\frac{v}{u}\, e^{v^2/2u^2} \qquad (11.31)$$

where *erf* is the error function. We can compare this mass ratio to the ratio of the mass (M) of the first stages of a rocket vehicle to the payload-carrying last stage mass (m), where v is the terminal velocity of the last stage, and v_e is the "exhaust velocity" of the fuel:

$$\frac{M}{m} = e^{v/v_e} - 1 \qquad (11.32)$$

which is exponential in v/v_e, whereas the rotating cable mass is exponential in the square of v/v_e. For a cable catapult (Sec. 11.3.3), the ratio of the tether mass to the payload mass goes only as the square of v/v_e, not the exponential of the square [Eq. (11.29)].

Because of the squared exponential growth of the rotating tether's mass, the maximum launch velocity attainable for practical launcher-to-payload mass ratios is three times the characteristic velocity of the cable material: 3 km/s for a 1-km/s Kevlar™ cable. A cable catapult using the same amount of cable material could conceptually give the payload a launch velocity of 30 times the cable's characteristic velocity, or 30 km/s. A rotating tether, however, is very simple, uses known materials and technology, and as shown below, can perform missions that are difficult and expensive to do with rockets.

A tether "Earth-lift" facility using the improved tether material Spectra™ has been designed to lift payloads out of the Earth's gravity well. The 300-km-long tethers would have a mass of only 7 tons, yet could provide greater than a 1 km/s Δv to a 10-ton payload. Two of these facilities could raise 10-ton payloads from LEO to escape. One tether facility would be placed in a circular 400-km orbit and another in a highly elliptical orbit (EEO) with a 4:1 period resonance. Payloads would be picked up from an Earth orbit at 150 km or less by the lower tether and tossed into an intermediate elliptical orbit with an orbital period twice that of the lower facility and half that of the upper facility. There, the higher tether would pick up the payloads and toss them to escape or to the Moon.

We can adjust the length, orbital altitude, and rotation rate of a rotating tether so the orbital altitude equals the tether arm's length and the tip speed matches the orbital velocity of the center of mass. Thus, a long rotating tether in orbit around the Moon or small airless planet can touch down to the surface six times an orbit, simultaneously dropping off and lifting up payloads weighing a reasonable fraction of the tether mass [Moravec, 1977]. Because the tether's tip speed is equal to the orbital velocity of its center, the tip of the tether is not moving with respect to the planetoid's surface when it touches down. By having the tip's grapple mechanism play out and pull in tether around the touchdown time, the stay time on the surface can be extended to many tens of seconds. At the Moon, the minimum mass of a Kevlar™ tether is 13 times the payload to be handled, or 130 tons for a 10-ton payload. The length of each arm of the rotating tether is 1/6th the diameter of the planetoid, which is also the orbital altitude of the tether's center of mass. For the Moon, this is 580 km, or a total length of 1160. For Mars, the mass ratio is 1100 for Kevlar™. Better materials can lower this figure, but no material known can construct a rotating touchdown tether that can operate around the Earth.

As shown in Fig. 11.18, combining the Earth-lift tether systems and the lunar-touchdown tether system results in a tether-transport system [Forward, 1991]. This system would have a total mass of only a few hundred tons but can move 10-ton payloads from LEO to the lunar surface and back again. By arranging things so an equal amount of mass flows in both directions, we can make this system self-powered. Bags of lunar dirt moving down the tether system into Earth's deep gravity well would be the "fuel" needed to move payloads from LEO to the surface of the Moon.

Pumped Tethers

By using an onboard power supply, we can *pump* a tether by contracting and extending its length. Because the tether is orbiting, it already has significant energy and angular momentum. The pumped tether can add and subtract energy from the total system with its power supply; it can add and subtract from the orbital angular momentum by changing its rotational angular momentum. Because the ends of the tether are in different parts of the Earth's gravitational field, a very long pumped tether can even couple to this gravitational field and therefore to the Earth's inertial mass. The capabilities gained by pumping a tether are numerous [Section 4.4.6 of Penzo, 1989], and we are not certain we have found all of them. One of the most bizarre is the ability of a pumped tether to "lift itself by its own bootstraps."

A spacecraft starting in a low circular orbit about Earth can use a power supply, a long tether, and a counterweight to "bootstrap" itself (along with the tether and the counterweight) up the gravity well and escape in less than a month without using propellant [Landis, 1991]. It works because, if two halves of a spacecraft (or a spacecraft and its expended booster) are extended on a long tether, the center-of-mass of the extended system shifts slightly downward from the original center-of-mass, and the orbital period decreases [Landis, 1991; Section 4.4.6 in Penzo, 1989]. This shift in the center-of-mass occurs because the Earth's gravity force causes an acceleration on the masses that varies as $1/r^2$, while the counteracting

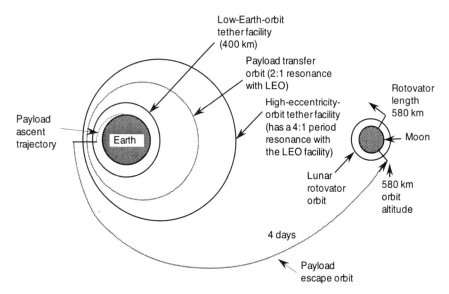

Fig. 11.18. Tether System for Transporting Payloads from LEO to the Moon's Surface. Two tether facilities can lift payloads from LEO to escape. A single facility at the Moon could capture this payload and lower it to the lunar surface.

centrifugal force due to orbital motion causes an acceleration that varies as r. For very long tethers, the two forces no longer exactly cancel at the two ends, so there is a residual, second-order force which must be balanced by a shift in the center-of-mass. When the tether is pulled in again, the center-of-mass of the combined system rises upward.

As Fig. 11.19 shows, by alternately extending and contracting the tether at proper points in the orbit, we can "pump" an initially circular orbit into a highly elliptical orbit. Theoretically, if the initial orbit is circular and at an altitude of greater than one Earth radius, the final orbit can be an escape parabola. Note that the angular momentum of the initial and final orbits are the same, so we do not have to supply angular momentum. The energy of the escape parabola is much greater than the energy of the initial circular orbit, so we must supply energy—either from an onboard power supply or by collecting externally supplied power. The final configuration has the vehicle, tether, and counterweight flying away from the Earth at some residual velocity, so it has some linear momentum. To conserve linear momentum, the tether has transferred linear momentum to the Earth by coupling to the Earth's tidal gravity fields through its extended length. Although the system appears to be "pulling itself up by its bootstraps," it is not. In effect, the tether is "climbing" out of the Earth's gravity well by coupling to the nonlinearities in the gravitational gradient fields or gravity tides.

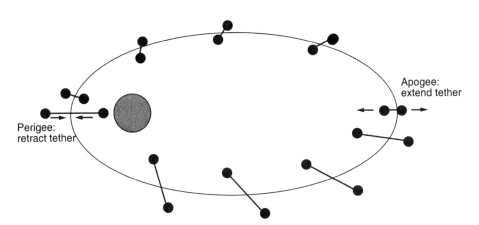

Fig. 11.19. "Bootstrap" Propulsion Using a Pumped Tether. By alternately extending and contracting a tether, we can change a circular orbit to an elliptical one. Angular momentum is unchanged, but the tether deployment mechanism adds energy to the orbit [Landis, 1991, courtesy of NASA-Lewis].

Propulsion Using an Electrodynamic Tether

A conducting tether can provide electric propulsion. In the electrodynamic-tether experiment attempted from the Space Shuttle in 1992, a conducting tether 2.5 mm in diameter and 20 km long was to be used to deploy a spacecraft upward from the Shuttle. The electrically positive spacecraft would collect electrons from the ionosphere and pass them down the conductive tether to the Shuttle, which would emit the electrons back into the ionospheric plasma with the help of an electron emitter [Section 1.1.1 in Penzo, 1989]. The motion of the conducting cable through the magnetic field of the Earth produces up to 5 kV of electrical potential between the two ends of the tether and up to 5 kW of power. The electrical energy, of course, comes from a decrease in the Shuttle's kinetic energy. If the Shuttle had an onboard power supply, we could pump current through the cable and therefore generate thrust through the cable's "pushing" against the Earth's magnetic field.

Lifetimes of Space Tethers

The lifetime of tethers in space is limited by space debris: micrometeorites from outer space and an increasing number of fragments produced by human activity [Sections 4.3.5 and 4.5.2 in Penzo, 1989]. Analysts have recently reviewed measurements of crater number versus size on skin panels of the Long Duration Experimental Facility, which spent six years in space [Forward, 1992]. They found that most space debris still comes from micrometeorites, but NASA's old space debris model needs to increase by a factor of 2.33. For single-strand space tethers with diameters greater than 1 mm, the lifetime is given by [Forward, 1992]:

$$t = \frac{1}{\pi\sigma D_0 L}\left(\frac{D}{D_0}\right)^{f-1} = \frac{385 \text{ km} \cdot \text{days}}{L}\left(\frac{D}{1 \text{ mm}}\right)^{2.46} \qquad (11.33)$$

where D = diameter of the tether (mm)
  L = length of the tether (m)
  D_0 = 1 mm (normalizing parameter)
  σ = 9.58 × 10^{-9} craters/s·m² (empirical space debris flux)
  f = 3.46 (an empirical parameter)

Using this equation, we can predict the lifetime of the electrodynamic tether with a diameter of 2.5 mm and a length of 20 km to be 183 days. For the follow-on aerodynamic experiment, using a tether with a diameter of 1.8 mm and a length of 120 km, the predicted $1/e$ lifetime is 14 days, or a probability of 7% that the tether would be cut in one day. Longer tethers would have even shorter lifetimes unless their diameter (and consequently their mass) were increased.

A fail-safe, multistrand tether overcomes this problem. It increases the single-strand lifetime by factors of 100 or more for a mass penalty of only 2 or 3 [Forward, 1992]. The lifetimes of this multistrand tether are very long, and the degradation is graceful and fail-safe. Thus, even for the improbable case of several cuts clustered together, these multistrand tethers should never need repair during their operational lifetime.

11.4 Interstellar Flight

The ultimate dream of those engaged in space propulsion is to be able to travel to the stars. In this section we show that going to the stars is extremely difficult but not impossible. To travel to the stars will take decades of time, gigawatts of power, kilograms of energy, and trillions of dollars. Yet, some technologies being developed for other purposes show promise of providing propulsion systems that will make rapid interstellar travel possible in the foreseeable future—if the world community decides to apply its energies and resources in that direction.

11.4.1 Relativistic Mechanics

Travel at significant fractions of the speed of light requires us to use equations for relativistic mechanics instead of classical mechanics [Feynman, 1963]. In relativistic mechanics, even the fundamental properties of time, length, and mass vary with the relative velocity of the observer. For example, the time (t) measured on a space vehicle moving at a velocity (v) relative to an observer is shorter than the time (t_0) measured by a stationary observer, by the relation

$$t = \sqrt{1 - \frac{v^2}{c^2}}\, t_0 \qquad (11.34)$$

where $c = 3 \times 10^8$ m/s is the speed of light. This equation results in the so-called "twin paradox," by which a traveling twin, after taking a round-trip journey at relativistic speeds, is younger on return than the stay-at-home twin. This effect is real. It has been experimentally measured to high accuracy not only with subatomic particles but also with macroscopic clock "twins," one of which traveled around the world at airline speeds of 1000 km/hr (~10^{-9} c) and "aged" measurably less than the clock twin that did not travel.

In the same manner, the length (ℓ) of a space vehicle moving along the direction of the velocity (v), as measured by a "stationary" observer, is shorter than the length (ℓ_o) measured when the vehicle is at rest by

$$\ell = \sqrt{1 - \frac{v^2}{c^2}}\, \ell_o \tag{11.35}$$

while the mass (m_v) of a moving space vehicle is greater than the rest mass (m_{vo}) of the vehicle by the relation:

$$m_v = \frac{m_{vo}}{\sqrt{1 - \frac{v^2}{c^2}}} \tag{11.36}$$

Another result of relativistic mechanics is that nothing can go faster than the speed of light. Suppose a vehicle is moving at a velocity (v) and shoots a projectile forward with a velocity (w) with respect to the moving vehicle. In this case the velocity (u) of the projectile as seen by the stationary observer is not $u = v + w$. Instead, it is

$$u = \frac{v + w}{1 + vw/c^2} \tag{11.37}$$

This equation always produces a velocity for u that is less than the speed of light.

Rocket Equation for Relativistic Velocities

The rocket equation for the mass ratio of a relativistic rocket [Ackeret, 1947] is different from the one for the classical mass ratio. In the classical derivation, a rocket with an initial mass (m_v) ejects an amount (dm) of expellant at an exhaust velocity with respect to the rocket of v_e, assumed to be a constant. In the system's center-of-mass, the rocket's resultant velocity is v and the expellant's velocity is u_e. For the relativistic derivations, the masses are replaced with their relativistic equivalents, which vary with the velocities v and u_e [Eq. (11.34)].

	Classical	Relativistic
body	m_v	$\dfrac{M_{vo}}{\sqrt{1-\dfrac{v^2}{c^2}}}$
exhaust	dm	$\dfrac{dm_o}{\sqrt{1-\dfrac{u_e^2}{c^2}}}$

Below, we compare the derivation of the classical rocket equation (left set of equations) with the relativistic derivation (right set of equations). From the law of conservation of mass (mass-energy in the relativistic case):

$$dm_v = -dm \qquad d\left[\frac{m_{vo}}{\sqrt{1-\dfrac{v^2}{c^2}}} \cdot c^2\right] = \frac{-dm_o}{\sqrt{1-\dfrac{u_e^2}{c^2}}} \cdot c^2 \qquad (11.38)$$

From the law of conservation of momentum:

$$d(m_v v) = u_e \cdot dm \qquad d\left[\frac{m_{vo}}{\sqrt{1-\dfrac{v^2}{c^2}}} \cdot v\right] = u_e \cdot \frac{dm_o}{\sqrt{1-\dfrac{u_e^2}{c^2}}} \qquad (11.39)$$

From the law of addition of velocities:

$$u_e = v_e - v \qquad u_e = \frac{v_e - v}{1 - \dfrac{v_e v}{c^2}} \qquad (11.40)$$

Expanding the derivatives and combining the above equations produces

$$\frac{dm_v}{m_v} = -\frac{dv}{v_e} \qquad \frac{dm_{vo}}{m_{vo}} = -\frac{dv}{v_e\left(1-\dfrac{v^2}{c^2}\right)} \qquad (11.41)$$

Integrating this result gives us

11.4 Interstellar Flight

$$\ln m_v = -\frac{v}{v_e} \qquad \ln m_{vo} = \frac{c}{2v_e}\ln\left[\frac{c+v}{c-v}\right] \qquad (11.42)$$

If the initial mass of the rocket is m_i, and the final mass is m_f, the rocket mass ratio (m_i/m_f) needed to reach the mission velocity $v = \Delta v$ is

$$\frac{m_i}{m_f} = e^{\Delta v/v_e} \qquad \frac{m_i}{m_f} = \left[\frac{1+\dfrac{\Delta v}{c}}{1-\dfrac{\Delta v}{c}}\right]^{c/2v_e} \qquad (11.43)$$

These are the classical and relativistic rocket equations. For the relativistic case, the maximum exhaust velocity for the expellant is given by

$$v_e = [e(2-e)]^{\frac{1}{2}} c \qquad (11.44)$$

where e is the fuel mass fraction converted into the expellant's kinetic energy.

11.4.2 Astronomical Data and Nearest Stellar Systems

The distances between stars is so large that a new unit of distance needs to be defined. A light-year is the distance that light travels in one year (3.16×10^7 s) at its speed of 3×10^8 m/s, or 1 light-year = 9.46×10^{15} m. This is 63,000 times the distance from the Sun to the Earth, which is one astronomical unit (1 AU = 1.50×10^{11} m). Within 20 light-years from the Sun, there are 59 stellar systems containing 81 visible stars. The nearest star system, Alpha Centauri, is 4.3 light-years from Earth. It is listed in Table 11.4, which gives prime targets for interstellar missions [Forward, 1976].

11.4.3 Time, Acceleration, Velocity, and Energy Requirements

The nearest star system is 4.3 light-years away, so even if we could rapidly accelerate a vehicle to near the speed of light, the vehicle would take a minimum of 4.3 years to arrive at Alpha Centauri and another 4.3 years to get information to Earth, either by radio or by returning. Thus, an interstellar mission takes at least 8.6 years, much longer than most of the interplanetary missions being carried out. The first real interstellar missions, probably by robotic probes, will require time to accelerate up to speed and will use sub-relativistic cruise speeds, so they will take many decades. But if a mission takes too long, a follow-on mission launched decades later with a better propulsion system may pass the old spacecraft. Thus, unless an interstellar mission can be completed in fewer than 50 years, it probably should not be started. Instead, the money should be invested in designing a better

Table 11.4. **Prime Targets for Interstellar Missions.** The table shows the key parameters of five of our closer star systems.

Stellar System	Distance (light years)	Remarks
Alpha Centauri	4.3	Closest system. Triple (G2, K5, M5). Component A similar to Sun (a G2 star).
Barnard's Star	6.0	Small, low luminosity M5 red dwarf. Next closest to solar system.
Sirius	8.7	Large, very bright A1 star with a white dwarf companion.
Epsilon Eridani	10.8	Single K2 star slightly smaller and cooler than the Sun. May have a planetary system similar to that of the solar system.
Tau Ceti	11.8	Single G8 star similar to the Sun. High probability of having a planetary system similar to that of the solar system.

propulsion system. As a result, we expect missions to the nearest stars to take between 8.6 and 50 years.

A requirement of 50 years for a mission produces acceleration and velocity requirements. If a vehicle accelerates at one Earth gravity (1 g) for one year (which would require a propulsion system better than any presently imagined), it would reach greater than 90% of the speed of light in a distance of only 0.5 light-years. It could coast for 3 years and then decelerate for one year to arrive in Alpha Centauri in about 5 years. This time is not much longer than the minimum travel time of 4.3 years. Thus, vehicle accelerations greater than 1 g do not significantly improve the mission time to the nearest stars. Accelerations less than 0.01 g result in mission times that are too long. A vehicle with an acceleration of 0.01 g takes 20 years to reach the Alpha Centauri half-way point and reaches only 20% of the speed of light. It then takes 20 years to decelerate. Add in the 4.3 years of communication time, and the mission time approaches the maximum time requirement of 50 years. Thus, missions to the nearest stars require accelerations between 0.01 and 1.0 g.

A time requirement of less than 50 years and an acceleration requirement of between 0.01 and 1.0 g produce requirements on the vehicle's cruise velocity. Tables 11.5 and 11.6 show the range of mission times of a one-way-flyby probe to the nearest star and a distant star, assuming an acceleration of 0.1 g to various maximum cruise velocities.

As a probe's coast velocity increases, the coast time decreases, but the acceleration periods lengthen. Increasing the coast velocity does not significantly improve the mission times to the nearest stars. These missions require cruise velocities between 0.1 and 0.5 times the speed of light (c).

We can conclude that, if the mission time is constrained to less than 50 years, a usable interstellar probe destined for the nearest stars needs to accelerate at 0.01 g–1.0 g (typically 0.1 g or 1 m/s^2) and have a velocity of 0.1–0.5 c (typically c/3 or

Table 11.5. Mission Times to Alpha Centauri at 4.3 light years. The table shows the variation in mission duration for various flight profiles. It would take 4.3 years after arrival to receive any data.

Maximum Velocity (c)	Acceleration Time-0.1 g (years)	Coast Time (years)	Mission Time (years)
1.00	10	0	14
0.50	5	6	15
0.40	4	9	17
0.30	3	13	20
0.20	2	21	27
0.10	1	43	48
0.05	0.5	85	90

Table 11.6. Mission Times to Tau Ceti at 11.8 light years. (Thirteen other star systems are closer.) This table shows the variation in mission duration for various flight profiles. It would take 12 years to receive a communication after arrival.

Maximum Velocity (c)	Acceleration Time-0.1 g (years)	Coast Time (years)	Mission Time (years)
1.00	10	6	28
0.50	5	21	38
0.40	4	28	44
0.30	3	40	55
0.20	2	60	74
0.10	1	120	133

10^8 m/s). At these speeds, relativistic effects are measurable but not large. For $v = c/3$, the time-dilation factor [Eq. (11.34)] is only 5.7%.

An interstellar probe requires a lot of energy. Assuming a flyby vehicle with a mass of 1 metric ton (1000 kg) and a coast velocity of $c/3$, the kinetic energy in the vehicle is 5×10^{18} J, about equal to the present energy consumed in the United States over three weeks. This amount of energy has a mass of 56 kg, so the vehicle's total mass is really 1056 kg, not 1000 kg. The actual energy requirements for the mission depend on the propulsion system's efficiency in converting input energy into kinetic energy of the vehicle.

With the minimum energy requirements determined, we can calculate the power requirements. Assuming an acceleration period of 3.3 years (producing a coast velocity of $c/3$ at an acceleration of 0.1 g), we must apply at least 50 GW of propulsion power to push 1000 kg to c/3 at 0.1 g. This is a remarkably low value. The Space Shuttle's three Main Engines operate at a power of 22 GW, so we have already built and flown engines that produce enough power and energy density for interstellar flight. If we use rockets, however, we must keep the rocket mass ratio reasonable. Thus, we must develop new rocket-engine concepts that have an exhaust velocity comparable to the required mission velocity of $c/3 = 10^8$ m/s or a specific impulse approaching 10^9 s. Otherwise, we must use propulsion systems without rockets (Sec. 11.3).

The "Impossibility" of Interstellar Flight

Papers have "proved" that interstellar flight is "impossible." These papers assumed that rockets are used for a round-trip manned mission to a star at 10 light-years or more away. The spacecraft is assumed to accelerate at 1 g until turnover, decelerate at 1 g, and then return in the same manner. Even assuming perfect matter-antimatter engines, when the mass ratio is calculated using the relativistic rocket equation [Eq. (11.43)], the amount of antimatter needed exceeds the mass of the Earth!

These papers are flawed not in their mathematics but in their assumptions. First, if we use nonrocket propulsion instead of a rocket, the spacecraft does not have to carry its expellant or energy source, the relativistic rocket equation does not apply, and the fuel mass required does not rise exponentially. Second, the assumption of constant acceleration at 1 g means that after one year, the rocket is close to the speed of light. Additional acceleration does not increase the speed or decrease the time of the mission (to the people back on Earth paying for the mission—the astronauts do benefit from the time-dilation effect); it only makes the vehicle and its load of fuel heavier and harder to push! It is better to optimize the mission so the acceleration and deceleration periods are only a year or so, with most of the mission carried out at a coast velocity somewhat below the speed of light. The energy and mass requirements are still high, but they are not "impossible."

11.4.4 Proposed Methods for Interstellar Flight

The presently conceived methods for achieving interstellar flight use many different technologies. A bibliography of all concepts is in Forward [1980] and its updates; Forward [1986] is a review of the more feasible concepts. Some concepts apply better to small flyby probes, whereas others are better for a round-trip manned mission. The first three discussed in the following sections are rockets that carry along their expellant, energy source, and engine. Any sort of rocket, even an antimatter rocket, has marginal performance for interstellar missions. Only nonrocket propulsion offers any prospects for travel to even the nearest stars. The most promising concepts involve some sort of beamed-power propulsion (Sec. 11.3.2). These nonrocket-propulsion systems keep the heavy parts of a vehicle (the expellant, the energy source, and the "engine" that puts the energy into the expellant) in the solar system. Because they are near the Sun, large amounts of mass are available, and we can maintain the energy source (usually the abundant sunlight) and the "engine" as the mission proceeds. The best technique seems to be beamed-laser propulsion.

Yet, beamed-laser propulsion is an inefficient way to put energy into a vehicle. At the start of the mission, most of the energy in the incident photons is still in the photons after they reflect from the sail. It is not until the vehicle velocity exceeds 0.5 c that the reflected photons are redshifted significantly, showing that much of the photon energy has gone into the vehicle. There must certainly be better and more energy-efficient methods to transport vehicles between the stars. Until they are found, however, these sections show that interstellar flight, although difficult, is not impossible.

Nuclear-Electric Rocket

The most advanced form of propulsion with high specific impulse available today is electric propulsion, and the electric power source with the highest energy density is a space nuclear reactor combined with some sort of thermal-to-electric generator. A nuclear-electric rocket consisting of a 1-MW_e nuclear reactor, powering a number of 12,500-s xenon ion thrusters, has been studied for an interstellar precursor mission out to 1000 AU* [Nock, 1987]. Such a system could reach a cruise velocity of 105 km/s (0.00035 c) at burnup of the nuclear fuel after 10 years, and would reach 1000 AU in 50 years. It would take 12,500 years to reach Alpha Centauri.

Nuclear-Pulse Rocket

The nuclear-pulse rocket described in Sec. 11.2.2 has been studied for its application to interstellar flight [Dyson, 1968]. Because the nuclear "pulse units" have a minimum size, the pusher plate and the vehicle would necessarily be large, with a payload of some 20,000 tons (enough to support a crew of hundreds). The total mass would be 400,000 tons, including a fuel supply of 300,000 fusion bombs massing about 1 ton each, resulting in a 4:1 mass ratio. The bombs would be exploded once every 3 s, accelerating the spacecraft at 1 g for 10 days to reach a cruise velocity of 10,000 km/s (c/30). At this speed, it would fly by Alpha Centauri in 130 years. To allow this ship to decelerate at the target star, it would need to have two stages, with the first stage weighing 160,000 tons. Although the nuclear-pulse rocket has a marginal performance as a starship, it could have been built and sent on its way twenty years ago.

---

* Remember, 1 AU is the average distance between the Earth and the Sun.

Antimatter Rockets

The antimatter rockets discussed in Sec. 11.2.3 have been evaluated for interstellar missions [Martin, 1982; Cassenti, 1984]. The mass ratio required is 5:1 up to a cruise velocity of 0.5 c, then rises slowly to 10:1 at 0.95 c. The amount of antimatter (m_a) needed for a vehicle of mass (m_v) carrying out a mission with velocity change (Δv) is given by Eq. (11.18). To send a 1-ton vehicle on a 48-year flyby mission to Alpha Centauri at a speed of one tenth the speed of light (0.1 c) (see Table 11.5), would require four tons of expellant and 9 kg of antimatter. The best exhaust velocity for this mission [Eq. (11.16)] is 0.063 c (19 Mm/s or a specific impulse of 190 million seconds). No such engine exists, although ideas have been published about various designs that use magnetic fields for reaction chambers and exhaust nozzles.

Interstellar Ramjet

One of the oldest concepts for interstellar transport (and the first of the non-rocket-propulsion systems) is the interstellar ramjet [Bussard, 1960 and entries in Forward (1980) and updates]. The *interstellar ramjet* consists of the vehicle carrying the payload, a fusion reactor, and a large scoop to collect hydrogen atoms from space. The hydrogen atoms are used as fuel in the fusion reactor, where the fusion energy is released and turned into kinetic energy of the reaction products, which are exhausted to provide thrust for the vehicle. Alternate versions merely collect the hydrogen atoms for use as expellant mass and heat the hydrogen either by onboard nuclear reactors, stored antimatter, beamed laser power, or beamed antimatter.

The ramjet is the only concept for interstellar transport that could reach ultrarelativistic speeds, because the faster it travels, the more fuel it gathers. Unfortunately, we have no workable designs for constructing a lightweight scoop that can collect enough hydrogen to achieve reasonable accelerations and cruise velocities. The main problem is that any magnetic-field configuration that is stronger near the "throat" of the scoop than at the perimeter repels the incoming charged particles rather than scooping them in. New ideas for scoops are needed before this concept can work.

Beamed-Pellet Propulsion

In the *beamed-pellet propulsion* system invented by Singer [see Forward, 1986], a long electromagnetic mass driver in the solar system accelerates small pellets toward the target star. Riding the pellet beam is the interstellar vehicle, which intercepts the pellets and reflects them back using a strong magnetic field. We do not need to guide the pellets to high absolute accuracy if we design the vehicle to be a self-adjusting "beam rider" and keep the angular variations between successive pellets small. Instead of having to hit a meter-sized vehicle at interstellar distances, the beam-guidance system needs only to ensure that the beam passes through the AU-sized planetary system of the target star. By rebounding the pellets from an expendable, unmanned, lead ship, we could decelerate the main vehicle at the target system.

Beamed-Microwave Propulsion

Beamed-microwave propulsion is suitable only for flyby interstellar missions using lightweight probes [Forward, 1986]. The short beaming distance of microwaves requires very high accelerations to reach near-relativistic speeds before the beam spreads too much. The basic structure of the vehicle is a large sail made of fine wire mesh. The payload consists of a large number of microcircuits embedded in the mesh at the junctions between the wires. As shown in Fig. 11.20, the mesh sail is pushed at high acceleration by a microwave beam formed by a large Fresnel zone-plate transmitter lens made of alternating sparse metal mesh rings and empty rings. The high acceleration allows the mesh to reach a coast velocity near that of light while it is still close to the transmitting lens.

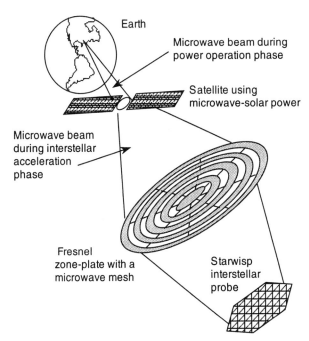

Fig. 11.20. Wire Mesh Interstellar Probe Being Pushed by a Microwave Beam. The Fresnel zone plate focuses a microwave beam onto the probe to provide thrust.

Upon arrival at the target star, the transmitter floods the star system with microwave energy. Using the wires in the mesh as microwave antennas, the microcircuits collect energy to power their optical detectors and logic circuits to form images of the planets in the system. The phase of the incident microwaves is sensed at each point of the mesh, and the phase information is used to form the mesh into a retrodirective, phased-array, microwave antenna that beams the images back to Earth. A minimal system would use a 1-km diameter mesh sail weighing 16 g and carrying 4 g of microcircuits. A 10-GW microwave beam would accelerate the mesh sail at 115 g's, reaching 0.2 c in a few days. Upon arrival at Alpha Centauri 21 years later, the mesh would collect 2 W of microwave power, which is enough to return high-resolution, television-frame-rate images as it flies through the system.

Beamed-Laser Propulsion

In *beamed-laser propulsion* [Forward, 1984], photon pressure from a solar-pumped laser array based in the solar system would push vehicles having large lightsails. The key element in the system is a final lens in the laser-beaming system that is 1000 km in diameter. With this size for the transmitting aperture, the spot size of the laser beam [Eq. (11.18)] would be 100 km in diameter (1/10th the size of the transmitting aperture) at Alpha Centauri's distance of 4.3 light-years and 1000 km at 44 light-years. Thus, all the transmitted laser power is available at the vehicle at any time during the mission, even over interstellar distances. The lens would be a Fresnel zone-plate consisting of rings of 1 µm plastic alternating with empty rings, with a mass estimated at 560,000 tons.

The diameter of the sail would be 3.6 km for an unmanned flyby probe using a 1-ton vehicle that is roughly one-third each of structure, sail, and payload. A laser power of 65 GW would give the sail an acceleration of 0.036 g's. If maintained for three years, this acceleration would produce a coast velocity of 0.11 c. The probe would fly through Alpha Centauri 40 years from launch.

A round-trip manned mission would involve larger-diameter sails and more massive vehicles that would require larger amounts of laser power. For a round trip to Epsilon Eridani, at 11 light-years, the diameter of the lightsail would be 1000 km, the same as the transmitter lens (a larger transmitter lens is not required). The lightsail would be divided into three nested, circular segments [Fig. 11.21]. The total vehicle mass would be 80,000 tons, including 3000 tons for the crew, their habitat, and their exploration vehicles. The lightsail would be accelerated at 0.3 g's by 43,000 TW of laser power. At this acceleration, the vehicle would reach 0.5 c in 1.6 years. The expedition would reach Epsilon Eridani in 20 years of Earth time and 17 years of crew time.

At 0.4 light-years from the target star, the 320-km rendezvous part of the sail detaches from the center of the lightsail and turns to face the 1000-km diameter ring sail that remains. The light from the solar system reflects from the ring sail, which acts as a mirror. The reflected light decelerates the smaller rendezvous sail

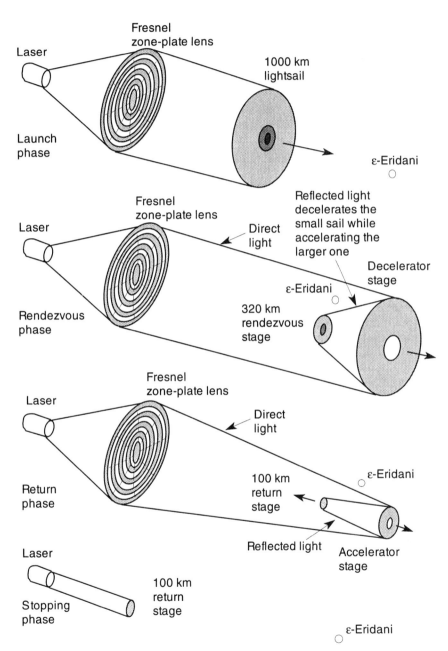

Fig. 11.21. Roundtrip Interstellar Flight Using Laser-Pushed Lightsails. By "staging" the lightsail at various points, we can do an interstellar return mission. (ε-Eridani refers to the star Epsilon Eridani.)

and brings it to a halt in the Epsilon Eridani system. The crew then explores the system for a few years using the lightsail as a solar sail.

For return, the 100-km diameter return sail detaches from the center and turns to face the 320-km diameter ring sail that remains. Laser light beamed from the solar system reflects from the ring sail onto the 100-km diameter return sail and accelerates it up to speed back toward the solar system. As the return sail approaches the solar system 20 Earth-years later, it is brought to a halt by a final burst of beamed laser power. Crew members have been away 51 years (5 years exploring) and have aged 46 years.

Beamed-laser propulsion seems to be the best available technique for interstellar travel because it uses known physics and known technology being developed for other purposes. Lightsails are being flight-tested for interplanetary travel. Researchers have demonstrated solar-pumped lasers and high-power, solar-electric sources for electrically powered lasers in space-power applications. Others have shown how to combine many laser beams into one clean coherent beam. Pointing that beam accurately has been demonstrated for space weapons. Interstellar, beamed-laser propulsion systems are larger than anything presently considered, but no new physics is involved. Beamed-power, laser-lightsail propulsion is energy inefficient because most of the energy in the incident laser beam remains in the reflected beam and does not transfer to the vehicle. We need better ways to transport payloads between the stars. But until that time, it is helpful to know at least one method conceptually allows round-trip, crewed travel over interstellar distances within a human lifetime.

References

Ackeret, J. 1947. On the Theory of Rockets. *Journal of the British Interplanetary Society.* 6: 116-123.

Augenstein, B. W., B. E. Bonner, F. E. Mills and M. M. Nieto, eds. 1988. *Antiproton Science and Technology.* Singapore: World Scientific.

Bekey, I. and P. A. Penzo. 1986. Tether Propulsion. *Aerospace America.* 24:40-43.

Bruckner, A. P., C. Knowlen, A. Hertzberg, and D. W. Bogdanoff. 1991. Operational Characteristics of the Thermally Choked Ram Accelerator. *Journal of Propulsion and Power.* 7:828-836.

Bussard, R. W. 1960. Galactic Matter and Interstellar Flight. *Astronautica Acta.* 6:179-194.

Cassenti, B. N. 1984. Optimization of Relativistic Antimatter Rockets. *Journal of the British Interplanetary Society.* 37:483-490.

Dyson, F. J. 1968. Interstellar Transport. *Physics Today.* October: 41-45.

Feynman, Richard P., Robert B. Leighton, and Matthew L. Sands. 1963. *The Feynman Lectures on Physics.* Reading, Mass: Addison-Wesley Publishing Co., Inc. [Volume 1, Chapters 15 and 16.]

Forward, Robert L. 1976. A Programme for Interstellar Exploration. *Journal of the British Interplanetary Society.* 29: 611-632.

Forward, Robert L., Eugene F. Mallove, Zbigniew Paprotny, Jurgen Lehmann and John Prytz. 1980. Interstellar Travel and Communication Bibliography. *Journal of the British Interplanetary Society.* 33:201-248; 1982 Update, 36: 311-329; 1984 Update, 37: 501-512; 1985 Update, 38: 127-136; 1986 Update, 40: 353-364.

Forward, Robert L. 1984. Roundtrip Interstellar Travel Using Laser-Pushed Lightsails. *Journal of Spacecraft and Rockets.* 21:187-195.

Forward, Robert L., Brice N. Cassenti, and David Miller. 1985. Cost Comparison of Chemical and Antihydrogen Propulsion Systems for High ΔV Missions. AIAA Paper 85-1455, AIAA/SAE/ASME/ASEE 21st Joint Propulsion Conference, 8-10 July 1985, Monterey, California.

Forward, Robert L. 1986. Feasibility of Interstellar Travel: A Review. *Journal of the British Interplanetary Society.* 39: 379-384.

Forward, Robert L. and Joel Davis. 1988. *Mirror Matter: Pioneering Antimatter Physics.* New York: John Wiley & Sons, Inc.

Forward, Robert L. 1990. Grey Solar Sails. *Journal of the Astronautical Sciences.* 38:161-185.

Forward, Robert L. 1991. 21st Century Space Propulsion. *Journal of Practical Applications in Space.* 2: 1-35.

Forward, Robert L. 1992. Failsafe Multistrand Tether Structures For Space Propulsion. AIAA Paper 92-3214 presented at the AIAA/SAE/ASME/ASEE 28th Joint Propulsion Conference, 6-8 July 1992, Nashville, Tennessee. See also: *Failsafe Multistrand Tether Structures for Space Propulsion.* Final Report by Forward Unlimited, P.O.Box 2783, Malibu, CA 90265 on Contract NAS8-39318 with NASA/MSFC, Huntsville, AL 35812, July 1992.

Friedman, L. 1988. *Starsailing: Solar Sails and Interstellar Travel.* John Wiley & Sons, New York.

Gabrielse, Gerald, et al. 1993. Extremely Cold Antiprotons for Antihydrogen Production, Proceedings of the Antihydrogen Workshop. Munich, Germany, 30-31 July, 1992. *Hyperfine Interactions.* 76:81–93.

Gerwin, R. A., G. J. Marklin, A. G. Sgro, A. H. Glasser. 1990. *Characterization of Plasma Flow Through Magnetic Nozzles.* Los Alamos National Laboratory. Final Technical Report AL-TR-89-092, February 1990, on Contract RPL: 69018 with Air Force Astronautics Laboratory, Edwards AFB, CA 93523, 1 May 1986 to 30 April 1987.

Haloulakos, V. E. and R. F. Bourque, 1989. *Fusion Propulsion Study.* AL-TR-89-005, Final Report on U.S. Air Force Contract F04611-87-C-0092. ASTIA Document No. A212935.

Hertzberg, A., A. P. Bruckner and C. Knowlen. 1991. Experimental Investigation of Ram Accelerator Propulsion Modes. *Shock Waves.* 1:17-25.

Hess, Harald F., et al. 1987. Magnetic Trapping of Spin-Polarized Atomic Hydrogen. *Physical Review Letters.* 59:672-675.

Kantrowitz, A. 1972. Propulsion to Orbit by Ground-Based Lasers. *Aeronautics and Astronautics.* 10:74-76.

Kare, J. 1989. Pulsed Laser Propulsion for Low. Cost, High Volume Launch to Orbit.

Lawrence Livermore National Laboratory preprint UCRL-101139, IAF Conference on Space Power, Cleveland, Ohio, 5 June 1989.

Keefer, D., A. Seghinasab, N. Wright, and Q. Zhang. 1990. Laser Propulsion Using Free Electron Lasers. 21st International Electric Propulsion Conference, Orlando, Florida, 18-20 July 1990.

Landis, G. A. and Hrach, F. J. 1991. Satellite Relocation by Tether Deployment. *Journal of Guidance and Control.* 14:214–216.

Laug, K. 1989. Solar Thermal Propulsion for Efficient Orbit Transfer. *Proceedings JANNAF Conference.* May 1989, Cleveland, Ohio. CPI Publication 515, Vol. 1, pp. 289–331.

Martin, Anthony R. and Alan Bond. 1979. Nuclear Pulse Propulsion: A Historical Review of an Advanced Propulsion Concept. *Journal of the British Interplanetary Society.* 32:283-310.

Martin, A. R., ed. 1982. Special Issue on Antimatter Propulsion. *Journal of the British Interplanetary Society.* 35: 387-424.

Moravec, H. 1977. A Non-Synchronous Orbital Skyhook. *Journal of the Astronautical Sciences.* 25:307-322.

McInnes, C. R. and J. F. L Simmons. 1989. Halo Orbits for Solar Sails—Dynamics and Applications. *ESA Journal.* 19:229-234.

Nock, K. T. 1987. TAU—A Mission to a Thousand Astronomical Units. Paper 87-1049 presented at the 19th AIAA/DGLR/JSASS International Electric Propulsion Conference, Colorado Springs, Colorado, 11-13 May.

Oates, G. C. 1988. *Aerothermodynamics of Gas Turbine and Rocket Propulsion. AIAA Education Series.* Washington, DC:American Institute of Aeronautics and Astronautics.

Paine, C. G. and G. M. Seidel. 1991. Magnetic Levitation of Condensed Hydrogen. *Reviews of Scientific Instruments.* 62:3022-3024.

Pecchioli, M. and G. Vulpetti. 1988. A Multi-Megawatt Antimatter Engine Design Concept for Earth-Space and Interplanetary Unmanned Flights. Paper 88-264 presented at the 39th Congress of the International Astronautical Federation, Bangalore, India, 8-15 October 1988.

Penzo, Paul A. 1985. Tethers for Mars Space Operations. Pg. 445–465. The Case for Mars II. McKay, C.P., editor. Vol. 62 Science and Technology Series, American Astronautical Society.

Penzo, Paul A. and H. L. Mayer. 1986. Tethers and Asteroids for Artificial Gravity Assist In the Solar System. *Journal of Spacecraft and Rockets.* 23:79-82.

Penzo, Paul A. and Paul W. Ammann. 1989. *Tethers In Space Handbook—Second Edition.* Office of Space Flight, NASA, Washington, DC 20546.

Seigel, A. E. Theory of High-Muzzle-Velocity Guns. *Interior Ballistics of Guns.* H. Krier and M. Summerfield, Eds. *Progress in Astronautics and Aeronautics.* American Institute of Aeronautics and Astronautics. 66:135-175

Shepherd, L. R. 1952. Interstellar Flight. *Journal of the British Interplanetary Society.* 11:149-167.

Snively, Leslie O. and Gerard K. O'Neill. 1983. Mass Driver III: Construction, Testing, and Comparison to Computer Simulation. *Space Manufacturing 1983*. Proceedings Space Manufacturing Conference held 9-12 May 1983, Princeton University, Princeton, New Jersey. *Advances in the Astronautical Sciences*. 53:391-401.

Solem, Johndale C. 1993. Medusa: Nuclear Explosive Propulsion for Interplanetary Travel. *Journal of the British Interplanetary Society*. 46: 21-26.

Sprik, R., J. T. M. Walraven and I. F. Silvera. 1983. Compression of Spin-Polarized Hydrogen to High Density. *Physical Review Letters*. 51:479-482.

Stwalley, William C., A. Marjatta Lyyra and Warren T. Zemke. 1991. Survey of Potential Novel High Energy Content Metastable Materials. pp. 1-5 *Proceedings of the High Energy Density Matter (HEDM) Conference*. 24-27 February 1991. Albuquerque, NM. M. E. Cordonnier, ed. Phillips Laboratory, Edwards AFB, CA 93523 PL-CP-91-3003, October, 1991.

U.S. Air Force, 1992. *Proceedings of the High Energy Density Materials Contractors Meeting*. Lancaster, CA 13-15 April 1992. See also: M.E. Cordonnier, ed. 1991. *Proceedings of the High Energy Density Matter (HEDM) Conference*. Albuquerque, NM 24-27 February 1991, report PL-CP-91-3003 dated October 1991; L. P. Davis and F. J. Wodarczyk, eds. 1990. *Proceedings of the High Energy Density Materials Contractors Conference*. Long Beach, CA 25-28 February 1990; and prior conferences back to 1987.

Wright, Jerome L. 1992. *Space Sailing*. Philadelphia: Gordon and Breach Science Publishers.

Zubrin, Robert M. 1993. The Use of Magnetic Sails to Escape From Low Earth Orbit. *Journal of the British Interplanetary Society*. 46:3-10.

Zubrin, Robert M. and Dana G. Andrews. 1991. Magnetic Sails and Interplanetary Travel. *Journal of Spacecraft and Rockets*. 28:197-203.

Related Bibliography

Belding, J. A., and W. B. Coley. 1973. Integral Rocket/Ramjet for Tactical Missiles, *Astronautics and Aeronautics*. 12:20-26.

Billig, F. S., P. J. Waltrup, and R. D. Stockbridge, Integral Rocket Dual Combustion Ramjets: A new Propulsion Concept. *AIAA Journal of Spacecraft*. 17:416-424.

Dugger G. L., ed. 1969. *Ramjets*. AIAA Selected Reprint Series. Vol. VI.

Ramjet and Ramrocket Propulsion Systems for Missiles. AGARD LS-136, 1984.

Ramjets and Ramrockets for Military Applications. AGARD CP-307, 1981.

Vaught C., Witt M., Netzer D. 1992. Investigation of Solid Fuel, Dual Mode Combustion Ramjets. *AIAA Journal of Propulsion and Power*. 8:1004-1011.

Weast, R. C. ed. 1978. CRC Handbook of Chemistry and Physics. 59th ed. Cleveland, OH:CRC Press.

Weinreich H. L. 1979. Overview of Propulsion Systems and Related Fluid Dynamic Aspects. AGARD - VKI Lecture Series No. 98 *Supplement on Missile Aerodynamics*.

Appendix A
Units and Conversion Factors[*]

The metric system of units, officially known as the *International System of Units*, or *SI*, is used throughout this book. By international agreement, the fundamental SI units of length, mass, and time are defined as follows (see National Bureau of Standards Special Publication 330, 1986):

The *meter* is the length of the path traveled by light in vacuum during a time interval of 1/299,792,458 of a second.

The *kilogram* is the mass of the international prototype of the kilogram.

The *second* is the duration of 9,192,631,770 periods of the radiation corresponding to the transition between two hyperfine levels of the ground state of the cesium-133 atom.

Additional base units in the SI system are the *ampere* for electric current, the *kelvin* for thermodynamic temperature, the *mole* for amount of substance, and the *candela* for luminous intensity. Mechtly [1977] provides an excellent summary of SI units for scientific and technical use.

We form the names of multiples and submultiples of SI units by applying the following prefixes:

Factor by which unit is multiplied	Prefix	Symbol
10^{18}	exa	E
10^{15}	peta	P
10^{12}	tera	T
10^{9}	giga	G
10^{6}	mega	M
10^{3}	kilo	k
10^{2}	hecto	h
10	deka	da
10^{-1}	deci	d
10^{-2}	centi	c
10^{-3}	milli	m
10^{-6}	micro	μ
10^{-9}	nano	n
10^{-12}	pico	p
10^{-15}	femto	f
10^{-18}	atto	a

---

[*] Much of the material in this appendix has been adapted, with permission, from *Space Mission Analysis and Design*, 1992. 2nd ed. Wiley J. Larson and James R. Wertz, eds., Torrance, California, and Norwell, MA: Microcosm, Inc. and Kluwer Academic Publishers.

Appendix A

For each quantity listed below, the SI unit and its abbreviation are in brackets. For convenience in computer use, most conversion factors are given to the greatest available accuracy. Note that some conversions are exact definitions and some (speed of light, astronomical unit) depend on the value of physical constants. All notes are on the last page of the list.

To convert from	To	Multiply by	Notes
Acceleration [meter/second^2, m/s^2]			
Gal (galileo)	m/s^2	1.0×10^{-2}	E
Inch/second^2	m/s^2	2.54×10^{-2}	E
Foot/second^2	m/s^2	3.048×10^{-1}	E
Free fall (standard), g	m/s^2	9.806 65	E
Angular Acceleration [radian/second^2, rad/s^2]			
Degrees/second^2	rad/s^2	$\pi/180$	E
		$\approx 1.745\,329\,251\,994\,329\,577 \times 10^{-2}$	
Revolutions/second^2, rev/s^2	rad/s^2	2π	E
		$\approx 6.283\,185\,307\,179\,586\,477$	
Revolutions/minute^2	rad/s^2	$\pi/1800$	E
		$\approx 1.745\,329\,251\,994\,329\,577 \times 10^{-3}$	
Revolutions/minute^2	deg/s^2	0.1	E
Radians/second^2, rad/s^2	deg/s^2	$180/\pi$	E
		$\approx 5.729\,577\,951\,308\,232\,088 \times 10^{1}$	
Revolutions/second^2, rev/s^2	deg/s^2	3.6×10^2	E
Angular Measure [radian, rad]			
Degree, deg	rad	$\pi/180$	E
		$\approx 1.745\,329\,251\,994\,329\,577 \times 10^{-2}$	
Minute (of arc)	rad	$\pi/10,800$	E
		$\approx 2.908\,882\,086\,657\,216 \times 10^{-4}$	
Radian	deg	$= 180/\pi$	E
		$\approx 5.729\,577\,951\,308\,232\,088 \times 10^{1}$	
Second (of arc)	rad	$\pi/648,000$	E
		$\approx 4.848\,136\,811\,095\,4 \times 10^{-6}$	
Angular Momentum [kilogram · meter^2/second, kg · m^2/s]			
Gram · cm^2/second	kg · m^2/s	1.0×10^{-7}	E
lbm · inch^2/second	kg · m^2/s	$2.926\,397 \times 10^{-4}$	
Slug · inch^2/second	kg · m^2/s	$9.415\,402 \times 10^{-3}$	
lbm · foot^2/second	kg · m^2/s	$4.214\,011 \times 10^{-2}$	
Inch · lbf · second	kg · m^2/s	$1.129\,848 \times 10^{-1}$	
Slug · foot^2/second = ft · lbf · second	kg · m^2/s	1.355 818	

To convert from	To	Multiply by	Notes
Angular Velocity [radian/second, rad/s]			
Degrees/second, deg/s	rad/s	$\pi/180$	E
		$\approx 1.745\ 329\ 251\ 994\ 329\ 577 \times 10^{-2}$	
Revolutions/minute, rpm	rad/s	$\pi/30$	E
		$\approx 1.047\ 197\ 551\ 196\ 597\ 746 \times 10^{-1}$	
Revolutions/second, rev/s	rad/s	2π	E
		$\approx 6.283\ 185\ 307\ 179\ 586\ 477$	
Revolutions/minute, rpm	deg/s	6.0	E
Radians/second, rad/s	deg/s	$180/\pi$	E
		$\approx 5.729\ 577\ 951\ 308\ 232\ 088 \times 10^{1}$	
Revolutions/second	deg/s	3.6×10^2	E
Area [meter^2, m^2]			
Foot^2, ft^2	m^2	$9.290\ 304 \times 10^{-2}$	E
Hectare	m^2	1.0×10^4	E
Inch^2, in^2	m^2	$6.451\ 6 \times 10^{-4}$	E
Mile^2 (U.S. statute)	m^2	$2.589\ 998 \times 10^6$	M
Yard^2, yd^2	m^2	$8.361\ 273\ 6 \times 10^{-1}$	E
(Nautical mile)^2	m^2	$3.429\ 904 \times 10^6$	E
Density [kilogram/meter^3, kg/m^3]			
g/cm^3	kg/m^3	1.00×10^3	E
lbm/inch^3	kg/m^3	$2.767\ 990\ 5 \times 10^4$	M
lbm/ft^3	kg/m^3	$1.601\ 846\ 3 \times 10^{1}$	M
Slug/ft^3	kg/m^3	$5.153\ 79 \times 10^2$	M
Electric Charge [coulomb, C]			
Abcoulomb	C	1.0×10^1	E
Faraday (based on C-12)	C	$9.648\ 70 \times 10^4$	M
Faraday (chemical)	C	$9.649\ 57 \times 10^4$	M
Faraday (physical)	C	$9.652\ 19 \times 10^4$	M
Statcoulomb	C	$3.335\ 640 \times 10^{-10}$	M
Electric Conductance [siemens, S]			
Abmho	S	1.0×10^9	E
Mho (Ω^{-1})	S	1.0	E
Electric Current [ampere, A]			
Abampere	A	1.0×10^1	E
Gilbert	A	$7.957\ 747\ 2 \times 10^{-1}$	M
Statampere	A	$3.335\ 640 \times 10^{-10}$	M
Electric Field Intensity [volt/meter $\equiv$ kilogram $\cdot$ meter $\cdot$ ampere^{-1} $\cdot$ second^{-3}, V/m $\equiv$ kg $\cdot$ m $\cdot$ A^{-1} $\cdot$ s^{-3}]			

To convert from	To	Multiply by	Notes
Electric Potential Difference [volt ≡ watt/ampere ≡ kilogram·meter^2·ampere^{-1}·second^{-3}, V ≡ W/A ≡ kg·m^2·A^{-1}·s^{-3}]			
Abvolt	V	1.0×10^{-8}	E
Statvolt	V	$2.997\,925 \times 10^{2}$	M
Electric Resistance [ohm ≡ volt/ampere ≡ kilogram·meter^2·ampere^{-2}·second^{-3}, Ω ≡ V/A ≡ kg·m^2·A^{-2}·s^{-3}]			
Abohm	Ω	1.0×10^{-9}	E
Statohm	Ω	$8.987\,554 \times 10^{11}$	M
Energy or Torque [joule ≡ newton·meter ≡ kilogram·meter^2/s^2, J ≡ N·m ≡ kg·m^2/s^2]			
British thermal unit, Btu (mean)	J	$1.055\,87 \times 10^{3}$	M
Calorie (mean), cal	J	$4.190\,02$	M
Kilocalorie (mean), kcal	J	$4.190\,02 \times 10^{3}$	M
Electron volt, eV	J	$1.602\,177\,33 \times 10^{-19}$	C
Erg ≡ gram·cm^2/s^2			
= pole·cm·oersted	J	1.0×10^{-7}	E
Foot poundal	J	$4.214\,011\,0 \times 10^{-2}$	M
Foot lbf = slug·foot^2/s^2	J	$1.355\,817\,9$	M
Kilowatt hour, kW·hr	J	3.60×10^{6}	E
Ton equivalent of TNT	J	4.184×10^{9}	E
Force [newton ≡ kilogram·meter/second^2, N ≡ kg·m/s^2]			
Dyne	N	1.0×10^{-5}	E
Poundal	N	$1.382\,549\,543\,76 \times 10^{-1}$	E
Ounce force (avoirdupois)	N	$2.780\,138\,5 \times 10^{-1}$	M
Pound force (avoirdupois), lbf ≡ slug·foot/s^2)	N	$4.448\,221\,615\,260\,5$	E
Illuminance [lux ≡ lumen/meter^2, lx ≡ lm/m^2]			
Footcandle	lx	$1.076\,391\,0 \times 10^{1}$	M
Phot	lx	1.00×10^{4}	M
Length [meter, m]			
Angstrom, Å	m	1.00×10^{-10}	E
Astronomical unit (IAU)	m	$1.495\,978\,70 \times 10^{11}$	M
Astronomical unit (radio)	m	$1.495\,978\,9 \times 10^{11}$	M
Earth equatorial radius, R_E	m	$6.378\,140 \times 10^{6}$	AA
Foot	m	3.048×10^{-1}	E
Inch	m	2.54×10^{-2}	E
Light year	m	$9.460\,55 \times 10^{15}$	M

To convert from	To	Multiply by	Notes
Micron, μm	m	1.0×10^{-6}	E
Mil (10^{-3} inch)	m	2.54×10^{-5}	E
Mile (U.S. statute)	m	$1.609\,3 \times 10^{3}$	M
Nautical mile (U.S.)	m	1.852×10^{3}	E
Parsec (IAU)	m	$3.085\,7 \times 10^{16}$	M
Solar radius	m	$6.960\,00 \times 10^{8}$	AA
Yard	m	9.144×10^{-1}	E

Luminance [candela/meter^2 ≡ cd/m^2]

Footlambert	cd/m^2	$3.426\,259$	M
Lambert	cd/m^2	$(1/\pi) \times 10^4 \approx 3.183\,098\,862 \times 10^{3}$	E
Stilb	cd/m^2	1.0×10^{4}	M

Magnetic Field Strength, H [ampere turn/meter, A/m]

Oersted (EMU)	A/m	$(1/4\pi) \times 10^{3}$ $\approx 7.957\,747\,154\,594\,766\,788 \times 10^{1}$	E, 1

Magnetic Flux [weber ≡ volt · s ≡ kilogram · meter^2 · ampere^{-1} · second^{-2}
$\quad$ **Wb ≡ V · s ≡ kg · m^2 · A^{-1} · s^{-2}]**

Maxwell (EMU)	Wb	1.0×10^{-8}	E
Unit pole	Wb	$1.256\,637 \times 10^{-7}$	M

Magnetic Induction, B [tesla ≡ weber/meter^2 ≡ kilogram · ampere^{-1} · second^{-2}
$\quad$ **T ≡ Wb/m^2 ≡ kg · A^{-1} · s^{-2}]**

Gamma (EMU)	T	1.0×10^{-9}	E, 1
Gauss (EMU)	T	1.0×10^{-4}	E, 1

Magnetic Dipole Moment [weber · meter ≡ kilogram · meter · ampere^{-1} · second^{-2}
$\quad$ **Wb · m ≡ kg · m · A^{-1} · s^{-2}]**

Pole · centimeter (EMU)	Wb · m	$4\pi \times 10^{-10}$ $\approx 1.256\,637\,061\,435\,917\,295 \times 10^{-9}$	E, 1
Gauss · centimeter^3 (Practical)	Wb · m	1.0×10^{-10}	E, 1

Magnetic Moment [ampere turn · meter^2 ≡ joule/tesla, A · m^2 ≡ J/T]

Abampere · centimeter^2 (EMU)	A · m^2	1.0×10^{-3}	E, 1
Ampere · centimeter^2	A · m^2	1.0×10^{-4}	E, 1

Mass [kilogram, kg]

Atomic unit (electron)	kg	$9.109\,389\,7 \times 10^{-31}$	C
Atomic mass unit (unified), amu	kg	$1.660\,540\,2 \times 10^{-27}$	C
Metric ton	kg	1.0×10^{3}	E
Ounce mass (avoirdupois)	kg	$2.834\,952\,312\,5 \times 10^{-2}$	E

To convert from	To	Multiply by	Notes
Pound mass, lbm (avoirdupois)	kg	$4.535\ 923\ 7 \times 10^{-1}$	E
Slug	kg	$1.459\ 390\ 29 \times 10^{1}$	M
Short ton (2000 lbm)	kg	$9.071\ 847\ 4 \times 10^{2}$	E
Solar mass	kg	1.991×10^{30}	H

Moment of Inertia [kilogram · meter^2, kg · m^2]

Gram · centimeter^2	kg · m^2	1.0×10^{-7}	E
lbm · inch^2	kg · m^2	$2.926\ 397 \times 10^{-4}$	
lbm · foot^2	kg · m^2	$4.214\ 011 \times 10^{-2}$	
Slug · inch^2	kg · m^2	$9.415\ 402 \times 10^{-3}$	
Inch · lbf · s^2	kg · m^2	$1.129\ 848 \times 10^{-1}$	
Slug · foot^2 = ft · lbf · s^2	kg · m^2	$1.355\ 818$	

Power [watt ≡ joule/second ≡ kilogram · meter^2/second^3, W ≡ J/s ≡ kg · m^2/s^3]

Foot lbf/second	W	$1.355\ 817$	M
Horsepower (550 ft · lbf/s)	W	$7.456\ 998 \times 10^{2}$	M
Horsepower (electrical)	W	7.46×10^{2}	E
Solar luminosity	W	3.826×10^{26}	W

Pressure or Stress [pascal ≡ newton/meter^2 ≡ kilogram · meter^{-1} · second^{-2}
$$Pa \equiv N/m^2 \equiv kg \cdot m^{-1} \cdot s^{-2}]$$

Atmosphere	Pa	$1.013\ 25 \times 10^{5}$	E
Bar	Pa	1.0×10^{5}	E
Centimeter of mercury (0° C)	Pa	$1.333\ 22 \times 10^{3}$	M
Dyne/centimeter^2	Pa	1.0×10^{-1}	E
Inch of mercury (32° F)	Pa	$3.386\ 389 \times 10^{3}$	M
lbf/foot^2	Pa	$4.788\ 025\ 8 \times 10^{1}$	M
lbf/inch^2, psi	Pa	$6.894\ 757\ 2 \times 10^{3}$	M
Torr (0° C)	Pa	$1.333\ 22 \times 10^{2}$	M

Solid Angle [steradian, sr]

Degree^2, deg^2	sr	$(\pi/180)^2$ ≈ $3.046\ 174\ 197\ 867\ 085\ 993 \times 10^{-4}$	E
Steradian	deg^2	$(180/\pi)^2$ ≈ $3.282\ 806\ 350\ 011\ 743\ 794 \times 10^{3}$	E

Specific Heat Capacity [joule · kilogram^{-1} · kelvin^{-1} ≡ meter^2 · second^2 · kelvin^{-1}
$$J \cdot kg^{-1} \cdot K^{-1} \equiv m^2 \cdot s^2 \cdot K^{-1}\ ]$$

cal · g^{-1} · K^{-1}	J · kg^{-1} · K^{-1}	$4.190\ 02 \times 10^{3}$	H
Btu · lbm^{-1} · °F^{-1}	J · kg^{-1} · K^{-1}	$4.190\ 02 \times 10^{3}$	H

Units and Conversion Factors

To convert from	To	Multiply by	Notes
Stress (see Pressure)			
Temperature [kelvin, K]			
Celsius, °C	K	$t_K = t_C + 273.15$	E
Fahrenheit, °F	K	$t_K = (5/9)(t_F + 459.67)$	E
Fahrenheit, °F	C	$t_C = (5/9)(t_F - 32.0)$	E
Thermal Conductivity [watt · meter^{-1} · kelvin^{-1} $\equiv$ kilogram · meter · second^3 · kelvin^{-1}			
W · m^{-1} · K^{-1} $\equiv$ kg · m · s^{-3} · K^{-1}]			
cal · cm^{-1} · s^{-1} · K^{-1}	W · m^{-1} · K^{-1}	$4.190\ 02 \times 10^2$	H
Btu · ft^{-1} · hr^{-1} · °F^{-1}	W · m^{-1} · K^{-1}	$1.732\ 067$	H
Time [second, s]			
Sidereal day, d_* (ref. = Υ)	s	$8.616\ 409\ 0 \times 10^4$ = 23^h 56^m 4.090 0^s	M
Ephemeris day, d_e	s	8.64×10^4	E
Ephemeris day, d_e	d_*	$1.002\ 737\ 9$	
Keplerian period of a satellite in low-Earth orbit	min	$1.658\ 669 \times 10^{-4} \times a^{3/2}$ (a in km)	W
Keplerian period of a satellite of the Sun	d_e	$3.652\ 569 \times 10^2 \times a^{3/2}$ (a in AU)	W
Tropical year (ref. = Υ)	s	$3.155\ 692\ 53 \times 10^7$	AA
Tropical year (ref. = Υ)	d_e	$3.652\ 421\ 91 \times 10^2$	AA
Sidereal year (ref. = fixed stars)	s_e	$3.155\ 814\ 98 \times 10^7$	AA
Sidereal year (ref. = fixed stars)	d_e	$3.652\ 563\ 63 \times 10^2$	AA
Calendar year (365 days)	s	$3.153\ 6 \times 10^7$	M
Julian century	d	$3.652\ 5 \times 10^4$	E
Gregorian calendar century	d	$3.652\ 425 \times 10^4$	E
Torque (see Energy)			
Velocity [meter/second, m/s]			
Foot/minute, ft/min	m/s	5.08×10^{-3}	E
Inch/second, ips	m/s	2.54×10^{-2}	E
Kilometer/hour, km/hr	m/s	$(3.6)^{-1} = 0.277777...$	E
Foot/second, fps or ft/s	m/s	3.048×10^{-1}	E
Miles/hour, mph	m/s	4.4704×10^{-1}	E
Knot (international)	m/s	$5.144\ 444\ 444 \times 10^{-1}$	M
Miles/minute	m/s	$2.682\ 24 \times 10^1$	E
Miles/second	m/s	$1.609\ 344 \times 10^3$	E
Astronomical unit/sidereal year	m/s	$4.740\ 388\ 554 \times 10^3$	W

To convert from	To	Multiply by	Notes
Velocity of light, c	m/s	$2.997\,924\,58 \times 10^8$	E
Viscosity [pascal · second ≡ kilogram · meter^{-1} · second^{-1}, Pa · s ≡ kg · m^{-1} · s^{-1}]			
Stoke	m²/s	1.0×10^{-4}	E
Ft² · s	m²/s	$9.290\,304 \times 10^{-2}$	E
lbm · ft^{-1} · s^{-1}	Pa · s	$1.488\,163\,9$	M
lbf · s/ft²	Pa · s	$4.788\,025\,8 \times 10^1$	M
Poise	Pa · s	1.0×10^{-1}	E
Poundal s/ft²	Pa · s	$1.488\,163\,9$	M
Slug · ft^{-1} · s^{-1}	Pa · s	$4.788\,025\,8 \times 10^1$	M
Rhe	(Pa · s)$^{-1}$	1.0×10^1	E
Volume [meter³, m³]			
Foot³	m³	$2.831\,684\,659\,2 \times 10^{-2}$	E
Gallon (U.S. liquid)	m³	$3.785\,411\,784 \times 10^{-3}$	E
Inch³	m³	$1.638\,706\,4 \times 10^{-5}$	E
Liter	m³	1.0×10^{-3}	E
Ounce (U.S. fluid)	m³	$2.957\,352\,956\,25 \times 10^{-5}$	E
Pint (U.S. liquid)	m³	$4.731\,764\,73 \times 10^{-4}$	E
Quart	m³	$9.463\,529\,46 \times 10^{-4}$	E
Yard³	m³	$7.645\,548\,579\,84 \times 10^{-1}$	E

NOTES:

E   (*Exact*) indicates that the conversion is exact by definition of the non-SI unit or that it is obtained from other exact conversions.
M   Values are those of Mechtly
W   Values are those of Wertz
AA Values are those of *Astronomical Almanac*
H   Values are those of Weast
(1) Take care in transforming magnetic units because the dimensions of magnetic quantities (**B**, **H**) depend on the system of units. Most of the conversions given here are between SI and EMU (electromagnetic). The following equations hold in both sets of units:

$$\mathbf{T} = \mathbf{m} \times \mathbf{B} = \mathbf{d} \times \mathbf{H}$$
$$\mathbf{B} = \mu \mathbf{H}$$
$$\mathbf{m} = I \mathbf{A} \text{ for a current loop in a plane}$$
$$\mathbf{d} = \mu \mathbf{m}$$

with the following definitions
$\mathbf{T} \equiv$ torque

B ≡ magnetic induction (commonly called "magnetic field")
H ≡ magnetic field strength or magnetic intensity
m ≡ magnetic moment
I ≡ current in loop
A ≡ vector normal to the plane of the current loop (in the direction of the angular velocity vector of the current about the center of the loop) with magnitude equal to the area of the loop
d ≡ magnetic dipole moment
μ ≡ magnetic permeability

The permeability of vacuum, μ_0, has the following values, by definition:

$\mu_0 \equiv$ 1 (dimensionless)   EMU

$\mu_0 \equiv 4\pi \times 10^{-7}$ N/A²   SI

Therefore, in electromagnetic units in vacuum, magnetic induction and magnetic field strength are equivalent and the magnetic moment and magnetic dipole moment are equivalent. For practical magnetostatics, space is a vacuum, but the spacecraft itself may have $\mu \neq \mu_0$.

References

Cohen, E. Richard and Taylor, B.N. *CODATA Bulletin No. 63*, Pergamon Press, Nov 1986.

Hagen, James B. and Boksenberg, A., eds. 1991. *Astronomical Almanac*. Washington, D.C.: U.S. Government Printing Office.

Mechtly, E. A. 1977. *The International System of Units*. Champaign, IL: Stipes Publishing Company.

Weast, R. C., ed. 1985. *CRC Handbook of Chemistry and Physics*. Boca Raton, FL: CRC Press.

Wertz, James R., ed. 1978. *Spacecraft Attitude Determination and Control*. Dordrecht, Holland: D. Reidel Publishing Company.

Appendix B
Thermochemical Data for Selected Propellants

In Chap. 4, we discuss much of the theory behind thermochemical analyses. To actually do these calculations, we usually must use sophisticated computer programs such as the *Lewis* code or the I_{sp} code (see Chap. 4 for references). As an alternative for conceptual design, the following charts represent the thermochemical data for some of the more popular propellant combinations.

Three approaches can be taken in evaluating thermochemical parameters (isentropic parameter [γ], molecular mass, and temperature of the combustion gases):

- Frozen flow—the equilibrium chemical reaction is assumed to occur very slowly, so the parameters do not change through the thrust chamber. This gives a conservatively low prediction for specific impulse (about 2%). All nozzle expansion ratios yield the same parameter values.

- Equilibrium flow—the reactions occur very quickly, so the flow is always in equilibrium through the thrust chamber. This approach gives optimistic results for specific impulse (about 3%) and the parameters depend on the nozzle expansion.

- Chemical kinetics—the actual chemical reaction rates are modeled. This can be the most accurate approach, but it is quite difficult to implement.

To give ourselves the most flexibility, we assume the frozen-flow approximation. This assumption allows us to use the following thermochemical data for any choice of nozzle expansion ratio and still come up with reasonably accurate thermochemical predictions. To develop the data given below, we assume a chamber pressure of 3.45 MPa (about 500 psi). Although the parameters change by a small amount for other chamber pressures, the change is quite small and is considered negligible for preliminary or conceptual design. For more detailed analysis, the equilibrium-flow or chemical-kinetics approach is best. In addition, remember that combustion efficiencies and nozzle efficiencies somewhat reduce the predicted values.

Table B.1 shows some of the basic properties of the propellants analyzed. Following that table is a series of four charts associated with each propellant combination. These charts show combustion flame temperature (T_f), exhaust-product molecular mass($\mathcal{M}$), exhaust-product isentropic parameter (γ), and vacuum specific impulse (I_{sp}) as a function of oxidizer-to-fuel ratio (O/F). The molecular mass is frozen in the combustion chamber (at Mach number = 0), and the isentropic parameter is frozen at the throat (Mach number = 1) to give us a bit more accuracy.

The charts showing vacuum specific impulse assume a chamber pressure of 3.45 MPa, an exit pressure of 13.8 kPa, and zero ambient pressure. The purpose of these charts is to show the O/F that gives the maximum specific impulse, as well as the variation of I_{sp} with O/F. Although the magnitude of specific impulse changes for different design conditions (chamber pressure, expansion conditions, ambient environment), the basic shape of the curve remains the same. For the curves involving hybrid propellants, we have included curve-fit equations. These equations allow us to model the effect of O/F shift on thermochemistry in computer simulations. For nonhybrids, we usually just pick a point from the curves for a fixed O/F.

Table B.1. Properties of Typical Liquid Propellants, Note: FP = freezing point at one atmosphere, BP = boiling point at one atmosphere, P_{vap} = vapor pressure, Al = aluminum alloy, SS = stainless steel, Ni = nickel alloy, Cu = copper (all possible alloys do not necessarily work).

Propellant		T_{FP} (K)	T_{BP} (K)	P_{vap} (Pa)	Density (kg/m^3)	Stability	Handling	Storage	Material Compatibility
Oxygen	O_2	54	90	5.07 MPa @ 154K	1142	good	good	cryogenic	Al, SS, Ni, Cu, Teflon, Kel-F
Hydrogen Peroxide	H_2O_2	267.4	419	345 @ 298K	1414	unstable at T>414K	burns skin flammable	decomposes at 1%/year	Al, SS, Ni, Kel-F
Nitrogen Tetroxide	N_2O_4	261	294	0.765 MPa @ 344K	1440	temp. dependent	burns skin toxic	good when dry	Al, SS, Ni, Teflon
Hydrogen	H_2	13.8	20.3	1.294 MPa @ 32.8K	71	flammable	flammable	cryogenic	Al, SS, Ni, Kel-F
RP-1	$CH_{1.97}$	229–291	445–537	2275 @ 344K	810	flammable	flammable	good	Al, steel, Ni, Cu, Teflon, Kel-F, Neoprene
Hydrazine	N_2H_4	274	386	19300 @ 344K	1010	toxic & flammable	toxic flammable	good	Al, SS, Teflon, Kel-F, polyethylene
Monomethyl Hydrazine	CH_3NH-NH_2	220	359	60657 @ 344K	878	toxic	toxic	good	Al, SS, Teflon, Kel-F, polyethylene
Unsymmetrical dimethyl hydrazine	$(CH_3)_2$N-NH_2	215	336	1.213 MPa @ 344K	789	toxic	toxic	good	Al, SS, Teflon, Kel-F
Fluorine	Fl	53	85	5.57 MPa @ 144K	1509	good	very toxic, flammable	cryogenic	Al, SS, Ni, brass
Poly Butadiene	C_2H_4	N/A	N/A	N/A	900 - 1000	excellent - solid	excellent - solid	excellent - solid	everything

Thermochemical Data for Selected Propellants

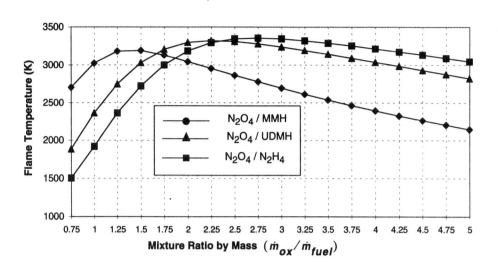

Fig. B.1. **Flame Temperature Versus Mixture Ratio.** Fuels are hydrazine (N_2H_4), mono-methyl hydrazine (MMH), unsymmetrical dimethyl hydrazine (UDMH). The oxidizer for each is nitrogen tetroxide (N_2O_4).

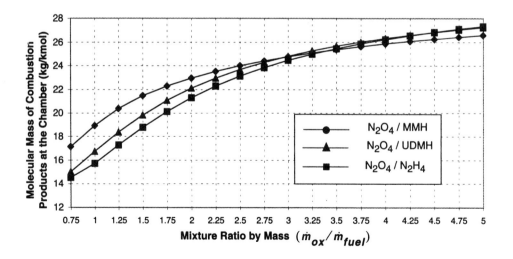

Fig. B.2. **Molecular Mass of Combustion Products Versus Mixture Ratio.** Fuels are hydrazine (N_2H_4), mono-methyl hydrazine (MMH), unsymmetrical dimethyl hydrazine (UDMH). The oxidizer for each is nitrogen tetroxide (N_2O_4).

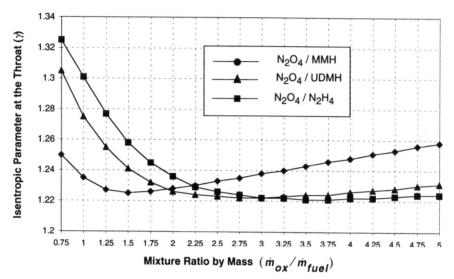

Fig. B.3. **Isentropic Parameter Versus Mixture Ratio.** Fuels are hydrazine (N_2H_4), mono-methyl hydrazine (MMH), unsymmetrical dimethyl hydrazine (UDMH). The oxidizer for each is nitrogen tetroxide (N_2O_4).

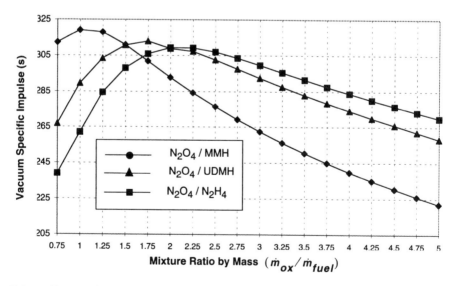

Fig. B.4. **Vacuum Specific Impulse Versus Mixture Ratio.** Fuels are hydrazine (N_2H_4), mono-methyl hydrazine (MMH), unsymmetrical dimethyl hydrazine (UDMH). The oxidizer for each is nitrogen tetroxide (N_2O_4).

Thermochemical Data for Selected Propellants

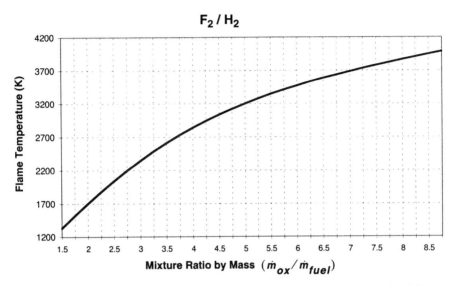

Fig. B.5. **Flame Temperature Versus Mixture Ratio.** The fuel is liquid hydrogen (H_2) and the oxidizer is liquid fluorine (F_2).

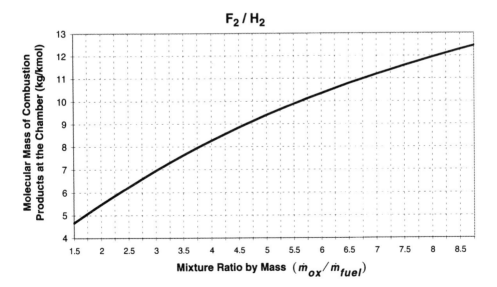

Fig. B.6. **Molecular Mass of Combustion Products Versus Mixture Ratio.** The fuel is liquid hydrogen (H_2) and the oxidizer is liquid fluorine (F_2).

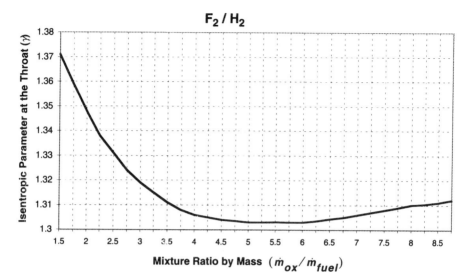

Fig. B.7. **Isentropic Parameter Versus Mixture Ratio.** The fuel is liquid hydrogen (H_2) and the oxidizer is liquid fluorine (F_2).

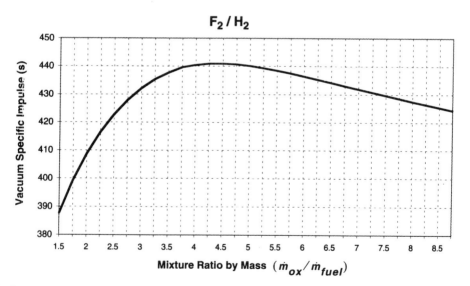

Fig. B.8. **Vacuum Specific Impulse Versus Mixture Ratio.** The fuel is liquid hydrogen (H_2) and the oxidizer is liquid fluorine (F_2).

Thermochemical Data for Selected Propellants

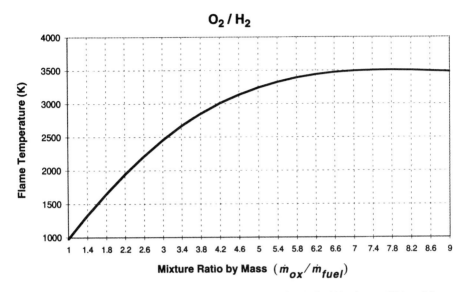

Fig. B.9. **Flame Temperature Versus Mixture Ratio.** The fuel is liquid hydrogen (H_2) and the oxidizer is liquid oxygen (O_2).

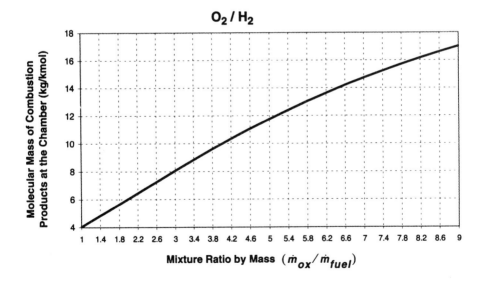

Fig. B.10. **Molecular Mass of Combustion Products Versus Mixture Ratio.** The fuel is liquid hydrogen (H_2) and the oxidizer is liquid oxygen (O_2).

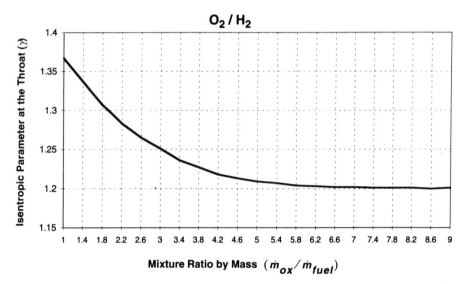

Fig. B.11. **Isentropic Parameter Versus Mixture Ratio.** The fuel is liquid hydrogen (H_2) and The oxidizer is liquid oxygen (O_2).

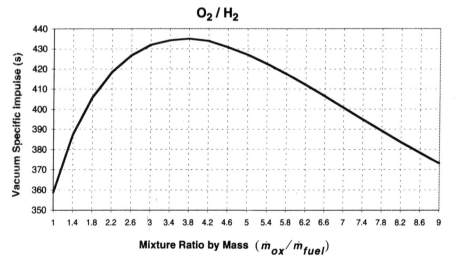

Fig. B.12. **Vacuum Specific Impulse Versus Mixture Ratio.** The fuel is liquid hydrogen (H_2) and the oxidizer is liquid oxygen (O_2).

Thermochemical Data for Selected Propellants

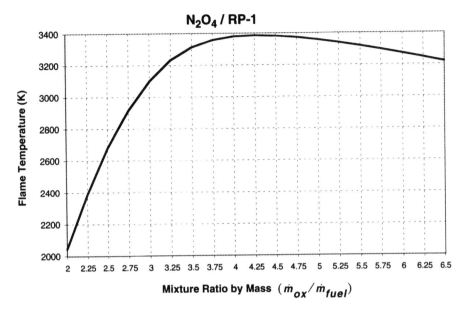

Fig. B.13. Flame Temperature Versus Mixture Ratio. The fuel is kerosene (RP-1) and the oxidizer is nitrogen tetroxide (N_2O_4).

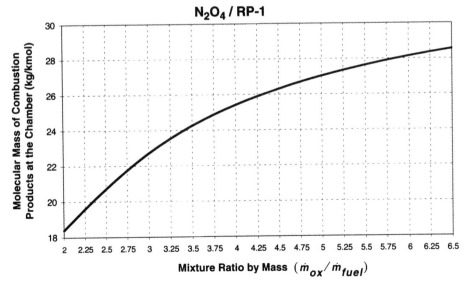

Fig. B.14. Molecular Mass of Combustion Products Versus Mixture Ratio. The fuel is kerosene (RP-1) and the oxidizer is nitrogen tetroxide (N_2O_4).

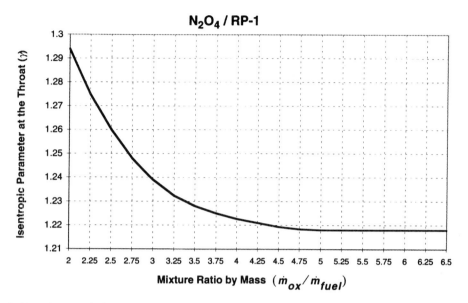

Fig. B.15. **Isentropic Parameter Versus Mixture Ratio.** The fuel is kerosene (RP-1) and the oxidizer is nitrogen tetroxide (N_2O_4).

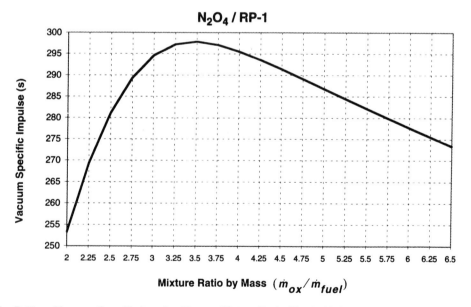

Fig. B.16. **Vacuum Specific Impulse Versus Mixture Ratio.** The fuel is kerosene (RP-1) and the oxidizer is nitrogen tetroxide (N_2O_4).

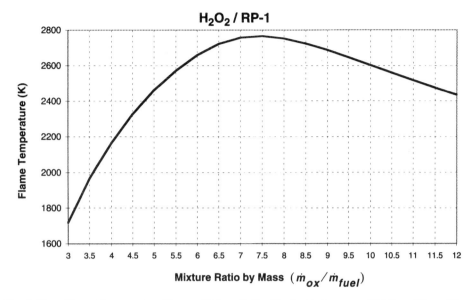

Fig. B.17. **Flame Temperature Versus Mixture Ratio.** The fuel is kerosene (RP-1) and the oxidizer is 90% hydrogen peroxide by mass (90% – H_2O_2 and 10% – H_2O).

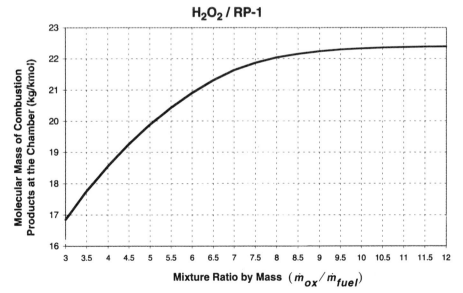

Fig. B.18. **Molecular Mass of Combustion Products Versus Mixture Ratio.** The fuel is kerosene (RP-1) and the oxidizer is 90% hydrogen peroxide by mass (90% – H_2O_2 and 10% – H_2O).

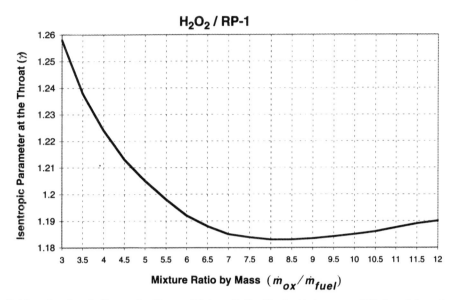

Fig. B.19. **Isentropic Parameter Versus Mixture Ratio.** The fuel is kerosene (RP-1) and the oxidizer is 90% hydrogen peroxide by mass (90% – H_2O_2 and 10% – H_2O).

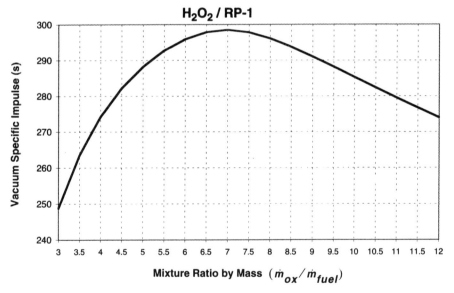

Fig. B.20. **Vacuum Specific Impulse Versus Mixture Ratio.** The fuel is kerosene (RP-1) and the oxidizer is 90% hydrogen peroxide by mass (90% – H_2O_2 and 10% – H_2O).

Thermochemical Data for Selected Propellants

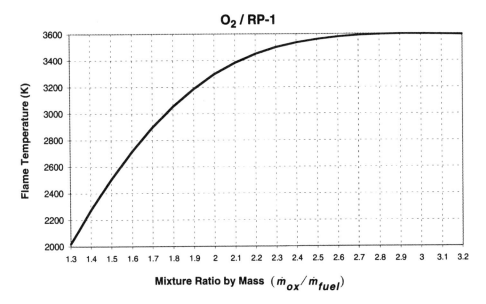

Fig. B.21. Flame Temperature Versus Mixture Ratio. The fuel is kerosene (RP-1) and the oxidizer is liquid oxygen (O_2).

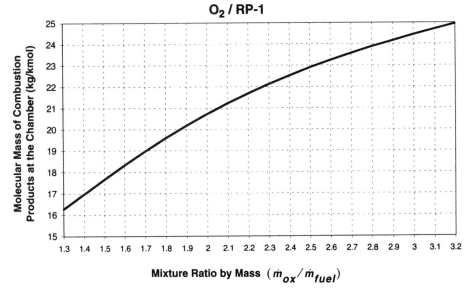

Fig. B.22. Molecular Mass of Combustion Products Versus Mixture Ratio. The fuel is kerosene (RP-1) and the oxidizer is liquid oxygen (O_2).

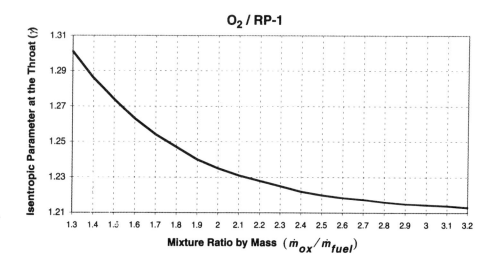

Fig. B.23. **Isentropic Parameter Versus Mixture Ratio.** The fuel is kerosene (RP-1) and the oxidizer is liquid oxygen (O_2).

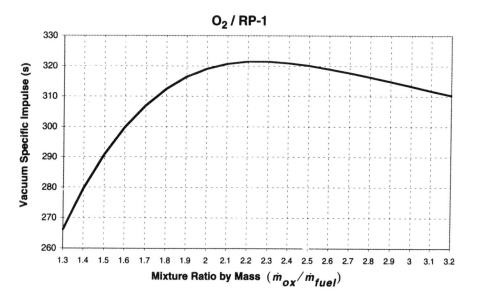

Fig. B.24. **Vacuum Specific Impulse Versus Mixture Ratio.** The fuel is kerosene (RP-1) and the oxidizer is liquid oxygen (O_2).

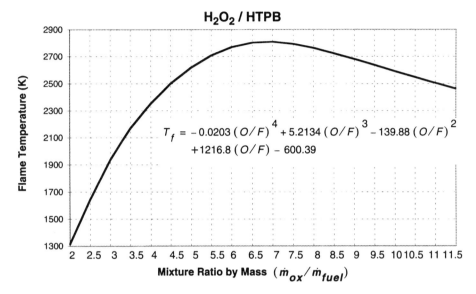

Fig. B.25. **Flame Temperature Versus Mixture Ratio.** The fuel is solid hydroxy terminated poly butadiene (HTPB) and the oxidizer is 90% hydrogen peroxide by mass (90% H_2O_2 and 10% H_2O). The equation shows a curve fit of this data.

$$T_f = -0.0203\,(O/F)^4 + 5.2134\,(O/F)^3 - 139.88\,(O/F)^2 + 1216.8\,(O/F) - 600.39$$

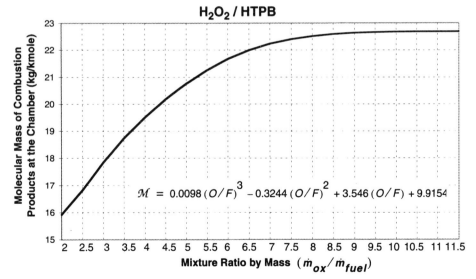

$$\mathcal{M} = 0.0098\,(O/F)^3 - 0.3244\,(O/F)^2 + 3.546\,(O/F) + 9.9154$$

Fig. B.26. **Molecular Mass of Combustion Products Versus Mixture Ratio.** The fuel is solid hydroxy terminated poly butadiene (HTPB) and the oxidizer is 90% hydrogen peroxide by mass (90% H_2O_2 and 10% H_2O). The equation shows a curve fit of this data.

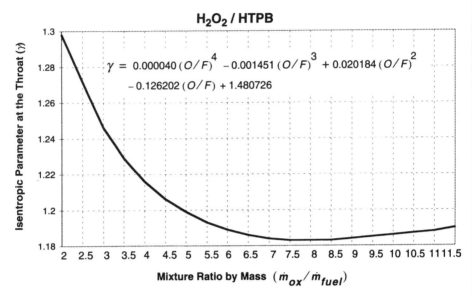

Fig. B.27. Isentropic Parameter Versus Mixture Ratio. The fuel is solid hydroxy terminated poly butadiene (HTPB) and the oxidizer is 90% hydrogen peroxide by mass (90% H_2O_2 and 10% H_2O). The equation shows a curve fit of the data.

$$\gamma = 0.000040\,(O/F)^4 - 0.001451\,(O/F)^3 + 0.020184\,(O/F)^2 - 0.126202\,(O/F) + 1.480726$$

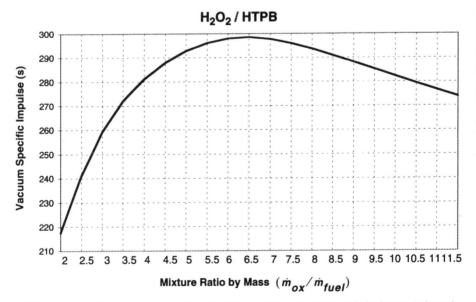

Fig. B.28. Vacuum Specific Impulse Versus Mixture Ratio. The fuel is solid hydroxy terminated poly butadiene (HTPB) and the oxidizer is 90% hydrogen peroxide by mass (90% H_2O_2 and 10% H_2O).

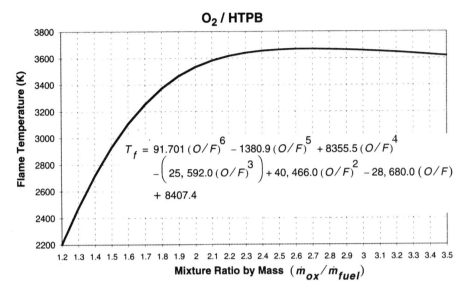

Fig. B.29. **Flame Temperature Versus Mixture Ratio.** The fuel is hydroxy terminated poly butadiene (HTPB) and the oxidizer is liquid oxygen (O_2). The equation shows a curve fit of the data.

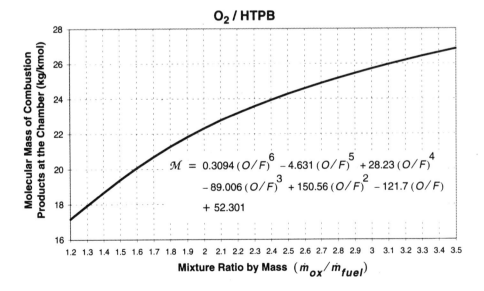

Fig. B.30. **Molecular Mass of Combustion Products Versus Mixture Ratio.** The fuel is hydroxy terminated poly butadiene (HTPB) and the oxidizer is liquid oxygen (O_2). The equation gives a curve fit of the data.

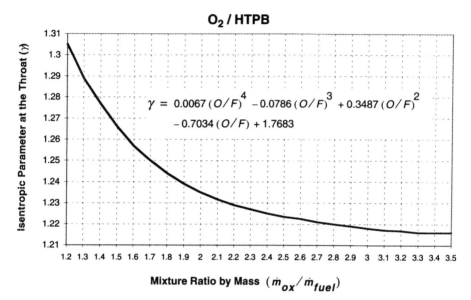

Fig. B.31. **Isentropic Parameter Versus Mixture Ratio.** The fuel is hydroxy terminated poly butadiene (HTPB) and the oxidizer is liquid oxygen (O_2).

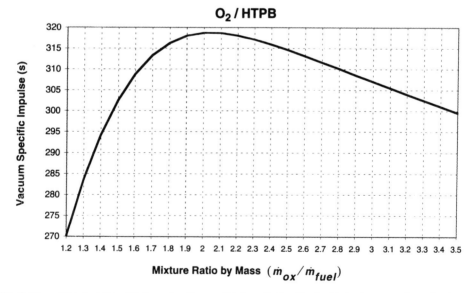

Fig. B.32. **Vacuum Specific Impulse Versus Mixture Ratio.** The fuel is hydroxy terminated poly butadiene (HTPB) and the oxidizer is liquid oxygen (O_2).

Thermochemical Data for Selected Propellants

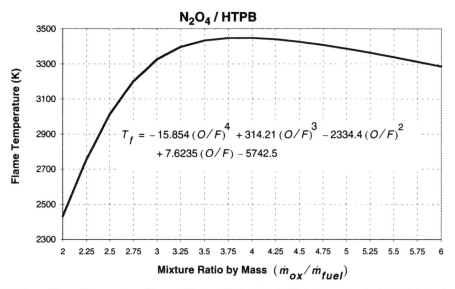

Fig. B.33. **Flame Temperature Versus Mixture Ratio.** The fuel is hydroxy terminated poly butadiene (HTPB), and the oxidizer is nitrogen tetroxide (N_2O_4). The equation gives a curve fit of the data.

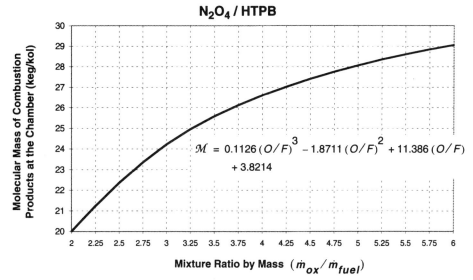

Fig. B.34. **Molecular Mass of Combustion Products Versus Mixture Ratio.** The fuel is hydroxy terminated poly butadiene (HTPB), and the oxidizer is nitrogen tetroxide (N_2O_4). The equation gives a curve fit of the data.

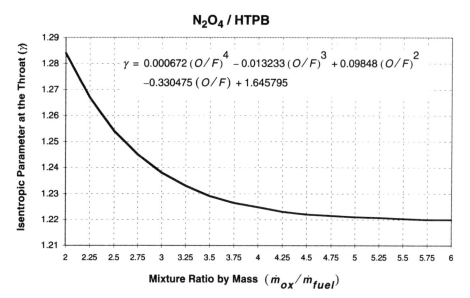

Fig. B.35. **Isentropic Parameter Versus Mixture Ratio.** The fuel is hydroxy terminated poly butadiene (HTPB), and the oxidizer is nitrogen tetroxide (N_2O_4). The equation gives a curve fit of the data.

$$\gamma = 0.000672\,(O/F)^4 - 0.013233\,(O/F)^3 + 0.09848\,(O/F)^2 - 0.330475\,(O/F) + 1.645795$$

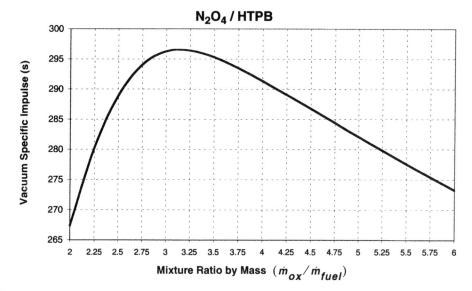

Fig. B.36. **Vacuum Specific Impulse Versus Mixture Ratio.** The fuel is hydroxy terminated poly butadiene (HTPB), and the oxidizer is nitrogen tetroxide (N_2O_4).

Appendix C
Launch Vehicles and Staging
Ronald W. Humble, U.S. Air Force Academy

Whenever we encounter missions requiring a large Δv, we run the risk of not being able to perform the missions with certain technologies. For example, from Table 2.10 we find that a typical launch Δv ranges from about 8.8 km/s to 9.3 km/s. In Sec. 1.1.5, we find there is a "not feasible" condition that gives us a relationship [Eq. (1.29)] between the mission Δv, average specific impulse (I_{sp}), and the inert-mass fraction (f_{inert}). Figure C.1 shows the regions that are and are not feasible for a launch mission using Eq. (1.29). The lower curve corresponds to the relation between inert-mass fraction and specific impulse for Δv = 8800 m/s. The upper curve corresponds to Δv = 9300 m/s. Specific impulses below the curve values for a particular Δv are not feasible. We have also overlaid discrete values for first stages of existing launch vehicles [Isakowitz, 1991]. Table C.1 lists the data for these stages. The systems above the line to the left of f_{inert} = 0.05 are the core first stages for the Titan vehicles, and the system above the line at f_{inert} = 0.088 is the Ariane 5 core stage. Of these possibilities, only the Titan-II and the Ariane 5 have enough Δv to get themselves to orbit. To be conservative, we have assumed the sea-level value for specific impulse for all of the "real" data. In some cases, Isakowitz does not specify the sea-level I_{sp}, so we simply reduce the vacuum I_{sp} by 5%.

There are at least two other considerations. First, the Titan-II and Ariane-5 can get themselves to orbit but without much payload. For example, using Eq. (1.20), we find the allowable payload is given by

$$m_{pay} = \frac{m_{prop}}{e^{\frac{\Delta v}{I_{sp} g_0}} - 1} - m_{inert} \tag{C.1}$$

We assume a conservative Δv = 8800 m/s and the inert mass is the difference between the gross mass and propellant mass in Table C.1 (this difference does not include a payload mounting structure or a fairing). If so, we can find the payload masses for Titan-II and Ariane-5 (payload masses for the other systems are negative):

- Titan-II = 858 kg
- Ariane-5 = 2,224 kg

The second consideration occurs when we are close to the feasible limit. As we approach this limit, our design space becomes very sensitive to small changes in key parameters. For example, Fig. 1.6 shows us that, for a given inert-mass fraction, as our specific impulse decreases, the slope of the mass curve gets steeper.

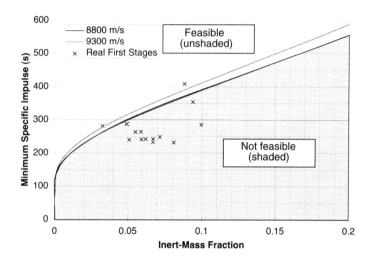

Fig. C.1. **Feasible Regions for Launch Systems.** The two curves shown here represent the minimum possible specific impulse, given a certain structural technology (f_{inert}), to perform a launch mission. Data for existing or historical (real) first stages is overlaid [Isakowitz,1991] and is listed in Table C.1. Several existing first-stage systems are feasible for a launch mission alone, based only on specific impulse and inert-mass fraction (other conditions may make these impractical or impossible).

This means that any small change or error in our design causes very large changes in the system. This situation is undesirable!

So, what can we do to resolve the dilemma of having technology—such as specific impulse or inert-mass fraction—that cannot do a large Δv mission? The obvious answer is to find or develop a solution that allows us to increase specific impulse or to decrease the inert-mass fraction. Looking again at Fig. 1.6, we find that, if we can increase the specific impulse of our propulsion system above 700 s, the mass curves become very flat and almost independent of structural technology (f_{inert}). Two technologies that can achieve this level of specific impulse at high thrust-to-weight ratios are nuclear fission (Chap. 8) and, perhaps, beamed-laser propulsion (Sec. 11.3.2).

Finding technology that can lower the inert-mass fraction can relieve us from a requirement for high specific impulse. This fact is also illustrated in Fig. 1.6, where we see that lower f_{inert}s shift our specific-impulse requirement, for a given initial mass, to a lower number. But existing systems are pretty good, and it is difficult to drastically improve structural technology. Having said this, we *can* drastically improve the "integrated inert-mass fraction" (the average mass fraction, integrated over a mission) by discarding inert mass as it becomes unnecessary. This approach is called *staging*. The basic philosophy behind staging is presented in Sec. 2.6.1.

Table C.1. Data on First Stages of Common Launch Vehicles. This is the basic data from Isakowitz [1991] used in Fig. C.1. Inert-mass fraction = (Gross Mass − Propellant Mass) / Gross Mass.

Stage	Propellant Mass (kg)	Gross Mass (kg)	Sea-Level I_{sp} (s)	f_{inert}
Atlas-E	112,900	121,000	233	0.067
Atlas-I	138,300	145,700	239.75	0.051
Atlas-II	155,900	165,700	240.75	0.059
Atlas-IIA	155,900	166,200	241.7	0.062
Atlas-IIAS	155,900	167,100	241.7	0.067
Delta	96,100	101,700	263.2	0.055
Titan-II	118,000	122,000	281	0.033
Titan-III	134,000	141,000	287	0.050
Titan-IV	155,000	163,000	287	0.049
Saturn S1-B	408,000	444,000	232	0.081
Saturn S1-C	2,080,000	2,210,000	264	0.059
Ariane-L33	233,000	251,000	248.5	0.072
Ariane-H150	155,000	170,000	409	0.088
Energia	820,000	905,000	354	0.094
Proton	410,200	455,600	285	0.100

Evaluating Staging

Having discussed the rationale for staging, how do we choose the number of stages, and how do we size the individual stages? In Sec. 2.6.1, we see that increasing the number of stages decreases the initial mass of our vehicle (Fig. 2.11). However, increasing the number of stages usually increases the cost of our system, if we have to design all of the stages from scratch. In fact, if we choose n stages, the cost for this system can be greater than n times the cost of a single stage. This claim assumes, of course, that doing a mission with a single stage is practical. The process outlined in Table C.2 allows us to size a vehicle with a number of stages.

To illustrate how we can evaluate staging, we look at several launch-vehicle systems as an example. We assume an average ascent $\Delta v = 9000$ m/s and a payload of 1 kg. The choice of payload mass allows us to normalize all of the other masses. This means we simply multiply all of the normalized masses by the payload mass to get the actual design mass. In summary:

- $\Delta v = 9000$ m/s
- payload mass $(m_{pay}) = 1$ kg

Table C.2. Sizing Process for Staged Vehicles. This process allows us to size individual stages and the entire vehicle.

Step	Comments
1. Choose the number of stages (n_{stage})	• Choose the minimum number of stages that is practical. • Choose different values for n_{stage} and compare the marginal differences.
2. Choose propellants for each stage	• These trades are discussed throughout the book.
3. Choose the inert-mass fraction for each stage	• Figs. 5.21, 5.22, and C.2 indicate reasonable choices. • There is a large dispersion in the numbers.
4. Allocate a fraction of Δv to each stage	• Let $f_1 \to f_{n_{stage}}$ be the fraction for each stage; 1 refers to the first stage, n_{stage} refers to the last stage. • $f_1 + f_2 + \ldots + f_{n_{stage}} = 1$ • $f_1 \Delta v_{tot} = \Delta v_1$ (Δv on first stage) $f_i \Delta v_{tot} = \Delta v_i$ (Δv on i-th stage) $f_{n_{stage}} \Delta v_{tot} = \Delta v_{n_{stage}}$ (Δv on last stage)
5. Size the stages and the vehicle	• We start at the uppermost stage and work back to the first stage. • The payload for each succeeding stage includes the previous stages and the actual payload for the mission.
6. Minimize the vehicle mass by optimizing the Δv fraction allotted to each stage	• We must vary f_1 through $f_{n_{stage}}$ to determine the combination that minimizes the vehicle's initial mass. • Usually requires a numerical iteration or optimizing algorithm which repeats steps 4 and 5 until we find a minimum initial mass of the vehicle.

Choose Propellants for Each Stage

The process for choosing the propellants for a particular stage is discussed throughout the rest of the book, where we have already discussed the usual considerations of specific impulse, handling, toxicity, and others.

But one point needs to be stressed. There is a common perception that the choice of propellants can be based on the density of the propellants. Further, this perception drives us to choose denser, and usually lower-specific-impulse, propellants (such as RP-1/LOx) for lower stages and less-dense, higher-specific-impulse propellants (such as H_2/LOx) for upper stages. We reason that higher-density propellants allow us a better (lower) inert-mass fraction, which leads us to a lighter first stage. Although this reasoning may be correct (depending on the mission and requirements), the overall vehicle mass usually increases above what is achievable with higher-performing propellants.

The Saturn family of launch vehicles used RP-1/LOx on the first stage and H_2/LOx on second and third stages. This approach is now universally accepted. However, keep in mind that these vehicles were huge because they were intended for the very large Δv mission of going from the Earth's surface to the Moon and back. If designers had made the first stage with H_2/LOx, it would have been too big to transport to the launch site and would have made vertical assembly and operation of the vehicle even more difficult than it was. Although the mix of propellants was appropriate for Saturn-V, it may not be appropriate for other missions.

If we look at the vehicles listed in Table C.1 and plot the inert-mass fraction versus the average propellant density, we get the result shown in Fig. C.2. We determine the average propellant density as follows:

- From the oxidizer-to-fuel ratio (O/F) for the individual systems (see Isakowitz [1991]), determine the mass of fuel and oxidizer based on the propellant mass [use Eqs. (5.29) and (5.30)]
- Determine the fuel and oxidizer volumes using Eqs. (5.31) and (5.32) and the density data given in Appendix B
- Add the volumes together to get the total volume
- Divide the total propellant mass by the total volume to determine the average propellant density

In Fig. C.2, the propellants on the left are LH_2/LOx (O/F range 5–6), the middle band contains RP-1/LOx systems (O/Fs about 2.25) and the right-hand band reflects values for hydrazine/N_2O_4 (O/Fs about 1.9). Clearly, inert-mass fraction decreases as propellant density increases, but large dispersions indicate other important factors are at play.

For our example problem, we look at several possibilities:

- The entire vehicle uses H_2/LOx, assuming 410 s I_{sp} for the first stages (slightly worse than the space value) and 435 s for all other stages (see Appendix B)
- The first stage uses RP-1, and the remaining stages use H_2/LOx, assuming a first stage I_{sp} of 290 s (slightly better than the sea-level value for the S-1C from Table C.1 or slightly worse than a space engine from Appendix B)
- The first stage uses hydrazine/N_2O_4, and the rest use H_2/LOx, assuming a first stage I_{sp} of 290 s (slightly worse than the vacuum value for Atlas (Table C.1) and worse than a space engine from Appendix B)
- All solid propellants, assuming 260 s for the first stage (slightly better than Scout at sea level—see Isakowitz [1991] or Chap. 6), and 290 s for all other stages (see Table 6.3)

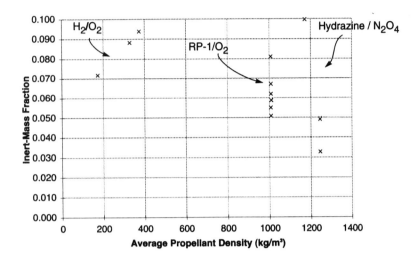

Fig. C.2. Inert-mass Fraction versus Average Propellant Density for the Vehicles Listed in Table C.1. As propellant density increases, inert-mass fraction decreases. But large dispersions indicate that other factors play a major role in these results. The density groupings indicated with text and arrow depend on the propellant combination used.

Choose the Inert-mass Fraction for Each Stage

Figures 5.29, 5.30, and C.2 show the trends in inert-mass fraction for liquid rockets. Table 6.2 and Figs. 6.9 and 6.10 show trends for solids. But the large dispersions in these figures are frustrating. For example, the inert-mass fraction for the Atlas family of vehicles ranges from 0.051 to 0.067 (see Table C.1). How can fractions vary by 30% for similar technology and propellants?

The dispersion in mass fractions depends on all of the design requirements and constraints that are part of any design. The Atlas-E is a simple vehicle that has no parallel stages and does not have much mass stacked on top. By contrast, the Atlas-IIAS first stage has solid rockets strapped to its side and has a large upper stage (Centaur) and payload on top. It makes sense that this more complex vehicle should have a larger mass fraction. When choosing an inert-mass fraction, we must consider complexity, plus propellant type and mass, and then decide how aggressive or conservative we want to be.

For our example, we choose the following inert-mass fractions:
Single stage to orbit

- H_2/LOx   = 0.075   (Fig. C.2)
- RP-1/LOx   = 0.055   (Fig. C.2)
- Hydrazine/N_2O_4 = 0.035   (Fig. C.2)
- Solids   = 0.080   (Table 6.3)

Multiple stages to orbit
- First stage, H_2/LOx = 0.095 (Fig. C-2 and Fig 5.29)
- First stage, RP-1/LOx = 0.070 (Fig. C-2 and Fig. 5.29)
- First stage, hydrazine/N_2O_4 = 0.050 (Fig. C-2 and Fig. 5.29)
- First stage, solid = 0.100 (Table 6.3)
- Others, H_2/LOx = 0.100 (Fig. 5.29)
- Others, RP-1/LOx = 0.085 (Fig. 5.29)
- Others, hydrazine/N_2O_4 = 0.075 (Fig. 5.29)
- Others, solid = 0.08 (Fig. 6.9)

Allocate a Fraction of Δv to Each Stage

How do we vary the proportions between stages? We want to divide up the Δv so the vehicle's total mass is minimized! We define f_i as the fraction of Δv allocated to the i-th stage. The constraint on f_i is that the sum of all of the fractions equals one. The Δv for each stage becomes

$$\Delta v_i = f_i \Delta v_{tot} \tag{C.2}$$

The best combination of these numbers minimizes the vehicle mass. In the special case of inert-mass fractions and specific impulses being equal for all stages, the fraction is

$$f_i = \frac{1}{n_{stage}} \tag{C.3}$$

For the more special case of a single-stage-to-orbit, the only fraction is $f_1 = 1$. For all other situations, we must rely on results of numerical analysis. We discuss this approach below.

Size the Stages and Vehicle

To size the vehicle, we start with the uppermost stage and work down the vehicle stack, stage by stage. Given the payload mass, Δv, specific impulse, and inert-mass fraction for each stage, we can determine the propellant mass, inert mass, and initial mass of that stage using Eqs. (1.27), (1.24), and (1.26) respectively. This initial mass then becomes the payload mass for the succeeding stage, and we repeat the analysis. As an example, consider a two-stage launch vehicle using all H_2/LOx propulsion. We assume the specific impulse and inert mass decisions as listed above. We also assume the Δv is divided, with 46% on the first stage and 54% on the second stage. (This assumption is justified by the analysis discussed in the next section.) If our payload mass is 1 kg, the numbers for the second (upper) stage are:

$$f_1 = 0.46 \rightarrow \Delta v_1 = 0.46\,(9000) = 4140 \text{ m/s}$$

$$f_2 = 0.54 \rightarrow \Delta v_2 = 0.54\,(9000) = 4860 \text{ m/s}$$

$$m_{prop_2} = \frac{m_{pay}\left[e^{\left(\frac{\Delta v_2}{I_{sp_2} g_0}\right)} - 1\right]\left(1 - f_{inert_2}\right)}{1 - f_{inert_2} e^{\left(\frac{\Delta v_2}{I_{sp_2} g_0}\right)}}$$

$$= \frac{(1)\left[e^{\frac{4860}{435\,(9.81)}} - 1\right](1 - 0.1)}{1 - 0.1 e^{\frac{4860}{435\,(9.81)}}} = 2.779 \text{ kg}$$

$$m_{inert_2} = \frac{f_{inert_2}}{1 - f_{inert_2}} m_{prop_2} = \frac{0.1}{1 - 0.1}(2.779) = 0.309 \text{ kg}$$

$$m_{i_2} = m_{pay} + m_{prop_2} + m_{inert_2} = 1 + 2.779 + 0.309 = 4.088 \text{ kg}$$

Now, for the first stage:

$$m_{prop_1} = \frac{(4.088)\left[e^{\frac{4140}{410\,(9.81)}} - 1\right](1 - 0.095)}{1 - 0.095 e^{\frac{4140}{410\,(9.81)}}} = 9.066 \text{ kg}$$

$$m_{inert_1} = \frac{0.095}{1 - 0.095}(9.066) = 0.952 \text{ kg}$$

$$m_i = 4.088 + 9.066 + 0.952 = 14.106 \text{ kg}$$

Optimize the Δv Fraction

So, how do we allocate Δv between stages? For a two-stage vehicle, we can vary one of the Δv fractions over the range from 0 to 1. If we choose to vary f_1, then f_2 is determined from the requirement that both numbers add up to 1. As we vary f_1, we can calculate the initial mass of the vehicle. If we plot the result of f_1 versus the initial mass, we can see the minimum value of initial mass, giving us our optimum distribution of Δv. The algorithm is as follows:

1. Choose a range of f_1 and divide this range into several increments that are Δf_1 apart
2. Let f_1 be the lowest value in the range of f_1
3. Let $f_2 = 1 - f_1$
4. Let $\Delta v_1 = f_1 \times \Delta v_{tot}$ and $\Delta v_2 = f_2 \times \Delta v_{tot}$
5. Calculate the initial mass of the vehicle with these Δv fractions
6. Let $f_1 = f_1 + \Delta f_1$, if we have not reached the end of our range
7. Go back to step 3

To illustrate this algorithm, we look at the example from above for the two-stage H_2/O_2 system. Figure C.3 shows how the initial mass varies as a function of f_1. The minimum initial vehicle mass is at $f_1 = 0.46$, as we used in our example above.

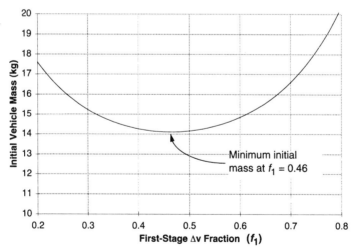

Fig. C.3. Two-Stage H_2/O_2 Vehicle Initial Mass versus First-Stage Δv Fraction. As we vary f_1 between 0.2 and 0.8, we see a minimum at $f_1 = 0.46$.

Doing this analysis for more than two stages is more difficult. We need to vary two Δvs over some range to find a minimum. For a three-stage system, we repeat the above algorithm for a range of f_2 values, choosing the minimum f_1 value for each f_2 (remember $f_3 = 1 - f_1 - f_2$). We can then plot the initial vehicle mass (each point being minimized for f_1) and choose the f_2 with the minimum initial-mass value.

Summary of Example Results

Let us now apply this analytical approach to our example problem. We start by looking at the single-stage-to-orbit problem. No optimizing is required because all of the Δv goes onto the only stage. Only the H_2/O_2 system and the hydrazine system turn out to be feasible for this mission, given our assumed numbers. The results are shown in Table C.3. The hydrogen-fueled vehicle is definitely lighter than the hydrazine-fueled vehicle.

Table C.3. **Results of the Single-Stage-to-Orbit Example.** Based on the assumed parameters, RP-1/O_2 and solids are not feasible. The H_2/O_2 system is lighter than the hydrazine/N_2O_4 system. Remember, we have normalized our vehicle masses by assuming a 1-kg payload. For other payloads, multiply these numbers by the payload mass to get actual mass.

	H_2/O_2	Hydrazine / N_2O_4
Specific impulse (s)	410	290
Inert-mass fraction	0.075	0.035
Propellant mass (kg)	26.06	127.04
Inert mass (kg)	2.11	4.61
Final mass (kg)	3.11	5.61
Initial mass (kg)	**29.17**	**132.64**
Mass of payload / initial mass	3.43 %	0.75 %
Minimum feasible I_{sp} [Eq. (1.29)]	354.2 s	273.66 s

Now, let us look at the four possibilities described above for two-stage vehicles. Table C.4 shows the results of the analysis. Notice that the vehicle using pure H_2/O_2 is substantially lighter than the vehicle with RP-1 fuel on the first stage and H_2 on the second stage. If we add up the mass for the first stage, we find that the all-H_2/O_2 vehicle has a first-stage mass of 10.018 kg, whereas the RP-1 first stage has a mass of 13.256. From Isakowitz [1991] we can deduce that typical stage densities are 256 kg/m^3 for H_2/O_2 (from the S-1C stage) and 655 kg/m^3 for RP-1/O_2 (from the Ariane-5 core stage). Using these numbers, we find that the volume of the H_2/O_2 stage is 0.04 m^3 and the volume of the RP-1/O_2 stage is 0.02 m^3. The H_2/O_2 stage is twice as big despite its being lighter. These numbers validate our previous discussion concerning why we would choose a lower specific impulse but denser propellant for a first stage, as was done for the Saturn-V.

Table C.4. Results of Analysis for Two-Stage Vehicles. The vehicle made up completely of propellants with high specific impulse outperforms all others. A two-stage, all-solid vehicle seems impractical. Remember, we have normalized our vehicle masses by assuming a 1-kg payload. For other payloads, multiply these numbers by the payload mass to get actual mass.

	All H_2/O_2	RP-1 and H_2	N_2H_4 and H_2	All Solids
Stage 1 - I_{sp} (s)	410	290	290	260
Stage 2 - I_{sp} (s)	435	435	435	290
Stage 1 - Inert-mass fraction	0.095	0.070	0.050	0.100
Stage 2 - Inert-mass fraction	0.100	0.100	0.100	0.080
Stage 1 - Δv (m/s)	4140	2610	2880	3780
Stage 2 - Δv (m/s)	4860	6390	6120	5220
Stage 1 - Propellant mass (kg)	9.066	12.328	12.558	63.179
Stage 1 - Inert mass (kg)	0.952	0.928	0.661	7.020
Stage 2 - Propellant mass (kg)	2.668	5.648	4.956	9.708
Stage 2 - Inert mass (kg)	0.296	0.628	0.551	0.844
Initial vehicle mass (kg)	**14.106**	**20.531**	**19.726**	**81.752**
Payload mass/Initial mass	7.1 %	4.9 %	5.1%	1.2 %

Performing similar analysis for three stages further lowers the masses of the vehicles. We find an initial mass for the all-H_2/O_2 vehicle of 12.312 kg and 47.356 kg for the all-solids vehicle. However, for both the RP-1 and hydrazine first-stage vehicles, we find that optimizing drives the first stage Δv to zero. This means that a two-stage vehicle using propellants with higher specific impulse is lighter than a three-stage vehicle using one stage with a low specific impulse.

The mass of the all-solid vehicle is still quite high compared to the one using liquids. This observation explains why existing vehicles, such as Scout and Pegasus, have so many stages.

Conclusions

We have shown why staging can be a valuable tool, presented an example of how staging can help in a launch mission (while hopefully dispelling some misconceptions), and shown how to size a vehicle. However, we have limited our discussion to fairly conventional approaches. It is very easy to quibble over the design numbers chosen here, but a sensitivity analysis shows that our basic conclusions do not change much if we vary specific impulse by 10 seconds or inert-mass fraction by a few percent.

We choose the launch mission as an example, but this type of analysis applies to any mission that requires a large Δv. Another example is transferring from low-Earth orbit to geostationary orbit. We typically use two stages for this mission—one stage for the perigee kick and another stage for the apogee kick in a Hohmann transfer. In the orbit-transfer example, the payoffs and sizing are a bit more straightforward because we are not trying to accelerate continuously. Other examples include lunar or planetary missions.

Many studies deal with optimizing missions by minimizing initial vehicle mass. But it is almost meaningless to minimize mass without including the cost of the minimization. As previously mentioned, doubling or tripling the number of stages may double or triple the cost. As designers, we would much rather try to get away with designing fewer stages. Each additional stage drastically increases the amount of work we must do and the probability that we might fail.

References

Isakowitz, Steven J. 1991. *International Reference Guide to Space Launch Systems.* Washington, DC: American Institute of Aeronautics and Astronautics.

Index

A

Ablative cooling *See also* **Cooling**
  analysis model   235
  definition of   202
  energy balance   239
  mass estimate   226
  schematic of   239
Absorption, neutron   472
Acceleration
  units and conversion factors   686
Acoustic velocity   97
  derivation of   97–100
Activation energy,
  in chemical kinetics   173
Adiabatic flow   95
Adiabatic process
  definition of   94
Advanced Solid Rocket Motor (ASRM)   297
Agena   191
Alfven critical speed   564
Alpha Centauri   671, 672, 673, 678
Alpha radiation
  definition of   468
Altitude
  and nozzle design   10
  vs. atmospheric density   46
  vs. specific impulse   208
  vs. thrust   114
Aluminum
  as hybrid rocket additive   404
  fuel for SRMs   324, 325
  properties of   270
Aluminum hydride
  fuel for SRMs   325
Amagat's Law   88
Ammonium nitrate
  as SRM oxidizer   326
  use as oxidizer   325
Ammonium perchlorate
  as SRM oxidizer   326
  use as oxidizer   325
Angular acceleration
  units and conversion factors   686
Angular measure
  units and conversion factors   686
Angular momentum
  Δv calculation for removing   612
  units and conversion factors   686
Angular velocity
  units and conversion factors   687
Antimatter propulsion   642–645
Antimatter rockets   645–647
  for interstellar travel   676
Antioxidant use in propellants   327
Apogee
  definition of   33
Apogee kick motor (AKM)   352
  example performance prediction   359
  preliminary design of   359
Applications
  of electric propulsion   512, 513
  of hybrid rockets   367, 370
  of rockets   4
  of solid rocket motors   295, 297
Arcjets   525, 553–559
  applied (solenoidal) magnetic field   527
  constricted-arc configuration   555, 556
  definition   509
  magnetoplasmadynamic (MPD)   526
  performance comparison   446
  system efficiency of   587
Area
  units and conversion factors   687
Area ratio
  nozzle   102, 204, 208
Area ratio, nozzle   103
Argument of perigee *See also* **Orbit elements**
  fixed in Molniya orbit   37, 43
  perturbations in   41–44
Ascending node   36
Astrodynamics   31–61
Astronomical data   671
Atmosphere
  density of   46
Atmosphere, sensible   64
Atomic structure   464
Attitude control   280–281
  definition of   2
  liquid rocket use   183
Attitude-control system
  mass sizing for   615
Avogadro's number   150

B

Ballistic coefficients	45
Ballistic missiles	
in liquid-rocket development	189
Ballistics	
in SRMs	331
of hybrid motors	371–403
Barnard's Star	672
Batteries	582
Beamed laser propulsion	678
Beamed laser thermal propulsion	654
Beamed microwave propulsion	677
Beamed pellet propulsion	676
Bearings	
arrangement for turbomachinery	266
Bell nozzles *See also* **Nozzles**	224
Bellows expulsion device	
configuration of	275
Beryllium	
fuel for SRMs	324, 325
Beryllium hydride	
fuel for SRMs	325
Beta radiation	
definition of	468
Binders	
in composite propellants	327
in solid propellants	325
Binding energy	
average	467
in nucleus	466
Bipropellants	
definition of	188
Blasius effect	381
Blowdown process	339
Blowdown system	
definition of	214
Blowing	
effect of	378
Boil-off, tank, description of	268
Boltzmann factor	550
Boron	
fuel for SRMs	325
Boundary layer	
in combustion	372
Boundary-layer cooling *See also* **Cooling**	
analysis model	237
definition of	203
energy balance	241
schematic of	241
Brahe, Tycho	32, 38
Burning rate	
data for hybrid rockets	382
data for solid fuels	330
derivation for hybrids	377
determining rate for SRMs	328
effect from throttling in hybrids	386
erosive	331
for propellants	327
in hybrid rockets	374–384
mesa burning	328
plateau burning	328
pressure dependence	329, 392, 395
St. Robert's Law	328
temperature dependence	330, 391
Burn-up, reactor core	477
Burst pressure	
definition of	269

C

c^* *See* **Characteristic exhaust velocity**	
Cable catapults	660
Case, or motor case	
materials for	310
pV/W estimate	312
sizing for SRMs	310–312
Castor IVA	298, 318
nozzle design	319
Catalyst beds	
for monopropellant decomposition	171
Catapult propulsion	656–662
cable	660
magnetic-coil accelerators	660
ram accelerators	658
Cathodes, hollow	579
Cavitation	
definition of	252
Centaur	191, 522, 605
CERMET engine	
configuration of	456
CERMET reactor	
configuration of	456
core sizing	488
description of	455
mass estimate for core	490
mass versus available power	490
propellant temperature limit	459

INDEX

volume versus available power 489
Chamber pressure *See* **Pressure levels**
Chamber, thrust
  configuring for LRPS 217–244
  liquid monopropellant 242–244
  LRPS mass estimate 226–229
Characteristic exhaust velocity (c^*) 112
Characteristic length
  definition of 219
  typical values 219, 220
Chemical bonds
  energies of 150–153
    table 152
Chemical equilibrium
  role in determining
    combustion products 161
Chemical kinetics 172–173
Chemical kinetics approach
  to thermochemical parameter
    evaluation 695
Chemical propulsion
  comparison with other
    technologies 446
Chemical reaction pressurant 213
Chemical reactions
  dissociation 156
  endothermic 156
  exothermic 156
  frozen-flow assumption 172
  kinetics assumption 172
  kinetics of 172–173
  products of 151
  reactants 151
  shifting equilibrium assumption 172
Child-Langmuir equation 578
Circle (conic section) 34
Circular orbit 35
Closed system
  definition of 79
Coefficient of thrust
  definition of 111
Cold-gas propulsion system
  definition of 3
  performance summary 11
  system diagram 182
Cold-gas thruster
  candidate gases for 141
  performance prediction 144–147
  system architecture for 138
  system diagram 182

thermodynamic design
  example 138–145
Combustion
  differences
    liquid, solid propellants 173–174
  efficiency
    of liquid rockets 206
  stability
    in liquid rockets 183
  thermally choked 658
Combustion chamber
  contraction ratio 221
  definition of 6, 187, 217
  flow Mach number 221
  sizing 218–222
Combustion chamber pressure
  See **Pressure levels**
Combustion chamber temperature
  calculating
    available heat method 168–172
Combustion efficiency
  in liquid rockets 206
Combustion ports
  in hybrid rockets 411
Combustion products
  determining 161–168
    equilibrium constant method 162
    minimization of free
      energy method 166–168
  key thermochemical parameters 176
  molecular mass 150, 176
Composite propellants 324
Composites
  properties of 270
Compton scattering
  definition of 472
Computer programs
  ballistic analysis 345
  for combustion analysis 173, 174
Conceptual design review (CoDR) 601
Conduction
  definition of 128
Configuration
  atomic nucleus 465
  bearing arrangements 266
  blowdown system 278
  CERMET engine and reactor 456
  cold-gas system 182
  common bulkhead tanks 270
  expander cycle 201

gas-generator	201
impulse turbine types	260
liquid rocket propulsion systems (LRPS)	
bipropellant	180
monopropellant	181
nuclear cross section model	473
nuclear reactor	450
nuclear rockets	449
particle-bed reactor	454
passive-expulsion devices	277
positive expulsion devices	275
pressure-regulated system	278
pump-fed liquid rocket	185
reaction turbine	262
shield, nuclear radiation	496
spherical propellant tanks	269
staged-combustion	201
tandem tanks	269
tank-pressure-fed liquid rocket	184
TDRS feed system	245
thrust chamber	218
turbomachinery arrangements	264
turbopump	248
Conic section	34
Conical nozzle	
for LRPS	223
performance of	117–119
Conservation of mass	
definition of	79
Constant pressure process	154
Constant volume process	154
Constants	
of motion	35–36
Contraction ratio	
combustion chamber	221
Control drums *See* **Control rods**	
Control mass approach	
in mass-transfer analysis	79
Control rods	
nuclear reactor	450, 451
Control volume	
first law for	122
Control volume approach	
in mass-transfer analysis	79–81
Controlled angle of attack	
in launch-vehicle steering	71
Convection	
definition of	130
Cooling	
choice of	202–203
for LRPS thrust chamber	234–242
pressure drop	209
structural	202
definition of	188
Copernicus, Nicolaus	32
Cost constraints	
for propulsion systems	26
mission requirements	20
Coulomb attraction	
description of	465
Critical design review (CDR)	601
Critical reactor	
definition of	444, 478
Cross sections, nuclear	472–475
configuration model	473
data for reactor materials	482
data for thermal neutrons	475
definition of	472, 473
Cure catalyst	
use of	326
Current	
units and conversion factors	687
Cycle, engine	
choice of	200–202
comparison of pressure levels	202
engine balance	211
liquid rocket propulsion systems (LRPS)	185
CZ-3	605

D

Dalton's Law	88
Darkening agent	
use in solid propellants	326
Debye length	549, 573
Decay heat	
definition of	469
Defect mass	
definition of	466
Delta launch vehicle *See* **Launch vehicles, Delta**	
Delta v (Δv)	
budget	24
for selected launch vehicles to low-Earth orbit	66
definition of	64–69
	17

Density	
of atmosphere vs. altitude	46
units and conversion factors	687
Diaphragm expulsion device	
configuration of	275
Discharge behavior, electrical	548–553
Dissociation reactions	156
Double-base propellants	323
Doublet *See* **Injectors, choosing type**	
Drag	
ballistic coefficient	45
coefficient	68
orbit perturbations from	44–46
Drag loss	67
Drift rate (in orbit)	55
Dynamic pressure	
in feed system	207
injector	209

E

Earth rotation	
effect on velocity budget	69
East-West drift	60
Eccentric anomaly	39
Eccentricity	32–36
vector	37
Effective exhaust velocity	
definition of	11, 110
Efficiency	
bell nozzle	225
liquid rocket combustion	206
nozzle	117–120
Electric charge	
units and conversion factors	687
Electric conductance	
units and conversion factors	687
Electric field intensity	
units and conversion factors	687
Electric potential difference	
units and conversion factors	688
Electric propulsion	
benefits of	521, 522
design process for	513–517
efficiency vs. exhaust speed	523
electric power system	510
electrical forces and currents in	535–545
elements of	510–511

mass calculations for	586–591
mission specification for	514–523
on-orbit use of	590
performance assessment	585–589
power conditioning for	583, 593
power conversion for	582
power sources	582–583
system evaluation	589–591
thermal management system	510, 584, 593
Electrical rocket propulsion system	
definition of	6
Electrodynamic tethered propulsion	667
Electromagnetic	
propulsion	526–532, 563–574
definition	510
performance summary	11
Electromagnetic units	693
Electrostatic	
propulsion	532–534, 575–581
definition	510
performance summary	11
Electrothermal propulsion	524–526
arcjet use in	553–559
definition	509
performance summary	11
Ellipse (conic section)	34
Energy	
activation, in chemical reactions	173
conservation of	82
internal	82, 153
specific	86
system	81
units and conversion factors	688
Energy equation (orbits)	35
Engine balance	211
Engine cycle *See* **Cycle**	
Engines	
definition of	6
Enrichment, uranium fuel	481
Enthalpy	
absolute	155
definition of	86, 154
relative	155
stagnation	100
static	100
table of	170
Entropy	
definition of	93, 166

E

Envelope constraints
  for propulsion systems — 27
Environment
  environmental impacts — 22
  flight limitations — 22
  operating — 25
  political, economic — 22
Ephemeris — 31
Epsilon Eridani — 672, 678
Equilibrium constant — 161
Equilibrium constant method
  for determining combustion products — 162–166
Equilibrium flow approach
  to thermochemical parameter evaluation — 695
Equilibrium state
  of chemical reactions — 162
Erosive burning
  in solid rockets — 331
Evaporation pressurant — 213
Exhaust nozzle
  definition of — 217
Exhaust pressure *See* Pressure levels
Exhaust velocity (exit velocity)
  effective — 11, 110
  for momentum thrust — 7
Exit pressure *See* Pressure levels
Expander cycle *See also* Cycle
  definition of — 200
  system configuration — 201
Expansion ratio of a rocket nozzle — 102, 204
Expulsion devices
  passive — 277
  positive — 275

F

Faraday's law — 540, 541
Fast reactors
  core sizing — 488
  definition of — 475
Fast-fission factor — 480
Fast-fission reactor
  CERMET type — 455
Figures of merit
  qualitative — 23
  quantitative — 23

  example of — 603
Filament-wound structures — 310
Film cooling *See also* Cooling
  analysis model — 237
  definition of — 203
  energy balance — 241
  schematic of — 241
Filters, description of — 246
Finite elements
  solid rocket ballistics — 345–348
  thermal analysis — 137
Fission
  definition of — 444
  fast
    use of moderator — 448
  nuclear
    energy distribution — 471
    process of — 469
  pulsed
    performance comparison — 446
  solid-core
    performance comparison — 446
  thermal
    use of moderator — 448
Flame temperature
  of a combustion gas — 176
  of selected propellants — 695–713
Flexible bearing (Flexseal) — 321
Flight readiness review (FRR) — 601
Flight-simulation programs
  for launch vehicles — 74–75
Flow
  isentropic — 95–107
  nozzles *See* Nozzles
  separated — 116
  supersonic — 97
Flow control
  definition of — 187
Flowfields, rocket
  multiphase — 171
Fluid ducting, description of — 245
Fluid flow
  characteristics of — 95
  overexpansion, underexpansion of — 115
  stagnation conditions — 100
Fluorine
  propellant properties — 696
Force
  units and conversion factors — 688
Fourier's Law — 128

Free energy *See also* Gibb's
 free energy
  use in combustion analysis 166
Free radicals 639
Friction coefficient
  blowing effect on 379
Frozen flow assumption 172, 695
Frozen-flow losses 523
Fuel elements, nuclear 449, 450
Fuel particle
  particle-bed reactor 454
Fuel *See* Propellants
Fuels, chemical
  high-energy density 636
Fusion
  inertial
    performance comparison 446
  magnetic
    performance comparison 446
  nuclear 467
  reactions 467

G

Gamma radiation
  attenuation by shielding 497
  definition of 468
Gaseous rocket *See* Cold-gas thruster
Gas-generator *See also* Cycle
  definition of 200
  system configuration 201
  system diagram 185
Gemini 191
General perturbations *See also*
  Orbits, Orbit perturbations 41
Geocentric Inertial Coordinates (GCI)
  in defining orbital elements 36–37
  inertial properties of 36
Gibbs free energy 167
Gimbal 180, 188
Global Positioning System (GPS)
  apogee kick motor in 299
Glow discharge 551
Glow-to-arc transitions 552
Goddard, Robert H. 189, 512
Grain, propellant
  in SRMs 296, 333
  volumetric loading 358
Graphite Epoxy Motor (GEM) 298, 299

Graphite *See* Composites
Gravity losses 67, 606
Gravity-turn maneuvers 512
  in launch-vehicle steering 69
Ground tracks
  repeating 58–60

H

H-10 605
H-2 Stg-2 605
Hall accelerators 540
Hall effect 531, 540, 570
Hall thrusters 531, 533, 570
  performance data for 574
Hard starts
  definition of 369
Head coefficient, description of 250
Head loss coefficient
  definition of 233
  typical values 233
Heat
  of fusion 160
  of reaction
    definition of 151
  of vaporization 160
Heat addition
  as energy transfer 120–127
  in constant area
    with steady flow 123–127
  in constant volume
    with no flow 120–121
  in steady flow 122
Heat of formation, standard 156
  table of 158
Heat transfer
  coefficients 133
  conductive 128–130
  convective 130–134
  in rocket propulsion 128–138
  numerical modeling of 137–138
    by finite differences 137
    by finite elements 137
  radiative 134–136
Heaters, description of 246
Helium (He)
  specific impulse performance 142
Hess's Law 157
HMX (Her Majesties' Explosive) 327

734 INDEX

Hohmann transfer orbit 47–52
  as most efficient two-burn transfer 47
  examples of using 606–609
  total required Δv 48
  transfer ellipse 47
Hohmann, Walter 47
Hoop stress
  mass estimate for liquid rockets 227
Hybrid rocket propulsion system
  definition of 5
  performance summary 11
Hybrid rockets
  advantages and
    disadvantages of 365–368
  aft mixing chamber use 425
  ballistics of 371–402
    parameter comparisons 389
    parameter variations 384–392
  burning rate 374–383
    alternate expressions for 383–384
    contours of 382
    data correlation for 397–401
    expression for 377
    kinetics effect on 395–397
    radiation effect on 392–395
  combustion chamber design 424, 437
  combustion process in 372, 373
  definition of 365
  design process for 403–405
  history of 368–371
  injectors for 424
  mass estimates for 427, 436
  nozzle design 426, 436
  performance estimate of 419–423, 433
    simulation algorithm for 421
  port configurations 411–419, 431
    volumetric efficiency 415
    web supports for 413
  pressure levels in 408, 429
  propellants for 404–408, 428
    requirements 410
  regression rates *See* Hybrid rockets,
    burning rate
  reverse hybrids 404
  system sizing 409–410, 431
  temperature sensitivity in 391–392
  thrust-vector control in 426
Hydrazine arcjets 617
Hydrazine *See also*
  Monopropellant systems 188
  decomposition of 242
  typical properties 696
Hydrocarbon fuels *See* **Propellants**
Hydrogen
  propellant properties 696
Hydrogen peroxide *See also*
  Monopropellant systems 188
  decomposition of 242
  typical properties 696
Hydroxy-terminated
  polybutadiene (HTPB) 327
Hyperbola (conic section) 34
Hypergolic propellants
  definition of 175, 187

I

Ideal expansion
  in rocket nozzles 113
Ideal gas
  definition of 87
  law of 87
Ideal rocket equation
  definition of 13
  derivation of 12–13
Igniters
  definition of 187
  in solid motors
    preliminary design of 356
  pyrogen 313
  sizing 313–314
Illuminance
  units and conversion factors 688
Impulse, total *See also*
  Specific impulse 16
Inclination 36
Inducer-inlet flow coefficient 250
Inductance
  definition 542
Inertial space 55
Inertial Upper Stage (IUS) 298, 605
Inert-mass fraction
  definition of 14
  for liquid rockets 215
  for reaction control systems 216
  for space engines 216
  in nuclear rockets 461
  purpose of 15
  vs. propellant mass 215

vs. stage number	215	**Isentropic spouting velocity**	
Injectors		definition of	258
choosing number and size	232	**Isp code**	168, 695
choosing type	231		
definition of	187, 217		
design of	229–234	**J**	
mass estimate	226		
pressure drop	209	**Jet Assisted Takeoff (JATO)**	190
typical numbers	234	**Jet engines**	
In-situ propellants		and rockets, differences between	6
nuclear stages and	448	**Jet vanes**	323
Instability, combustion *See*		**Juno**	189
Combustion, stability		**Jupiter**	189
Insulation			
cryogenic tanks	273	**K**	
in SRMs	314–316		
Interceptors		**Kepler, Johannes**	32, 38
nuclear stage	447	**Keplerian orbits** *See also*	
Interconnect plumbing and		Two-body equations	
components		of motion	32–40
definition of	187	**Kilogram, definition of**	685
Internal ballistics analysis, in		**Kinetics assumption**	
solid rocket motors *See*		for combustion prediction	172
Solid rocket motors (SRM),			
ballistics prediction		**L**	
International System of			
Units (SI)	685–692	L^*, *See* Characteristic length	
Inter-regenerative cooling		L', *See* Characteristic length	
definition of	235	**L-7/EPS**	605
Interstellar ramjet	676	**Launch azimuth**	57
Interstellar travel	668–680	**Launch vehicles**	
astronomical data and	671	Ariane	
methods for	674–680	basic data on	717
possibility of	674	Δv budget	66
requirements for	671	solid rocket motor use	297
targets for	672	Atlas	189
Ion engines	532, 534, 575–581	basic data on	717
case study for	592–595	Δv budget	66
definition	510	development history	190
electron-bombardment	577, 579	solid rocket motor use	297
non-neutral fields in	550	Delta	
performance values for	582	basic data on	717
Ionization processes, in plasmas	545–548	Δv budget	66
Isentropic flow	95	solid rocket motor use	297, 298
Isentropic parameter		Energia	
of a combustion gas	176	basic data on	717
of selected propellants	695–714	general aspects of design	61
Isentropic process		H-2	
definition of	94	solid rocket motor use	297
Isentropic relations	95		

liquid rocket use	183
Long March	
solid rocket motor use	297
nuclear	446
Pegasus	300
solid rocket motor use	297, 298
Proton	
basic data on	717
Saturn	189
basic data on	717
Δv budget	66
development history	190
propellant choices	719
Scout	
solid rocket motor use	297
Space Shuttle *See* Space Shuttle	
staging	62, 715–726
Δv allocation	721, 723
inert-mass fractions for	720
sizing process for	718, 721
steering	69–74
controlled angle of attack	71
gravity turn	69
linear-tangent	71–73
thrust-to-weight ratios	195
Titan	189
basic data on	717
Δv budget	66
development history	190
solid rocket motor use	297, 298
thrust vector control in	301, 323
trajectory simulation programs	74–75
Launch windows	55–58
Launch-site selection	24
Length, units and conversion factors	688
Lewis code	168, 695
Lewis dot structure	152
Lewis number	375
Librating tethers	663
Light-gas guns	658
driver stage	658
launcher stage	658
Linear-tangent steering	
in launch vehicles	71–74
Liquid Injection TVC (LITVC)	321, 426
Liquid oxygen *See* **Oxygen**	
Liquid propellants *See* **Propellants**	
Liquid rocket propulsion systems (LRPS)	179–291
advantages of	181, 183

applications of	183
definition of	4
design process for	192–193
disadvantages of	183
engine sizing	195–199
flow control	
bipropellant	180
cold-gas thruster	182
monopropellant	181
history of	189–191
hoop stress	
mass estimate	227
interconnect plumbing and components	
bipropellant	180
cold-gas thruster	182
monopropellant	181
mass estimate	226
monopropellant	
ignition of	187
performance summary	11
preliminary design decisions for	194–217
propellant feed system	
bipropellant	180
cold-gas thruster	182
monopropellant	181
propellant storage	
bipropellant	180
cold-gas thruster	182
monopropellant	181
structural mounts	
bipropellant	180
cold-gas thruster	182
mass estimate	281
monopropellant	181
system diagram	
bipropellant	180
monopropellant	181
tank pressurization	
bipropellant	180
cold-gas thruster	182
monopropellant	181
thrust chamber	
bipropellant	180
thruster	
monopropellant	181
thrust-to-weight ratios	195–199
thrust-vector control	
bipropellant	180

INDEX

Local sidereal time (LST)   57
Lorentz force   535
Low-thrust maneuvers   27
LRPS *See* Liquid rocket propulsion systems
LST (Local sidereal time)   57
Luminance
    units and conversion factors   689
Lumped-parameter methods
    for ballistics prediction in solid motors   334
    and changes in burn surface or throat area   340
    steady-state   334
    unsteady conditions   337
    for nuclear reactor sizing   476

M

Mach number
    at nozzle throat   104
    of a fluid flow   97, 100
Magnesium
    fuel for SRMs   325
Magnetic dipole moment   693
    units and conversion factors   689
Magnetic field
    calculation of   542–545
Magnetic field strength   693
    units and conversion factors   689
Magnetic flux
    convection of   540–541
    units and conversion factors   689
Magnetic induction   693
    units and conversion factors   689
Magnetic moment   693
    units and conversion factors   689
Magnetic pressure   542
Magnetic Reynolds number   541
Magnetic tension   542
Magnetic-coil accelerators   660
Magnetoplasmadynamic (MPD) arcjets   526
    performance comparison   446
Magsail (magnetic loop sail)   652
Mars mission
    nuclear stage   447
    radiation exposure   447
Marxman's formula
    for blowing effect   379
Mass
    jet-specific   519, 588
    units and conversion factors   689
Mass drivers   660
Mass estimate
    cylindrical tanks   271
    liquid rocket propulsion systems
        ablative material   226
        analytical method for   228
        injectors   226
        nozzles   226
        thrust chamber   226
        typical fractions   227
    nuclear reactor core   488
    of nuclear reactor vs. available power   490
    propellant tank for electric propulsion   588
    pV/W method for tanks   272
    spherical tanks   270
    stored-gas pressurant systems   279
    structural mounts for liquid rockets   281
    turbomachinery   266
Mass fraction
    in electric propulsion systems   589
    of a gas   89
    payload   520
Mass fraction *See* Inert-mass fraction
Mass transfer
    in rockets   78–81
        control volume approach   79
        system approach   79
Materials
    for LRPS thrust chambers   229
    for SRM nozzles   317
    high-temperature properties   129
    refractory   128
    structural properties   270
    structural properties for SRM cases   310
Maximum expected operating pressure (MEOP)   311
Mean anomaly   38–40
Mean motion   39–40
Meter, definition of   685
Metric (SI) units (*See* International System of Units)
Metric prefixes   685

Minimization of free energy method
  for determining
      combustion products   166–168
Mission concepts
  developing alternative   23–24
Mission requirements
  cost constraints   20
  defining   20
  example of   602
  hard and soft   23
  schedule   20
  technical risk   20
Mixture ratio *See* **Oxidizer-to-fuel ratio**
Moderators
  in nuclear reactors   448
  in nuclear rockets   450
Mole
  definition of   89
Mole fraction
  definition of   89
Molecular mass, exhaust product
  of selected propellants   695–713
Molniya orbits *See also* **Orbits**
  parameters   43
Moment of inertia
  units and conversion factors   690
Momentum equation   108
Momentum thrust
  equation for   7
Monomethyl hydrazine
  typical properties   696
Monopropellant systems
  definition of   4
  operating sequence   243
  performance of   188
  sizing of   196
  system diagram   181
  thrust chamber   242
Monopropellants
  definition of   188, 242
Motor cases
  mass estimates of   311
  materials for   310
  preliminary design of   354
Motors
  definition of   6
Multiplication factor, nuclear   479

N

NERVA (Nuclear Engine for Rocket Vehicle Applications)
  fuel element   453
  mass estimate, reactor core   490
  mass versus available power   490
  program description   452
  propellant temperature limit   459
  reactor description   452
  reactor sizing   476–486
  reactor sizing with reflector   485–486
  reactor sizing without reflector   481–485
  volume versus available power   489
Net Positive Suction Head (NPSH)
  definition of   252
Neutral burning motors   332
Neutron production
  effectiveness factor   480
Neutrons
  attenuation by shielding   497
  energy of   475–476
  fast   475
  free   465
  thermal   475
    cross section data   475
Newton, Isaac   32
Newton's Second Law   7, 107
Nitrogen tetroxide
  typical properties   696
Nitronium perchlorate
  as SRM oxidizer   326
Nodal vector   37
Nonleakage probability
  for neutrons   480
North-South drift *See also*
  Geosynchronous orbit   60
Nozzle exit velocity
  determination of   106
Nozzles
  bell   223, 224
  cooling of *See* Cooling
  deLaval   223
  exit pressure   205
  expansion ratio   102, 204
  in hybrid rockets   426
  in liquid rockets   188
  in nuclear rockets   449
  in solid motors
    preliminary design of   356

INDEX

sizing of	316–320
mass estimate	226
nonaxial flow in	117
overexpansion, underexpansion	115
Rao	223
separation of flow	116
sizing	222–225
thrust ratio in	120

Nuclear fission *See* **Fission, nuclear**

Nuclear fusion
in rockets	640
inertial-confinement	641
magnetic-confinement	641

Nuclear physics
atomic structure	464
binding energy	466
defect mass	466
definition of	464
modified four-factor formula	478

Nuclear reactors
CERMET	455
comparison of	455, 457
core burn-up	477
core sizing	477
critical, supercritical, subcritical	478
definition of	478
fast	475
materials	492
NERVA	452
Particle-bed	452
power requirements	461, 462
propellant temperatures	459
sizing	463–493
thermal	475
thermal hydraulics	491–492

Nuclear rocket propulsion
systems	443–507
advantages of	444
configuration of	448–451
definition of	5
design process for	455–459
operation of	448–451
performance summary	11

Nuclear rockets
applications of	457
configuration of	449
electric	
and interstellar travel	675
operation of	498
pressure levels	462

propellant flow rates	461
shadow shields for	496
sizing	461

Nuclear-pulse rockets
for interstellar travel	675

Nucleons
definition of	465

O

Oberth, Hermann	189, 512

Oblateness of the Earth
effect on orbits	59
Ohm's law, generalized	539, 570

Open system
definition of	79

Orbit elements 36–38
argument of perigee	37
eccentricity	36
eccentricity vector	37
inclination	36
nodal vector	37
right ascension of ascending node	36
semi-major axis	36
time since perigee passage	37
variations	
secular and periodic	40

Orbit insertion
definition of	2
liquid rocket use	183

Orbit maintenance 58–61
definition of	2
liquid rocket use	183

Orbit maneuvering
coplanar orbit transfers	47–52
liquid rocket use	183
one-tangent-burn	50
spiral transfer	52

Orbit period
equations for	38

Orbit perturbations 40–47
atmospheric drag effects	44
due to nonspherical Earth	42–44
due to third-body interventions	41
general perturbations	41
long-period variations	40
secular variations	40
short-period variations	40
solar radiation effects	46

special perturbations	41
Orbit rendezvous	54–55
Orbit transfer *See also*	
Hohmann transfer orbit	47–55
plane change	52–54
Orbital maneuvering	47–55
Orbital-transfer vehicle	
parametric cost modeling for	590
upper-stage motor preliminary design of	352–361
Orbital-transfer vehicles	
nuclear	447
Orbits *See also* **Orbit perturbations, Sun-synchronous orbit, Molniya orbit, Geosynchronous orbit, Orbit transfer, Stationkeeping, Constellation**	
circular velocity	35
elements	36–38
equations of motion	32–36
escape velocity	36
Hohmann transfer orbit (*See also* Hohmann transfer orbit)	47
Keplerian orbits	32–40
maneuvers	47–55
Molniya orbits	43
ORBUS 21	298, 322
ORBUS 6E	298, 301, 306
Overexpanded nozzle	115
Oxidizer *See* **Propellants**	
Oxidizer-to-fuel ratio	
choice of, in liquid rockets	199
definition of	176
of selected propellants	695–714
Oxygen	
propellant properties	696

P

Pair production	
definition of	472
PAM-DII *See* **Payload Assist Module**	
Parabola (conic section)	34
Particle-bed reactor	
advantages	453
configuration of	454
engine	454
fuel element	453, 454
fuel particle	454
mass estimate for core	490
mass versus available power	490
propellant temperature limit	459
reactor description	452
sizing	486–488
volume versus available power	489
Particles	
delayed, prompt definition of	470
Particles, collections of charged	
dynamics of	536
Particles, single charged	
dynamics of	535
Passive expulsion devices	274
Passive propellant expulsion device	
configuration of	277
Path dependence	
for work and energy	83
Payload Assist Module (PAM)	298
Pegasus launch vehicle *See* **Launch vehicles, Pegasus**	
Perfect gas	
calorically perfect	92
definition of	87
law of	87
properties of	94
Performance	
Isp vs. expansion ratio	208
of liquid rocket thrust chambers	208
parameters for	25
Performance requirements	26
Perigee	
definition of	33
Perturbations, orbit	40–47
Phasing orbit	54
Photoelectric-effect	
definition of	472
Pintle *See* **Injectors**	
Pintles *See also* **Injectors, Thrust vector control**	323
Piston expulsion device	
configuration of	275
Piston pumps	
application of	186
Plasma	
calculating electric field in	538–540
definition	538
ionization of	545–548
microinstabilities in	547

Plasma propulsion	
definition	510
Plasticizer	
use of	326
Poison *See* **Control rods**	
Polar boss	
definition of	295
design of	313
example	355
function	296
Polar equation of a conic section	34
Polybutadiene	
typical properties	696
Positive column	548
Positive expulsion devices	273–275
Potassium nitrate	
as SRM oxidizer	326
Potassium perchlorate	
as SRM oxidizer	326
Power	
units and conversion factors	690
Power balance	211
PPT *See* **Thrusters, pulsed-plasma**	
Prandtl number	375
Preliminary design decisions	
definition of	192
for nuclear rockets	459–463
Preliminary design review (PDR)	601
Preliminary requirements review (PRR)	600
Pressurant	
chemical reaction	213
evaporation	213
sizing process for	279
stored gas	213
type vs. tank volume	214
Pressurant systems	
advantages and disadvantages	276
design and sizing	275–280
stored-gas	
use and sizing	278
Pressure	
units and conversion factors	690
Pressure feed	
definition of	185
system diagram	184
Pressure levels	
chamber pressure	203
cooling loss	209
determination of	203–214
dynamic pressure	207
effect on engine size	204
effect on LRPS performance	207
feed system pressure drop	209
gas-generator	205
ideal expansion	205
in nuclear rockets	462
injector, pressure drop in	209
nozzle exit/exhaust	205
nozzle expansion ratio	203
pressurant system	213
propellant tank	209
pump-pressure-feed	205
tank-pressure-feed	204
Pressure regulator, description of	246
Pressure vessels	
in nuclear rockets	450
Processes	
design for LRPS	192–193
designing and sizing pumps	253
designing and sizing turbines	259
designing propulsion systems	19–30
designing SRMs	305
designing turbomachinery	251
hybrid rocket design	403
nozzle sizing for LRPS	222–225
sizing pressurant	279
summarizing pump requirements	252
Progressive burning motors	332
Propellant burn surface area	332
Propellant density	
in solid propellants	332
Propellant expulsion	
devices for	273–275
passive devices	274
Propellant feed system	
channel sizing	207, 244–246
configuring	244–246
dynamic pressure	207
in liquid rockets	184–186
in nuclear rockets	449
pressure loss	209
pump-pressure feed	
pressure levels	205
system diagram	185
sizing	244–246
tank-pressure feed	
pressure levels	204
system diagram	184

Propellant grain
  in solid rocket motors   296
Propellant handling system
  definition of   184
Propellant mass fraction   309
  definition of   14, 306
  trends in   308
Propellant storage
  definition of   186
  expulsion of propellants   186
Propellants
  definition of   2
Propellants, chemical
  characteristics of   174–176
  chemical reactivity of   175
  choosing   199–200
  for hybrid rockets   404
  for nuclear rockets   460
  heat of formation   175
  heat transfer properties   175
  hypergolic   187
  in nuclear reactors   462
  liquid
    typical properties   696
  solid   323–331
    antioxidant use   327
    binders   325
    bonding agents in   327
    burning rate of   327–331
    composite   324, 330
    cure catalyst for   326
    darkening agent in   326
    density of   332
    double-base   323
    erosive burning of   331
    fuels   324
    mesa burning   328
    minor ingredients   326
    oxidizers   325
    plasticizer use   326
    plateau burning   328
    solids loading   325
    temperature sensitivity   330
  specific gravity of   175
  thermochemical data for   695–714
  viscosity of   175
Propulsion systems *See also*
  specific technologies, such as
  Nuclear rocket propulsion,
  Hybrid rocket propulsion
  defining requirements for   24–28
  design process for   19–30
  design process table   21
  developing preliminary designs   29
  developing selection criteria   22
  functions of   2
  performance summary   11
Propulsion technologies
  table of functions   4
Pulse mode
  definition of   181
Pulsed-plasma thrusters (PPT)   559–563
Pump cycle *See* Cycle
Pump feed
  definition of   185
  system diagram   185
Pumped tethers   665
Pumps
  efficiency of   255
  maximum pressure rise   254
  requirements for   250, 252
  sizing   253, 254

Q

Quasistatic process
  definition of   95

R

Radial reflector
  in nuclear reactors   448
Radiation
  attenuation, materials for   497
  dosage limits   495
  dosages   447, 493
  Mars mission exposure   447
Radiation cooling *See also* Cooling
  analysis model   236
  definition of   202
  energy balance   240
  schematic of   240
Radiation shields
  in nuclear rockets   449, 493–497
  sizing   493–497
Radioactivity
  alpha radiation   468
  beta radiation   468
  gamma radiation   468

natural	468
Ram accelerators	658–660
combustion modes	658
Ramjets	
interstellar	676
Ramrockets	
dual-mode	633
liquid-fuel	632
performance of	634–635
solid propellant gas-generator	632
solid-fuel	633
Rate constant, in chemical kinetics	173
Rate law, in chemical kinetics	173
Reactor pressure vessel	
in nuclear reactors	448
Redesigned Solid Rocket Motors (RSRM)	297
Redstone	189, 190
Reflectors	
in nuclear rockets	450
Refractory materials	
for rocket applications	128
Regenerative cooling	
in rocket nozzles	133
Regenerative cooling *See also* **Cooling**	
analysis model	235
definition of	203
energy balance	238
in rocket nozzles	132
schematic of	238
Regressive burning motors	332
Relativistic mechanics	668
Resistance	
units and conversion factors	688
Resistojets	524, 534, 553
definition	509
performance comparison	446
Resonance escape probability	
of neutrons	479
Restrictors	332
Reversible process	
definition of	94
Reynolds analogy	375
Reynolds number, magnetic	541
Richardson equation	551
Richardson's constant	551
Right ascension of ascending node	36, 41
Rocket equation	307
antimatter	645
for relativistic velocities	669
Rockets	
air-augmented	631–635
and jet engines, differences between	6
antimatter	645–647
definition and basic elements	2
electric	636
nuclear-fusion	640
nuclear-pulse	641, 675
Rolling diaphragm expulsion device	
configuration of	275
Rotor	
definition of	258
RP-1	
typical properties	696

S

Scale height (atmosphere)	
atmospheric density	45
Scattering, neutron	472
Schematics *See* **Configuration**	
Schottky effect	551
Second, definition of	685
Sectoral terms (geopotential)	42
Semi-major axis	36
Separated flow	116
SERT I, SERT II *See* **Space Electric Rocket Test**	
Shadow shields	
configuration	496
Shifting equilibrium assumption	172
Shock wave, separated flow	116
SI *See* **International System of Units**	
Similarity relationships	
discussion of	249–250
Single-stage-to-orbit (SSTO) launch vehicle	
parametric analysis for	26
requirements for	25
Sirius	672
Sizing	
catalyst beds	242
CERMET reactor core	488
channels, feed system	244
combustion chamber	218–222
exhaust nozzle	222–225
fast reactor core	488
feed system channels	207

hybrid rocket systems	409–410, 431
igniters	313
internal insulation in SRMs	314
liquid propellant tanks	214–217
mass estimates for hybrid rockets	426
monopropellant thrust chamber	242
motor cases	310–312
NERVA-type reactors	481
nuclear reactor	
core volume	477
nuclear reactors	463–493
nuclear rockets	461
particle-bed reactors	486
propellant tanks	268
pumps	253
radiation shields	493–497
staged vehicles	718
stored-gas pressurant systems	278
tanks for electric propulsion	588
thrust chamber	217, 226–229
thrust vector control systems	320–323
turbines	257

Slotted grains
treatment of	348

Sodium nitrate
as SRM oxidizer	326

Solar arrays
mission constraints for	594
representative values for	584

Solar propulsion
performance comparison	446
solar lightsail	649–651
solar thermal	648

Solar wind   652

Solid angle
units and conversion factors	690

Solid Performance Program (SPP)   174

Solid rocket motors (SRM)
ballistics prediction	331–351
lumped-parameter methods	
See also Lumped-	
parameter methods	334–341
preliminary design example	357
spatial variation methods *See*	
also Spatial variation	
methods for ballistics	
prediction in	
solid motors	342–348
design process for	305
dimensions	302
efficiency calculation	351
fuels for	325
igniters	313
preliminary design of	357
internal insulation	
preliminary design of	356
safety factor	315
sizing of	314–316
major components	295
mass summary	303
motor case sizing	310–312
nozzles	
materials for	316–317
preliminary design of	356
sizing of	316–320
performance summary	304
specific impulse	351
thrust vector control systems	
sizing of *See also*	
Thrust vector control	320–323
vacuum thrust calculation	351

Solid rocket propulsion system
definition of	5
performance summary	11

Space Electric Rocket Test (SERT I, SERT II)   513

Space Shuttle   674
Δv budget	66
liquid rocket engine	189, 191
Redesigned Solid Rocket	
Motor (RSRM)	318
nozzle design	318
solid rocket motor use	297

Space storables
definition of	189

Spatial variation methods for ballistics prediction in solid motors
ballistic element methods	345
single-element model	342–345
treatment of slotted grains	348

Special perturbations *See also*
Orbits, Orbit perturbations   41

Specific angular momentum   36

Specific gas constant
derivation of	90

Specific heat
and energy transfer	91–93
definition of	91
nuclear propellants	460

ratio of	93	Pressurant systems	
Specific heat capacity		**Strength**	
units and conversion factors	690	for tank materials	270
Specific impulse		**Stress**	
and expansion ratio	208	units and conversion factors	690
comparison of technologies	446	**Strong nuclear force**	
definition and equation for	10	definition of	465
of selected propellants	695–714	**Structural cooling** See **Cooling**	
Specific mass		**Structural mass fraction**	
comparison of technologies	446	equation for	14
Specific mechanical energy	35	**Structural materials**	
Specific speed, description of	249	for propellant tanks See	
SPT See **Thrusters, stationary plasma**		Tanks, propellant	
St. Robert's Law	328	**Structural mounts**	
Staged-combustion See also **Cycle**		in liquid rockets	188
definition of	200	in nuclear rockets	449
system configuration	201	**Subcritical reactor**	
Staging		definition of	478
choosing propellants for	718	**Suction-specific speed**	249
definition of	716	**Sufficiently inertial**	
delta-V allocation	721, 723	coordinate frame	36
evaluating	717	**Sun-synchronous orbits**	
inert-mass fractions for	720	determining inclination	43
sizing	721	**Supercritical reactor**	
sizing process for	718	definition of	478
Stagnation conditions	100–102	**Supersonic flow**	97–107
Star 48B	298, 301, 318	**Surface tension**	
nozzle design	320	defining experiment	277
State symbols		for passive propellant expulsion	274
representations of	150	**System approach**	
Stationkeeping	591	to mass-transfer analysis	79
Stationkeeping See also			
Orbit maintenance	60	**T**	
Δv calculation for	612		
delta-v requirements	522	**Tank pressurization** See also	
electric propulsion use	512, 591	**Pressurant systems**	275
mass comparisons for	591	definition of	186
Statite	651	pressurant gases for	186
Stator		**Tanks, propellant**	
definition of	258	cylindrical	
Steel		mass estimate	271
properties of	270	in nuclear rockets	449
Steering loss	68	pressure level	209
Stefan-Boltzmann constant	135	sizing of	214–217, 268
Stoichiometric coefficients		spherical	
definition of	150	mass estimate	270
Stoichiometric length		structural materials for	270
definition of	387	tandem	
Stored-gas pressurant	213	configuration of	269
Stored-gas systems See			

Tau Ceti	672, 673	schematic of	218
TDK-DM	605	sizing	217–244
TDRS *See* Tracking and Data Relay Satellite		Thrust equation, derivation of	109–110
		Thrust history	27
Technology risk level		solid rocket example	28
for propulsion systems	27	Thrust skirt	
Techroll seal	321	design of	312
Telstar IV	513, 554	example	355
arcjet thruster on	555	function	296
Temperature		Thrust vector control (TVC)	280–281
of isentropic flow	95–102	definition of	188
stagnation	101	Flexible bearing (Flexseal)	321
units and conversion factors	691	in hybrid rockets	426
Tesseral terms (geopotential)	42	jet vanes and pintles	323
Tethered propulsion	662–668	Liquid Injection TVC	321
electrodynamic	667	Techroll seal	321
librating and rotating tethers	663	typical control authority	281
lifetimes in space	667	Thrust, maximum	
pumped	665	conditions for	113
vertically stabilized	662	Thrusters	
Thermal conductivity		augmented-hydrazine	513, 553
units and conversion factors	691	cold-gas	
Thermal hydraulics		thermodynamic design	
in nuclear reactors	491	example of	138–145
Thermal neutrons *See* Neutrons, thermal		efficiency comparison of	585
		electric	
Thermal reactors		types of	524
definition of	475	electrostatic	533
Thermal utilization factor		Hall	531, 570
of neutrons	479	magnetoplasmadynamic	
Thermochemistry		performance comparison	446
nuclear rocket	459–460	magnetoplasmadynamic (MPD)	531, 563–570
Thermodynamics		applied-field	568–570
first law	82, 153	efficiency of	587
for a control volume	86	performance data for	572
second law	166	self-field	563–568
Theta-pinch engines	529	microwave-heated	526
Thor	189	pulsed inductive	529, 530, 534
Throat		efficiency of	587
area for a given mass flow	113	pulsed-plasma (PPT)	
heating	135		513, 529, 534, 559–563
Mach number	104	stationary-plasma (SPT)	
Thrust chamber *See also* Combustion chamber			531, 534, 570–575
		Thrust-to-weight	
cooling analysis	234–241	comparison of technologies	446
cooling of	202	Thrust-to-weight ratio	
definition of	187	definition of	17
generalized view	78	for launch vehicles	18
magnetic	647		
mass estimate	226		

Thrust-to-weight ratios	
for orbit-transfer vehicles	18
Time	
units and conversion factors	691
Time-of-flight	54
in an elliptical orbit	38–40
Titan launch vehicle *See*	
Launch vehicles, Titan	
Titanium	
fuel for SRMs	325
properties of	270
Torque, units and conversion factors	688
Total impulse	
equation for	16
Toxicity of propellants	174, 324
Tracking and Data Relay Satellite (TDRS)	
feed system schematic	245
Trajectory *See also* **Orbits**	31
Transducers, description of	246
Transfer Orbit Stage (TOS)	298, 605
Transient Performance Program (TPP)	173
Transtage	605
Triplet *See* **Injectors, choosing type**	
Tripropellants	
advantages of	188
True anomaly	37, 39
Tsiolkovsky, Konstantin	189
Turbines	
allowable speeds for materials	261
efficiency	263
impulse	
configuration of	260
pressure-compounded	
configuration of	260
reaction	
configuration of	262
requirements for	257
sizing	257
process table for	259
stage	
definition of	258
velocity-compounded	
configuration of	260
Turbomachinery	
bearing arrangements	266
configuring	264
design of	247–268
design process	251

mass estimate	266
Turbopump cycle *See* **Cycle**	
Turbopumps *See also* **Turbomachinery**	
configuring	264
cycles	185
design of	247–268
hot-gas source	185, 186
in nuclear rockets	449
TVC *See* **Thrust vector control**	
Two-body equations of motion	
See also **Orbits**	34–36
orbit maintenance	58–60
perturbations	42

U

Ullage, tank	268
Underexpanded nozzle	115
Unsymmetrical dimethyl hydrazine	
typical properties	696
Upper stage motor	
preliminary design of	352–361

V

V-2 rockets	189, 190
Valves	
check	246
control	246
drain	245
isolation	246
on/off	246
tank fill	245
Van Allen belts	512
effect on solar cells	522
Van der Waals effects	87
Velocity	
acoustic	97
ideal propulsive	65
needed for orbit injection	65
nozzle exit	106
units and conversion factors	691
Velocity budget *See* Δv **budget**	
Vernal equinox	57
Vertically stabilized tethers	662
Viking	191
Viscosity	
units and conversion factors	692

Volume
    units and conversion factors   692
Volumetric loading efficiency   312

W

Wall thickness
    liquid rocket propulsion systems
        mass estimate for   227
Web distance   335
Web fraction   357
Work function, in electrode material   551

X

X-15 rocket   190

Z

Zirconium
    fuel for SRMs   325
Zonal coefficients   42
Z-pinch engines   529